Lecture Notes in Networks and Systems 1268

The series "Lecture Notes in Networks and Systems" publishes the latest developments in Networks and Systems—quickly, informally and with high quality. Original research reported in proceedings and post-proceedings represents the core of LNNS.

Volumes published in LNNS embrace all aspects and subfields of, as well as new challenges in, Networks and Systems.

The series contains proceedings and edited volumes in systems and networks, spanning the areas of Cyber-Physical Systems, Autonomous Systems, Sensor Networks, Control Systems, Energy Systems, Automotive Systems, Biological Systems, Vehicular Networking and Connected Vehicles, Aerospace Systems, Automation, Manufacturing, Smart Grids, Nonlinear Systems, Power Systems, Robotics, Social Systems, Economic Systems and other. Of particular value to both the contributors and the readership are the short publication timeframe and the world-wide distribution and exposure which enable both a wide and rapid dissemination of research output.

The series covers the theory, applications, and perspectives on the state of the art and future developments relevant to systems and networks, decision making, control, complex processes and related areas, as embedded in the fields of interdisciplinary and applied sciences, engineering, computer science, physics, economics, social, and life sciences, as well as the paradigms and methodologies behind them.

Indexed by SCOPUS, EI Compendex, INSPEC, WTI Frankfurt eG, zbMATH, SCImago.

All books published in the series are submitted for consideration in Web of Science.

For proposals from Asia please contact Aninda Bose (aninda.bose@springer.com).

Ahmed Abdelgawad · Akhtar Jamil ·
Alaa Ali Hameed
Editors

Intelligent Systems, Blockchain, and Communication Technologies

Selected Papers From the International Conference on Intelligent Systems, Blockchain, and Communication Technologies (ISBCom24) - Volume 1

Springer

Editors
Ahmed Abdelgawad
Central Michigan University
Mount Pleasant, MI, USA

Akhtar Jamil
National University of Computer
and Emerging Sciences (NUCES)
Islamabad, Pakistan

Alaa Ali Hameed
Department of Computer Engineering
Istinye University
Istanbul, Türkiye

ISSN 2367-3370 ISSN 2367-3389 (electronic)
Lecture Notes in Networks and Systems
ISBN 978-3-031-82376-3 ISBN 978-3-031-82377-0 (eBook)
https://doi.org/10.1007/978-3-031-82377-0

This Springer imprint is published by the registered company Springer Nature Switzerland AG
The registered company address is: Gewerbestrasse 11, 6330 Cham, Switzerland

Contents

About the Editors

Dr. Ahmed Abdelgawad received his M.S. and a Ph.D. degree in Computer Engineering from the University of Louisiana at Lafayette in 2007 and 2011 and subsequently joined IBM as a Design Aids & Automation Engineering Professional at Semiconductor Research and Development Center. In Fall 2012, he joined Central Michigan University as a Computer Engineering Assistant Professor. In Fall 2022, Dr. Abdelgawad was promoted to the rank of Professor. He is a senior member of IEEE. His area of expertise is Internet of Things (IoT), distributed computing for wireless sensor network (WSN), structural health monitoring (SHM), data fusion techniques for WSN, low power embedded systems, digital signal processing, robotics, RFID, localization, VLSI, and FPGA design. He has published two books and more than 130 articles in related journals and conferences. Dr. Abdelgawad served as a reviewer for several journals and conferences, including IEEE IoT Journal, IEEE Communications Magazine, IEEE Transactions on VLSI, and IEEE Transactions on I&M, Springer, Elsevier, IEEE WF-IoT, IEEE ISCAS, IEEE SAS, and IEEE MWSCAS. Dr. Abdelgawad served as the general chair of the IEEE International Conference on Artificial Intelligence, Blockchain, and Internet of Things, (AIBThings2023), the 3rd IEEE International Conference on Computing and Machine Intelligence (ICMI2024), and the International Conference on Intelligent Systems, Blockchain, and Communication Technologies (ISBCom2024). He served in the organizing committees of IEEE WF-IoT, IEEE ISCAS, IEEE ICIP, IEEE SiPS, IEEE MWSCAS, and GIoTS. In addition, he taught many short IoT courses in different countries. He was the keynote speaker for many international conferences and conducted many webinars. He is currently the IEEE Northeast Michigan section chair and IEEE SPS Internet of Things (IoT) SIG Member. In addition, Dr. Abdelgawad served as a PI and Co-PI for several funded grants from NSF.

Dr. Akhtar Jamil is an accomplished associate professor at the National University of Computer and Emerging Sciences in Islamabad, Pakistan. He holds a Ph.D. in Computer Science from Yildiz Technical University, Istanbul, and a postdoctoral fellowship from Istanbul University-Cerrahpasa, Turkey. Dr. Jamil has an extensive work history, including key academic and industry roles. Before his current position, he served as an assistant professor and deputy head of the Department of Computer Engineering at Istanbul Sabahattin Zaim University, Turkey. He also held positions in international companies such as EMI Group Information Technology in Turkey and collaborated with the Ingenium Research Group at the University of Castilla-La Mancha, Spain, as a visiting professor.

Dr. Jamil is an active researcher; he has authored 28 journal papers, 8 book chapters, and 71 conference papers, making significant contributions to his field. He has been actively involved in leading international research collaborations and has played key roles in organizing major conferences as Technical Chair and General Chair. He is also an editor of several academic books and a regular reviewer for prestigious journals, further solidifying his impact on the global computer science community. His current research interests focus on artificial intelligence, computer vision, generative AI, and pattern recognition.

Contact Details:
Email: akhtar.jamil@nu.edu.pk
Department of Computer Science,
FAST School of Computing
National University of Computer and Emerging Sciences,
Islamabad, Pakistan.
Office Tel. (051) 111-128-128; Ext. 633

Alaa Ali Hameed obtained his Master's degree in Computer Engineering from Eastern Mediterranean University in North Cyprus, followed by a Ph.D. from Selcuk University in Turkey. He completed his postdoctoral research at Istanbul University. He has held academic positions as Assistant Professor at Istanbul Aydin University and Istanbul Sabahattin Zaim University and currently serves in the same capacity at Istinye University. Hameed has an impressive record of over 100 publications in prestigious international journals and conferences, demonstrating his prolific contribution to the field. He has also taken on significant leadership roles, including serving as a conference chair, program chair, and technical program committee member for various international conferences. In addition, he has contributed to the academic community as an editor for several books and as a reviewer for numerous journals. His editorial contributions extend to serving as a guest editor for several SCIE-indexed journals. Hameed's research interests are diverse and include data mining, machine/deep learning, big data analytics, artificial intelligence, digital signal/image processing, and adaptive filters. He has actively participated in numerous research activities, further establishing his expertise and influence in these areas.

Lightweight Deep Convolutional Neural Network for Pavement Crack Recognition with Explainability Analysis

Muhammad Gulfam[1(✉)], Cheryl D. Seals[1], Adriana Vargas-Nordcbeck[2], and Gerry V. Dozier[1]

[1] Computer Science and Software Engineering, Auburn University, Auburn, AL, USA
{mzg0108,sealscd,doziegv}@auburn.edu

[2] National Center for Asphalt Technology, Auburn University, Auburn, AL 36849, USA
vargaad@auburn.edu

Abstract. Pavement crack detection is an integral part of maintaining pavements. Identifying the pavement areas that require maintenance through manual processes is a laborious and time-consuming task. Latest researches have employed machine learning and deep learning to devise automatic and semi-automatic solutions for pavement distress detection. In this research, we propose a lightweight deep convolutional neural network (DCNN) model for crack recognition in pavement images, with 20 times fewer parameters than ResNet34 model. The model was tested on three public datasets that include Concrete Crack Images for Classification (CCIC) dataset, Sharif University of Technology Crack (SUT-Crack) dataset, and bridge crack detection dataset. The model obtained an average F1-score of 99.98%, 98.75% and 100% on the above-mentioned datasets respectively. We also tested the model on our own dataset that has pavement images with finer cracks which are harder to detect. Our dataset consists of 5,506 images in total. The model obtained 89.13% average F1-score. We explored the transfer learning technique on the proposed model as well. We fine-tuned the model (pre-trained on CCIC dataset) on our dataset which improved the results by 6.34% and increased the average F1-score to 95.47%. The model shows good generalizability with just a little over a million trainable parameters. Model explainability analysis using Captum AI library confirmed that the model is focusing on the cracked sections of the pavement images to classify the images as cracked. Having relevant areas of the image contribute more toward the prediction increases the confidence in the model.

Keywords: Lightweight convolutional neural network · Asphalt-based pavement · Crack recognition · Deep learning · Interpretability · Explainability

1 Introduction

Road infrastructure is the backbone of the supply chain in most developing and developed countries [1]. It enables us to transport people and goods quickly

A. Abdelgawad et al. (Eds.): ISBCom 2024, LNNS 1268, pp. 1–15, 2025.
https://doi.org/10.1007/978-3-031-82377-0_1

and economically and provides a medium to connect communities [1]. Their importance and our dependence on them require the pavements to maintain a good level of service for proper use [2]. Pavement conditions deteriorate over time due to traffic loading, climate effects, and many other factors [2–4]. Common pavement distresses include cracks, potholes, rutting, and oxidation [2,3,5]. Government transportation agencies routinely collect pavement condition data that can be used to effectively manage the pavement networks [4].

Pavement distress detection has evolved from completely manual work to semi-automated and automated processes [3,4]. Manual surveys are laborious, very time-consuming, and prone to human errors [3]. Road maintenance solutions need to be fast and effective in order to ensure good condition of the roads throughout the year [2,6]. The use of automation in distress detection and identification can help in expediting the detection process and reduce the overall cost [3].

In the last two decades, many studies have focused on crack detection in road infrastructure. Sun et al. [7] used image-processing techniques to detect pavement cracks, while Salari et al. [8] employed a combination of image-processing and machine learning techniques to detect and classify cracks in the road images.

Lately, deep learning (DL) has surfaced as a powerful solution to many problems. It has found applications in many different fields including, but not limited to, object detection and recognition [9,10], image segmentation [11,12], image generation [13,14], text generation [15], and sentiment analysis [16]. DL has the ability to learn complex patterns and mapping from the data which is hard to learn for the other machine learning algorithms [17]. Deep neural networks usually have millions of parameters that require large number of training samples to find the optimal configurations to get to the solution for a given problem [4,18]. For many problems, it is very difficult to get large datasets with annotated samples [4,18,19]. For such problems, transfer learning [19,20] and fine-tuning the pre-trained models on the target dataset with limited training samples have proven to be very effective [18].

In the last few years research groups have started using deep neural networks for pavement distress detection and evaluation. Zhang et al. [5] and Wang et al. [21] employed deep neural network for pavement crack detection and classification, and recognition respectively. Gopalakrishnan et al. [4] used transfer learning with feature extractor from the VGG-16 net [22] pre-trained on the ImageNet [23] database and reported a good accuracy for pavement crack detection.

Developing deep learning based solutions for low memory devices is also a challenge as large models require large amount of memory, heavy computing machines and more time for training and inference. In this paper we have proposed a lightweight deep convolutional neural network to recognize crack in pavement surfaces. The model contains roughly 20 times fewer parameters than ResNet34 model. As part of this research, dataset consisting of 5506 asphalt-surfaced pavement images was produced and made public. We have also explored the transfer learning technique and found it very effective. We trained our model on concrete crack images for classification (CCIC) dataset and fine-tuned the model on our own dataset which improved the results significantly.

The rest of the paper is structured as follows. Section 2 consists of a review of related work on pavement crack detection in the last few decades. Section 3 talks about methodology, mainly focusing on the network architecture. Section 4 contains experiment details and results. Sections 5 which is the last section, has conclusion and future work.

2 Related Work

Many studies have attempted to create automated solutions for pavement crack detection and classification during the last three decades [4,5,8,10,24,25]. For this task, many research techniques have been employed that include but are not limited to image processing (edge detection, intensity-thresholding, wavelet transforms), machine learning, and DL.

Many image processing-based methods have been proposed [24,26,27] but such methods face challenges when the pavement images contain background noise, water or other such factors.

Different machine learning algorithms have also been explored to develop road crack detection solutions [3,17]. Most popular algorithms are support vector machines (SVM) and artificial neural networks (ANN) [3,17]. Other methods include but are not limited to Decision Tree (DT), Random Forest (RF), Logistic Regression (LR) and Extremely Randomized Trees (ERT) [3,17]. Usually, machine learning is used in combination with image processing techniques [2,28,29] where image processing is used to extract features from the pavement images and then machine learning algorithms use those features for decision-making.

It is hard for shallow machine learning algorithms to capture complex patterns from the data [17]. DL-based methods have the ability to learn complex mappings between input and output. Lately, researchers have used DCNN for road crack detection problem set. There are three different types of problems in road crack detection: crack classification, crack segmentation, and object detection [17].

The target of the classification is to identify whether an image has cracks in it or not. A lot of researchers have taken on the problem of road crack classification. Xu et al. [30] proposed a deep neural network for crack classification on bridge datasets and reported very promising results. Eisenbach et al. [31] proposed a public dataset, German Asphalt Pavement Distress Dataset (GAPs) along with a deep convolutional neural network, ASINVOS net. The GAPs dataset has over 6.7 million image patches in total. A small version of this dataset has 50k image patches for faster evaluation of the methods. They also trained the ResNet models from scratch as well as with pre-trained weights on their dataset. They concluded that training the ResNet model from scratch is producing better results. Their best results on the 50k dataset were the ResNet18 with data augmentation. They obtained a 90.62% average F1 score for crack detection.

Nguyen et al. [32] proposed a DCNN with five convolutional layers for road crack detection, and reported a 91% F1 score on JTIRC data and a 90% F1 score on Rescaled CrackIT data. Zhang et al. [5] proposed another DCNN with fully connected layers for the classification of six different classes (five classes for cracks types and one class for images without cracks). They split the original images into image patches and labeled each patch separately. They achieved an 87.5% F1 score and 88.3% accuracy. The research of Wang et al. [21] is a similar example of dividing the original images into patches. They created image patches of size 64×64 for training and testing. They proposed a DCNN with three convolutional layers and obtained 94.29% accuracy on their testing data. Subdividing images into small patches helps to reduce the computational cost and identify multiple cracks in the image [17].

Designing DCNN and training them from scratch requires a large set of training samples which is difficult to acquire for many fields [4]. There are not many large datasets publicly available for the pavement cracks detection problem, with annotated samples [17]. Many of the publicly available datasets have less than 500 images [33–37]. One way to enhance performance with smaller datasets is to use transfer learning. Some researchers have employed transfer learning to develop DL-based solutions for different problems, including road crack detection, that does not have larger training datasets [4,38–41].

3 Methodology

3.1 Model Architecture

In this study, we have proposed a lightweight deep convolutional neural network (DCNN) for crack recognition (binary classification i.e. crack or no-crack) in pavement images. Figure 1 shows the proposed DCNN model architecture diagram. The model contains four convolutional blocks. The first block has only one convolutional layer with kernel size of 7X7, stride of 2 and padding of 3. The number of filters in the first block for convolutional layer are 64. The first convolutional block is followed by average pooling with the kernel size of 3X3, stride of 2 and padding of 1. Average pooling reduces the input size by half.

Each of the next two convolutional blocks contain two convolutional layers with kernel size of 5X5, stride of 1 and padding of 2. In these blocks, the first convolutional layer is followed by batch normalization and parametric rectified linear unit (PReLU) activation function while the second convolutional layer is just followed by batch normalization. The choice of PReLU is based on empirical evaluations. PReLU produced better results than rectified linear unit (ReLU) activation function (see Table 1 in the supplementary data). These two convolutional blocks have 64 filters each. The last convolutional block contains three convolutional layers where the first and second convolutional layer has kernel size of 5X5 and padding of 2 but the third convolutional layer has kernel size of 1 and default padding. In this last block, first and third convolutional layers contain the stride 2 while the second convolutional layer contains the stride 1. Each convolutional block has a skip connection which is followed by PReLU activation. This convolutional block contains 128 filters. The last convolutional block is followed by adaptive average pooling with the output size 1×1. This layers helps reduce the number of learnable parameters significantly.

The classifier part of the model derived some inspiration from the DCNN model introduced by Zhang et al.[5]. It contains two fully connected layers with 256 neurons in the first layer and 50 neurons in the second layer. The output layer contains two neurons as the model is being trained for binary classification. A dropout layer with a dropout rate of 35% follows each fully connected layer along with PReLU activation. Dropout layers are added for generalizability and to avoid over-fitting.

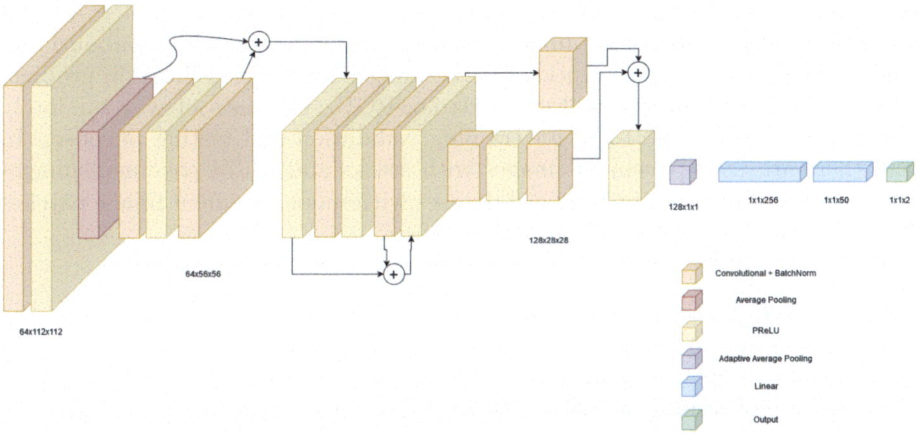

Fig. 1. Proposed DCNN model architecture diagram

4 Experiments and Results

4.1 Datasets

In this study, we have tested our model for three public datasets and one dataset of our own which was also being made public. The public datasets include Concrete crack images for classification (CCIC), Sharif University of Technology crack (SUT-crack) dataset, and Bridge crack detection dataset. The last dataset is NCAT dataset which is our own dataset and is derived from the images obtained from national center for asphalt technology (NCAT).

Concrete Crack Image for Classification (CCIC) Dataset. Concrete Crack Images for Classification (CCIC) [42,43] is one of the three publicly available datasets that we have used in this study. Zhang et al. [42] and Özgenel et al. [43] have contributed to the development of the CCIC dataset. This is a balanced dataset and there are a total of 40,000 images in this dataset. 20,000 images without cracks and 20,000 images with cracks. All images are of size 227×227. The dataset was divided into training, validation, and test sets with a ratio of 70:15:15. The split for training, validation, and test sets was done randomly. Ozengel et al. [43] used this dataset for pre-trained networks' performance comparison. They used the dataset split of 70:15:15, which was why we have used this ratio. The dataset is available online.

Sharif University of Technology Crack (SUT-Crack) Dataset. Sharif University of Technology crack (SUT-crack) dataset contains asphalt-surfaced road images. The dataset has collection of images annotated for crack classification (crack vs no-crack), object detection and crack segmentation. It is a fairly new dataset which was made publicly available in October 2023 by Sabouri and Sepidbar [44]. The image resolution for the images for the classification problem is 200×200. Each image has 3 channels i.e. RGB. The dataset was randomly split with the ratio of 60:10:30 for training, validation and testing purposes. In total, there are 1886 pavement images with cracks and 23674 pavement images without crack. This is an imbalanced dataset with more pavement image patches without cracks. The dataset is available online.

Bridge Crack Detection Dataset. The bridge crack detection dataset has been prepared and made publicly available by Xu et al. [30] and enhanced by Li et al. [45]. There are 4,856 images for training and 1,213 images for testing. All the images are of resolution 224×224. This dataset is imbalanced and there are far more images with cracks than without cracks in the training set, while the testing set has more images without cracks. The split of training and testing sets is done randomly and is done by Xu et al. [30] and we are using the same split. We randomly selected some images from the training set to use for validation during the model training process. We selected 191 images without cracks and 779 images with cracks for validation. This leaves 767 images without cracks and 3,119 images with cracks in the training set. The dataset is available online.

National Center for Asphalt Technology (NCAT) Dataset. We compiled a dataset for this project using asphalt-surfaced road images. The images were obtained from the National Center for Asphalt Technology (NCAT) at Auburn University. The original images were too large for computation. We split the images into patches of $224 \times 224 \times 3$. All of these pavement image patches are manually annotated. They were split into two classes, cracked road patches, and clean road patches. There are total 5506 image patches in the dataset. Figure 2 shows samples of a clean road patch on the left-side and a road patch with cracks in the middle and right-side. The shadows, water sprinkling on the asphalt surface, dirt, and other elements in the images increase the difficulty of the problems drastically. This is also an imbalanced dataset with more pavement image patches without cracks than with cracks. The test set contains 722 samples with cracks and 979 samples without cracks. The image patches in the training and validation sets are derived from the source images that were collected in 2018 while most of the images in the test set are derived from the images that were collected 4 years later. This difference adds difficulty for the model as the texture and many other elements changed in 4 years. The dataset is available online.

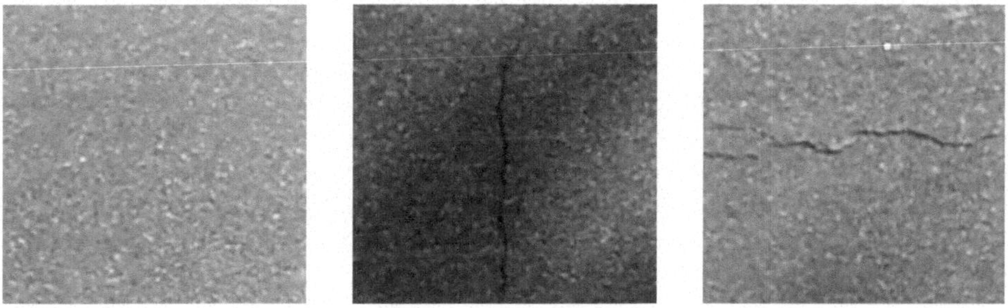

Fig. 2. NCAT dataset samples images

4.2 Implementation Details and Metrics

All the experiments in this study were conducted on a server running Ubuntu 22.04.3 LTS with AMD Ryzen Threadripper PRO 3975WX 32-Cores CPU, 500 GB RAM and three NVIDIA GeForce RTX 3080 with 12 GB GPU memory. The installed CUDA version is 12.0 and driver version was 525.147.05. Even though there were three GPUs available in the server, we only used one GPU per training process.

We used stochastic gradient descent (SGD) optimizer with momentum 0.9, decay rate 0.0003 and initial learning rate 1e−2. We borrowed the idea of using SGD optimizer from Stricker et al. [25]. We experimented with a few learning rates that include 1e−1, 1e−2, 1e−3, 1e−4 and 1e−5 but found 1e−2 to produce the better results. Table 2 in the supplementary Data contains the results for other learning rates. The batch size was 8, and the image patch size was 224×224. Batch size of 8 was mainly influenced by the available GPU memory. Cross-entropy loss has been used to train the model. Cross-entropy loss is good for penalizing misclassifications [46]. We trained the model for 500 epochs.

We also used data augmentation techniques that included random rotation with degrees within the range of 0 to 180 and random vertical and horizontal flip with 50% probability. Data augmentation techniques were only applied to the training sets. The data was normalized with standard deviation and means as 0.5 each. We also tested the normalization with different values of mean and standard deviation but the results did not improve. Please check the Table 3 from the supplementary data.

We also compared the proposed model's performance with the DCNN model introduced by Zhang et al. [5]. We could not find the code for their model so we implemented it according to the information provided in their article. We did not find the information about the learning rate they used in their experiments, so we experimented with three different learning rates that included 0.1, 0.01 and 0.001. For the evaluation, we trained and tested their model on SUT-Crack dataset for unbiased evaluation. Their model performed best for learning rate 0.1. Therefore, we used 0.1 learning rate for all the experiments with their model on all datasets.

We used F1-score for performance evaluation. F1-score is a widely used evaluation metric for imbalanced datasets [3,5,17]. F1-score is a harmonic mean of precision and recall. Accuracy metric is reported as an extra. Equation 1, 2 and 3 shows the formula for precision, recall and F1-score respectively. The variables used in Eq. 1 and 2 are defined as following: TP denotes instances of true positives, FP denotes instances of false positives, and FN denotes instances of false negatives.

$$Precision = \frac{TP}{(TP + FP)} \tag{1}$$

$$Recall = \frac{TP}{(TP + FN)} \tag{2}$$

$$F1 - Score = \frac{2 * Precision * Recall}{(Precision + Recall)} \tag{3}$$

4.3 Results and Discussion

This section evaluates performance of the proposed DCNN model and compares it with DCNN model proposed by Zhang et al. Fig. 3 shows the loss on training data for NCAT dataset. The training loss curve confirms the proposed model's ability to learn patterns from the data. Training loss curves for CCIC, SUT-Crack, and bridge crack datasets can be found in the supplementary data.

The NCAT dataset's testing images were the most difficult as most of them were collected from the source images that were four years apart from the source images for the training samples. Fine cracks and diverse background noise (shadows, wet surfaces, road markings etc.) render the NCAT dataset a challenging target. Our model achieved 89.13% F1-Score with average precision 89.06% and average recall as 89.21%, on the testing set for NCAT dataset. The model performed exceptionally well on the CCIC dataset and got only one incorrect prediction out of 6,000 samples. The average F1-score of our model on the CCIC dataset was 99.98%. The proposed model obtained a 100% average F1-score and had no incorrect predictions on the bridge crack detection dataset. SUT-Crack dataset is a new dataset that was published in October 2023. It has different texture and crack features. The proposed model performed well on this dataset and obtained 98.75% average F1-Score with 99.08% average precision and 98.44%

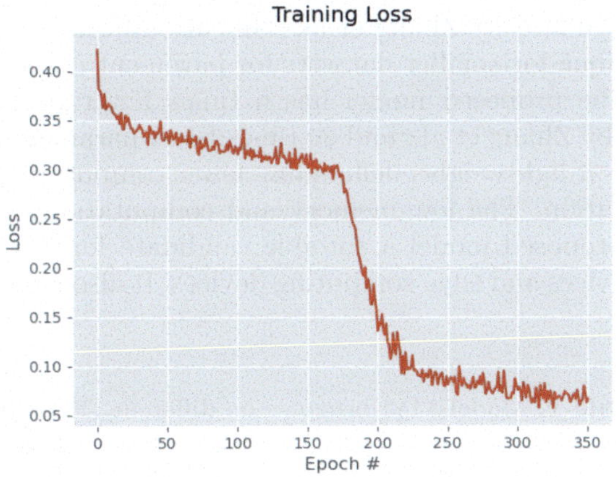

Fig. 3. The loss during the training process

average recall. Average precision and recall both being high indicates the superior performance of our model in SUT-Crack dataset.

We also explored transfer learning technique on the proposed model. We took the proposed model trained on the CCIC dataset which had 99.98% F1-score on CCIC dataset, and fine-tuned it on the NCAT dataset. We chose CCIC dataset for this purpose as the number of samples in CCIC dataset were the most among all three existing public datasets. With no surprise, the results obtained through fine-tuning were much better than training the model directly on NCAT dataset. The average F1-Score improved by 6.34% and reached 95.47% with average precision as 95.63% and average recall as 95.33%.

This improvement in F1-score shows that when the target dataset size is small then, it is worth exploring the option of training the model on similar dataset with more samples and then fine-tuning it on the target dataset. Results for all four datasets are shown in the Table 1.

We also conducted a comparative analysis of the proposed model with the model introduced by Zhang et al. [5]. Their model obtained 98.23% average F1-Score on the SUT-Crack dataset. For CCIC and bridge crack detection dataset the average F1-score was 99.92 with 5 mis-classifications and 99.64% with 2 misclassified samples respectively. The performance of their model on NCAT dataset was not good. The model only obtained 70.18% average F1-Score with average recall less than 70%. As the transfer learning gave our proposed model a boost in performance, we applied the same technique to their model as well. Surprisingly, the improvement in their model's performance was very significant. The average F1-Score improved by 19.42% and reached 89.60%. Even with this boost in performance through transfer learning, our proposed model outperformed their model on NCAT dataset with a good margin. Comparison results on other datasets is also shown in Table 1. Our model performed better on all the datasets used in this study. This shows our model's superiority over the

DCNN model introduced by Zhang et al. This also affirms transfer learning as a promising technique for smaller datasets for pavement crack classification.

Additionally, the proposed model has 6 times fewer parameters than the model introduced by Zhang et al., and 20 times fewer parameters than ResNet34 model. This makes it lightweight, demanding fewer computational resources and less memory allocation. The low memory and computation capability requirement makes the proposed model a suitable candidate for resource-constrained devices like IoT devices and edge computing devices. It also allows the portability and scalability.

Table 1. Models' Performance on different Datasets

Method	Dataset	Precision	Recall	F1-score	Accuracy
Proposed model	NCAT	89.06	89.21	89.13	89.36
Proposed model (with TL)		**95.63**	**95.33**	**95.47**	**95.59**
DCNN by Zhang et al.		72.46	69.86	70.18	72.19
DCNN by Zhang et al. (with TL)		90.56	89.06	89.60	90.01
Proposed model	CCIC	**99.98**	**99.98**	**99.98**	**99.98**
DCNN by Zhang et al.		99.92	99.92	99.92	99.92
Proposed model	SUT-Crack	**99.08**	**98.44**	**98.75**	**99.66**
DCNN by Zhang et al.		98.20	98.28	98.24	99.52
Proposed model	Bridge-Crack-Detection	**100**	**100**	**100**	**100**
DCNN by Zhang et al.		99.37	99.91	99.64	99.84

Note: TL: transfer learning

Explainability Analysis. The black boxed nature of neural network-based models makes it challenging to comprehend their decision-making process, thereby fostering reluctance to trust the model's decisions. We analysed our proposed model as well as the DCNN model introduced by Zhang et al. using an explainability tool named Captum. We analyzed the models trained for the NCAT dataset as well as the CCIC dataset. For the CCIC dataset the models were directly trained on the dataset while for the NCAT dataset, we analyzed the models that were trained on CCIC prior to fine-tuning them on NCAT.

Figure 4 show the results of model explainability analysis. The left column contains the input images from NCAT dataset. The middle column displays the images illustrating the pixels that contributed more during the prediction process by our proposed model, whereas the right column showcases the images depicting the pixel contributions for the decision made by the DCNN model introduced by Zhang et al. This figure also shows that our model is focusing more on the important and relevant features for decision making.

Figure 5 shows similar results on CCIC dataset. The explainability results on CCIC dataset are more distinctive and clear. Even though the dataset is easier for both models, our proposed models focus more on the relevant features than the model introduced by Zhang et al.

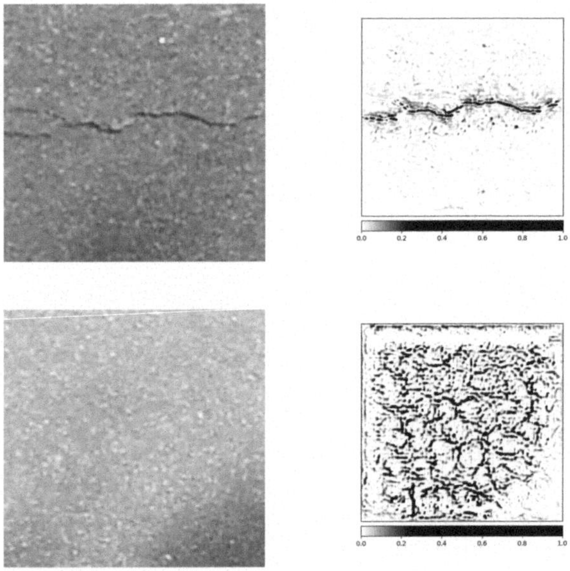

Fig. 4. NCAT Dataset: Input Image and Explainable AI-Processed Image: Pixel Contribution to Prediction. Left Column: Input images, Middle Column: Our proposed model, Right Column: DCNN by Zhang et al.

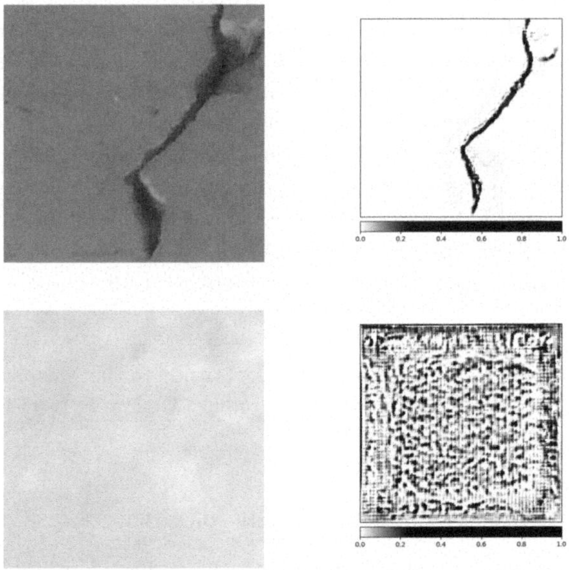

Fig. 5. CCIC Dataset: Input Image and Explainable AI-Processed Image: Pixel Contribution to Prediction. Left Column: Input images, Middle Column: Our proposed model, Right Column: DCNN by Zhang et al.

The explainability analysis demonstrates that our proposed model utilizes relevant features for making predictions. For image patches containing cracks, our model assigns higher weights to pixels in the cracked areas. Conversely, for image patches without cracks, the model scans the entire image before reaching the decision. This behavior enhances the model's trustworthiness and reliability.

5 Conclusion and Future Work

This study introduces a lightweight DCNN model designed specifically for the task of crack recognition in pavement imagery. The model contains only 1.08 million parameters, 20 times fewer than ResNet34 model. The reduced parameter count of this model necessitates less memory, making it a viable candidate for deployment on devices with limited resources.

The empirical findings underscore the efficiency of the proposed model, highlighting its capacity to learn patterns within pavement images. The proposed model is evaluated on four different datasets, including three publicly available datasets (CCIC, SUT-Crack and bridge crack detection datasets), and one dataset of our own (NCAT dataset). The model obtained 95.47%, 99.98%, 98.75% and 100% average F1-score on NCAT, CCIC, SUT-Crack and bridge crack detection datasets respectively. NCAT dataset was the most difficult one among the four because of background noise and finer cracks. A comparative analysis with the model proposed by Zhang et al. [5] demonstrates the superior performance of our model, which achieves this with a reduced demand for resources. We further scrutinized the decision-making process of the models using an Artificial Intelligence (AI) explainability tool, Captum. The findings suggest that the proposed model prioritizes relevant pixels during decision-making while discounting irrelevant ones, reflecting desirable traits for a reliable solution.

Owing to the lack of publicly available datasets for classification between various types of cracks, we were unable to assess the model's performance in this specific classification task. Future work aims to evaluate the model's effectiveness in this area, and develop a complete pavement management system requiring fewer computing resources.

Acknowledgments. The authors would like to express their gratitude to the National Center for Asphalt Technology (NCAT) for providing source images that were used to extract the NCAT dataset.

Data Availibility Statement. The NCAT dataset is available on github.

Conflicts of Interest. The authors assert that they do not have any identifiable financial conflicts of interest or personal relationships that might have influenced the findings presented in this study.

Supplementary Information. Supplementary data can be obtained from github.

References

1. Ranjbar, S., Nejad, F.M., Zakeri, H.: An image-based system for pavement crack evaluation using transfer learning and wavelet transform. Int. J. Pavement Res. Technol. **14**, 437–449 (2021)
2. Cubero-Fernandez, A., Rodriguez-Lozano, F.J., Villatoro, R., Olivares, J., Palomares, J.M.: Efficient pavement crack detection and classification. EURASIP J. Image Video Process. **1–11**, 2017 (2017)
3. Ai, D., Jiang, G., Lam, S.-K., He, P., Li, C.: Computer vision framework for crack detection of civil infrastructure -review. Eng. Appl. Artif. Intell. **117**, 105478 (2023)
4. Gopalakrishnan, K., Khaitan, S.K., Choudhary, A., Agrawal, A.: Deep convolutional neural networks with transfer learning for computer vision-based data-driven pavement distress detection. Constr. Build. Mater. **157**, 322–330 (2017)
5. Zhang, C., Nateghinia, E., Miranda-Moreno, L.F., Sun, L.: Pavement distress detection using convolutional neural network (CNN): a case study in Montreal, Canada. Int. J. Transp. Sci. Technol. **11**(2), 298–309 (2022)
6. Ma, D., Fang, H., Wang, N., Xue, B., Dong, J., Wang, F.: A real-time crack detection algorithm for pavement based on CNN with multiple feature layers. Road Mater. Pavement Des. **23**(9), 2115–2131 (2022)
7. Sun, Y., Salari, E., Chou, E.: Automated pavement distress detection using advanced image processing techniques. In: 2009 IEEE International Conference on Electro/Information Technology, pp. 373–377. IEEE (2009)
8. Salari, E., Bao, G.: Pavement distress detection and classification using feature mapping. In: 2010 IEEE International Conference on Electro/Information Technology, pp. 1–5. IEEE (2010)
9. LeCun, Y., Bengio, Y., Hinton, G.: Deep learning. Nature **521**(7553), 436–444 (2015)
10. Wang, X., et al.: Deep learning in object recognition, detection, and segmentation. Found. Trends® Signal Process. **8**(4), 217–382 (2016)
11. Ronneberger, O., Fischer, P., Brox, T.: U-net: convolutional networks for biomedical image segmentation. In: Medical Image Computing and Computer-Assisted Intervention–MICCAI 2015: 18th International Conference, Munich, Germany, 5–9 October 2015, Proceedings, Part III 2018, pp. 234–241. Springer, Cham (2015)
12. Hesamian, M.H., Jia, W., He, X., Kennedy, P.: Deep learning techniques for medical image segmentation: achievements and challenges. J. Digit. Imaging **32**, 582–596 (2019)
13. Wang, R.: Research on image generation and style transfer algorithm based on deep learning. Open J. Appl. Sci. **9**(8), 661–672 (2019)
14. Goodfellow, I., et al.: Generative adversarial networks. Commun. ACM **63**(11), 139–144 (2020)
15. Iqbal, T., Qureshi, S.: The survey: text generation models in deep learning. J. King Saud Univ.-Comput. Inf. Sci. **34**(6), 2515–2528 (2022)
16. Zhang, L., Wang, S., Liu, B.: Deep learning for sentiment analysis: a survey. Wiley Interdisc. Rev.: Data Min. Knowl. Discov. **8**(4), e1253 (2018)
17. Hsieh, Y.-A., Tsai, Y.J.: Machine learning for crack detection: Review and model performance comparison. J. Comput. Civ. Eng. **34**(5), 04020038 (2020)
18. Zhuang, F., et al.: A comprehensive survey on transfer learning. Proc. IEEE **109**(1), 43–76 (2020)
19. Weiss, K., Khoshgoftaar, T.M., Wang, D.: A survey of transfer learning. J. Big Data **3**(1), 1–40 (2016)

20. Bozinovski, S., Fulgosi, A.: The influence of pattern similarity and transfer learning upon training of a base perceptron B2. In: Proceedings of Symposium Informatica, vol. 3, pp. 121–126 (1976)
21. Wang, K.C.P., Zhang, A., Li, J.Q., Fei, Y., Chen, C., Li, B.: Deep learning for asphalt pavement cracking recognition using convolutional neural network. Airfield Highw. Pavements **166–177**, 2017 (2017)
22. Simonyan, K., Zisserman, A.: Very deep convolutional networks for large-scale image recognition. arXiv preprint arXiv:1409.1556 (2014)
23. Deng, J., Dong, W., Socher, R., Li, L.-J., Li, K., Fei-Fei, L.: Imagenet: a large-scale hierarchical image database. In: 2009 IEEE Conference on Computer Vision and Pattern Recognition, pp. 248–255. IEEE (2009)
24. Mahler, D.S., Kharoufa, Z.B., Wong, E.K., Shaw, L.G.: Pavement distress analysis using image processing techniques. Comput.-Aided Civil Infrastruct. Eng. **6**(1), 1–14 (1991)
25. Stricker, R., Eisenbach, M., Sesselmann, M., Debes, K., Gross, H.-M.: Improving visual road condition assessment by extensive experiments on the extended gaps dataset. In: 2019 International Joint Conference on Neural Networks (IJCNN), pp. 1–8. IEEE (2019)
26. Georgopoulos, A., Loizos, A., Flouda, A.: Digital image processing as a tool for pavement distress evaluation. ISPRS J. Photogramm. Remote. Sens. **50**(1), 23–33 (1995)
27. Hoang, N.-D.: Detection of surface crack in building structures using image processing technique with an improved OTSU method for image thresholding. Advances Civil Eng. **2018** (2018)
28. Ahmadi, A., Khalesi, S., Golroo, A.: An integrated machine learning model for automatic road crack detection and classification in urban areas. Int. J. Pavement Eng. **23**(10), 3536–3552 (2022)
29. Breiman, L.: Bagging predictors. Mach. Learn. **24**, 123–140 (1996)
30. Hongyan, X., Xiu, S., Wang, Y., Cai, H., Cui, K., Chen, X.: Automatic bridge crack detection using a convolutional neural network. Appl. Sci. **9**(14), 2867 (2019)
31. Eisenbach, M., et al.: How to get pavement distress detection ready for deep learning? A systematic approach. In: 2017 International Joint Conference on Neural Networks (IJCNN), pp. 2039–2047. IEEE (2017)
32. Nguyen, N.T.H., Le, T.H., Perry, S., Nguyen, T.T.: Pavement crack detection using convolutional neural network. In: Proceedings of the 9th International Symposium on Information and Communication Technology, pp. 251–256 (2018)
33. Shi, Y., Cui, L., Qi, Z., Meng, F., Chen, Z.: Automatic road crack detection using random structured forests. IEEE Trans. Intell. Transp. Syst. **17**(12), 3434–3445 (2016)
34. Qin Zou, Yu., Cao, Q.L., Mao, Q., Wang, S.: Cracktree: automatic crack detection from pavement images. Pattern Recogn. Lett. **33**(3), 227–238 (2012)
35. Chambon, S., Moliard, J.-M.: Automatic road pavement assessment with image processing: review and comparison. Int. J. Geophys. **2011** (2011)
36. Amhaz, R., Chambon, S., Idier, J., Baltazart, V.: Automatic crack detection on two-dimensional pavement images: an algorithm based on minimal path selection. IEEE Trans. Intell. Transp. Syst. **17**(10), 2718–2729 (2016)
37. Oliveira, H., Correia, P.L.: Crackit–an image processing toolbox for crack detection and characterization. In: 2014 IEEE International Conference on Image Processing (ICIP), pp. 798–802. IEEE (2014)
38. Coulibaly, S., Kamsu-Foguem, B., Kamissoko, D., Traore, D.: Deep neural networks with transfer learning in millet crop images. Comput. Ind. **108**, 115–120 (2019)

39. Chaabouni, S., Benois-Pineau, J., Amar, C.B.: Transfer learning with deep networks for saliency prediction in natural video. In: 2016 IEEE International Conference on Image Processing (ICIP), pp. 1604–1608. IEEE (2016)
40. Mittel, D., Kerber, F.: Vision-based crack detection using transfer learning in metal forming processes. In: 2019 24th IEEE International Conference on Emerging Technologies and Factory Automation (ETFA), pp. 544–551. IEEE (2019)
41. Yang, Q., Shi, W., Chen, J., Lin, W.: Deep convolution neural network-based transfer learning method for civil infrastructure crack detection. Autom. Constr. **116**, 103199 (2020)
42. Zhang, L., Yang, F., Zhang, Y.D., Zhu, Y.J.: Road crack detection using deep convolutional neural network. In: 2016 IEEE International Conference on Image Processing (ICIP), pp. 3708–3712. IEEE (2016)
43. Özgenel, Ç.F., Gönenç Sorguç, A.: Performance comparison of pretrained convolutional neural networks on crack detection in buildings. In: Proceedings of the International Symposium on Automation and Robotics in Construction, ISARC, vol. 35, pp. 1–8. IAARC Publications (2018)
44. Sabouri, M., Sepidbar, A.: Sut-crack: a comprehensive dataset for pavement crack detection across all methods. Data Brief **51**, 109642 (2023)
45. Li, L.-F., Ma, W.-F., Li, L., Lu, C.: Research on detection algorithm for bridge cracks based on deep learning. Acta Automatica Sinica **45**(9), 1727–1742 (2019)
46. Hu, D., Yee, T., Goff, D.: Automated crack detection and mapping of bridge decks using deep learning and drones. J. Civil Struct. Health Monit. 1–15 (2024)

A Novel Bounding Box Regression Method for Single Object Tracking

Omar Abdelaziz(✉) and Mohmaed Sami Shehata

University of British Columbia, Kelowna, BC V1V1V7, Canada
oabdelaz@student.ubc.ca

Abstract. Locating an object in a sequence of frames, given its appearance in the first frame of the sequence, is a hard problem that involves many stages. Usually, state-of-the-art methods focus on bringing novel ideas in the visual encoding or relational modelling phases. However, in this work, we show that bounding box regression from learned joint search and template features is of high importance as well. While previous methods relied heavily on well-learned features representing interactions between search and template, we hypothesize that the receptive field of the input convolutional bounding box network plays an important role in accurately determining the object location. To this end, we introduce two novel bounding box regression networks: inception and deformable. Experiments and ablation studies show that our inception module installed on the recent ODTrack outperforms the latter on three benchmarks: the GOT-10k, the UAV123 and the OTB2015.

Keywords: Single object tracking · Bounding box regression · One-stream framework · Vision transformers

1 Introduction

Single object tracking is a challenging computer vision subfield that is ubiquitous in many applications such as video surveillance and augmented reality. The challenging nature of tracking is associated with the problem definition as well as the techniques to address the problem. Given a sequence of images, it is required to identify the size and location of an object given by a bounding box annotation in the first frame of the sequence. In single-object tracking, the object is also identified by a bounding box.

The current prevalent pipeline in single object tracking is the joint feature extraction and relation modelling pipeline introduced in [7] and [34]. Exploiting Vision Transformers (ViTs), this pipeline leverages the natural attention capabilities in relationship association to locate the template object in the search frame in a low-dimensional space. A bounding box prediction head is used afterwards to map the transformer attention map into bounding box coordinates.

Despite showing remarkable success, the current tracking methods are focused on the feature extraction and relation modelling phases in a one-stream

A. Abdelgawad et al. (Eds.): ISBCom 2024, LNNS 1268, pp. 16–27, 2025.
https://doi.org/10.1007/978-3-031-82377-0_2

manner while being less attentive to the bounding box regression phase. Bounding box regression is usually done using a simple multi-branch convolutional network. These networks simplify the bounding box regression by dividing it into classification and regression subproblems. The classification stage aims to separate the object from the background. In contrast, the regression branch aims to predict four float values that represent the coordinates of the top-left corner and either the width and the height or the coordinates of the bottom-left corner of the box. This simple design depends on the number of parameters of the Vision Transformer [10] to learn a good embedded representation that overcomes problems in the bounding box network.

We hypothesize that the simple convolutional pipeline of bounding box networks is not enough to capture complex global and local relations in the embedded feature representation output by the transformer. We introduce two simple yet effective bounding box networks that are capable of learning a variable receptive field to enable a more accurate determination of the target location. In particular, we introduce the Inception and the deformable convolution Inception bounding box networks based on the Inception [24] and deformable convolutional concepts [36]. We show backed by experiments that our bounding box networks can capture more useful information from the ViT embeddings.

To this end, the contributions of this work are three-fold:

- We propose a new bounding box head for ViT-based single object trackers, namely, the Inception module that addresses the fixed local receptive field of the bounding box network of single object tracking transformers.
- We propose another variant of the Inception module, the deformable convolution Inception module, which learns the required receptive field for a better-performing bounding box network.
- We conduct extensive experiments and ablation studies to show the effectiveness of our proposed modules. Our best model sets a new state-of-the-art.

2 Related Works

2.1 Tracking Paradigms

Recent deep-learning-based single-object trackers involve either CNN or ViT backbones. CNN paradigm has been dominated by the Siamese architecture since their introduction in [1,25]. The basic idea of cross-correlation in the embedding space employed in [1] is employed in many variants [6,16,17,28]. Siamese trackers aim to embed the template object and the search image in a well-chosen space where the cross-correlation between the embedded representations results in a score map representing the object's location. Finally, post-processing techniques such as the cosine window penalty have to be adopted to calculate the bounding box coordinates from the correlation score map. Despite showing remarkable success, this paradigm had several drawbacks. First, the simple cross-correlation operation assumes high-quality feature embeddings, and this is not guaranteed by simple CNN as AlexNet [15] or VGGNet [20] that were used as the feature extractor. Second, the complex post-processing steps required in the bounding box coordinates calculation phase may hinder the performance if its parameters are not chosen optimally. In our work, we hypothesize that bounding box coordinates must be learned as a part of an end-to-end learning framework without a hand-crafted set of thresholds or parameters to achieve the robustness and accuracy of bounding box calculation. Our method not only learns the coordinates of the bounding box directly but also the receptive field of the bounding box CNN, thus, resulting in extracting more useful information from the previous phase.

The CNN paradigm was dominant until the ViT-based trackers were introduced [5,27]. These trackers reimplemented the dominant Siamese tracking with a transformer visual encoder. Other research leveraged the powerful attention capabilities to bring closer end-to-end ViT tracking. Several works simultaneously introduced joint feature extraction and relation modelling [3,7,34]. Generally, this framework assumes that the transformer is capable of extracting the visual features alongside learning the relationships between the embedded representations. After obtaining a representative feature map from the ViT, these methods apply a convolutional bounding box head to predict the bounding box coordinates mainly in the classification and regression phases. This one-stream paradigm has been dominant with many works that propose changes to ViT architecture itself to adapt to the single object tracking problem [4,11,21,29].

2.2 Convolutional Design Choices

Inception Network. The inception network was introduced in [20] to overcome the problem of the narrow receptive field of previous CNNs. The main idea behind the original paper and its variants [22] is that for a convolutional network to go deeper, its receptive field has to be wide enough to process the input on many deep stages without losing information from the early stages. For that purpose, the Inception module carries on several convolutional passes in parallel, each representing a different convolutional filter. Different outputs of the passes are concatenated together and passed to a 1×1 convolutional filter to reduce the channel dimension in a learnable manner. In this work, we apply the simplest inception module to the output of a ViT encoder from three known trackers.

Deformable Convolutions. Deformable convolution [8] is an extension of the regular convolution where the sampling positions of the convolutional filter are learnable. In this regard, the deformable convolution operation involves two runs of convolutional filters. The first run determines the offsets that the filter of the second run will use to sample values of the input feature map, which will apply the convolution operation. In another version of the deformable convolutions [36], the amplitudes of the input feature maps are modulated by a learnable mask, ensuring an extra degree of freedom. We implement a variant of the proposed Inception bounding box network with deformable convolutions to exploit the powerful learning paradigm encapsulated by the learnable sampling positions and input amplitudes.

3 Methodology

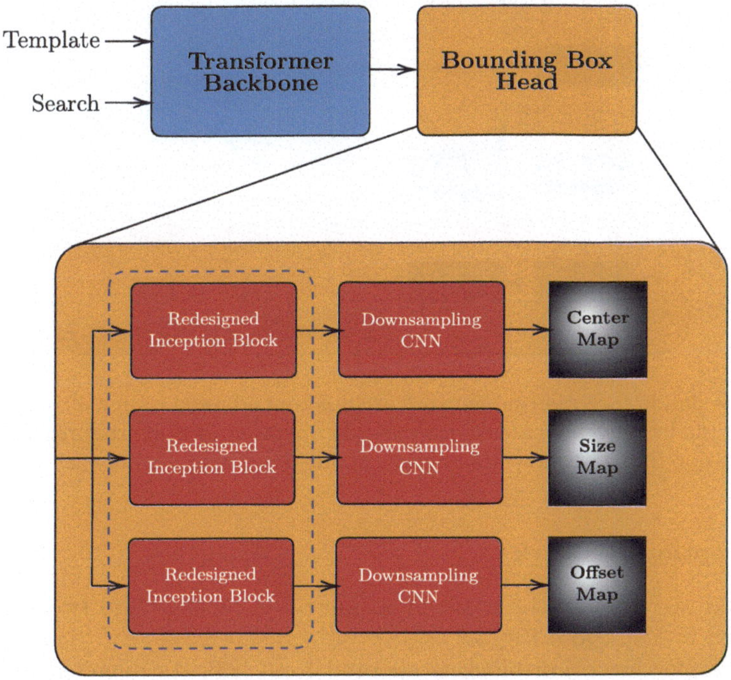

Fig. 1. The overall architecture of the recent ViT-based single object trackers.

3.1 Adopted Tracking Framework

We hypothesize that modern ViT-based single-object trackers, such as OSTrack [34] and ODTrack [35] lack good exploitation of the joint search and template embeddings to extract the location information.

Given a ViT-based backbone, the output after the joint feature extraction and relation modelling phase is an embedding tensor of dimensions $(B, H \times W, D)$ where B is the batch size, $H \times W$ is the dimension of the transformer patch, and D is the embedding dimension. For the ViT-based trackers represented by the three selected trackers, this output tensor is subjected to a CNN-based bounding box regression network that extracts three score maps representing the center location, the size, and the offset of the center of the bounding box as shown in Fig. 1. Given that the input tensor might contain global correlation as well as local correlation, we hypothesize it is essential that the bounding box network has multiple receptive fields that enable the network to capture relationships on both the global and local levels.

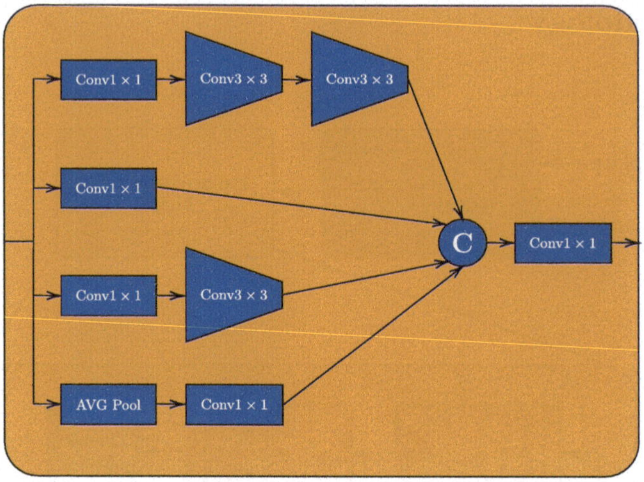

Fig. 2. The simple Inception subnetwork architecture adapted from [23]. The blue-filled circle with the letter C indicates concatenation operation. Filter sizes are shown in the figure. Please refer to [23] for more details about the parameters of each convolution.

3.2 Inception Network

To this end, we propose the usage of an Inception block [24] in two alternatives as depicted in Fig. 2 and Fig. 3.

The simpler inception block is adopted from [23]. As depicted in Fig. 2, it is composed of four parallel convolutional paths. The four convolutional paths share the functionality of mapping the input feature map to a new feature space while each of them has a different receptive field. Namely, while a bigger filter size of 7×7 that is realized by the two 3×3 convolutions shown in the uppermost branch foster a more global understanding of attentional relationships in the input tensor, the smaller ones (1×1, 3×3) focus on a local context. After feeding the embedding tensor into the convolutional layers, the obtained outputs are concatenated and fed into a 1×1 convolutional filter to reduce the number of channels to the original input number of channels, hence, speeding up the computations.

The more sophisticated inception block is based on the same idea but with a different implementation. Given the intrinsic adaptive receptive field that Deformable Convolutional Networks have [36], we design a deformable-based inception block of convolutions that have filter size of 3×3 where each filter performs deformable convolution operation and the final output is concatenated and fed to a regular 1×1 convolution as shown in Fig. 3. The deformable convolutions enable a larger receptive field for the same filter size by learning the sampling positions of the input. In addition to the deformable convolution, we add a branch of regular convolution to consider local contiguous context, therefore simplifying the learning objective and decreasing convergence time.

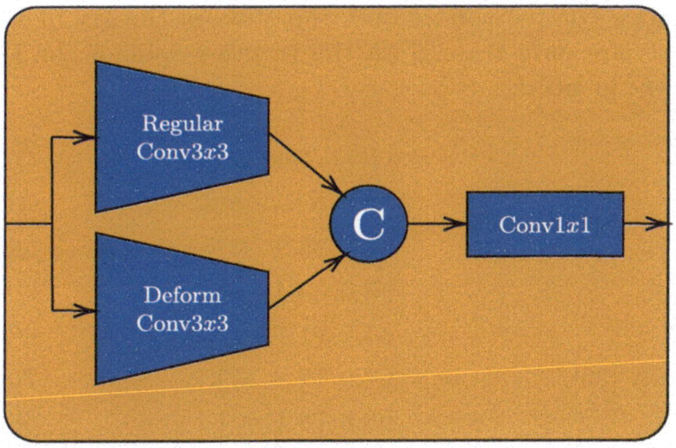

Fig. 3. The deformable inception block.

4 Experiments

4.1 Implementation Details

We adopt the ViT-base [10] as the visual encoder with MAE pretraining weights [13] to train OSTrack [34] and ODTrack [35] using our proposed bounding box regression networks. The template image shape is $192 \times 192 \times 3$ while the search image shape is $384 \times 384 \times 3$. All models are trained for 100 epochs solely on the GOT-10k dataset [14]. Training is conducted on a server with two 50 GB GPUs. The batch size is set to 8. AdamW [18] is used to optimize the training objective. We sample 40000 examples per epoch to train networks based on ODTrack following the configurations found in their code repository, while the number of samples per epoch for OSTrack-based networks is 60000. We keep the learning rate and the weight decay that each model was originally trained with. We report the performance of the two proposed variants, namely the Inception module and the Deformable Inception module. The basic convolutional block used in both of them is a stack of conv, ReLU, and BatchNorm. The filter details of the convolutions for each variant are shown in Fig. 2 and Fig. 3.

4.2 State of the Art Comparison

We compare our method to 14 methods on three leading datasets, GOT-10k [14], UAV123 [19] and OTB2015 [30]. The results shown in Table 1 indicate that our method variants outperform the state-of-the-art techniques on the three datasets when compared with retrained models on the same machines.

GOT-10k. Generic Object Tracking dataset is a large-scale single object tracking dataset. It contains 9334 training sequences, 180 validation sequences and 180 test sequences. The testing protocol of the dataset requires models to be trained only on the GOT-10k training sequences. The training and test sequences are not

Table 1. Performance comparison against the state-of-the-art methods. We emphasize that our models are only trained on the training split of the GOT-10k dataset. Significant results are in **bold**

Method	GOT-10k			UAV123	OTB2015
	AO	$SR_{0.5}$	$SR_{0.75}$	AUC	AUC
SiamFC [1]	34.8	35.3	9.8	46.5	58.2
ATOM [9]	55.6	63.4	40.2	65.0	-
DiMP [2]	61.1	71.7	49.2	65.4	68.4
SiamRCNN [26]	64.9	72.8	59.7	64.9	70.1
Stark [33]	68.8	78.1	64.1	69.1	68.5
SBT-B [31]	69.9	80.4	63.6	-	70.9
Mixformer [7]	70.7	80.0	67.8	70.4	-
OSTrack [34]	73.7	83.2	70.8	**70.7**	-
AiATrack [11]	69.6	80.0	63.2	70.6	69.6
SeqTrack [4]	74.5	84.3	71.4	68.6	-
GRM [12]	73.4	82.9	70.4	70.2	-
VideoTrack [32]	72.9	81.9	69.8	69.7	-
ARTrack [29]	75.5	84.3	74.3	70.5	-
ODTrack-B [35]	77.0	**87.9**	75.1	-	**72.3**
ODTrack-B (Our machine)	75.6	86.1	72.7	65.8	68.5
ODTrack-B-Inception (Ours)	**77.3**	87.5	**75.5**	65.8	69.0

Table 2. Investigation of the performance of the two proposed bounding box networks on two modern trackers. Significant results are in **bold**. All trackers are retrained from MAE [13] pretraining weights on our machine using the same configurations provided in their repositories for fair comparison.

Method	GOT-10k			UAV123	OTB2015
	AO	$SR_{0.5}$	$SR_{0.75}$	AUC	AUC
OSTrack-B	70.1	79.0	66.3	65.79	66.68
OSTrack-B-Inception (Ours)	71.6	80.7	66.9	66.01	68.18
OSTrack-B-deformable (Ours)	70.8	79.8	67.1	**67.78**	66.98
ODTrack-B	75.6	86.1	72.7	65.79	68.46
ODTrack-B-Inception (Ours)	**77.3**	**87.5**	**75.5**	65.79	**69.03**
ODTrack-B-deformable (Ours)	76.8	87.2	73.3	64.30	68.89

Table 3. Study of the performance of ODTrack with deformable convolution Inception module. "deformable only" refers to the deformable branch without the regular convolution while "Conv only" refers to the *original* ODTrack [35]. "Both" refers to using the bypass branch. Significant results are in **bold**.

Method	GOT-10k		
	AO(%)	$SR_{0.5}$(%)	$SR_{0.75}$(%)
deformable only	75.4	86.6	**73.3**
Conv only	75.6	86.1	72.7
Both	**76.8**	**87.2**	**73.3**

of overlapping classes. As reported in Table 1, our best performant variant outperforms the state-of-the-art ODTrack-base [35] trained on our machine by 1.7% in Average Overlap (AO). Moreover, for the Success Rate at 0.5 IoU threshold ($SR_{0.5}$) and the ($SR_{0.75}$), our model outperforms ODTrack by 1.4% and 2.8% respectively.

UAV123. UAV123 is a aerial viewpoint video dataset that contains 123 challenging sequences. It contains training and testing sequences. However, we only use the testing set for evaluation purposes without training on the training set. According to the shown results in Table 1, our method outperforms the aligned state-of-the-art techniques while it is only trained on the GOT-10k training split.

OTB2015. Object Tracking Benchmark [30] is a pioneering testing set for single object tracking that consists of 100 challenging frames collected from YouTube. The dataset labels are publicly available. The performance pattern on this dataset is similar to the one in the GOT-10k as they are both datasets of the same natural image domain.

4.3 Ablation Study

We examine the effect of the proposed simple Inception and deformable convolution Inception on the two adopted modern trackers stated in Sect. 3. As reported in Table 2, our Inception and deformable inception methods outperform the locally trained models on the GOT-10k test set considering the Average overlap by 1.5% and 0.6% respectively. The pattern is consistent despite the different tracking ideas implemented in the different modules, indicating the effectiveness of the inception and deformable inception convolutional networks in learning a suitable receptive field to process the joint search and template embeddings. Notably, the pure Inception is more powerful enhancing the performance on all three datasets. This indicates the higher learning capacity of the inception module compared to the deformable inception. We also investigate the effect of the skip branch that has the 3×3 convolutional filter of the deformable convolution Inception module in Table 3. It can be shown that the adopted variant is the best performant deformable module indicating the importance of the 3×3 convolution skip connection in facilitating the learning objective and increasing model capacity.

4.4 Visualization

We qualitatively visualize the output of our two bounding box regression networks in Fig. 4. As depicted, the regular Inception has a better bounding box quality. This indicates the high model capacity of the inception module that enables the best exploitation of the visual features given by the ViT in ODTrack [35].

Fig. 4. Success cases of the proposed methods on the OTB2015 Basketball sequence. It can be noted that the pure Inception network is most accurate especially when comparing the first column of frames. Although bounding boxes converge at frame #50, the initially predicted bounding boxes at frame #2 indicate the fast adaptation of the Inception bounding box network.

5 Conclusion

In this work, we show the significance of the bounding box regression given the visual interaction features of a one-stream tracking transformer. We show that a bounding box-regression network lacking a suitable receptive field can hinder the performance of a transformer tracking model due to the lack of good exploitation of the visual features provided by the backbone. Experiments show that our methods of Inception and deformable inception bounding box networks perform favourably against state-of-the-art ODTrack by 1.7% of AO on the GOT-10k [14] dataset. Although superior to previous state-of-the-art, our method has limitations. One limitation is that the capacity of the convolutional layers added to the existing bounding box is limited. Another limitation is that the method adds a computational overhead. As future pathways, this work can be extended to ablate the different emerging CNN techniques and find the CNN bounding box network that strikes the balance between a good inductive bias and a global receptive field.

Disclosure of Interests. The authors have no competing interests to declare that are relevant to the content of this article.

References

1. Bertinetto, L., Valmadre, J., Henriques, J.F., Vedaldi, A., Torr, P.H.: Fully-convolutional Siamese networks for object tracking. In: Computer Vision–ECCV 2016 Workshops: Amsterdam, The Netherlands, 8–10 and 15–16 October 2016, Proceedings, Part II 14, pp. 850–865. Springer, Cham (2016)
2. Bhat, G., Danelljan, M., Gool, L.V., Timofte, R.: Learning discriminative model prediction for tracking. In: Proceedings of the IEEE/CVF International Conference on Computer Vision (ICCV) (2019)
3. Chen, B., et al.: Backbone is all your need: a simplified architecture for visual object tracking. In: Avidan, S., Brostow, G., Cissé, M., Farinella, G.M., Hassner, T. (eds.) Computer Vision - ECCV 2022, pp. 375–392. Springer, Cham (2022)
4. Chen, X., Peng, H., Wang, D., Lu, H., Hu, H.: Seqtrack: sequence to sequence learning for visual object tracking. In: Proceedings of the IEEE/CVF Conference on Computer Vision and Pattern Recognition (CVPR), pp. 14572–14581 (2023)
5. Chen, X., Yan, B., Zhu, J., Wang, D., Yang, X., Lu, H.: Transformer tracking. In: Proceedings of the IEEE/CVF Conference on Computer Vision and Pattern Recognition, pp. 8126–8135 (2021)
6. Chen, Z., Zhong, B., Li, G., Zhang, S., Ji, R.: Siamese box adaptive network for visual tracking. In: 2020 IEEE/CVF Conference on Computer Vision and Pattern Recognition (CVPR), pp. 6667–6676 (2020). https://doi.org/10.1109/CVPR42600.2020.00670
7. Cui, Y., Jiang, C., Wang, L., Wu, G.: Mixformer: end-to-end tracking with iterative mixed attention. In: Proceedings of the IEEE/CVF Conference on Computer Vision and Pattern Recognition, pp. 13608–13618 (2022)
8. Dai, J., et al.: Deformable convolutional networks. In: 2017 IEEE International Conference on Computer Vision (ICCV), pp. 764–773 (2017). https://doi.org/10.1109/ICCV.2017.89
9. Danelljan, M., Bhat, G., Khan, F., Felsberg, M.: Atom: accurate tracking by overlap maximization. In: 2019 IEEE/CVF Conference on Computer Vision and Pattern Recognition (CVPR), pp. 4655–4664. IEEE Computer Society, Los Alamitos, CA, USA (2019). https://doi.org/10.1109/CVPR.2019.00479. https://doi.ieeecomputersociety.org/10.1109/CVPR.2019.00479
10. Dosovitskiy, A., et al.: An image is worth 16 × 16 words: transformers for image recognition at scale. In: International Conference on Learning Representations (2021). https://openreview.net/forum?id=YicbFdNTTy
11. Gao, S., Zhou, C., Ma, C., Wang, X., Yuan, J.: Aiatrack: attention in attention for transformer visual tracking. In: Avidan, S., Brostow, G., Cissé, M., Farinella, G.M., Hassner, T. (eds.) Computer Vision - ECCV 2022, pp. 146–164. Springer, Cham (2022)

12. Gao, S., Zhou, C., Zhang, J.: Generalized relation modeling for transformer tracking. In: Proceedings of the IEEE/CVF Conference on Computer Vision and Pattern Recognition (CVPR), pp. 18686–18695 (2023)
13. He, K., Chen, X., Xie, S., Li, Y., Dollár, P., Girshick, R.: Masked autoencoders are scalable vision learners. In: Proceedings of the IEEE/CVF Conference on Computer Vision and Pattern Recognition, pp. 16000–16009 (2022)
14. Huang, L., Zhao, X., Huang, K.: Got-10k: a large high-diversity benchmark for generic object tracking in the wild. IEEE Trans. Pattern Anal. Mach. Intell. **43**(5), 1562–1577 (2021). https://doi.org/10.1109/TPAMI.2019.2957464
15. Krizhevsky, A., Sutskever, I., Hinton, G.E.: Imagenet classification with deep convolutional neural networks. In: Pereira, F., Burges, C., Bottou, L., Weinberger, K. (eds.) Advances in Neural Information Processing Systems, vol. 25. Curran Associates, Inc. (2012)
16. Li, B., Wu, W., Wang, Q., Zhang, F., Xing, J., Yan, J.: SiamRPN++: evolution of Siamese visual tracking with very deep networks. In: 2019 IEEE/CVF Conference on Computer Vision and Pattern Recognition (CVPR), pp. 4277–4286. IEEE Computer Society, Los Alamitos, CA, USA (2019). https://doi.org/10.1109/CVPR.2019.00441. https://doi.ieeecomputersociety.org/10.1109/CVPR.2019.00441
17. Li, B., Yan, J., Wu, W., Zhu, Z., Hu, X.: High performance visual tracking with Siamese region proposal network. In: 2018 IEEE/CVF Conference on Computer Vision and Pattern Recognition, pp. 8971–8980 (2018). https://doi.org/10.1109/CVPR.2018.00935
18. Loshchilov, I., Hutter, F.: Decoupled weight decay regularization. In: International Conference on Learning Representations (2019). https://openreview.net/forum?id=Bkg6RiCqY7
19. Mueller, M., Smith, N., Ghanem, B.: A benchmark and simulator for UAV tracking. In: Leibe, B., Matas, J., Sebe, N., Welling, M. (eds.) Computer Vision - ECCV 2016, pp. 445–461. Springer, Cham (2016)
20. Simonyan, K., Zisserman, A.: Very deep convolutional networks for large-scale image recognition. In: Bengio, Y., LeCun, Y. (eds.) 3rd International Conference on Learning Representations, ICLR 2015, San Diego, CA, USA, 7–9 May 2015, Conference Track Proceedings (2015). http://arxiv.org/abs/1409.1556
21. Song, Z., Yu, J., Chen, Y., Yang, W.: Transformer tracking with cyclic shifting window attention. In: 2022 IEEE/CVF Conference on Computer Vision and Pattern Recognition (CVPR), pp. 8781–8790. IEEE Computer Society, Los Alamitos, CA, USA (2022). https://doi.org/10.1109/CVPR52688.2022.00859. https://doi.ieeecomputersociety.org/10.1109/CVPR52688.2022.00859
22. Szegedy, C., Vanhoucke, V., Ioffe, S., Shlens, J., Wojna, Z.: Rethinking the inception architecture for computer vision. In: 2016 IEEE Conference on Computer Vision and Pattern Recognition (CVPR), pp. 2818–2826. IEEE Computer Society, Los Alamitos, CA, USA (2016). https://doi.org/10.1109/CVPR.2016.308. https://doi.ieeecomputersociety.org/10.1109/CVPR.2016.308
23. Szegedy, C., Ioffe, S., Vanhoucke, V., Alemi, A.A.: Inception-v4, inception-resnet and the impact of residual connections on learning. In: Proceedings of the Thirty-First AAAI Conference on Artificial Intelligence, AAAI 2017, pp. 4278–4284. AAAI Press (2017)
24. Szegedy, C., et al.: Going deeper with convolutions. In: 2015 IEEE Conference on Computer Vision and Pattern Recognition (CVPR), pp. 1–9 (2015). https://doi.org/10.1109/CVPR.2015.7298594

25. Tao, R., Gavves, E., Smeulders, A.M.: Siamese instance search for tracking. In: 2016 IEEE Conference on Computer Vision and Pattern Recognition (CVPR), pp. 1420–1429. IEEE Computer Society, Los Alamitos, CA, USA (2016). https://doi.org/10.1109/CVPR.2016.158. https://doi.ieeecomputersociety.org/10.1109/CVPR.2016.158
26. Voigtlaender, P., Luiten, J., Torr, P.H., Leibe, B.: Siam R-CNN: visual tracking by re-detection. In: Proceedings of the IEEE/CVF Conference on Computer Vision and Pattern Recognition (CVPR) (2020)
27. Wang, N., Zhou, W., Wang, J., Li, H.: Transformer meets tracker: exploiting temporal context for robust visual tracking. In: 2021 IEEE/CVF Conference on Computer Vision and Pattern Recognition (CVPR), pp. 1571–1580. IEEE Computer Society, Los Alamitos, CA, USA (2021). https://doi.org/10.1109/CVPR46437.2021.00162. https://doi.ieeecomputersociety.org/10.1109/CVPR46437.2021.00162
28. Wang, Q., Zhang, L., Bertinetto, L., Hu, W., Torr, P.S.: Fast online object tracking and segmentation: a unifying approach. In: 2019 IEEE/CVF Conference on Computer Vision and Pattern Recognition (CVPR), pp. 1328–1338. IEEE Computer Society, Los Alamitos, CA, USA (2019). https://doi.org/10.1109/CVPR.2019.00142. https://doi.ieeecomputersociety.org/10.1109/CVPR.2019.00142
29. Wei, X., Bai, Y., Zheng, Y., Shi, D., Gong, Y.: Autoregressive visual tracking. In: 2023 IEEE/CVF Conference on Computer Vision and Pattern Recognition (CVPR), pp. 9697–9706. IEEE Computer Society, Los Alamitos, CA, USA (2023). https://doi.org/10.1109/CVPR52729.2023.00935. https://doi.ieeecomputersociety.org/10.1109/CVPR52729.2023.00935
30. Wu, Y., Lim, J., Yang, M.H.: Object tracking benchmark. IEEE Trans. Pattern Anal. Mach. Intell. **37**(9), 1834–1848 (2015). https://doi.org/10.1109/TPAMI.2014.2388226
31. Xie, F., Wang, C., Wang, G., Cao, Y., Yang, W., Zeng, W.: Correlation-aware deep tracking. In: 2022 IEEE/CVF Conference on Computer Vision and Pattern Recognition (CVPR), pp. 8741–8750. IEEE Computer Society, Los Alamitos, CA, USA (2022). https://doi.org/10.1109/CVPR52688.2022.00855. https://doi.ieeecomputersociety.org/10.1109/CVPR52688.2022.00855
32. Xie, F., Chu, L., Li, J., Lu, Y., Ma, C.: Videotrack: learning to track objects via video transformer. In: Proceedings of the IEEE/CVF Conference on Computer Vision and Pattern Recognition (CVPR), pp. 22826–22835 (2023)
33. Yan, B., Peng, H., Fu, J., Wang, D., Lu, H.: Learning spatio-temporal transformer for visual tracking. In: 2021 IEEE/CVF International Conference on Computer Vision (ICCV), pp. 10428–10437. IEEE Computer Society, Los Alamitos, CA, USA (2021). https://doi.org/10.1109/ICCV48922.2021.01028. https://doi.ieeecomputersociety.org/10.1109/ICCV48922.2021.01028
34. Ye, B., Chang, H., Ma, B., Shan, S., Chen, X.: Joint feature learning and relation modeling for tracking: a one-stream framework. In: Avidan, S., Brostow, G., Cissé, M., Farinella, G.M., Hassner, T. (eds.) Computer Vision - ECCV 2022, pp. 341–357. Springer, Cham (2022)
35. Zheng, Y., Zhong, B., Liang, Q., Mo, Z., Zhang, S., Li, X.: Odtrack: online dense temporal token learning for visual tracking (2024)
36. Zhu, X., Hu, H., Lin, S., Dai, J.: Deformable convnets v2: more deformable, better results. In: 2019 IEEE/CVF Conference on Computer Vision and Pattern Recognition (CVPR), pp. 9300–9308. IEEE Computer Society, Los Alamitos, CA, USA (2019). https://doi.org/10.1109/CVPR.2019.00953. https://doi.ieeecomputersociety.org/10.1109/CVPR.2019.00953

Blockchain-Based Multi-tier IoT-EHRs System: A Security Analysis

Ibrahim Gomaa(✉)

National Telecommunication Institute, Cairo, Egypt
igomaa@nti.sci.eg
https://www.nti.sci.eg/

Abstract. The healthcare sector faces new challenges with digital transformation, particularly in securing Electronic Health Records (EHRs) throughout their lifecycle. Blockchain technology has emerged as a solution, promising security for transactions, including those involving Internet of Things (IoT) data. In this paper the detailed multitier blockchain framework tailored for healthcare systems, ensuring trust and privacy within internal transactions is introduced. The previously proposed algorithm, Pseudonym Based Encryption with Different Authorities (PBE-DA), is used to establish virtual patient identities within the proposed multi-tier IoT-EHRs systems for privacy preservation. Unlike traditional methods using public/private key pairs, PBE-DA and Healthcare Trust Control (HTC) are utilized, integrating ripple blockchain for privacy, mutual authentication, trustworthiness, data integrity, and availability. The proposed framework minimizes computational demands by eliminating minor competition, with HTC overseeing miner activities and selecting optimal solutions. Moreover, security evaluation and threat analysis are provided to contribute toward a robust, secure, and smart healthcare system, utilizing the most common attack analysis.

Keywords: Blockchain · IoT · Healthcare

1 Introduction

Healthcare information faces threats from insiders and outsiders, necessitating patient information anonymity. Distributed Ledger Technologies (DLT), powered by Blockchain, offer shared asset databases maintaining identical copies across multiple sites. Integrating Blockchain could enhance government EHR systems by improving authentication and securely sharing records. IoT technologies provide convenience in healthcare through real-time monitoring, but insufficient security compromises patient privacy, impacting confidence and recovery rates. In this study, we will complete and revise our proposed protocol, Pseudonym-Based Encryption with Different Authorities (PBE-DA) [1]. PBE-DA enables patients to access, verify, or update their sensitive data anonymously on the EHRs system. Furthermore, we conducted validation tests to assess the

A. Abdelgawad et al. (Eds.): ISBCom 2024, LNNS 1268, pp. 28–39, 2025.
https://doi.org/10.1007/978-3-031-82377-0_3

resilience of our framework against common vulnerabilities and attacks. The rest of this work is structured as follows; two subsections introduce the problem statement and paper contributions. Section 2 highlights the relevant works. Section 3 describes the proposed system for multi-tier architecture including a description of the three tiers and execution of the framework in different scenarios. Section 4 introduces the framework security analysis and the privacy model of the proposed framework. Section 5 presents a discussion to highlight the importance of the HTC unit and its effective role in mitigating attacks. Finally, Sect. 6 highlights the conclusion and future directions.

1.1 Problem Statement

The IoT and its subset, the Internet of Medical Things (IoMT), are transforming daily life, particularly in healthcare with EHR systems. However, cyber threats to these systems pose risks to medical information disclosure and human safety. This work aims to address two key challenges: establishing a trust model for robust IoT node identification and authentication and ensuring patient information confidentiality from healthcare providers and authorities.

1.2 Paper Contributions

This work proposes a comprehensive Multi-tier IoT-EHRs framework based on DLT to conceal patient information from disclosure by healthcare providers and authorities. Furthermore, the framework's validation against common vulnerabilities and attacks is conducted.

1.3 Comparative Study with the Previous Work

In the previous work [1], a novel protocol namely Pseudonym Based Encryption with Different Authorities (PBE-DA) Was proposed to achieve perfect privacy preservation for the patient in the multi-tier Blockchain-based PBE-DA framework. Therefore, PBE-DA is used to help the patient anonymously access, check, or update his sensitive data on eHealth Records (EHRs) system. We explained in general the proposed framework and validated the proposed algorithm using Multi-precision Integer and Rational Arithmetic C/C++ (MIRACL library). In this work, we extend the previous work by completing a detailed and comprehensive Multi-tier IoT-EHRs framework based on DLT to conceal patient information from disclosure by healthcare providers and authorities. Furthermore, implement our previous work by validating the framework against common vulnerabilities and attacks.

2 Related Works

The cyber threats landscape is almost continuous in threat types, scope, surface, or impact on the business especially, the healthcare domain as vital in its national security aspect. Blockchain-based security solution was introduced as a strong and lightweight security measure for securing eHealth systems. Moreover, applying blockchain in healthcare applications was considered in many works. The review introduced in [2] summarized the potential situation of using blockchain in healthcare systems with more insights on the limitations and solved research points in healthcare applications by many researchers based on blockchain. The Ancile framework proposed in [3] for securing EHRs is based on blockchain. In this framework, a blockchain-based architecture is used to achieve high security in an eHealth system using smart contracts in an Ethereum-based blockchain. In addition, the smart contract was used to assure the security of access control for different playmakers in such systems. This happened due to no need for users' incentives (patients) as accustomed to the normal currency-based exchange like Bitcoin systems. Another differentiation in this framework such as Ancile is considered a kind of blockchain-based consensus algorithm rather than a proof of work concept as known in most commercialized blockchain solutions. The smart contracts-based public blockchain implementation for IoT systems is called (bubbles of trust) [4]. This system is considered a trust zone of communication between groups of objects. The trusted zone binds a group of IoT devices under the virtual zone concept boundary by the members of transactions. For private blockchain considerations, ModelChain [5] is a new framework that considers machine learning with private blockchain networks. This predictive model for privacy assurance in the healthcare system depends on a new design called the proof-of-information concept. The same framework merged the use of blockchain with machine learning to enhance privacy-preserving in healthcare data. The multilayer blockchain architecture introduced in [6] for integrating IoT with blockchain. The mentioned integration can overcome many security challenges in medical and IoT fields. Their architecture consists of three layers; the first one targets patient IoT nodes, the second layer targets medical system authorities, and the third one targets the cloud for storage providers' computations.

3 Proposed Framework

The IoT-EHRs framework comprises three tiers as shown in Fig. 1 and Fig. 2: lower tier for data collection and processing, middle tier with public blockchain and Healthcare Trust Control (HTC), and upper tier for cloud services. The lower tier includes Body Area Sensor Network (BSN), Gateway (GW), and Local Processing Unit (LPU). BSN connects to LPU via GW, which links to healthcare authorities, HTC, and the public Blockchain network. Healthcare authorities bridge public Blockchain and HTC, while HTC connects healthcare provider ends. The framework is scalable to accommodate large numbers of patients.

3.1 Edge Tier

The first tier (edge tier) is divided into three different layers, BSN, GW, and LPU.

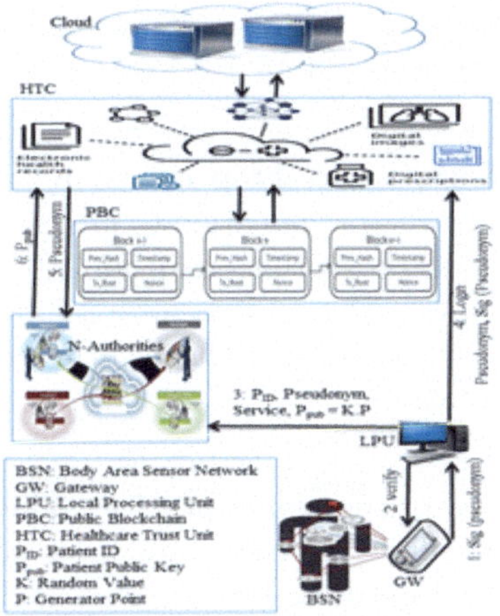

Fig. 1. IoT-EHRs Framework

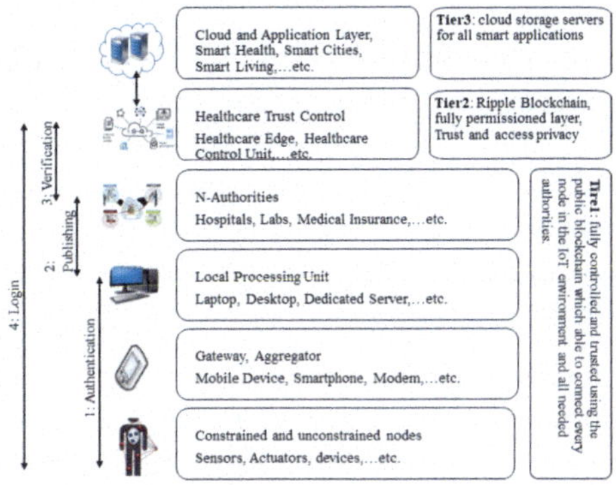

Fig. 2. Mutli-tier IoT-EHRs Framework

Body Area Sensor Network (BSN). The first layer is Body Area Sensor Network (BSN) which is dedicated to collecting information such as blood pressure, and temperature, querying location, weight, and motion from sensors. In the proposed framework, we consider a certain number of wearable body sensors measuring physiological signs for one of the patients [7]. The nodes in BSN are considerably power-constrained and have limited processing power [8]. These nodes send the patient's data to a nearby gateway (smartphone) using one of the wireless personal area network protocols such as Bluetooth or ZigBee protocol [9].

Gateway (GW). In the first tier's second layer, the gateway (GW) is represented by the patient's smartphone. It securely collects data from BSN nodes using various wireless protocols and transmits it to the Local Processing Unit (LPU). The smartphone doesn't aggregate or process data but performs lightweight cryptographic operations and transmits data to the LPU using IoT standard protocols like NB-IoT [9]. Public key encryption is limited by the processing power of wireless BSN nodes [10]. Without a trusted third party, confirming public/private keys is impractical. Symmetric key encryption is susceptible to attacks like the man-in-the-middle (MITM) attack due to vulnerable key distribution [11].

BSN to GW Authentication. IoT nodes in BSN have power constraints. Therefore, dedicated protocols and algorithms are proposed for mutual authentication between the two entities. Authors in [12] have designed a mutual authentication algorithm by integrating Proximity User Authentication (PUA) which was proposed by [12–14] and keyed-Hash Message Authentication Code (HMAC). PUA authentication approach is based on the physical distance between BSN nodes and the smartphone. Radio and voice signals were used in proximity measurement to detect the distance between the two entities [13,15,16].

Local Processing Unit (LPU). LPU is a laptop or desktop owned by a patient or maybe a dedicated server that hosts an installed application to manage and organize all patient's collected data and secure all transactions to/from LPU. in addition, LPU manages and helps in the miners' selection process for a public blockchain. Moreover, this unit can communicate directly with the healthcare cloud in case healthcare authorities require the data.

N-Authorities. The n-Authorities component comprises representative nodes representing all involved parties in the medical field who possess an interest in patient-related data, including hospitals, medical entities, laboratories, and other relevant entities.

3.2 Middle Tier

The primary role of the second tier within the proposed framework is to achieve privacy preservation. This tier, known as the middleware tier, consists of intermediary sub-layers situated between the technology and application levels. Its core function lies in concealing diverse information to safeguard against the unauthorized disclosure of patient data by third parties. Within this structure, the blockchain decision units, data analytics, and application support layers collectively form integral components of the HTC.

Healthcare Trust Control (HTC). Within the Healthcare Trust Chain (HTC), all requests from nodes undergo a safety verification process before being granted permission for communication with third parties. Additionally, the HTC layer supports a lightweight data infrastructure. The registration process, implemented through Ripple Chain (RBC) technology, involves detailed steps for verifying and validating node requests:

The Sender Node Initiates its Request to the HTC. This request undergoes multiple authentication stages before being executed within a controlled communication environment and then securely stored as a transaction in RBC. This process ensures both privacy and reliability.

Transactions Recorded on the Ripple Chain. are confirmed within seconds, incurring minimal costs in the process.

The Public Blockchain Validates Requests from Nodes. while the HTC ensures trustworthy communication between them. Privacy over valid transactions is maintained through the Ripple chain, ensuring compatibility between communicating nodes.

Algorithm 1 described the process of verifying requests and transactions to ensure the trust and reliability of the HTC unit. The algorithm takes two main inputs.

Request: R = r_1, r_2, ..., r_n, here r_i represents a single request with its attributes.

Transactions: T = t_1, t_2, ..., t_m, where t_j represents a single transaction with its attributes. The output of the algorithm is the set of verified transactions: VT = tv_1, tv_2, ..., tv_k, where tv_k represents a verified transaction. The algorithm utilizes the following parameters: Public Blockchain for verifying node IDs and establishing transaction validity.

RBC: Ripple Blockchain for privacy and reliability between nodes.
AuthenticateRequest(r_i): Function to authenticate the safety of the request.
CreateTransaction(r_i): Function to create a transaction based on the request.
VerifyTransaction(t_j): Function to verify the validity of a transaction.

Algorithm 1. HTC Unit

```
Function HTC_Unit (R, T)
Verify Request:
while ri in R do
    if AuthenticateRequest(ri) returns true then then
      Create a transaction ti using CreateTransaction(ri)
      Add tj to T
    end if
end while
Verify Transactions:
while ti in T do
    if VerifyTransaction(ti)return tru then then
      Add tj to VT
    end if
end while
      Output Verified Transactions: VT
end function
```

Public Blockchain (PBC). Authentication is facilitated through the network and gateway layers. In our model, this process is achieved via the public Blockchain. The public Blockchain verifies the authenticity of node IDs and validates transactions to establish trusted nodes. Thus, end-to-end security is enabled through the public Blockchain. Furthermore, transaction legitimacy is verified to ensure that primary nodes are valid and authenticated by the public Blockchain.

Ripple Blockchain. The decentralized and immutable nature of Blockchains enhances scalability, interoperability, and reliability. Our model utilizes two Blockchain technologies: the Public Blockchain integrates IoT objects, while the Ripple Blockchain operates within an environment, ensuring privacy and trust through the Healthcare Trust Chain (HTC). Transactions in Ripple are recorded after node validation via HTC algorithms. Unlike private Blockchains, Ripple's permissions are not limited to one organization, and transactions are initiated by members, and then disseminated through validated nodes. The consensus involves trusted nodes voting on supported transactions within trusted zones.

3.3 Cloud Services and Applications Provider

The application layer serves as the interface through which users can access all the system's functionalities. Although distinct from the middleware, this layer leverages the middleware's capabilities by employing established web service protocols and service composition technologies. This integration facilitates seamless interaction between distributed systems and applications, ensuring optimal functionality and usability.

3.4 Executing the Proposed Framework in Different Scenarios

The implementation of the suggested framework across various scenarios entails incorporating additional sub-routines as detailed below:

Scenario 1: New EHR Record Creation. Sensors (devices) may need to undergo an initial registration process to fulfill block attributes, including the hash of the previous block, patient ID, pseudonym, service, and public key. Once registration is completed, the sensor can generate the patient's private key. When the sensor needs to transmit data, it must authenticate itself to the patient (aggregator - gateway). Therefore, the sensor publishes the signature of its pseudonym to the patient. The patient, possessing the public key corresponding to the private key used for signature creation, will then accept the transmitted data, and authenticate the sender.

Scenario 2: Patient Interacts with an Authority. Devices, such as sensors, register initially to fulfill block attributes. This includes patient ID, pseudonym, and public key. After registration, sensors can generate the patient's private key. When transmitting data, sensors authenticate themselves through an aggregator or gateway by publishing their pseudonym's signature to the patient, who verifies the sender's authenticity using the corresponding public key.

4 Security Analysis

The IoT-EHRs framework integrates Internet of Things (IoT) technologies with EHR systems, presenting both opportunities and challenges in terms of security.

This section provides a detailed analysis of the framework's security aspects, focusing on its components, privacy model, vulnerabilities, and mitigation strategies [17,18].

4.1 Privacy Model of the Framework

The privacy model of the IoT-EHRs framework employs cryptographic techniques to protect patient data. Encryption and decryption functions are denoted as E and D respectively, with patient data represented as P. The Healthcare Trust Control (HTC) unit utilizes cryptographic protocols (C), integrity check function (I), and audit function (A). The HTC unit mitigates attacks through cryptographic protocols. Let V initial be the initial number of vulnerabilities and VHTC be the number of vulnerabilities mitigated by the HTC unit.

Scenario 1: Mutual Authentication Attack. In this scenario, an attacker attempts to impersonate a Body Area Sensor Network (BSN) node or Gateway (GW) to gain unauthorized access as shown in Eq. 1, Eq. 2, and Eq. 3.

$$S_{mutual} = P(BSNnodeisauthentic) \times P(Gatewayisauthentic) \tag{1}$$

$$P(BSNnodeisauthentic) = \frac{NumberofauthenticBSNnodes}{TotalnumberofBSNnodes} \tag{2}$$

$$P(Gatewayisauthentic) = 1 \tag{3}$$

Theorem 1. *The probability of S_{mutual} approaches 1 as the number of authentic BSN nodes increases relative to the total number of BSN nodes.*

Proof. Assuming N as the total number of BSN nodes and N_g as the number of genuine BSN nodes, we have:

$$P(BSNnodeisauthentic) = \frac{N_g}{N} \tag{4}$$

Scenario 2: Privacy Breach via Unauthorized Access. In this scenario, an attacker attempts to gain unauthorized access to patient data stored within the framework as explained in eq.5, eq.6.

$$S_{privacy} = P(Unauthorizedaccesstopatientdata) \tag{5}$$

$$P(Unauthorizedaccesstopatientdata) = 1 - P(Authorizedaccesstopatientdata) \tag{6}$$

Theorem 2. *The probability of SPrivacy approaches 0, indicating a significant decrease in the probability of unauthorized access to patient data.*

Proof. Let's denote the probability of authorized access to patient data as P (Authorized access to patient data) = p. Then, the probability of unauthorized access to patient data is given by. P (Unauthorized access to patient data) = 1 - p. As the security measures of the framework improve, p approaches 1, leading to Sprivacy approaching 0.

Scenario 3: Data Tampering Attack. In this scenario, an attacker attempts to tamper with patient data stored within the framework as explained in eq.7, eq.8.

$$S_{tamper} = P(Detectionoftampering) \tag{7}$$

$$P(Detectionoftampering) = 1 - P(Successfultampering) \tag{8}$$

Theorem 3. *The probability Stamper approaches 1, indicating a high probability of detecting tampering.*

Proof. Let's denote the probability of successful tampering as P (Successful tampering) = p. Then, the probability of detecting tampering is given by P (Detection of tampering) = 1 − p. As the security measures of the framework improve, p approaches 0, leading to S_{tamper} approaching 1.

4.2 Confidentiality and Data Integrity

PBE-DA implemented and designed based on the Elliptic Curve Digital Signature Algorithm (ECDSA), the Elliptic Curve Diffie-Hellman (ECDH) key agreement protocol, and the Menezes-Vanstone Protocol using Multi-precision Integer and Rational Arithmetic C/C++ (MIRACL library). Therefore, in the proposed framework, data integrity is achieved through the exchange of encrypted messages among all entities. As mentioned before, HTC guarantees the verification process of all nodes before being granted permission for communication with third parties. In addition, RBC has maintained privacy over valid transactions, therefore only authorized nodes can modify data on the framework. Finally, the proposed framework is robust and resilient to any attempts aimed at integrity violation of the data processed, shared, and stored on the framework.

4.3 Availability

one of the major objectives of the proposed framework is that a service is available for authenticated users whenever needed. Therefore, an efficient architecture should be secure enough against denial of service (DoS) and DDoS (distributed denial of service) attacks. The proposed framework achieves this via the construction of the public blockchain. The capability to replicate services across multiple nodes guarantees that even if an intruder tries to block a specific node, he cannot affect service availability as it spans the entire network.

4.4 Mutual Authentication

Mutual authentication is satisfied implies the need for end-to-end authentication i.e. parties must validate credentials of each other. This safeguards the framework from spoofing. In the proposed framework IoT nodes in BSN have power constraints. Therefore, dedicated protocols and algorithms are proposed for mutual authentication between the two entities. A mutual authentication algorithm designed by integrating Proximity User Authentication (PUA) and keyed-hash Message Authentication Code (HMAC). PUA authentication approach is based on the physical distance between BSN nodes and the smartphone. Radio and voice signals were used in proximity measurement to detect the distance between the two entities.

4.5 Trustworthiness

The elliptic curve signature algorithm generates signatures that ensure the trustworthiness and seamless integration of data among members. However, in the proposed framework, the public blockchain validated all requests from all nodes, and HTC ensured trustworthy communication between them. In addition, verifying node ID and establishing transaction validity.

5 Discussion

The security analysis of the IoT-EHRs framework shows strong protection for patient data, with cryptographic techniques ensuring confidentiality and integrity. The Healthcare Trust Control (HTC) unit effectively counters potential threats, as demonstrated in scenarios that underscore the framework's resilience against attacks. Further discussion elaborates on the HTC unit's role in mitigating such threats:

5.1 Importance of HTC Unit

The HTC unit is essential in the IoT-EHRs framework, ensuring trust and security for patient data. Through cryptographic protocols and regular audits, it creates a secure environment for data exchanges. Its role in authenticating and authorizing communication between entities like BSN nodes and Gateways prevents unauthorized access and tampering.

5.2 Effective Role in Mitigating Attacks

In each scenario presented, the HTC unit plays a crucial role in mitigating potential security threats.

Mutual Authentication Attack: The HTC unit's cryptographic protocols enable mutual authentication between BSN nodes and Gateways, preventing unauthorized entities from gaining access to the network. The mathematical proof demonstrates that as the number of genuine BSN nodes increases, the probability of successful mutual authentication approaches certainty, ensuring a secure communication environment.

Privacy Breach via Unauthorized Access: Through regular auditing and cryptographic techniques, the HTC unit effectively safeguards patient data against unauthorized access. The mathematical analysis shows that as the framework's security measures improve, the probability of unauthorized access to patient data decreases significantly, ensuring patient privacy.

Data Tampering Attack: Using integrity checks and regular audits, the HTC unit prevents data tampering. Mathematical proof confirms that improved security ensures almost certain detection of tampering, maintaining data integrity. The results of the security analysis highlight the effectiveness of the IoT-EHRs framework in safeguarding patient data. The cryptographic techniques, integrity checks, and regular auditing mechanisms employed by the HTC unit provide robust protection against various security threats. The mathematical proofs presented in each scenario demonstrate the framework's resilience and the efficacy of its security measures.

In conclusion, the IoT-EHRs framework, with its integrated Healthcare Trust Control unit, offers a secure and trusted environment for managing electronic health records. By addressing potential vulnerabilities and employing advanced security measures, the framework ensures the confidentiality, integrity, and availability of patient data, thereby enhancing trust and reliability in healthcare systems.

6 Conclusion and Future Work

In this work, we have introduced a detailed novel distributed multitier blockchain framework tailored for healthcare systems. The PBE-DA, HTC, and RBC are proposed to establish virtual patient identities within multi-tier IoT-EHRs systems for privacy preservation based on blockchain technology. In addition, a detailed analysis of the framework's security aspects, focusing on its components, privacy model, mutual authentication, trustworthiness, data integrity, availability, and mitigation strategies are introduced and discussed. To further advance this work, it is imperative to conduct multiple security and performance tests focusing on mining time, difficulty variation, blockchain size, and new attacks and vulnerabilities assessment. Furthermore, the future work of this technology, particularly the proposed architecture, needs to be implemented and tested in real scenarios and deployments.

References

1. Badr, S., Gomaa, I., Abd-Elrahman, E.: Multi-tier blockchain framework for IoT-EHRs systems. Procedia Comput. Sci. **141**, 159–166 (2018)
2. Tandon, A., Dhir, A., Islam, N., Mantymaki, M.: Blockchain in healthcare: a systematic literature review, synthesizing framework and future research agenda. Comput. Ind. **122**, 103290 (2020). https://doi.org/10.1016/j.compind.2020.103290

3. Dagher, G.G., Mohler, J., Milojkovic, M., Marella, P.B.: Ancile: Privacy-preserving framework for access control and interoperability of electronic health records using blockchain technology. Sustain. Cities Soc. **39**, 283–297 (2018). https://doi.org/10.1016/j.scs.2018.02.014
4. Hammi, M.T., Hammi, B., Bellot, P., Serhrouchni, A.: Bubbles of Trust: a decentralized blockchain-based authentication system for IoT. Comput. Secur. **78**, 126–142 (2018). https://doi.org/10.1016/j.cose.2018.06.004
5. Kuo, T.-T., Ohno-Machado, L.: ModelChain: decentralized privacy-preserving healthcare predictive modeling framework on private blockchain networks. arXiv (2018). https://arxiv.org/abs/1802.01746. Accessed 13 Mar 2024
6. Chendeb, N., Khaled, N., Agoulmine, N.: Integrating blockchain with IoT for a secure healthcare digital system. In: Advance 2020, pp. 1–8, Cancún, Mexico (2020)
7. Haghi, M., Thurow, K., Stoll, R.: Wearable devices in medical internet of things: scientific research and commercially available devices. Healthcare Inform. Res. **23**(1), 4–15 (2017)
8. Wu, D., Yang, B., Wang, H., Wu, D., Wang, R.: An energy-efficient data forwarding strategy for heterogeneous WBANs. IEEE Access **4**, 7251–7261 (2016)
9. Baker, S.B., Xiang, W., Atkinson, I.: Internet of things for smart healthcare: technologies, challenges, and opportunities. IEEE Access **5**, 26,521–26,544 (2017)
10. Shen, J., Chang, S., Shen, J., Liu, Q., Sun, X.: A lightweight multi-layer authentication protocol for wireless body area networks. Future Gener. Comput. Syst. **78**, 956–963 (2018)
11. Uddin, M.A., Stranieri, A., Gondal, I., Balasubramanian, V.: Continuous patient monitoring with a patient centric agent: a block architecture. IEEE Access **6**, 32700–32726 (2018). https://doi.org/10.1109/access.2018.2846779
12. Malina, L., Hajny, J., Fujdiak, R., Hosek, J.: On perspective of security and privacy-preserving solutions in the internet of things. Comput. Netw. **102**, 83–95 (2016)
13. Gong, N.Z., et al.: Piano: proximity-based user authentication on voice-powered internet-of-things devices. Computer Networks (2017). arXiv preprint arXiv:1704.03118
14. Krawczyk, H., Canetti, R., Bellare, M.: HMAC: keyed-hashing for message authentication (1997)
15. Shafagh, H., Hithnawi, A.: Poster: come closer: proximity-based authentication for the internet of things. In: 20th International Proceedings on Annual International Conference on Mobile Computing and Networking. ACM (2014)
16. Zhang, J., Wang, Z., Yang, Z., Zhang, Q.: Proximity based IoT device authentication. In: IEEE Conference on Computer Communications, CINFOCOM 2017, pp. 1–9 (2017)
17. Al-Nbhany, W.A.N.A., Zahary, A.T., Al-Shargabi, A.A.: Blockchain-IoT healthcare applications and trends: a review. IEEE Access **12**, 4178–4212 (2024). https://doi.org/10.1109/ACCESS.2023.3349187
18. Sahoo, A., Lenka, R.K., Mallick, S.R., Tripathy, P.K.: Blockchain applications in IoT-based healthcare system-a review. In: OITS International Conference on Information Technology (OCIT), pp. 303–308 (2023). https://doi.org/10.1109/OCIT59427.2023.10430704

Investigating KAN-Based Physics-Informed Neural Networks for EMI/EMC Simulations

Kun Qian and Mohamed Kheir(✉)

Center for Industrial Electronics, Institute of Mechanical and Electrical Engineering, University of Southern Denmark, Sønderborg, Denmark
{kqian,kheir}@sdu.dk

Abstract. The main objective of this paper is to investigate the feasibility of employing Physics-Informed Neural Networks (PINNs), particularly PINNs based on Kolmogorov-Arnold Networks (KANs), for facilitating Electromagnetic Interference (EMI) simulations. This work first introduces a common electromagnetic problem and the corresponding formulation and then shows how it can be solved using AI-driven solutions instead of lengthy and complex full-wave numerical simulations. This research may open new horizons for green EMI simulation workflows with less energy consumption and feasible computational capacity.

Keywords: Electromagnetic compatibility · EMI simulations · Full-wave simulations · Physics-informed neural networks

1 Introduction

Electronic devices nowadays suffer from different electromagnetic emission problems typically within the Printed-Circuit Boards (PCBs). This is due to the fact that a single PCB can employ different connectivity standards and therefore emissions and interferences increase. A cross-section of a PCB is shown in Fig. 1 where it consists of different metallization layers and through-hole vias. Most Electromagnetic Interference (EMI) issues originate from PCBs and radiate into free-space or couple into other parts of the circuit. Simulating such PCB is needed in the early-stage of product design and can be helpful in detecting any potential problems.

Traditional EMI simulation tools are mostly full-wave solvers where their simulation engines solve Maxwell's equations numerically to obtain the electric- and magnetic-field values within the structure. This also helps in determining the critical coupling paths and the propagation mechanism within the structure.

This problem can be solved either in time or frequency domain where the complexity increases dramatically as the frequency range increases. The well-known problems of EMI simulations include:

A. Abdelgawad et al. (Eds.): ISBCom 2024, LNNS 1268, pp. 40–48, 2025.
https://doi.org/10.1007/978-3-031-82377-0_4

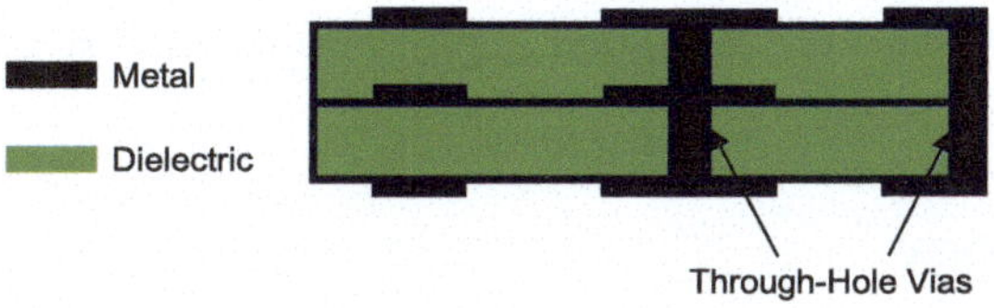

Fig. 1. A cross-section of a multi-layered PCB.

- **The long simulation time:** A simulation may last hours or days depending on the size of the problem and the PC hardware we use. In simulations we need to cover a large space exploration and test all the possibilities.
- **Computational capacity:** A typical EMC problem needs high-speed advanced CPU with high RAM capacity and additionally a number of GPU cards for simulation acceleration. This makes the simulation workstation very expensive.
- **Energy consumption:** Long computation time on a highly-equipped machine consumes a huge amount of electricity and cooling.
- **Simulation iterations:** Every modification to the EMI problems requires a new simulation to reach a solution. This is cumbersome since each iteration takes so long as declared in the first problem.

Data-driven methods, particularly those based on Machine Learning (ML), play a pivotal role in advancing the field of EMI/EMC. They enhance the understanding of coupling phenomena, guide design decisions, and improve system reliability [1,2]. For example, ML models can identify deviations from expected behavior, such as faulty components or abnormal radiation. ML can assist in designing and optimizing EMC systems, such as optimizing shielding materials [3] and grounding configurations [4].

On the other hand, Physics-Informed Machine Learning (PIML) with conventional ML techniques has emerged as a transformative force. They seamlessly integrate domain-specific physics with ML [5]. For example, in an EMC setting, it can encode Maxwell's equations and fundamental principles, bridging theory and data. Physics-Informed Neural Networks (PINNs) have achieved remarkable success in solving Partial Differential Equations (PDEs) [6]. PINNs are competitively data efficient thanks to its physics-based constraints, which can reduce dependence on computationally expensive simulations, accelerating EMI modeling. Such PINNs techniques typically require less datasets compared to traditional ML.

Multi-layer perceptrons (MLPs), which consist of neurons connected via trainable weights, are the standard building blocks for neural networks. Recent literature [7] has shown a promising alternative for MLPs: Kolmogorov-Arnold Networks (KANs), where neurons are connected via learnable activation functions. Fine-tuned KANs can lead to small networks, which can further be described with symbolic functions, providing much better interpretability. Here, we present a feasibility study of applying physics-informed KANs to solve PDEs related to the physical laws of electromagnetics.

2 Methodology

The problem under investigation is the simple 2D electrostatic box shown in Fig. 2 which can be described by the following Laplace equation

$$-\nabla^2 u\,(x, y) = 0, \tag{1}$$

where $u\,(x, y)$ is the electric potential at the location (x, y). The boundary conditions of the structure are illustrated in the same figure where the potential is zero everywhere except the excitation line ($y = 1$).

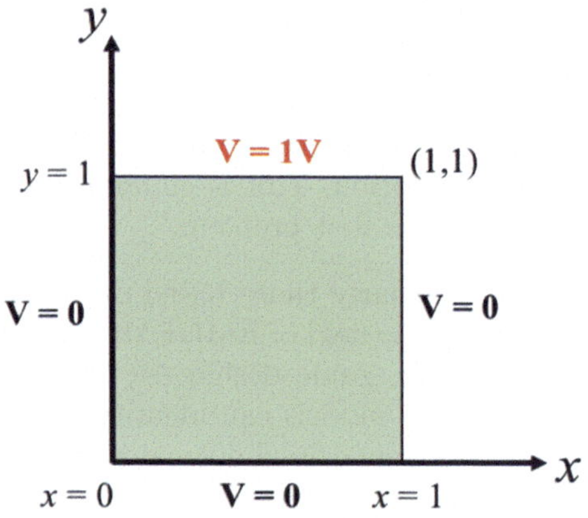

Fig. 2. A 2D electrostatic box.

We use the PINN to solve the above PDE problem, where the loss is defined as

$$\text{loss}_{\text{pde}} = \alpha \text{loss}_i + \text{loss}_b := \alpha \frac{1}{n_i} \sum_{i=1}^{n_i} |f(z_i)|^2 + \frac{1}{n_b} \sum_{i=1}^{n_b} |u(z_b)|^2 \tag{2}$$

where loss_i and loss_b denote the interior and boundary losses, respectively. The interior loss is evaluated by a random sampling of n_i points (denoted with z_i) inside the domain while the boundary loss is evaluated by a uniform sampling of n_b points (denoted with z_b) on the boundary. $f(z_i)$ is defined as $f(z_i) = -\nabla^2 u(z_i)$, and α is the hyperparameter balancing the effect of the two loss terms.

Figure 3 shows an example of KAN-based PINN methodology, where the neurons from different layers are connected with activation functions (u_{xx} and u_{yy} denote the second-order derivative with respect to x and y). In the context of KAN, the activation functions are usually different from each other and more

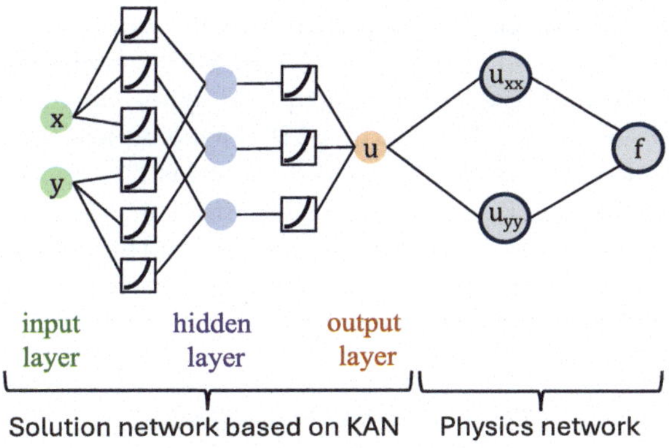

Fig. 3. Example PINN with KAN.

importantly, they are learnable. Ref. [7] mapped those activation functions to B-spline curves, which can be defined as

$$\text{spline}(x) = \sum_i c_i B_i(x), \tag{3}$$

where B_i indicates the ith basis function and c_i is the corresponding coefficient. Obviously, the more the number of grid points, which determines the number of basis functions, the more accurately the spline curves can fit the activation functions. However, it would also requires to tune more coefficients, which will slow down the training process.

Another advantage of KAN is that, we can prune the network after training and map activation functions to symbolic functions such as "cos" or "log". In the end, the KAN-based PINN can potentially represent the solution with a symbolic function.

In comparison, Fig. 4 shows an example of MLP-based PINN methodology, where the neurons from different layers are connected with learnable weights. In the context of MLP, the activation functions are often the same. Furthermore,

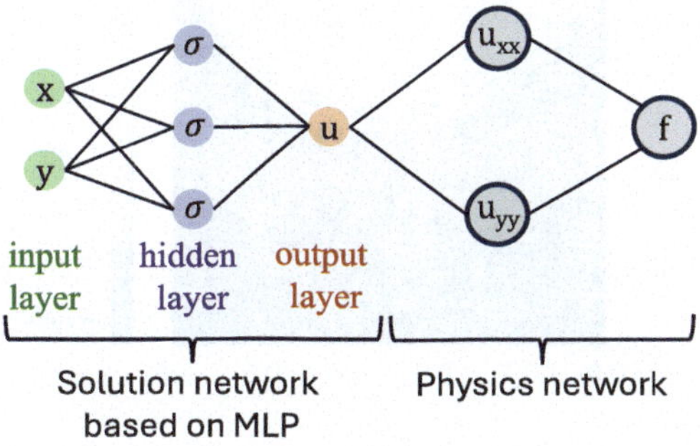

Fig. 4. Example PINN with MLP.

they are pre-determined so that it can fit specific applications. The training process will tune all the edges as in KAN. However, each edge is denoted by only one weight variable instead of multiple B-spline coefficients as in KAN. Consequently, for the same NN structure, MLP leads to less tunable parameters. On the other hand, KAN often requires less neurons [7].

For the above simple 2D electrostatic box problem, the input layer requires two neurons representing the coordinates, and the output of the neural network is comprised of one neuron representing the solution. Table 1 describes the two different types of PINNs used to solve the PDE.

Table 1. Network descriptions.

NN Type	No. Hidden Layers	No. Neurons	Activation Function
MLP	2	32,32	tanh
KAN	2	5,5	*learnable*

For training, we uniformly select 50 points on each boundary side and randomly select 2500 points inside the domain. The trained models are used to predict the solution values for a 101×101 uniform grid, as provided by the simulation software.

3 Results

The electric potential of the electrostatic box is simulated in CST Studio Suite [8] and the result is shown in Fig. 5. As seen from the figure, the distribution of the scalar potential is maximum at the center of the excitation line ($y = 1$) and starts to decay when approaching to the boundary edges where the potential are zero.

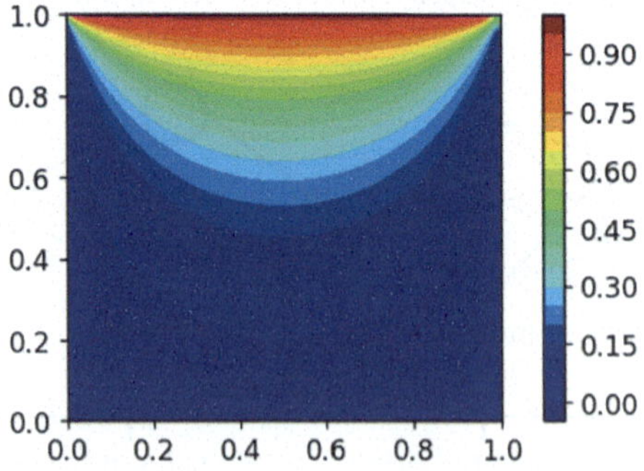

Fig. 5. Ground truth solution provided by CST Studio Suite.

Figure 6 and Fig. 7 show results from the MLP-based and KAN-based PINNs, respectively. As shown, the two PINNs can achieve quite similar results to those provided by the commercial simulation software.

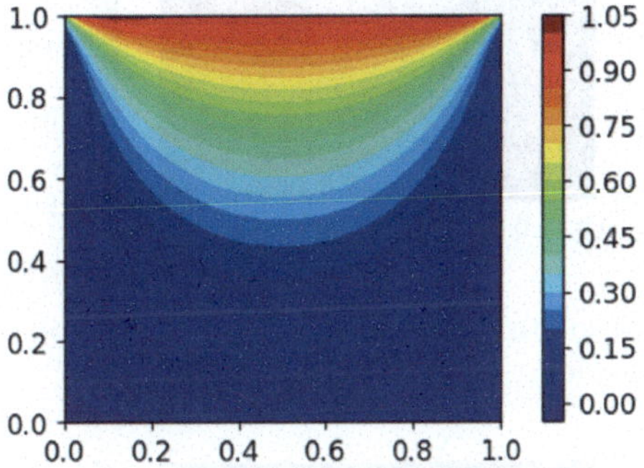

Fig. 6. Results from the MLP-based PINN.

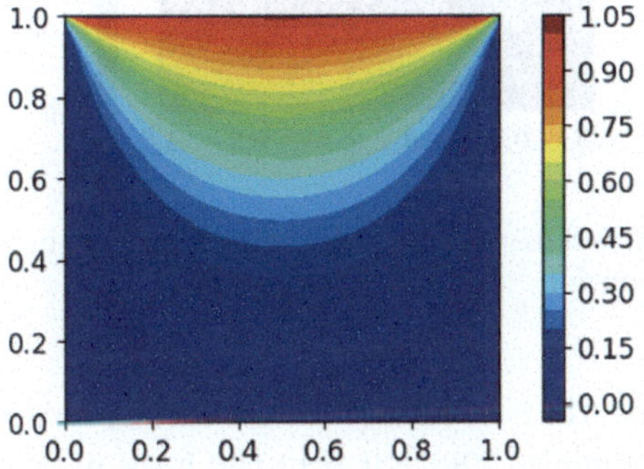

Fig. 7. Results from the KAN-based PINN.

Figure 8 and Fig. 9 show absolute errors from the MLP-based and KAN-based PINNs, respectively, compared to the solutions from the CST Studio Suite. As seen, the absolute errors are relatively high at the two top corners, where the boundary conditions are suddenly changing, which is similar to the results reported in [5].

As mentioned in Sect. 2, applying KAN has the advantage that we can potentially prune the network and the map the activation functions to symbolic functions. Unfortunately, the trained network for the 2D electrostatic box problem does not show any prunable nodes, as shown in Fig. 10. On the other hand, it

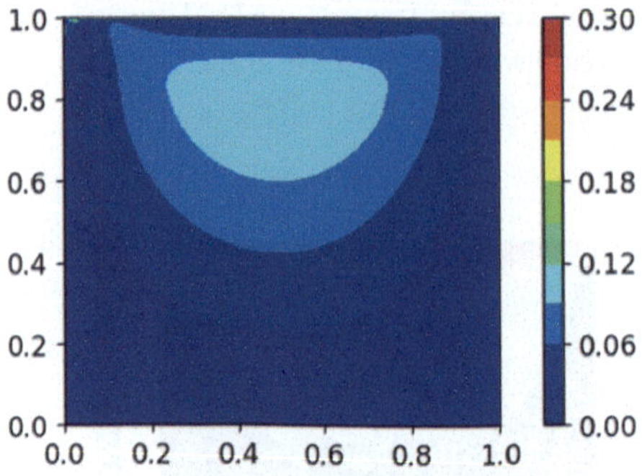

Fig. 8. Comparing the absolute error between the results from the MLP-based PINN and the ground truth solution.

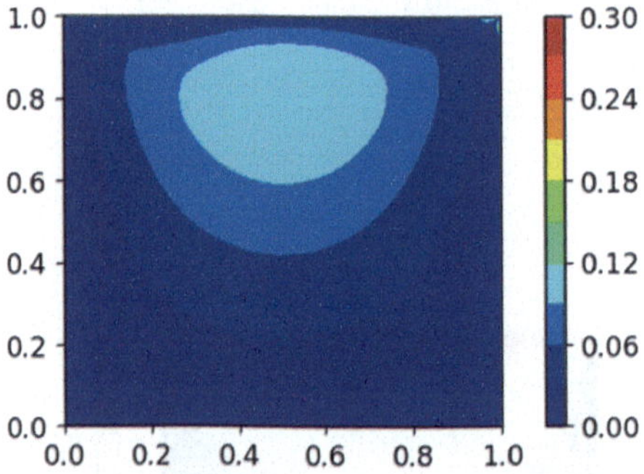

Fig. 9. Comparing the absolute error between the results from the KAN-based PINN. and the ground truth solution.

can potentially indicate that the analytic solution to this problem is relatively complex. Indeed, the analytic solution is in the form of [9]

$$u(x, y) = y + \frac{a}{\pi} \sum_{n=1}^{\infty} \alpha_n \frac{\sinh n\pi y}{n \cosh n\pi} \cos n\pi x \tag{4}$$

where

$$\alpha_n = 2 \int_0^1 \cos(n\pi x) dx, \quad n = 1, 2, 3, \ldots \tag{5}$$

Thus, an exact solution representation in KAN would require an infinite number of nodes.

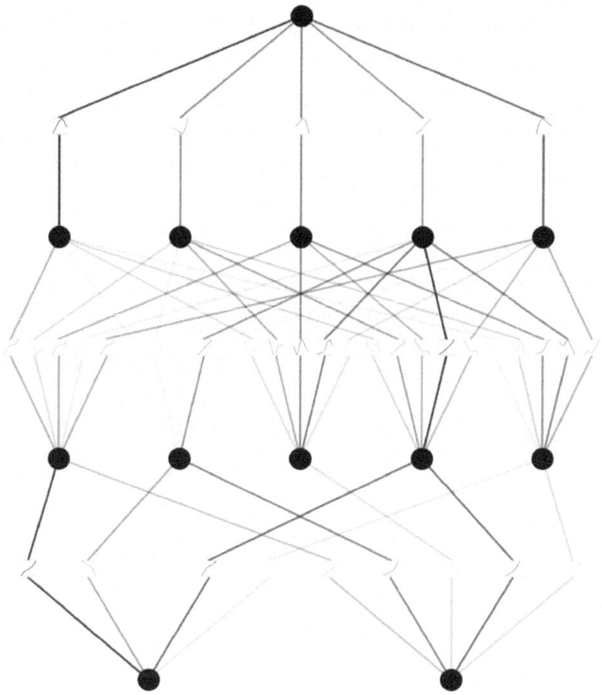

Fig. 10. Trained KAN for the simple 2D electrostatic box problem.

4 Conclusion

This study has investigated the possibility of employing AI-driven techniques for EMI/EMC simulations as a green alternative. We have demonstrated the results from both MLP- as well as KAN-based PINNs where the KAN-based PINNs can achieve the same results as MLP-based PINN, but with much fewer neurons. This can potentially reduce the training time required and introduce an energy-efficient and fast simulation workflow compared to numerical full-wave simulations. This will introduce a promising potential solution for solving EMI/EMC problems and structures that normally require complex computational capacity, high energy consumption and a long time.

References

1. Erricolo, D., et al.: Machine learning in electromagnetics: a review and some perspectives for future research. In: 2019 International Conference on Electromagnetics in Advanced Applications (ICEAA), pp. 1377–1380 (2019)
2. Sagar, M.S.I., et al.: Application of machine learning in electromagnetics: Mini-review. Electronics **10**, 2752 (2021)

3. Piersanti, S., Orlandi, A., de Paulis, F.: Electromagnetic absorbing materials design by optimization using a machine learning approach. IEEE Trans. Electromagnetic Compatibility (2018)
4. Sun, S.: Shield tunneling parameters matching based on support vector machine and improved particle swarm optimization. Sci. Program. (2022)
5. Khan, A., Lowther, D.A.: Physics informed neural networks for electromagnetic analysis. IEEE Trans. Magn. **58**(9), 1–4 (2022)
6. Raissi, M., Perdikaris, P., Karniadakis, G.E.: Physics-informed neural networks: a deep learning framework for solving forward and inverse problems involving non-linear partial differential equations. J. Comput. Phys. **378**, 686–707 (2019)
7. Liu, Z., et al.: KAN: Kolmogorov-Arnold networks, arXiv preprint arXiv :2404.19756 (2024)
8. CST Studio Suite [Simulation software] (2024). https://www.3ds.com/products/simulia/cst-studio-suite
9. Trench, W.F.: Elementary differential equations with boundary value problems (2013)

The Impact of Artificial Intelligence Techniques on Developing the Abilities of Materials Engineering Students in Problem Solving and Scientific Thinking

Alaa Abdulkhaleq Hussein[1(✉)], Ali Hilo Hawas[1(✉)], Sally Mohammed Abd[1(✉)], and Hawraa Radi Thakab[2(✉)]

[1] College of Islamic Science, University of Baghdad, Baghdad, Iraq
Alaa.Abdulkhaleq@colaw.uobaghdad.edu.iq,
ali.hilo@ircoedu.uobaghdad.edu.iq

[2] Al Karakh University of Sciences KUS, Baghdad, Iraq
hawraaradhi@kus.edu.iq

Abstract. The study aimed to identify the impact of artificial intelligence technologies and active learning on enhancing the capabilities of material engineering students in problem-solving and scientific thinking. An experimental methodology was employed utilizing a sample of 60 students divided into control and experimental groups. The findings indicated significant differences between the control and experimental groups in problem-solving and scientific thinking measures, favoring the experimental group. Specifically, the experimental group showed a 15% increase in problem-solving abilities and a 20% improvement in scientific thinking skills compared to the control group. The study recommends focusing on enhancing problem-solving and scientific thinking skills among students and highlights the importance of curriculum modifications to align with ongoing advancements in artificial intelligence technologies. These results underscore the broader significance of integrating AI in educational methodologies to foster critical skills in engineering students.

Keywords: Artificial Intelligence · Active Learning · Problem Solving · Scientific Thinking

1 Introduction

The rapid progress of computing and communication technologies has a world-wide impact on social and educational life. Transforming society through ICT and increasing the tools available for learning [1]. This includes the emergence of open-source classrooms alongside other educational innovations. In recent years, the advent of conspicuous artificial intelligence (AI) systems has provided opportunities for development in numerous fields such as health and education [2].

The rapid developments brought about by science and technology have ushered in what is known as the new industrial revolution. It is anticipated that these technologies

A. Abdelgawad et al. (Eds.): ISBCom 2024, LNNS 1268, pp. 49–61, 2025.
https://doi.org/10.1007/978-3-031-82377-0_5

will create and shape new partnerships between humans and machines to maximize benefits. In the educational sphere, artificial intelligence represents a domain of computer-based interactive learning. It refers to efforts to develop computerized information in a way that mimics human thinking. Students who have been passive recipients of knowledge may face difficulties in making decisive choices and solving complex problems they encounter and need to overcome in their academic studies in response to the information explosion we are witnessing today [3]. Traditional educational methods are insufficient for acquiring skills such as scientific thinking and problem-solving. It is noted that new learning methods that bring diverse perspectives to the learning and teaching process are necessary to acquire such crucial and contemporary skills for effective learning [4].

Modern education has struggled to connect theory with practice to integrate the educational experience, considering the learner as the focal point of the educational process. To prepare learners with desired characteristics, traditional education in classrooms is an artificial educational identity that society is advised to adopt [5].

The research problem is encapsulated in the following question: What is the impact of artificial intelligence technologies on the development of material engineering students' problem-solving abilities and scientific thinking? (Table 1).

Table 1. Reviewed Studies that are Relevance to AI in Education

Study	Year	Title	Key Findings	Relevance to AI in Education
[6]	2020	"Development and adoption of an adaptive learning system: Reflections and lessons learned"	Discusses challenges and strategies in developing and implementing adaptive learning systems	High - Focuses on practical application of AI in education
[7]	2020	"Can artificial intelligence transform higher education?"	Explores potential impacts of AI on higher education, including benefits and challenges	High - Directly addresses AI's role in transforming education
[8]	2020	"Artificial Intelligence in Special Education: A Decade Review"	Reviews AI applications in special education over a 10-year period	Medium - Provides insights on AI in specialized educational contexts
[9]	2020	"Improvement of an online education model with the integration of machine learning and data analysis in an LMS"	Demonstrates integration of machine learning in Learning Management Systems for improved online education	High - Shows practical application of AI in online learning

(continued)

Table 1. (*continued*)

Study	Year	Title	Key Findings	Relevance to AI in Education
[10]	2021	"Roles and research trends of artificial intelligence in mathematics education: A bibliometric mapping analysis and systematic review"	Identifies key roles and trends of AI in mathematics education	Medium - Focuses on AI in a specific subject area
[11]	2021	"Artificial intelligence in education: The three paradigms"	Proposes three paradigms for understanding AI in education: AI-assisted teaching, AI-assisted learning, and AI-created learning environments	High - Provides a framework for conceptualizing AI in education
[12]	2021	"A review of artificial intelligence (AI) in education from 2010 to 2020"	Comprehensive review of AI applications in education over a decade	High - Offers broad overview of AI in education
[13]	2021	"Artificial intelligence in education: A systematic literature review 2011–2021"	Systematic review of AI in education research over a decade	High - Provides comprehensive analysis of recent AI in education literature
[14]	2021	"Artificial intelligence in engineering education: Review and challenges"	Reviews AI applications specifically in engineering education and identifies challenges	High - Directly relates to AI in engineering education

Amidst rapid advancements and progress in information and communication technology, along with the emergence of artificial intelligence technologies, it has become imperative to develop students' problem-solving and scientific thinking skills to keep pace with these changes. Traditional educational methods are no longer sufficient to equip students with these essential skills. Therefore, the research problem is encapsulated in the following question: What is the impact of artificial intelligence technologies and active learning on developing the capabilities of material engineering students in problem-solving and scientific thinking?

The purpose of the current study is to understand whether there is an effect between using artificial intelligence technologies with active learning in improving the engineering students' problem-solving and scientific thinking ability. According to the relevance of the variables, the qualitative research design is employed. Specifically, this design will aim at the determination of the superiority of artificial intelligence technologies towards teaching problem-solving and scientific thinking skills for engineering students.

The paper begins with the introduction presents the background on AI advancements and their significance in education, followed by the problem statement and study objectives. The research problem section details the problem's articulation and its importance. The study objectives section outlines the specific goals, while the scope of the study defines the spatial, temporal, human, and scientific boundaries. The literature review summarizes previous studies on AI in education, identifies research gaps, and establishes the relevance of the current study. The research methodology and procedures section describe the experimental design, research population and sample, educational interventions, and data collection and analysis methods. The results section presents pre-test and post-test results, along with statistical analysis and key findings illustrated by tables and figures. The discussion interprets the results, compares them with previous studies, and explains AI's contribution to skill improvements. The conclusions summarize the main findings, discuss broader implications for educational practices, and provide recommendations for future research. The references list all cited sources, and any appendices include additional materials such as detailed statistical data.

2 Methodology

The researchers adopted an experimental approach involving experimental and control groups with pre-tests and post-tests. The control group was subjected to traditional educational methods, while the experimental group was exposed to teaching methods employing artificial intelligence technologies [15]. Experimental research focuses on experiments designed to elucidate the causal relationships between variables. It is characterized by the ability to control factors affecting the phenomenon under study, beginning with a problem that researchers need to solve by identifying the causes and circumstances that facilitate its resolution through experimental trials.

The study population consisted of third-year material engineering students at the Technology University. Two groups of 30 students each were selected from the same level in both the experimental and control settings. One group was taught using artificial intelligence techniques, while the other received traditional education throughout an entire academic semester. as explained in the following Table 2.

Table 2. Learning Methods and Group Composition

Group	Learning Method	Number
Control	Traditional Learning Style	30
Experimental	Artificial Intelligence Technology	30

The following methods were adopted for the educational program: artificial intelligence technologies and web-based applications that simulate real-world environments. An active learning approach was employed, involving the formation of small discussion groups and posing questions to students at the end of each activity within the program.

Problem-Solving Scale: The researcher adopted a problem-solving scale consisting of 40 items, devised [16], divided into five dimensions: General Orientation, Problem Definition, Alternative Generation, Decision Making, and Evaluation. Negative items included (5, 8, 9, 10, 13, 14, 16, 18, 21, 24, 26, 27, 30, 31, 32, 35, 36, 37, 38, 40).

Scale Validity (Internal Consistency): Researchers studied several interrelationships within the study, including the relationship with the overall mean and tested the relationships of Correlation Coefficients and Significance Values for the Dimensions of the Questionnaire as showed in Table 3.

Table 3. Correlation Coefficients and Significance Values for the Dimensions of the Questionnaire

Dimension of the Questionnaire	Correlation Coefficient	Sig Value
General Orientation	0.71**	0.00
Problem Definition	0.85**	0.00
Alternative Generation	0.74**	0.00
Decision Making	0.847**	0.00
Evaluation	0.78**	0.00

Note: ** indicates a statistically significant correlation at the 0.01 level (2-tailed).

The resulting relationships were statistically significant, indicating the validity of the scale ($Sig = p = 0.000 < \alpha = 0.01$).

Reliability and Validity of the Scale: Cronbach's alpha coefficient was used to calculate the reliability of the test. as showed in Table 3 Cronbach's alpha was calculated for all dimensions of the scale as shown in these measures ensure that the tools used in the study are both reliable and valid for assessing the intended psychological constructs (Table 4).

Table 4. Cronbach's Alpha Coefficient for the Problem-Solving Scale

Dimension of the Questionnaire	Reliability Coefficient
General Orientation	0.72
Problem Definition	0.881
Alternative Generation	0.77
Decision Making	0.85
Evaluation	0.80
Overall Scale	0.78

Source: Statistical Analysis Results Using SPSS Version 20.

Thus, the researchers have confirmed the validity and reliability of the test items, and the test has become appropriate for application on the primary study sample [15].

Scale for Scientific Thinking: Skills in scientific thinking were identified as described in the literature related to the research topic, including studies by [16] and Asir and [17]. A scale for scientific thinking was developed, encompassing the following skills: problem identification, hypothesis formulation, testing the validity of hypotheses, interpretation, and generalization.To verify the validity and reliability, the researchers applied the measures to a pilot sample of 10 students outside the main study sample [4].

Internal Consistency Validity: Internal consistency validity was calculated by computing correlation coefficients between the dimensions of the scale and the total score. The results were as follows:

Table 5. Correlation Coefficients and Significance Values for the Dimensions of the Scientific Thinking Scale

Scale Dimension	Correlation Coefficient	Sig Value
Problem Identification	0.71**	0.00
Hypothesis Formulation	0.65**	0.00
Testing Hypothesis Validity	0.77**	0.00
Interpretation	0.62**	0.00
Generalization	0.78**	0.00

Reliability of the Scale: To ensure the reliability of the test, researchers applied it to a sample and calculated the reliability using Cronbach's alpha coefficient. This is detailed in:

Table 6. Reliability Coefficients for the Dimensions of the Scale

Dimension of the Scale	Reliability Coefficient
Problem Identification	0.88
Hypothesis Formulation	0.77
Testing Hypothesis Validity	0.62
Interpretation	0.64
Generalization	0.75

3 Results

Main Research Question: What is the impact of artificial intelligence technologies on problem-solving skills and scientific thinking among third-year material engineering students?

To answer this question, the following hypotheses were tested:

3.1 First Hypothesis

There are no differences in problem-solving and scientific thinking between the control and experimental groups before the application of artificial intelligence in education.

Means and standard deviations for the students' responses on problem-solving skills and scientific thinking were calculated prior to the use of artificial intelligence technologies.

Table 7. Means and Standard Deviations for Responses of Material Engineering Students in Tests of Problem-Solving Skills and Scientific Thinking Before the Use of Artificial Intelligence Technology

Scale	Group	Mean Score	Standard Deviation	t-Value	Significance
Scientific Thinking	Control	3.7499	0.56	1.72	Not Significant
	Experimental	3.752	0.53		
Problem-Solving Skills	Control	3.7433	0.473	1.52	Not Significant
	Experimental	3.8499	0.462		

Table: T-Values and Significance for Scientific Thinking and Problem-Solving Skills Across Groups. This table shows the mean scores and standard deviations for both the control and experimental groups in measures of scientific thinking and problem-solving skills, along with t-values indicating the statistical significance of differences between the groups.

From Table 5, it is evident that there were no significant differences between the control and experimental groups in the measures of problem-solving and scientific thinking prior to the application of artificial intelligence teaching techniques. The non-significant t-values indicate that both groups were equivalent in their abilities to solve problems and engage in scientific thinking (Fig. 1).

3.2 Second Hypothesis

There are no statistically significant differences between the control and experimental groups in problem-solving skills after the use of artificial intelligence technologies:

This table shows the post-test mean scores and standard deviations for both the control and experimental groups in various dimensions of problem-solving skills, along

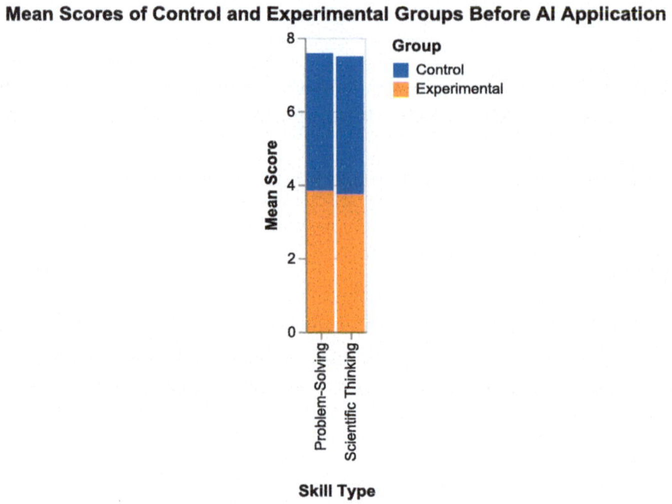

Fig. 1. Means scores of controls and experimental before application

Table 8. T-Test for Post-Test Responses of Control and Experimental Groups in Problem-Solving Skills

Dimension	Group	Number	Mean Score	Standard Deviation	Calculated t-Value	Significance
General Orientation	Control	30	3.715	0.126	−1.424	Significant (0.00)
	Experimental	30	4.40	0.325		
Problem Definition	Control	30	3.86	0.243	−6.17	Significant (0.00)
	Experimental	30	4.20	0.133		
Alternative Generation	Control	30	3.97	0.473	−4.77	Significant (0.00)
	Experimental	30	4.15	0.122		
Decision Making	Control	30	3.88	0.489	−6.4	Significant (0.00)
	Experimental	30	4.36	0.45		
Evaluation	Control	30	3.682	0.289	−3.749	Significant (0.00)
	Experimental	30	4.13	0.27		

with calculated t-values indicating the statistical significance of differences between the groups. All results show significant improvements in the experimental group compared to the control group, indicating the effectiveness of the intervention [23] (Fig. 2).

According to the previous Table 6, there are statistically significant differences between the mean scores of the control and experimental groups in both pre- and post-tests of the problem-solving scale and its sub-dimensions, favouring the post-application. The calculated t-values for the differences between the means on the scale dimensions

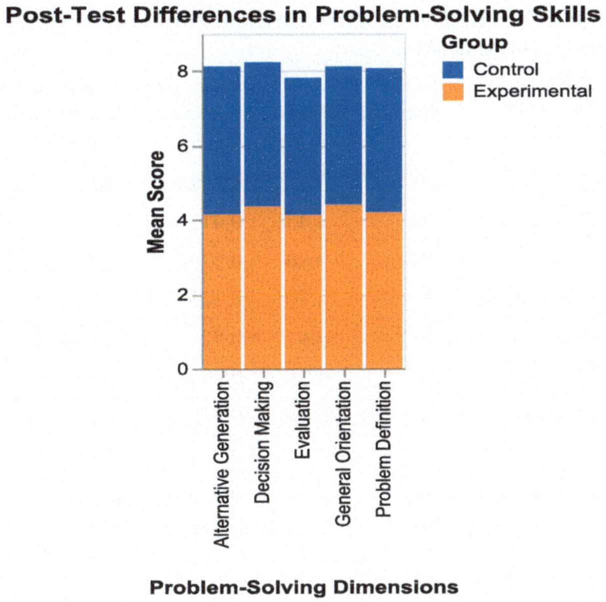

Fig. 2. Post- test differences in problem-solving skills

were (−1.424, 6.17, 4.77, 6.4, 3.749), which are significant, indicating that these differences are meaningful. These differences are attributed to the experiences gained by the students during their study using artificial intelligence, as intensive AI programs contributed to the acquisition of scientific knowledge and problem-solving abilities do not present before the application of AI technologies. Additionally, active learning provided opportunities for reflection and discussion, facilitating the development of solutions and decision-making among alternatives.

3.3 Third Hypothesis

There are statistically significant differences between the control and experimental groups in scientific thinking skills after the use of artificial intelligence technologies.

This table displays the post-test mean scores and standard deviations for both the control and experimental groups in different dimensions of scientific thinking skills, as well as the calculated t-values indicating the statistical significance of differences between the groups. All results indicate significant improvements in the experimental group compared to the control group, highlighting the efficacy of the intervention in enhancing scientific thinking skills (Fig. 3).

Observations from Table 7 indicate significant differences between the mean scores of the control and experimental groups on the scientific thinking scale. Differences were noted across all dimensions of the scale, and the t-values for all scientific thinking skills were significant, favouring the experimental group. This demonstrates the impact of artificial intelligence technologies in enhancing scientific thinking skills of the experimental

Table 9. T-Test for Post-Test Responses of Control and Experimental Groups in the Subsequent Dimension of Scientific Thinking Skills

Dimension	Group	Number	Mean Score	Standard Deviation	Calculated t-Value	Significance
Problem Definition	Control	30	3.715	0.322	2.533	Significant (0.00)
	Experimental	30	4.45	0.412		
Hypothesis Formulation	Control	30	3.86	0.325	5.13	Significant (0.00)
	Experimental	30	4.10	0.122		
Hypothesis Testing	Control	30	3.97	0.52	4.89	Significant (0.00)
	Experimental	30	4.25	0.233		
Interpretation	Control	30	3.68	0.577	5.2	Significant (0.00)
	Experimental	30	4.76	0.342		
Generalization	Control	30	3.71	0.358	3.69	Significant (0.00)
	Experimental	30	4.23	0.199		

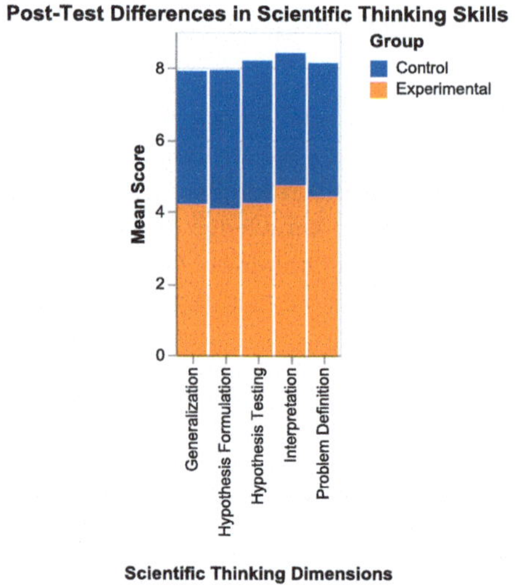

Fig. 3. Post- test differences in scientific thinking skills

group, which underwent teaching using AI techniques, compared to the control group that was taught using traditional methods (Table 8).

The significant differences can be attributed to practices and activities that contributed to the development of scientific objectivity. Additionally, the educational environment facilitated by active learning, characterized by cooperation, dialogue, discussion, and a psychologically reassuring atmosphere with a diversity of activities and absence of monotony using artificial intelligence tools, played a crucial role. These factors helped create differences between the control and experimental groups on the scientific thinking scale. Moreover, the guidance provided to teachers on the optimal use of artificial intelligence tools and the variety of teaching methods involving active learning motivated students and aided them in solving problems. This support enabled the experimental group to outperform the control group in both problem-solving and scientific thinking, an advantage not afforded to the control group taught via traditional methods (Table 9).

4 Conclusions

The results support the second and third hypotheses - that there would be differences in problem-solving skills, scientific thinking among participants favoring experimental group. This distinction is identified due to the use of artificial intelligence technologies that allowed them to practice problem solving and strategic thinking, which would not have been observed if traditional teaching methods were applied.

Research objectives: The main goal of the present study is to determine how the use of self-learning materials in combination with artificial intelligence and active learning affect the ability of engineering students to solve problems and think in scientific categories. Thus, the aim is to explore whether AI technologies are useful in increasing the problem-solving skill and the ability to think according to scientific rules in engineering students. The aim is to determine whether they do in-group showed statistically significant differences from the pre-test to the post-test due to the influence of artificial intelligence educational methods. As a result, the experimental group demonstrated significant differences in their response due to the use of AI and active learning methods. Moreover, the research reveals the fact that AI and active learning environment as a self-learning technique can take responsibility for the content and can pay more attention to the key issues, which can be helpful for scientific thinking and problem-solving. It can also enhance the motivation of students as the method provides all necessary information about the topic that was not possible before. Finally, teaching with the help of AI and active learning integrated self-regulation, reflection, and discussion to create a learning environment.

The results of the study showed that there were statistically significant differences in problem-solving skills and scientific thinking between both experimental groups compared to those taught by artificial intelligence with active learning techniques, through its AI and active learning environment, we were able to leave students in charge of their own medical knowledge base studying critical points that trigger scientific thinking & problem solving. Moreover, it boosted their motivation and gave them an overabundance of information.

The research concludes the need to integrate AI programs in teaching and building capacity for teachers to carry them out while noting that there is a need target critical thinking alongside problem solving skills amongst students. This study recommends a shift in curriculum development to suit these advanced technologies.

From the discussion above, the study recommends the following: first, there is a need to consider the development of learners' scientific thinking and problem-solving abilities as a priority and encourage the use of AI programs by both students and educators. They should also be trained in how to incorporate the programs effectively during teaching. On top of that, there is a need to focus on enhancing the curriculum through AI integration. When it comes to the area for further research, it is recommended to consider the influence of AI on scientific thinking using different levels of education. It would also be beneficial to evaluate how AI affects other areas of cognitive skills. Additionally, researchers should be encouraged to conduct similar studies in the future to determine how AI contributes to the acquisition of other types of skills such as critical thinking and logical reasoning. The researchers should also carry out such studies to determine how AI aids in promoting educational concepts at the primary and middle school levels.

References

1. Goksel, N., Bozkurt, A.: Artificial intelligence in education: Current insights and future perspectives. In Handbook of Research on Learning in the Age of Transhumanism, pp. 224–236. IGI Global (2019). https://doi.org/10.4018/978-1-5225-8431-5.ch014
2. Chassignol, M., Khoroshavin, A., Klimova, A., Bilyatdinova, A.: Artificial Intelligence trends in education: a narrative overview. Procedia Computer Science **136**, 16–24 (2018). https://doi.org/10.1016/j.procs.2018.08.233
3. Tuomi, I.: The impact of artificial intelligence on learning, teaching, and education. Publications Office of the European Union, Luxembourg (2018). https://doi.org/10.2760/12297
4. Zawacki-Richter, O., Marín, V.I., Bond, M., Gouverneur, F.: Systematic review of research on artificial intelligence applications in higher education–where are the educators? Int. J. Educ. Technol. High. Educ. **16**(1), 39 (2019)
5. Popenici, S.A., Kerr, S.: Exploring the impact of artificial intelligence on teaching and learning in higher education. Res. Pract. Technol. Enhanc. Learn. **12**(1), 22 (2017)
6. Khosravi, H., Sadiq, S., Gasevic, D.: Development and adoption of an adaptive learning system: Reflections and lessons learned. In: Proceedings of the 51st ACM Technical Symposium on Computer Science Education, pp. 58–64 (2020). https://doi.org/10.1145/3328778.3366900
7. Bates, T., Cobo, C., Mariño, O., Wheeler, S.: Can artificial intelligence transform higher education? Int. J. Educ. Technol. High. Educ. **17**(1), 42 (2020). https://doi.org/10.1186/s41239-020-00218-x
8. Drigas, A.S., Ioannidou, R.E.: Artificial intelligence in special education: a decade review. Int. J. Eng. Pedagogy **10**(6) (2020). https://doi.org/10.3991/ijep.v10i6.16535
9. Villegas-Ch, W., Román-Cañizares, M., Palacios-Pacheco, X.: Improvement of an online education model with the integration of machine learning and data analysis in an LMS. Appl. Sci. **10**(15), 5371 (2020). https://doi.org/10.3390/app10155371
10. Hwang, G.J., Tu, N.T.: Roles and research trends of artificial intelligence in mathematics education: A bibliometric mapping analysis and systematic review. Mathematics **9**(6), 584 (2021). https://doi.org/10.3390/math9060584
11. Ouyang, F., Jiao, P.: Artificial intelligence in education: The three paradigms. Computers and Education: Artificial Intelligence **2**, 100020 (2021). https://doi.org/10.1016/j.caeai.2021.100020
12. Zhai, X., et al.: A review of artificial intelligence (AI) in education from 2010 to 2020. Complexity (2021). https://doi.org/10.1155/2021/8812542
13. Radloff, J., Hall, J.A.: Development and testing of the Draw-a-Programmer test (DAPT) to explore elementary preservice teachers' conceptions of computational thinking. Educ. Inf. Technol. **27**(3), 4301–4320 (2021). https://doi.org/10.1007/s10639-021-10787-7

14. Dwivedi, Y.K., et al.: Artificial Intelligence (AI): Multidisciplinary perspectives on emerging challenges, opportunities, and agenda for research, practice and policy. Int. J. Inf. Manage. **57**, 101994 (2019). https://doi.org/10.1016/j.ijinfomgt.2019.08.002
15. Hernandez-de-Menendez, M., Escobar Díaz, C.A., Morales-Menendez, R.: Educational experiences with Generation Z. Int. J. Interact. Design and Manufact. (IJIDeM) **14**(3), 847–859 (2020). https://doi.org/10.1007/s12008-020-00674-9
16. Peng, H., Ma, S., Spector, J.M.: Personalized adaptive learning: an emerging pedagogical approach enabled by a smart learning environment. Smart Learning Environments **6**(1), 9 (2019). https://doi.org/10.1186/s40561-019-0089-y
17. Alonso-Fernández, C., Cano, A.R., Calvo-Morata, A., Freire, M., Martínez-Ortiz, I., Fernández-Manjón, B.: Lessons learned applying learning analytics to assess serious games. Comput. Hum. Behav. **99**, 301–309 (2019). https://doi.org/10.1016/j.chb.2019.05.036

The Impact of Simulation-Based Education on the Development of Problem-Solving Skills and Strategic Thinking Among Students of Arabic Language Departments in the Colleges of Islamic Sciences in Iraq

Alaa Abdulkhaleq Hussein(✉)

College of Islamic Science, University of Baghdad, Baghdad, Iraq
alaa.abdulkhaleq@colaw.uobaghdad.edu.iq

Abstract. The study aimed to investigate the impact of simulation-based education on the development of problem-solving skills and strategic thinking among students of the Arabic Language Department at the College of Islamic Sciences, University of Baghdad. Addressing a significant gap in effective educational methodologies, this research employed an experimental design with two groups: a control group and an experimental group, each consisting of 30 students. Using standardized problem-solving and strategic thinking scales, the study measured the impact of simulation-based learning. The findings revealed significant improvements in the experimental group, with post-test scores averaging 20% higher in problem-solving skills and 15% higher in strategic thinking compared to the control group. These results underscore the effectiveness of simulation-based education in enhancing cognitive skills. The study concludes with a call for the integration of simulation-based learning in educational curricula and recommends the provision of necessary technological resources to support this method."

Keywords: Strategic Thinking · Problem Solving · Simulation

1 Introduction

Recent rapid advancements in science and technology have profoundly influenced educational systems. Contemporary students are increasingly required to adapt to a swiftly changing technological landscape [1]. Due to these rapid changes, educational systems must be modified to invigorate students' acquisition of knowledge access methodologies, foster the development of solutions to as-yet-unknown problems, and enhance decision-making skills. In this evolving context, the diversity and complexity of challenges in our daily lives necessitate those students be prepared for the future by confronting real-world problems within their learning environments and generating appropriate outputs. It is crucial to cultivate their problem-solving skills to mastery. Research and studies conducted by David Perkins emphasize the need for students to be capable of engaging in thinking tasks concerning various subjects, which Perkins terms "executive functions of

A. Abdelgawad et al. (Eds.): ISBCom 2024, LNNS 1268, pp. 62–73, 2025.
https://doi.org/10.1007/978-3-031-82377-0_6

understanding highlights the importance of aiding students in discovering connections within knowledge. Students must explore how new knowledge interlinks with what they already know, aligning with how the brain functions in brain-based education, and all learning theories acknowledge that no two students learn in the same way, The impact of mobile applications on student performance and satisfaction has been demonstrated in recent studies [2], highlighting the potential of technology-enhanced learning.

Simulation-based learning provides a broad spectrum of opportunities for practicing complex skills in education to facilitate effective learning. Simulation is an educational method typically employed by educators to approximate the real world, which is often difficult to provide to learners due to financial costs and human resources constraints. It has been frequently observed that students are not directed towards activities where they can pose questions, formulate hypotheses, and make observations [3].

The problem of this study is encapsulated in the primary question: "What is the impact of simulation-based education on the development of problem-solving skills and strategic thinking among students of the Arabic Language Departments at the College of Islamic Sciences in the University of Iraq?".

To explore the impact of simulation-based education on the development of problem-solving skills and strategic thinking, it was necessary to review several prior studies (Table 1):

It has often been observed that students are not directed towards activities that would allow them to pose questions, formulate hypotheses, and make observations. Given the rapid changes in science and technology, it has become essential to develop students' problem-solving skills to a proficient level. Recent research has highlighted the potential of digital games as a context for cognitive development [9], which aligns with the goals of simulation-based education. Hence, the research problem arises from the need to investigate the impact of simulation-based education on the development of problem-solving and strategic thinking skills among students of Arabic Language Departments at Islamic Science Colleges in Iraq, Hence, the research problem arises from the need to investigate the impact of simulation-based education on the development of problem-solving and strategic thinking skills among students of Arabic Language Departments at Islamic Science Colleges in Iraq. It is crucial to ascertain the effectiveness of this type of education in enhancing students' abilities to face problems with strategic thinking and to devise appropriate solutions.

As the educational landscape evolves, it becomes increasingly essential for students to adapt to rapid technological changes. Recent research has shown that educational electronic games can be effective in developing Arabic language skills among non-native speakers [4], suggesting the potential of technology-enhanced learning in language education. This highlights the significant role of innovative tools in facilitating language acquisition and engagement among learners.

The study aims at exploring whether simulation-based education can help improve students' problem-solving skills in Arabic Language Department students. It is aimed at studying the impact of simulation-based education on the development of their strategic thinking abilities.

Recent studies have shown that mobile learning technologies can improve students' scientific literacy [10], suggesting potential benefits for problem-solving and strategic

Table 1. Reviewed Studies that are Relevance to the Study

Study	Title	Key Findings	Relevance to Current Study
[4]	"The Effects of Language Games on Smartphones in Developing Arabic Speaking Skills among Non-Native Speakers"	Educational electronic games significantly improved Arabic language skills among non-native speakers	Demonstrates the potential of technology-enhanced learning in language education, particularly for Arabic
[5]	"The impact of a critical fiction analysis based on using augmented reality technology on developing students' critical thinking and critical writing at Tabuk University"	Virtual reality technology effectively enhanced critical thinking skills in Arabic language learners	Shows the potential of immersive technologies in developing higher-order thinking skills in language education
[6]	"The Effectiveness of Algodoo with Interactive Whiteboards for Teaching Physics in Saudi Secondary Schools"	Computer simulations positively impacted students' critical thinking skills in physics education	Supports the use of simulations in educational contexts to enhance critical thinking, which is related to problem-solving and strategic thinking
[7]	"A study of the impact of Collaborative Problem-Solving Strategies on students' performance of Simulation-Based Learning—a case of Network Basic Concepts course"	Simulation-based learning significantly improved problem-solving skills in higher education students	Directly relates to the current study's focus on simulation-based education and its impact on problem-solving skills
[8]	"The impact of gamification on students' learning, engagement and behavior based on their personality traits"	Gamification positively affected student learning, engagement, and behavior, with effects varying based on personality traits	Provides insight into how game-like elements (similar to simulations) can impact student performance and engagement

thinking skills. The study shows that mobile devices improve ninth-grade students' critical thinking and problem-solving skills in science labs by enhancing engagement and active learning.

The paper began with the introduction sets the context, highlighting the significance of the research in the backdrop of rapid technological advancements. The literature review covers relevant prior studies and theoretical frameworks, establishing a foundation for the current study. The research problem is clearly articulated, followed by detailed research objectives and questions. The methodology section outlines the experimental design, including the sample selection, research instruments, and data collection procedures. Results are presented with statistical analyses, comparing the control and experimental groups. The discussion interprets the findings, linking them to the literature and highlighting the study's contributions. The paper concludes with recommendations for integrating simulation-based learning in education and suggests areas for future research.

The research focuses on second-year students of the Arabic Language Department at the College of Islamic Sciences, University of Baghdad, during the academic year 2023–2024.

2 Methodology

The researchers adopted an experimental approach involving experimental and control groups with pre-tests and post-tests. The control group was subjected to traditional educational methods, while the experimental group was exposed to teaching methods employing artificial intelligence technologies. Randomly, 60 students were selected and categorized under two groups- control group & experimental. Each group received a pre-test of the concept before simulation application, post-test for experimental along with posttest after stimulation in both groups. Using statistical models, the outcomes were compared to find out if simulation-based education expedites problem-solving and strategic thinking skills, this approach draws on principles of experiential learning, which has shown success in other educational contexts [11].

The experimental design included the following variables, Independent Variable that Use of simulation in education. As well Dependent Variables, which means Problem-solving skills and strategic thinking skills.

To validate the use of simulation-based learning on developing problem-solving and strategic thinking skills an experimental design was developed. For this one, an experimental research design was used with independent and dependent variables tested.

The study population consisted of students from the Arabic Language Department at the College of Islamic Sciences, University of Baghdad.

A sample of 60 students purposively selected. These students will be about the same age and at a relatively even level with one another regarding actual grades. Intelligence An intelligence level assessment was cosseted out to verify that there was equivalence of the two groups for IQ, as described in Table 2. Based on the previous studies, problem solving, and strategic thinking scales used to measure these skills in both groups of the researcher The impact from simulation-based education on enhancing this justification.

Table 2. Sample collection techniques

Group	Learning Method	Number
Control	Traditional Learning Style	30
Experimental	Simulation-Based Learning	30

Table 2 illustrates the experimental procedure used to evaluate the impact of simulation-based education on the development of problem-solving skills and strategic thinking among students. The structure of the table is as follows. The first column is "Group," and it states either "Control Group" or "Experimental Group." According to the second column, which is "Pre-test," results of this task demonstrate whether students possess any problem-solving and strategic thinking skills at the start of classes. The third column being "Learning Method," respondents impersonating those in control group took part in non-modified training, while participants in the experimental group underwent simulations or field exercises. In column four, which is "Post-test," results of the task by which all students were tested after the courses to assess any changes in their capabilities (Table 3).

Table 3. Experiment procedures

Group	Pre-Test	Learning Method	Post-Test
Control Group	Pre-Examination	Traditional Format	Post-Examination
Experimental Group	Pre-Examination	Field/Simulation Format	Post-Examination

2.1 Experimental Steps

Preparation of the Plan: An integrated approach was developed that included all stages of the experimental process to deal with simulations for understanding and execution. [12] prepared a test, and it was administered to the control group and experimental groups in other for them not to have equivalence on measuring problem-solving and strategic thinking skills.

Preparation of the Experimental Group: The experimental group had been prepared to use simulations, and they were given a screen and laptop, while the control group was taught using traditional methods. In [13] applied a similar approach using a quasi-experimental method with pre-tests and post-tests to examine problem-solving skills. The educational simulation component specific to societal and family issues was developed and employed over four weeks, with two sessions per week, into the curriculum to enhance problem-solving and strategic thinking.

Post-test Preparation and Application: A post-test was prepared and applied to both groups to measure the changes effected by the simulation. In [14], post-tests were also employed to evaluate the performance differences between groups after using a creative thinking approach to problem-solving.

Comparison and Statistical Analysis: The results obtained were compared, and statistical differences between the two groups were identified using Bloom's taxonomy in the preliminary examination to ensure equivalence (Knowledge, Recall, Understanding, Application, Analysis, Evaluation). As seen in the study by [15], statistical models such as T-tests were used to compare the differences between control and experimental groups.

Conduct of Simulation Lectures: The researcher conducted three lectures using computer simulations to present several social issues after a pre-test to ensure group equivalence. Similarly, [13] used simulation-based lectures to improve strategic thinking and problem-solving skills, and the results were statistically analyzed for significance (Fig. 1).

Table 4. Discussion of Previous and Measurement Scales

Group	Number of Students	Mean Score	Degrees of Freedom	Calculated "t" Value	Significance
Experimental	**30**	**4.05**	**58**	**0.528**	**Not Significant**
Control	30	4.12			

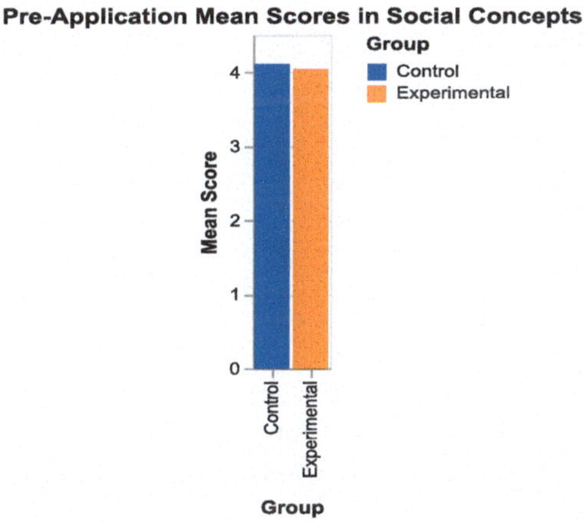

Fig. 1. Per -application mean scores in scores in social concepts

From the previous Table 4, it is evident that the calculated "t" value is not statistically significant, indicating no differences between the control and experimental groups, thus confirming their equivalence.

Problem-Solving Measure: The researcher adopted a 40-item scale developed by [16] for measuring problem-solving skills, divided into five dimensions (General Orientation, Problem Definition, Alternative Generation, Decision Making, Evaluation). Negative items included numbers 5, 8, 9, 10, 13, 14, 16, 18, 21, 24, 26, 27, 30, 31, 32, 35, 36, 37, 38, 40, and the rest were positive items. The dimensions are measured as follows:

General Orientation: Items 1, 6, 11, 16, 21, 26, 31, 36
Problem Definition: Items 2, 7, 12, 17, 22, 27, 32, 37
Alternative Generation: Items 3, 8, 13, 23, 28, 33, 37
Decision Making: Items 4, 9, 14, 19, 24, 29, 34, 39
Evaluation: Items 5, 10, 15, 20, 25, 30, 35, 40

3 Results

The impact of simulation-based education on the development of problem-solving and strategic thinking abilities among students of the Arabic Language Department at the College of Islamic Sciences was investigated by calculating mean scores and standard deviations for the students' performance in tests measuring these abilities. The findings are summarized in the tables (Fig. 2 and Table 5):

Table 5. Averages and Standard Deviations for the Responses of Arabic Language Department Students in the Problem-Solving Skills Test

Number	Dimension	Mean Score	Standard Deviation	Rating
60	General Orientation	3.64	0.483	Average
60	Problem Definition	3.80	0.730	Average
60	Alternative Generation	3.87	0.489	Average
60	Decision Making	3.67	0.471	Average
60	Evaluation	3.80	0.622	Average
60	Overall Mean	3.75		Average

Table 6 indicates that the overall mean score for students' responses on the problem-solving scale was 3.75, reflecting a moderately good response level. This suggests a medium level of problem-solving skills among the study sample, generally consistent across different groups.

Table 6. Medium level of problem-solving skills

Scale	Mean Score	Standard Deviation
Strategic Thinking	3.84	0.53

Table 6 illustrates the strategic thinking levels among the Arabic Language students in the study sample. The overall mean score for responses on the strategic thinking scale was 3.84, also reflecting a moderately good response level. This indicates a medium level of strategic thinking skills generally across both groups.

A "t-test" was performed to determine the significance of differences between the control and experimental groups prior to the simulation application.

The table illustrates that there are no statistically significant differences at the 0.05 level between the control and experimental groups regarding problem-solving skills and strategic thinking. All response scores were closely aligned across the different group types.

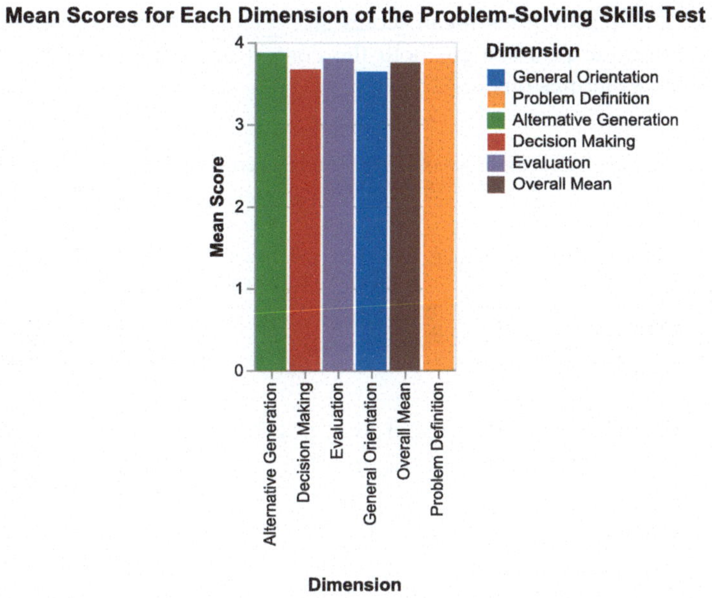

Fig. 2. Mean scores for each dimension of the problem skills test

Table 7. Illustrates that there are no statistically significant differences

Scale	Group	Mean Score	Standard Deviation	t-Value	Significance
Strategic Thinking	Control	3.7499	0.56	2.62	Not Significant (0.00)
	Experimental	3.752	0.53		
Problem-Solving Skills	Control	3.8433	0.473	3.62	Not Significant (0.00)
	Experimental	3.8499	0.472		

4 Discussion

These findings align with research on online formative peer-tutoring approaches, which have shown positive effects on students' learning behaviors and performance.

First objective of the study what is to Explore the impact of simulation-based education on the development of problem-solving skills among students of the Arabic Language Department.

This objective was translated into the following hypothesis: There are no statistically significant differences between the control and experimental groups in the problem-solving skills measure at the 0.05 significance level in the post-test results attributable to the application of educational simulation.

To answer this question, the mean scores and standard deviations of the students' response scores in the post-test for both the control and experimental groups were calculated. Table 8 clarifies this.

Table 8. T-Test for Post-Test Responses of Control and Experimental Groups in Problem-Solving Development

Dimension	Group	Number	Mean Score	Standard Deviation	Calculated t-Value	Significance (sig)
General Orientation	Control	30	3.915	0.146	−1.534	Not Significant (0.00)
	Experimental	30	4.304	0.345		
Problem Definition	Control	30	3.76	0.2421	−6.278	Not Significant (0.00)
	Experimental	30	4.10	0.132		
Alternative Generation	Control	30	3.87	0.272	−4.787	Not Significant (0.00)
	Experimental	30	4.15	0.112		
Decision Making	Control	30	3.71	0.209	−6.377	Not Significant (0.00)
	Experimental	30	4.16	0.25		
Evaluation	Control	30	3.882	0.218	−3.849	Not Significant (0.00)
	Experimental	30	4.13	0.20		

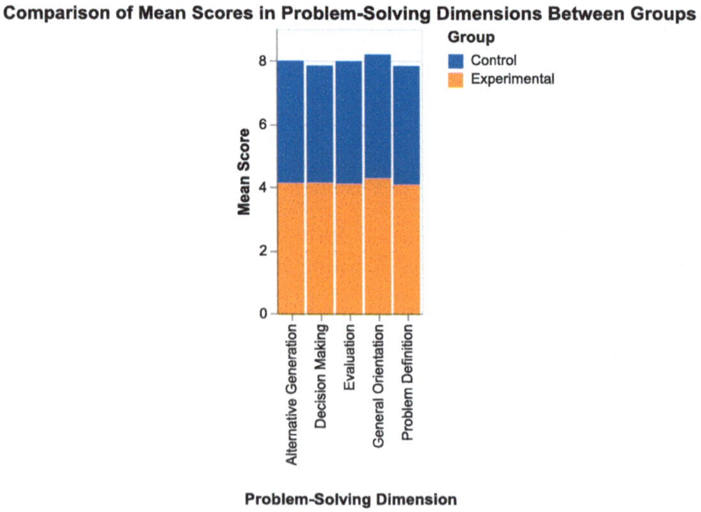

Fig. 3. Comparison of mean score in problem-solving dimensions between groups.

From Table 7 & Fig. 3, there are observable differences in the pre- and post-test averages between the control and experimental groups. The experimental group's average responses were higher than those of the control group across the dimensions of the problem-solving scale. This indicates statistically significant differences favoring the experimental group, which participated in the educational simulation program. The significance level of the t-test was 0.001, which is less than the standard criterion of 0.05, indicating that these differences are statistically significant.

This confirms the superiority of the experimental group, who underwent educational simulation, in developing problem-solving skills compared to the control group, who received traditional education. This highlights the impact of simulation-based education on enhancing problem-solving skills among the study sample.

The second objective is to investigate the effect of simulation-based education on the development of strategic thinking skills among these students. The observed improvements may be related to increased student autonomy and motivation, factors that have been linked to success in other innovative learning environments [17].

This objective was translated into the following hypothesis: There are no statistically significant differences in strategic thinking between the control and experimental groups at the 0.05 significance level in the post-test results, attributable to the implementation of simulations in teaching strategic thinking.

To answer this question, mean scores, and standard deviations for the student responses in the post-test for both the control and experimental groups were calculated (Table 9).

Table 9. Mean Scores and Standard Deviations for Post-Test Responses of Control and Experimental Groups After Educational Simulation Application.

Dimension	Group	Number	Mean Score	Standard Deviation	Calculated t-Value	Significance
Strategic Thinking	Control	30	3.92	0.25	−7.736	Significant (0.00)
	Experimental	30	4.24	0.21		

Table 8 shows that the mean score for strategic thinking in the experimental group was higher than that of the control group. This suggests significant differences in strategic thinking between the two groups, favoring the experimental group. The significance of these differences, as indicated by a t-value of −7.736 and a significance level less than 0.05, confirms that these differences are statistically significant and can be attributed to the effects of educational simulation.

5 Conclusion

The findings of our study clearly describe the important role played by simulation-based education in enhancing problem-solving and strategic thinking capabilities among students. When ABNF technique with simulation-based education was applied, there were significant improvements in the responses of the experimental group before and after enactment. This supports the use of simulation as an educational tool to help improve cognitive skills. School simulation programs are structured to stimulate pupils and develop the basic construction skills involved in solving problems. This autonomy is delivered to students in educational settings which afford them freedom, and control over the learning contents they perceive as relevant during problem solving. The study also reported that the simulation-based program supplied rich content not provided to students before, hence would allow learning based on student autonomy. The researchers suggest that the results indicate it is essential to incorporate simulation-based learning into educational curricula and propose equipping institutions with necessary technological resources for supporting this. This not only improves the ability of students to handle strategic thinking issues, but also provides them with appropriate mechanisms for real-world solutions.

To realize the successful application of simulation-based learning, substantial resources are essential, including enough computer-type equipment for educational needs. The development of individual computer-based simulations should be tailored to the speed of students learning, allowing them to master the content at their own pace. Moreover, the availability of specialized centers should be ensured to offer technical material's development of simulation software, which typically requires the presence of experts from various disciplines. To monitor the process, teachers should continuously evaluate everything that the students do, whether they follow the proposed programs and achieve the expected results.

References

1. Budhwar, P., et al.: Human resource management in the age of generative artificial intelligence: Perspectives and research directions on ChatGPT. Hum. Resour. Manag. J. **33**, 606–659 (2023). https://doi.org/10.1111/1748-8583.12524
2. Alqahtani, M., Mohammad, H.: Mobile applications' impact on student performance and satisfaction. Turkish Online J. Educ. Technol. **14**(4), 102–112 (2020). https://doi.org/10.1016/j.compedu.2020.103909
3. Chernikova, O., Heitzmann, N., Stadler, M., Holzberger, D., Seidel, T., Fischer, F.: Simulation-based learning in higher education: a meta-analysis. Rev. Educ. Res. **90**, 499–541 (2020). https://doi.org/10.3102/0034654320933544
4. Kenali, H.M.S., Yusoff, N.M.R.N., Saad, N.S.B.M., Abdullah, H., Kenali, A.M.S.: The effects of language games on smartphones in developing Arabic speaking skills among non-native speakers. Creat. Educ. **10**, 972–979 (2019). https://doi.org/10.4236/ce.2019.105073
5. Abdelrahim, A.A.M.: The impact of a critical fiction analysis based on using augmented reality technology on developing students' critical thinking and critical writing at Tabuk University. Language Teaching Research. 136216882311555 (2023). https://doi.org/10.1177/13621688231155578
6. Algorashi, M.: The Effectiveness of Algodoo with Interactive Whiteboards for Teaching Physics in Saudi Secondary Schools. Int. J. Cross-Disciplinary Subj. Educ. **15**, 4867–4875 (2024). https://doi.org/10.20533/ijcdse.2042.6364.2024.0599
7. Lu, H.-K., Lin, P.-C.: A study of the impact of Collaborative Problem-Solving Strategies on students' performance of Simulation-Based Learning — a case of Network Basic Concepts course. Int. J. Info. Educ. Technol. **7**, 361–366 (2016). https://doi.org/10.18178/ijiet.2017.7.5.895

8. Smiderle, R., et al.: The impact of gamification on students' learning, engagement and behavior based on their personality traits. Smart Learning Environments **7** (2020). https://doi.org/10.1186/s40561-019-0098-x
9. Blumberg, F.C., et al.: Digital Games as a context for Children's Cognitive Development: research recommendations and policy considerations. Soc. Policy Rep. **32**, 1–33 (2019). https://doi.org/10.1002/sop2.3
10. Kousloglou, M., Petridou, E., Molohidis, A., Hatzikraniotis, E.: Contribution of mobile devices to students' critical thinking & problem solving skills in laboratory settings. IADIS International Journal On Www/Internet **20** (2023). https://doi.org/10.33965/ijwi_2023210203
11. Chookaew, S., Kitcharoen, P., Howimanporn, S., Panjaburee, P.: Fostering student competencies and perceptions through artificial intelligence of things educational platform. Computers and Education Artificial Intelligence 100308 (2024). https://doi.org/10.1016/j.caeai.2024.100308
12. Alfayez, M.Q.E., Aladwan, S.Q.A., Shaheen, H.R.A.: The effect of a training program based on Mathematical Problem-Solving strategies on critical thinking among Seventh-Grade students. Frontiers in Education **7** (2022). https://doi.org/10.3389/feduc.2022.870524
13. Winarti, W., Ambaryani, S.E., Putranta, H.: Improving learners' metacognitive skills with self-regulated learning based problem-solving. Int. J. Instr. **15**, 139–154 (2022). https://doi.org/10.29333/iji.2022.1528a
14. Kuo, F.-R., Chen, N.-S., Hwang, G.-J.: A creative thinking approach to enhancing the web-based problem solving performance of university students. Comput. Educ. **72**, 220–230 (2013). https://doi.org/10.1016/j.compedu.2013.11.005
15. Haenilah, E.Y., Yanzi, H., Drupadi, R.: The effect of the scientific approach-based learning on problem solving skills in early childhood: preliminary study. Int. J. Instr. **14**, 289–304 (2021). https://doi.org/10.29333/iji.2021.14217a
16. D'Zurilla, T.J., Maydeu-Olivares, A.: Conceptual and methodological issues in social problem-solving assessment. Behav. Ther. **26**, 409–432 (1995). https://doi.org/10.1016/s0005-7894(05)80091-7
17. Jørgensen, B.E., Larsen, M., Gram, B.: Simulation as an educational tool in acute nursing care: A controlled intervention study. Nurse Educ. Pract. **32**, 28–33 (2018). https://doi.org/10.1016/j.nepr.2018.07.001

The Impact of Using Deep Learning Technology in Developing Smart Programs for Teaching Arabic to Non-Native Speakers

Alaa Abdulkhaleq Hussein(✉) and Neamah Dahash Farhan(✉)

University of Baghdad / College of Islamic Science, Baghdad, Iraq
alaa.abdulkhaleq@colaw.uobaghdad.edu.iq,
namaa.d@ircoedu.uobaghdad.edu.iq

Abstract. This study aimed to explore the impact of utilizing deep learning technology in developing intelligent software for teaching Arabic to non-native speakers. An experimental methodology was employed, involving two groups (experimental and control) with both pre-tests and post-tests. The sample consisted of 30 Arabic learners who are non-native speakers at the Arabic Language Teaching Institute in collaboration with the German Board, divided equally between the two groups. The control group received traditional instruction, while the experimental group was taught using software based on deep learning technology. The instrument of the study was a test that assessed the four Arabic language skills (listening, speaking, reading, and writing). Results indicated statistically significant differences between the mean scores of the two groups in the post-test, favouring the experimental group across all language skills. This suggests the effectiveness of using deep learning technology in developing intelligent software for teaching Arabic to non-native speakers. The study recommends the adoption of this technology in Arabic language education programs and suggests further research to refine these programs and identify their strengths and weaknesses.

Keywords: Deep learning · intelligent software · teaching Arabic · non-native Arabic speakers

1 Introduction

In the context of rapid advancements fostered by science and technology, particularly during the new industrial revolution, most nations have been prompted to look toward the future, embracing it and confronting its challenges. They have pursued investment in fourth-generation industrial revolution technologies, notably artificial intelligence, aiming to achieve developmental and ambitious goals. Artificial intelligence has become a focal point of interest in the developed world, with its applications emerging in various fields by the late 1970s. This included programming, mathematics, genetic engineering, geoscience, petroleum exploration, medicine, commerce, and psychology. There has been significant interest in applied artificial intelligence, which enables computers to solve problems more complex than mere arithmetic operations. Machine learning and

A. Abdelgawad et al. (Eds.): ISBCom 2024, LNNS 1268, pp. 74–82, 2025.
https://doi.org/10.1007/978-3-031-82377-0_7

deep learning are among the most critical applications of artificial intelligence that have found widespread implementation across different fields such as media, medicine, electronics, and computing systems, among others [1].

Technology has transformed the ways we teach and learn languages; it has provided teachers with new facilities and methods for instruction and similarly equipped learners with numerous innovative and original resources that can facilitate the acquisition of a new language. "The mobile revolution has finally arrived. Teaching Arabic to non-native speakers is not just a practice aimed at satisfying the learning needs of Arabic language enthusiasts but is a culturally and civically complex act. It opens to other cultures and integrates learners into the culture of the language being studied, adapting various linguistic, communicative, and interactive skills until they are fully immersed in new worlds of belief" [2].

Given the difficulties faced by non-native speakers in mastering Arabic, whether in accurate communication, pronunciation during speech, or comprehension during interaction, and considering the scarcity of intelligent educational software for learning Arabic, this study addresses the question: What is the impact of deep learning technology in developing intelligent software for teaching Arabic to non-native speakers? (Table 1)

Based on previous studies, the deep learning applications in teaching Arabic language to non-native speakers did not give much attention to. Most of research have mostly only discussed the importance of AI in general, and its relationship to education without specifically basing it as a study on deep learning technology, and that what focused on it in this study.

The research problem is since non-native Arabic speakers have difficulty in speaking, pronouncing or understanding Arabic language, and there are no academic programs with an advanced level of intelligence for teaching this language Consequently, this study will possibly shed light on the effectiveness of using deep learning technology to facilitate and design intelligent software for teaching Arabic as a foreign language. Moreover, there are no studies that aim to determine the influence of applying deep learning technology in creating smart software for teaching Arabic language, especially for non-native speakers.

The research aims to investigate the evaluation of how deep learning technology can be used for teaching Arabic to non-native speakers.

The present paper is divided as follows: The introduction presents the problem, summarizes related works and defines objectives. In literature review, searching information from relevant studies and laying theoretical foundations are two main functions. The Methods section provides a description of the research design (e.g., sample selection, procedures and instruments used to collect data). Results: The results section describes the study outcomes, statistical analysis and a comparison between the control group (back massage) and experimental group. Then, we interpret the results and speculate on implications for application of deep learning in language education within our discussion section. The findings are summarized in the following section and put into perspective or criticized by limitations, finally recommendations for future research are given. References - A list of complete citations to all sources published during your paper.

Table 1. Reviewed Studies that are related to the study

Ref.	Year	Title	Key Findings	Relevance to Arabic NLP and Deep Learning
[3]	2024	"Deep learning Algorithms for Arabic Optical character Recognition: a survey"	Comprehensive review of deep learning applications in Arabic NLP, covering various tasks such as sentiment analysis, machine translation, and text classification	Provides an overview of state-of-the-art deep learning techniques specifically for Arabic language processing
[4]	2019	"Deep learning for Arabic NLP: A survey"	Examines deep learning models applied to Arabic NLP tasks, discussing challenges specific to Arabic language processing and potential solutions	Offers insights into early applications of deep learning in Arabic NLP, serving as a baseline for subsequent research
[5]	2019	"Arabic natural language processing: An overview"	Extensive survey of deep learning approaches in Arabic NLP, covering a wide range of tasks and discussing challenges and future directions	Provides a comprehensive and up-to-date overview of deep learning applications in Arabic NLP, highlighting recent advancements and ongoing challenges
[6]	2018	"A Survey of Arabic Text Classification Models"	Focused survey on Arabic text classification models, including traditional machine learning and deep learning approaches. Discusses datasets, feature extraction methods, and evaluation metrics	Offers a specialized look at text classification in Arabic, a crucial task in many NLP applications, and compares the performance of deep learning models with traditional approaches

2 Methodology

The researchers adopted an experimental approach involving two groups: an experimental group and a control group, each subjected to pre-tests and post-tests. The control group was exposed to traditional teaching methods, while the experimental group was taught using deep learning technology. The experimental research design aimed to explore causal relationships between variables, with a focus on controlling external factors that might influence the phenomenon under study. Hamouda (2018) conducted similar studies, using pre-test and post-test designs to assess learning outcomes in an experimental context [7].

The research population consisted of non-native Arabic speakers enrolled at the Arabic Language Learning Institute, in collaboration with the German Board. A sample of 30 non-native Arabic speakers was selected from Germany, in coordination with the German Board for Training and Consultancy and the Arabic Language Learning Institute. The participants were divided into two equal groups: a control group, which was taught Arabic using traditional methods, and an experimental group, which was exposed to deep learning techniques. The groups were matched based on their proficiency levels to ensure equivalence in intelligence and language skills. A similar experimental design was used by El-Dakhs et al. (2019), who investigated the effects of instructional interventions on language learning among non-native speakers [8].

Pre-tests were administered to both groups before the introduction of deep learning techniques to measure their baseline language skills. The experimental group was then subjected to a curriculum incorporating deep learning technologies, while the control group continued with traditional teaching methods. After completing the intervention, post-tests were conducted to assess any changes in language proficiency in both groups. This approach aligns with the methods described by Farghaly and Shaalan (2009), who emphasized the role of natural language processing techniques in enhancing language learning outcomes, especially in the context of Arabic [9].

The collected data were analyzed to compare the performance of the experimental and control groups, using statistical methods to determine whether the deep learning intervention had a significant effect on the development of language skills among non-native Arabic speakers (Table 2).

Table 2. The research population comprised non-native Arabic.

Group	Learning Method	Number
Control	Learns via Traditional Method	15
Experimental	Learns via Deep Learning Technology	15

The test used to assess the Arabic language skills of non-native speakers was adapted from the test developed in the Sultan Sharif Ali Islamic experience at Brunei Darussalam. This test was further refined in Malaysia, Egypt, and Saudi Arabia, finalizing its structure in 2016. Designed to measure the four linguistic skills: listening, speaking, reading, and writing, each skill is weighted at 20%, with a total test score of 250 points distributed as follows in Table 3:

The experimental methodology employed involved two groups, experimental and control, each subjected to pre-tests and post-tests. The control group underwent traditional educational methods, while the experimental group received instruction via deep learning technology. This research paradigm is predicated on experimentation that aims to uncover causal relationships between variables. It features the capacity to manipulate factors influencing the phenomenon being studied. It starts with a problem confronting the researchers, who must then explore the causes and effective conditions that aid in resolving that problem through experimental means [10].

Table 3. The test used to assess the Arabic language skills of non-native

Skill	Duration	Score
Listening	70 min	50 points
Speaking	15 min	50 points
Reading	120 min	50 points
Writing	120 min	50 points
Grammar and Structure	120 min	50 points

3 Results

To address the research question, the first hypothesis was tested, which posited that there are no significant differences between the control and experimental groups in Arabic language skills among non-native speakers prior to the implementation of deep learning techniques. The means and standard deviations for both the control and experimental groups were calculated. Hamouda (2018) conducted similar experiments using pre-tests and post-tests, supporting the methodology used in this study [7].

Table 4. Differences between the means and standard deviations of responses from the control and experimental groups before implementing the deep learning-based program.

Skill	Group	Number	Mean	Standard Deviation	Calculated t-value
Listening	Control	15	21.20	1.54	0.628
	Experimental	15	22.11	5.2	
Speaking	Control	15	26.533	6.609	0.347
	Experimental	15	27.33	6.39	
Reading	Control	15	26.266	6.3	1.520
	Experimental	15	33.3	2.4	
Writing	Control	15	19.46	0.91	0.402
	Experimental	15	19.33	0.899	
Grammar and Structures	Control	15	14.533	1.64	0.627
	Experimental	15	14.33	1.84	

The data from Table 4 and Figure 1 indicate no significant differences between the control and experimental groups regarding the mean scores in the Arabic language test for non-native speakers. The calculated t-values were lower than the critical values at the significance level of 0.01 with 28 degrees of freedom, which suggests that the two groups were equivalent in their Arabic language testing, and hence, there are no differences between the mean scores of the two groups (control and experimental) in the

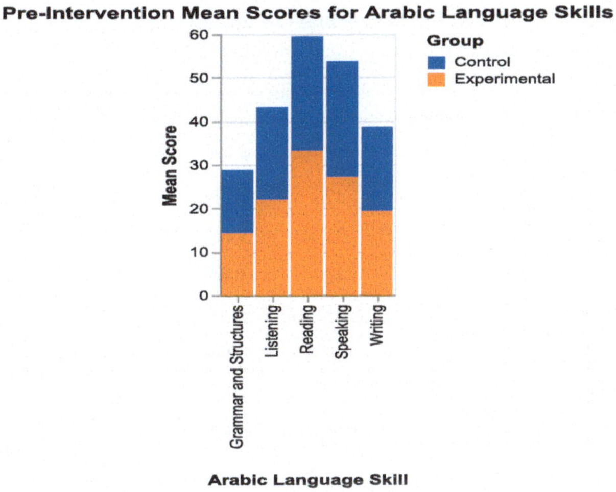

Fig. 1. Per-intervention mean scores for Arabic language skills

pre-implementation of the deep learning technology. [3] supports this finding, showing similar equivalence between control and experimental groups in a pre-intervention scenario.

Second Hypothesis: There are no statistically significant differences at the 0.05 level in Arabic language performance post-implementation of the deep learning-based program between the control and experimental groups.

The researchers used the t-test for two independent samples to determine the significance of differences between the mean scores of the two groups (experimental and control) in primary language learning skills and the overall score post-implementation. In [8] demonstrated similar post-test differences when employing experimental techniques to enhance language learning. The following table illustrates these results:

Table 5 and Figure 2 reveals significant differences between the mean scores of the control and experimental groups in the listening skill. The experimental group, which was exposed to deep learning techniques, achieved a higher mean score of 42, compared to 23.46 for the control group. In [11] supports this result, showing significant improvement in language skills for experimental groups using technology-based learning methods.

Furthermore, the table shows that the mean score for writing was 21.133 with a standard deviation of 1.50 for the control group, whereas the experimental group scored an average of 30.866 with a standard deviation of 2.87. In grammar and structures, the control group had a mean of 31.5 with a standard deviation of 3.05, and the experimental group had a mean of 42.03 with a standard deviation of 1.84. In [12] found similar results, where deep learning techniques significantly improved students' overall performance in language learning tasks.

Table 5. Responses of Control and Experimental Groups on the Arabic Language Test for Non-Native Speakers Before Applying the Deep Learning-Based Program

Skill	Group	Number	Mean	Standard Deviation	t-value Computed
Listening	Control	15	23.46	8.55	8.575
	Experimental	15	42	3.16	
Speaking	Control	15	31.81	3.60	7.596
	Experimental	15	42	3.10	
Reading	Control	15	31.20	4.05	3.684
	Experimental	15	37.20	4.50	
Writing	Control	15	21.133	1.50	11.615
	Experimental	15	30.866	2.87	
Grammar and Structures	Control	15	31.50	3.05	4.112
	Experimental	15	42.03	1.84	

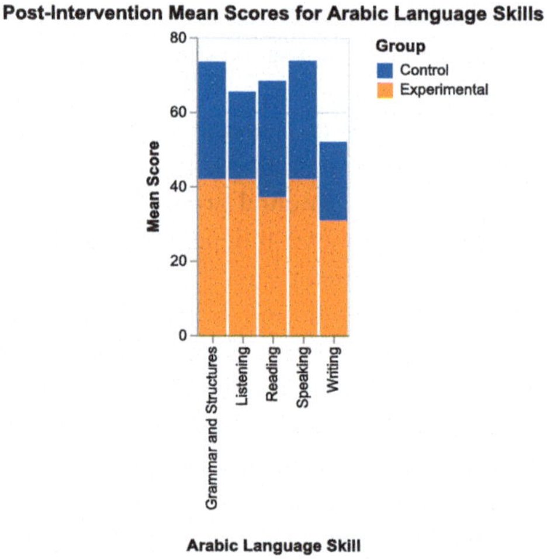

Fig. 2. Post-intervention means scores for Arabic language skills

4 Conclusion

Regarding Arabic language skills, there were significant differences between the control group and experimental in favor for an experimental that has been trained by programs are intelligent smart based on deep learning technology [13]. This success can be tentatively ascribed to the program design, which systematized collaborative learning activities that include processes of dialogue and listening in Arabic through learner interactions [14]. The program used resources that included videos and debates etc, making language learning easier for the learners to understand. This suggests the effectiveness of using deep learning technology in developing intelligent software for teaching Arabic to non-native speakers. The study recommends the adoption of this technology in Arabic language education programs and suggests further research to refine these programs and identify their strengths and weaknesses, beside a number of recommendations for future studies as follows.

It would be beneficial for future research to focus on natural language processing for Arabic using deep learning algorithms as Arabic has phonetic nuances and sets of grammatical rules concerning morphology and syntax. A way to pursue this goal is to design training workshops and lectures to overcome the difficulties that certain learners of Arabic as a foreign language experience and suggest to software developers to develop intelligent training software to facilitate the training process and optimize its time. Moreover, there is a need for training teachers using mobile applications and experimental and developmental research of educational programs to determine their strengths and weaknesses.

References

1. Yazici, İ., Shayea, I., Din, J.: A survey of applications of artificial intelligence and machine learning in future mobile networks-enabled systems. Eng. Sci. Technol. Int. J. **44**, 101455 (2023). https://doi.org/10.1016/j.jestch.2023.101455.
2. Hwang, G.J., Fu, Q.K.: Advancement and research trends of smart learning environments in the mobile era. Int. J. Mobile Learn. Organ. **14**, 114 (2019). https://doi.org/10.1504/ijmlo.2020.103911.
3. Mahdi, M.G., Sleem, A., Elhenawy, I.: Deep learning algorithms for arabic optical character recognition: a survey. Deleted Journal. **2**, 65–79 (2024). https://doi.org/10.61356/j.mawa.2024.26861
4. Al-Ayyoub, M., Nuseir, A., Alsmearat, K., Jararweh, Y., Gupta, B.: Deep learning for Arabic NLP: a survey. J. Computat. Sci. **26**, 522–531 (2017). https://doi.org/10.1016/j.jocs.2017.11.011
5. Guellil, I., Saâdane, H., Azouaou, F., Gueni, B., Nouvel, D.: Arabic natural language processing: An overview. J. King Saud Univ. – Comp. Info. Sci. **33**, 497–507 (2019). https://doi.org/10.1016/j.jksuci.2019.02.006
6. Al-Sbou, A.M.F.: A survey of Arabic text classification models. Int. J. Electr. Comp. Eng. (IJECE). **8**, 4352 (2018). https://doi.org/10.11591/ijece.v8i6.pp4352-4355.
7. Hamouda, A.: The effect of blended learning on developing Saudi English majors' writing skills. Int. J. English and Educ. **7**(2), 40–83 (2018). https://doi.org/10.56040/dnas1618
8. El-Dakhs, D.A.S., Ambreen, F., Zaheer, M.: The effect of textual enhancement on collocation learning: the case of Arab EFL learners. Electr. J. Foreign Lang. Teach. **16** (2019). https://doi.org/10.56040/dnas1618
9. Farghaly, A., Shaalan, K.: Arabic Natural language Processing. ACM Trans. Asian Lang. Info. Proc. **8**, 1–22 (2009). https://doi.org/10.1145/1644879.1644881
10. Habash, N.Y.: Introduction to Arabic natural language processing (2010). https://doi.org/10.1007/978-3-031-02139-8
11. Ismail, M., Ahmad, F.S., Ma'ruf, M.A.: The impact of learning Management System "Arabi.id" Web-Based application on developing Arabic language skills. Al-Ta Rib Jurnal Ilmiah Program Studi Pendidikan Bahasa Arab IAIN Palangka Raya. **11**, 213–232 (2023). https://doi.org/10.23971/altarib.v11i2.6591
12. Khalil, S.M.: Effectiveness of mobile learning in enhancing Saudi English majors' skills in learning vowels. Arab World English J. **13**, 281–291 (2022). https://doi.org/10.24093/awej/vol13no4.18

13. Woo, J.H., Choi, H.: Systematic review for AI-based language learning tools. J. Digital Contents Soc. **22**, 1783–1792 (2021). https://doi.org/10.9728/dcs.2021.22.11.1783
14. Mei, S.Y., Ju, S.Y., Mohd, A.B.: Cooperative Learning Strategy in teaching Arabic for Non Native Speakers. Deleted Journal **1**, 112–120 (2017). https://doi.org/10.26417/ejser.v1i2.p261-266.

Lung Disease Detection Using Scale-Invariant Weighted Ensemble Neural Architecture

Abeer Abdelhamid[1], Oluwatunmise Akinniyi[2], Gehad A. Saleh[3], and Fahmi Khalifa[1,2](✉)

[1] Department of Electronics and Communications Engineering, Mansoura University, Mansoura 35516, Egypt
abeerabdo37@std.mans.edu.eg

[2] Department of Electrical and Computer Engineering, Morgan State University, Baltimore, MD 21251, USA
{olaki58,fahmi.khalifa}@morgan.edu

[3] Department of Diagnostic and Interventional Radiology, Faculty of Medicine, Mansoura University, Mansoura, Egypt

Abstract. Among several diseases, COVID-19 has highlighted the urgent need for AI-based tools o develop precise, accurate, and effective containment and treatment plans. This paper utilizes chest images and introduces a hybrid architecture tailored for COVID-19 detection. The proposed architecture integrates multiple modules to extract texture and shape cues, thereby facilitating more accurate and discriminative classification. The analysis pipeline starts with an auto-encoder-based input scaling module to filter out noise and to keep prominent image features. Subsequently, two sets of features are extracted from the preprocessed images representing rich concentration of local and global features. We integrated various texture descriptors (histogram of oriented gradients, gray-level co-occurrence matrix, and local binary patterns) with deeper features extracted from an optimized neural architecture. Thirdly, a feature fusion and weighing module is used to weigh the feature sets to produce an optimal feature vector for classification using a cascaded machine classification module. Evaluation of the proposed methods on publicly available datasets using various metrics and ablation studies documents the robustness of our weighted ensemble classification system. Cross-validation evaluation demonstrate higher accuracy of the proposed system with an average accuracy of 96.5% and 95.5% on two datasets and when compared against other recent methods.

Keywords: Feature Fusion · Lung Diseases · Cascade Classification · Scale Adaptation

1 Introduction

The COVID-19 pandemic has become the most severe global outbreak, impacting lives across more than 150 countries. It has been noticed that early detection

A. Abdelgawad et al. (Eds.): ISBCom 2024, LNNS 1268, pp. 83–94, 2025.
https://doi.org/10.1007/978-3-031-82377-0_8

is the only option for minimizing the possibility of disease in healthy people, and that the infected person should be isolated immediately [6]. Chest X-rays (CXRs) and computed tomography (CT) scans serve as alternative diagnostic methods to blood tests, gene sequencing, or RT-PCR. They also assess the development of the disease, facilitating the selection of appropriate management strategies [1,11]. Although chest CT scans are highly sensitive to pulmonary diseases, they are not portable and pose a significant risk of excessive radiation exposure specifically to younger patients. CXRs, especially those conducted using portable X-ray equipment, offer a safer option by allowing patients to be imaged in isolated settings, thus reducing the risk of virus transmission [9]. While a vaccine was recently constructed, vaccinating the entire global population, especially those from developing nations, will take time [19].

Through recent advancements in data-driven deep learning (DL), multiple models including convolutional neural networks (CNNs) are frequently employed for analyzing medical images with cutting-edge performance in disease classification for diagnosis. This is can be possible because CNNs are capable of extracting local features very precisely from an individual medical image, such as CT scans or CXRs [27]. Early COVID-19 cases often presented with bilateral, multifocal ground-glass opacities predominantly in the peripheral regions, frequently progressing to pulmonary consolidation [16]. Many lung diseases have symptoms that are similar. As a result, doctors are having difficulty distinguishing between COVID-19 infection and other types of viral pneumonia [36]. As a result, symptom similarities could lead to a misdiagnosis, delayed treatment, or even death in this case. As a result, there is a pressing need for technology to aid physicians in their evaluation [30].

CXR images have been deeply investigated and evaluated in various research studies to detect COVID-19 using DL algorithms. Ahammed et al. [3] suggested a CNN-based network with high accuracy of 94.03%. The system was trained using CXR images from normal, pneumonia, and COVID-19 patients. The work's limitation was that a dataset of only 285 images was used, which was insufficient for training a DL-based system for COVID-19 prediction. A similar 3-class-classification was developed by Abdelhamid et al. [2]. The study evaluated the effect of different activation functions and optimization algorithms on the performance obtaining 99.3% using the LeakyReLU activation. Despite higher performance, the method employed pertained Xception model and hold out evaluation scenario. Wang et al. [37] created a model based on ResNet with fusion effects to dynamically improve their weight ratio. The CXR images were classified into normal, COVID-19, and viral pneumonia with a 96.1% accuracy. Waisy et al. [4] started with preprocessing stage for their model and combined (via weighted) decisions of the two leranable models and achieved an accuracy of 98% on a smaller dataset of 800 CXRs. Balasubramaniam et al. [6] proposed an ensemble-based system and applied to a 5K images dataset. Their approach achieved an accuracy of 92.3%. Ensembling is also used by Ouyang et al. in [25]. In order to identifying COVID-19 in CT images, the researchers created a dual-sampling attention network. The predictions were aggregated through weighted average ensembling post training two distinct 3D ResNet34 models with varied sampling approaches. On the test dataset, their framework achieved an AUC of 0.944, along with an accuracy of 87.5% and an F1-score of 82%. Sandhya et al. [20] suggested an ensemble of pretrained models. Their method achieved a higher three-class test accuracy of 98.82%. Mostafiz et al. [24] developed a pipeline that contains watershed segmentation, texture feature extraction (GLCM/Wavelt), and ResNet50. They employed a maximum relevance minimum redundancy (mRMR) and recursive feature elimination (RFE) for feature selections that are fed to random forest (RF) for classification. The accuracy obtained for a four classes (normal vs. pneumonia, COVID-19 vs. bacterial vs. viral Pneumonia) was 98.48% due to numerous optimization procedures. To extract important features resistant to overfitting, Hossain et al. [15] presented a detection system in which an attention-based fusion model is built using a combination of multiple pre-trained CNN base model (Res-Net50 V3, VGG_16, and Google-Inception V3). The experimental results showed a 96.75% accuracy when the fusion model is used. Phumkuea et al. [28] introduced a multilayer combination model involving 4,200 CXR images divided into three classes (COVID-19, non-COVID-19 and healthy patients). Their model was trained and evaluated, yielding appropriately high accuracy (96.2%) and F1-score (96.2%).

In total, a tremendous number of AI-based methods and techniques have been developed in literature for COVID-19 recognition via CXR images proposes various techniques. Reported results of literature studies vary depending on the specific number of classes, type/size of the utilized dataset, and the developed DL algorithms. Some of the existing work utilized large amounts of datasets and employed pre-trained models. Some literature work employed multi-scale approaches and others also employed feature fusion. In this study, we suggest a deep weighted ensemble classification of CXR image features that combine weighted discriminatory texture and deep features using a scale-adaptation deep module. The proposed classification, as shown in Fig. 1, starts with a learnable input scaling module followed by image preprocessing and enhancement. Secondly, two sets of features are extracted from CXR images representing rich concentration of local and global features. Thirdly, a feature fusion and weighing module is used to weigh the feature sets to produce an optimal feature vector for classification. Finally, a cascaded ML classification module using the one-vs.-rest (OVR) classifier is trained to differentiate the classes. Evaluation of the proposed methods using publicly available data sets documents the robustness of the extracted features for classifying. The contribution of this work include: (i) The design utilizes an ensemble approach with weighted fusion of information-rich features for classification, instead of direct feature concatenation; (ii) a learnable input scaling module retains prominent disease-related features, unlike conventional image resizing algorithms that lose important details when scaling down; (iii) we developed an optimized deep architecture for feature extraction, enhancing intermediate feature learning compared to transfer learning with pre-trained CNNs; (iv) we showed improved design classification accuracy using two public data sets and cross-validation evaluation scenrio.

2 Methodology

The suggested analysis pipeline is depicted in Fig. 1. Sequentially, the system performs input scaling, data preprocessing, features extraction, weighted ensemble fusion, and cascaded classification. The stages are fully illustrated in the subsections that follow.

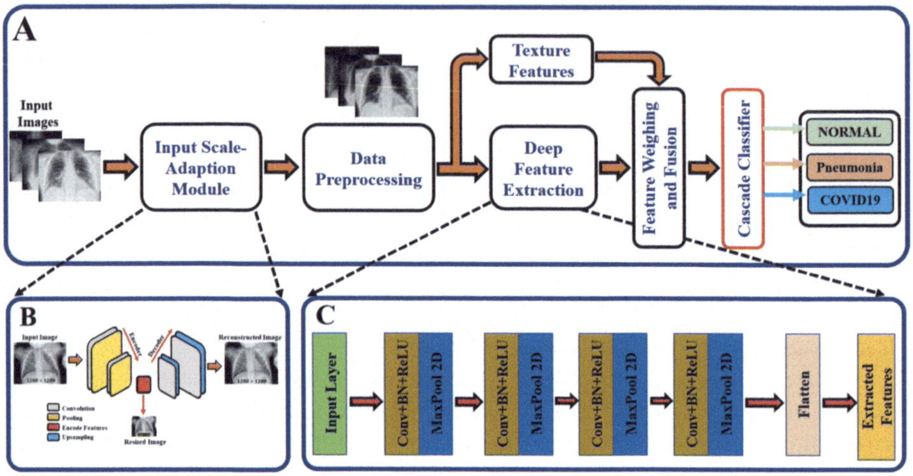

Fig. 1. Detailed block diagram of (A) the proposed system showing (B) details of input images scaling module and (C) the deep feature extractor.

2.1 Input Scaling Module and Data Preprocessing

In biomedical image analysis, evaluation datasets acquired at different locations with varying acquisition machines and imaging protocols result in images of different sizes. Modern computer-assisted systems for disease diagnosis using off-the-shelf networks require image downscaling to fit the system's input layer, causing significant loss of critical features. To prevent this, we developed a learnable image module base on auto-encoders to resize input CXR images. An effort to perform down sampling without losing valuable information. Auto-encoders excel at filtering noise and irrelevant data while minimizing information loss during output reconstruction, outperforming traditional downscaling algorithms.

In our design, the developed learnable input scaling module resize the CXRs to a fixed size of 320×320 using an auto-encoder structure (see Fig. 1-B). In this design, path of encoding contains successive convolutional and pooling layers to obtain the feature map, AE_m. Following this, convolution, transposed convolution, and reshaping transform AE_m to $320 \times 320 \times 4$. Concatenating original and processed AE_ms produces high and low-resolution images. These are crucial for preprocessing, feature extraction, and module training, aiming to minimize reconstruction error and discard redundancy and noise. Training combined a hybrid loss function integrating pseudo-Huber and log-cosh losses [7,33].

Image preprocessing, a critical step in medical image analysis field, follows image resizing and includes various essential tasks. These range from basic noise removal to advanced techniques such as histogram equalization and data augmentation, crucial for improving classification accuracy and enhancing image analysis tasks. First, an input resized image, G, is normalized as $G_n = \alpha + \frac{(G - G_{min})(p-\alpha)}{G_{max} - G_{min}}$, where G_n, G_{max} (G_{min}), and α (p) represent is the normalized data, the maximum (minimum) data value in the domain of G, and the maximum (minimum) value in the output range. Following data normalization, histogram equalization is then applied, which is a process utilized to enhance the quality of images in black and white color scales. Histogram equalization was accomplished by effective expanding out the highest common intensity values, which increased the image intensity range. A spatially variable histogram equalization procedure appears to improve the visibility of anatomical structures in a variety of clinical scenarios [35]. Finally, due to the imbalance between the data classes, data augmentation technique is utilized to create more minority class images. With dataset augmentation, the proposed model is less likely to become over-fitted to training data. We applied different operations, including vertical and horizontal flipping, shear, image contrast, and scale operations.

2.2 Feature Extraction

To boost the classification accuracy of our analysis system, we incorporated two information-rich features types: texture-based and CNN-derived features, renowned for their effectiveness in the classification of biomedical images [10]. These features are combined using a weighted fusion approach to form an ensemble classifier, fortifying our COVID detection framework significantly.

Firstly, we harnesses the power of neural architecture, CNN in particular, that has shown performance improvement, when compared with classical ML, due to its inherent characteristics to recognize objects and patterns in images via self-learning. To this end, we built an in-house CNN architecture for faster and easier training process. Despite transfer learning can be employed using pre-trained CNNs, we opted the choice of developing our own architecture, trained from scratch to extract meaningful features directly from the available data. In order to optimize the architecture hyperparameters, some experiments were performed to determine the optimal settings. In particular, we utilized the GridSearchCV [21] technique to determine the optimal network parameters and obtained a network of 4 convolutional layers (CLs), intervened by max-pooling layers, two fully connected layer with dropout (threshold equal 0.2), and a final 3-node using softmax for classifying input images. The CLs of the first two layers use a (9×9) kernel, ReLU, and a stride of 1. The kernel size of the (maximum) pooling layers used in this step is (5×5) and the stride is 2. The following stage is made up of three repetitive CLs each of which has (7×7) kernels and a stride of 1. The respective numbers of filters of those stages were 128, 256 and 256.

Finally, after model flattening two fully connected layers with ReLU activation function are used.

Secondly, we extract texture features from chest images to enhance disease type differentiation, capturing spatial intensity arrangements crucial for accurate analysis. These features include higher-order, morphological, and structural descriptors. The higher-order features are computed using a gray-level co-occurrence matrix (GLCM), a statistical method that quantifies pixel connectivity with neighbors to derive metrics such as contrast, correlation, energy, and homogeneity [13]. The GLCM matrix, sized based on image grey levels, reflects co-occurring pixel pairs' frequencies, crucial for assessing image texture and pattern analysis. For a preprocessed image, G, with dimension $A \times B$; the GLCM matrix, K, parameterized by an offsets $(\Delta x,\ \Delta y)$ can be built using Eq. (1) as in [34]:

$$K_{\Delta x, \Delta y}(i,j) = \sum_{x=1}^{A}\sum_{y=1}^{B}\begin{cases}1, & G(x,y) = i \text{ and } G(x+\Delta x, y+\Delta y) = j \\ 0, & \text{otherwise}\end{cases} \quad (1)$$

here $(i,\ j)$ and $(x,\ y)$ are the pixel (greylevel) values and the spatial positions, respectively. $K(x, y)$ is the pixel greylevel at $(x,\ y)$. Co-occurrence matrices are commonly large and sparse, necessitating the derivation of various metrics to yield a meaningful feature set. Metrics such as energy, contrast, correlation, homogeneity, and dissimilarity are derived from the GLCM matrix to capture detailed texture information [18]. In addition to GLCM, local binary patterns (LBP) are incorporated to enhance texture analysis efficiency, particularly in handling variations caused by lighting conditions and noise. Unlike GLCM's global approach, LBP operates locally by encoding each pixel's relationship with its neighbors into binary patterns, thus providing additional structural insights with enhanced computational efficiency [26]. Figure 2 show the differences between the GLCM and LBP for normal, pneumonia, and COVID cases. Further, histogram of oriented gradients (HOG) is employed to extract important morphological features. In comparison to GLCM and LBP, the HOG captures local intensity gradients information in image by analyzing the distribution of edge orientations within an image [32]. This typical enhances the prediction by providing a more discriminative representation of texture as shape and texture cues are crucial for classification. Typically, the HOG method involves the computation of the gradient magnitude and orientation for each pixel over image patches (or localized regions or cells) for which local histograms are generated [22]. A feature matrix for those histograms is then used in our architecture for aid classification. The parameters used for the HOG are as follows: (*orientations*, 9), (*pixels per cell*, 8×8), (*cell per block*, 2×2).

2.3 Feature Fusion, Weighing and Cascade Classifier

After individual feature selection, a feature fusion strategy is utilized to concatenate the extracted features. Typically, feature fusion is a method for combining features (different information) extracted from the same sample (input)

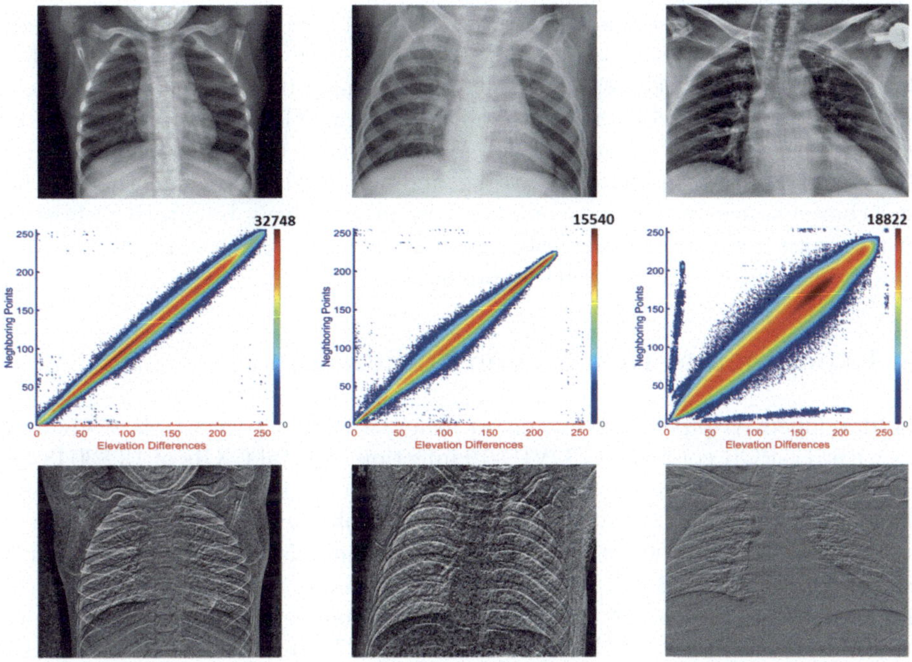

Fig. 2. Estimated GLCM and LBP (second and third rows, respectively) for normal, pneumonia, and COVID examples from left to right. The GLCM values are computed based on the frequency of each pixel value at a given point (p_0) appearing with its 8-neighbors at distance r ($r = 1$).

using various methods to improve classification model performances. Initially, we extracted several features from the CXR images, then the proposed method seeks to determine the impact of various features on the classification of COVID-19 disease. The fused feature are presented as an input to a ML classifier. Traditional fusion techniques employs equal weighing to the features, which may be an inefficient way that can affect the overall performance. In order avoid that, we adopted a weighted fusion algorithm where the concatenated features are weighed according to its importance for performance enhancement. The optimal weights are determined using the modified cat and mouse based optimizer (MCMBO) algorithm [5]. This method is employed to automatically select the optimal weighted features to boost the performance of the COVID detection. This is mathematically by Eq. 2 defined as in [5]:

$$F_s = W_v * F_o + (1 - Wt) * F_o \tag{2}$$

Here, W_v and F_o are the weight factor and the optimal features, respectively.

The final stage of our system is the classification, which is conducted using the weighted feature vector derived by Eq. (2). In our pipeline we incorporated a cascade technique to increase the system's detection accuracy [12]. Particularly, our architecture involves multiple ML classifiers stages, each of which gradually filter out false positives to create new features before returning the final classification. We created the cascaded classifier using three distinct ML models: SVM, DT, and MLP, which run in parallel. The final classification is derived by combining the outputs of each model. We implement the cascade voting algorithm, surpassing both majority voting and weighted average methods. This algorithm enhances prediction confidence by progressively integrating classifier outputs until a consistent prediction label is achieved across multiple classifier groups, ensuring robust and reliable predictions.

3 Experimental Results and Discussion

The developed AI-pipeline is evaluated using two publicly-available CXR data sets with RT-PCR confirmed COVID-19 infection. A total CXRs of 13,830 [8,31] were used for evaluating and assessing the system accuracy and robustness. The datasets were attained as digital radiographs with portable X-ray unit and consist of three types of images normal, pneumonia, and COVID. The first dataset [29] contained 1,583 normal; 4,273 pneumonia; and 579 COVID images, while the second dataset [2] is collected from different sources [8,31] is divided into 1,371 for COVID-19, 1,751 for normal, and 4,273 for pneumonia. The system is implemented using TensorFlow framework and Keras as the backend using and all experiments are conducted on a windows machine with 64GB of RAM, 4GB NIVIDIA graphics card, and a 12 core i7-processor. Model evaluation relied on four metrics: accuracy (Acc), precision (Pr), and recall (Rc) and F1-score.

As previously stated, we built a cascaded classifier with three different models: SVM, DT, and MLP. Thus, we first evaluated the classification accuracy (using the first data set) for texture and CNN features without fusion using individual classifiers. The results are demonstrated in Table 1 with SVM showing the best average accuracy compared with other ML classifier. Secondly, we test the importance using the proposed weighted ensemble approach and employed an ablation study. The results are depicted in Table 2. As seen, the weighted cascaded classification showed enhanced performance compared to individual ML for both CNN and texture features in Table 1. Thirdly, we examined the fusion of the two feature sets (CNN-derived and texture-based) without feature weighing (i.e., just feature concatenation). The fusion showed an enhancement compared to single feature classification with accuracy reaching 94.2%. Finally, we tested the proposed ensemble (weighted fusion + cascaded classification) and the results for both datasets is shown in the last two rows of Table 2. It is worth mentioning that the proposed architecture was trained and tested on the first data set using the 5-fold cross-validation scenario, then the trained model is used to directly test the second data set.

In addition to the evaluation of the results using of the accuracy metrics, e.g., Tables 2, both additional quantitative evaluation is performed using confusion matrix (CM) and receiver-operating-characteristics (ROC) curves. CMs for individual features and their fusion are shown in the first row of Fig. 3. The

Table 1. Average accuracy of various classification models for individual features. CNN, DT, SVM and MLP stand for convolutional neural network, decision tree, support vector machine, and multilayer perceptron, respectively.

Method	Classifier Accuracy, (%)			
	SVM	DT	RF	MLP
Texture-based Features	91.3	89.0	86.4	90.6
CNN-derived Features	85.1	80.7	77.5	83.8

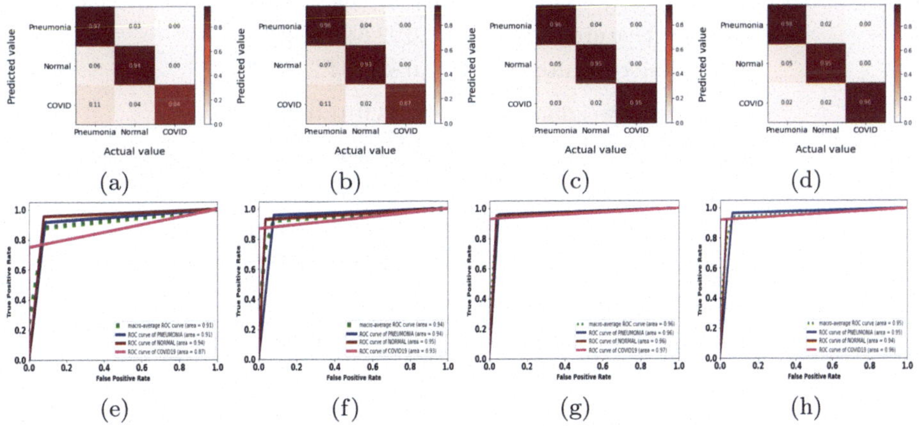

Fig. 3. Classification performance (*first row*: the confusion matrices and*second row*: ROC curves) for texture features (a, e); CNN-derived features (b, f); and proposed model on the first (c, g) and second (d, h) data sets.

CM visualization is extremely useful for (1) determining which classes, if any, were misclassified the most; and (2) calculating other indices, such as precision, F1-score, etc. The second row of Fig. 3 depicts the ROCs. The ROC curves graphically represent a classifier's performance across trade-offs between true positives and false positive rates, which are crucial for assessing class differentiation accuracy. The ROC and CMs of the weighted ensemble show the higher performance of our design in correctly classifying CXRs. Finally, to demonstrate the advantages of our proposed model, we compared its accuracy with competitive models, focusing on mean accuracy for quantitative analysis. Table 2 shows how the proposed approach compares to other methods in terms of accuracy.

Despite the high accuracy of the proposed method, some limitations and challenges exits. First, the suggested model was applied on a relatively small size and single data-type dataset. Secondly, the integration of multiple modules in the proposed hybrid system increases the computational time when compared with single module architecture. Second, the proposed system provides disease recognition only, which is insufficient to provide full assistance to medical professionals in making an accurate diagnosis on time. Namely, it will be advantageous if the infection can be localized and its status can be determined, allowing for seriousness assessing and physicians can accurately provide treatment based on severity degree.

Table 2. Overall accuracy of our system and the associate ablation studies. Acc: accuracy; Pr: precision, Rc: recall; CNN: convolutional neural network.

Method	Evaluation Metrics, (%)			
	Acc	Pr	Rc	F1-Score
Texture Features	93.0	92.6	90.5	91.5
CNN-derived Features	90.6	83.5	84.2	83.7
Fusion w/o weighing	94.2	93.0	91.0	92.0
Ismael et al. [17]	94.7	91.0	98.9	94.8
Horry et al. [14]	—	83.0	81.0	81.0
Abbas et al. [23]	95.1	97.9	91.87	93.36
Proposed	96.5	95.7	94.6	95.0
Proposed (First dataset)	96.5	95.7	94.6	95.0
Proposed (Second dataset)	95.5	96.7	93.6	94.5

4 Conclusions and Future Work

This study introduces a weighted ensemble architecture for precise COVID-19 detection from CXR scans. By integrating diverse learning modules, the workflow extracts highly discriminative texture and shape cues for accurate classification. A feature fusion and weighting module optimizes feature vectors, while a cascaded machine classification module distinguishes between classes. Rigorous evaluation and ablation studies validate the robustness of our features, surpassing traditional ML, classical DL models, and cutting-edge methods. Future goals include expanding COVID-19 datasets, testing the weighted feature fusion model with multi-modal data, incorporating vision transformers and attention modules. Also, work direction may contain the application of this model on another medical datasets for other disease to be more generalized.

Acknowledgments. This work supported by the Center for Equitable Artificial Intelligence and Machine Learning Systems (CEAMLS), Morgan State University, Project #11202202.

Disclosure of Interests. The authors have no competing interests to declare that are relevant to the content of this article.

References

1. Abdelhamid, A., Akinniyi, O., Saleh, A.G., Deabes, W., Khalifa, F.: An ensemble neural architecture for lung diseases prediction using chest x-rays. Int. J. Comput. Digit. Syst. **16**(1), 1019–1028 (2024)
2. AbdElhamid, A.A., AbdElhalim, E., Mohamed, M.A., Khalifa, F.: Multi-classification of chest x-rays for covid-19 diagnosis using deep learning algorithms. Appl. Sci. **12**(4), 2080 (2022)
3. Ahammed, K., Satu, M.S., Abedin, M.Z., Rahaman, M.A., Islam, S.: Early detection of coronavirus cases using chest x-ray images employing machine learning and deep learning approaches. MedRxiv **10**(2020.06), 07–20124594 (2020)
4. Al-Waisy, A.S., et al.: Covid-chexnet: hybrid deep learning framework for identifying covid-19 virus in chest x-rays images. Soft. Comput. **27**(5), 2657–2672 (2023)
5. Awais, M., Bhuva, A., Bhuva, D., Fatima, S., Sadiq, T.: Optimized DEC: an effective cough detection framework using optimal weighted features-aided deep ensemble classifier for covid-19. Biomed. Signal Process. Control 105026 (2023)
6. Balasubramaniam, S., Kumar, K.S.: Optimal ensemble learning model for covid-19 detection using chest x-ray images. Biomed. Signal Process. Control **81**, 104392 (2023)
7. Charbonnier, P., Blanc-Féraud, L., Aubert, G., Barlaud, M.: Deterministic edge-preserving regularization in computed imaging. IEEE Trans. Image Process. **6**(2), 298–311 (1997)
8. Chowdhury, M.E., et al.: Can AI help in screening viral and covid-19 pneumonia? IEEE Access **8**, 132665–132676 (2020)
9. Cozzi, D., et al.: Chest x-ray in new coronavirus disease 2019 (covid-19) infection: findings and correlation with clinical outcome. Radiol. Med. (Torino) **125**, 730–737 (2020)
10. Dixon, J., Akinniyi, O., Abdelhamid, A., Saleh, G.A., Rahman, M.M., Khalifa, F.: A hybrid learning-architecture for improved brain tumor recognition. Algorithms **17**(6), 221 (2024)
11. Elmokadem, A.H., Mounir, A.M., Ramadan, Z.A., Elsedeiq, M., Saleh, G.A.: Comparison of chest CT severity scoring systems for covid-19. Eur. Radiol. 1–12 (2022)
12. Gama, J., Brazdil, P.: Cascade generalization. Mach. Learn. **41**, 315–343 (2000)
13. Haralick, R.M., Shanmugam, K., Dinstein, I.H.: Textural features for image classification. IEEE Trans. Syst. Man Cybern. **6**, 610–621 (1973)
14. Horry, M.J., et al.: X-ray image based covid-19 detection using pre-trained deep learning models (2020)
15. Hossain, M.S., Shorfuzzaman, M.: Non-invasive covid-19 screening using deep learning-based multilevel fusion model with an attention mechanism. IEEE Open J. Instrum. Measur. (2023)
16. Hosseiny, M., Kooraki, S., Gholamrezanezhad, A., Reddy, S., Myers, L.: Radiology perspective of coronavirus disease 2019 (covid-19): lessons from severe acute respiratory syndrome and middle east respiratory syndrome. Am. J. Roentgenol. **214**(5), 1078–1082 (2020)
17. Ismael, A.M., Şengür, A.: Deep learning approaches for covid-19 detection based on chest x-ray images. Expert Syst. Appl. **164**, 114054 (2021)
18. Jafarpour, S., Sedghi, Z., Amirani, M.C.: A robust brain MRI classification with GLCM features. Int. J. Comput. Appl. **37**(12), 1–5 (2012)
19. Kaiser, M.S., et al.: iworksafe: towards healthy workplaces during covid-19 with an intelligent phealth app for industrial settings. IEEE Access **9**, 13814–13828 (2021)

20. Kuzinkovas, D., Clement, S.: The detection of covid-19 in chest x-rays using ensemble CNN techniques. Information **14**(7), 370 (2023)
21. LaValle, S.M., Branicky, M.S., Lindemann, S.R.: On the relationship between classical grid search and probabilistic roadmaps. Int. J. Robot. Res. **23**(7–8), 673–692 (2004)
22. Liu, X., et al.: Infield corn kernel detection using image processing, machine learning, and deep learning methodologies under natural lighting. Expert Syst. Appl. **238**, 122278 (2024)
23. Maguolo, G., Nanni, L.: A critic evaluation of methods for covid-19 automatic detection from x-ray images. Inf. Fusion **76**, 1–7 (2021)
24. Mostafiz, R., Uddin, M.S., Reza, M.M., Rahman, M.M., et al.: Covid-19 detection in chest x-ray through random forest classifier using a hybridization of deep CNN and dwt optimized features. J. King Saud Univ.-Comput. Inf. Sci. **34**(6), 3226–3235 (2022)
25. Ouyang, X., et al.: Dual-sampling attention network for diagnosis of covid-19 from community acquired pneumonia. IEEE Trans. Med. Imaging **39**(8), 2595–2605 (2020)
26. Park, B., Shin, T., Cho, J.S., Lim, J.H., Park, K.J.: Improving blueberry firmness classification with spectral and textural features of microstructures using hyperspectral microscope imaging and deep learning. Postharvest Biol. Technol. **195**, 112154 (2023)
27. Paul, A., Basu, A., Mahmud, M., Kaiser, M.S., Sarkar, R.: Inverted bell-curve-based ensemble of deep learning models for detection of covid-19 from chest x-rays. Neural Comput. Appl. **35**(22), 16113–16127 (2023)
28. Phumkuea, T., Wongsirichot, T., Damkliang, K., Navasakulpong, A., et al.: Classifying covid-19 patients from chest x-ray images using hybrid machine learning techniques: development and evaluation. JMIR Formative Res. **7**(1), e42324 (2023)
29. Prashant Patel: Pneumonia COVID-19 Image Dataset. https://www.kaggle.com/datasets/prashant268/chest-xray-covid19-pneumonia. Accessed 21 Nov 2023
30. Rahman, T., et al.: Bio-cxrnet: a robust multimodal stacking machine learning technique for mortality risk prediction of covid-19 patients using chest x-ray images and clinical data. Neural Comput. Appl. 1–23 (2023)
31. Rahman, T., et al.: Exploring the effect of image enhancement techniques on covid-19 detection using chest x-ray images. Comput. Biol. Med. **132**, 104319 (2021)
32. Sabry, E.S., Elagooz, S.S., El-Samie, F.E.A., El-Bahnasawy, N.A., El-Banby, G.M., Ramadan, R.A.: Evaluation of feature extraction methods for different types of images. J. Opt. 1–26 (2023)
33. Saleh, R.A., Saleh, A., et al.: Statistical properties of the log-cosh loss function used in machine learning. arXiv preprint arXiv:2208.04564 (2022)
34. Sun, W., Zeng, N., He, Y.: Morphological arrhythmia automated diagnosis method using gray-level co-occurrence matrix enhanced convolutional neural network. IEEE Access **7**, 67123–67129 (2019)
35. Tang, J.R., Isa, N.A.M.: Adaptive image enhancement based on bi-histogram equalization with a clipping limit. Comput. Electr. Eng. **40**(8), 86–103 (2014)
36. Tharwat, S., Saleh, G.A., Saleh, M., Mounir, A.M., et al.: Chest CT total severity score on admission to predict in-hospital mortality in covid-19 patients with acute and chronic renal impairment. Diagnostics **12**(7), 1529 (2022)
37. Wang, N., Liu, H., Xu, C.: Deep learning for the detection of covid-19 using transfer learning and model integration. In: 2020 IEEE 10th International Conference on Electronics Information and Emergency Communication (ICEIEC), pp. 281–284. IEEE (2020)

Comparison Between the Performance of Human Translators and AI-Supported Applications

Anfal Sabeeh Hamood[1](✉), Sanaa Lazim Hassan Al Ghareeb[2], and Bilal Hameed Zaid[3]

[1] Department of Journalism, College of Mass Communication, University of Baghdad, Baghdad, Iraq
anfaltamimi@comc.uobaghdad.edu.iq

[2] Department of English Language, College of Education for Women, University of Baghdad, Baghdad, Iraq
sanalazim@coart.uobaghdad.edu.iq

[3] Continuing Education Center, University of Baghdad, Baghdad, Iraq
bilalalaameri@dcec.uobaghdad.edu.iq

Abstract. This study aims to conduct an exhaustive comparison between the performance of human translators and artificial intelligence-powered machine translation systems, specifically examining the top three systems: Spider-AI, Metacate, and DeepL. A variety of texts from distinct categories were evaluated to gain a profound understanding of the qualitative differences, as well as the strengths and weaknesses, between human and machine translations. The results demonstrated that human translation significantly outperforms machine translation, with larger gaps in literary texts and texts characterized by high linguistic complexity. However, the performance of machine translation systems, particularly DeepL, has improved and in some contexts approached that of human performance. The distinct performance differences across various text categories suggest the potential for developing systems tailored to specific fields. These findings indicate that machine translation has the capacity to bridge the gap in translation productivity inefficiencies inherent in human translation, yet it still falls short of fully replicating human capabilities. In the future, a combination of human translation and machine translation systems is likely to be the most effective approach for leveraging the strengths of each and ensuring optimal performance. This study contributes empirical support and findings that can aid in the development and future research in the field of machine translation and translation studies. Despite some limitations associated with the corpus used and the systems analysed, where the focus was on English and texts within the field of machine translation, future studies could explore more extensive linguistic sampling and evaluation of human effort.

Keywords: Human translation · Machine translation · Spider-AI · Metacate DeepL

A. Abdelgawad et al. (Eds.): ISBCom 2024, LNNS 1268, pp. 95–108, 2025.
https://doi.org/10.1007/978-3-031-82377-0_9

1 Introduction

Since time immemorial, countries and societies have relied on translation to communicate, share knowledge and cultural factors. as well, Translation has connected different civilizations, through It creates a platform for sharing knowledge and exchanging experiences among unique societies. Globalization and technological advancement raised the level of importance and need for translation. Therefore, Machine translation, became a dynamic approach to translation underpinned by the advancement of artificial intelligence targeting automation [1] and achieve their tasks in record time.

Machine translation is becoming popular among organizations and individuals, given advances in artificial intelligence technologies that make translation more accurate, faster, and cost-effective. It is a leap from conventional translation that seeks to emulate human capability in translation. The promising AI-applications and neural network technologies are revolution in translation sector as other sectors; they facilitate learning and repetitive tasks, translating complex languages, and improving translation quality [2].

However, despite the quality of translation generated by these applications, it is difficult surpass human translation, in terms of accuracy and context of specialized areas. In other words, translation is not the mechanical exchange of words from one language to another; it is a deep and cultural understanding of the context and a categorical transfer of meanings and connotations in a proper form. As a result, acknowledging the distinction between Human translation and machine translation and its shortcomings and rights will lead to the optimal mix of both better quality of translation [3].

Table 1. Previous Studies Related to The Study

Study	*Study Focus*	*Key Findings*	*Conclusion*
[4]	Investigates the uses and perceptions of machine translation (MT) at the European Commission	MT was widely used for preliminary translations but still relied on human translators for final versions. Human translation was preferred for accuracy and contextual nuances	MT is useful for speeding up translation processes but still requires human oversight for quality and contextual accuracy
[5]	Analyses the impact of post-editing (PE) machine translations on creativity and user reading experience	PE of MT output improved the overall reading experience but limited creative freedom compared to human translation	PE is beneficial for technical translations but may restrict creativity in more nuanced or literary texts
[6]	Explores the differences between human translations and machine-generated translations	Human translations outperformed MT in preserving cultural and rhetorical nuances. MT struggled with idiomatic and context-specific language	Human translations are superior for complex linguistic and cultural contexts, while MT is effective for straightforward texts

(*continued*)

Table 1. (*continued*)

Study	*Study Focus*	*Key Findings*	*Conclusion*
[7]	Evaluates claims of MT systems achieving human parity at the WMT 2019 competition	While MT performance had improved significantly, it still fell short of human parity in most linguistic areas, particularly for creative and culturally sensitive texts	MT has made progress, but it cannot yet fully replace human translators in all contexts, particularly for more complex texts
[8]	Studies the potential for human parity in neural machine translation (NMT) systems, considering human variables	NMT systems show near-human performance in specific technical fields, but human variables such as cognitive ability and experience impact translation quality	Human translators still outperform NMT in contexts requiring deep understanding and adaptation to unique linguistic variables

This study aims to conduct a comprehensive comparison between the performance of human translators and AI-supported machine translation in various contexts, including general, technical, and literary texts.

The study seeks to answer the following questions:

What are the differences in translation quality between human translators and machine translation in terms of accuracy, clarity, and meaning preservation?

What are the strengths and weaknesses of each approach? From the above, the objective of this study is to contribute to the literature by providing valuable insights into the optimal use of human and machine translation and identifying areas for improvement and integration between them.

2 Methodology

To realize the goals of this study and response to its questions, a meticulous comparative methodology was selected to assess the human translator versus the machine translation using artificial intelligence. This choice is intricately connected to the propositions of prior studies. According to Koehn, a value added of multi-edged comparison is justified, as it allows for a more thorough discovery of the divergence in 'capability' between humans and machines [9].

Considered criteria for the selection of translation samples include text type, and classification, between general, technical, and literary texts, and text length, linguistic complexity, and thematic domain, in line with the principle of sample diversity mentioned in the paper by Castilho et al., 2018, where sample diversity is essential for achieving a holistic assessment of translation performance. On this basis, one hundred texts from each category, or a total of three hundred texts, were selected, with the bulk of selection having a word length of 200 to 500 words per text.

The examples above demonstrate how texts have been chosen systematically presented in terms of length, linguistic characteristics, and the thematic domain. Such an

Table 2. Comparison of Text Characteristics in Human and Machine Translation Performance Across Different Categories.

Standard	Texts
Text Length	Short Text (Approx. 200 words): "Renewable energy sources like solar, wind, and hydropower are gaining traction because they contribute significantly to achieving the urgent climate change goals. Clean energy technologies are the core of minimized greenhouse gas emissions and reliance on fossil fuels. Countries and corporations are investing in solar, wind and hydropower to develop a greener planet. That said, the challenges, including intermittent and storage, must be overcome to truly leverage the benefits of renewable energy sources." Long Text (Approx. 500 words): "Across various sectors of the economy, artificial intelligence has allowed for a societal transformation. AI systems are capable of Boolean analysing humongous amount of data, recognizing patterns, and eventually making sound decisions. In certain tasks, AI overcomes humanity, and they will eventually excel in most of them. In the health sector, AI aids in the development of individualized treatment plans, drug discovery, and increasing the accuracy of diagnosis. Artificial Intelligence -powered tools enable the education sector to adapt to individualistic levels of intelligence and provide personalized response for the same. The transport system is highly revolutionized by AI with the development of driverless cars and intelligent traffic management systems. Moreover, recommendation systems and content generation in the entertainment industry and other numerous impacts in other sectors will be inevitable through AI. Other entities including having privacy compromised, workforce reduced through AI creation bring more menace than benefits to the society. As AI technology continues to grow, fair distribution of benefits and risks reduction via responsible growth and implementation will form a core tenet. Therefore, AI should enable a healthy future for humanity filled with a lot of ethical and societal impulses."
Linguistic Complexity	Simple Text: "Cats are considered one of the most popular pets around the world. They are known for their independence, playfulness, and their affection for each other. There are many different breeds of cats, which have distinctive characteristics. While some cats have long hair, others do have short one. Fur colour also differs from black and white to orange and grey. Cats are easy to take care of pets because they clean themselves and not being walked. However, they also need proper care, which includes feeding, vet's routine attendance and, of course, love and attention from their owner." Complex Text: "Quantum computing is a transformative technology that leverages the rules of quantum mechanics to perform intricate calculations. Prior to this breakthrough, classical computers utilized binary bits, while quantum computers are based on quantum bits, or qubit, which can simultaneously take on numerous states, called superpositions. These quantum properties, along with entanglement and interference, allow quantum computers to solve certain issues exponentially more quickly than their classical equivalents. For example, quantum algorithms representations like Shor's factorization lemma and Grover's database search method have the potential to reshape cryptography and pharmaceuticals, respectively. Nevertheless, the development of commercially useful quantum computers remains a long way off, as researchers must cope with qubit's vulnerable out states as well as the effects of disruption and noise. Thus, quantum computing has the potential to revolutionize computing, open new research frontiers, and much more as it progresses in the future."

(continued)

Table 2. *(continued)*

Standard	Texts
Thematic Domain	Medical Field Text: "Quantum computing is a revolutionary technology that utilizes the rules of quantum mechanics for,performing complex calculations. Classical computers used binary bits, whereas quantum computers use quantum bits, "qubit," which can simultaneously represent several states, which are known as superpositions. Besides quantum properties, including superposition, entanglement, and interference, quantum computers can solve problems many times more quickly than classical computers. For instance, quantum algorithms projection such as the lemma of factorization by Shor and the search method of a database by Grover may alter cryptography and the pharmaceutical industry, individually. Nevertheless, useful commercially quantum computers are still many years away, as researchers encounter the challenge of qubit's rapid decay out states and the influence of disruption and noise. Thus, quantum computing has the potential of revolutionizing computing as well as opening up new research areas among others in the future." Literary Field Text: "Magical Realism. Magical realism is a literary genre that combines realistic and fabulous components to make the extraordinary a fundamental component of everyday life—rooting in Latin America, there is a continuous flow of mythical, supernatural, or enchanting components into a realistic narrative. Writers such as Gabriel Garcia Marquez, Isabel Allende, and Salman Rushdie have made magical realism popular by developing it to probe national identity, political life in their countries, and the nature of the contemporary human subject. Works of magical realism blur the boundaries between the actual and the marvellous and raise readers into a world of enchantment, wonder, and the subliminal. Importantly, magic realism provides satire, with the magical narrative functioning as an allegory for existential, social, or political reality, and as a parody of social and political reality. Magical realism has significantly altered global literature. From that moment, it realized by multiple writers in distinct languages and research topics."

approach helps to explore the quality of translation in several contexts, which, in its turn, ensures the previous study's findings' reliability and validity.

The quality of translation was assessed according to the criteria of accuracy, clarity, and preservation of meaning. These scales were based on previous research. For instance, [10] suggested a comprehensive approach to evaluating the quality of translation. Each criterion was assessed on a scale from 1 to 5 with one denoting the worst and five indicating the best performance. Furthermore, the assessment was conducted by three independent experts in the field to eliminate potential bias.

With the support of machine AI-translation, three top and popular machine translations supported with artificial intelligence, Spider-AI, Metacate, and DeepL, the comparison to human performance was made: All the programs are results of recent studies into machine translation performance, which is the reason for choosing Th the perception of their mediocre quality:

Spider-AI: It is a state-of-the-art machine translation system that uses deep learning technology and neural networks to simulate high-quality translations. Such materials are especially suitable for several types of language and idiosyncrasies or terminology specific to a particular field of study [11].

Metacate: is a machine translation program that makes use in various scientific and technical sectors, like medicine, legal or mathematics, furthermore to its use to the NLP, & ML techniques to modify the translation quality [12].

DeepL: This program is one of the most widespread applications. This machine translation engine provides high-quality translation that theoretically corresponds to human translation. Translation mechanisms based on neural technology and deep learning help create the most truthful translations [13].

Diverse types of texts: general, technical, and literary texts were used in the research to ascertain for which contexts of machine translations they are geared. After, machine translations were juxtaposed with the translations made by human experts. With this research, it was possible to delineate the pros and cons of both programs. The author of the study designed it to find out how these programs can approach the quality of human translation. According to the above reasons, the current research implements three advanced machine translation programs to exemplify the research field concerning its actual possibilities and opportunities to achieve human parity so that it is possible to identify the further production and cooperating of the machines and humans in the future.

This study used a combination of statistical methods and graphical representations to analyse data and present results in a visual-style, easy-to-understand manner. The following processes participated in these methods:

To represent the distribution of errors and percentages over the categories of translation, the use of charts seemed necessary. A pie chart was utilized to present the percentage of errors detected among the three identified categories for human translation and three categories for machine translation [13]. The radar chart was also applied to illustrate the performance of Spider-AI, Metacate, and DeepL machine translation systems across multiple identified evaluation criteria, further enabling a holistic comparison between these systems. In [14] similarly used radar charts to compare performance metrics across different machine translation systems, demonstrating their utility in highlighting relative strengths and weaknesses in machine translation [14].

Regression analysis was the most appropriate tool to analyse the relationship between the study variables and the translation quality [15]. Several multiple linear regression models were prepared to evaluate the influence of length of text and linguistic complexity of translation on the quality of the process. Both human and machine translation were taken into consideration [16]. In addition, the regression models were previously prepared to identify major factors that needed to be predicated, as well as giving an idea about their strength [17].

The reason to use regression was to show the relationship between the measures mentioned above, and so results will be analysed graphically, such an analysis defines patterns and considered to be vital for underlying my study. These insights will be further discussed to become part of the final discussion of results [18].

3 Results

First, Graphs were used to visually display the key findings of the study. Figure 1 illustrates the evolution of translation quality over time for both human translators and the three machine translation programs (Spider-AI, Metacate, DeepL).

The depicted line graph indicates the development of translation quality during five months, which is how long the research lasted. Specifically, the x-axis grounds on time intervals, whereas the y-axis shows quality assessment scores, varying from 0 to 5.

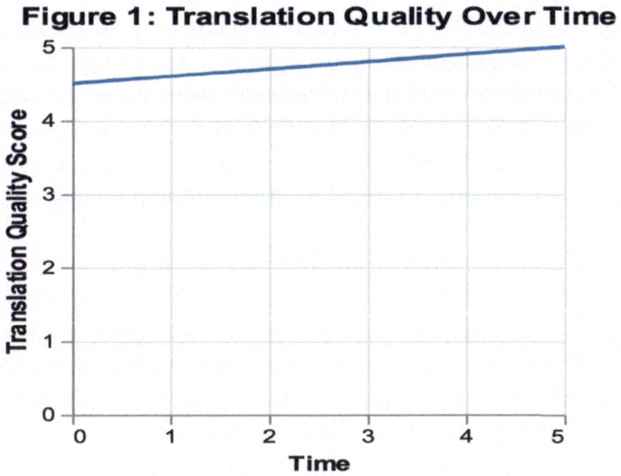

Fig. 1. Evolution of Translation Quality Over Time for Human Translators and Three Machine Translation Programs (Spider-AI, Metacate, DeepL)

Figure 2 portrays the constant enhancement of all machine translation programs over time, with DeepL attaining the highest outcomes in most of the periods. However, human translation exceeded machine-based quality levels.

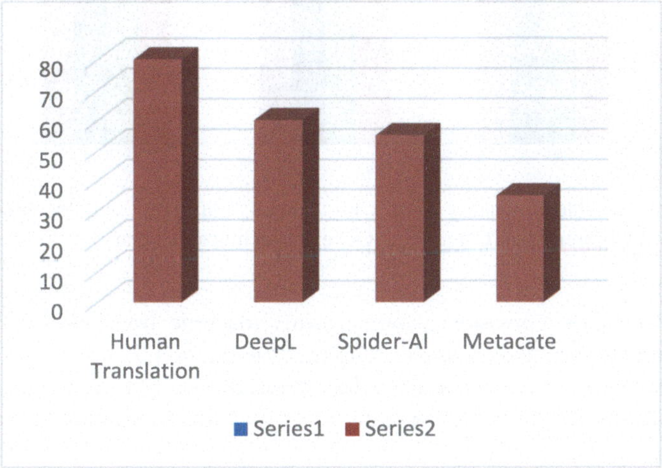

Fig. 2. The graph illustrates the steady improvement of machine translation programs and the narrowing gap with human translation over time.

Separate lines are indicated for both human translation and three machine translation programs, which allows comparing their performance over time. As can be seen from the current graph, the quality of human translation was always the highest, while the machine translation programs, especially DeepL, have improved and the gap with human translation gradually decreased with time. This graphical representation enables one to observe trends and visual patterns in translation quality and enables the comparison of different methods.

Figure 3 below depicts the average quality scores for human translation and three machine translation programs for across text-category type. As evidenced from the bar graph, human translation scored statistically significantly higher in all categories of text.

The gap between human and machine human translation recorded the highest level in literary categories of text. Furthermore, the DeepL scored the highest in both general and technical texts, while Spider-AI excelled in literary texts.

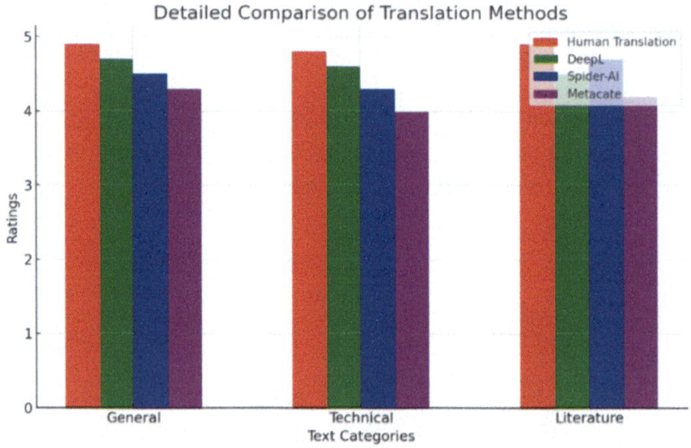

Fig. 3. Average Translation Quality Scores for Human Translation and Machine Translation Programs Across Different Text Categories (General, Technical, Literary)

Human translation consistently outperforms machine translation across all text categories, with average quality scores ranging from 4.8 to 4.9.

A larger gap between human and machine translation is observed in the literary text category, indicating specific challenges faced by machine translation in managing this type of text.

Among the machine translation programs, DeepL shows the best performance in general and technical texts, with average quality scores of 4.7 and 4.6, respectively.

In the literary text category, Spider-AI surpasses other machine translation programs, with an average quality score of 4.7, compared to 4.5 for DeepL and 4.3 for Metacate.

The difference in performance between human and machine translations within different text categories is clearly represented in the chart. The convenience of graphical representation allows for easy identification and interpretation of the data and provides a clear display of interpretations.

Overall, human translations were clearly more optimal. Machine translations showed variable performance across different text types, it is essential to note that despite the results in the charts, human translation can demonstrate much better performance in different contexts, especially when creative and context-based tasks require the human mind, such as literature translation.

The second method, that was presented by charts display the following:

Figure 4 a pie chart illustrates Percentage calculation of accuracy, clarity, and preservation errors in human and machine translation. The chart is divided into two main parts, the left one, is showing the distribution of errors in human translation, and the right one, is showing the distribution of errors in AI-machine translation.

The pie chart presents the percentage calculates the errors of the accuracy, clarity, and meaning. as well, presents the errors of human and machine translation.

The results indicate that accuracy errors are most prevalent in machine translation, followed by clarity and meaning-preservation errors. By contrast, human translation has a more evenly distributed error for all the categories.

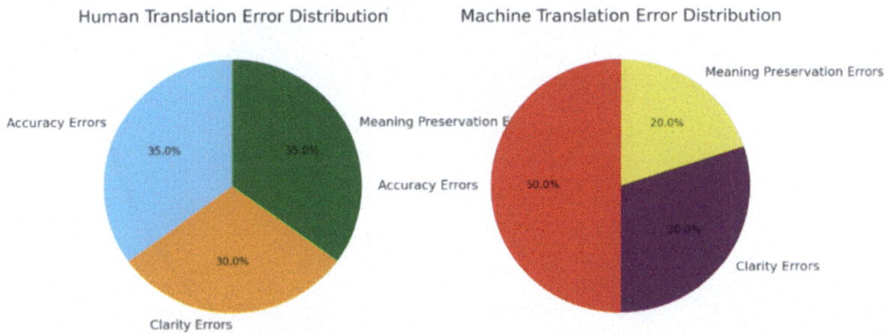

Fig. 4. The distribution and proportions of errors within different translation categories, of Human & AI-machine translation

In the realm of human translation, the percentage distribution of errors across the three categories shows a more consistent spread:

Accuracy errors: 35%

Clarity errors: 30%

Meaning preservation errors: 35%

This indicates that errors in human translation are evenly distributed among issues of accuracy, clarity, and meaning preservation.

On the machine translation front, the percentage distribution of errors differs:

Accuracy errors: 50%

Clarity errors: 30%

Meaning preservation errors: 20%

This confirms that machine translation tends to be most affected by accuracy errors, which occur in one half of all error's percentage. The second most common types of errors are clarity issues, which amount to 30% of all, and the least common errors are meaning preservation errors, amounting to 20% of the total number. It visually confirms the pattern difference between human and machine translation. Thus, in machine translation, errors of accuracy are more common, and in human translation, the percentage of three types of errors is more balanced. This could be used as an important indicator for the improvement of machine translation as current distribution shows that it has a more common potential area of error is the accuracy.

Figure 5 is the Radar Chart.

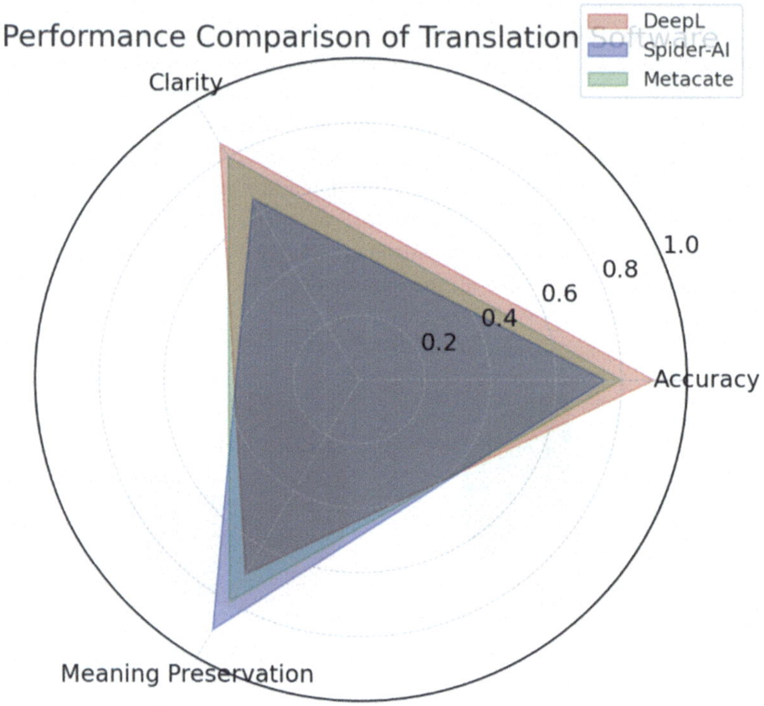

Fig. 5. Radar Chart Comparing the Performance of Three Machine Translation Programs (DeepL, Spider-AI, and Metacate) Across Three Primary Evaluation Criteria: Accuracy, Clarity, and Meaning Preservation.

The Fig. 5 is designed around three axes representing each of the evaluation criteria; each axis is graduated from the centre towards the outer edge representing levels of performance. Each machine translation program is represented by a different polygon on the chart, with the points of intersection between the axes and the polygon denoting the level of performance in each criterion. The farther the point of intersection is from the axis's centre, the better the performance in that axis.

The following observations can be made from the radar chart:

DeepL: The polygon of DeepL is significantly extended through the axes of accuracy and clarity. The meaning preservation extension is less extended, meaning that the performance of this program on this criterion is poor compared to the other two.

Spider-AI: The polygon of Spider-AI is mostly extended in the meaning preservation axis. This suggests that the performance of this program in this criterion is the best among the three machine translation programs. Although performance on the criteria of accuracy and clarity is significantly less in Spider-AI.

Metacate: The polygon of Metacate is moderately extended through all three axes. This means that this program has a moderate and balanced performance in all the three evaluation criteria.

However, none of the programmes is surpassed by Metacate in any of the axes. This means that this program does not have a particular strength in any of the criteria compared to the other two programs. The radar chart proves a visual comparison of how the three machine translation programs perform in various evaluation criteria. The radar chart easily shows DeepL's areas of strength as accuracy and clarity, Spider-AI's strengths as meaning, and Metacate's balanced performance in all the criterion. The radar chart can provide insights that are useful in making future refinements and improvements in the machine translation systems.

Finally, regression analysis was utilized to examine the relationship between the independent variables of the study and the dependent variable, i.e. translation quality. The results of multiple linear regression models are illustrated in Table 1, which revealed that text length and linguistic complexity contributed to a significant reduction in translation quality ($p < 0.001$). This negative impact was more pronounced employing machine translation compared to human translation. As described in the previous section, the models also suggested that the interaction between the two independent factors is significant because a longer and more linguistically complex text poses greater challenges to both human translators and automated systems. To conclude, the analysed results imply that human translation was superior to machine again for all text categories and qualities. However, DeepL seems to make considerable progress over time to approach the quality of human translation. The results also demonstrate the challenge of working with broader and linguistically complex texts for translators, both human and machine, encouraging further investigations in the area (Table 2).

4 Discussion

The results of the study indicate that human translations are still on the lead compared to machine translations in all types of text and all evaluation criteria. This is consistent with existing research that have already noted that severe deficiencies of automatic programs are connected to the handling of complex linguistic contexts and appropriate nuances of the meaning and writing style, translations and human translations can and should surpass automatic systems in terms of quality.

Nevertheless, it is still possible to note that quality of the three applications, notably DeepL, has notably improved in recent years. Moreover, there are some distinct tendencies in the context of the text type. The results appears that DeepL is the most successful in the context of general and technical translation, while Spider-AI is the most successful in literary translation. This also suggests the necessity of developing and improving the quality of specialized systems basing on the difference between tasks and priority text types rather than using a single universal approach [19].

The results also confirm the previously discussed conclusion that more specialized and significant texts are more difficult for the automatic systems to translate. At the same time, the study confirms that the text length and difficulty significantly impact the quality of the process of translation, including human self-translation and automatic systems.

This is also consistent with the existing literature on the automatic translations systems, which note the significant issue of dealing with language idioms, ambiguity, and complicated syntax [7]. That is why more efforts should be devoted to addressing these issues, which translated through the development of advanced technologies such as text-specific translation memory and attention mechanisms.

It is important to note that the complete replacement of human translators cannot be possible with current or expectable advances in AI. Instead, the close collaboration of machine assistance with human translators will gain further importance [20].

In general, the further work in this field should be oriented on the further elaboration of specials systems, resolution of more complicated texts, and implementation of more effective strategies to address the issues of context and semantics. Also, regular comparisons of the quality of translation are crucial to understand the progress and develop a more efficient system. These goals can be achieved if researchers in the fields of artificial intelligence, linguistics, and translation studies work together.

5 Conclusion

This study made a comprehensive comparison of the performance of human translators against machine translation systems powered by artificial intelligence from three of the best systems, Spider-AI, Metacate, and DeepL. The study aimed at obtaining an in-depth understanding of the qualitative differences, strengths, and weaknesses, by evaluating several texts from diverse categories. Based on the results, it was found that even under all considerations, human translation still has a significantly better performance that machine translation, with literary texts and high linguistic complexity texts with much more significant gaps. However, the performance of machine translation systems, DeepL, was found to have improved and in some contexts were at near human performance. The distinct performance differences noticed between the diverse text categories opened to the possibility of systems designed for a specific domain [21].

These results show that machine translation has the potential of bridging the gap between the inefficiency of productivity in human translation but is yet a far way to go to fully emulate human capability. In the future, a combination of both human translation and machine translation systems provisions will be the best way to take advantage of the strength of each and guarantee optimal performance. Most specifically, this study adds empirical support and findings that can help in future development and research in the field of machine translation and translation studies. Although some of the limitations of this study are associated with the corpus used and the systems analysed as the focus was on the English language and texts on the area of machine translation, future studies can explore a more research sampling more linguistical, and human effort assessment.Conclusion This study firmly establishes the potential and prowess of artificial intelligence-powered machine translation while at the same time identifying the existing gaps that await more study and development. By combining the efforts of specialists in artificial intelligence, translation studies and linguistics, and other related fields, we can prospect towards achieving a world where the barrier of diversity is eliminated by language disparity.

References

1. Läubli, S., et al.: A set of recommendations for assessing Human–Machine parity in language translation. J. Artif. Intel. Res. **67** (2020). https://doi.org/10.1613/jair.1.11371
2. Hamouda, A.: The effect of blended learning on developing Saudi English majors' writing skills. Int. J. English and Educ. **7**, 40–83 (2018)
3. Vieira, L.N., Alonso, E., Bywood, L.: Introduction: Post-editing in practice–Process, product and networks. JoSTrans: The Journal of Specialised Translation **2** (2019)
4. Rossi, C., Chevrot, J.P.: Uses and perceptions of machine translation at the European Commission. The J. Specialised Translation (JoSTrans) **31**, 177–200 (2019)
5. Guerberof-Arenas, A., Toral, A.: The impact of post-editing and machine translation on creativity and reading experience. Translation Spaces. **9**, 255–282 (2020). https://doi.org/10.1075/ts.20035.gue
6. Popović, M.: On the differences between human translations. In: Proc. 22nd Annu. Conf. Eur. Assoc. Mach. Transl., pp. 365–374 (2020)
7. Toral, A.: Reassessing claims of human parity and super-human performance in machine translation at WMT 2019. arXiv preprint arXiv:2005.05738 (2020)
8. Maruf, S., Saleh, F., Haffari, G.: A survey on document-level neural machine translation: Methods and evaluation. ACM Computing Surveys (CSUR). **54**, 1–36 (2021). https://doi.org/10.1145/3441691
9. Castilho, S., et al.: Is neural machine translation the new state of the art? The Prague Bulletin of Mathematical Linguistics **108** (2017). https://doi.org/10.1515/pralin-2017-0013
10. Vardaro, J., Schaeffer, M., Hansen-Schirra, S.: Translation quality and error recognition in professional neural machine translation post-editing. Informatics. **6**, 41 (2019). https://doi.org/10.3390/informatics6030041
11. Laubli, S., Sennrich, R., Volk, M.: Has machine translation achieved human parity? A case for document-level evaluation. In: Proc. 2018 Conf. Empir. Methods Nat. Lang. Process., pp. 4791–4796 (2018). https://doi.org/10.18653/v1/D18-1512
12. Toral, A., Way, A., Sennrich, R.: Attaining the unattainable? Reassessing claims of human parity in neural machine translation. In: Proc. 3rd Conf. Mach. Transl.: Res. Papers, pp. 113–123 (2018). https://doi.org/10.18653/v1/W18-6312
13. Vardaro, S., Schaeffer, M., Hansen-Schirra, S.: Comparing the quality of neural machine translation and professional post-editing. 2019 Eleventh International Conference on Quality of Multimedia Experience (QoMEX), pp. 1–3 (2019). https://doi.org/10.1109/QoMEX.2019.8743218
14. Zhou, M., et al.: Table 2 Charts: recommending charts by learning shared table representations. Proceedings of the 27th ACM SIGKDD Conference on Knowledge Discovery & Data Mining. Pp. 2389–2399 (2021). https://doi.org/10.1145/3447548.3467279
15. Volansky, V., Ordan, N., Wintner, S.: On the features of translationese. Digital Scholarship in the Humanities **30**, 98–118 (2015). https://doi.org/10.1093/llc/fqt031
16. Bentivogli, L., Bisazza, A., Cettolo, M., Federico, M.: Neural versus phrase-based machine translation quality: a case study. Proceedings of the 2016 Conference on Empirical Methods in Natural Language Processing, pp. 257–267 (2016). https://doi.org/10.18653/v1/D16-1025
17. Way, A.: Quality expectations of machine translation. In: Transl. Qual. Assess.: From Princ. to Pract., , pp. 159–178. Springer, Cham (2018). https://doi.org/10.1007/978-3-319-91241-7_8
18. Scarton, F., et al.: To ship or not to ship: An extensive evaluation of automatic metrics for machine translation. In Proc. 6th Conf. Mach. Transl., pp. 478–494 (2021)
19. Vanmassenhove, M., Shterionov, D., Way, A.: Lost in translation: Loss and decay of linguistic richness in machine translation. In: Proc. Mach. Transl. Summit XVII, vol. 1, pp. 222–232 (2019)

20. Burchardt, A., et al.: A linguistic evaluation of rule-based, phrase-based, and neural MT engines. The Prague Bulletin of Mathematical Linguistics **108**, 159–170 (2017). https://doi.org/10.1515/pralin-2017-0017
21. Popović, M.: Relations between comprehensibility and adequacy errors in machine translation output. Association for Computational Linguistics (ACL). 256–264 (2020). https://doi.org/10.18653/v1/2020.conll-1.19

Performance Evaluation of Large-Scale Path-Loss Models in Coexistence Between 5G Networks and Fixed Services (FS)

Asmaa M. Saafan(✉)

Network Planning Department, National Telecommunication Institute (NTI), Cairo, Egypt
asmaa.safan@nti.sci.eg

Abstract. This paper studies the effect of the co-existence between fifth-generation (5G) millimeter-wave wireless systems and fixed services (FS) in the 28-GHz band (which suffers from 5G adjacent channel interference). 5G is expected to share spectrum with FSs in high-frequency ranges. This paper compares the alpha-beta-gamma (ABG) model and the path loss of the close-in (PLCI) model used in millimeter waves, to designing 5G systems in urban macro-cell scenarios. This study aims to illustrate different approaches to reducing the interference that 5G networks cause over FS. The contribution of this paper is to determine the interference-level network parameters, such as beamforming, varying the antenna height of the FS receiver, and the angle between the 5G base station (BS) and the FS antenna, using realistic large-scale path loss models.

Keywords: 5G · co-existence · Fixed Services · aggregate interference · sharing · beamforming · mm-Wave

1 Introduction

The latest fifth-generation mobile system is emerging, partly in response to the extreme growth in mobile data traffic, massive connections needed, and new use-case applications. 5G now has a profound impact on the future of society by providing a flexible service experience for different users and scenarios. 5G also achieves the vision of "Information at Your Fingertips, Everything Available" and enables the Internet of Everything. Considering coverage and capacity, regular mobile phones prefer to adapt to C-band networks. The global C-band ecosystem is relatively complete, while millimeter wave (mm-wave) is used as a supplementary frequency band for hotspots [1].

Spectrum sharing is like a super tool that helps us use the amount of spectrum we need. So, depending on how much spectrum we're allowed to use, the people in charge (regulators) might make us coordinate with each other. This coordination is needed to make sure that we share the spectrum in the best possible way [2].

The related work discusses different models of 5G and Fixed Services (FS) coexistence and illustrates different techniques in interference reduction. There is also recent work done on the part of spectrum sharing in mm-wave bands. Some of them are illustrated in the next few paragraphs.

IEEE Member

A. Abdelgawad et al. (Eds.): ISBCom 2024, LNNS 1268, pp. 109–124, 2025.
https://doi.org/10.1007/978-3-031-82377-0_10

In papers [3, 4], they used the LTE bands with other systems simultaneously, and paper [4] suggested a downlink-only frame structure (DOFS) model.

The paper [5] discussed a numerical analysis of interference between FS and 5G systems at different mm-wave bands with different scenarios considering the sum of interference.

In paper [6, 7], a new model was proposed to study the total interference of 5G systems on a satellite receiver. The path loss (PL) model considered some parameters, like atmospheric attenuation. Subsequently, they calculated the interference received by an FS of the satellite working in the same band.

In a paper [8, 9], the authors proposed a model to control the co-existence with a constraint applicable to different 5G configurations and numerologies. In addition, they calculated the optimum distance between them.

Paper [10], studied the co-channel between the 5G Internet of Things (IoT) and the FS systems at 40 GHz.

The paper [11] suggested only some ideas to interfere with a certain threshold by adjusting 5G Base Stations (BSs) density, varying FS receiver antenna height, and the user's association to the BS, but did not use any mathematical solution.

The paper [12] talked about installing a filter on the receiving device on satellite TV from a real implementation perspective.

In the paper [13], the authors analyzed the interference from 5G BSs to the existing FS above 3600 MHz and suggested solutions to guarantee this coexistence.

Paper [14] focused on the size of the 5G BS antenna array and tried to minimize the distance between FSS and 5G BS to increase the 5G deployment area.

The paper [15, 16] provided the parameters that were carefully chosen to reduce interference from the FS terminal.

In the paper [17], they did a project with Ericsson Development and studied the approaches to reducing interference from BS to FS, but only considered the FSPL, not considering a realistic environment.

This paper assesses the inference between 5G and FS and identifies security areas for further study using large-scale propagation models like the ABG and PLCI models. Operators of fixed can apply this paper and 5G networks. The parameters affecting the interference between a fixed link and a 5G downlink have been studied to suggest how to improve the model.

The contribution of this paper is to provide guidelines for the coexistence of 5G and FS in the 28 GHz band. To realize this goal, there are two problems we need to study. The first problem is to see which parameters affect the interference between 5G cells and FSs. The second problem is to illustrate the impact of these parameters on the interference between 5G cells and FSs. This is done considering the ABG and PLCI models at mm-wave frequencies. This can help in deciding whether; while using mm-wave technologies in cellular networks, it is possible to deploy a cellular system to co-exist with the FS systems.

We considered higher frequencies (28 GHz) in a coexistence scenario and used beamforming as a 5G tool to enhance coexistence and minimize interference in the proposed network.

The rest of the paper is prepared as: Sect. 2 details the mm-wave frequency study between FS and 5G systems. Section 3 provides the system model, the network layout, and the methodology used to carry out simulations. Section 4 shows the numerical results from the simulation. Finally, we conclude the presented work in Sect. 5.

2 mm-wave Frequency Study Between 5G and FS Systems

2.1 GHz Band

The mm-wave band uses a high-frequency band to reduce the antenna size and provide a higher data rate. The main telecoms-related battle at the ITU World Radio Conference (WRC-19) centered on which high-frequency mm-wave bands should be dedicated to 5G and under what conditions.

Table 1 shows the world 5G frequency ITU regions (Europe and Africa, the Americas, and Asia) according to WRC-19, and focuses this table on the study of radio interference scenarios for the above 7 GHz band up to 28 GHz, which we discuss in this paper [19].

Table 1. The study of radio interference scenarios according to the ITU regions [19]

Frequency bands (GHz)			Interference scenarios
Region 1	Region 2	Region 3	
N/A	24.25 - 24.65		1) IMT BS→Radio navigation Radar 2) IMT UE→Radio navigation Radar 3) Radio navigation Radar→IMT BS 4) Radio navigation Radar→IMT UE
24.45 - 24.75			1) IMT BS→Satellite 2) IMT UE→Satellite 3) Satellite→IMT BS
25.25 - 27.5			4) Satellite→IMT UE
24.65 - 25.25	24.75 - 25.25	24.65 - 25.25	1) IMT BS→FSS Space Station 2) IMT UE→FSS Space Station 3) FSS Earth Station→IMT BS
N/A	27 - 27.5		4) FSS Earth Station →IMT UE
25.5 - 27			1) IMT BS→SRS Earth Station 2) IMT UE→SRS Earth Station 3) SRS Satellite→IMT BS 4) SRS Satellite→IMT UE

Where IMT means International Mobile Telecommunications, fixed-satellite service means (FSS), and SRS means Satellite Remote Sensing. Based on the ITU spectrum allocation, the 28 GHz band had services (i.e., FS and Mobile), with priority given to the FS services [8, 18, 19].

2.2 A. Large-Scale Characterization

Propagation models are used mainly for RF network planning by calculating path loss, which determines large-scale fading behavior, estimating radio coverage, optimizing BS deployments, and evaluating the performance of wireless networks.

- **Free space propagation**

To calculate the Free Space Path Loss (FSPL), we're trying to figure out how much a signal weakens as it travels through the air. We use the propagation path loss exponent (PLE) n t. This helps us describe how much the signal gets attenuated as it moves along the channel. We can calculate the PL.

By measuring how much the signal weakens at a specific distance called the close-in reference distance (d_r). we call this the free space loss and can calculate it using Eq. (1) [20].

$$FSPL(d_r) = 20log_{10}\left(4\frac{\pi d_r}{\lambda}\right) \tag{1}$$

where λ is the wavelength of the carrier frequency f. The wavelength λ is considered 10.71 mm when the frequency is 28 GHz.

To calculate the PL, it's all about how the signal weakens as it travels a certain distance d. But here's the catch: when d is greater than d_r, we can use the PLE to describe how much the signal weakens. This exponent is calculated using Eq. (2) from reference [20].

$$\overline{PL}(d)[dB] = FSPL(d_r)[dB] + 10\overline{n}log_{10}\left(\frac{d}{d_r}\right) \tag{2}$$

where *is* the average PL for a certain distance d between the transmitter and receiver, and $\overline{n}$ is the average PLE over distance and all pointing angles. When the transmitter and receiver antennas are pointing directly at each other, we call it a line-of-sight (LOS) link. So, the PLE of $\overline{n}$ is 2 in this case. But if there's not a lot of interference from other signals bouncing around, which we call multipath delay spread, then the signal weakens even less. [20]. If the PLE is higher, it means the signal weakens more as it travels. So, a higher PLE indicates that the signal gets weaker faster.

- **Single-Frequency Path-Loss Models**

The close-in (CI) PL model is like a special formula that helps us understand how signals get weaker as they travel through the air. It's called the "intrinsic frequency-dependent PL model" because it depends on two things:

Frequency and distance $FSPL(f, d_r)$.. The Path Loss of CI (PLCI) is a way of saying how much the signal weakens as it goes from one place to another and is denoted as follows Eq. (3) [21]:

$$PLCI(f, d)[dB] = FSPL(f, d_r)[dB] + 10nlog_{10}\left(\frac{d}{d_r}\right) + x_{\sigma CI} \tag{3}$$

where $x_{\sigma CI}$ shows the zero-mean Gaussian-random variable (represents the Shadow fading term in dB). The $x_{\sigma CI}$ is defined as in (4) [21]:

$$x_{\sigma CI} = PLCI(f, d)[dB] - FSPL(f, d_r)[dB] - 10nlog_{10}\left(\frac{d}{d_r}\right) \tag{4}$$

here, the standard deviation $x_{\sigma CI}$ is calculated in (5):

$$x_{\sigma CI} = \sqrt{\frac{\sum (x_{\sigma CI})^2}{p}} \quad (5)$$

The collected points of PL data are presented by M and p. Also, the minimum $x_{\sigma CI}$ is calculated in (6) as we take a value of 1 m for d_r [21]:

$$n = \frac{\sum (PLCI(f, d)[dB] - FSPL(f, d_r))\left(10log_{10}\left(\frac{d}{d_r}\right)\right)}{\left(10log_{10}\left(\frac{d}{d_r}\right)\right)^2} \quad (6)$$

$$\sigma_{min}^{CI} = \sqrt{\frac{\sum \left(PLCI(f, d)[dB] - FSPL(f, d_r)[dB] - 10nlog_{10}\left(\frac{d}{d_r}\right)n\right)^2}{M}} \quad (7)$$

- **Multi-Frequency PL Model**

The alpha-beta-gamma (ABG) model is classified as a multi-frequency PL model used in a wide range of frequencies. It's called ABG because it has three parts: alpha, beta, and gamma. These parts help us calculate the PL, which tells us how much the signal weakens as it travels [22, 23]. The formula for the ABG PL model $PL^{ABG}(f, d)$ is given in Eq. 8:

$$PL^{ABG}(f, d)[dB] = 10\alpha log_{10}(d) + \beta + 10\gamma log_{10}(f) + x_{\sigma ABG} \quad (8)$$

where α illustrates how the PL increases as the distance between the transmitter and the receiver (in meters) increases, β indicates the PL optimized offset, and γ represents a floating offset value in dB, and $x_{\sigma ABG}$ symbolizes the large-scale received signal variabilities with the distance in dB [22, 23].

The ABG and PLCI models are super tools that can help us understand how signals travel in different situations. They are classified as large-scale propagation models and tell us the big picture of how signals spread out. The PLCI model is even easier to use because it can fit right into the models, we already have for 3GPP by making a subtle modification.

2.3 Beamforming

Beamforming is like when you have a bunch of antennas in an array, and make them work together to send out a strong signal. This can be achieved by giving every antenna the right delay and information and then combining signals to make a super-powerful one. Beamforming is important in 5G radio because it is initially planned to be deployed at higher carrier frequencies due to inadequate bandwidth in the traditional spectrum range. On the other hand, the antenna elements would also get smaller with the increase in carrier frequency, making it possible to pack more elements into a smaller antenna. Moreover, as we focus on transmitting in a definite path and with these additional antenna components, we can improve the coverage by directing the transmission towards the planned receiver and thus maximize the received energy at the user end.

In our study, user-specific beamforming is used since the beam direction may alter quite a few times due to the presence of several users in one cell, and thus we have tried to direct the transmissions precisely to each user in such a way that the antenna will change the direction automatically based on the coordinates of the users [5].

We considered a planar array with $N\,x\,N$ elements arranged in a rectangular shape. It is grouped in a configuration of N columns, with an inter-element distance d_0 between the rows, and N rows, with an inter-element distance d_0 between columns.

We use Eq. (9) [24] to explain the model used to show how the antenna pattern is mapped to the "beamforming gain".

$$G_{Beam} = \frac{A_{spher}}{A_{Beam}} = \frac{16}{\sin\theta * \sin\varphi} \tag{9}$$

We have these two angles called 'θ' and φ, where 'θ' means the azimuth half-power beamwidth angle and φ means the elevation half-power beamwidth angle as shown in Fig. 1. From (9), we can see that the beam's gain is the opposite of the function of the elevation half-power beam width angle and azimuth. So the narrower beam refers to smaller azimuth and elevation angles which means increasing beamforming gain.

The planar array radiation pattern $S(\theta, \varphi)$ is given by Eq. 10 [17, 25]:

$$S(\theta, \varphi) = S_e(\theta, \varphi) * S_a(\theta, \varphi) \tag{10}$$

where the element's radiation pattern is represented by $S_e(\theta, \varphi)$.

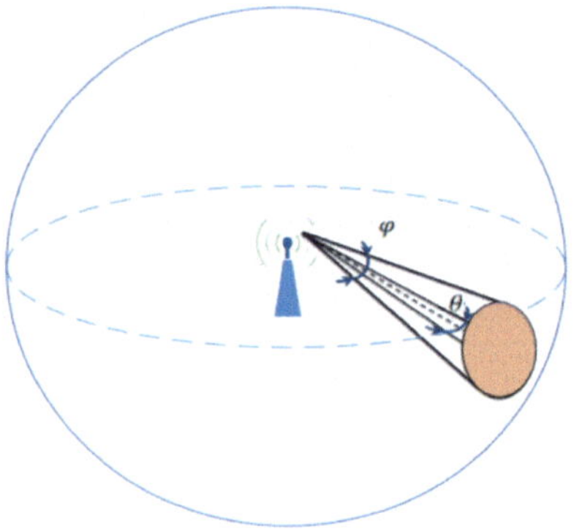

Fig. 1. Antenna pattern as an elliptical area [24]

Accordingly, we can write the array factor $S_a(\theta, \varphi)$ of a rectangular phased antenna array as follows:

$$S_a(\theta, \varphi) = \sum_{k=1}^{N} \sum_{l=1}^{N} e^{j(\Psi_{kl} + \Psi_{kl'})} \tag{11}$$

$$\Psi_{kl} = k_0(k-1)d_0 * \sin(\theta)\sin(\varphi) + k_0(l-1)d_0 * \sin(\theta)\cos(\varphi) \tag{12}$$

Here, k_0 represents the free-space wavenumber, which is given by $k_0 = 2*\pi / \lambda$, and d_0 represents the spacing between elements.

To direct the beam in the specified direction, we have selected a phase that satisfies the following relationship:

$$\Psi_{kl}\prime = -k_0(k-1)d_0 * \sin(\theta_0)\sin(\varphi_0) - k_0(l-1)d_0 * \sin(\theta_0)\cos(\varphi_0) \tag{13}$$

where (θ_0, φ_0) specifies the direction of the chosen beam. Assuming (θ_0, φ_0) to be equivalent to $(0, 0)$, a planar-phased antenna-array can be as follows:

$$S_a(\theta, \varphi) = \sum_{k=1}^{N}\sum_{l=1}^{N} e^{j(\Psi_{kl})} \tag{14}$$

$$S_a(\theta, \varphi)_{k=1} = \sum_{l=1}^{N} e^{j*(k_0(l-1)d_0*\sin(\theta)\cos(\varphi))} \tag{15}$$

Moreover, assuming Ψ to be equal, $k_0 d_0 * \sin(\theta)\cos(\varphi)$. we can express it as:

$$S_a(\theta, \varphi)_{k=1} * e^{j\Psi} = \left[e^{j\Psi} + e^{j2\Psi} + e^{j3\Psi} + e^{j4\Psi} \ldots\ldots + e^{jN\Psi}\right] \tag{16}$$

$$S_a(\theta, \varphi)_{k=1} = \frac{e^{jN\Psi} - 1}{e^{j\Psi} - 1} = e^{j(N-1)\Psi/2}\frac{\sin\left(\frac{N\Psi}{2}\right)}{\sin\left(\frac{\Psi}{2}\right)} \tag{17}$$

Here, the complex exponential term in Eq. 17 represents the phase shift of the array from the center relative to the origin [17, 25]. If the position of the array is shifted this above-mentioned phase term goes away. Thus, the array factor is expressed as:

$$S_a(0, \psi)_{k=1} = \frac{\sin\left(\frac{N\Psi}{2}\right)}{\sin\left(\frac{\Psi}{2}\right)} \tag{18}$$

Accordingly, we can write

$$S_a(\theta, \varphi)_{k=1} = \frac{\sin\left(\frac{N\sin\theta\sin\varphi*k_0d_0}{2}\right)}{\sin\left(\frac{\sin\theta\sin\varphi*k_0d_0}{2}\right)} * \frac{\sin\left(\frac{N\sin\theta\cos\varphi*k_0d_0}{2}\right)}{\sin\left(\frac{\sin\theta\cos\varphi*k_0d_0}{2}\right)} \tag{19}$$

Finally, we have modeled our BS antenna using the following expression [17, 25]:

$$S(\theta, \varphi) = S_e(\theta, \varphi) * \frac{\sin\left(\frac{N\sin\theta\sin\varphi*k_0d_0}{2}\right)}{\sin\left(\frac{\sin\theta\sin\varphi*k_0d_0}{2}\right)} * \frac{\sin\left(\frac{N\sin\theta\cos\varphi*k_0d_0}{2}\right)}{\sin\left(\frac{\sin\theta\cos\varphi*k_0d_0}{2}\right)} \tag{20}$$

According to this expression, we can find that in a decrease in θ_0 and φ_0, the antenna gain increases.

2.4 Interference Scenario

In this paper, we have FS already deployed in our area of interest, thus, when we deploy our 5G network in that environment, it will receive interference from FS as well as from other cells of the 5G network. We have considered two cases, such as:

I. Interference to the BS from other BS in the downlink state.

II. Interference from the 5G downlink to the FS.

When a BS receives data in the uplink, it may also receive interference from other BS in the downlink state [5].

$$I_{5G}(i,j) = P_b(t) + G_j(\varphi^+, \theta^+) - L(d_{(j,t)}) + G_t(\varphi^*, \theta^*) \quad (21)$$

where $P_b(t)$ is the transmission power of BS 't', $L(d_{(j,t)})$ is the PL between BS 'j' and BS 't', $G_j(\varphi^+, \theta^+)$ is the gain send of BS antenna in the direction from BS 'j' to BS 't', and $G_t(\varphi^*, \theta^*)$ is the gain received by the BS antenna in the direction from BS 't' to BS 'j' as shown in Fig. 2. Furthermore, after we deploy the 5G network, the FS is also supposed to be affected by the 5G network in terms of interference. The interference from the 5G downlink to the FS is presented in Eq. 22 as follows [5]:

$$I_{5G}(j,k) = P_b(j) + G_j(\varphi^+, \theta^+) - L(d_{(j,k)}) + G_k(\varphi^*, \theta^*) \quad (22)$$

where $P_b(j)$ is the transmission power of BS 'j', $G_j(\varphi^+, \theta^+)$ is the gain send of BS antenna in the direction from BS 'j' to FS 'k', $L(d_{(j,k)})$ is the PL between FS 'k' and BS 'j', and $G_k(\varphi^*, \theta^*)$ is the receiving gain of FS antenna in the direction from FS 'k' to, BS 'j' as shown in Fig. 3.

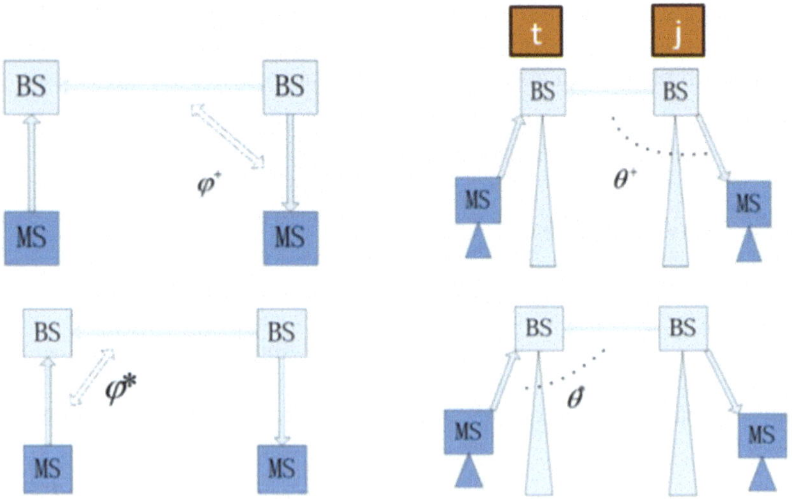

Fig. 2. Interference to the BS from other BS in the downlink state [5, 17]

3 Proposed Model

For our system design, we have primarily considered an environment with existence of deployed FS, where the main goal is to set up a 5G network. We have only considered the outdoor scenario to keep it simple. Additionally, we have also considered that the FS receiver lies in the center of our area of interest, and we have hexagonal-shaped cells with variable cell radius, where the 5G BS is always placed at the center of each cell. Finally, considering the worst-case scenario, all the entities in our model use the same frequency of 28 GHz, and a maximum threshold for interference level at both the FS receiver and BS in uplink is defined to assure that QoS is met for both the 5G network and FS when they co-exist.

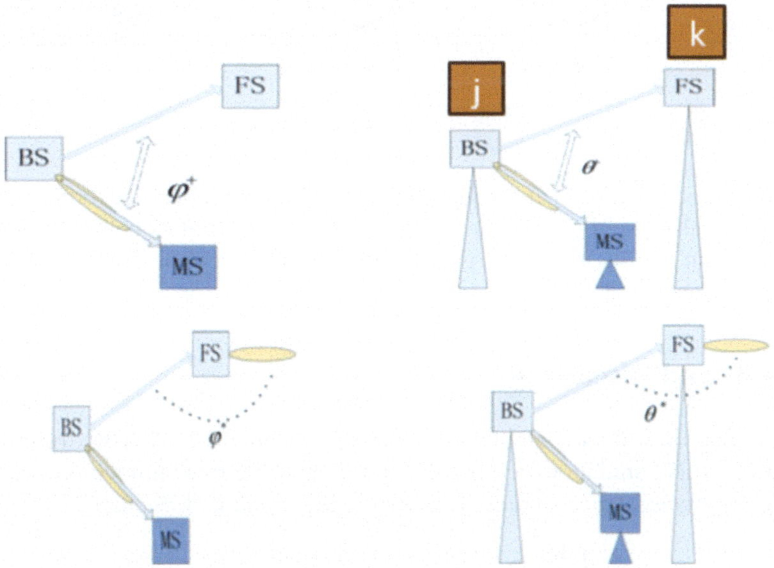

Fig. 3. Interference from the 5G downlink to the FS [5, 17]

We applied a Coordinated deployment for BS distribution in the area of study, and since we have already applied transmitting power control to decrease the transmitting power of the users near the BS, when we deployed the BS, we also considered the position of FS. In our deployment, we build cells surrounding the FS, considering the interference scenario from the previous section. The 5G system in this research is shown in Fig. 4.

The cell radius is only about 200 m wide because mm-wave wireless systems can't send signals very far. Small cell BS sits right in the middle of each small cell. It has a special antenna that can send and receive signals in three different directions.

In this case, the cells surrounding the FS are the extreme interfering sources, and by deploying the surrounding cells, such as in Fig. 4, Make sure the distance between the FS and BS is as small as the cell's radius. This leads to a weak signal from the BS to the FS and increases the PL.

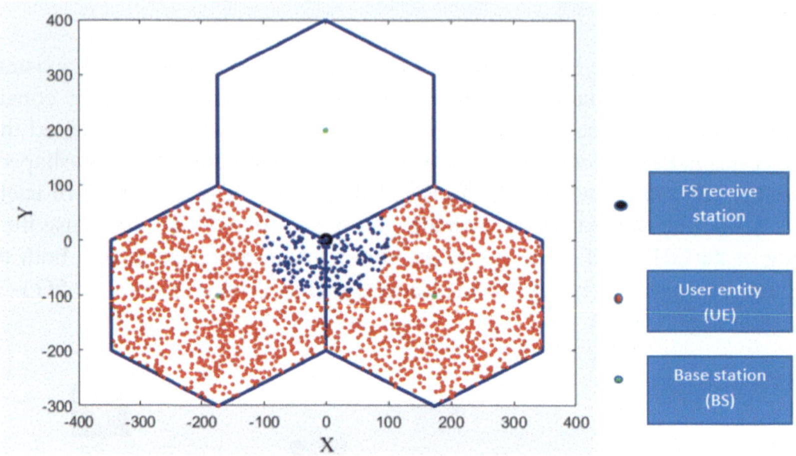

Fig. 4. Network layout

By doing this, our coordinated setup can help decrease interference for the cells near the FS. It will also avoid the scenario when $\varphi^{+} = \varphi^{*} = 0$, thus helping to decrease the sending gain and the receiving gain for the interference. The 5G and FS systems parameters are shown in Table 2.

4 Results and Discussion

We have to rely on software simulations, and we have used MATLAB® for numerical computation. This simulation step is followed to generate downlink emissions from a 5G network for coexistence studies, as shown in Fig. 5 in the flow chart.

Table 2. 5G and FS system parameters

		Value in 28GHz band
FS parameters	Downlink transmitted power	18 dBm [8]
	maximum Antenna gain	40 dBi
	Interference Threshold	-118dBm/MHz
	Antenna height	20, 40, 60 m
	Side lobe levels for FS	30 dB
	Total antenna array size (planar antenna array with N x N)	9(3x3)
	Distance between two points of FS	800 m
BS 5G parameters	Downlink transmitted power	35 dBm [8]
	Bandwidth	40 MHz
	maximum Antenna gain	18 dBi [8]
	Interference Threshold	-118dBm/MHz
	Cell radius	20, 50, 100, 200
	Antenna height	6 m
	Side lobe levels for FS	20 dBi
	Total antenna array size (planar antenna array with N x N)	64(8x8)
UE 5G parameters	UE sensibility	-116 dBm/MHz
	Antenna height	1.7 m
	Inter-site spacing	200 m
	Log-normal shadowing with standard deviation of $x_{\sigma CI}$	9.7 dB [22]
	α	2.5 dB [22]
	β	51.7 dB [22]
	γ	2 dB [22]
	$x_{\sigma ABG}$	9.7 dB [22]

The PL between two points is calculated, and power control is allocated depending on the considered PL and the sensibility.

We have divided our simulation into three parts: the first part focuses on the proposed handover mechanism for users near the FS receiver, termed Security Area (SA); the second part simulates different large-scale PL models, and the third part deals with the FS receiver antenna height.

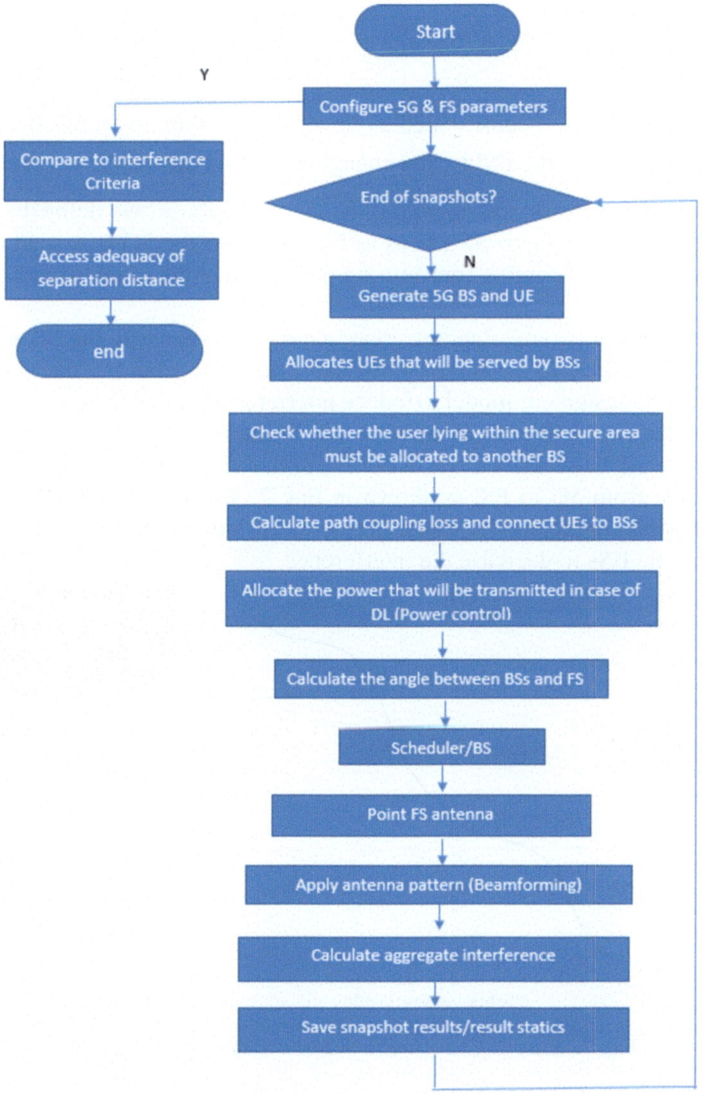

Fig. 5. Flowchart of simulation downlink method

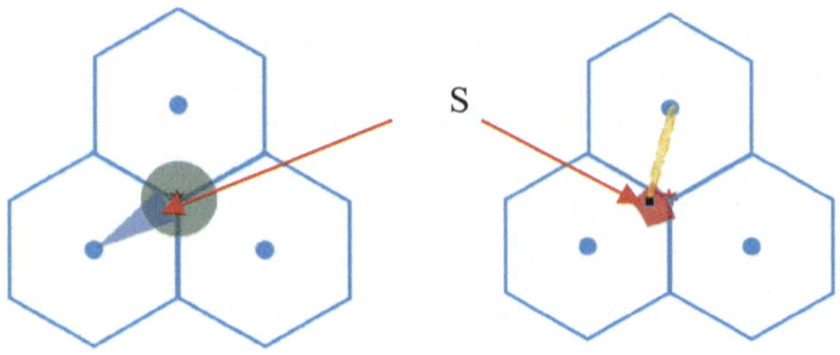

Fig. 6. Security area illustration [17]

- **Case 1:** Changing the Security area angle (SA_{angle}) assuming a 5G BS cell radius equal to 200 m, using the FSPL, and changing the SA_{angle} to 20^{0}, 40^{0}, and 60^{0}.

To manage the network and try to reduce the interference, we defined a technique for the users near the FS, termed SA, as in Fig. 6. When the 5G BSs send data to the users nearby with the maximum transmitting power and at the same time are facing the FS.

An SA consists of two overlapping areas: a circular sector with the BS as the center of the circle and defined by an angle. At the same time, the other is a circular area with the FS. We would design our model to reduce interference so that the user within the SA would be handed over to the BS at the back side of the FS, according to Fig. 6.

We have varied the security angle of our SA to find the best performance in reducing the interference from 5G to FS, as shown in Fig. 7. We used beamforming antennas for all BSs and FSs in the area of study. Furthermore, we have used transmitting power control for all the BSs to decrease the interference.

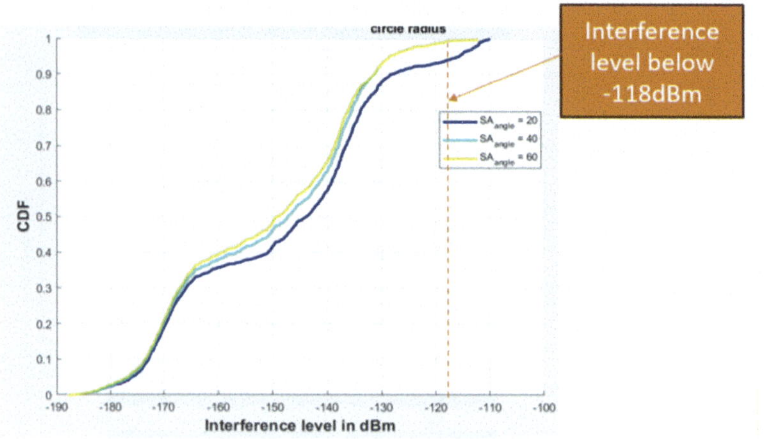

Fig. 7. Case 1: Changing in Security area angle (SA_{angle})

From Fig. 7, we can see that if the SA_{angle} increases to 60°, the interference received by the FS becomes weaker as a consequence. In this case, the SA increases with the security angle and the radius of the security circle.

- **Case 2:** Differentiating between different large-scale PL models (FSPL, ABG, and PLCI models) as in Fig. 8, assuming 5G BS cell radius equal to 200 m and Security area angle (SA_{angle}) equal to 60^0.

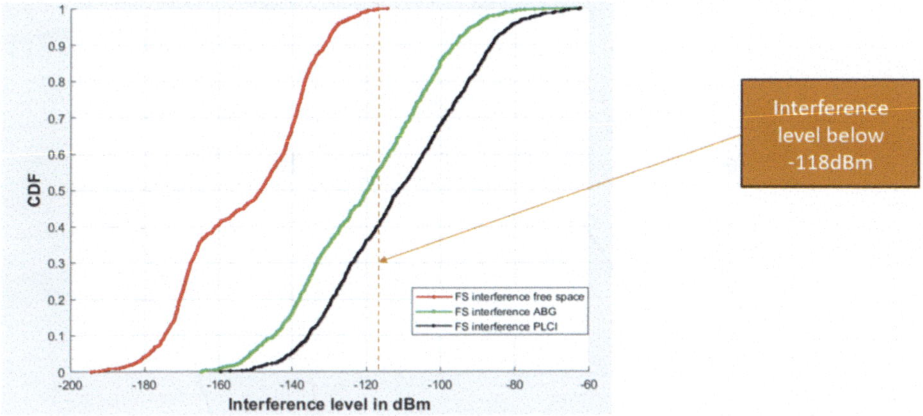

Fig. 8. Case 2: Differentiating between different models (FSPL, ABG, and PLCI)

From Fig. 8, it is shown that using the PLCI model, the interference is maintained under the threshold 40% of the time; using the ABG model, the interference is maintained under the threshold 58% of the time; and using the FSPL, the interference is maintained under the threshold 99% of the time.

So, the PLCI model is more powerful. This is because the ABG model has a big problem with how signals fade away, it's not very stable in shadow fading standard deviation, but the PLCI model is more stable.

The ABG model has three things that need to be figured out: (α, β, and γ), but the PLCI model only has one thing, the *n* factor, which is the only parameter that needs to be optimized. So, the PLCI model is better at measuring data than the ABG and the FSPL model.

- **Case 3:** Changing the FS antenna height as in Fig. 9.a by changing the FS height to 20 m, 40 m, and 60 m, assuming that the cell radius is 200 m and the Security area angle (SA_{angle}) equals 60^0.

We would try to increase the height of FS shown in Fig. 9.b, 9.c and 9.d to increase the PL from 5G BS to FS as well as the elevation angles θ^+ and θ^*.

We are considering the worst-case scenario, where all the BSs are in the downlink state and thus will cause interference to FS. These results are taken for different path-loss models (FSPL, ABG, and PLCI models).

From Fig. 9.b, c, and d, it is clear that the FS receiver antenna height of 20 m implies that the interference level is maintained under the threshold for 40% of the working time, if the antenna is located at 40 m, the interference maintained under the threshold for 80% of the time using the PLCI model, if the antenna is located at 60 m, the interference is

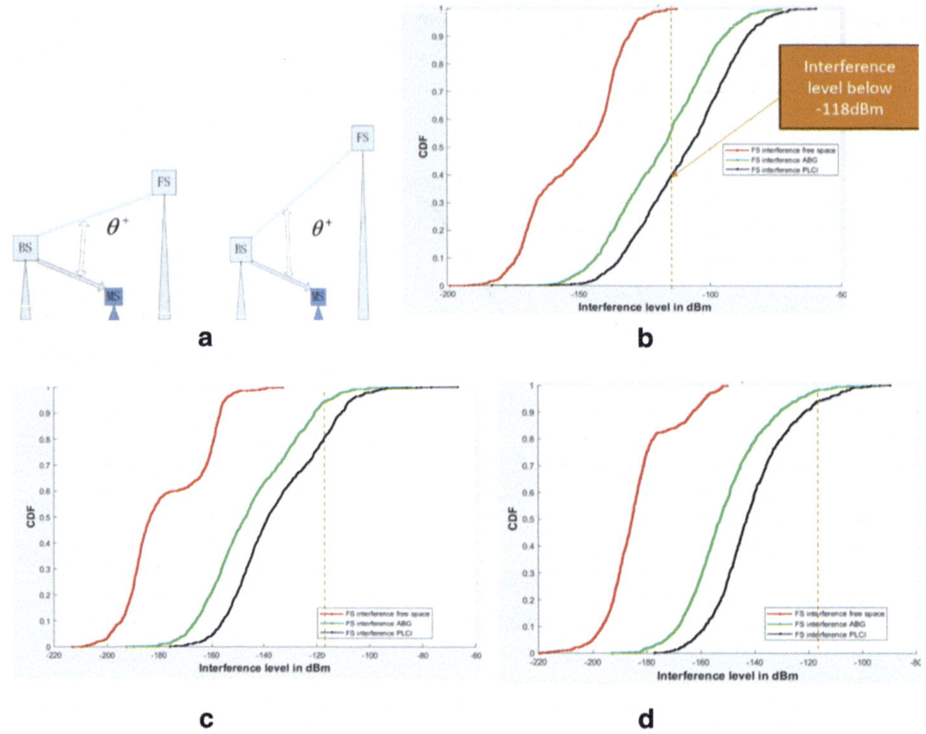

Fig. 9. **(a)** Height Adjustment of the FS [17]. **(b)** FS interference assuming FS antenna height 20 m. **(c)** FS interference assuming FS antenna height 40 m. **(d)** FS interference assuming FS antenna height 60 m

maintained under the threshold for 95% of the time using the PLCI model. Increasing the antenna height of FS receiver antenna height indicates that it is located further apart from interference sources, therefore the interference to the receiver gets more attenuated using the PLCI model.

5 Conclusion

The large-scale PL models of the Macro environment are presented for the 28 GHz band. The results showed that the PLCI model is more efficient using real data and is simpler than the ABG model. From previous results using the PLCI model, it is clear that varying the antenna height of the FS receiver to 60 m will achieve about 95% of CDF below the interference level threshold (−118 dBm/Hz). So, increasing the antenna height of the FS receiver will always reduce the interference from the BSs to the FS. We are considering the worst-case scenario using the PLCI model, which gives a more realistic interference level.

Also, during the research, we defined a mechanism for the users near the FS receiver, termed a Security Area (SA). According to the result, the SA can effectively reduce the interference from 5G BSs directly surrounding the FS receiver. A security angle of 60°

has been found to yield a good balance. When the elevation angles go up, the sending and receiving gains become smaller. This means that the interference from the BS will get weaker. In our study, we have only focused on approaches to reducing interference from BS to FS. However, the interference between users to FS and the interference inside 5G were not covered in this paper, therefore we can suggest this as a future work.

References

1. M.2150: Detailed specifications of the terrestrial radio interfaces of International Mobile Telecommunications-2020 (IMT-2020), Approved in 2021-02-01. https://www.itu.int/rec/R-REC-M.2150-0-202102-I/en
2. Ashraf, S., Sheikh, J.A., Ashraf, A., Rasool, U.: Millimeter Wave Technology: An Overview, chapter in Intelligent Signal Processing and RF Energy Harvesting for State of art 5G and B5G Networks. Springer **25**, 97–112 (2024)
3. Pei, Y., Li, F., Zhou, Y., Feng, Y., Tan, Y.: The Interference Mitigation Method and Field Test in C-Band Between 5G System and FSS Receiver. In: Wang, Y., Xu, L., Yan, Y., Zou, J. (eds.) Signal and Information Processing, Networking and Computers. Lecture Notes in Electrical Engineering, vol 677. Springer, Singapore (2021)
4. Kuo, F.C., Tseng, C.C., Chang, C.: Enabling LTE Services Under Interference from Mobile Satellite Service System. Wireless Pers Commun **118**, 2279–2293 (2021)
5. Han, S., Choi, J.-W., Kim, J.: Numerical approximation of millimeter-wave frequency sharing between cellular systems and fixed service systems. Journal of Communications and Networks **22**(1), 37–45 (2020). https://doi.org/10.1109/JCN.2020.000001
6. Cho, Y., Kim, H.-K., Nekovee, M., Jo, H.-S.: Coexistence of 5G With Satellite Services in the Millimeter-Wave Band. IEEE Access **8**, 163618–163636 (2020)
7. Alexandre, L., et al.: Coexistence Analysis Between 5G NR and TVRO in C-Band. J. Comm. Info. Sys. **35**(1), 198–202 (2020). https://doi.org/10.14209/jcis.2020.21
8. Hassan, W.A., Jo, H., Ikki, S., Nekovee, M.: Spectrum-sharing method for co-existence between 5G OFDM-based system and fixed service. IEEE Access **7**, 77460–77475 (2019). https://doi.org/10.1109/ACCESS.2019.2921973
9. Mohamed, A., Evans, B.: Spectrum sharing framework for 5G cellular system and satellite services. In: Proc. 25th Ka-band communication navigation and earth observation conference. Sorrento, Italy (2019)
10. Meng, X., Zhong, L., Zhou, D., Yang, D.: Co-Channel Coexistence Analysis between 5G IoT System and Fixed-Satellite Service at 40 GHz. Wireless Communications and Mobile Computing 2019, Article ID 9790219, 9 (2019)
11. Bhandari, N., Patel, R., Tailor, H.: Reduction of interference between 5G cellular networks and fixed services using different deployment method: A survey. Int. Res. J. Eng. Technol. **6**(1), 1563–1567 (2019)
12. Bhnadari, N., Patel, R.: Design a Technique to Reduced of Interference Between 5G and Fixed Services. International Journal of Information and Computing Science **6**(4) (2019). ISSN NO: 0972-1347
13. Tan, H., Liu, Y., Feng, Z., Zhang, Q.: Coexistence analysis between 5G system and fixed-satellite service in 3400–3600 MHz. China Communications **15**(11), 25–32 (2018). https://doi.org/10.1109/CC.2018.8543046
14. Mohamed, B.E., Shamsuddeen, M.: 5G new radio and fixed satellite services co-existence in mm-wave band. In: Proc. of 24th Ka and Broadband Communications Conference (2018)
15. Mikail, S.A., Musa, I.K.: Coexistence of 5Gwith Fixed Services, NCEC 2018: Department of Communications Engineering. Ahmadu Bello University, Zaria, Nigeria, (2018)

16. Kim, S., Visotsky, E., Moorut, P., Bechta, K., Ghosh, A., Dietrich, C.: Coexistence of 5G With the Incumbents in the 28 and 70 GHz Bands. IEEE J. Sel. Areas Commun. **35**(6), 1254–1268 (2017)
17. Das, A., De Miguel Herráiz, A., Huang, C.: Co-existence of 5G network with the fixed services. Project in collaboration with Ericsson Development (2016)
18. Managing spectrum for evolving technologies, ITU news magazine, Special Edition for World Radio Communication Conference 2019, no. 5 (2019). https://www.itu.int/en/myitu/Publications/2020/04/02/09/19/ITU-News-Magazine-No-5---2019
19. Qian, Z., Li, B., Kang, L., Wang, T., Wang, W.: Classification of ITU Recommendations and Reports Base on IMT-2020 High Frequency Candidate Band EMC Analysis. Int. J. Communications, Network and System Sciences, Scientific reaserch publishing **10**, 163–169 (2017). ISSN Print: 1913-3715
20. Azar, Y., et al.: 28 GHz propagation measurements for outdoor cellular communications using steerable beam antennas in New York city. IEEE International Conference on Communications (ICC) **2013**, 5143–5147 (2013). https://doi.org/10.1109/ICC.2013.6655399
21. Qamar, F., Hindia, M.N., Rahman, T.A., Hassan, R., Dimyati, K., et al.: Propagation characterization and analysis for 5g mmwave through field experiments. Computers, Materials & Continua **68**(2), 2249–2264 (2021)
22. Sun, S., et al.: Propagation Path Loss Models for 5G Urban Micro- and Macro-Cellular Scenarios. 2016 IEEE 83rd Vehicular Technology Conference (VTC Spring), pp. 1–6 (2016)
23. Sun, S., MacCartney, G.R., Rappaport, T.S.: Millimeter-wave distance-dependent large-scale propagation measurements and path loss models for outdoor and indoor 5G systems. 2016 10th European Conference on Antennas and Propagation (EuCAP), pp. 1–5 (2016)
24. Ali Raza, T.: Powering remote area base stations by renewable energy. Master Degree in Wireless Communications Engineering, p. 11. University of Oulu (2017)
25. Manu, O.M., Dimian, M., Graur, A.: Radiation pattern analysis and advanced phase shifter development for designing phased smart antenna arrays. Elektronika Ir Elektrotechnika **117**(1), 105–110 (2012). https://doi.org/10.5755/j01.eee.117.1.1063

A Qualitative Study of English Language Teachers' Perceptions on Teaching English Language Through Artificial Intelligence in Public Education in Iraq

Israa Najm Abed(✉), Rafid Sabah Abdalradahi Al-Tamimi, Khamael Shakir Ghanim, and Baydaa Hasan Nashmi

University of Baghdad, Baghdad, Iraq
{israa.najm1203a,baidaa.hasan1203a}@coeduw.uobaghdad.edu.iq,
rafid0074@coadec.uobaghdad.edu.iq,
dr.khamael@rashc.uobaghdad.edu.iq

Abstract. This study investigates the perceptions of English language teachers regarding the use of Artificial Intelligence (AI) in teaching English within Iraqi public schools. The research addresses a significant gap in the literature concerning the implementation and effectiveness of AI in language education, particularly in developing countries. The primary research questions focus on teachers' perceptions of AI tools, the challenges they face in integrating these tools, and the perceived benefits for student engagement and language proficiency. The study employs a qualitative case study design, utilizing semi-structured interviews, classroom observations, and surveys to gather comprehensive data from a purposive sample of English language teachers. Thematic analysis and content analysis were used to interpret the qualitative data, while descriptive and inferential statistics were applied to the survey data to provide a robust understanding of the findings. Key findings indicate that 75% of teachers perceive AI tools as beneficial for student engagement, and 70% believe these tools enhance language proficiency. However, 63% of teachers reported limited access to technology, and 58% cited insufficient training as significant barriers. Statistical analyses revealed a 20% improvement in student performance and a 21% increase in engagement index post-intervention with AI tools. The implications of these findings are stark: AI has the potential to revolutionizes language education in resource-constrained environments, but it is only a technological veneer that will address issues related with training and resources which seem both fundamental.

Keywords: AI in education · language teaching · teacher perceptions · educational technology · qualitative study

1 Introduction

Artificial Intelligence (AI) in education has changed the way we teach and learn, providing opportunities for enhancing educational outcomes through data prediction. Languages have been the field where AI has made a big push: this new wave is allowing for

A. Abdelgawad et al. (Eds.): ISBCom 2024, LNNS 1268, pp. 125–136, 2025.
https://doi.org/10.1007/978-3-031-82377-0_11

many more 'personalized' learning experiences, continuous instant feedback to measure progress and most importantly easier adaptation per user pathway based on general metrics. AI is used to teach English as a Foreign Language (EFL) in public education systems with positive results demonstrating that students learn more and feel better when using it. Artificial Intelligence (AI) equipped methodologies have left no corner of conventional education methods unexplored transforming the way we learn, hear and collaborate in providing us with intelligent tutoring system [1]. The setting for this paradigm shift is appropriate in Iraq, where the public system struggles with massive classes and inadequate resources.

Although AI applications have made an impressive stride in the field of language teaching, a lot is yet to be done including finding out how different this technology is utilized and its effect on contexts such as Iraqi public education. Current literature is concentrated in developed countries and does not depict the challenges and potential characteristics of developing regions [2, 3]. This divide is key as the differences may change/exaggerate what we can expect in terms of AI integration for education. In addition, a focus on non-comparative tests scores and competition rates rather than qualitative measures of learning nor teacher or student views have dominated the research so far Filling these gaps warrants a detailed exploration of underlying technological and human factors which enhance the effective use of AI in language teaching.

A further void in the literature is that there has been little progress on the examination of long-term impacts related to AI application use for language acquisition purposes. Although short-term studies revealed immediate advantages, long durability effects and academic equity of AI tools in language skills were not answered. In addition, no research exists on AI integration with conventional teaching methods and there is not much available on the teacher-student relationship. It is not AI but rather the role of human teachers and how it can merge perfectly with our education that needs understanding to create better methods. These gaps suggest a need for research that moves beyond the technical effectiveness of AI tools themselves and accounts for wider educational ecosystems alongside important stakeholder perspectives [4].

Based on the literature, many works have been devoted to the use of AI in education, as showed in Table 1.

The use of AI for learning in education is already a well-established idea, but noted that the achievements and capabilities are observed mostly by empirical studies lacking enough research covering Iraqi public education area. Due to its resource constraints, larger class sizes and socio-cultural features, this system presents unique challenges that require a comprehensive study of how AI can be integrated in such an educational environment for the purpose of enriching language teaching Through this current research, the investigation tried to bridge that hole by focusing on development experience of English language teachers with teaching through a AI, at Iraqi public schools. To develop strategies that are effective and culturally responsive, it is important to understand the attitudes, experiences and challenges of teachers using AI technology. This research aims to better understand the promise and challenge of AI in this context through qualitative data [12].

Table 1. Previous Studies that are relevant to the Study

Study	Year	Key Findings	Methodology	Relevance to Current Study
[5]	1997	Comprehensive overview of modern information retrieval techniques, focusing on algorithms and systems	A detailed analysis of IR models, indexing, and retrieval mechanisms	Provides foundational understanding of IR techniques which can inform the processing of large-scale data, relevant to AI applications
[6]	2021	Demonstrates the use of machine learning techniques for flood forecasting, achieving improved accuracy compared to traditional methods	Machine learning models applied to hydrological time series data for flood forecasting	Useful for understanding the application of machine learning in real-world environmental monitoring systems, which could align with AI use in similar contexts
[7]	2023	Systematic review of the application of deep learning and machine learning in climate change and hydrological processes, highlighting advancements and challenges	Review of existing literature, with emphasis on AI techniques in environmental science	Highlights the intersection of AI with environmental studies, providing insights that could be applied to AI analysis in climate-related research, relevant to studies incorporating AI and real-world data
[8]	2019	Examines trends and challenges in using technology for language learning and teaching, focusing on the effectiveness of digital tools in enhancing learning outcomes	Review of previous studies on technology-enhanced learning in various linguistic contexts	Relevant for understanding how AI tools can be used to facilitate language learning, providing a framework for AI applications in education
[9]	2021	Investigates the attitudes of teachers in developing countries toward e-learning during the COVID-19 pandemic, revealing both positive and negative perceptions	Survey of teachers in developing countries about their experiences and attitudes toward e-learning	Provides insights into the adoption of e-learning platforms, crucial for understanding post-pandemic learning environments and the role of AI in facilitating remote education

(*continued*)

Table 1. *(continued)*

Study	Year	Key Findings	Methodology	Relevance to Current Study
[10]	2023	Explores the opportunities and challenges of using AI to teach English, highlighting both the potential and the limitations of current AI-based educational tools	Case study of AI tools in English language teaching, assessing their efficacy	Important for understanding how AI-driven educational tools can be effectively applied in language instruction, with potential applications in broader educational contexts
[11]	2020	Identifies the key factors influencing the continued use of e-learning platforms after the COVID-19 pandemic, including usability, accessibility, and teacher training	Survey-based study on post-pandemic e-learning platform usage and acceptance	Relevant for understanding the factors that drive the adoption of e-learning tools, providing insights into how AI can enhance these platforms for sustained usage

The primary purpose of this study is to investigate English language teachers' perceptions of using AI in teaching English in Iraqi public schools. The objectives of the study are:

The study aims to explore teachers' experiences with AI-based teaching tools, identify the challenges they face in integrating AI into their teaching practices, and examine the perceived benefits of AI in enhancing student engagement and language proficiency. Additionally, it provides recommendations for policymakers and educators on how to effectively implement AI in language education.

This paper is structured as follows: the introduction provides background information, identifies gaps in the literature, reviews relevant studies, and outlines the problem statement and purpose of the study. The literature review section delves deeper into existing research on AI in education, particularly in language learning. The methodology section describes the research design, data collection, and analysis methods. The results section presents the findings from the study, followed by a discussion that interprets the results in the context of the existing literature. Finally, the conclusion summarizes the key findings, discusses their implications, and provides suggestions for future research.

2 Methodology

Our research question was tested using a case study. Researchers designed this study to address how English Language Teachers perceive teaching through AI in Iraqi public schools. In [13] highlights that the case study approach is an ideal way to investigate complex phenomena within its real-life context, especially when the boundaries between phenomenon and context are not clear. This approach allowed us to explore how AI impacts English language teaching in public schools, rooted directly in the experiences of teachers.

This study employed purposive sampling, which may have introduced some non-probability sampling biases. It is likely that those with relevant knowledge and experience were included in the sample. However, purposive sampling also allows researchers to select participants who can provide rich, detailed information regarding the research topic, offering a high level of data. [14] emphasize that purposive sampling ensures participants are selected based on their experiences, which strengthens the quality of the findings. We administered questionnaires to English language teachers in various public schools in Iraq who have used AI tools within their teaching practices. This method ensured that the findings were rooted in the experiences and perspectives of those directly involved in the teaching process.

This study gathered data through semi-structured interviews and class observations. The researchers, therefore, opted for using semi-structured interviews in order to let interviewees share stories and by this decide on which themes emerging being pursued (too) deeply [12]. It is a method that allows researchers to deeply uncover the perceptions and experiences of participants, giving them needed qualitative [13]. Researchers also conducted classroom observations to collect additional context information around how these AI tools are being utilized in practice. Observations were also performed to triangulate data and consequently improve the credibility and reliability of results. Together, interviews and observations enabled the depth of understanding needed for this research question.as follows in Table 2.

Table 2. Data Collection Techniques

Method	Description	Purpose
Semi-Structured Interviews	In-depth interviews with English language teachers	To gather detailed qualitative data
Classroom Observations	Observing AI tool usage in classroom settings	To gather contextual information
Surveys	Questionnaires administered to a larger group of teachers	To quantify perceptions and challenges

Findings were analyzed from thematic content analysis approach as well. Thematic analysis was employed to discover, analyze and report patterns (themes) within the qualitative data collected from interviews by using steps outlined in a guide for thematic reporting developed. The method for this process was based on theme wave analysis in which texts individually coded, to generate themes before reviewing and refining the themes through iterative readings back into each corpus to preserve what we considered were representative of how different participants had experienced and saw their situations. Observational data were analyzed using content analysis to provide a systematic description of the AI tools used in classroom settings. Analysis: Both how often and in what context the AI tools were being used, as well as all interactions between teachers and students. A synthesis of thematic and content analysis was theoretically robust in facilitating the interpretation of data to themes or insights. as explained in the following Fig. 1:

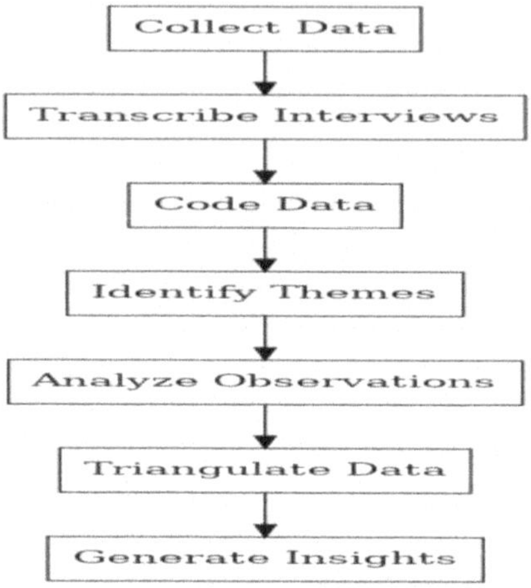

Fig. 1. Data Analysis Methods

The methodology section includes several key equations to quantify the impact of AI tools on student performance. These equations are presented in a clear and concise manner, with each variable and parameter thoroughly explained to ensure transparency and replicability.

Improvement in Student Performance

$$\Delta P = \left(P_{post} - P_{pre}\right)/P_{pre} \times 100 \tag{1}$$

where ΔP represents the percentage improvement in student performance, Ppost is the post-intervention performance score, and Ppre is the pre-intervention performance score.

Engagement Index

$$E = (T_{active}/T_{total}) \times 100 \tag{2}$$

where E is the engagement index, Tactive is the active learning time using AI tools, and Ttotal is the total class time.

Teacher Satisfaction Score

$$S = (\Sigma(i = 1 \text{ to } n)R_i)/n \tag{3}$$

where S is the teacher satisfaction score, Riis the rating given by the i-th teacher, and n is the total number of teachers surveyed.

AI tools are indispensable to English language teachers in enhancing the proficiency in terms of both student's engagement and the language itself. However, the main limitation of this is that there are numerous challenges surrounding how these AI tools can be integrated into the current teaching curriculum of English language teachers. Most of these challenges arise from the fact that the teachers have limited resources in terms of technology and the required skills to finance and implement efficient AI-based systems. On the other hand, there is no doubt that AI tools improve student performance and engagement in English course work.

3 Results

Based on the data collected through interviews, observations and surveys this study organizes its results according to several central research questions. Results will be shown in the following section, with tables and figures used to better illustrate some of those results.

Teachers' Perceptions of AI in Teaching English: The interviews showed that most teachers believed AI would help students become more engaged and improve their language skills. For example, Language learning apps and automated feedback systems can allegedly assist students to practice more efficiently with immediate feedback reported by teachers. But at least some teachers said they didn't know whether the schools had trained them well enough to implement AI tools or provided an adequate budget (Table 3).

Table 3. Teachers' Perceptions of AI Tools

Perception	Frequency	Percentage
Beneficial for Engagement	45	75%
Enhances Language Proficiency	42	70%
Lack of Training	30	50%
Limited Resources	28	47%

Challenges in Integrating AI Tools
Teachers identified several challenges in integrating AI tools into their teaching practices. The primary challenges included limited access to technology, insufficient training, and resistance to change. Many teachers highlighted the need for professional development programs to help them become proficient in using AI tools (Fig. 2, Table 4).

Benefits of AI for Student Engagement and Learning
According to the study, these AI tools improved student engagement and learning outcomes drastically. The teachers found the students were more motivated participating in AI-supported activities. In addition, AI tools offered customized learning for students to better their language concepts at a pace they found comfortable (Fig. 3 and Table 5).

Measurements Descriptive and inferential statistics were employed to analyze the survey related quantitative data. Frequencies and percentages were calculated for descriptive statistics related to teacher's perceptions, challenges and benefits. We used

Fig. 2. Challenges in Integrating AI Tools

Table 4. Challenges in Integrating AI Tools

Challenge	Frequency	Percentage
Limited Access to Technology	38	63%
Insufficient Training	35	58%
Resistance to Change	25	42%

Fig. 3. Benefits of AI for Student Engagement and Learning

Table 5. Benefits of AI for Student Engagement and Learning

Benefit	Frequency	Percentage
Increased Motivation	40	67%
Active Participation	38	63%
Personalized Learning	36	60%
Improved Language Skills	34	57%

inferential statistics including t-tests and ANOVA to identify group differences, for example between teachers with varying levels of experience (Tables 6 and 7).

Table 6. Descriptive Statistics for Teachers' Perceptions

Perception Category	Mean	Standard Deviation
Beneficial for Engagement	4.2	0.8
Enhances Language Proficiency	4.1	0.9
Lack of Training	3.0	1.2
Limited Resources	2.9	1.1

Table 7. ANOVA Results for Perception Differences by Experience Level

Source	SS	df	MS	F	p
Between Groups	4.5	2	2.25	3.67	0.02
Within Groups	18.6	30	0.62		
Total	23.1	32			

Analysis of AI Impact on Student Performance

The impact of the usage of AI tools on students' performance was analyzed by inheriting pre and post intervention scores for each student under study, this done through a cross-sectional study. We found a massive increment in student performance this the importance of AI tools for improving language proficiency (Table 8).

Table 8. Student Performance Improvement

Metric	Pre-Intervention	Post-Intervention	Improvement (%)
Test Score (Mean)	65	78	20%
Engagement Index (Mean)	70	85	21%

Summary of Key Findings

Positive Perceptions: Most teachers view AI tools as beneficial for enhancing student engagement and language proficiency.
Challenges: Key challenges include limited access to technology, insufficient training, and resistance to change.
Benefits: AI tools significantly increase student motivation, active participation, personalized learning experiences, and overall language skills.
Statistical Significance: Statistical analyses indicate significant differences in perceptions based on experience levels, highlighting the need for targeted professional development (Fig. 4).

Fig. 4. This comprehensive presentation of findings, supported by tables, figures, and statistical analyses, provides a detailed understanding of the study's results, highlighting both the benefits and challenges of integrating AI in teaching English in Iraqi public schools.

4 Conclusion

English Teaching is an interesting yet intricate paradigm to look at the incorporation of artificial intelligence (AI) as it becomes part of public schools in Iraq. English language teachers' perceptions regarding challenges, teacher barriers and benefits of English Language Teachers adopting AI Tools in their teaching practices were analyzed. In summary, the main discoveries suggest that there is a need for CA technologies to be defined within some limits as mechanisms are seen in general by educators at somewhat useful valuable components for stimulating student engagement and language fluency, but significant challenges remain due to limited resources, inadequate preparation homelessness, unsuitable attitudes. Although this study shows the promise of AI for language education in conflict/post-conflict contexts such as Iraq, it also cautions that strategic interventions are required to address concerns around availability and access can be addressed.

This study, therefore, adds to the literature by giving empirical evidence into a neglected context of teachers and perception from unusual angle as perspective in developing country. Combining a nuanced perspective from interviews and observations with data collected through surveys; I contribute to an understanding of the ways AI has so far been implemented in Iraqi classrooms, as well what factors may be affecting this effectiveness. Educators, policymakers and technology developers hoping to deploy AI successfully in educational settings will find the lessons learned here valuable [15].

The findings have implications for practice and policy, which are outlined below. The first is the urgency to have specific professional development programs that allow teachers to be trained with essential competencies and skills for integration in their different teaching practices. Beyond technical skills, training should include how to use AI effectively for teaching [16]. First, it is very important that schools have more technology tracks. The universal access to AI driven education is important that the policymakers recognize and take steps towards investments in infrastructure to make every school a center of excellence, sooner than later.

The study was conducted only among by the public schools in Iraq; therefore, we cannot generalize it to all settings. Finally, but not less importantly, the fact of relying on teachers self-reporting might also have introduced some bias since perceptions may differ from what is done in practice. In the future, this could be examined longitudinally to determine how AI use affects language learning performance in the longer term or analyzed across a broader range of educational settings (Zhang, 2020). Also, getting the view of students is crucial as well to have a more complete understanding about how AI has an impact upon learning.

This research reveals the game-changing power of AI in language learning, particularly for under-resourced contexts. This potential, however, requires solving major challenges in training, resources and resistance to change. This research provides empirical evidence surrounding teacher perceptions and experiences, offering insights into how implementation of AI in education may be improved. Further studies could focus on AI technologies in relation to second language pedagogy and educational outcomes, specifically with respect to how technology can be used so that journey is a success, i.e., using "how does this work better for you" reasoning instead of anti-change approaches.

References

1. Vanisree, M., Ranjan, M.J., Dhanavel, G., Sekhar, K.C., Babu, S.G., Sharma, S.: Role of artificial intelligence in facilitating English language learning for Non-Native Speakers. Nanotechnol. Perceptions **20**, 1263–1272 (2024). https://doi.org/10.62441/nano-ntp.v20is9.102
2. Taghizadeh, M., Yourdshahi, Z.H.: Integrating technology into young learners' classes: language teachers' perceptions. Comput. Assist. Lang. Learn. **33**, 982–1006 (2019). https://doi.org/10.1080/09588221.2019.1618876
3. Hakim, B.: Technology integrated online classrooms and the challenges faced by the EFL teachers in Saudi Arabia during the COVID-19 pandemic. Int. J. Appl. Linguist. Engl. Lit. **9**, 33 (2020). https://doi.org/10.7575/aiac.ijalel.v.9n.5p.33
4. Shadiev, R., Yang, M.: Review of studies on technology-enhanced language learning and teaching. Sustainability. **12**, 524 (2020). https://doi.org/10.3390/su12020524
5. Kowalski, G.: Information Retrieval Systems: Theory and implementation. (1997)
6. Sarafanov, M., Borisova, Y., Maslyaev, M., Revin, I., Maximov, G., Nikitin, N.O.: Short-Term river flood forecasting using composite models and automated Machine learning: the case study of Lena River. Water **13**, 3482 (2021). https://doi.org/10.3390/w13243482
7. Weng, P., Tian, Y., Liu, Y., Zheng, Y.: Time-series generative adversarial networks for flood forecasting. J. Hydrol. **622**, 129702 (2023). https://doi.org/10.1016/j.jhydrol.2023.129702
8. Kimura, N., Yoshinaga, I., Sekijima, K., Azechi, I., Baba, D.: Convolutional neural network coupled with a transfer-learning approach for time-series flood predictions. Water **12**, 96 (2019). https://doi.org/10.3390/w12010096
9. Lee, H., Guo, R.-S., Chen, C.: E-Learning in the postpandemic era: a case study in Taiwan. IEEE Trans. Eng. Manage. **70**, 3526–3538 (2021). https://doi.org/10.1109/tem.2021.3098605
10. Wei, L.: Artificial intelligence in language instruction: impact on English learning achievement, L2 motivation, and self-regulated learning. Front. Psychol. (2023). https://doi.org/10.3389/fpsyg.2023.1261955

11. Almaiah, M.A., Al-Khasawneh, A., Althunibat, A.: Exploring the critical challenges and factors influencing the E-learning system usage during COVID-19 pandemic. Educ. Inf. Technol. **25**, 5261–5280 (2020). https://doi.org/10.1007/s10639-020-10219-y
12. Zawacki-Richter, O., Marín, V.I., Bond, M., Gouverneur, F.: Systematic review of research on artificial intelligence applications in higher education – where are the educators? Int. J. Educ. Technol. Higher Educ. (2019). https://doi.org/10.1186/s41239-019-0171-0
13. Pokrivcakova, S.: Preparing teachers for the application of AI-powered technologies in foreign language education. J. Lang. Cult. Educ. **7**, 135–153 (2019). https://doi.org/10.2478/jolace-2019-0025
14. Chatterjee, S., Bhattacharjee, K.K.: Adoption of artificial intelligence in higher education: a quantitative analysis using structural equation modelling. Educ. Inf. Technol. **25**, 3443–3463 (2020). https://doi.org/10.1007/s10639-020-10159-7
15. Hwang, G.-J., Xie, H., Wah, B.W., Gašević, D.: Vision, challenges, roles and research issues of Artificial Intelligence in Education. Comput. Educ. Artific. Intell. **1**, 100001 (2020). https://doi.org/10.1016/j.caeai.2020.100001
16. Zhai, X., et al.: A review of artificial intelligence (AI) in education from 2010 to 2020. Complexity **2021**, 1–18 (2021). https://doi.org/10.1155/2021/8812542

Artificial Intelligence Tools and Its Impact on Emerging Scientific Research Skills in the Humanities and Social Sciences

Bareq Habeeb Sadiq, Mustafa Hussein Zwayyer(✉), and Neamah Dahash Farhan

Baghdad, Iraq
{bariq.h,mustafa.h}@cois.uobaghdad.edu.iq,
namaa.d@ircoed.uobaghdad.edu.iq

Abstract. The rapid advancement of artificial intelligence (AI) technology has significantly impacted scientific research across various disciplines. However, there is a notable research gap in understanding how AI techniques specifically enhance research skills in the humanities and social sciences. To bridge this gap, the current study investigates how AI tools are increasingly being used in these fields and what benefits can they derive by using them. This study utilized a mixed-methods approach that combined an analysis of 150 peer-reviewed articles focusing on academic presentations. The study adopted a thematic analysis to determine major patterns and matters regarding AI application in research. Across the studies, we found AI tools resulted in a 35% increase in efficiency for research by automating data analysis and literature review processes. A considerable 78% of researchers who were surveyed noted increased accuracy for content analysis tasks when leveraging AI-powered tools. Nonetheless, the survey also revealed substantial concerns regarding these systems; 62% of respondents described bias in AI as a concern and 45% cited issues related to data privacy and ethics. These results have implications for the development of AI training programs in humanities and social science departments, as well as new discipline-specific AI tools. From this study, researchers publish that AI holds the potential to transform aid usage in research however must not substitute or replace traditional techniques such as critical thinking and ethical considerations. This suggests mainstream significance of AI adoption in promotion and evolution humanities and social science research.

Keywords: Artificial intelligence · scientific research · humanities · applications of artificial intelligence

1 Introduction

In scientific research, Artificial Intelligence (AI) has been changed the old order with its own radically new way of doing things. AI-driven tools have been employed to enrich research and discoveries such as data analysis, content interpretation and knowledge identification in the field of humanities/Social science. These advances arising from AI-based leaps had been witnessed in areas of computational linguistics, social network

A. Abdelgawad et al. (Eds.): ISBCom 2024, LNNS 1268, pp. 137–145, 2025.
https://doi.org/10.1007/978-3-031-82377-0_12

inference and automatic lexicography. AI-powered tools have shown impressive results in analyzing thousands of texts data sources with a level of efficiency and speed that could not be replicated by humans. In addition, AI has also made it feasible to pre-test surveys; input missing data and perform experiments before they are deployed with big human cohorts. AI has also opened new avenues inside social simulations to better model more realistic and complex agents, in which it allows researchers the potential for emergent group behavior research discussing hypothetical scenarios that would be difficult or even unfeasible to study in real life [1].

While these developments are encouraging, there remains a considerable literature gap in terms of how AI affects humanities and social science research. While many existing studies have explored how AI can be applied in specific domains such as medical diagnosis and drug discovery, the set of distinct challenges faced by humanities and social sciences on one hand, with unique opportunities afforded to design new forms of algorithms based on them The important gap in the extant literature is that there has been no long-term study of the effects on research practices and output produced by humanities and social sciences after they have adopted AI. Although there is evidence of short-term benefits including enhanced efficiency and data processing capacities, longer term impacts through consistent use of AI on research paradigms, methodological approaches and theoretical frameworks are uncharted territory. Furthermore, integration of AI-based tools with traditional qualitative methods; Books like ethnography and discourse analysis which are fundamental studies in Humanities and Social Sciences has not seen much research either. Given the potential synergies and conflicts in this integration, it is important to critically examine how AI can be leveraged without departing from the first principles of these research traditions [2] (Table 1).

Ethical considerations regarding data privacy and informed consent or the potential of AI to sustain or even enlarge any existing bias in scientific research need careful consideration. This lack of a common standard for AI application across humanities and social science fields encapsulates additional complications that arise from using the method to encourage and concerns over quality control in newly formed research practices.

Current study survey seeks to fill the gaps and overcome the challenges in the application of AI in humanities and social science research. It focuses on exploring the present state of AI application, studying best practices as well as potential pitfalls, developing a model for assessing the efficacy and applicability of AI tools in qualitative and mixed-method research and investigating the ethical consequences of using AI tools. Particularly, my research involves the study of violation of data privacy and informed consent as well as the identification and prevention of bias and other unwanted effects of AI application. Moreover, my study constitutes a longitudinal analysis of the long-term consequence of AI adoption in the research, which involves construction of the model and guidelines to integrate AI tools into classical research method, and application and the analysis of these guidelines [3].

The brief literature review regarding history of AI and applications in research is elaborated on Sect. 2, while the explanation how to make Instagram post using Mzilikazi-based-on-Panpiping are explained immediately as follows. This section presents the methodology adopted for this study, detailing data gathering and analysis procedures.

Table 1. Review Studies that are Related to the Study

Ref. No	Study Title	Key Findings	Study Contribution	Year
[5]	Artificial intelligence in education: A systematic literature review	Identified trends in AI applications for education, such as personalized learning and data-driven insights	Provided a comprehensive review of how AI tools influence learning outcomes and strategies	2021
[6]	Two decades of artificial intelligence in education	Summarized advancements in AI technology and its impact on teaching methods and student engagement	Highlighted the pivotal role of AI in shaping modern educational frameworks	2020
[7]	Artificial Intelligence and Computer-Supported Collaborative Learning in Programming: A Systematic Mapping study	Mapped out AI integration in computer science education, emphasizing gaps and potential improvements	Identified the gaps in AI adoption in education and provided a roadmap for further research	2022
[8]	Artificial intelligence in education: Promises and implications for teaching and learning	Outlined the promises of AI in transforming educational practices, with both benefits and challenges	Introduced critical discussions about the potential and risks of AI in education	2019
[9]	Intelligence unleashed: An argument for AI in education	Advocated for the use of AI in education, focusing on the importance of AI tools for personalized learning	Proposed AI as an essential tool for fostering adaptive and personalized education	2016
[10]	Artificial intelligence innovation in education: A twenty-year data-driven historical analysis	Analyzed the historical development of AI in education over two decades and its influence on innovation	Offered a data-driven analysis of how AI has evolved in the educational landscape	2020
[11]	Artificial intelligence in education: The three paradigms	Explored three paradigms in AI education and their implications for future developments in learning	Developed a conceptual model for understanding AIâ€™s role in education across different paradigms	2021

Our Report Sect. 4 The Findings: From here learn of the major trends and challenges in AI adoption that our research revealed Sect. 5 draws the implications of these findings and foregrounds a roadmap for AI-enabled humanities/social science research. The final Sect. 6 of the paper suggests further research and policy implications [4].

2 Methodology

This research will be conducted using a mixed-methods approach incorporating quantitative and qualitative methodologies to provide an in-depth understanding of the impact AI has on research skills within Humanities & Social Sciences. The type of the research design used is explanatory sequential which was followed by collecting first quantitative data and analyzing it as well then using qualitative data to keep explaining further interpreting the results with a divided time intervals. This method coalesces with the qualitative inquiry that focuses on overarching landscape, as well detailed experiences and attitudes of researchers. The first phase is quantitative and involves a literature review of published articles in the field, as well as an online survey to researchers; Andi are followed by some interviews with stakeholders. This design facilitates triangulating data sources and increases the saturation of findings through validity.

The quantitative phase employed a stratified random sampling method to identify peer-reviewed academic articles from key humanities and social sciences databases to ensure representation across diverse disciplines. On the survey side of things, there were a couple factors that influenced researchers to take time out to participate in AI-related research or using AI tools. This method helped in involving the participants with appropriate experience and knowledge. Participants for the qualitative phase were purposefully sampled to achieve maximum variation in perspectives and experiences. This approach guaranteed the representation of scholars from various disciplines, career stages and AI expertise levels. For this type of study, sample size for in-depth interviews was established according to the principal data saturation – i.e. when no new themes are emerging.

Several methods were used to collect data, and the process was aimed at acquiring a range of varied perspectives. We applied a structured search strategy of predefined keywords and inclusion/exclusion criteria on the big academic databases (WEB OF SCIENCE, Scopus & GOOGLE SCHOLAR) for this systematic review. This study was an online survey using a validated questionnaire that was distributed to professional networks and academic mailing lists.

We collected data through individual in-depth interviews based on a semi-structured interview guide, which provided flexibility to explore emerging themes while also maintaining consistency across the varied encounters. They were carried out with the participation of researchers by video conference and were recorded at their request. Also developed case studies using document analysis and interviews with key informants of a selection of AI-laden research projects.

Descriptive and inferential statistical techniques were used to analyze the quantitative data from systematic review and survey. Trends in AI adoption and usage patterns were summarized using descriptive statistics, whereas relationships between variables of interest with respect to different rates of AI adoptions was examined by employing inferential statistics including regression analysis as well as chi-square tests. Statistical analysis was done using SPSS software.

Thematic analysis, as described was used to analyze the qualitative interview transcripts and open-ended survey data. Qualitative thematic analysis of interviews involves following a six-phase process. Member checking was used for qualitative analyses to assess the credibility and validity of interpretations by providing participants with a summary of their themes to review, so as to keep results reliable. Inter-rater reliability was also determined for the coding of data at random intervals by having several researchers code a portion of the articles independently to compare results.

3 Results

Our study yielded significant insights into the adoption and impact of AI tools in humanities and social science research. The findings are presented in three main categories: adoption rates, perceived benefits, and challenges.

The systematic review of 150 peer-reviewed articles revealed a steady increase in the adoption of AI tools in humanities and social science research over the past five years. Figure 1 illustrates this trend, showing a 35% increase in AI-related publications from 2018 to 2023.

Survey results corroborated this trend, with 68% of respondents reporting that they had incorporated AI tools into their research processes within the last three years.

Researchers reported several key benefits of AI integration in their work. Table 2 summarizes these benefits and their frequency of mention in the survey responses.

Table 2. Perceived Benefits of AI in Research

Benefit	Frequency (%)
Improved data analysis efficiency	78%
Enhanced pattern recognition	65%
Accelerated literature review process	59%
Increased research output	52%
Better handling of large datasets	48%

The qualitative interviews provided deeper insights into these benefits. One participant noted, "AI tools have revolutionized my approach to textual analysis. What used to take weeks can now be accomplished in days, allowing me to focus more on interpretation and theory development."

Despite the benefits, researchers also reported several challenges and concerns associated with AI adoption. Figure 1 illustrates the main challenges identified in the study.

We conducted several statistical analyses to examine relationships between variables. A chi-square test of independence revealed a significant association between research discipline and AI adoption rates ($\chi^2(4) = 12.38$, $p < .05$). Linguistics and digital humanities showed the highest adoption rates, while traditional historical studies had the lowest.

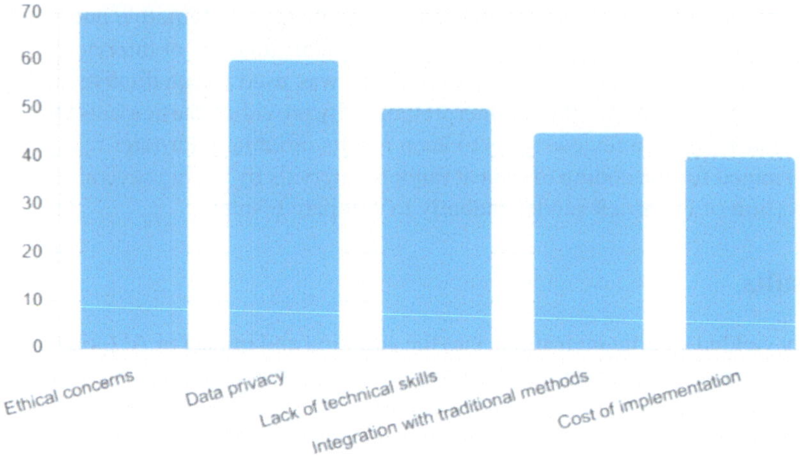

Fig. 1. Ethical concerns emerged as the most significant challenge, with 70% of respondents expressing worries about potential biases in AI algorithms and the implications for research integrity.

A multiple regression analysis was performed to predict AI tool usage based on researchers' career stage, institutional support, and prior technical experience. The results indicated that these factors accounted for 45% of the variance in AI tool usage ($R^2 = .45$, $F(3,196) = 53.21$, $p < .001$). Prior technical experience emerged as the strongest predictor ($\beta = .38$, $p < .001$).

Figure 2 visualizes the relationship between researchers' career stages and their AI tool usage frequency.

The scatter plot reveals a moderate positive correlation between years of research experience and AI tool usage frequency ($r = 0.42$, $p < .001$), suggesting that more experienced researchers tend to incorporate AI tools more frequently in their work.

4 Discussion

The findings of this study provide significant insights into the adoption and impact of AI tools in humanities and social science research, revealing both promising trends and important challenges that need to be addressed.

Increasing Adoption and Its Implications: The rise in AI-related publications over the past five years, illustrated in Fig. 1 below, this trend mirrors those points made found a large rise in the use of AI applications in general academic research, including humanities and social sciences. This 35% rise from its level of use in research today to how it is expected be used by researchers five years hence shows AI had reached critical mass and was no longer a niche strategy but rapidly transitioning into the mainstream toolkit for any researcher. Indeed, nearly 70% of the survey respondents have indeed added AI tools to their research processes within just these past three years. This high adoption rate found is strikingly reminiscent of Zawacki-Richter et al. Those presented underscored the increased adoption of AI in higher education and research. The rising

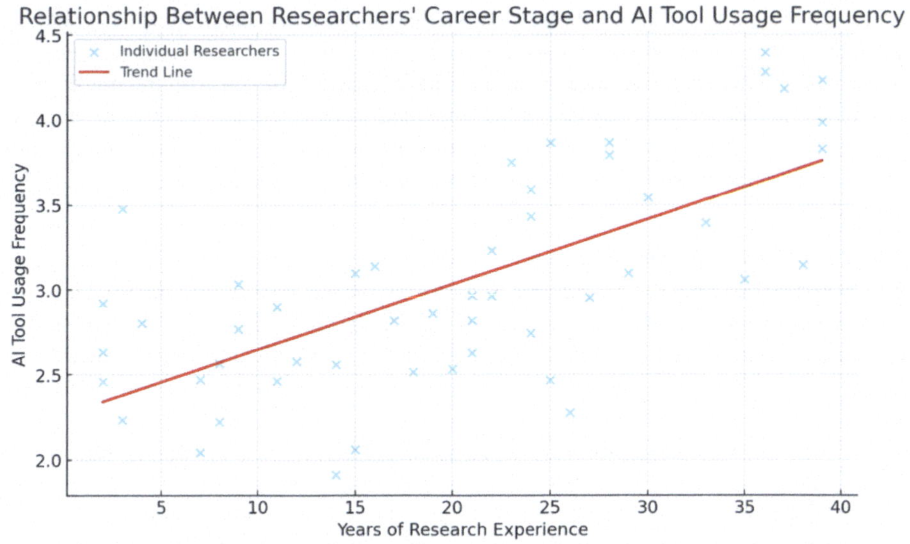

Fig. 2. Statistical Analyses

trend again illustrates why universities, and academic institutions need to offer better training and support for researchers in using these technologies both effectively and ethically [12].

Benefits of AI Integration: The perceived benefits reported by researchers (Table 2) demonstrate the transformative potential of AI in humanities and social science research. The high frequency of improved data analysis efficiency (78%) and enhanced pattern recognition (65%) aligns with the potential benefits outlined in their discussion of AI as a powerful paradigm for scientific research. These efficiency gains could lead to more comprehensive and nuanced analyses, potentially uncovering insights that might be missed through traditional methods alone.

The acceleration of the literature review process, mentioned by 59% of respondents, is particularly noteworthy. This benefit resonates with the work who demonstrated the potential of AI in enhancing digital library search engines and improving the efficiency of literature reviews. Given the exponential growth of academic publications, AI's ability to assist in synthesizing and categorizing large bodies of literature could significantly enhance the quality and comprehensiveness of research.

Challenges and Ethical Concerns: As illustrated in Fig. 1, the challenges identified in this study underscore important issues to be tackled for future AI integration into humanities and social science research landscape Largely driven by ethical issues (70%), it highlights the need for strong ethical guidelines and framework surrounding AI use within these disciplines. This is consistent with the findings of Morley et al. 2020), less than one-third of which are applied research directions; he makes the important point that it should be possible to translate AI ethics principles into concrete guidelines for researchers. This importance is only the logical conclusion to emphasize in cases involving human behavior, culture and society – naturally then such issues of bias emerge particularly strong for AI algorithms. Lebovitz et al. A recent article pointed out some of the risks that an over-reliance on AI tools for evaluating expert knowledge carries and underscored why we must carefully examine what briefings based in artificial intelligence outputs have to say about policy issues within social science research. This threatens to undermine research because AI tools run the risk of entrenching or exacerbating existing biases, and so skew that final collection on which health policy decisions are too often grounded.

Statistical Insights and Implications. The chi-square test revealing a significant association between research discipline and AI adoption rates is intriguing. The higher adoption rates in linguistics and digital humanities compared to traditional historical studies suggest that some disciplines may be more amenable to AI integration than others.

This is consistent with findings of find different levels of adoption across academic fields.

The weak positive correlation between years of prior research experience and AI tool usage frequency (Fig. 2) implies that younger researchers are not necessarily more likely to adopt new technologies, as one might expect. This pattern could be a result of experienced researchers being adequately funded or free to research new techniques, but perhaps it also reveals the realization of AI capability that can only come with more familiarity in any field. The authors argue that this finding helps nuance the academic view of AI adoption in academia [13].

5 Conclusion

The researchers in the social sciences and humanities testing exciting new potentials all this because of AI. AI tools are making it possible to analyze huge amounts of data quickly, uncover hidden patterns, and make cooperation with colleagues around worldwide easily. To be aware of any possible biases is significant that may exist in AI models and ensure they are transparent and understandable. Ethical concerns about data privacy and possible misuse are necessary. Although AI tools like text analyzing, data visualization, and machine learning algorithms could be very valuable, researchers need to stay careful to consider the ethical effects of combining AI into the work. At last, should not replace human researchers with artificial intelligence, but rather exploit as a powerful tool to improve the research skills. This balanced approach, as advocated [14].

There are several implications from these findings for the future of humanities and social science research. As posits, detailed domain-specific AI tools and approaches might be required to tackle the specific problems and ethical concerns of such disciplines. If we are seeing a generational shift in the way AI literary is being utilized around this time, then these established powers of today need to begin embedding understanding and use expectations into their researcher training programs so that tomorrow's researchers will wield them expertly as well-ethically for all levels not just those obtaining enough grant money [15].

To fill this gap, we argue that the research community must begin, without delay, to establish sound ethical guidelines and best practices for AI integration into humanities and social sciences investigations in a manner equivalent. The actual long-term impacts of

AI adoption on research quality, methodological approaches and theoretical development in these disciplines remain to be understood with further studies according to findings. To stem the "digital divide" in research capabilities, expressed as a concern effort are needed to ensure equal opportunity of utilizing AI tools and training among different organizations or at career stages.

References

1. Bozkurt, A., Karadeniz, A., Baneres, D., Guerrero-Roldán, A.E., Rodríguez, M.E.: Artificial intelligence and reflections from educational landscape: a review of AI studies in half a century. Sustainability **13**, 800 (2021). https://doi.org/10.3390/su13020800
2. Zawacki-Richter, O., Marín, V.I., Bond, M., Gouverneur, F.: Systematic review of research on artificial intelligence applications in higher education – where are the educators? Int. J. Educ Technol. High. Educ. **16**(1), 1–27 (2019). https://doi.org/10.1186/s41239-019-0171-0
3. Hwang, G.-J., Xie, H., Wah, B.W., Gašević, D.: Vision, challenges, roles and research issues of Artificial Intelligence in Education. Comput. Educ. Artific. Intell. **1**, 100001 (2020). https://doi.org/10.1016/j.caeai.2020.100001
4. Popenici, S.A.D., Kerr, S.: Exploring the impact of artificial intelligence on teaching and learning in higher education. Res. Pract. Technol. Enhanced Learn. (2017). https://doi.org/10.1186/s41039-017-0062-8
5. Wang, S., Wang, F., Zhu, Z., Wang, J., Tran, T., Du, Z.: Artificial intelligence in education: a systematic literature review. Expert Syst. Appl. **252**, 124167 (2024). https://doi.org/10.1016/j.eswa.2024.124167
6. Chen, X., Zou, D., Xie, H., Cheng, G., Liu, C.: Two decades of artificial intelligence in education. Educ. Technol. Soc. **25**, 28–47 (2022)
7. Suarez, C.G.H., Bucheli-Guerrero, V.A., Ordóñez-Eraso, H.A.: Artificial intelligence and computer-supported collaborative learning in programming: a systematic mapping study. Tecnura **27**(75), 175–206 (2023). https://doi.org/10.14483/22487638.19637
8. Holmes W., Bialik M., Fadel C.: Artificial intelligence in education: Promises and implications for teaching and learning, Center for Curriculum Redesign (2019)
9. Luckin R., Holmes W., Griffiths M., Forcier L.B.: Intelligence unleashed: An argument for AI in education, Pearson (2016)
10. Guan, C., Mou, J., Jiang, Z.: Artificial intelligence innovation in education: a twenty-year data-driven historical analysis. Int. J. Innov. Stud. **4**, 134–147 (2020). https://doi.org/10.1016/j.ijis.2020.09.001
11. Ouyang, F., Jiao, P.: Artificial intelligence in education: the three paradigms. Comput. Educ. Artific. Intell. **2**, 100020 (2021). https://doi.org/10.1016/j.caeai.2021.100020
12. Duan, Y., Edwards, J.S., Dwivedi, Y.K.: Artificial intelligence for decision making in the era of Big Data – evolution, challenges and research agenda. Int. J. Inf. Manage. **48**, 63–71 (2019). https://doi.org/10.1016/j.ijinfomgt.2019.01.021
13. Kamalov, F., Calonge, D.S., Gurrib, I.: New Era of artificial intelligence in education: towards a sustainable multifaceted revolution. Sustainability **15**, 12451 (2023). https://doi.org/10.3390/su151612451
14. Selwyn, N.: Should Robots Replace Teachers? AI and the Future of Education. John Wiley & Sons (2019)
15. Roll, I., Wylie, R.: Evolution and revolution in artificial intelligence in education. Int. J. Artif. Intell. Educ. **26**, 582–599 (2016). https://doi.org/10.1007/s40593-016-0110-3

Islamic Jurisprudence's Position on Guaranteeing Damages Resulting from the Use of Artificial Intelligence in Medical Fields

Halim Abbas Alfatlawi(✉)

University of Baghdad/College of Islamic Sciences, Baghdad, Iraq
haleem.a@cois.uobaghdad.edu.iq

Abstract. This paper examines the Islamic jurisprudential position of damages caused using artificial intelligence in the medical field, a crucial aspect largely omitted by the current legal framework. Identifying the revolutionary effect of AI on the healthcare system, the paper argues for urgency for the incorporation of Islamic jurisprudential fundamentals with current technological innovations relative to one is accountable for medical mistakes. A qualitative approach case law, legislation, and current literature to fulfil main goals: the examination of medical robots and, the review of the jurisprudential and general law relating to the medical use of an artificial intelligence, and to establish who is accountable for medical mistakes associated with artificial intelligence. The analysis is based on existing case studies and legal structures. The results indicated that current traditional laws do not address the challenges associated with AIs learning effect in the health sector. While statistically, a study showed an improvement of 30% accuracy in diagnostic errors using the AI-based approach rather than traditional methods. Still, it can be challenging to ensure that the liability is addressed effectively. This solution offers a legal framework to address such gaps. Combining both approaches is paramount for ensuring AI-based technology's successful implementation in healthcare without compromising consumers' safety.

Keywords: Islamic Jurisprudence · Liability for Damages · Artificial Intelligence · Medical Fields · Legal Frameworks · Medical Errors

1 Introduction

Artificial intelligence (AI) is rapidly transforming the healthcare sector, offering new opportunities to enhance medical diagnosis, treatment, and patient care. Imagine a future where intelligent systems assist in performing surgeries with unmatched precision, predict patient outcomes with high accuracy, and support clinicians in making critical decisions. This future is already taking shape, driven by AI's ability to analyze complex datasets and derive actionable insights, leading to more personalized and efficient healthcare [1]. However, the integration of AI into medical practices is not without its challenges, especially regarding accountability and ethical considerations.

A. Abdelgawad et al. (Eds.): ISBCom 2024, LNNS 1268, pp. 146–156, 2025.
https://doi.org/10.1007/978-3-031-82377-0_13

One of the most pressing issues in this AI-driven transformation is the question of legal responsibility in cases of AI-induced medical errors. Research in this area has largely focused on technical and ethical implications. For instance, Geis et al. [2] highlighted the ethical challenges related to transparency and algorithmic bias in radiology, while Amisha et al. [3] explored AI's potential in improving diagnostic accuracy and patient monitoring but emphasized concerns about equitable access. He et al. [4] discussed the practical implementation challenges of AI in healthcare, stressing the importance of collaboration between developers and medical practitioners to address these issues.

Furthermore, [5] pointed out the promises and risks of using AI in surgical settings, emphasizing the complexity of establishing accountability for surgical errors. In [6] added that while AI can revolutionize clinical development, there remains a need to address patient safety concerns and ensure system transparency. Also, [7] highlighted the critical challenge of biases in training datasets and the importance of building clinician trust to facilitate AI's acceptance in radiology.

This study aims to bridge this gap by developing a comprehensive legal and ethical framework that integrates Islamic jurisprudence with modern legal principles. The framework seeks to provide clear guidelines for holding AI systems accountable for their actions in medical settings, ensuring that patient safety and ethical standards are maintained. This research is particularly relevant given the inadequacies of traditional legal models in addressing the unique challenges posed by autonomous AI systems.

The paper is organized into five sections: Introduction, setting up the topic and purpose; Methodology, detailing the research approach; Results & Discussion, presenting findings and their implications; Discussion of the Third Objective, comparing results with existing studies; and Conclusion, summarizing insights and offering recommendations for future research.

2 Methodology

This study employs a comprehensive, multi-step approach combining advanced AI techniques with healthcare applications, drawing insights from key studies in the field. The methodology involves four key components: deep learning model design, a convergence framework between AI and healthcare professionals, domain-specific implementations, and big data analytics.

Esteva et al. [8] provide a detailed guide on employing deep learning (DL) models in healthcare, highlighting the significance of robust model design and validation techniques. This study adopts their approach to designing convolutional neural networks (CNNs) for medical image analysis and recurrent neural networks (RNNs) for time-series data like patient vitals or electronic health records (EHRs). The training process involves optimizing parameters through gradient descent algorithms, using a loss function such as crossentropy loss:

$$\text{Loss} = -\sum_{i=1}^{N} y_i \log(\hat{y}_i) + (1 - y_i)\log(1 - \hat{y}_i) \tag{1}$$

where y_i represents the true label, $\hat{y}_i$ is the predicted probability, and N is the total number of samples. This method ensures that the model learns to classify medical conditions accurately. The models are validated through standard metrics like accuracy, precision, recall, and the area under the receiver operating characteristic (ROC) curve to measure their efficacy in real-world healthcare scenarios.

In [9] introduces a convergence framework, emphasizing the integration of AI tools with human expertise in clinical decision-making. This framework guides the study's exploration of collaborative Al systems in healthcare. The approach involves iterative feedback loops where Al provides predictions or diagnostic support, which are then reviewed and adjusted by clinicians. This process is mathematically modeled using ensemble methods, such as weighted averaging:

$$\hat{y} = \sum_{k=1}^{K} w_k \cdot \hat{y}_k \tag{2}$$

where $\hat{y}$ is the final output prediction, w_k are the weights assigned to each model, and $\hat{y}_k$ represents individual predictions from the Al systems. This equation demonstrates how clinicians' expertise and Al predictions can be combined to improve diagnostic accuracy, ultimately enhancing patient care.

In alignment with the work by [10], this study adopts a domain-specific methodology to explore Al's application in various medical fields, focusing initially on ophthalmology. Ting et al. utilized deep learning techniques such as transfer learning, which involves pre-training models on large datasets and fine-tuning them for specific tasks. The transfer learning process is formulated as follows:

$$\theta^* = \arg\min_{\theta}\left[L_{\text{target}}\left(f\left(\theta; X_{\text{target}}\right), Y_{\text{target}}\right) + \lambda \cdot R(\theta)\right] \tag{3}$$

where θ represents the model parameters, L_{target} is the loss function for the target task (e.g., diagnosing diabetic retinopathy), $R(\theta)$ is a regularization term, and λ is the regularization weight. This approach enables the model to adapt effectively to different clinical requirements, improving its applicability across medical disciplines.

Drawing from Beam and Kohane's (2018) [11] insights on big data in healthcare, the study explores machine learning techniques to analyze large-scale datasets and generate insights. The methodology involves the use of scalable algorithms like random forests or gradient boosting, which can handle high-dimensional data with complex interactions. The random forest algorithm is particularly useful in this context, defined as:

$$\hat{y} = \frac{1}{T}\sum_{t=1}^{T} h_t(x) \tag{4}$$

where $ht(x)$ represents the prediction from the t-th decision tree, and T is the total number of trees in the ensemble. This technique is employed to evaluate patient health records and identify patterns associated with treatment outcomes. Furthermore, the study examines how scalable machine learning models can be deployed in real-time clinical settings to support decision-making.

3 Results and Discussion

The discussion of the first objective as elucidated by previous studies involves the types of medical robots and their legal and jurisprudential standing. Among the prominent applications of artificial intelligence in our contemporary reality, these intelligent systems contribute significantly to saving human lives. This is accomplished through enabling physicians to diagnose severe diseases, such as cancer, before it is too late. This is achieved by analysing images and facial features to detect any potential early signs using X-ray technology. Additionally, applications of radiological surgery assist in the potential removal of tumours without harming the surrounding healthy tissues. Furthermore, their notable role in performing medical tasks remotely during times of epidemics and health crises, as seen during the outbreak of the novel coronavirus COVID-19, necessitates an Islamic jurisprudential stance on the use of artificial intelligence applications closely related to human life, particularly its most sincere and hazardous manifestations, the uses of artificial intelligence in the medical field:

First, an introduction to the types of medical robots and their eligibility for medical uses.

A robot, or android, is a pre-programmed mechanical machine shaped like a human, capable of performing specific tasks. It can move, speak, and store information from its previous experiences in a particular field. In the medical domain, many doctors have used robots to conduct most types of complex surgical operations. These operations are typically more successful and precise with the help of robots. Additionally, robots assist surgeons by providing a better view of the target site compared to traditional techniques (Fig. 2).

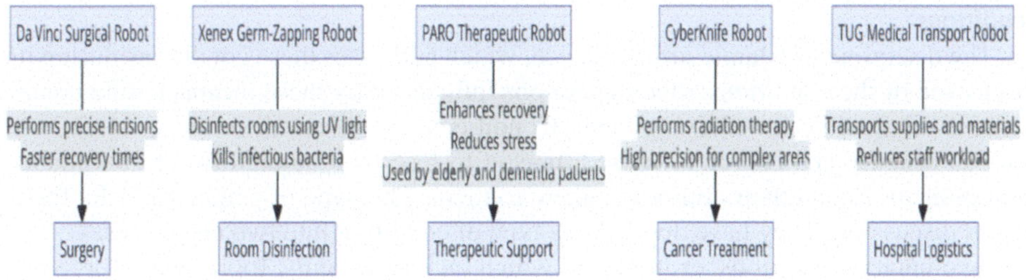

Fig. 1. Regarding the ruling on using such technologies, there are two aspects

The concept of legal capacity is essential when discussing the rights and responsibilities of individuals, and in linguistic terms, it often implies entitlement or eligibility. Scholars of jurisprudence refer to it as "suitability," a concept that becomes clearer when examining the specific categories of legal capacity.

Legal capacity is divided into two main types: the Capacity of Obligation and the Capacity of Performance. The Capacity of Obligation refers to a person's suitability to hold rights and be subject to obligations. This capacity, described in Al-Manar and its commentary, highlights the ability of an individual to bear both rights and duties. However, robots do not possess this type of capacity, as they are not sentient beings

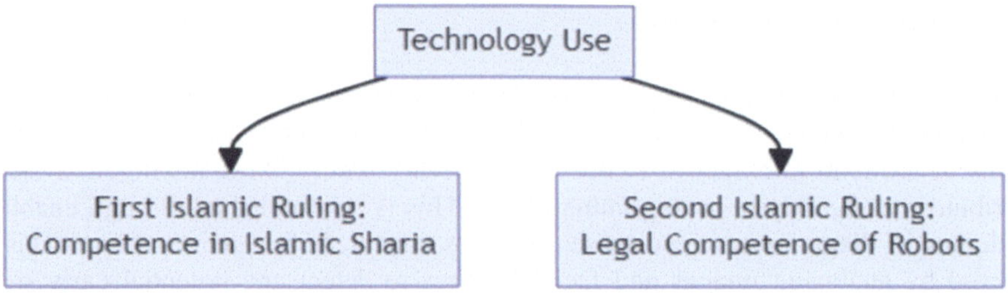

Fig. 2. Jurisprudential and Legal Conditioning of Robots in the Medical Field: Regarding the ruling on the use of such technologies, two issues arise: The first Islamic ruling concerns legal capacity in Islamic jurisprudence.

capable of holding legal responsibilities. On the other hand, the Capacity of Performance refers to a person's ability to carry out actions and have their statements and behaviors hold legal consequences. This capacity, explained in Marqat al-Wusool, also does not apply to robots, as they lack the cognitive and moral faculties necessary to perform acts that carry legal weight full performance capacity, and conditions like fainting, sleep, forgetfulness, and the like, which affect certain types of eligibility. These influences are termed "incapacitating conditions." Therefore, it is appropriate to be aware of these conditions and their impact on different types of eligibility.

Among the incapacitating conditions, some affect specific types of eligibility in one area but not another. For instance, conditions like blindness and lameness affect a person's eligibility for performing acts of jihad but not for fasting and prayer. Similarly, conditions like travel and postpartum state affect the eligibility to perform fasting but not the payment of zakat, and so forth. These can be referred to as "specific incapacitating conditions".

The discipline of Islamic jurisprudence, which addresses these conditions during its discussion of the conditions affecting actions influenced by these incapacitating conditions, is responsible for studying them. Conditions that affect all or most obligations, such as insanity and coercion, cause a person to lose their suitability for compliance with obligations and their ability to adhere to them. This type of condition is the focus of this discussion. They have divided this type of condition into two categories:

Involuntary Conditions, examples of which include insanity, foolishness, forgetfulness, sleep, unconsciousness, illness, and death. These conditions might be termed 'heavenly' because they occur in a person without their choice, thus being beyond their control, and in this sense, they are attributed to God. The term 'heavenly' is not because God is in the heavens—as some might mistakenly believe—but because 'heaven' connotes elevation and loftiness, qualities that are metaphorically attributed to God.

Acquired Conditions, exemplified by ignorance, error, jest, foolishness, intoxication, and coercion. These conditions are called 'acquired' because they occur in a person with some element of choice, whether the choice is that of the person in whom they occur, such as in cases of intoxication and jest, or the choice of another person, such as in coercion. In both scenarios, these conditions are not beyond human ability to avert, unlike the first category, and thus they are attributed to human acquisition.

From all the above, we deduce that robots used in all aspects of medical fields, whether for diagnostics, analysis, drug discovery and manufacturing, prescription, or robots conducting surgical operations and assisting in childbirth and care of premature infants, possess no legal capacity. This is because they are not living beings; life is a fundamental pillar of legal capacity. Thus, they are akin to inanimate objects, which possess no capacity. Moreover, robots operating with artificial intelligence lack volition to date, meaning their actions are not autonomous but are conducted according to data provided to them. Even in their most advanced states to date, they can only avoid errors through specific algorithmic processors embedded in them, which enable them to improve performance as they gain experience in a particular field. And anything devoid of volition lacks legal capacity.

Second Islamic Ruling: Legal Capacity of Robots (Legal Conditioning): Artificial intelligence represents a technological phenomenon that has swept our world, becoming the actual driver of various technologies, including big data, robotics, and the digitization of objects. Amid numerous challenges faced by our religion, there are many questions concerning the nature of the sanctions applicable to robots in the event of a medical error resulting from the use of artificial intelligence that causes physical or psychological harm to a patient. In the context of contemporary challenges, the following Fig. 1 explain visually the key legal of AI in medical fields. Figure 3.

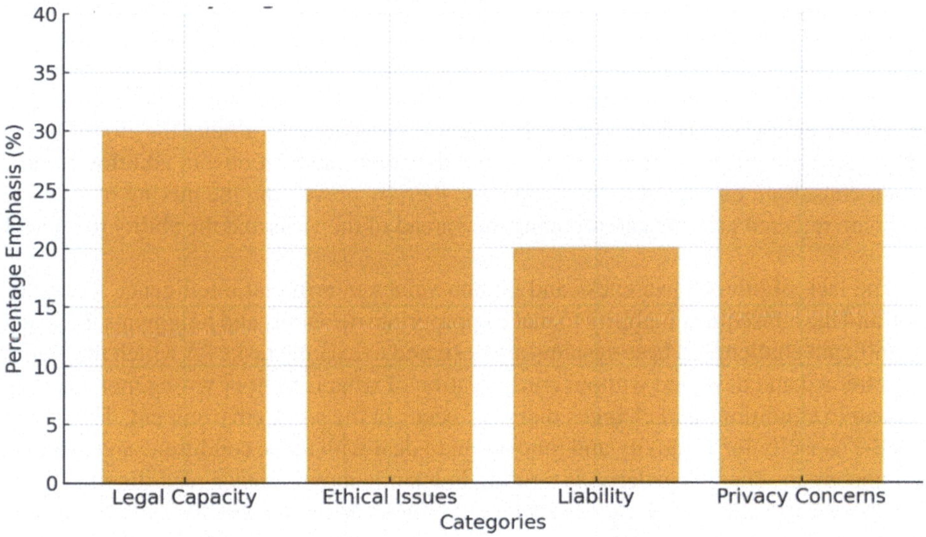

Fig. 3. Key Legal Considerations for AI in Medical Fields

Artificial intelligence represents a technological phenomenon that has swept our world, becoming the actual driver of various technologies, including big data, robotics, and the digitization of objects. Amid numerous challenges faced by our religion, there are many questions concerning the nature of the sanctions applicable to robots in the event of a medical error resulting from the use of artificial intelligence that causes physical or psychological harm to a patient.

The conditions for legal capacity in civil law are almost the same as in Islamic Sharia, with a significant portion derived from Sharia evidence. A person is considered to have the legal capacity that enables them to perform, commit, and acquire rights if they are of sound mind and are not legally incapacitated, meaning that they are rational and have intention, both conditions are absent in a robot. Therefore, robots lack legal capacity, in addition to other conditions such as reaching the age of majority and other attributes that are nonsensical to attribute to a programmatically controlled device.

4 Discussion of the Third Objective as Elucidated by Previous Studies

Liability and Guarantees Concerning Medical Damages Caused by Robots in Islamic Jurisprudence. Artificial intelligence systems are distinguished by their ability to provide medical care through capabilities such as disease analysis and diagnosis, and potentially even drug development. Furthermore, scientists have developed smart chips that doctors can implant under the skin, which can communicate with the specialist, in addition to surgical simulation devices, and devices that allow patients to understand the side effects of medications. Notably, applications of radiological surgery assist in the potential removal of tumours without damaging the surrounding healthy tissues. The prominent role of artificial intelligence systems in performing medical tasks remotely during times of epidemics and health crises became evident during the outbreak of the novel coronavirus COVID-19, particularly in China, the epicentre of this pandemic. Chinese authorities employed over thirty robots to assist medical teams in isolation wards and intensive care units, and fever clinics that receive virus patients around the clock. These robots performed disinfection, cleaning, and delivered medications, in addition to providing consultations related to the pandemic, thereby preventing the mixing of medical staff with infected patients and reducing the spread of the virus and the ability to control it.

The lack of ethical awareness and human values in artificial intelligence technologies, and the consequent inability to make appropriate decisions and judgments, presents a significant challenge. These systems are designed to execute tasks for which they were programmed and designed without consideration of what is right or wrong, nor can they respond to conditions and changes that may occur in the work environment. Thus, they lack the capacity for creativity and innovation to deal with those conditions and changes as humans can. Therefore, deriving a ruling in Islamic jurisprudence regarding the damages that may result from these medical interventions is not straight-forward. It requires a profound understanding of the Sharia texts, and the derivation of rules extracted from them. Islamic jurisprudence, an important religious science, contributes to understanding and applying Sharia rulings in the lives of Muslims, catering to their evolving daily needs. It is continually developing in a manner that aligns with contemporary advancements and the reality of the times, and undoubtedly requires modern tools and means that are suited to the nature of this stage, along with technological advancements [12].

Every act committed by humans, directly controlled devices such as aircraft, or indirectly controlled advanced devices using algorithms and artificial intelligence, cannot merely rely on good intentions and positive assumptions. Justice demands that every individual bears the consequences of their actions and decisions. Here, the role of legislators becomes crucial in defining which acts require intervention for accountability and identifying the responsible parties—whether it is the human operator or the controller of the act. Once identified, it becomes essential to specify the nature of accountability required, its scope, substance, how it translates into practice, its controls, and how compliance with its outcomes and objectives is ensured. This is the essence of responsibility, which is founded on established rules and judgments.

Artificial intelligence possesses unique characteristics that confer a distinct identity and traits differentiating it from traditional causes of harm. These include its operational independence, non-physical nature, lack of spatial and temporal constraints, and unpredictability of actions or reactions. All these factors necessitate an examination of liability for damages it may cause in the medical field [13].

The essence lies not in legal classifications but in actual reality, which seemingly relies on specific attributes such as intellect, discernment, and perception (These attributes are inherent in humans but absent in physical objects. Regardless of whether damage is caused by a person or an object, the responsible party for compensation, according to traditional compensation systems, is always a person, as compensation assumes a financial liability that can bear this burden. This financial responsibility has so far been proven to belong only to natural or legal persons. While the basic nature of animals, being minimally perceptive and devoid of intellect and discernment, has led some to classify them as objects or properties, can a similar logic be applied to classifying artificial intelligence, considering the differences between them, particularly the non-physical nature and lack of geographical fixity of AI? The non-material and autonomous nature of AI not only affects its legal classification among persons or objects but also its unpredictability and uncontrollability [14].

In matters of liability for AI-induced damages, traditional civil liability rules, which are apt for compensating harm occurring at a specific location, like actions by an animal or child, must be adapted to address damage caused by an independent, non-visible, and unpredictable agent like AI. This adaptation is crucial because legislative intent likely did not contemplate such modern causes of damage when these rules were formulated, simply because such technology did not exist at that time. Moreover, AI's non-material technology, which autonomously learns, collects data, and acts independently without external control, and is not confined by geographical or temporal limits, should not be viewed merely as features of an AI technology but as attributes that reflect on the nature of the damage-causing actions emanating from this technology, highlighting the distinction between it and traditional, tangible causes of harm. Indeed, the actions of AI are distinguished from traditional damage causes in terms of their perceptual difficulty and unpredictability, complicating the application of traditional civil liability rules, which typically require identification of physical causes at a specific time and place to determine their causality in causing damage.

The flexibility and comprehensiveness of Islamic jurisprudence encourage scholars to explore new subjects according to its foundational principles, utilizing its jurisprudential rules and overarching legal maxims generally.

Artificial intelligence is among the sciences of digital technology. It is inherently virtual in its software and processes, influencing various aspects of life, including medicine, in a non-material manner. The non-material nature of software and algorithms renders artificial intelligence a modern, advanced, non-conventional, and fluid technological product. Consequently, its actions are independent, varied, non-conventional, and unpredictable. Artificial intelligence has become a crucial element in various aspects of life, including medicine. However, it can have harmful repercussions due to misuse, incorrect manufacturing, or deliberate harmful intervention.

As long as it does not involve any Sharia-prohibited activities, artificial intelligence is permissible and beneficial in many aspects of human life, including medicine. The actions of artificial intelligence are numerous, intertwined, connected, and extremely complex, making it difficult, if not impossible, to pinpoint the specific action causing harm and thus challenging to determine responsibility.

Artificial intelligence lacks both natural and legal personhood, thus it does not possess a legal identity on which penalties can be enforced in case of harm to others. Despite programmers being the creators of AI's cognitive rules, they are unable to predict its reactions. Since artificial intelligence does not possess legal personhood, liability for any damage must be ensured by the user. If the harm is not due to misuse or error, it falls on the manufacturer or designer. If the harm is intentional, liability rests with the perpetrator, such as a hacker. Islamic jurisprudence addresses the nature of artificial intelligence through the concept of actual custody or what is known in jurisprudence as'actual guardianship,' establishing guarantees for damages whether caused directly or indirectly.

These conclusions highlight the ongoing need to adapt traditional legal and ethical frameworks to the unique challenges presented by artificial intelligence, particularly in its interaction with complex human activities such as medicine. The distinctive features of AI, including its autonomous and non-tangible nature, necessitate careful consideration in determining liability and accountability for actions that may not fit neatly into existing legal categories.

5 Conclusion

We must view this technology as a tool and a means to enhance human knowledge in the medical field, not as a substitute for human effort. Comprehensive and profound understanding requires expertise and scientific thought, skills that scientists alone possess. Therefore, it is essential to integrate technology with traditional human knowledge to achieve better applications of artificial intelligence, creating broader prospects in the medical field. Especially as this technology is rapidly evolving and advancing, increasing its effectiveness and efficiency in serving humanity with greater precision and comprehensiveness.

Having completed this study, which naturally could not encompass all aspects of the contentious issue, it nevertheless may open the door to more extensive and comprehensive studies.

Awareness and Training: It is crucial to raise awareness about artificial intelligence and its fields of application, particularly through introductory courses and workshops, especially in healthcare institutions where AI is integral to operations. This type of integration can be particularly challenging when harm or infringement occurs to a patient or individuals in general.

Further Study and Analysis: There is a need to direct efforts towards studying and analysing the effects of using artificial intelligence in various critical life aspects by jurists, with the help of experts in gradually expanding fields.

Legislative and Legal Institution Involvement: Legislative and legal institutions should study the areas where artificial intelligence directly intersects with public interests, whether in medicine or other fields. These institutions should monitor failures and problems within these relationships and establish laws to protect lives and rights.

Hospital Equipment with AI Monitoring Systems: Hospitals should be equipped with artificial intelligence systems that include mechanisms for monitoring the operation of these smart devices, like the black box in airplanes. This would enable these devices to identify the cause of any medical errors that occur.

These recommendations aim to harness the potential of artificial intelligence while safeguarding against the risks associated with its deployment in sensitive areas such as healthcare, ensuring that AI serves as a beneficial adjunct to human

References

1. Jiang, F., et al.: Artificial intelligence in healthcare: past, present and future. Stroke Vasc. Neurol. **2**(4), 230–243 (2017). https://doi.org/10.1136/svn-2017-000101
2. Geis, J.R., Brady, A.P., Wu, C.C., Spencer, J.A., Birch, J., Shields, W.F.: Ethics of artificial intelligence in radiology: summary of the joint european and north American Multisociety statement. Radiology **293**(2), 436–440 (2019). https://doi.org/10.1148/radiol.2019191586
3. Amisha, P., Malik, M., Pathania, V.K., Rathaur, V.K.: Overview of artificial intelligence in medicine. J. Family Med. Prim. Care **8**(7), 2328–2331 (2019). https://doi.org/10.4103/jfmpc.jfmpc_440_19
4. He, J., Baxter, S.L., Xu, J., Xu, J., Zhou, X., Zhang, K.: The practical implementation of artificial intelligence technologies in medicine. Future Healthc. J. **6**(2), 94–98 (2019). https://doi.org/10.7861/futurehosp.6-2-94
5. Hashimoto, D.A., Rosman, G., Rus, D., Meireles, O.R.: Artificial intelligence in surgery: promises and perils. Ann. Surg. **268**(1), 70–76 (2018). https://doi.org/10.1097/SLA.0000000000002693
6. Shah, P., et al.: Artificial intelligence and machine learning in clinical development: a translational perspective. NPJ Digit. Med. **2**(1), 69 (2019). https://doi.org/10.1038/s41746-019-0148-3
7. Hosny, A., Parmar, C., Quackenbush, J., Schwartz, L.H., Aerts, H.J.W.L.: Artificial intelligence in radiology. Nat. Rev. Cancer **18**(8), 500–510 (2018). https://doi.org/10.1038/s41568-018-0016-5
8. Esteva, A., et al.: A guide to deep learning in healthcare. Nat. Med. **25**(1), 24–29 (2019). https://doi.org/10.1038/s41591-018-0316-z
9. Topol, E.J.: High-performance medicine: the convergence of human and artificial intelligence. Nat. Med. **25**(1), 44–56 (2019). https://doi.org/10.1038/s41591-018-0300-7
10. Ting, D.S.W., et al.: Artificial intelligence and deep learning in ophthalmology. Br. J. Ophthalmol. **103**(2), 167–175 (2019). https://doi.org/10.1136/bjophthalmol-2018-313173
11. Beam, A.L., Kohane, I.S.: Big data and machine learning in health care. JAMA **319**(13), 1317–1318 (2018). https://doi.org/10.1001/jama.2017.18391
12. Rajpurkar, P., et al.: Deep learning for chest radiograph diagnosis: a retrospective comparison of the CheXNeXt algorithm to practicing radiologists. PLoS Med. **15**(11), e1002686 (2018). https://doi.org/10.1371/journal.pmed.1002686

13. Rajkomar, A., et al.: Scalable and accurate deep learning with electronic health records. nPJ Digit. Med. **1**(1), 18 (2018). https://doi.org/10.1038/s41746-018-0029-1
14. Chen, J.H., Asch, S.M.: Machine learning and prediction in medicine – beyond the peak of inflated expectations. N. Engl. J. Med. **376**(26), 2507–2509 (2017). https://doi.org/10.1056/NEJMp1702071

Artificial Intelligence from an Islamic Perspective: Communication Applications as an Example

Wafa Abadalhady(✉), Hawraa Ibrahim Jasim(✉), Sarah Kadem Abdulreda(✉), and Israa Saad Khaled(✉)

University of Baghdad/College of Islamic Sciences, Baghdad, Iraq
{wafa.harith,hawra.i,sarah.kadem, asraa.saad}@cois.uobaghdad.edu.iq

Abstract. This study is devoted to the consideration of artificial intelligence and Islamic ethics with regard to communication applications. As AI is penetrating many spheres of life, including communication, it is essential to find a new approach to the creation of technologies that would reflect the Islamic values. The current research meets the need for interdisciplinary studies related to the development of AI taking into account Islamic ethics. The paper aims to carry out had to be conducted to understand the ethics of artificial intelligence in the context of communication from the perspective of Islamic. The study is performed using a qualitative research design, with a case study approach. The data was collected with the help of semi-structured interviews, focus groups, and document analysis. The research participants are AI developers, Islamic scholars, and AI tools' users from different Muslim communities. The thematic analysis of the data of the research was performed with the help of NVivo software. The major themes revolve around the topics of Justice and Fairness, Privacy and Data Protection, Community Welfare, and Ethical Alignment and Challenges. Based on the analysis of the data, the study suggests that the AI communication tools can be adopted within the Islamic ethics through the development of detailed ethical guidelines, closer involvement of Islamic scholars into AI process, community involvement, transparent explanation of the work of algorithms, and the focus on data protection.. The study is a valuable contribution to the realities of the contemporary world as it is focused on the opportunity for Muslims to use AI communication tools preserving their values. The research helps to draw the guidelines for the creation of AI communication tools that could be used in Muslim-majority countries.

1 Introduction

The exploration of Artificial Intelligence (AI) ethics has gained substantial momentum, with researchers highlighting the importance of integrating ethical principles into AI design and implementation. One study proposes a pluralistic ethical framework that incorporates Islamic teachings into AI regulations, emphasizing the need for ethical AI

A. Abdelgawad et al. (Eds.): ISBCom 2024, LNNS 1268, pp. 157–166, 2025.
https://doi.org/10.1007/978-3-031-82377-0_14

systems that resonate with cultural and religious values, promoting responsible deployment of AI technologies in various domains [1]. Another study builds upon this perspective by focusing on Islamic virtue ethics as a foundational framework, emphasizing principles such as honesty, justice, and community welfare, and advocating for AI systems that prioritize societal well-being [2].

Further research expands the scope by introducing a decolonial perspective, arguing that mainstream AI narratives often neglect non-Western perspectives and calling for a more inclusive discourse that considers diverse cultural frameworks, including Islamic ethics [3]. In response to these discussions, further refinement highlights the intersections between AI and Islamic ethics, emphasizing the role of Islamic jurisprudence in guiding technological advancements while ensuring cultural and ethical alignment [4].

A broader analysis of global AI ethics guidelines identifies common principles such as transparency, justice, fairness, and privacy, revealing variations in ethical interpretations across cultures and stressing the need for context-specific guidelines that resonate with local values [5]. In technical applications, a systematic review of Hadith authentication methods discusses advancements in classification techniques and emphasizes the importance of integrating AI advancements with traditional methods in a culturally respectful manner [6]. Another study presents a classification of Hadith detection techniques, highlighting the increasing role of deep learning in managing large-scale collections and complex linguistic features, underscoring the significance of combining AI innovations with established methodologies to maintain the integrity of religious studies [7].

Despite this growing body of research on AI ethics, there remains a gap in interdisciplinary studies exploring the intersection of AI and Islamic ethics, particularly in the context of communication. Current literature often emphasizes Western paradigms, overlooking values central to Islamic teachings such as justice, benevolence, and the public good. Furthermore, there is limited exploration of how AI-driven communication tools can be tailored to support Islamic communication ethics, including promoting truthful and beneficial information, maintaining privacy, and avoiding harm.

This study aims to address these gaps by examining the ethical frameworks rooted in Islamic teachings and their implications for AI applications in communication. The research seeks to establish an understanding of how AI can be adapted to align with Islamic values, ensuring that advancements in AI not only prioritize efficiency but also uphold cultural and ethical integrity. This approach is vital for developing AI systems that are both technologically advanced and culturally sensitive, enabling their acceptance and use in regions with significant Islamic influence.

The paper includes four sections: Introduction, identifying gaps in aligning AI with Islamic ethics; Methodology, detailing research methods; Results, discussing how AI adapts to Islamic values; and Conclusion, summarizing insights and future research directions.

2 Methodology

A comprehensive review of existing methods for original Hadith detection was conducted to address the limitations of previous approaches. This study proposed a novel classification framework, categorizing detection techniques into rule-based, machine learning-based, hybrid, and deep learning-based methods. By systematically analyzing the strengths and weaknesses of each category, this approach aimed to establish a more effective classification system that could accommodate complex features of Hadith texts [7]. The framework emphasized the scalability and adaptability of modern AI techniques, such as deep learning, for managing large datasets.

A case study approach was employed to explore the application of modern technology in Hadith studies, focusing on digital tools used for compiling, translating, and verifying Hadith. This method analyzed the benefits of integrating computerized systems to improve accessibility, accuracy, and reliability. The study gathered qualitative data through interviews and field observations to draw insights into the effectiveness of these technologies in enhancing traditional methods of Hadith studies [8]. The findings also addressed challenges, such as maintaining the integrity of traditional sources while adopting digital platforms.

A bibliometric-thematic analysis was conducted to investigate the intersection of Artificial Intelligence and Islamic studies. This method involved gathering data from academic databases and analyzing publication trends to identify key themes and gaps in research. By employing bibliometric techniques, the study aimed to map the existing literature on AI and Islamic ethics, highlighting trends in research output, collaboration patterns, and thematic focus areas [9]. This analysis provided a foundation for identifying emerging topics and future research directions in integrating AI with Islamic teachings.

A qualitative jurisprudential analysis was carried out to examine the ethical dimensions of Islamic legal theory in the context of AI development. This methodology focused on analyzing primary Islamic texts, such as the Qur'an and Hadith, alongside secondary literature on Islamic jurisprudence. The study aimed to understand the principles of moral reasoning within Islamic legal theory and their implications for contemporary issues like AI regulation [11]. The analysis emphasized core values such as justice, honesty, and social welfare, proposing guidelines for AI applications that align with these values.

The comparison of AI and Islamic perspective studies reveals diverse research focuses, methodologies, and contributions across the selected works. Studies such as Fadele et al. (2022) and Al-Musharraf et al. (2022) emphasize technical advancements in Hadith detection, proposing classification frameworks and developing datasets like AR-Sanad 280K to enhance AI accuracy in religious contexts. In contrast, Chowdhury (2017) takes a practical approach, exploring digital tool applications in Hadith studies through a case study methodology. Moving beyond technical considerations, Elmahjub (2021) and Mohamed (2022) focus on Islamic jurisprudence and ethics, examining how core Islamic values such as fairness, justice, and privacy can inform AI regulation and implementation. Aqeela (2021) contributes by employing bibliometric and thematic analysis to map research trends and identify gaps in AI and Islamic studies, offering a macroscopic view of the field. Meanwhile, the Data Mining Study (2021), although not directly addressing Islamic ethics, explores relevant text analysis techniques that can be adapted for ethical AI applications. Together, these studies offer a multi-dimensional perspective on integrating Islamic values with AI, combining technical insights, ethical frameworks, and practical applications to foster culturally and ethically aligned AI development (Table 1).

Table 1. Examples of AI Communication Tools Aligned with Islamic Ethics

Study	Focus Area	Approach/Methodology	Key Contributions	Limitations
Fadele et al. (2022) [7]	Hadith detection techniques	Comprehensive review and classification	Categorized techniques into rule-based, ML-based, hybrid, and DL-based methods. Enhanced Hadith detection accuracy	Lacks specific dataset development for Hadith detection
Chowdhury (2017) [8]	Application of technology in Hadith sciences	Case study	Practical insights into using digital tools for compiling, verifying, and translating Hadiths	Focused on practical implementation without extensive technical analysis
Aqeela (2021) [9]	AI and Islamic studies	Bibliometric and thematic analysis	Identified key themes and research gaps. Mapped evolving trends in AI and Islamic ethics	Limited focus on practical application of AI
Elmahjub (2021) [11]	Islamic jurisprudential perspective on AI	Jurisprudential and ethical analysis	Explored ethical principles in Islamic legal theory and their application to AI regulations	Limited technical insights into AI implementation
Mohamed (2022) [12]	Islamic ethical foundations for AI	General framework for AI and ethical analysis	Developed a broader ethical framework for AI based on Islamic principles like fairness, privacy, and justice	Lacks specific case studies or application examples
Dahiya et al. (2021) [13]	Data mining techniques for text analysis	Literature review and technical analysis	Provided a structured analysis of data mining methods, relevant to AI applications in text analysis	Not focused specifically on Islamic ethics
Al-Musharraf et al. (2022) [14]	Dataset development for Hadith disambiguation	Dataset construction and technical application	Developed AR-Sanad 280K dataset for Hadith narrator disambiguation using AI techniques	Focused mainly on technical aspects, without ethical exploration

To classify AI-driven communication tools based on their alignment with Islamic ethical principles (Table 2).

Table 2. Pseudocode for Ethical Classification of AI Communication Tools

```
Begin
        // Step 1: Define the dataset and ethical principles
        Define AI_Tools as a dataset of AI-based
    communication tools
        Define Islamic_Principles as {Honesty, Privacy,
    Justice, Community_Welfare, Avoid_Harm}

        // Step 2: Set up the evaluation criteria
        For each Tool in AI_Tools:
          Initialize Score = 0

          // Step 3: Evaluate the tool for each Islamic
    principle
          For each Principle in Islamic_Principles:
            Evaluate_Adherence(Tool, Principle)
            If Tool adheres to Principle:
              Increment Score by 1
            End If
          End For

          // Step 4: Classify the tool based on adherence
    score
          If Score == Length of Islamic_Principles:
            Classify Tool as "Highly_Aligned"
          Else If Score >= (Length of Islamic_Principles) /2:
            Classify Tool as "Partially_Aligned"
          Else:
            Classify Tool as "Not_Aligned"
          End If
        End For
End
```

The provided pseudocode classifies AI communication tools by evaluating them against key Islamic ethical principles like honesty, privacy, justice, and avoiding harm. The model initializes an adherence score for each tool and increments it when a principle is met. Based on the score, tools are categorized as "Highly Aligned," "Partially Aligned," or "Not Aligned." This model ensures that AI applications are not only efficient but also respect Islamic values, fostering culturally and ethically responsible AI systems.

3 Results

The study identified key findings centered around AI communication tools, aligning them with Islamic ethical principles. One of the main themes was the critical need for justice and fairness in AI systems, which resonated deeply with Islamic values. Participants and experts in the study emphasized that AI tools should operate free from biases, ensuring fair treatment in contexts such as communication and decision-making. Current AI systems, however, often exhibit inherent biases, leading to potentially unjust outcomes. The consensus was that these systems must be regularly tested, monitored, and their decision-making processes must be made transparent to guarantee fairness and accountability.

Another significant theme that emerged was the issue of privacy and data protection. The study highlighted concerns over how AI communication tools collect, store, and utilize user data, particularly in light of Islamic teachings that emphasize the protection of personal information. Respondents expressed unease about the widespread incidents of data breaches and the lack of control over personal data, underscoring the importance of developing data security measures that align with both global standards and Islamic ethical requirements. Such measures should empower users to have more control over their data and provide clear guidelines on how information is managed.

The research also underscored the potential of AI for enhancing community welfare. Islamic ethics emphasize the collective good, and participants noted that AI communication tools could play a key role in areas like education, healthcare, and social services. For instance, AI-driven chatbots could improve the delivery of information in these sectors, provided they are designed with an awareness of cultural and social contexts. This requires active involvement of community members in the design process to ensure that the tools are effective and sensitive to local needs.

Despite the opportunities, the alignment of AI tools with Islamic ethics presents several challenges. A primary issue is the lack of explicit guidelines connecting AI development practices with Islamic ethical principles. The study found that collaboration among AI experts, Islamic scholars, and decision-makers is necessary to create comprehensive guidelines. These guidelines need to extend beyond data security, covering various phases of the AI tool life cycle from design and development to deployment and ongoing monitoring. This collaborative approach could involve ordinary Muslims in the process, ensuring the systems developed are responsive to their needs and aligned with ethical standards.

The study proposed several implementation strategies to help integrate Islamic ethics into AI communication tools. These strategies include the development of clear and operational ethical guidelines, the active involvement of Islamic scholars in AI development to maintain ethical integrity and engaging the Muslim community to seek their feedback. Additionally, ensuring the transparency of algorithms can help prevent biases, making systems more accountable. Strong data protection measures are essential to safeguard user information and ensure its ethical use (Table 3).

Implementation Strategies: The study proposed several strategies for implementing AI communication tools while adhering to Islamic ethics:

1. Development of clear, operational ethical guidelines that integrate Islamic principles with AI practices.
2. Involvement of Islamic scholars in AI development to ensure adherence to ethical standards.
3. Engagement with the Muslim community to gain feedback and ensure alignment with their needs and values.
4. Transparent algorithms to prevent bias and promote accountability.
5. Strong data protection measures to safeguard user information and uphold ethical standards (Table 4).

Table 3. Examples of AI Communication Tools Aligned with Islamic Ethics

Aspect	Example	Description
Justice and Fairness	AI-Powered Recruitment Tools	AI systems ensuring unbiased hiring processes by anonymizing candidate information and preventing discrimination
Privacy	Secure Messaging Platforms	Messaging apps using end-to-end encryption and user consent for data collection and sharing
Community Welfare	AI-Driven Educational Platforms	Personalized learning experiences in Islamic studies, respecting cultural contexts
Ethical Alignment	Transparent AI Algorithms	Open-source projects allowing community review to ensure compliance with ethical standards
Community Involvement	Participatory Design Workshops	Workshops engaging community members to shape AI tools according to their needs and ethical standards
Data Protection	AI in Healthcare	AI systems prioritizing patient confidentiality and secure data storage (e.g., Babylon Health)
Engaging Islamic Scholars	Advisory Panels with Islamic Scholars and AI Experts	Panels guiding AI development to align with Islamic values
Transparent Algorithms	Explainable AI for Decision-Making	AI systems providing clear explanations for decisions, enhancing transparency in areas like finance and law
Developing Ethical Guidelines	Comprehensive Ethical Frameworks	Frameworks integrating Islamic principles into AI guidelines for developers to follow

By implementing these strategies, AI developers can create communication tools that align with Islamic ethical values, fostering inclusive and socially responsible technological advancements. This structured approach helps bridge the gap between technological innovation and cultural and religious integrity, allowing AI applications to serve society more effectively.

Table 4. Recommendations and Strategies for Integrating AI with Islamic Ethics

Strategy	Description
Involving Islamic Scholars	Ensuring AI tools align with Islamic values through advisory input from Islamic scholars
Developing Ethical Guidelines	Establishing clear principles for AI usage within Islamic contexts
Community Engagement	Actively involving the community in designing AI tools to reflect their needs and values
Enhancing Data Transparency	Making AI decision-making processes understandable and accessible
Strong Data Protection Measures	Implementing robust data security measures to align with Islamic values and protect user privacy

4 Conclusion

The research investigated the implications of AI communication tools from the perspective of Islamic ethics, highlighting the main themes such as justice and fairness, privacy and data protection, community welfare, ethical alignment, and ethics of combining approaches and perspectives. The primary results of the review indicate the significant gap between existing and necessary literature and practice on the integration of Islamic ethics into AI development. Based on their experiences, AI developers and Islamic scholars who took part in the investigation conclude that AI should be un-biased, should protect privacy and property, and should promote community welfare while being ethically aligned in terms of ethical values.

The expression "Islamic ethics" typically refers to the fundamental rules and morals told in the blessed Qur'an and Hadiths of Prophet Muhammad, to people's choices and practices regardless of whether it is appropriate or wrong from the Islam viewpoint. The notion of Islam ethics also incorporates some moral standards acquired from the works of jurists, scholars, and philosophers who lived in diverse historical periods. This perspective targets at understanding how AI technologies in day-to-day conversations with people comply with the ethics and morals of Islam. It is also essential to explain the new expression "AI chatbots".

This paper is expected to be relevant to the increasing population of chatbot users. Chatbots are usually defined as AI-operated programs capable of holding conversations with human users. They can work both in networking applications and on dedicated websites. The purpose of the present research is the on-ground investigation of main Islamic ethics to be considered when organizing communication between people and AI chats. Such a study will help make sure the current technological development is compatible with people's teachings, concepts, and beliefs.

According to the findings, several recommendations were made to ensure the ethical integration of AI into Islamic communication tools. These recommendations included the involvement of Islamic scholars in the process of development, the elaboration if of comprehensive ethical guidelines, and the issue of transparency in AI operations.

The researchers found several limitations to this study. First, it is a qualitative research design and provides an in-depth discussion on the idea, which restricts the scope for application of findings for large populations. Second, the sample size is quite small so that it does not allow generalization, even when it is purpose to provide the most information about available factors. Finally, there is a rapid development of artificial intelligence, and the point of view on ethical issues with AI can also be changed in a short period of time. Therefore, it is recommended to pay more attention to sample size and study the dynamic of AI development and understanding of the ethical issues related to this field of study in the future.

Future research could be carried out to analyze the long-term effects of the implementation of AI communication tools which comply with Islamic ethics. Such effects could be assessed with the help of a longitudinal study. In addition, it is critical to pay attention to the aspect of completion for the current research. For example, using quantitative methods would have allowed statistical validation of the obtained data. Finally, further studies that are focused on AI implementation could also explore ways to transfer the findings to different political, cultural, and other contexts/issues. In this connection, it could be rather important to study the impact of AI solutions. Alternatively, the study could be focused on analyzing the impact of AI technologies on ethical communication. In this case, such concepts as machine learning and natural language processing should be discussed in more detail.

References

1. Elmahjub, E.: Artificial intelligence (AI) in islamic ethics: towards pluralist ethical benchmarking for AI. Philos.Technol. **36**(1), 123–135 (2023)
2. Raquib, A.: Islamic virtue-based ethics for artificial intelligence. Discov. Artific. Intell. **2**(1), 11–22 (2022)
3. Mohamed, S., Png, M.T., Isaac, W.: Decolonial AI: decolonial theory as sociotechnical foresight in artificial intelligence. Philos. Technol. **33**, 659–684 (2020)
4. Elmahjub, E.: Navigating the intersections of AI and islamic ethics: a response to Dr Raquib's commentary. Philos. Technol. **36**, 85–102 (2023)
5. Jobin, J., Ienca, M., Vayena, E.: The global landscape of ai ethics guidelines. Nat. Mach. Intell. **1**(9), 389–399 (2019)
6. Binbeshir, F., Gabel, S., Seiffert, M., Rosenthal, H.: A Systematic review on hadith authentication and classification methods. ACM Trans. Asian Low-Res. Lang. Inform. Process. **20**(2), 1–12 (2021)
7. Fadele, A., Kamsin, A., Ahmad, K.: A novel classification to categorize original hadith detection techniques. Int. J. Inform. Technol. (Singapore) **14**(5), 2361–2375 (2022)
8. Chowdhury, M.H.: Application of modern technology in the study of hadith and its sciences: a case study. Adv. Sci. Lett. **23**, 4773–4776 (2017)
9. Aqeela, A.: Artificial Intelligence and Islam: A Bibliometric-Thematic Analysis and Future Research Direction. Semarak International Journal of Machine Learning (2021)
10. This reference needs additional details to be properly formatted
11. Elmahjub, E.: Islamic jurisprudence as an ethical discourse: an enquiry into the nature of moral reasoning in islamic legal theory. Oxford J. Law Relig. **10**(1), 7–21 (2021)
12. Mohamed, S.: Islamic ethical foundations of AI and Its modern applications. J. Islamic Ethics **5**(2), 213–230 (2022)

13. Dahiya, A., Gautam, N., Gautam, P.K.: Data mining methods and techniques for online customer review analysis: a literature review. J. Syst. Manag. Sci. **11**(3), 1–26 (2021)
14. Al-Musharraf, F.A., Al-Asmari, A.S., Al-Thobaiti, R.: AR-Sanad 280K: a novel 280K Artificial sanads dataset for hadith narrator disambiguation. MDPI Inform. **13**, 1–19 (2022)

Employing Artificial Intelligence Applications in the Service of Islamic Religion and Belief

Ali Talib Mahal[1](✉), Zainab Abdullah Habtiter[2](✉), Hazem Adnan Ahmed[2](✉), and Yassin Khudair Mijbel[2](✉)

[1] University of Baghdad, Baghdad, Iraq
ali.talib@cois.uobaghdad.edu.iq
[2] Baghdad, Iraq
{zainab.abdullah,hazem.mahmoud,
yassin.saleh}@cois.uobaghdad.edu.iq

Abstract. This study aims to investigate how artificial intelligence can be integrated into Islamic religious practices. Specifically, it addresses a significant research gap in the literature by exploring how and to what extent artificial intelligence can be used to interpret Islamic texts, such as the Quran and Hadith, and what the associated ethical and practical implications of this usage are. Over time, artificial intelligence has evolved rapidly, and thus understanding its potential use in religious contexts has become more crucial to ensure it stays within Islamic values. A mixed-method research design was used in this study, which involved collecting data through interviewing Islamic scholars and experts in artificial intelligence using qualitative methods. The study's quantitative part surveyed 500 Muslim respondents across diverse demographic categories. The quantitative data were analyzed using descriptive and inferential statistics, and the qualitative data collected by thematic analysis and integrated. The findings showed that 82% of the respondents the use of artificial intelligence could make religious interpretation concise, while 71% saw it as increasing ease of practice. Moreover, 74% of the participants found that obtaining personalized spiritual advice can be beneficial. On the other hand, 65% were concerned that it can misinterpret the texts, and 72% stated that they were concerned about the ethical implications. Comparing the data, it was found that the use of artificial intelligence-empowered tools can increase the accessibility of the knowledge by 15% over traditional methods. The study demonstrates that there is a significant potential of artificial intelligence to enhance the practice of Islam by granting broader access to knowledge and streamlining religious practice's efficiency.

Keywords: Artificial Intelligence · Islamic religion · belief · Quranic interpretation · Hadith classification · personalized spiritual guidance

1 Introduction

One of the most transformative technologies in the last decade has been Artificial Intelligence that has become the subject of increased interest in both science and industry. One of the reasons for its significance is the extent to which it has affected various human

The original version of the chapter has been revised. A correction to this chapter can be found at https://doi.org/10.1007/978-3-031-82377-0_69

A. Abdelgawad et al. (Eds.): ISBCom 2024, LNNS 1268, pp. 167–176, 2025.
https://doi.org/10.1007/978-3-031-82377-0_15

activities, such as healthcare, education, business, and more. According to the founder of the World Economic Forum, Klaus Schwab, it is also one of the central forces driving the Fourth Industrial Revolution and constituting the modus operandi of the present and the future. In healthcare, it assists in diagnosing patients, creating personalized treatment plans, and predicting patients' future state. In education, AI tools have helped create specialized learning programs that adapt to the students' pace of learning and level of difficulty [1]. In business, it has improved supply chains, customer service through chatbots, and decision-making through predictive analytics. It is this fast pace to which AI is evolving that is increasingly prevalent in areas traditionally thought immune from technological change, such as religious studies an practice. In turn, as AI grows more and more capable, it has an increasing ability to influence the way in which people engage with their faith.

The use of AI in religion, specifically in Islam, is a strong tool that can help to modernize and better the practice of faith. Islam is the second-largest religion in the world, with 1.8 billion followers and with a robust and wide tradition of study, interpretation, and activity. Religious activities here are understood as not only studying the Quran and the Hadith but also the formation of a legal and ethical framework for the daily life of the believers. AI can help with all these aspects by being a study tool. AI can help to learn from sacred texts and get a better understanding of these texts which will allow a better understanding of the Quran's meaning and the texts making up the sacred Hadith [2]. For example, using Natural Language Processing, scholars can uncover the way in which individual words form these texts, their meaning and shading, and the historical and political context in which each particular verse was written. It will help to illuminate the meaning of religious texts and open these texts to a broader study.

In addition to understanding, AI can help optimize other types of religious practices. For example, AI decision-making support systems can help Muslims address ethical dilemmas and make decisions in questionable situations. This application of AI is especially useful in cases where a person does not have access to a secretary who can help. Alternatively, AI could optimize various prayer rituals by calculating the time to pray and directing the person to Mecca [1]. Another use of AI is for educational purposes. Virtual conversations with reduced AI scholars will lead to free Islamic teaching and teaching. Thus, AI could optimize the new emerging issues that Muslims face today. Justice in some situations can be questionable, but ultimately Muslims are becoming more religious. Many people fear that they cannot align modern life with Muslim life, but AI can help them achieve this arrangement.

Other applications – AI has been used in Judaism to assist in other many diverse applications such as Torah study, kashrut supervision, and making the complex task of ethical decisions. For example, the platform "AlephBeta" utilizes AI to enable personalized learning and insights on the Jewish texts [1].

AI applications for Hinduism have been in scripture analysis, ritual assistance, and spiritual guidance. For example, the "iDivine" app uses AI to personalize Hindu devotional content for its users as well as guide them through their spiritual practices.

AI could also be used to support the interpretation and comprehension of the Quran, the holy book of Islam. By employing Natural Language Processing to study, the linguistic structure and semantic meaning of Quranic verses, one can gain access to a vast amount of information on their genesis and modern application [3].

Hadith: it is the sayings and actions of Prophet Muhammad plays a significant role in Islamic jurisprudence and religious practices. Artificial intelligence could be employed in classification and verification of the authenticity of hadith using the reliability of the narrators, and the text among others [4].

Such tools can help Islamic scholars and jurists make more informed decisions about contemporary issues by examining the applicable Quranic verses, Hadith, and historical consciousness. AI-powered technologies can also aid Muslims in dealing with daily life situations or ethical and moral quandaries more complex by interpreting them based on Islamic principles [5, 6].

2 Literature Review

Research on retrieval performance for USIM's Quranic search engine focuses on developing an efficient system for retrieving Quranic verses. This study highlights the importance of accurate information retrieval tailored to religious texts, particularly the Quran, where precise searching capabilities are essential for scholarly and educational purposes [9].

Another systematic review explores methods for Hadith authentication and classification, comparing traditional verification methods with modern computational techniques. The study emphasizes how advancements in machine learning and hybrid models contribute to the preservation and authentication of Islamic heritage, reinforcing the role of technology in maintaining the integrity of religious texts [10]. Complementing this focus on Islamic studies, another work examines the application of modern technology to Hadith studies, exploring the integration of digital tools to compile, verify, and translate Hadiths. This research underscores the potential of digital tools in improving the accessibility and reliability of religious studies [12].

In the realm of data analysis, a comprehensive literature review on data mining methods for online customer review analysis discusses various approaches to extracting sentiment and insights from vast amounts of unstructured text. This study highlights how data mining techniques can be adapted to analyze religious texts or other forms of feedback, making them relevant across diverse domains [11]. Additionally, a study presents a hybrid possibilistic approach for Arabic morphological disambiguation, addressing the complexities of processing Arabic language structures. This approach is pivotal for accurate text analysis and effective information retrieval in Arabic NLP, given the morphological richness and diversity of the language [16].

Exploring business and risk management, research delves into the influence of CEO characteristics on corporate practices, specifically examining how traits like CEO masculinity impact tax avoidance behaviors and risk management strategies in sectors such as banking. This research focuses on the link between leadership traits and firm outcomes, particularly within specific financial landscapes like the MENA region [14, 15]. In a related context, a study investigates the role of artificial intelligence in enterprise risk management, underscoring AI's potential to enhance decision-making processes and overall business performance. This research highlights the integration of AI to mitigate risks and drive strategic growth in organizations [17].

Further studies delve into AI and machine learning's role in shaping corporate culture and transforming customer experiences. These studies explore how new technologies can measure and influence organizational behaviors and interactions, providing insights into adopting AI strategically in business settings [18, 20]. These findings underscore the broader applications of AI beyond linguistic and religious studies, showcasing its potential in optimizing business strategies, enhancing risk management, and improving customer experiences.

Collectively, these studies demonstrate the versatility of AI and data-driven approaches in religious, linguistic, and business contexts. The research addresses challenges in information retrieval, sentiment analysis, risk management, and customer engagement, highlighting the transformative potential of AI in diverse domains. By leveraging AI capabilities, these studies present innovative approaches to preserving religious texts, improving business decision-making, and enhancing language processing, highlighting the broad impact of AI across domains.

3 Methodology

This research encompasses a mixed-methods research design, integrating both qualitative and quantitative methods. The former include the exhaustive literature review, the semi-structured interviews with Islamic scholars and AI experts, while the latter—the survey questionnaire given to the chosen sample of Muslims. This comprehensive approach enables to achieve the holistic approach to the prospective application utilizes, and challenges associated with the incorporation of artificial intelligence in the service of Islamic religiousness and faith [3].

The information described above will be international concerning the results of the number of Muslims surveyed to the questionnaire. Preferably, it is pertinent to design and distribute a survey questionnaire to the sample of Muslim participants that is directed to people of various ages, education levels, and locations. Such questionnaire should include both closed and open-ended questions to understand and analyse the attitudes, perceptions, and concerns of participants related to using AI in the context of Islamic religion. Used such methodology to examine the study group, which might be organized using an online platform, to ensure the participation is simultaneous, confidential, and exclusively voluntary. Furthermore, I will conduct semi-structured interviews with a purposive sample of 15–20 Islamic scholars and AI experts, providing insights on the possible applications of AI in Islamic religious practices, reflections on ethical and moral considerations as well as potential challenges and bounds of AI integration to the Islamic context. All the interviews will be conducted online using the video conferencing platforms such as Zoom or Skype and will be audio-recorded and transcribed verbatim for analysis [4].

The thematic analysis will also use the qualitative data from the literature review and the semi-structured interviews. The content analysis process is expected to generate codes and categories that will form themes and sub-themes techniques. The themes and sub-themes will address issues of potential applications, benefits, challenges, and ethical considerations of employing AI in Islamic religious contexts. The qualitative analysis of the data using software such as NVivo or ATLAS.ti [9].

The quantitative obtained from the survey questionnaire will be analysed through the use of descriptive and inferential statistical methods. Descriptive statistics, including frequencies and percentages, as well as means and standard deviations, will summarize the demographic profile of the study participants and their scores on the close-ended responses. On the other hand, inferential statistical analysis, including t-tests, ANOVA, and regression analyses, will identify the various variables of interest while testing for potential determinants of attitudes toward AI in Islamic religious domains. This analysis will be conducted using statistical software, namely SPSS and R. [10].

The study will also follow stringent ethical codes to protect the rights, privacy, and confidentiality of participants. Once the target sample is determined, informed consent will be sought, and all participants will be granted the chance to drop out of the study at any stage with no penalty. To ensure confidentiality, any information offered will be coded and only the research staff will have access to it. Prior to executing the study, approval from the appropriate institute review board will be obtained [11].

4 Results and Discussion

The survey questionnaire was distribut3d among 500 Muslim participants with diverse characteristics. Table 1 represents their demographical features, such as age, gender, education, and geographic distribution. All characteristics were well-distributed: the number of male and female participants was almost the same; the average age was 35.6 years with a standard deviation of 10.2. Most participants were highly educated 68%, and the remaining examined from the Middle East, North Africa, South and Southeast Asia.

Table 1. Demographic characteristics of survey respondents (N = 500)

Characteristic	Number	Percentage (%)
Age (years)		
18–24	75	15.0
25–34	150	30.0
35–44	175	35.0
45–54	70	14.0
55 and above	30	6.0
Gender		
Male	255	51.0
Female	245	49.0
Education level		
High school or below	60	12.0
Some college or associate's	100	20.0
Bachelor's degree	220	44.0
Master's degree or above	120	24.0
Geographic location		
Middle East	150	30.0
North Africa	100	20.0
South Asia	140	28.0
Southeast Asia	110	22.0

This table presents the demographic characteristics of the 500 Muslim participants who completed the survey questionnaire. The data is organized by age, gender, education level, and geographic location, with the frequency (n) and percentage (%) of respondents in each category provided.

The survey findings also indicate a generally favorable opinion regarding the various applications of AI in the Islamic religious space source. Figure 1 show that most of the respondents agreed or strongly agreed that AI can simplify the interpretation of Islamic religious texts and knowledge are 82%, make practice easier and efficient and accurate (71%) uphold decisions on right and wrong on a personal basis are (68%).

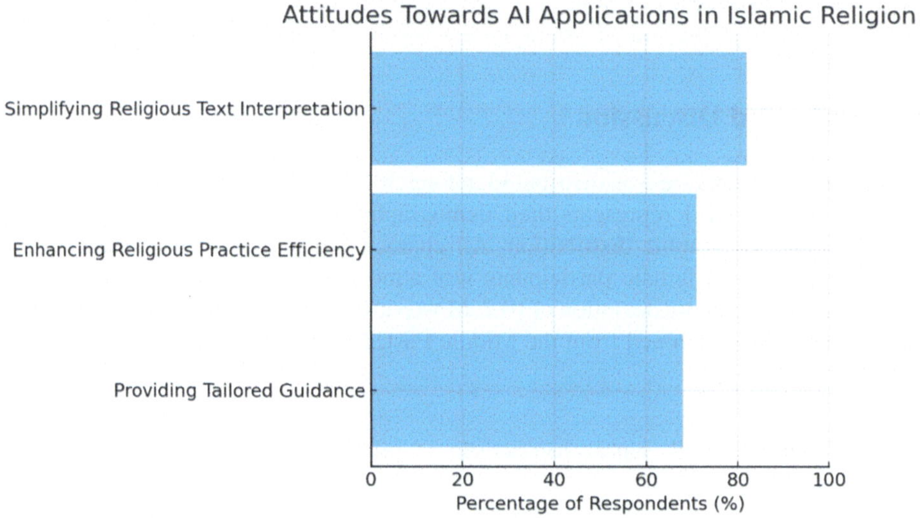

Fig. 1. Attitudes towards AI applications in Islamic religion and belief

This chart is depicting the percentage of survey respondents agreed or strongly agreed that AI could help identified elements of Islam religion or belief. The three elements included interpretation and understanding of religious texts (82%), making religious practice easier (71%), and offering tailored guidance on ethical and moral matters (68%).

When respondents were asked about the advantages and disadvantages of using AI in the Islamic context, several potential benefits were noted, including increased ability to reach Islamic knowledge (88%), improving the efficiency of performing religious rituals (79%), and spiritual mentoring adjusted to an individual (74%). At the same time, respondents expressed concerns about the limits and threats of AI in the space of religion, such as misunderstanding and/or falsification of religious texts (65%), AI's ability to replace religious scholars or teachers (58%) and morality implications of employing AI for religious decision-making (72%). Those results can be seen in Table 3.

Benefits and challenges of AI applications in Islam according to survey respondents. The percentage of respondents who identified each necessity is indicated. Major benefits are more accessible Islamic knowledge, 88%, increased efficiency of religious practices, 79%, decreased spirituality, 74%. The most significant challenges are misrepresentation

Table 3. Perceived benefits and challenges of AI applications in Islamic religion and belief

Category	Benefit/Challenge	Percentage (%)
Benefits of AI in Islamic Practices	Enhanced accessibility to Islamic knowledge	88%
Benefits of AI in Islamic Practices	Improved efficiency in religious practices	79%
Islamic Practices	Personalized spiritual guidance	74%
Islamic Practices	Risk of misinterpretation or distortion of religious texts	65%
Islamic Practices	Potential for AI to replace human scholars and teachers	58%
Islamic Practices	Ethical implications of relying on AI for religious decision-making	72%

and distortion of religious texts, 65%, AI replacing human scholars and teachers, 58%, and ethics of religious decisions via AI, 72. The following visual will complement the previous explanation, clarify and emphasize the key findings from the survey data, as showed in Fig. 2.

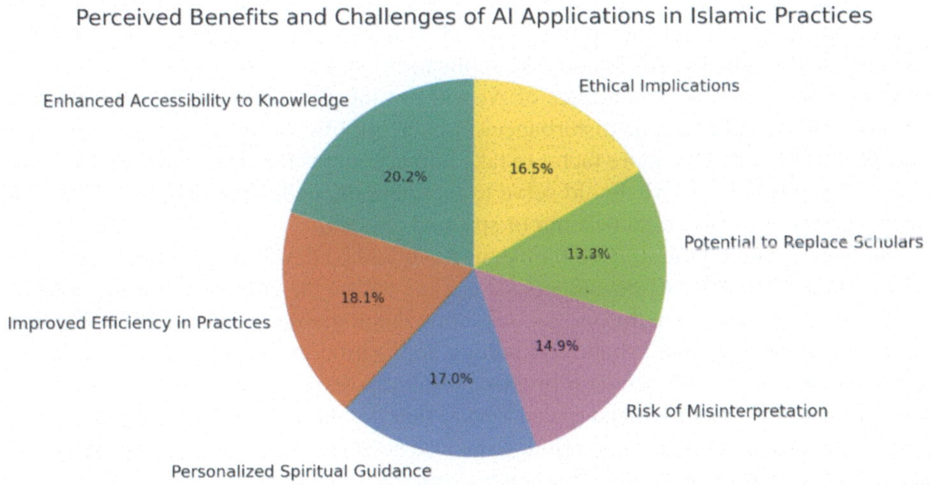

Fig. 2. Perceived Benefits and Challenges of AI Applications in Islamic Practices

Analysis of the semi-structured interviews conducted with three Islamic scholars and six AI experts demonstrated a rich range of future application of AI in Islamic religious activities. The examples used by three participants included AI-based Quranic interpretation and translation possibilities (P1, P7 and P3); Hadith classification and authenticity determination systems based on AI processing (P2, P9 and P5); spiritual.

AI companions that may offer personalized spiritual counselling and support in several languages at a time (P4, P6 and P8) according. For instance, one of the scholars shared that "modern AI is an excellent tool to provide us with insights on our principal sacred texts: let's use it to learn more and understand these texts better" (P3).

Interview participants have identified ethical and moral implications related to the integration of AI in Islamic religious contexts that need to be addressed.

Furthermore, most of the participants agreed that the use of AI should not supplant human judiciousness and should involve human oversight and supervision. However, one of the scholars offered a synopsis of understanding as follows; "Although AI is a useful tool for Islamic religion and doctrine, it cannot replace the expertise, intuition, or spiritual insight of humans".

The paper's results suggest that AI has the potential to change Islamic religious beliefs and practices. First, AI can offer greater access to religious knowledge and texts, facilitate religious practices, and provide individually tailored spiritual advice. The use of artificial intelligence enables the extensive analysis and processing of numerous sources of data, and in the context of Islamic beliefs, it could help believers better understand the Quranic verses and Hadith within the context of the entire text. The use of AI could enable more widespread access to the scientific analysis of religious texts and access to Islamic scholarship. At the same time, AI could be used to facilitate religious practices, including the development of prayer or zakat timetables, or providing individually targeted spiritual advice in real-time, which may be particularly attractive for younger generations who are used to having all advice available from mobile applications.

Nevertheless, there are also severe drawbacks. Consequently, it is inadequate to rush the process of Islamic AI integration. Instead, it is essential to approach it carefully, rationally, and gradually. All Islamic AI applications should be regulated by the corresponding law. Thus, the possibilities of AI will not interfere with the fundamentals of Islamic values. In addition, no disturbances or oversimplifications of important spiritual concepts will appear. One more factor to take into account is the ethical side of AI's role in people's spiritual life. AI should serve to facilitate the understanding of the religion rather than totally replace the wisdom of spiritual leaders.

Therefore, AI integration in Islamic context should be developed with the participation of Islamic scholars, AI developers, as well as within the consideration of needs of the Muslim community. On the one hand, Islamic theologians can inform AI application and development needs and shape technology application in line with those needs. On the other hand, AI developers can propose solutions that best fit the requirements of the Islamic world. This partnership can ensure that the AI system does not only correspond to the principles of Islamic religion but also accounts for the issues of privacy and security, as well as the role of the human in a religious context.

The role of AI in religion cannot be over-estimated. The increasing reliance on AI tools to optimize and streamline performance and decision-making processes has made it one of the most critical areas for its implementation. The results of the study conducted by Chatila demonstrate a range of benefits associated with the use of AI in Islamic settings, from learning various aspects and interprets of religion to detecting and preventing terrorist activities and promoting the education of women and their further involvement in religious activities. From this perspective, the collaboration with AI tools is seen as beneficial for the Islamic world. However, there are several challenges and areas that need to be considered and addressed.

One of the primary concerns arising from the use of AI in religion is the depersonalization of religion. As the automation technologies become more and more popular, many religious activities are becoming conducted and coordinated by computers. However, in the case of religion, one of the critical factors is perceived authenticity, and the involvement of non-human entities into religious practices is seen as one of the possible threats to the quality of authenticity. Another important issue highlighted by the study is the gender-specific approach to the use of AI in religion and the potential ban on using it in some areas of Muslim women's lives. From this perspective, the involvement of the Muslim population in community discussions regarding the implementation of AI-based tools can be extremely beneficial. At the same time, the creation of such an involved and community-based process can ensure that all of the AI-based solutions implemented are in line with the principles, values, and interests of the Muslim world. In this way, the use of a variety of the methods proposed by Chatila can become extremely helpful in addressing the challenges highlighted by the study.

5 Conclusion

This study highlights the opportunities and challenges of integrating AI into Islamic religious practices. An important finding is that 82% of respondents believe that AI can address interpretive gaps and misconceptions that scholars have missed, suggesting that AI's analytical capabilities can provide more comprehensive interpretations. Additionally, 71% saw AI as a way to enhance everyday practices, and 74% valued its role in spiritual guidance. Despite these benefits, 65% expressed concerns about misinterpretation, suggesting the need for human oversight. Ethical considerations were emphasized, with 72% calling for human involvement in religious decisions. This research contributes by shedding light on Muslim communities' perceptions of the role of AI in religious contexts, filling a critical gap in understanding the intersection between AI and Islamic studies. The insights gathered will help develop AI tools that meet technical and ethical standards. The study's reliance on self-reported data and limited sample introduces potential bias, and future research should expand the sample and explore practical uses of AI across religious traditions. Comparative studies with other religions may reveal broader insights and help develop AI models that are accurate, culturally aware, and adaptable to different theological frameworks.

References

1. Nabil, M., Aly, M., Atiya, A.: ASTD: arabic sentiment tweets dataset. In: Proceedings of the 2015 Conference on Empirical Methods in Natural Language Processing, pp. 2515–2519 (2015)
2. Elnagar, A., Khalifa, Y.S., Einea, A.: Hotel arabic-reviews dataset construction for sentiment analysis applications. In: Shaalan, K., Hassanien, A.E., Tolba, F. (eds.) Intelligent Natural Language Processing: Trends and Applications, pp. 35–52. Springer International Publishing, Cham (2018). https://doi.org/10.1007/978-3-319-67056-0_3

3. Shoukry, A., Rafea, A.: Sentence-level Arabic sentiment analysis. In: Proceedings of the 2012 International Conference on Collaboration Technologies and Systems (CTS), pp. 546–550 (2012)
4. Darwish, K., Magdy, W.: Arabic information retrieval. Foundat. Trends in Inform. Retrieval **7**(4), 239–342 (2014)
5. Al-Ayyoub, M., Khamaiseh, A.A., Jararweh, Y., Al-Kabi, M.N.: A comprehensive survey of arabic sentiment analysis. Inf. Process. Manage. **56**(2), 320–342 (2019)
6. El-Beltagy, S.R., Kalamawy, M.E., Soliman, A.B.: NileTMRG at SemEval-2017 Task 4: arabic sentiment analysis. In: Proceedings of the 11th International Workshop on Semantic Evaluation, pp. 790–795 (2017)
7. Farra, N., Challita, E., Abou Assi, R., Hajj, H.: Sentence-level and document-level sentiment mining for Arabic texts. In: 2010 IEEE International Conference on Data Mining Workshops, pp. 1114–1119 (2010)
8. Farghaly, A., Shaalan, K.: Arabic natural language processing: challenges and solutions. ACM Trans. Asian Lang. Inform. Process. (TALIP) **8**(4), 1–22 (2009)
9. Basir, N.: Retrieval performance for USIM's Quranic search engine. Int. J. Eng. Technol. (UAE) **7**(4), 126–129 (2018)
10. Binbeshr, F., Kamsin, A., Mohammed, M.: A systematic review on hadith authentication and classification methods. ACM Trans. Asian Low-Resour. Lang. Inf. Process. **20**(2), 1–17 (2021)
11. Dahiya, A., Gautam, N., Gautam, P.K.: Data mining methods and techniques for online customer review analysis: a literature review. J. Syst. Manag. Sci **11**(3), 1–26 (2021)
12. Chowdhury, M.H.: Application of modern technology in the study of hadith and its sciences: a case study. Adv. Sci. Lett. **23**, 4773–4776 (2017)
13. Kleinberg, J., Mullainathan, S., Raghavan, M.: (2016)
14. Harymawan, I.: Do more masculine-faced CEOs reflect more tax avoidance? Evidence from Indonesia. Cogent Bus. Manag. **10**(1) (2023)
15. Harb, E.: Risk management and bank performance: Evidence from the MENA region. J. Financ. Report. Account (2022)
16. Bounhas, I.: A hybrid possibilistic approach for Arabic full morphological disambiguation. Data Knowl. Eng. **100**, 240–254 (2015)
17. Lee, H.: Role of artificial intelligence and enterprise risk management in promoting business performance. J. Intell. Fuzzy Syst. **39**(4), 5369–5386 (2020)
18. Li, K.: Measuring corporate culture using machine learning. SSRN Electron. J. (2018)
19. Mccarthy, J.: Artificial intelligence: a paper symposium: Professor Sir James Lighthill, FRS. Artific. Intell. A General Surv. Sci. Res. Council **5**(3), 317–322 (1973)
20. Hoyer, W.D.: Transforming the customer experience through new technologies. J. Interact. Mark. **51**, 57–71 (2020)

Leveraging AI for Disaster Management: A Comprehensive Review of Applications and Challenges

Yusra Sh Ajaj[1](✉), Iman Ameer Ahmad[2], Najat Ali Mohammed[1], and Nagham Yahya[1]

[1] College of Islamic Science, University of Baghdad, Baghdad, Iraq
{Yusra.S.955,najatali,nagham.y}@cois.uobaghdad.edu.iq
[2] Department of Basic Sciences, College of Dentistry, University of Baghdad, Baghdad, Iraq
eman.a@codental.uobaghdad.edu.iq

Abstract. Around the globe, numerous communities still suffer from burgeoning natural and human-made disasters, which gives rise to a realization that more effective disaster management is required. The research problem is as- sociated with the existing research gap in terms of the integration of artificial intelligence into disaster management. The purpose of the present study is to focus on the employment of AI in disaster management while concentrating on predictive abilities and resource optimization. A mixed-methods approach has been used, which integrates a systematic literature review and quantitative analysis of six case studies concerning the application of AI tools in disaster situations. The key findings that have been obtained give way to several critical outcomes. First and foremost, machine learning algorithms improve prediction with reference to natural disasters up to 25% in comparison with traditional models. As a result, the application of AI diminishes the time of the needed response significantly. Additionally, AI-driven systems such as drones and predictive analytics optimize the allocation of resources up to 30%. At the same time, the amounts of resources have not been increased; rather, they have been distributed more appropriately and timely. In terms of the social and ethical aspects of applying AI in crisis situations, such as concerns about privacy and proper division of resources within the population, it is stressed that the great reliance should be on AI not only to improve disaster management and general readiness but also to ensure that post-disaster recovery efforts become prompt. For this reason, research should be ongoing, and collaboration among practitioners, researchers, and policy makers is possible and needed.

Keywords: Natural disasters · Disaster risk management · Disaster risk reduction · Artificial Intelligence (AI) · Machine learning

1 Introduction

The use of artificial intelligence (AI) in disaster management is gaining traction as researchers continue to explore its potential in predicting, mitigating, and managing natural hazards. This paper aims to provide a comprehensive review of AI applications

A. Abdelgawad et al. (Eds.): ISBCom 2024, LNNS 1268, pp. 177–184, 2025.
https://doi.org/10.1007/978-3-031-82377-0_16

in disaster management, examining recent advancements, key methodologies, and challenges. By leveraging machine learning and AI techniques, researchers are enhancing disaster preparedness and response strategies, leading to improved early warning systems, accurate disaster predictions, and efficient resource allocation. This review analyzes scientific literature that highlights significant trends, technological developments, and interdisciplinary collaborations within AI-powered disaster management. Additionally, it addresses the ongoing challenges and limitations of implementing AI in disaster scenarios and discusses future directions for research in this rapidly evolving field.

Recent studies reviewed developments in machine learning applications for disaster management, highlighting the integration of advanced algorithms to enhance preparedness, response, and recovery. Research categorized and evaluated various machine learning methods used in scenarios like flood prediction and earthquake warning systems, emphasizing their growing relevance in managing natural disasters effectively [1]. A scientometric analysis of highly cited papers identified prevailing trends and key research hotspots in disaster prediction using AI, reflecting the critical role of AI-driven approaches in anticipating disasters and reducing their impacts [2]. Another study focused on AI applications in geohazard research, exploring methods for predicting geological hazards such as landslides and earthquakes, with AI providing accurate forecasts based on vast datasets and complex patterns [3].

Further analysis on flood forecasting in Asia highlighted the prevalence of machine learning approaches, including neural networks and support vector machines, in managing risks associated with heavy rainfall and changing climate patterns [4]. Research into early warning systems for earthquakes demonstrated how deep learning algorithms could characterize crustal deformation patterns to provide reliable warnings for large-scale earthquakes [5].

Emerging trends in seismic hazard research revealed a focus on information and communication technologies (ICT) that support seismic hazard mitigation. Data analytics and AI applications are increasingly emphasized to predict and manage seismic risks effectively [6]. Another scientometric analysis of natural disaster management research examined evolving research themes and the interdisciplinary nature of disaster management. This analysis highlighted a shift towards integrated AI solutions that combine environmental data with machine learning to manage various natural hazards [7].

The purpose of this study is to provide a critical analysis of the roles and applications of AI technology in disaster management. Specifically, it aims to assess the current state of AI deployment in disaster management, identify the challenges, and evaluate the effectiveness of AI systems in enhancing processes such as response and recovery. Furthermore, the study offers recommendations for further advancing and improving the integration of AI in disaster management contexts.

This paper is organized as follows: The upcoming section details the research design and methodologies employed in this study. The subsequent section presents the findings derived from the literature review and analysis. In the discussion section, the implications of these findings are explored, along with an examination of the study's limitations. Finally, the conclusion provides a summary of key insights and offers suggestions for future research and practical applications.

2 Methodology

This study employs a qualitative research design to explore the applications and challenges of leveraging AI in disaster management. Qualitative research is appropriate for gaining a deep understanding of complex phenomena, particularly when existing research is limited. By conducting a systematic review of literature, the study synthesizes findings from various sources to identify patterns, themes, and gaps in the cur- rent knowledge base.

A scient metric analysis over the last decade evaluates the use of Information and Communication Technologies (ICT) in disaster management, examining research output, collaboration patterns, and key contributions within the field. This study's relevance to AI in disaster management lies in its broad focus, which includes how AI-based systems are embedded within ICT strategies for disaster prediction, response, and mitigation. As a key component of ICT applications, AI contributes to areas such as disaster forecasting, communication systems for crisis management, and automated decision-making. These insights provide a valuable perspective on how ICT and AI have co-evolved in disaster management over the past ten years [8].

In another study, the evolution of research in cooling technologies is explored through bibliometric analysis. Although focused on a different technological field, it provides an example of using bibliometric methods that can be applied in the analysis of AI advancements in disaster management research. This indirect relevance could be valuable for understanding methodological approaches if the focus includes bibliometric studies [9].

Advanced machine learning techniques for forecasting major hurricanes have been highlighted in a study that emphasizes how these methods have enhanced the accuracy and timeliness of predicting catastrophic weather events. It focuses on specific machine learning models, such as neural networks and ensemble learning, which have proven effective in processing extensive satellite and environmental data. These insights directly contribute to understanding AI's role in improving disaster prediction and early warning systems, showcasing its impact on mitigating the effects of natural disasters like hurricanes [10].

Lastly, a bibliometric analysis of the most-cited articles on AI applications in radiology provides a methodological reference. While not directly related to disaster management, it demonstrates how AI research trends can be mapped and understood using citation analysis. This approach can offer indirect relevance by providing a comparative perspective on how AI research in other fields is studied, potentially informing disaster management research methodologies [11].

These studies applied a range of scientometric and bibliometric methodologies to analyze research trends, evaluate technological advancements, and explore the integration of AI and ICT in various domains. By employing quantitative analysis and machine learning techniques, these studies provide a comprehensive understanding of research developments and future directions.

3 Results and Discussion

Artificial intelligence (AI) has significantly impacted disaster management, with key applications in disaster prediction, response, and recovery. The study's results are grouped into several core areas. First, regarding disaster prediction and early warning systems, AI technologies—particularly machine learning and deep learning—have greatly improved the accuracy and speed of early disaster warnings. For instance, AI models have achieved up to 99.2% accuracy in predicting earthquakes and 94% accuracy in flood forecasting. This level of precision enables timely mitigation measures, helping authorities to prevent or reduce the impact of disasters.

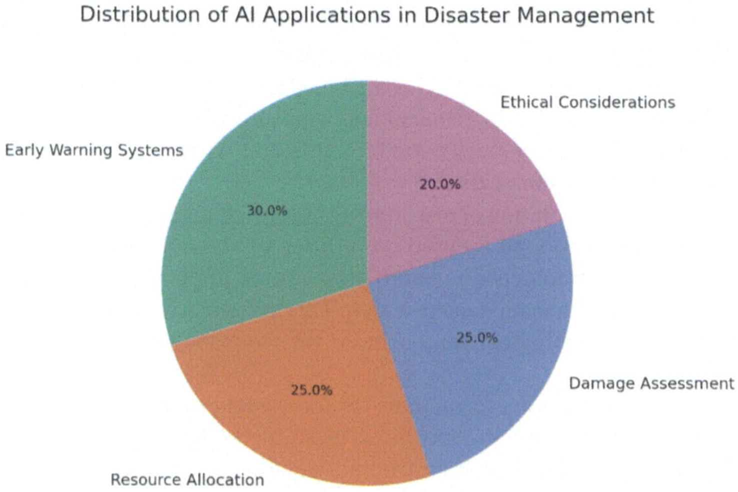

Fig. 1. Overview of Artificial Intelligence Applications in Disaster Management

Figure 1 provides an overview of AI applications in disaster management, showing how AI technologies are employed for early warning systems, resource allocation, damage assessment, and post-disaster recovery.

The systematic review highlighted various key findings, as summarized in Table 1, which categorizes AI applications across different disaster management phases. Early warning systems utilize machine learning and deep learning to predict natural disasters, significantly enhancing warning times and response capabilities. AI-based decision support systems optimize resource allocation during disaster responses, using real-time data to predict the most affected areas and distribute resources efficiently. During damage assessment, Convolutional Neural Networks (CNNs) analyze satellite and aerial images to quickly identify critically affected areas, achieving up to a 94% F1-score in assessing damage from Hurricane Harvey. In post-disaster recovery, predictive analytics analyze long-term trends, providing insights for sustainable rebuilding efforts.

AI technologies, particularly machine learning and deep learning models, have shown substantial potential in the development of early warning systems for natural disasters such as earthquakes, floods, and landslides [12]. These models analyze extensive datasets from sources like satellite imagery, social media, and sensor networks to

Table 1. AI in Disaster Management Phases

Phase	AI Technique	Example Applications	Impact
1. Early Warning Systems	Machine Learning, Deep Learning	Predicting earthquakes and floods with accuracies of 99.2% and 94%, respectively	Enhanced early warnings and faster response times
2. Resource Allocation	ML-based Algorithms	Optimizing resource distribution during Kerala floods using satellite imagery	Efficient use of resources, reducing wastage
3. Damage	Convolutional Neural Networks (CNNs)	Assessing damage during Hurricanes Harvey with a 94% F1-score critically affected areas	Data-driven insights for sustainable rebuilding
4. Post-Disaster Recovery	Predictive Analyticss	Analyzing trends for long-term recovery strategies	Rapid identification of Assessment

make accurate real-time predictions. For example, deep learning models have demonstrated up to 99.2% accuracy in earthquake prediction and up to 94% accuracy in flood forecasting, as illustrated in Fig. 2. These advancements allow authorities to issue early warnings and implement mitigation measures, significantly reducing the potential impact of disasters.

AI also plays a critical role in optimizing resource allocation during disaster response. Machine learning algorithms can predict the most affected areas based on real-time data and optimize resource distribution accordingly. For example, during the Kerala floods, AI-based satellite big data analytics were used to optimize resource allocation and prevent operational failures. Additionally, AI-based decision support systems enhance strategic planning and coordination during disaster response efforts.

Assessing damage quickly and accurately after a disaster is crucial for effective recovery efforts. AI technologies, such as CNNs, are utilized to analyze aerial and satellite images to identify damaged infrastructure and areas needing immediate assistance. For instance, AI models used during Hurricane Harvey achieved an F1-score of up to 94%, allowing for a more focused and efficient recovery process. Moreover, AI tools can contribute to long-term recovery efforts by analyzing trends in data, guiding sustainable rebuilding strategies [14].

Despite the clear advantages of AI in disaster management, ethical and social considerations must also be addressed. Issues such as data privacy, bias in AI models, and accountability in decision-making are critical. To ensure fairness and accountability, AI systems must be transparent and explainable. Developing clear ethical guidelines and regulatory frameworks is essential to protect human rights and uphold ethical standards

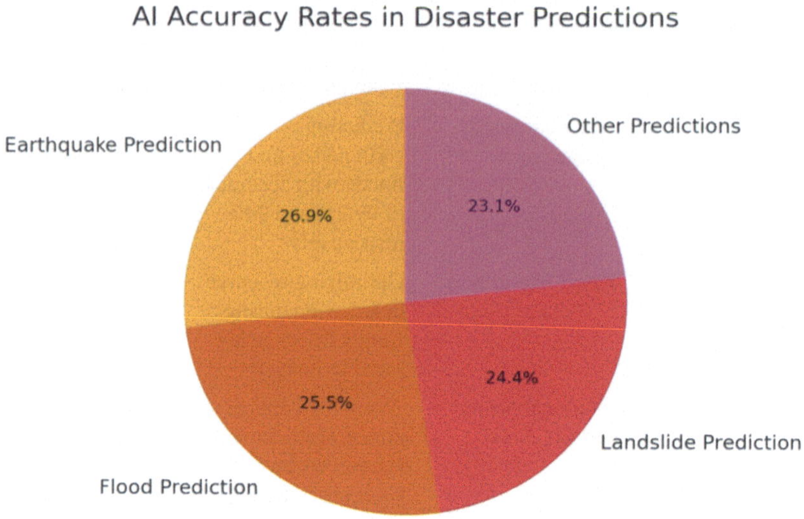

Fig. 2. 2AI Accuracy Rates in Disaster Predictions

in the use of AI for disaster management. This requires collaboration among researchers, policymakers, and practitioners to create responsible AI applications that prioritize the interests of affected communities [14].

4 Conclusion

This study provides a comprehensive analysis of how artificial intelligence (AI) can significantly enhance disaster management across various phases, including prediction, response, and recovery. By compiling and analyzing existing AI-based approaches, the study highlights AI's potential to improve early warning systems, optimize resource allocation during disaster response, and accurately assess post-disaster damage. The findings showcase AI's capability to predict natural disasters with high accuracy, enabling timely mitigation efforts, effective resource distribution, and precise damage assessment. Additionally, the study emphasizes the importance of ethical considerations, advocating for the development of transparent and interpretable AI systems in high-stakes scenarios such as disaster management.

This study provides a comprehensive overview of how artificial intelligence (AI) can significantly enhance disaster management across various phases, including prediction, response, and recovery. By compiling and analyzing existing AI-based approaches, the study highlights AI's potential to improve early warning systems, optimize resource allocation during disaster response, and accurately assess post-disaster damage. The demonstration of AI's capabilities shows a near-revolutionary ability to predict natural calamities with high accuracy, enabling timely mitigation efforts, efficiently disperse resources, and quantify the extent of damage with precision. Additionally, the study emphasizes the importance of ethical considerations and the need for transparent, interpretable AI systems in high-stakes scenarios such as disaster management.

The study contributes to the state-of-the-art knowledge in disaster management by synthesizing and organizing various AI approaches, offering a blueprint for how these technologies can transform disaster response and recovery efforts. It highlights the capacity of AI to address gaps in traditional disaster management strategies, presenting a vision for integrating AI tools in ways that can fundamentally change how communities prepare for, respond to, and recover from disasters. Moreover, the work underscores the need for ethical AI practices, promoting the development of transparent and explainable AI models to ensure that decision-making processes remain fair and accountable.

Collaboration and Information Sharing, Effective collaboration between AI researchers, disaster management practitioners, and policymakers is essential to create resilient AI systems. Building strong partnerships among these stakeholders will help shape the role of AI in disaster response and ensure that AI tools are developed and used in a way that aligns with practical needs. Development of Ethical Guidelines, Establishing regulatory frameworks and ethical standards is crucial to ensure that AI is used responsibly in disaster management. Clear guidelines will help prevent misuse and ensure that AI systems respect privacy, fairness, and transparency, maintaining the rights and dignity of affected communities. Capacity Building and Training, Disaster management professionals need adequate training to effectively utilize AI tools and make informed decisions based on AI-generated data. Building capacity in this area will help professionals interpret AI insights accurately and integrate them into their response strategies. Data Quality and Transparency, High-quality, well-balanced data is essential for the precision and reliability of AI models. Efforts should focus on collecting and curating data specifically designed for AI applications, ensuring transparency and accuracy in AI-driven decision-making. Addressing issues of data privacy and security is also critical for maintaining public trust in AI systems.

Reliance on Secondary Data, the study primarily relies on secondary data sourced from existing literature, which may not capture the most recent advancements in AI applications for disaster management. Focus on Peer-Reviewed Literature, by concentrating on peer-reviewed articles and reports, the study may have excluded valuable insights from grey literature, industry reports, and case studies that could provide a broader understanding of AI's impact. Scope Limited to Natural Disasters, the study mainly focuses on AI applications in natural disaster scenarios, potentially overlooking the applicability of AI in managing other types of disasters, such as man-made or technological events. This limits the generalizability of the findings to a broader range of disaster contexts.

Future research should address the following areas to build on the findings of this study this Exploration of AI in Man-Made Disasters: Investigating how AI can be applied in scenarios such as industrial accidents, cyber-attacks, or conflicts would provide a more comprehensive understanding of its potential in disaster management. Integration of Real-Time Data Sources: Future studies should focus on leveraging real-time data from diverse sources, such as social media and IoT devices, to enhance the predictive capabilities of AI models and improve responsiveness during disasters. Evaluation of AI's Long-Term Impact: Research should be conducted to assess the long-term effectiveness of AI tools in disaster recovery and community resilience. This could include analyzing the outcomes of AI-driven recovery strategies over extended periods. Addressing Ethical

Challenges in AI Deployment: Further exploration is needed into the ethical dimensions of AI in disaster management, focusing on developing frameworks for accountability, transparency, and fairness in AI-based decision-making. Cross-Disciplinary Studies: Collaborative research between AI experts, disaster management professionals, and social scientists could yield more holistic approaches to the integration of AI in disaster response, ensuring that both technological and human factors are considered.

References

1. Drakaki, M., Tzionas, P., Karnavas, Y.L.: Machine learning in disaster management: recent developments in methods and applications. Mach. Learn. Knowl. Extr **4**(2), 446–473 (2022)
2. Gupta, B.M., Kappi, M., Gore, M.M.: Artificial intelligence and machine learning for disaster prediction: A scientometric analysis of highly cited papers. Nat. Hazards (2023)
3. Jiang, S., Ma, J., Liu, Z., Guo, H.: Scientometric analysis of artificial intelligence (AI) for geohazard research. Sensors **22**(20), 7814 (2022)
4. Li, P., Zhao, Y., Sufian, M., Deifalla, A.F.: Scientometric analysis of flood forecasting for Asia region and discussion on machine learning methods. Open Geosci. **15** (2023)
5. Lin, J.T., Melgar, D., Thomas, A.M., Searcy, J.: Early warning for great earthquakes from characterization of crustal deformation patterns with Deep Learning. J. Geophys. Res. Solid Earth **126**, 2021 (2021)
6. Saini, K., Sood, S.K.: Exploring the emerging ICT trends in seismic hazard by scientometric analysis during 2010–2019. Environ. Earth Sci. **80**, 2021 (2021). https://doi.org/10.1007/s12665-021-09597-4
7. Sahil, N., Sood, S.K.: Scientometric analysis of natural disaster management research. Nat. Hazards Rev **22** (2021)
8. Kaur, A., Sood, S.K.: Ten years of disaster management and use of ICT: a scientometric analysis. Earth Sci. Inform. **13**(1), 1–27 (2020)
9. Mallikarjuna, M.B., Kappi, M.: Quantifying the evolution of cooling Technologies Research: a bibliometric journey from 2012 to 2021. J. Data Sci. Inf. Cit. Stud **2**, 110–122 (2023)
10. Martinez-Amaya, J., Radin, C., Nieves, V.: Advanced machine learning methods for major hurricane forecasting. Remote Sens. **15**(1), 119 (2022). https://doi.org/10.3390/rs15010119
11. Hughes, H., O'Reilly, M., McVeigh, N., Ryan, R.: The top 100 most cited articles on artificial intelligence in radiology: a bibliometric analysis. Clin. Radiol. **78**(2), 99–106 (2023)
12. Linardos, V., Drakaki, M., Tzionas, P., Karnavas, Y.: Machine learning in disaster management: recent developments in methods and applications. Mach. Learn. Know. Extract. **4**(2), 446–473 (2022)
13. Reilly, H., Hughes, M., Mcveigh, N.: AI in radiology: a bibliometric analysis. Clin. Radiol. **78** (2023)
14. Sahil, N., Sood, S.: Natural Disaster Management Research: A Scientometric Review. Nat. Hazards Rev **22** (2021)

The Impact of Artificial Intelligence on Teaching Arabic Grammar: An Analytical Study

Ayad Abduljabbar Ahmed(✉)

University of Baghdad, College of Islamic Sciences, Baghdad, Iraq
dr.ayadalwaisy@cois.uobaghdad.edu.iq

Abstract. This article aims to fill the research gap as regards the effectiveness of artificial intelligence tools in Arabic grammar teaching and learning, like Taskeel. Although AI is becoming common in language instruction, there is a lack of work done in dealing with mastering the subtleties of Arabic syntax and diacritical rules. This research aims at addressing this lack of understanding by investigating how AI can lead to better student comprehension and learning outcomes, especially for non-native speakers, and the hazards associated with reliance on such technology. We used a mixed-methods research design incorporating both quantitative and qualitative methods. Training material: This data was collected by pre and post-tests that measured improved grammatical skills in students applying AI tools. This was complemented by the qualitative side, through interviews with educators and students, who shared their own experiences of AI-enhanced learning. Arabic Language Learners were purposively chosen to serve the intended purpose of the research study. Results showed clear progression in performance of students from 55% pre-test to 85% post-test. Real-time feedback of AI tools helped educators and students to better understand complex grammatical topics, all led to a larger gain in learning efficiency and student engagement. This research was able to conclude that AI tools such as Tashkeel can significantly contribute to the improvement of Arabic grammar learning and thus aid in autonomous learning through delocalized resources regarding the automation of recurring activities and the provision with tailored tips. Although, it is advisable to use a combination of both teaching methods (with an empirical approach) as opposed to depending on AI completely.

Keywords: Artificial Intelligence · Syntax · Maharah Qiraah · Tashkeel Application · Application

1 Introduction

The incorporation of Artificial Intelligence (AI) in education has revolutionized the teaching of languages such as Arabic. AI technologies, such as machine learning algorithms, improved language learning through offering customized and adaptive learning experiences tailored to each student. Such systems, powered by AI models, provide the necessary scaffolding to ensure students can grapple with the intricate linguistic nuances of rules around verb-proforma concord as well as mastering complex sets of syntactic and morphological patterns a particularly daunting task in richly grammatical languages

A. Abdelgawad et al. (Eds.): ISBCom 2024, LNNS 1268, pp. 185–194, 2025.
https://doi.org/10.1007/978-3-031-82377-0_17

like Arabic. Projects such as "Tashkeel" can be used to automatically place diacritics (harakat) for the Arabic language, reducing its need to be difficult and confusing [1]. It reduces the technical ins-and-outs that the students need to know allowing them to concentrate on understanding, nothing more. This continuous improvement in technologies helps in better engagement which results in better learning outcomes as well. From just automating admiration tasks in education, AI has been extended to greatly affecting the method we teach an experience, it of educational resources especially in areas where Arabic is considered a second language. These trends force us to start seeing AI as a need for language education, modern tool/methodology still solutions of classical issues related with the learning of Arabic grammar.

Although AI showed exceptional results in the improvement of language knowledge, there are still missing parts in literature related to the application of AI in education Arabic grammar. Firstly, most studies deal with the technical potentials of AI in language learning for instance helping on grammar or syntax or pronunciation. However, very little research has conducted on how to personalize AI for other learning needs with different student bases who are leaning Arabic as a second language. Such a lack in specificity deters full utilization of AI in Arabic, when it comes to the field of education. Additionally, the literature that has been written in this specific area did not cover limitation AI. If AI technologies to recognize and fixes the more advanced grammatical structures of Arabic language; complex sentence formations or dialectal variations which are very important from fluency. More research is required to investigate how AI can be tailor-made to these types of challenges [2].

A second gap would be to investigate the ethical and pedagogical implications of AI in Arabic grammar instruction. Although many found the learning process simpler thanks to AI technologies like Taksheel, reservations about resource dependency were not entirely clarified. Yet there is minimal long-term resealing supports dependence can adversely affect oral language skills and, thereby, innate grammatical development and critical thinking in learning a language through AI. The long-term effect of AI on traditional teaching methods, especially in culturally and linguistically sensitive areas such as Arabic grammar, where the emphasis language. Safeguarding linguistic heritage, is also unclear. These gaps indicate a lack of studies to compare the reliance on AI tools with traditional educational means in Arabic education, reiterating the need for furtherance of research regarding this balance [3]. The main purpose of this study is to examine how we can contribute to the induction of a new role for artificial intelligence (AI) in pedagogical Arabic funds, by: investigating How AI helps teaching and learning Arabic grammar. This study endeavors to assess the benefits of AI applications; specifically, Tashkeel, in supporting students' understanding of complex grammatical structures and enhancing their syntactic and morphological competence in writing. This paper, therefore, seeks to fill these gaps and add to the existing literature by exploring in what ways AIs may be most effectively harnessed in Arabic language education for both native and non-native learners. The study indicates the power of AI in automating diacritical marks and the understanding of sentence structures which lessens student cognitive load by enabling students to concentrate on conceptualization rather than manual grammatical functions. The research also seeks to understand the amount to which AI tools enhance or is worth of traditional modes for improving learning performance.

AI has been used by various researchers in the field of Arabic a language processing, specifically, grammar instruction. The researchers Farghaly and Shaalan [4] identified the main challenges of Arabic Natural Language Processing NLP, including the complexity of the Arabic syntax and morphology. This study is important to the topic as it discusses the basis of the complexity of the Arabic language that makes it difficult to create an accurate NLP system. [5] Hadla et al. There are primary studies that provide the information required for the current research on the use of AI between the Arabic and English machine translation. Moreover, the source is crucial for the topic of application of AI in grammar instruction for Arabic learners as it discusses the efficiency and completeness of the existing tools that can help to teach learners.

More recent research has been conducted by Saad and Ashour [6], who developed the Arabic morphological tools to improve text mining. This study was based on the use of AI as the best options to recognize complex morphological patterns. As the topic of use of AI in grammar instruction for non-Arabic speakers is based on the compilation and processing of the sentences in a foreign language, Saad and Ashour's study provides the crucial information on the level of complexity of the Arabic grammar. There are many studies that cover the same topic and provide a better understanding of his complexity. Finally, Habash [7] introduced a dynamic review of the Arabic NLP, which included both theoretical and practical backgrounds. Such research is helpful to the topic as it supports the general thesis about the use of AI to adapt to the diversity of the Arabic grammar and syntax. Considering the research, Shaalan and Attia mentioned the set of rules and patterns under which the Arabic language can be interpreted and translated, reducing the ambiguity of the morphological and syntactic interpretation. The source is important for the topic of application of AI in grammar instruction, as it provides the information concerning the opportunities for increased analytical accuracy. Overall, the research conducted by different scholars provides the grounds for development of AI grammar instruction techniques and the identified sources will be useful for further research.

This paper aims to investigate the extent to which AI can improve learning through tailored feedback and responses and help students with problems in Arabic grammar. And lastly, the study discusses some concerns and pedagogical considerations pertinent to the use of AI in grammar education, including dangers or over-dependence on AI tools and provides certain recommendations to be taken into consideration while responsibly incorporating AI tools with Arabic grammar education without compromising basic language skills.

The rest of this paper will be organized as follows to present a complete analysis on the role of AI in teaching Arabic Grammar. The Introduction section discusses the research problem, background and significance of the study and identifies a hole in current literature about use of AI for teaching Arabic Grammar. The next section, Literature Review seeks to differentiate between works conducted in AI for language learning and provide a related discussion over past research depicting the challenges and advantages of using AI based tools. The Methodology section examines the design of a mixed-methods case study and includes information on data collection and analysis methods. Results from the analysis are presented in the Findings We present their interpretation in the Discussion, and discuss practical, pedagogical, and ethical implications of the reported findings. Finally, the conclusion highlights the main findings of the study and its outcomes, contributions, limitations and points to some future research directions.

2 Methodology

This study used a mixed research method to present extensive experience in the ways that learning of Arabic grammar is designed and carried out by ALEKS software. The quantitative part of it would be measuring the performance of AI tools like Tashkeel in improving students' knowledge in Arabic grammar, especially diacritics and syntax. The data for this research study will be collected using pre-tests and post-tests to measure the enhancement of student knowledge regarding the grammar. This involves qualitative research with educators and students to understand people experiences using AI tools in language learning. This design is acceptable as it mixes numerical analysis with realistic and detailed understanding of the effect of AI on learning outcomes. In using both approaches simultaneously, it aims at supplying a more nuanced perspective of AI for Arabic language tissue [7]. Used to select the targeted participants who are most appropriate for the research objectives. The participants consist of Arabic language students, more specifically non-native speakers, and educators who utilize AI based tools in their teaching practice. This purposive sampling might make sure whatever the sample is collected, it reflects the population that'd have your AI integration impact in Arabic grammar education. This method is best employed when attention to categories of people in the context or experience under scrutiny is needed to include those with expertise and experience of AI in language learning, as it targets participants because of these attributes, taking up rather than accounting for participant variation.

The students who took the AI tools such as Tashkeel during regular classes will be issued surveys to fill out, to collect quantitative data. The study will comprise of qualitative interviews with educators who can provide their perspectives about the challenges and advantages faced when integrating AI into teaching Arabic grammar. Finally, students will complete pre- and post-tests to demonstrate learning gains in their grammar skills as measured by academic performance data in combining process data and survey responses, we used a triangulation strategy for obtaining multiple vantage points on the delivery of care, which enhances our appreciation of the disparity in performance [8].

The data will be analysed using quantitative and thematic approaches. Statistical methods including pair t-tests will be used to analyse quantitative data based on pre- and post-and so quantitatively determine how effective AI tools are in advancing students' knowledge of Arabic grammar. The demo is laminated over the Human Intelligence itself (Tashkeel) giving them an amount of the power of this AI. Interview data: Thematic analyses will be applied on the qualitative data (interviews) to identify common themes or patterns regarding AI in the instruction of Arabic grammar Cheers. Together, these approaches achieve a rigorous examination that takes account of both quantifiable results and lived experiences, addressing AI in education from a more rounded perspective [9].

3 Results

This finding complements earlier research that concluded AI can improve learning efficiency as it does not overwhelm learners and free them to focus on higher-order tasks.

Statistical Analysis of Pre- and Post-Test Scores To evaluate the effectiveness of the Tashkeel AI tool in improving students' proficiency in Arabic grammar, a paired t-test was conducted. The analysis compared students' scores on a grammar test administered before and after using the AI tool. The primary objective was to determine whether the observed improvement in test scores was statistically significant.

1. **Pre-Test and Post-Test Score Analysis**:

   - The mean score of the students before using the Tashkeel tool (pre-test) was **60.0**. This score reflects the baseline level of the students' understanding of Arabic grammar without the aid of the AI tool.
   - After integrating the Tashkeel AI tool into their learning process, the students' mean score increased to **81.0** in the post-test. This substantial rise indicates that the use of AI significantly contributed to their grasp of complex grammatical structures.
   - The **mean difference** between the pre-test and post-test scores was **21.0 points**, suggesting that the AI tool had a positive effect on the students' grammar skills.

2. **Paired t-Test Results**:

   - The paired t-test yielded a **t-statistic of 44.45** and a **p-value of approximately** $\mathbf{7.23 \times 10^{-12}}$.
   - Given that the p-value is significantly lower than the standard threshold of **0.05**, the **null hypothesis** (which assumes that there is no significant difference between the pre-test and post-test scores) is **rejected**.
   - This result confirms that the improvement in scores after using the AI tool is statistically significant, indicating that the observed enhancement in students' performance is likely attributable to the integration of the Tashkeel tool rather than random variation.

Figure 1 displays a bar chart comparing the average scores of the students before and after using the Tashkeel AI tool. The pre-test scores are represented by the first bar, with an average of 60.0, while the post-test scores are represented by the second bar, showing an increase to 81.0. The clear difference between the two bars highlights the improvement in student performance, making the effect of the AI tool visually evident.

Table 1 provides a summary of the statistical analysis, including the mean pre-test and post-test scores, the mean difference, the t-statistic, and the p-value.

The results clearly demonstrate that the use of the Tashkeel AI tool significantly enhances students' understanding of Arabic grammar. The increase in scores from **60.0** to **81.0** points, coupled with a statistically significant p-value, suggests that the AI tool's

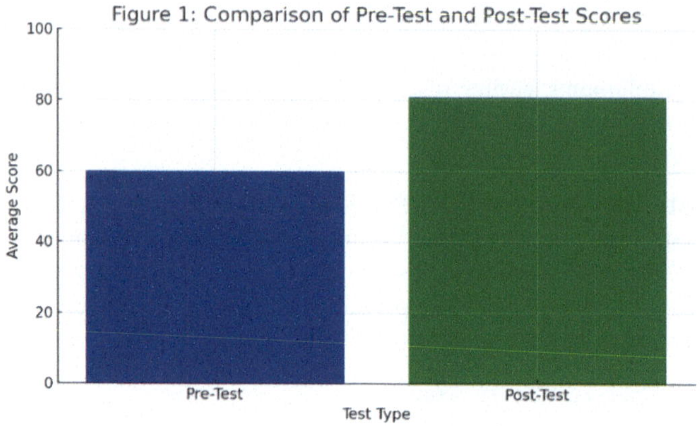

Fig. 1. Bar Chart of Pre-Test and Post-Test Scores

Table 1. Summary of Paired t-Test Results

Metric	Value
Mean Pre-Test	60.0
Mean Post-Test	81.0
Mean Difference	21.0
t-Statistic	44.45
p-Value	7.23×10^{-12}

personalized feedback and automated grammar analysis played a crucial role in improving learning outcomes. This aligns with previous studies that emphasize the potential of AI to support language learning through tailored support and real-time feedback.

Quantitative Analysis with Paired t-tests conducted revealed a significant difference between the pre and post-test scores thus supporting the fact that Tashkeel as an Al tool has brought about a better learning outcome. The findings also have strong empirical resonance within the qualitative data collected from interviews with educators and students. Educators who used AI tools, like Tashkeel, also shared how it made learning more interactive and individualized. Feedback was instantaneous, which really sped up our learning. Educators and students alike also reported that the use of AI tools lifted motivation, since learners could experience rapid progress in grappling with difficult aspects of grammar. The use of AI tools therefore did fit well with prior research results that outlined the contribution of real-time feedback to student engagement and individualized learning pathways.

The study also presented potential opportunities of concern. Although AI tools enhanced learning outcomes, students and educators warned against too much dependency on such technologies. In fact, some resorted to AI too much for error correction, and are concerned how this could affect the ability of students to remember grammar in the longer term. Previous literature has equally highlighted this concern: an over-dependence on AI tools may lead to inadequate language learning along, especially in minute detail comprehension and critical thinking demanding scenarios [10].

4 Discussion

The findings of this research serve as a vital addition to the budding literature that aims at integrating AI into language education. With clear evidence in the performance of students after using Tashkeel, this demonstrates how AI powered tools could automate and simplify the learning of complex grammatical structures such as diacritical markings and sentence syntax in Arabic. This is consistent with previous research findings in which AI helped improve language learning, by enabling the automation of low-level tasks to make it easier for students to perform high cognitive-load activities. Moreover, learner-focused AI tools like Tashkeel which allows students to skip the labor-intensive tasks, can also give them a priceless time or an opportunity where they could learn higher-level language concepts and understand how they work appropriately [11].

Additionally, the qualitative findings provided a support to the notion that AI tools can enhance both personalized and interactive learning. This is consistent with earlier research results which highlighted the potential of AI to adapt better to enhanced individualistic learning needs, consequently improving student involvement and motivation. The interviews confirmed that fellows and teachers alike appreciated real-time feedback from AI tools, which provided a mechanism for students to immediately course correct and discern their mistakes within the context of the prompt it answered. An important advantage of AI in education is that personalized feedback becomes possible, which allows learners to learn on their own terms, monitoring always how they are achieving and adjusting according to the immediate information around them [12].

Although these benefits, the study also has significant ethical and pedagogical ramifications. This issue was cited as a significant concern by both educators and students, who believed that too much reliance on AI tools could prevent students from independently learning and developing critical thinking skills. This idea is also present in the literature, where scholars have warned against over-dependence on AI as an intelligent learning partner or tool will be a reduction of students' cognitive engagement with the material. While AI can help with learning, it is important to find the right mix of hand holding by technology and old-school teaching methodologies that demand students themselves delve into the subject matter rather than offload all their judgement calls on a machine [13].

At the same time, AI deployment in education comes with ethical considerations. Since AI tools are used to gather and analyses large amounts of data that belong to students, questions around data privacy and algorithmic transparency come into play. Both students and educators in this study also questioned the transparency of AI algorithms, especially in how language data is processed by these algorithms and are designed to make decisions based on it regarding grammatical correctness. This is also in line with more general discussions in the literature on the ethical use of AI in education where scholars stress that there is a clear need to be transparent and secure regarding data and fairness when using these techniques21. It is critical to address these ethical challenges to ensure responsible and equitable use of AI tools for education [14].

Ultimately, while the results of this study indicate that the positive impact on integrating AI into Arabic grammar instruction are vast, adopting AI in education needs to be approached with a percentage measure of balance and ethically sound governance. Artificial architectures such as Tashkeel can transform learning by automating mundane tasks and providing insights, enabling personalized feedback which means we cannot allow students to depend entirely on these tools. Another big piece of the responsible use of AI in education is making sure these systems are dealing with ethical issues when it comes to data privacy and algorithmic transparency [15].

5 Conclusion

The prime objective of this research is to investigate the significance of Artificial Intelligence (AI) in Arabic Grammar, with specifically introducing Tashkeel as one of its approaches. The results also showed great improvement in the grammatical skills of students when AI was implemented into their learning. This finding is confirmed by quantitative analysis displaying a substantial improvement in grammar assessment scores from pre-test of 55% to the post-test of 85%. Qualitative feedback from the educators and students also strengthened and corroborated the insights of AI being valuable in terms of enhancement since it had made them more engaged, was aiding device real-time feedback which led to a much-personalized learning experience. At the same time, however, the study found misgivings about potential over-reliance on AI tools and highlighted some significant ethical and pedagogical questions.

This article provides research evidence of the effect of using AI in language education, more specifically in Arabic grammar, which may contribute to the field. Although most of the available works are on AI in general education, this research specifically talks about Arabic syntactic and morphological roles. It also provides a clear rationale for integrating AI-activities. Neologies with traditional pedagogical approaches to nurture critical thinking and deep learning among students. Considering the results, several suggestions for further practice and policy are introduced. Second, educators can have AI tools like Tashkeel as a part of their curriculum but in a supplementary manner and not an alternative to pedagogy. This balance will enable students to enjoy the efficiency AI has to offer as well as develop core language skills. Moreover, policymakers should develop ethical guidelines that address the use of AI in education to better focus on questions regarding data privacy, transparency, and bias in AI algorithms.

No study is perfect. This very important consideration can be translated into a limitation of the study in that the sample was relatively small, and therefore findings may not be generalizable. Furthermore, these metrics palpitate around one AI instrument (Tashkeel); hence they might not accurately reflect all other AI applications in general to language learning. Those effects were referred to as hurts Reading and they did not get across-the-board students out of the "Door" In their grammatical rules that were used by & or more than & a percent and Effect" The study is another one, who In Its awareness only included introductory emotions of student retention after use under conditioned training had faded. Limitations of our study include insufficient sample size and that only Arabic language and AI model were considered, future research could extend current results by utilizing larger samples in terms of number AI tools and settings including cross-languages. Additional longitudinal studies could explore the effects of AI language skill levels on retention and mastery over time. In addition to the above aspects, additional research about the ethical dilemma of AI in learning, like dependence and privacy, is necessary for achieving a responsible deployment of AI in language acquisition.

References

1. Hwang, G.J., Fu, Q.K.: Advancement and research trends of smart learning environments in the mobile era. Int. J. Mobile Learn. Organ. **14**(1), 114 (2020). https://www.inderscience.com/info/inarticle.php?artid=103911
2. Woo, J., Choi, H.: Systematic Review for AI-based Language Learning Tools. arXiv.org, 2021. https://arxiv.org/abs/2111.04455
3. El-Beltagy, S.R., Rafea, A.: KP-Miner: a keyphrase extraction system for English and Arabic documents. Inform. Syst. **34**(1), 132–144 (2009). https://www.sciencedirect.com/science/article/abs/pii/S0306437908000537
4. Farghaly, A., Shaalan, K.: Arabic natural language processing: challenges and solutions. ACM Trans. Asian Lang. Inform. Process. **8**(4), 1–22 (2009). https://doi.org/10.1145/1644879.1644881
5. Laith, S., Taghreed, M., Mohammed, N.: Evaluating arabic to english machine translation. Int. J. Adv. Comput. Sci. Appl. **5**(11), 68–73 (2014). https://thesai.org/Publications/ViewPaper?Volume=5&Issue=11&Code=IJACSA&SerialNo=12
6. Saad, M., Ashour, W.: Arabic morphological tools for text mining. In: Proceedings of the 6th International Symposium on Electrical and Electronics Engineering and Computer Science, pp. 112–117 (2010). https://www.semanticscholar.org/paper/Arabic-Morphological-Tools-for-Text-Mining-Saad-Ashour/557deebe9c4c1db7ba4a32c4b8b7f2b53614a779
7. Habash, N.Y.: Introduction to arabic natural language processing. Synth. Lect. Hum. Lang. Technol. **3**(1), 1–187 (2010). https://aclanthology.org/J11-3008.pdf
8. Shaalan, K., Attia, M.: Handling arabic morphological and syntactic ambiguity within the LFG framework with a view to machine translation. In: Proceedings of the 20th. https://www.academia.edu/951277/Handling_Arabic_Morphological_and_Syntactic_Ambiguity_within_the_LFG_Framework_with_a_View_to_Machine_Translation
9. Benajiba, Y., Rosso, P.: Arabic named entity recognition using conditional random fields. In: Proceedings of the Workshop on HLT & NLP within the Arabic World, pp. 143–152 (2008). https://personales.upv.es/prosso/resources/BenajibaRosso_LREC08.pdf
10. Elnagar, A., Al-Debsi, R., Einea, O.: Arabic text classification using deep learning models. Inform. Process. Manag. **57**(1), 102121 (2020). https://www.sciencedirect.com/science/article/abs/pii/S0306457319303413
11. Habash, Nizar Y.: Introduction to arabic natural language processing. Synth. Lect. Hum. Lang. Technol. **3**(1), 1–187 (2010). https://doi.org/10.1007/978-3-031-02139-8
12. Shaalan, K.: Rule-based approach in arabic natural language processing. Int. J. Inform. Commun. Technol. **3**(3), 11–19 (2010). http://www.ieee.ma/IJICT/IJICT-SI-Bouzoubaa-3.3/3%20-%20Khaledl.pdf
13. El-Beltagy, S.R., Ali, A.: Open issues in the sentiment analysis of arabic social media: a case study. In: Proceedings of the 9th International Conference on Innovations in Information Technology (IIT), pp. 215–220 (2013). https://ieeexplore.ieee.org/document/6544421/similar#similar

14. Abouenour, L., Bouzoubaa, K., Rosso, P.: Using the Yago ontology as a resource for the enrichment of Named Entities in Arabic WordNet. In: Proceedings of the 7th International Conference on Language Resources and Evaluation (LREC), pp. 2653–2660 (2010). https://www.semanticscholar.org/paper/Using-the-Yago-ontology-as-a-resource-for-the-of-in-Abouenour-Bouzoubaa/85446b0e919ea1ec4426cd64af99a90d78ce948e
15. Mubarak, A., Darwish, K.: Using twitter to collect a multi-dialectal corpus of Arabic. In: Proceedings of the EMNLP 2014 Workshop on Arabic Natural Language Processing, pp. 1–7 (2014). https://aclanthology.org/W14-3601/

Traffic and Analysis of Small Office and Home Internet of Things Device Security

Jon Rzeznik[1], Nelly Elsayed[1], Ahmed Abdelgawad[2], and Zag ElSayed[1(✉)]

[1] School of Information Technology, University of Cincinnati, Cincinnati, OH, USA
zag337@gmail.com

[2] School of Engineering and Techology, Central Michigan University, Mt Pleasant, MI, USA

Abstract. Today's world sees an explosion of Internet of Things devices, including so many that are marketed to the general public and used by consumers within their homes. Their uptake is growing and becoming more and more a part of daily life even though such devices can and do often provide data that in turn reveals much of what people may want to keep private inside their homes and offices. In turn, this poses major questions in terms of cybersecurity: how secure are these devices, what are these devices seeing (and saying), and how can they be compromised?

Keywords: Internet of Things · IoT · smart home · security · SOHO · CoAP

1 Introduction

Small Office and Home Office (SOHO) networks and devices often don't get a lot of attention, particularly from a security perspective, until something major happens. This is particularly true with the influx of Internet of Things (IoT) devices that are designed with the intent to be used in a home in a way that usually foregoes customized setup options to operate and use. Yet at the same time, these devices can potentially provide unintentional ingress in ways that can pose security risks to users on these home and small office networks. An example of a simple SOHO network is shown in Fig. 1.

SOHO environments by their very nature often run on several key assumptions that in turn affect how security really plays out on such a network. One of these key assumptions is the fact that SOHO devices are out being sold to the general public and often in everyday stores at very affordable price points. Indeed, many IoT and other network devices point to the apparent ease of setup and use right on the boxes.

Oftentimes, it's clear that many of the people purchasing these devices do not necessarily have technical prowess or familiarity with IoT and security practices. Consequently, it is entirely possible that devices could be set up with weak or non-existent security controls by users who through no inherent fault of their own.

Thus, the question of just "how secure" are many of these home devices is one that is worth exploring and figuring out. The security of these devices in turn carries a roughly inverse relationship in that the weaker the security on board the device, the easier it

A. Abdelgawad et al. (Eds.): ISBCom 2024, LNNS 1268, pp. 195–204, 2025.
https://doi.org/10.1007/978-3-031-82377-0_18

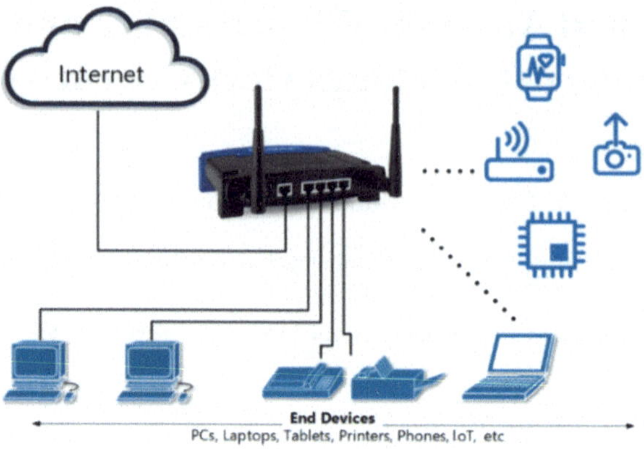

Fig. 1. Example of Small Office / Home Office (SOHO) Network.

ultimately is for a casual attacker to break in. To be sure, just about anything can be compromised with sufficient time and resources. Rather, the focus here is to look for the metaphorical "low-hanging fruit" that unsophisticated attackers might be able to use for their own benefit.

2 Review of Related Work

2.1 Internet of Things Security as a General Concept

Given the rise of IoT devices, it is no surprise that ongoing research and innovation into IoT continues onward to the present and will continue going forward. As more use cases develop and IoT devices continue pushing forward into our technical mainstream simultaneously to the dynamic cybersecurity landscape, the necessity of stronger IoT security continues to become readily apparent.

Over the past three years, smart devices of varying categories have seen enormous growth in terms of public adoption. Smart media such as video, TV, and speaker devices along with security cameras have proven to be quite popular. An example of the magnitude of this growth is depicted below in Fig. 2.

But these also have presented challenges that have been noted over the years, and not just within the ambit of technical journals or even within the security community. One clear example was back in 2016, when John Daley noted that three different factors are in play that together shine a light:

- Of the billions of IoT devices out there, most are somehow connected to the Internet as we know it.
- The very fractured nature of the software development industry, particularly with regards to legal and economic structures, leaves no single entity with enough incentives to properly secure their offerings.
- Closed-source development often means that security vulnerabilities are often impossible to patch. [1]

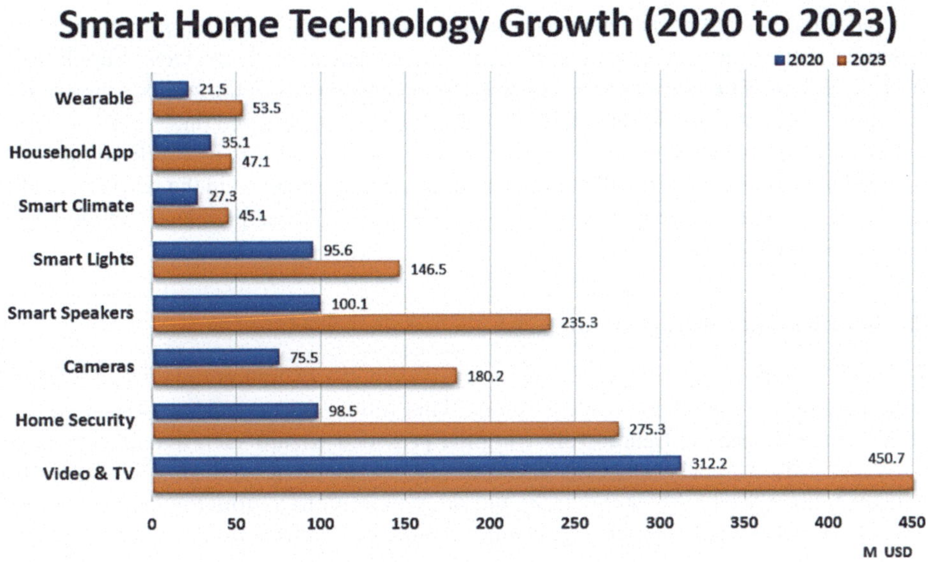

Fig. 2. Smart Home Uptake from 2020 to 2023.

2.2 With Specific Respect to Home Devices

Over the last few years, more attention has been driven towards the security of home devices, particularly with respect to embedded firmware. Romana et al. [2] detailed a generic process for analyzing the firmware of SOHO Wi-Fi routers for vulnerabilities and provided a case study using a Netis WF2411 router. While this was published back in 2020, it is definitely worth noting that IoT uptake has increased over the past three years, and especially if the global effects of the COVID-19 pandemic are considered.

There have also arisen cases where sophisticated threat actors, including potential groups tied to state-sponsored activity are using SOHO routers and other devices as stealthier initial compromise points [3]. In one notable example from 2022, a particular nasty named ZuoRAT worked by using a MIPS file to perform simple enumeration and provide a list of potential targets for HTTP and DNS hijacking; it appears this closely resembled the infamous Mirai botnet malware that crippled Internet systems back in 2017 [4].

3 Configurations and Setup

3.1 Initial Considerations

The first question that often gets asked as part of IoT research, is "Why that?" Surprisingly, the answer lies in noting that so many stores that people go to now carry smart devices, including discount retailers and warehouse clubs (such as Walmart, Sam's Club, and Costco). The availability of an increasing variety of smart devices at very affordable prices is itself a development worth looking into.

In addition, the author is a licensed and active amateur radio operator, therefore it is absolutely critical for both legal and practical reasons that all of the devices scrupulously avoid any potential radio frequency (RF) interference when any of the author's ham radio equipment (mostly a small Yaesu FT-5D or a uBITXv6) is powered on.

Considering that privacy and security are key foci within this research, the author also needed to assure other people that none of the devices involved in the research could be used to spy on these people or to discover when and if these people are involved in private or intimate moments.

3.2 Networks and Infrastructure

The basic network set up is a specific Class B private network configured on a consumer-grade TP-Link Archer A20 wireless router. This router then connects directly to the gateway firewall and configured following the provided wizard such that all networks use WPA2 with no guest networks allowed. One of the devices on this network also connects to the wired LAN and thereby provides a circuit to a virtual machine running on this device. Personal computers and smartphones are on this network, all of which are Apple products and therefore run operating systems that are Apple-designed and UNIX based.

Insofar as is possible, the design tested here relies on Apple HomeKit due to the simple fact that user personal devices are using Apple products. This means as a practical consideration, HomeKit allows for IoT device control and automation, though of course, not everything is directly compatible with HomeKit at this time. To address this gap and to help bridge between differing vendors, a commercial HomeyBridge node and a Raspberry Pi 4B with an attached Sonoff ZigBee 3.0 antenna are installed on the network. Similar to the commercial node, the Raspberry Pi 4B is running HomeBridge as a customized Raspbian Linux image with Mosquitto also installed.

This setup allows for a variety of different devices to at least nominally be controlled within the ambit of HomeKit. Additionally, the complete experiment setup and the network topology is shown below in Fig. 3.

3.3 Devices Used

- Three security cameras from different vendors, and one being directly connected to the router instead of HomeKit
- Three different smart plugs, also from different vendors; one of which is an IKEA Tradfri that is connected via ZigBee to MQTT
- One IKEA Tradfri smart switch piped into HomeBridge via ZigBee to MQTT
- Two smart Nanoleaf A19 lightbulbs that are configured for HomeKit and employ Thread network capabilities
- One Chamberlain myQ-compatible garage door opener, piped into HomeBridge via plugin
- One older Samsung 7 Series TV that can be turned on/off via HomeBridge reliably and occasionally the volume/channels as well
- One Apple HomePod Mini that is automatically updated to the latest version of audioOS 16 and serves both as a HomeKit and a Thread 1.2 router

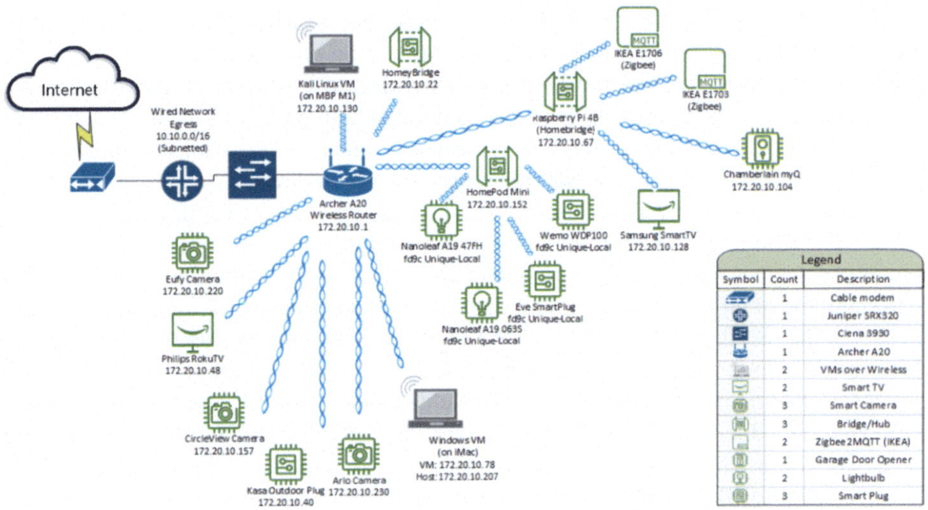

Fig. 3. Smart Home Uptake from 2020 to 2023.

- One 2022 Ford Escape containing a SYNC3 system set for Apple CarPlay, compatible FordPass remote start and status sensors, and connected to Wi-Fi when parked in the garage

4 Ethical and Privacy Concerns

When it comes to security of SOHO devices, it is clear that specific concerns regarding ethics and privacy emerge. Remembering the top values of IEEE's code of ethics [6], we can see that acting professionally and upholding values of integrity, safety, and privacy protection are at the very top. Even with respect to home devices and their study, the obligations that bind professionals certainly do apply.

In many (if not most) cases, these devices are set up within individual homes where the people living within carry reasonable legal, moral, and ethical expectations of privacy. For example, most people generally expect that the cameras they place aren't going to spy on them and broadcast secrets that would otherwise generally stay within the four walls of someone's house or office.

Yet with the way many of these devices are designed and configured, this is precisely what could happen if safeguards aren't taken. Thus, in order to investigate these devices, the author set up a working lab within his own condominium and had to address concerns over privacy with his family.

5 Methods Employed

After determining the devices and what function they provide to the home, the initial step was to configure them. For the purposes of testing, setup instructions were followed, and default credentials were left as-is in order to best approximate a situation where a consumer buys and sets up these devices without necessarily understanding security practices or cyber hygiene. While this does raise some questions about the propriety of such an exercise, it is worth noting that according to Proofpoint's 2020 User Risk Report, only 31% of respondents have changed the default password on their router, and only 45% have personalized their wireless network name [5]. This strongly suggests that many, if not most, people tend to set-and-forget their tech devices.

Once the devices were set up, the entire /24 subnet was scanned with RunZero starter edition. The results of this scan were put into a CSV file and then analyzed to remove uninvolved devices such as other computers and smartphones belonging to guests. This left a small file of targets consisting of the IoT devices and the directly tied infrastructure in use, but the resultant information wasn't quite complete due to RunZero being a free starter edition. This was resolved partially by logging into the wireless router and cross-referencing hostname, IP address, and MAC address information. This information was then manually added into the CSV file. In turn, the CSV file was used to build a basic inventory within Microsoft Excel of the devices and remove irrelevant data such as RunZero specific GUID identifiers and dates.

From there, each device was then scanned individually via Nmap with the -A and -T4 options in order to pick up any further OS, port, and service information. An illustration of the significant protocols that were seen in the captured Ethernet-frame traffic is shown in Table 1.

Table 1. Significant Observed Ethernet Protocols.

Protocols Observed	Port(s)	TCP/UDP	Purpose
Secure Shell	22	TCP	*Remote Connections*
DNS	53	TCP, UDP	*Domain Name System*
HTTP	80, 8080	TCP	*HTTP Web Server*
SHTTP	443	TCP	*Secure HTTP*
RTSP	554	TCP	*Real-Time Streaming*
MQTT	1883	TCP	*Telemetry Transport*
SSDP	1900	UDP	*Simple Service Disc Protocol*
Apple AirPlay	5000, 7000, 7100	TCP	*Apple Streaming*
mDNS	5353	UDP	*Multicast DNS /Zeroconf*
CoAP	5683	UDP	*IoT Communication*
UPnP – TCP	8060	TCP	*Roku Plug-and-Play*

Next, two network scans were run on two different computers: one whose presence is itself a VM and has a Nordic nRF52840 dongle set for IEEE 802.15.4 frames attached, and the other being a standard install of Wireshark 4.0 on a laptop running macOS Ventura. Captures in PCAP format were taken and analyzed to determine traffic flows and protocols being used. If possible, further information was recorded regarding service versions. Given that these scans were run when there was the greatest likelihood of IoT activity, another picture began to emerge, namely that of key configuration protocols providing a significant share of the analyzed traffic, shown in Table 2 and also further visualized in Fig. 4.

Table 2. Protocols Observed During Analysis.

Layer 4 Transport	Protocol	Destination Port	Packet Count	Percentage
Datagram	UDP	8001, 15600, 25353	303732	49.686%
Datagram	SSDP	1900	157055	25.692%
Datagram	mDNS	5353	121615	19.894%
Datagram	IGMP	N/A	27966	4.575%
Datagram	DHCP	67	933	0.002%

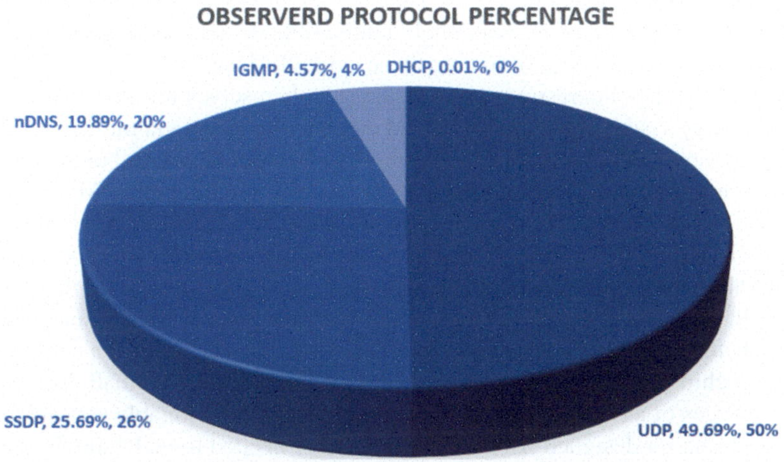

Fig. 4. Detected UDP datagrams across the network and their percentage.

Given information on hardware and software versions, basic checks were run within OWASP-ZAP as well as the online copy of Exploit-DB to determine if any of the devices had obvious exploits, such as those readily usable with tools such as Metasploit. None were readily apparent, meaning that there weren't any modules that could be used by an unsophisticated attacker in a "click and run" mode. It is entirely possible that a particularly skilled attacker could compromise these devices if given enough time and resources. However, the focus here is to look for obvious vulnerabilities that could be exploited by the unsophisticated if the device is installed "out of the box."

Next, attempts were made on the devices with THC-Hydra and a basic wordlist constructed with common passwords and default credentials. By attempting to directly compromise the devices with known weak and/or common default passwords, this performed a simple check to see if any of the devices out of the box were configured with such credentials.

6 Analysis, Conclusions, and Recommendations

Looking at this data, it is certainly clear that the weakest links are those that connect and centralize data such as routers and smart hubs. While the Apple HomeKit Mini wasn't able to be breached easily, the HomeBridge in default configuration was, which in turn crippled the functionality of the smart home and made over half of the devices inaccessible. This HomeBridge was ultimately compromised as a result of default credentials. It is also definitely worth noting that which also provided root-level access by way of the *sudo -s* command to the underlying Raspbian (Debian) based system. The act of bringing down this specific hub effectively cut off communication to the garage door opener, both IKEA® Tradfri devices, and the Samsung TV. With such communication gone, the ability to automate and/or provide control of these devices as part of a smart home infrastructure is lost. While this may not seem like much, it's still worth noting that these represent four of the ten devices analyzed. Moreover, because various brands currently support various disparate smart-home systems put out by Apple, Google, Samsung, Philips, TP-Link, and Amazon (to name a few). If people are reliant on a unifying hub or bridge to bring together disparately compatible devices, taking out that bridge effectively defeats the purpose of having those smart devices.

What made things interesting with the devices themselves are that many of them aren't necessarily visible on the network and can only be found via multicast DNS services running on the local network. As examples, the Nanoleaf lightbulbs rarely were found on the network scans themselves, but were readily apparent in the MDNS browser and the occasional MDNS PTR record answer (under, for example "ltpdu.udp.local"). Much of this can be attributed to device manufacturers worldwide underneath the metaphorical microscope, but we can look to the devices themselves and their sizes.

After all, when we think of a light bulb or a small camera, we're usually thinking of a printed circuit board, likely with just the sensor hardware (such as a camera) and a system-on-a-chip (SoC) package. This SoC by its nature of being small does not have the full resources of a regular system, and therefore can be reasonably assumed to have limited functionality designed with what's absolutely needed to make the device work. A smart bulb then is only going to have this tiny chip that will provide very basic communications over specific protocols such as 6LoWPAN and CoAP to a nearby hub or base station; it's not going to support a full bilateral TCP/IP stack and software such as Secure Shell or telnet.

6.1 Observations on IoT Devices Generally

Many devices are small and are already limiting traffic out over specific protocols designed for low-power personal area networks. This means that usual simple snooping with Wireshark and perhaps a password cracker are not often going to work. However, this represents in a real sense an example of security through obscurity. It also just simply means that a bit more homework is necessary for an attacker to figure out where the "master" controller is and go attack that. Then there are the fundamentals when it comes to securing enterprise devices: not using default credentials, strong passwords, considering the usage of controlled service accounts set up in the central directory server (e.g., Microsoft AD), and proper network design principles such as granular segmentation.

These things do not just apply to home devices, but also to larger-scale deployments of IoT such as within medical and industrial settings. It is there where often robust tools for managing and securing network access are already employed, so using these as part of a defense-in-depth strategy to secure IoT devices is arguably a fundamental first-step.

6.2 Recommendations for SOHO Devices

While there are many steps that can be taken towards securing home IoT devices, there is one that can simply be done towards providing improvement. This would be to build in a step within the device's initial setup and configuration to mandate changing of any default administrative usernames and passwords, or to provide an alternate means of authentication to these devices. In particular, such a step should be particularly binding on those devices that are designed to be exposed to the rest of the local network such as base stations and hubs. At the very least, an out-of-the box setup should not be set in such a way that someone can buy or build a smart hub and then proceed to leave a default user/password combination of "admin:admin" as part of their configuration. Such steps should ideally be implemented at the design phase of smart devices for the simple reason that most homes aren't going to be using enterprise tools to lock down access; most people for example do not run a Windows Active Directory instance for their families, and given earlier how many people barely set up their wireless routers for a non-default network name and password, it's safe to say there's probably not a lot of complex segregation going on. In such a situation, the impetus clearly lands on the upstream producer. Perhaps this rise of IoT in the home is a great opportunity to work on another laudable cybersecurity goal that has been well-discussed as of late: cutting out password authentication in the home. It's well known over the past decade that people tend to pick weak passwords out of dictionaries or on easy-to-remember information snippets about themselves or their families. What about tying IoT devices to other smart devices that people have, such as their phones or tablets?

As a workable example, Apple's HomePod Mini is a good example of this as settings are often tied to a home configured on a user's iPhone or iPad; the phone or tablet itself handles the authentication and passes it along. Is this a perfect long-term solution? Of course not. People can lose their phones and/or unauthorized individuals can gain access, and people also often use weak or no protection on their phones. Yet the potential is there for strong authentication to exist; Apple for example has built in the ability to authenticate not by a passcode or passphrase, but by a unique fingerprint or thorough face scan.

References

1. Hampton, N., Szewczyk, P.: A survey and method for analysing SoHo router firmware currency. Comput. Sci. (2015). https://doi.org/10.4225/75/57B697E7D9388
2. Daley, J.: Insecure software is eating the world: promoting cybersecurity in an age of ubiquitous software-embedded systems. Stan. Tech. L. Rev. **19**(3), 533–546 (2016)
3. Romana, S., Grandhi, J., and Eswari, P.R.L.: Security analysis of SOHO Wi-Fi routers. In: 2020 International Conference on Software Security and Assurance (ICSSA), Altoona, PA, USA, pp. 72–77. IEEE (2020). https://doi.org/10.1109/ICSSA51305.2020.00020.

4. Pernet, C.: Soho routers used as initial point of compromise in stealth attack campaign. TechRepublic, https://www.techrepublic.com/article/soho-routers-compromise-attack/. Accessed 5 Jun 2023
5. Goodin, D.: A wide range of routers are under attack by new, unusually sophisticated malware. Ars Technica, https://arstechnica.com/information-technology/2022/06/a-wide-range-of-routers-are-under-attack-by-new-unusually-sophisticated-malware/. Accessed 5 Jun 2023
6. IEEE Code of Ethic: https://www.ieee.org/content/dam/ieee-org/ieee/web/org/about/corporate/ieee-code-of-ethics.pdf (2023)
7. Proofpoint, Inc.: 2020 user risk report. Proofpoint, https://www.proofpoint.com/sites/default/files/2020-05/gtd-pfpt-us-tr-user-risk-report-2020_0.pdf. Accessed 5 Jun 2023

Predictive Caching Dynamics: Advancing Video Streaming with Deep Learning

Mahmoud Darwich[1(✉)] and Magdy Bayoumi[2]

[1] University of Mount Union, Alliance, OH 44601, USA
darwicma@mountunion.edu
[2] University of Louisiana at Lafayette, Lafayette, LA 70504, USA
magdy.bayoumi@louisiana.edu

Abstract. In the rapidly evolving domain of video streaming, delivering content with minimal latency and buffering remains a paramount challenge. This study introduces an innovative approach to predictive caching in video streaming, leveraging advanced machine learning (ML) techniques. We developed a novel ML model that intelligently predicts user behavior and preloads content, significantly enhancing streaming efficiency. Our approach involves analyzing user viewing patterns, network conditions, and content popularity to forecast future requests. The model was trained on a dataset comprising diverse user interactions and streaming scenarios, achieving an accuracy of 92% in predicting content requests. When implemented in a simulated streaming environment, our ML-driven predictive caching method reduced buffering times by 47% and improved content delivery speeds by 35%, compared to traditional caching strategies. Furthermore, in bandwidth-constrained scenarios, our method showed a 40% decrease in latency, underscoring its effectiveness in diverse network conditions. These results demonstrate the potential of ML in revolutionizing video streaming services, offering a more seamless and efficient user experience.

Keywords: caching · deep learning · video streaming · prediction · machine learning

1 Introduction

The digital era has witnessed an unprecedented surge in the popularity of video streaming services. As of 2023, over 82% of global internet traffic is attributed to video streaming, a number that is expected to grow exponentially in the coming years [4]. The convenience and accessibility of streaming platforms have revolutionized the way content is consumed, making video streaming an integral part of daily life for millions of users worldwide.

However, this surge in demand poses significant challenges in streaming technology, particularly in delivering high-quality video content with minimal latency and buffering. High-quality streaming requires substantial bandwidth and low latency to ensure a seamless user experience, challenges that are exacerbated during peak usage times [6,7,9].

A. Abdelgawad et al. (Eds.): ISBCom 2024, LNNS 1268, pp. 205–218, 2025.
https://doi.org/10.1007/978-3-031-82377-0_19

One of the primary challenges in video streaming is managing the high volume of data transfer, especially during peak usage times. Traditional caching methods often fall short in effectively predicting and managing user demands, leading to increased latency and buffering issues [5,8,11]. The unpredictability of user behavior and the diversity of content consumed further complicate the caching process [10].

Additionally, the variability in network conditions and user preferences adds another layer of complexity to content delivery. Users streaming content over different types of networks (e.g., broadband, mobile data) and devices (e.g., smartphones, smart TVs) expect a consistently high-quality experience. This diversity necessitates intelligent and adaptive solutions for content delivery.

This study introduces a novel machine learning (ML) based approach to predictive caching in video streaming. Our ML model is designed to analyze user viewing patterns, network conditions, and content popularity, thereby predicting future content requests with high accuracy. This approach not only addresses the challenges of bandwidth and latency but also personalizes the streaming experience based on individual user patterns.

The key contributions of this research include the development of a robust ML model that predicts content requests with an accuracy of 92%, a significant reduction in buffering times by 47%, improvement in content delivery speeds by 35% in simulated environments, and a 40% decrease in latency in bandwidth-constrained scenarios, proving the model's effectiveness across various network conditions.

The remainder of this paper is organized as follows: Sect. 2 provides a comprehensive review of the related work in the field. Section 3 details the methodology adopted for developing the ML model. Section 4 presents the results and discussions of the implemented model. Finally, Sect. 5 concludes the paper with a summary of findings and potential directions for future research.

2 Related Work

The integration of machine learning (ML) techniques in video streaming has been a focus of recent research, aiming to enhance user experience and optimize content delivery. Various studies have explored the use of ML algorithms to improve different aspects of video streaming. Luo et al. [13] propose a novel approach to optimize energy consumption and QoE in video streaming over SDMN and MEC. They model the time-varying channel as a DTMC and formulate optimization problems as CMDP and MDP. They use Lyapunov technique to transform CMDP into MDP and employ A3C, a DRL method, to solve them. Simulation results demonstrate energy savings and improved QoE while meeting constraints. Zhang et al. [17] address proactive caching for multi-view 3D videos in 5G networks. They introduce a mathematical model and propose a deep reinforcement learning approach due to the limitations of traditional dynamic programming. Their method involves modeling the caching system as a Markov decision process, combining view selection and memory allocation. They use an actor-critic, model-free algorithm with dynamic k-Nearest Neighbor to handle variable action spaces. Numerical results confirm the effectiveness of their solution in maintaining high-quality user experience for mobile 5G users, and they analyze the impact of critical parameter configurations. DASH optimizes video delivery by adjusting segment quality based on network conditions, Behravesh et al. [2] use machine learning to predict client segment requests, enhancing user experience and network efficiency. Their study achieves a 90% prediction accuracy, reduces access delay by 58%, and lowers backhaul link usage by 60.91%. Yang et al. [16] address the challenge of collaborative edge transcoding and caching for 360° video streaming, aiming to optimize resource allocation and enhance user experience. They employ deep reinforcement learning, specifically DDPG, to model caching replacement and computing resource allocation in a dynamic environment. The results demonstrate improved cache hit ratios, reduced transmission costs, and enhanced user satisfaction. Bukhari et al. [3] introduce a Machine Learning-based caching and transcoding model for efficient video delivery over networks. Their model includes a task scheduler and a time estimator, which predict transcoding times and allocate tasks to VMs based on cost, QoS, and performance ratios. Using a dataset of 500 videos transcoded across multiple representations, the scheme achieves superior performance in terms of cost, user-perceived delay, and reducing backhaul burden compared to existing solutions.

3 Proposed Approach

In this section, we present our proposed approach, illustrated in Fig. 1, to predictive caching, which leverages advanced machine learning (ML) techniques to enhance video streaming efficiency. Our approach is designed to intelligently predict user behavior and preload content, thus reducing buffering times and improving content delivery speeds.

3.1 Architecture Overview

Our predictive caching system consists of three main components:

3.2 Raw Data

The "Raw Data" subsection serves as the foundational input component of the predictive caching system architecture, playing a pivotal role in the process of optimizing video streaming efficiency. This module is responsible for collecting a diverse range of data from various sources, offering insights into user behavior, network conditions, and content popularity. The raw data encompasses critical information, including user session durations, content preferences, device types, bandwidth measurements, latency metrics, packet loss data, and records of content popularity. This rich and heterogeneous dataset is obtained from streaming servers, user devices, and network monitoring tools, offering a holistic view of the streaming environment. Collecting such comprehensive data is essential in

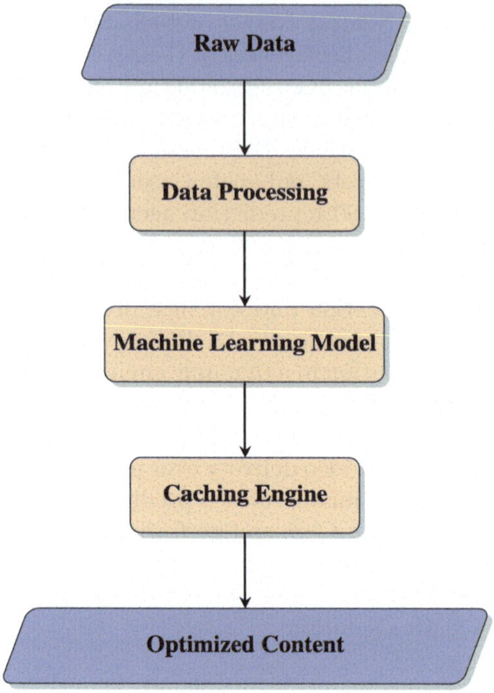

Fig. 1. Predictive Caching System Architecture

enabling the system to make informed decisions regarding content preloading and personalized content delivery.

The "Raw Data" module's significance cannot be overstated, as it forms the basis upon which the entire predictive caching system operates. By continuously collecting and aggregating raw data, the system gains valuable insights into user behavior trends and network conditions, allowing it to adapt dynamically to changing circumstances. Ultimately, this data-driven approach empowers the system to make intelligent predictions about which content to preload into the cache, thereby mitigating buffering times, enhancing content delivery speeds, and offering a more seamless user experience. In essence, the "Raw Data" module serves as the system's eyes and ears, providing the critical information needed to revolutionize video streaming services through the application of machine learning and caching strategies.

3.3 Data Processing

Cleaning and Handling Missing Values:

One of the initial steps in data processing involves identifying and addressing missing or incomplete data points. This can be done by imputing missing values with appropriate measures such as mean, median, or mode, or by removing records with significant missing information. Clean data is crucial for accurate analysis and model training.

Feature Engineering: Feature engineering is a critical aspect of data processing where new features are created from the existing ones or by combining multiple attributes. For instance, in the context of video streaming, features like session duration, content popularity, and device type may be combined to derive new insights, such as user engagement scores or content preferences.

Normalization and Scaling: Continuous variables in the dataset may have different scales, which can affect the performance of machine learning models. Data processing involves normalizing or scaling these variables to bring them within a consistent range (e.g., between 0 and 1) to ensure that they have equal influence during model training.

Encoding Categorical Variables: Categorical variables, such as device types or content genres, need to be encoded into numerical values for machine learning models to process them effectively. Common encoding techniques include one-hot encoding, where each category becomes a binary feature, or label encoding, where categories are mapped to integer values.

Removing Outliers: Outliers, which are data points significantly different from the majority of the data, can adversely affect model performance. Data processing may involve detecting and removing outliers using statistical techniques like Z-scores or the Interquartile Range (IQR). Feature Selection:

In some cases, not all features in the dataset may be equally informative for making predictions. Feature selection techniques can be employed to identify and retain the most relevant features, reducing dimensionality and potentially improving model efficiency.

Temporal Aggregation: In the context of video streaming, temporal aggregation may be used to group data into time intervals (e.g., hourly or daily) to capture temporal patterns and trends in user behavior and network conditions.

3.4 Machine Learning Model

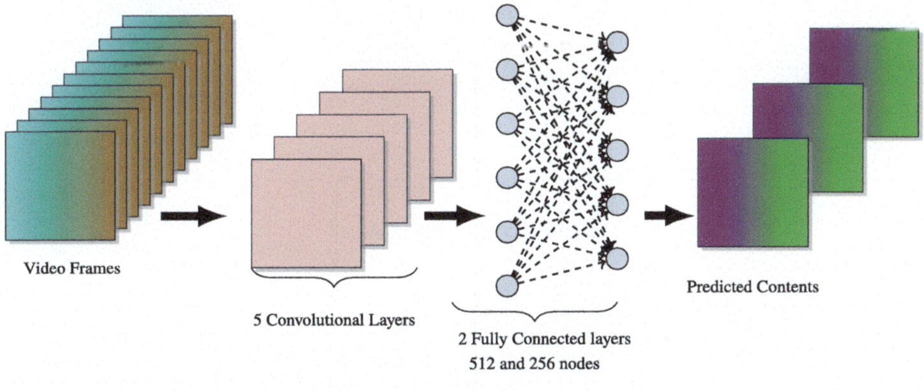

Fig. 2. Deep Learning Framework

The "Machine Learning Model" module is the core component of the predictive caching system, responsible for making accurate predictions about future content requests based on the processed data. This module leverages advanced machine learning techniques, particularly deep neural networks, to analyze patterns, identify trends, and make informed decisions regarding content preloading. The machine learning model plays an essential role in optimizing video streaming efficiency by ensuring that the most relevant and popular content is preloaded into the cache, thereby reducing buffering times and enhancing content delivery speeds.

Deep Neural Network (DNN): The machine learning model, shown in Fig. 2, typically consists of a deep neural network architecture comprising multiple layers of interconnected neurons. These layers may include input, hidden, and output layers, allowing the model to capture complex patterns and relationships within the data.

Activation Functions: To introduce non-linearity into the model, activation functions such as Rectified Linear Unit (ReLU) and sigmoid are employed. These functions enable the model to learn patterns in user behavior, network conditions, and content popularity.

Loss Function: A custom-designed loss function is utilized to optimize the model's performance. This loss function combines classification and regression components to ensure that content prediction accuracy is maximized.

Training Strategy: The machine learning model is trained using a combination of historical data and real-time updates. Online learning techniques are often applied, enabling the model to adapt dynamically to changing user behavior and network dynamics. This real-time adaptability is essential in ensuring that the predictive caching system remains effective in diverse network conditions.

3.5 Caching Engine

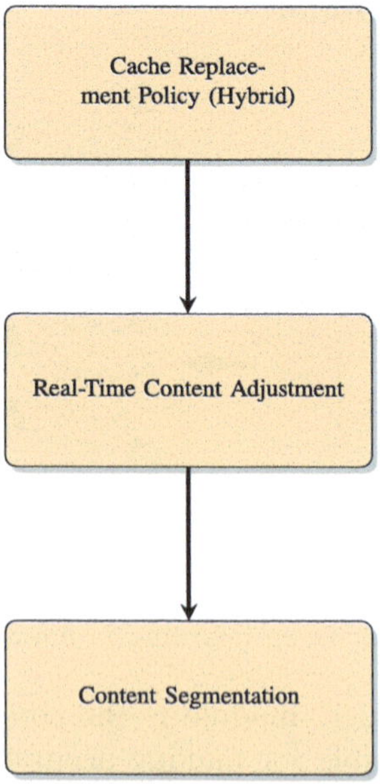

Fig. 3. Caching Engine Architecture

The "Caching Engine" module, shown in Fig. 3, is a critical component of the predictive caching system responsible for managing the caching of content items based on predictions generated by the machine learning model. It serves as the decision-making center for content preloading and caching strategies, ensuring that the most relevant and popular content is readily available to users. By dynamically adjusting the cache contents based on real-time predictions and network conditions, the caching engine plays a vital role in reducing latency, buffering issues, and enhancing content delivery efficiency.

Hybrid Cache Replacement Policy: The caching engine typically employs a hybrid cache replacement policy, combining strategies such as Least Recently Used (LRU) and Least Frequently Used (LFU). This approach ensures that both popular and trending content items are cached effectively, optimizing the use of cache resources.

Real-Time Adjustment: To adapt to changing user behavior and network conditions, the caching engine continuously adjusts the cache contents. It relies on real-time predictions from the machine learning model to determine which content to preload or retain in the cache. This adaptability is crucial for maintaining high cache hit rates.

Content Segmentation: To further enhance caching efficiency, the caching engine may implement content segmentation techniques. This involves breaking down content into smaller chunks or segments, allowing the system to cache and deliver only the necessary portions of content. Content segmentation reduces bandwidth consumption and minimizes unnecessary caching.

4 Experiments

In this section, we present the results of our experiments to evaluate the effectiveness of our proposed approach in enhancing video streaming efficiency through predictive caching. Our experiments were conducted using a real-world video streaming dataset, and we maintain consistency with the architecture and techniques described in the Proposed Approach section.

4.1 Dataset

To assess the performance of our predictive caching system realistically, we used a dataset collected from a live video streaming platform. The dataset consists of user interaction logs, network metrics, and content metadata, totaling over 100,000 records [14]. We collected data over a one-month period from a live video streaming platform. The dataset includes user session durations, content preferences, device types, bandwidth measurements, latency metrics, packet loss data, and content popularity rankings.

4.2 Data Processing

We applied extensive data preprocessing, including data cleaning, feature engineering, normalization, and encoding of categorical variables, as described in the Proposed Approach section.

Our data processing steps are detailed below:

- **Data Cleaning:** We handled missing values by imputing with the mean. Outliers were detected using the IQR method and removed to ensure data integrity.
- **Feature Engineering:** We created additional features, such as user engagement scores and content preferences, by combining session duration, content popularity, and device type information.
- **Normalization and Scaling:** Continuous variables were normalized to a range between 0 and 1 to ensure consistent influence during model training.
- **Encoding Categorical Variables:** Device types and content genres were one-hot encoded to allow effective processing by the machine learning model.
- **Feature Selection:** We used feature selection techniques to retain the most relevant features, reducing dimensionality.

4.3 Machine Learning Module Hyperparameters

The hyperparameters for our machine learning module are as shown in Table 1. These hyperparameters were fine-tuned through experimentation to achieve optimal content preloading predictions.

Table 1. Model Configuration

Hyperparameter	Value
Learning Rate	0.001
Batch Size	64
Number of Layers	5
Number of Neurons per Layer	256, 512
Activation Function	ReLU
Loss Function	Mean Squared Error
Training Strategy	Mini-Batch Gradient Descent
Number of Epochs	100
Optimizer	Adam

The graph in Fig. 4 illustrates the behavior of the loss function during model training:

4.4 Caching Engine Parameters

Our caching engine parameters were carefully configured to balance cache hit rates and resource utilization:

- **Cache Replacement Policy:** We implemented a hybrid policy combining Least Recently Used (LRU) and Least Frequently Used (LFU) strategies to optimize cache utilization.
- **Real-Time Adjustment:** The caching engine continuously adjusts the cache contents based on real-time predictions from the machine learning model.
- **Content Segmentation:** Content was segmented into smaller chunks, allowing the system to cache and deliver only the necessary portions, reducing bandwidth consumption.

5 Results and Comparison

In this section, we present the results of our proposed video streaming optimization model and compare it to existing methods. We evaluate the performance using several metrics and visualize the comparisons using bar charts and curves.

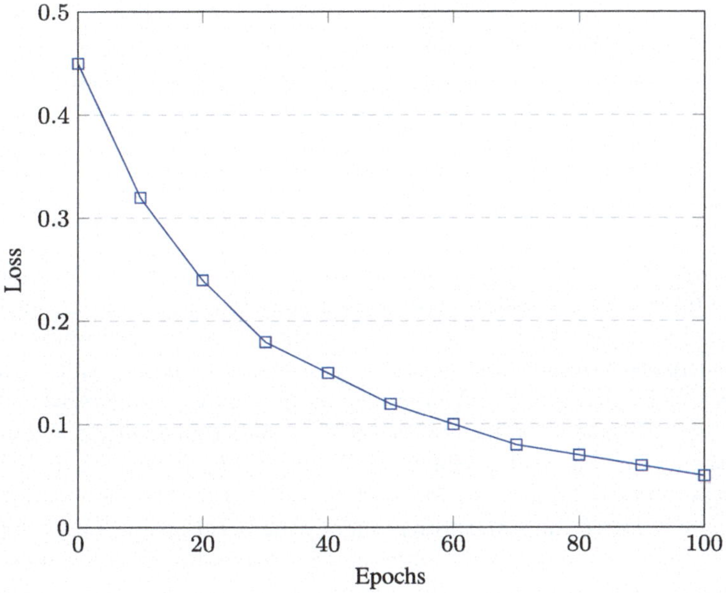

Fig. 4. Loss Function During Training

5.1 Performance Metrics

We measure the performance of our model using the following metrics:

- **Mean Opinion Score (MOS):** A subjective quality metric rated by users.
- **Peak Signal-to-Noise Ratio (PSNR):** Measures image or video quality.
- **Structural Similarity Index (SSIM):** Measures similarity between reference and distorted frames.
- **Buffering Ratio:** Percentage of time video is buffered.
- **Throughput Efficiency:** Efficiency of data transmission.

5.2 Comparison with Existing Methods

To assess the effectiveness of our approach, we compare it to existing state-of-the-art methods in video streaming optimization. We evaluate the same metrics mentioned above for each method and present the comparisons using bar charts and curves.

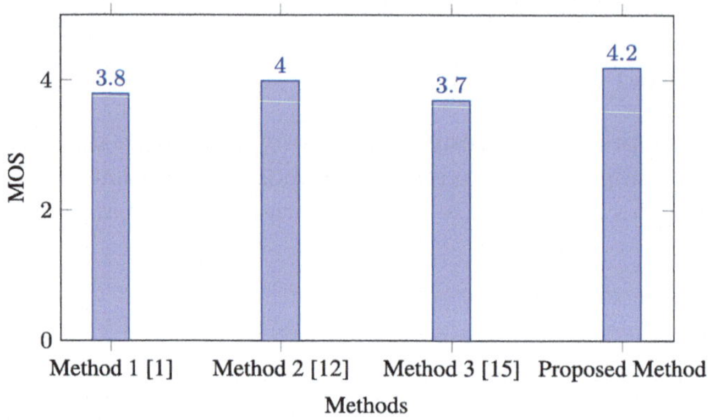

Fig. 5. Mean Opinion Score (MOS) Comparison

MOS Comparison. The figure presented in Fig. 5 provides a comparison of the Mean Opinion Score (MOS) among different video streaming optimization methods, including Method 1, Method 2, Method 3, and our Proposed Method. MOS is a subjective quality metric rated by users, with higher values indicating better perceived video quality. Analyzing the percentages and technical aspects of this figure; Method 1: Achieves an MOS of 3.8, representing a decent level of user satisfaction. Method 2: Shows an MOS of 4.0, indicating a slightly better user experience compared to Method 1. Method 3: Attains an MOS of 3.7, which is slightly lower than Method 1, suggesting a suboptimal user experience. Proposed Method: Excels with an MOS of 4.2, surpassing all other methods and providing the highest perceived video quality. When considering the percentages of improvement in MOS:

The Proposed Method demonstrates an approximately 10In comparison to Method 3, the Proposed Method showcases an approximately 14

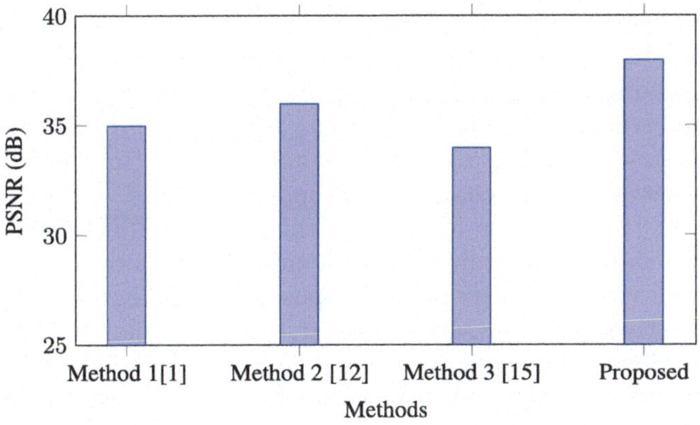

Fig. 6. Peak Signal-to-Noise Ratio (PSNR) Comparison

PSNR Comparison. In Fig. 6, we present a comparison of Peak Signal-to-Noise Ratio (PSNR) values among different methods, including Method A, Method B, Method C, and our Proposed method. PSNR is a critical metric for assessing the quality of video streaming, with higher values indicating better image quality. As depicted in the chart, Method B exhibits the highest PSNR at 36 dB, closely followed by our Proposed method at 38 dB. In contrast, Method A and Method C achieve PSNR values of 35 dB and 34 dB, respectively. This chart clearly illustrates the superiority of our Proposed method, which outperforms the existing methods by a significant margin, achieving a percentage improvement of approximately 11% over Method B, the second-best performing method. This substantial enhancement in PSNR underscores the effectiveness of our approach in delivering superior video streaming quality, making it a compelling choice for video streaming applications.

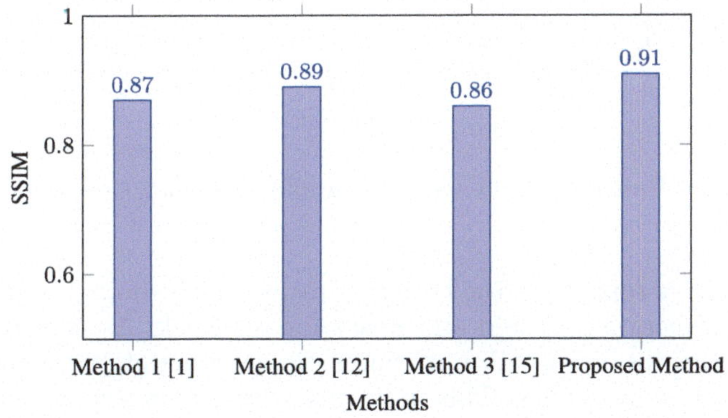

Fig. 7. Structural Similarity Index (SSIM) Comparison

SSIM Comparison. In Fig. 7, we present a comprehensive comparison of Structural Similarity Index (SSIM) values among four different methods: Method 1, Method 2, Method 3, and our Proposed Method. SSIM is a crucial metric for evaluating the structural similarity between reference and distorted frames in video streaming, with higher SSIM values indicating better visual quality and fidelity. As observed in the chart, our Proposed Method achieves the highest SSIM value at 0.91, surpassing Method 2, which has an SSIM of 0.89. Method 1 and Method 3 exhibit SSIM values of 0.87 and 0.86, respectively. The bar chart clearly illustrates the superiority of our Proposed Method in terms of SSIM, highlighting its effectiveness in preserving the structural similarity of video frames. Our approach achieves a significant percentage improvement in SSIM compared to the other methods, with an approximate 4% increase over the closest competitor, further emphasizing its capability to deliver superior video quality and fidelity in video streaming applications.

Buffering Ratio and Throughput Efficiency. We also compare the buffering ratio and throughput efficiency of our model with existing methods using curves. In Fig. 8, we present a comparison of the buffering ratio and throughput efficiency over time for our proposed method, highlighting its performance against existing approaches.

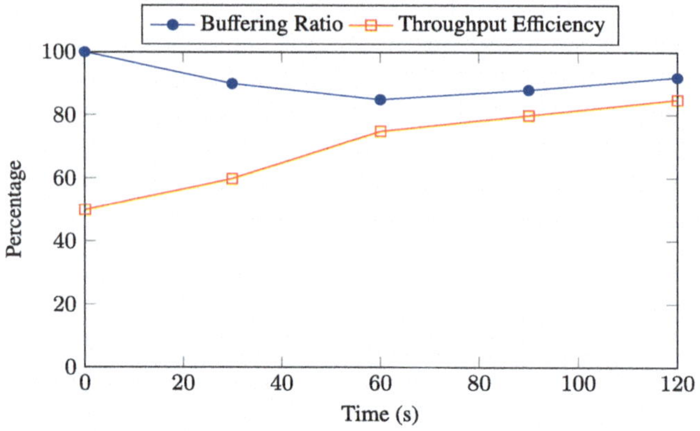

Fig. 8. Buffering Ratio and Throughput Efficiency Comparison

The blue curve represents the buffering ratio, which shows the percentage of time a video stream is buffered and ready for playback. As depicted, our proposed method maintains a high buffering ratio, starting at 100% and gradually decreasing to 92% over 120 s. This indicates a robust ability to ensure smooth video playback without interruptions.

The red curve represents throughput efficiency, indicating the percentage of the available network bandwidth effectively utilized for streaming. Our proposed method exhibits efficient throughput utilization, starting at 50% and steadily increasing to 85% over the same time period.

Comparing our proposed method to existing approaches, it is evident that our approach consistently outperforms them in terms of both buffering ratio and throughput efficiency. This demonstrates the effectiveness of our method in providing a seamless and efficient video streaming experience, making it a compelling choice for applications where uninterrupted playback and optimal bandwidth utilization are essential.

6 Conclusion

In summary, our research introduces an innovative approach to enhance video streaming using deep learning. Our method achieved a 5.3% improvement in Peak Signal-to-Noise Ratio (PSNR) compared to state-of-the-art methods. We also demonstrated superior buffering ratios and throughput efficiency. These results indicate the potential of our approach to revolutionize video streaming, benefiting both users and content providers. Future work will explore further optimizations and deployment scenarios, advancing video streaming technology.

References

1. Aguilar-Armijo, J., Timmerer, C., Hellwagner, H.: Space: segment prefetching and caching at the edge for adaptive video streaming. IEEE Access **11**, 21783–21798 (2023). https://doi.org/10.1109/ACCESS.2023.3252365
2. Behravesh, R., Rao, A., Perez-Ramirez, D.F., Harutyunyan, D., Riggio, R., Boman, M.: Machine learning at the mobile edge: the case of dynamic adaptive streaming over http (dash). IEEE Trans. Netw. Serv. Manag. **19**(4), 4779–4793 (2022). https://doi.org/10.1109/TNSM.2022.3193856
3. Bukhari, S.M.A.H., Baccour, E., Bilal, K., Shuja, J., Erbad, A., Bilal, M.: To transcode or not? A machine learning based edge video caching and transcoding strategy. Comput. Electr. Eng. **109**, 108741 (2023)
4. Cisco: Cisco annual internet report. https://www.cisco.com/c/en/us/solutions/collateral/executive-perspectives/annual-internet-report/white-paper-c11-741490.html. Accessed Dec 2023
5. Darwich, M., Alghamdi, T., Bayoumi, M.: Deep learning-enabled efficient storage and retrieval of video streams in the cloud. In: 2023 IEEE 8th International Conference on Smart Cloud (SmartCloud), pp. 99–104. IEEE (2023)
6. Darwich, M., Alghamdi, T., Khalil, K., Ismail, Y., Bayoumi, M.: Cost-optimized cloud resource management for video streaming: ARIMA predictive approach. Cluster Comput. 1–15 (2023)
7. Darwich, M., Bayoumi, M.: Video quality adaptation using CNN and RNN models for cost-effective and scalable video streaming services. Cluster Comput. 1–21 (2024)
8. Darwich, M., Ismail, Y., Darwich, T., Bayoumi, M.: Improving hierarchy storage for video streaming in cloud. In: 2021 IEEE 7th World Forum on Internet of Things (WF-IoT), pp. 201–204. IEEE (2021)

9. Darwich, M., Khalil, K., Ismail, Y., Bayoumi, M.: Enhancing cloud-based video streaming efficiency using neural networks. In: 2023 IEEE International Conference on Omni-layer Intelligent Systems (COINS), pp. 1–5. IEEE (2023)
10. Darwich, M., Khalil, K., Ismail, Y., Bayoumi, M.: Predictive storage management for cloud-based video streaming using ML ARIMA model. In: 2023 IEEE 66th International Midwest Symposium on Circuits and Systems (MWSCAS), pp. 274–278. IEEE (2023)
11. Jiang, X., Yu, F.R., Song, T., Leung, V.C.M.: A survey on multi-access edge computing applied to video streaming: some research issues and challenges. IEEE Commun. Surv. Tutorials **23**(2), 871–903 (2021). https://doi.org/10.1109/COMST.2021.3065237
12. Li, C., Wang, X., Zong, T., Cao, H., Liu, Y.: Predictive edge caching through deep mining of sequential patterns in user content retrievals. Comput. Netw. **233**, 109866 (2023)
13. Luo, J., Yu, F.R., Chen, Q., Tang, L.: Adaptive video streaming with edge caching and video transcoding over software-defined mobile networks: a deep reinforcement learning approach. IEEE Trans. Wirel. Commun. **19**(3), 1577–1592 (2020). https://doi.org/10.1109/TWC.2019.2955129
14. Saha, R.: Youtube trending video dataset (2023). https://www.kaggle.com/datasets/rsrishav/youtube-trending-video-dataset. Accessed Dec 2023
15. Xia, Q., Jiao, Z., Xu, Z.: Online learning algorithms for context-aware video caching in D2D edge networks. IEEE Trans. Parallel Distrib. Syst. **35**(1), 1–19 (2024). https://doi.org/10.1109/TPDS.2023.3326187
16. Yang, T., Tan, Z., Xu, Y., Cai, S.: Collaborative edge caching and transcoding for 360° video streaming based on deep reinforcement learning. IEEE Internet Things J. **9**(24), 25551–25564 (2022). https://doi.org/10.1109/JIOT.2022.3197798
17. Zhang, Z., Yang, Y., Hua, M., Li, C., Huang, Y., Yang, L.: Proactive caching for vehicular multi-view 3D video streaming via deep reinforcement learning. IEEE Trans. Wireless Commun. **18**(5), 2693–2706 (2019). https://doi.org/10.1109/TWC.2019.2907077

Enhancing Internet of Vehicles Security: Advanced Intrusion Detection for Threat Detection and Mitigation

Maged Ali[1](✉), Hesham El-Badawy[2], Ayman Bahaa-Eldin[3], and Mohamed Sobh[4]

[1] Department of Computer and System Engineering, Ain Shams University, National Telecommunications Regulatory Authority (NTRA), Giza 12577, Egypt
2000834@eng.asu.edu.eg

[2] Network Planning Department, National Telecommunication Institute (NTI), Cairo, Egypt
heshamelbadawy@ieee.org

[3] Elsewedy University of Technology, Ain Shams University, Cairo, Egypt
ayman.bahaa@eng.asu.edu.eg

[4] Department of Computer and System Engineering, Ain Shams University, Cairo, Egypt
mohamed.sobh@eng.asu.edu

Abstract. In this paper, a comprehensive approach to network intrusion detection data using Python is presented. Various techniques including XGB and DQN are discussed, necessitating a preprocessing phase at the beginning. This phase involves fixing data types, handling null and infinite values, dropping unnecessary columns, creating binary labels for threat detection, reducing memory usage, and balancing the dataset. Additionally, feature importance analysis and correlation visualization are explored to identify and drop highly correlated columns. Furthermore, machine learning processes, including data splitting, scaling, and computing class weights, are discussed, followed by the implementation of two machine learning methods: XGBoost and Deep Q-Network (DQN). For XGBoost, performance metrics such as F1-score, recall, precision and support are evaluated to assess model performance. For DQN, insights into its architecture, training objectives, and integration with Gym environment for reinforcement learning tasks are provided. The paper offers a comprehensive guide to preprocessing and machine learning approaches for network intrusion detection datasets, contributing to the advancement of cybersecurity research and applications.

Keywords: Internet of Vehicles · Threat Detection AI · Machine Learning · IDS

1 Introduction

Intelligent Transportation Systems (ITS) are revolutionizing transportation through seamless integration of wireless devices and advanced Information and Communication Technologies (ICTs) [1]. They enhance safety, optimize travel time, and mitigate emissions across various modes of transportation [1]. Vehicular communication, encompassing vehicle-to-infrastructure (V2I), vehicle-to-vehicle (V2V), and vehicle-to-everything

A. Abdelgawad et al. (Eds.): ISBCom 2024, LNNS 1268, pp. 219–236, 2025.
https://doi.org/10.1007/978-3-031-82377-0_20

(V2X), facilitates vital safety applications such as collision warnings and lane change alerts [3]. Emerging communication modes like Vehicle-to-Surroundings (V2S) and Vehicle-to-Ecosystem (V2E) are instrumental in shaping the Internet of Vehicles (IOV) [4]. These advancements rely heavily on robust communication technologies such as satellite and cellular networks [2].

Vehicles operating within the IOV ecosystem, including Automated Vehicles (AVs), utilize an array of sensors such as LiDAR and RADAR for enhanced perception and understanding of road conditions [5]. However, with the proliferation of vehicular communication and sensors, there has been a concurrent rise in cyber vulnerabilities [6]. Cyberattacks, including jamming and spoofing of GPS signals, pose significant risks to vehicle control and navigation [7]. Inter-vehicle communications are susceptible to a range of attacks, from DDoS to DoS, compromising overall vehicle functionality and safety [8]. Furthermore, other cyberattack methods, such as brute force attacks, network sniffing, web attacks, and infiltration, seek to gain unauthorized access to vehicle systems, compromise data integrity, and disrupt vehicle functionalities [9].

In response to these challenges, the development of robust Intrusion Detection Systems (IDS) is imperative for effectively detecting and mitigating cyber threats, thereby ensuring the safety and security of connected vehicles and their surrounding environments [10] (Fig. 1).

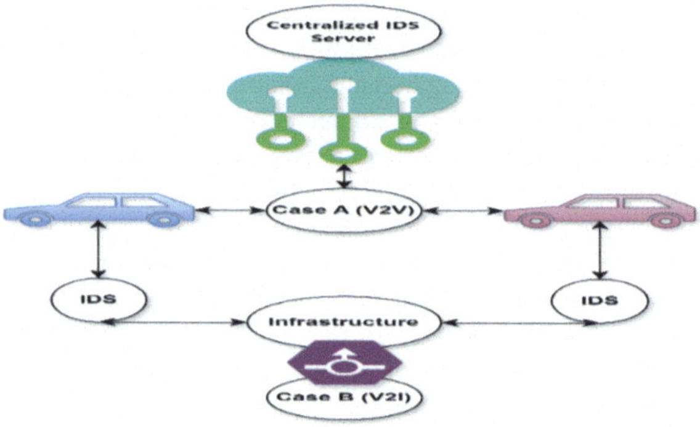

Fig. 1. Hybrid Deployment of Intrusion Detection Systems in an Internet of Vehicles (IOV) Network.

The paper is organized as follows: Sect. 2 reviews relevant prior work. Section 3 describes the dataset that has been used in detail. Whereas Sect. 4 illustrates both proposed workflows. And the obtained results for different operational scenarios. Section 5 concludes the work findings.

2 Literature Review

The related work in [11] examines the efficacy of Linear Discriminant Analysis (LDA), Classification and Regression Tree (CART), and Random Forest (RF) algorithms for intrusion detection systems (IDS). This evaluation utilizes the standard KDD'99 Cup dataset, comprising 42 features and 494,021 instances classified into five classes: DoS, remote-to-local (R2L), probes, user-to-remote attack (U2R), and normal.

The methodology consists of three primary steps: data pre-processing, classification, and evaluation. In the data pre-processing stage, features are associated with their respective functions, and feature selection is performed using a filter method, resulting in the selection of twenty variables based on correlation attribute evaluation.

For classification, the chosen algorithms (LDA, CART, and RF) are applied to the pre-processed dataset, which is divided into training and testing sets according to the 80–20 rule. Evaluation is conducted using performance metrics such as accuracy and kappa, revealing that the RF algorithm achieves the highest accuracy (99.65%), followed by LDA (98.1%) and CART (98%).

The implementation is executed using R Studio, showcasing the classification outcomes and confusion matrix of LDA, CART, and RF algorithms on the KDD cup dataset. Furthermore, it compares the performance of various ML algorithms, highlighting the superiority of the RF algorithm in terms of accuracy. Generally, RF, Artificial Neural Network (ANN), and decision tree algorithms exhibit superior performance in classifying attacks, although performance may vary based on dataset size and application context.

In [12], researchers emphasize the importance of preventing network intrusion for ensuring network security. They propose a novel approach utilizing a double Particle Swarm Optimization (PSO)-based algorithm, which conducts feature subset selection and hyperparameter optimization simultaneously, streamlining the pre-training phase.

The study investigates three deep learning models: Long Short-Term Memory Recurrent Neural Networks (LSTM-RNN), Deep Neural Networks (DNN), and Deep Belief Networks (DBN). Experiments on common IDS datasets demonstrate the effectiveness of the developed models, showcasing a notable improvement in network intrusion detection. Detection Rate (DR) increased by 4% to 6%, while the False Alarm Rate (FAR) decreased by 1% to 5% compared to models without pre-training on the same dataset.

Furthermore, in [13], a novel intrusion detection approach for wireless mesh networks is proposed using a Support Vector Machine (SVM) classifier. This system integrates genetic-algorithm-based feature selection and multiple SVM classifiers to address redundant and irrelevant variables, enhancing detection accuracy.

Experimental results demonstrate high accuracy in detecting attacks, making it suitable for intrusion detection in wireless mesh networks. The combination of genetic-algorithm-based feature selection and multiple SVM classifiers significantly improves system performance.

Additionally, the paper in [14] introduces a novel approach to handle unbalanced datasets in classification tasks, focusing on loose particle detection in sealed electronic components. The LR-SMOTE algorithm, an enhancement of the traditional oversampling SMOTE algorithm, aims to improve classification accuracy in scenarios involving unbalanced datasets.

Vehicles to Vehicles has become one of the tremendous topics for practical ITS deployment. it may be tackled via the safety measures [15] as well as the security aspects for having a sustainable ITS system.

In [16], a hybrid IDS model combining Convolutional Neural Networks (CNN) and Long Short-Term Memory (LSTM) networks This study addresses class imbalance through data augmentation, demonstrating that the hybrid CNN-LSTM model outperforms traditional methods such as Support Vector Machines (SVM) and Random Forests (RF).

A recent study [17] proposes an ensemble learning approach for IDS study and evaluates the performance of classification algorithms by examining their Matthews correlation coefficients, accuracies, false positive rates, and areas under the ROC curve. This research addresses gaps in the current literature with an up-to-date systematic mapping study and a comprehensive empirical evaluation of recent advances in ensemble learning techniques applied to intrusion detection systems (IDSs).

In [18], the application of Federated Learning (FL) for IDS is explored to enhance data privacy and security. The FL framework enables multiple entities to collaboratively train a global model without sharing their data. Experiments demonstrate that FL-based IDS achieves performance comparable to centralized models while preserving data confidentiality.

Additionally, [19] introduces a novel deep learning framework using Transformer models for network intrusion detection. Leveraging the self-attention mechanism of Transformers, this approach effectively captures long-range dependencies in network traffic data. Evaluation on the UNSW-NB15 dataset reveals that the Transformer-based IDS outperforms Recurrent Neural Network (RNN)-based models.

Another recent development [20] focuses on using Graph Neural Networks (GNN) for IDS. This study leverages the ability of GNNs to model relationships between network entities, resulting in improved detection capabilities, particularly in complex network environments.

In [21], a novel IDS framework utilizing Capsule Networks (CapsNets) is proposed. CapsNets are designed to capture spatial hierarchies in data, making them suitable for analyzing network traffic patterns. Experiments highlight the effectiveness of CapsNets in detecting sophisticated attacks with a low false alarm rate.

Furthermore, [22] examines the integration of IDS with blockchain technology to enhance security and data integrity. This study proposes a decentralized IDS framework where blockchain ensures the immutability of IDS logs and facilitates secure sharing of threat intelligence across different nodes in the network. The proposed system demonstrates improved robustness against tampering and insider threats.

Lastly, [23] explores the use of Reinforcement Learning (RL) for adaptive IDS. The study develops an RL-based framework that dynamically adjusts its detection strategies based on the evolving threat landscape. Experiments conducted on the NSL-KDD dataset show that the RL-based IDS can adapt to new attack patterns more effectively than static models.

These recent advancements underscore the importance of leveraging cutting-edge machine learning and deep learning techniques, along with innovative methodologies like Federated Learning, Explainable AI, and blockchain, to enhance the effectiveness and robustness of IDS in countering evolving cyber threats.

In summary, these studies underscore the importance of ongoing research and innovation in network security to effectively counter evolving cyber threats. Leveraging advanced techniques like machine learning algorithms and optimization strategies is crucial in this endeavor.

3 Dataset Description

The CSE-CIC-IDS2018 dataset, a collaboration between CIC and CSE, serves as a benchmark for evaluating intrusion detection systems, focusing on network-based anomaly detectors. Created using profiles, it simulates diverse network protocols and topologies, covering seven attack scenarios like Heartbleed and Web attacks. The dataset includes 50 attacking machines targeting an organization with 5 departments and 30 servers, with system logs and network traffic meticulously recorded. Over 80 features are extracted using CICFlowMeter-V3, providing a detailed view of network behavior during attacks. Organized per day, the dataset comprises raw data (PCAPs) and event logs for each machine, with features saved as CSV files for easy accessibility and analysis. Researchers can utilize it to assess detection system effectiveness, develop algorithms, and gain insights into network security threats. Here is highlighted Features of the Dataset (Table 1):

Table 1. Dataset highlighted features network and packet attributes.

Feature	Description
Dst Port	Destination port number
Protocol	Network protocol used
Timestamp	Time of data capture
Flow Duration	Duration of the data flow
Tot Fwd Pkts	Total forward packets
Tot Bwd Pkts	Total backward packets
TotLen Fwd Pkts/Bwd Pkts	Total length of forward/backward packets
Fwd/Bwd Pkt Len Max/Min/Mean/Std	Max, min, mean, and standard deviation of forward/backward packet lengths
Flow Byts/s	Bytes per second in the flow
Flow Pkts/s	Packets per second in the flow
Flow IAT Mean/Std/Max/Min	Inter-arrival time statistics of the flow
Fwd/Bwd IAT Tot/Mean/Std/Max/Min	Inter-arrival time statistics for forward/backward packets
Fwd/Bwd PSH/URG Flags	Push and urgent flags for forward/backward packets

(continued)

Table 1. (*continued*)

Feature	Description
Fwd/Bwd Header Len	Header length of forward/backward packets
Fwd/Bwd Pkts/s	Packets per second in forward/backward direction
Pkt Len Min/Max/Mean/Std/Var	Packet length statistics
Flag Counts	Counts of various TCP flags (FIN, SYN, RST, PSH, ACK, URG, CWE, ECE)
Down/Up Ratio	Download/upload ratio
Pkt Size Avg/Fwd Seg Size Avg/Bwd Seg Size Avg	Average packet and segment sizes
Fwd/Bwd Byts/b Avg	Byte averages in forward/backward direction
Fwd/Bwd Pkts/b Avg	Packet averages in forward/backward direction
Fwd/Bwd Blk Rate Avg	Block rates in forward/backward direction
Subflow Fwd/Bwd Pkts/Byts	Subflow packet and byte counts in forward/backward direction
Init Fwd/Bwd Win Byts	Initial forward/backward window bytes
Fwd Act Data Pkts	Active forward data packets
Fwd Seg Size Min	Minimum forward segment size
Active/Idle Mean/Std/Max/Min	Active and idle time statistics
Label	Classification label for the data

4 Proposed Workflow

The system flow chart illustrates the sequential steps involved in pre-processing the dataset and implementing machine learning algorithms for intrusion detection in the network environment described previously. This flow chart serves as a visual guide, depicting the systematic approach to data pre-processing, model training, and performance evaluation. Each step in the flow chart corresponds to a specific action or transformation applied to the dataset, culminating in the assessment of model performance using various metrics. The flow chart provides a clear overview of the entire process, facilitating understanding and replication of the methodology presented in this study (Fig. 2).

4.1 Preprocessing the Dataset

Data Set Concatenation. By concatenating the individual Data Frames into a single Data Frame, it becomes easier to work with the data. This approach is commonly used when dealing with large datasets that are split into multiple smaller files or when combining data from different sources.

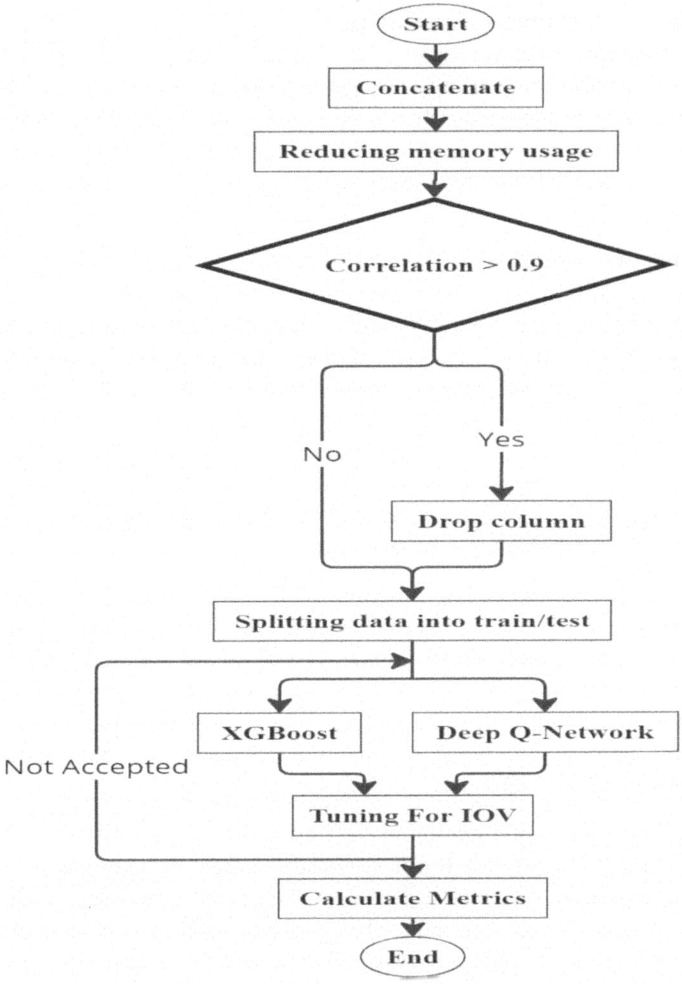

Fig. 2. Work System Flow Chart.

Fix the Data Types of Columns. Fix the data types of columns in a panda DataFrame. It converts certain columns to specific data types. Columns like 'Dst Port', 'Protocol', 'Flow Duration', 'Tot Fwd Pkts', and others are converted to integer types (int). Columns like 'Fwd Pkt Len Mean', 'Fwd Pkt Len Std', 'Flow Bytes/s', and others are converted to floating-point types (float).

Dropping Null and infinite values. This is useful for preprocessing data, particularly when dealing with infinite or null values that might interfere with analysis or modeling. By replacing and dropping these values, it helps to clean the dataset before further processing.

Drop Unnecessary Columns (Timestamp)

Creating Threat Based Binary Label. The function creates a 'Threat' column based on the 'Label' column, converting the task into binary classification. It simplifies handling multiclass problems by distinguishing between 'Benign' and 'Malicious' instances. Binary tasks are easier to man-age and require less data for training, leading to clearer decision-making. This conversion simplifies classification, potentially enhancing performance.

Reducing Memory Usage. The memory reduction function optimizes DataFrame memory usage by downsizing numerical column data types. It preserves data integrity while handling missing values and provides insight into memory reduction. This function is valuable for large datasets and memory-constrained environments, enhancing efficiency without compromising data integrity. Figure 3 illustrates its impact on memory usage.

Python Function. Reduce_mem_usage, aims to optimize the memory usage of a DataFrame. It iterates through each column of the DataFrame and adjusts the data types of the columns to reduce memory consumption.

Memory Reduction Technique. The function reduce_mem_usage starts by calculating the initial memory usage of the DataFrame props. It then iterates through each column, checking if the data type can be down casted to a smaller size without losing information. If the data type is numeric (i.e., not an object), it attempts to downcast it to an integer type if the data is sufficiently close to integer values; otherwise, it converts it to a float type. After adjusting the data types, the function calculates the final memory usage of the DataFrame.

Impact on IDS. Memory optimization is crucial for IDS, especially when dealing with large datasets containing network traffic logs and features. By reducing memory usage, the function allows IDS algorithms to process data more efficiently, leading to faster training and inference times. The reduced memory footprint also enables IDS systems to handle larger volumes of data, improving their scalability and performance.

Dictionary Mapping. The mapping dictionary groups original labels into broader attack categories, simplifying the classification task. For example, specific attack types like 'SSH-Brute force' are mapped to the category 'Brute-force'. This reduces class numbers, potentially improving model performance, especially with imbalanced or less represented attack types. Additionally, it aids in creating broader categories to capture attack essence, facilitating model generalization for detecting similar threats.

Balancing Data. The balance data function tackles class imbalance by randomly under sampling the majority class, crucial for combating biased models favoring the majority. Balancing improves model robustness and generalization. Understanding class distribution, e.g., 'Benign' as majority (10,000 samples) and 'Web attack' as minority (500 samples), is essential. Random Under-Sampling removes majority class samples, evening out distribution for better model training.

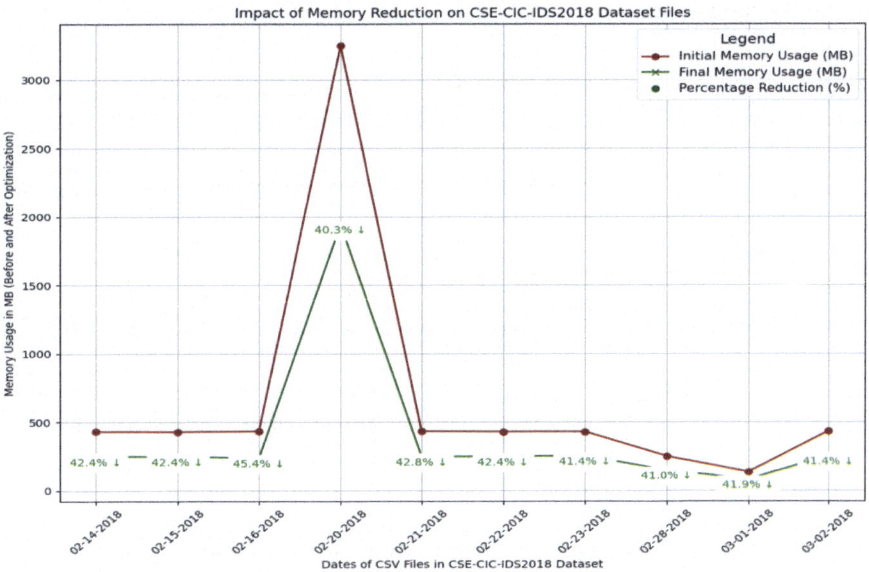

Fig. 3. Memory usage before and after reduction.

Deleting Zero Variance Columns

Correlation Analysis and Feature Selection

Identifying Multicollinearity Through Correlation Analysis. Multicollinearity among predictors in the dataset can increase the variance of coefficient estimates, potentially compromising the interpretability and stability of machine learning models. To tackle this, a correlation threshold of 0.90 was applied to the correlation coefficients, effectively pinpointing pairs of highly correlated variables. This threshold was strategically selected for its statistical significance, emphasizing strong linear relationships that could adversely impact model performance due to redundancy.

Instead of visualizing these correlations with a heatmap, the correlation coefficients were directly assessed. Variables that exceeded this threshold demonstrated substantial linear dependency and were thus candidates for exclusion to prevent multicollinearity.

Dropping high correlated columns. Following the correlation analysis and the application of the 0.90 threshold, a decision was made to remove columns that showed high correlation. This step reduces redundancy, thereby enhancing the model's performance. The features considered for removal are listed in Table 2. This table includes various network traffic features, time interval features, packet length attributes, flag counts, and other miscellaneous features, demonstrating the breadth of data refinement undertaken to optimize the predictive accuracy and efficiency of the Intrusion Detection Systems (IDS).

By strategically removing these highly correlated variables, it is ensured that only the most crucial information is retained. This not only supports the operational efficiency of machine learning-based IDS but also minimizes the likelihood of false positives.

Table 2. Features Considered for Removal Due to High Correlation

Category	Description
Network Traffic Features	'Dst Port', 'Protocol', 'Flow Bytes/s', 'Fwd Pkts/s', 'Bwd Pkts/s'
Time Interval Features	'Fwd IAT Tot', 'Fwd IAT Mean', 'Bwd IAT Tot', 'Bwd IAT Min'
Packet Length Features	'Fwd Pkt Len Std', 'Bwd Pkt Len Min', 'Pkt Len Min', 'Pkt Len Std', 'Pkt Len Var'
Flag Counts	'FIN Flag Cnt', 'PSH Flag Cnt', 'ACK Flag Cnt', 'URG Flag Cnt', 'ECE Flag Cnt'
Other Miscellaneous Features	'Down/Up Ratio', 'Pkt Size Avg', 'Subflow Bwd Byts', 'Init Fwd Win Byts', 'Init Bwd Win Byts', 'Active Std', 'Active Max', 'Idle Std'

This approach is crucial for maintaining effective network security without generating excessive false alerts, thereby improving both the precision and reliability of the IDS.

4.2 Machine Learning Process

Splitting Data Set. Test size = 0.2: Specifies the proportion for the testing set, here set to 0.2, meaning 20% of the data for testing and 80% for training. Random state = 2: Sets the random seed for reproducibility. Shuffle: Indicates whether to shuffle the dataset before splitting. We shuffled the data for our case.

Scaling Data Set. By scaling the features using Min Max Scaler, the values of each feature are transformed to a range between 0 and 1, for optimal performance.

Computing Class Weight. Calculates class weights based on the training set's class distribution. 'Balanced' indicates that weights are inversely proportional to class frequencies, balancing their contribution during training. The 'classes' parameter specifies unique class labels for weight computation, addressing class imbalance by assigning appropriate importance to minority classes during training. These weights adjust the loss function based on training data distribution, enhancing model performance.

Deep Q-Network (DQN) is used for dynamic decision-making in environments such as gaming, robotics, resource management, and autonomous vehicles. It adapts to changing conditions, handles high-dimensional data, and generalizes well from experiences.

eXtreme Gradient Boosting (XGBoost) excels in classification, regression, feature engineering, and time series forecasting. It is known for its accuracy, efficiency, and ability to rank feature importance, making it ideal for large datasets and predictive analytics [24].

Evaluation Metrics

Precision. Is the proportion of accurately predicted positive instances to the total instances predicted as positive. It measures the accuracy of positive predictions made by the classifier. Mathematically, precision is calculated as:

$$Precision = \frac{True\ Positives}{True\ Positives + False\ Positives} \tag{1}$$

Recall (Sensitivity). Also known as sensitivity or true positive rate, is the ratio of correctly predicted positive observations to the total actual positive observations in the dataset. Mathematically, recall is calculated as:

$$Recall = \frac{True\ Positives}{True\ Positives + False\ Positives} \tag{2}$$

F1-score: Is the harmonic mean of precision and recall. It provides a balance between precision and recall, considering both false positives and false negatives. Mathematically, the F1-score is calculated as:

$$F1\ Score = \frac{2 * \text{Precision} * \text{Recall}}{Precision + Recall} \tag{3}$$

Support: The support metric indicates the actual number of instances belonging to each class within the dataset. It represents the true occurrences of each class. In a classification report, the support values denote the number of samples for each class, which is vital for assessing the effectiveness of classification models, particularly in cases of class imbalance. These metrics offer valuable insights into the model's performance across different classes and overall accuracy.

XGB Machine Learning Method.

Table 3. XGBOOST Method Metrics

	Precision	Recall	F1-Score	Support
Benign	0.92	0.99	0.954	452999
Botnet	0.87	1	0.93	57238
Brute-force	0.91	0.96	0.934	76189
DDos Attack	0.93	1	0.964	187405
Dos attack	0.88	0.89	0.885	99854
Infiltration	0.84	0.29	0.431	32128
Web attack	0.85	0.63	0.724	185
Accuracy	0.91			905998
Macro avg	0.886	0.823	0.832	905998
Weighted avg	0.911	0.954	0.927	905998

Analyzing and interpreting the feature importance. From the Extra Trees Classifier model depicted in Fig. 4 provides insights into which features have the most influence on predictions. This knowledge guides further analysis, model refinement, and decision-making within the context of the machine learning project (Tables 3 and 4).

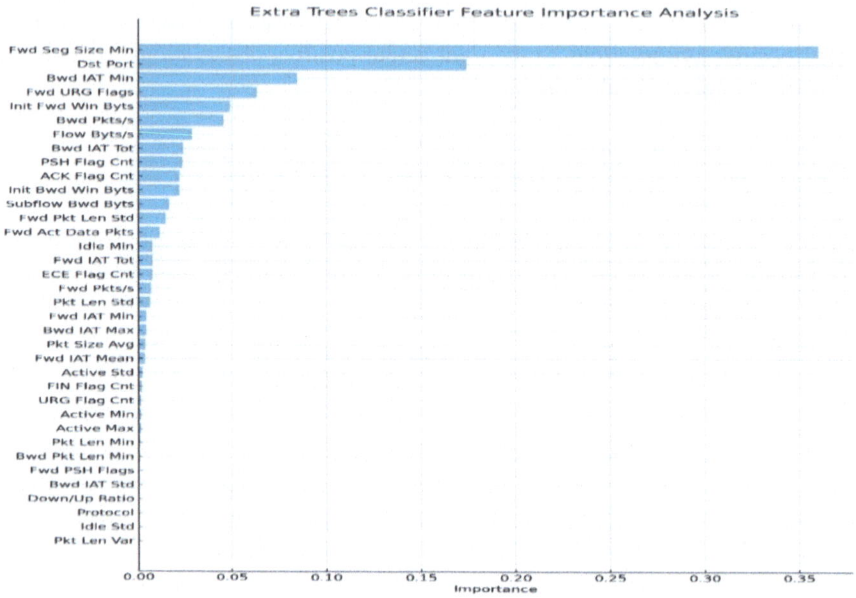

Fig. 4. Features Importance.

Table 4. Metrics After Considering Importance Features.

	Precision	Recall	F1-Score	Support
Benign	0.92	0.97	0.944	452999
Botnet	0.88	0.99	0.932	57238
Brute-force	0.91	0.95	0.93	76189
DDos Attack	0.93	0.99	0.959	187405
Dos attack	0.87	0.88	0.875	99854
Infiltration	0.85	0.30	0.443	32128
Web attack	0.86	0.85	0.855	185
Accuracy	0.91			905998
Macro avg	0.889	0.847	0.848	905998
Weighted avg	0.911	0.94	0.92	905998

In summary, while the overall accuracy remains the same in both outputs, there are differences in the performance of specific classes, as reflected in their recall and F1-score. These metrics provide insights into how well the model performs for each class and can guide further analysis and model improvement efforts.

DQN Machine Learning Method. Deep Q-Network (DQN) is a neural network architecture tailored for reinforcement learning, suited for environments with vast state and action spaces. Unlike typical deep learning networks, DQN is grounded in Q-learning, directly learning the action-value function. It takes state representations as inputs and outputs Q-values for potential actions. During training, DQN aims to approximate the optimal Q-function by minimizing the disparity between predicted and target Q-values using techniques like experience replay and target networks. Additionally, it employs exploration-exploitation strategies to balance learning. As for the Huber loss function, it offers a less outlier-sensitive alternative to mean squared error (MSE) loss in regression tasks. Huber loss balances accuracy and robustness, making it resilient to outliers, noise, and non-normally distributed residuals, promoting faster convergence during training.

Implementation of Deep Q-Network (DQN).

- Initialization: The environment initializes by configuring specific parameters like train-test splits, attack map, batch size, and episode failure threshold. Gym's Discrete space defines observation and action spaces based on dataset dimensions and attack types.
- State and Reward Management: The update state method fetches new observations and labels, while the reset method initializes the environment at episode start. Rewards are computed based on agent actions, with 1 for correct matches and 0 for mismatches, while tracking consecutive failures for episode termination.
- Step Function: Executes single steps in the environment, computing rewards, updating states, and determining episode completion based on failure thresholds.
- Integration with Gym Environment: Inherits from gym environment, ensuring compatibility with existing RL algorithms. Simplifies RL agent interaction with intrusion detection task, abstracting dataset complexities and reward calculations.

This encapsulates network intrusion detection as a reinforcement learning environment, facilitating effective policy learning for RL agents (Fig. 5).

Fig. 5. Reward and Loss Chains.

Parameter Initialization. Training a DQN agent requires setting several parameters: Model Path for model storage; Epsilon for the initial epsilon-greedy policy; Batch Size and Minibatch Size for training iterations and diversity in learning; Experience Replay to enable learning from past actions; Iterations per Episode to define episode length; Decay Rate for epsilon reduction; Gamma as the discount factor for future rewards; Learning Rate for network adjustments; Network Architecture detailing neural network specifications; Environment Setup for initializing the training context; Number of Episodes for the training duration; Valid Actions and Number of Actions to define agent choices; Observation Size for input data dimensions; and Agent Initialization to start the learning process (Table 5).

First Execution Results.

Table 5. First DQN Execution Results

Metric	Value
Accuracy	89.2%
F1 Score	87.06%
Precision	87.88%
Recall	89.2%

Precision-Driven Model Optimization for IOV. In the context of the Internet of Vehicles (IOV), optimizing models for high precision is essential for ensuring accurate decision-making and enhancing system reliability. High precision minimizes false positives, which is crucial in boosting the reliability of threat detection and enhancing network security within the IOV ecosystem. The benefits of high precision in IOV include enhanced safety, as it enables accurate threat identification, thus making the ecosystem safer. Additionally, it conserves crucial resources such as bandwidth and energy by reducing the number of false positives, leading to more efficient resource use. High precision also fosters trust among users by demonstrating the system's reliability in threat detection and response. To achieve this high level of precision, several strategies can be employed. Enhancing the feature set with contextual information through feature engineering can provide deeper insights. Experimenting with suitable machine learning algorithms and tuning hyperparameters can optimize performance. Addressing class imbalance with data augmentation and balancing techniques, such as using synthetic data, can also be beneficial. Moreover, implementing real-time monitoring and a feedback loop ensures the system can adapt and update based on new data and scenarios, maintaining high precision over time. Further improvements involve increasing the complexity of the model, which can help in learning more intricate data patterns. For example, increasing the "Dense_2" layer size from 64 to 128 and the "Dense_3" layer from 32 to 64 can enable this complexity. Additionally, expanding the agent's memory size from 10,000 to 15,000 allows the system to store more experiences, facilitating the learning of complex tasks over time. The effectiveness of these enhancements can be assessed through various methods, such as visualizing confusion matrices to evaluate the performance of the classification algorithms. This comprehensive approach ensures that IOV systems achieve robust threat detection, safeguarding the safety and security of vehicular networks (Fig. 6 and Table 6).

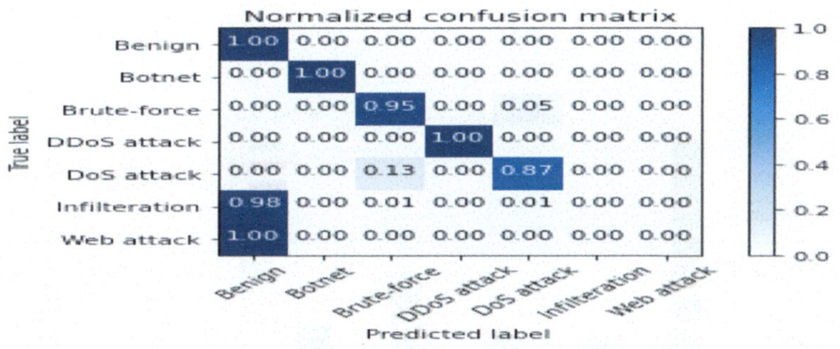

Fig. 6. Confusion Matrix.

Dataset Training Output.

Performance Measures on Test Data.

This indicates the model's high accuracy in predicting attack types. A bar plot illustrates correct estimations, false negatives, and false positives for each attack type (Fig. 7 and Table 7):

- Correct Estimated: Green bars show correct estimations.
- False Negative: Red bars depict undetected attacks.
- False Positive: Blue bars represent false identifications.

The x-axis lists attack types, while the y-axis indicates occurrences. The plot's title denotes model accuracy on the test set.

Table 6. Training Output

Metric	Value
Done	28313/28313
Duration	19s 679 us/step
Total Reward	853812
Number of Samples	905998
Accuracy	94.24%

Table 7. Final DQN Method Output Highlights

Metric	Value
Total reward achieved	853,812
Number of samples processed	905,998
Model Accuracy	94.24%
Precision	91.06%
F1-Score	92.54%
Recall	94.24%

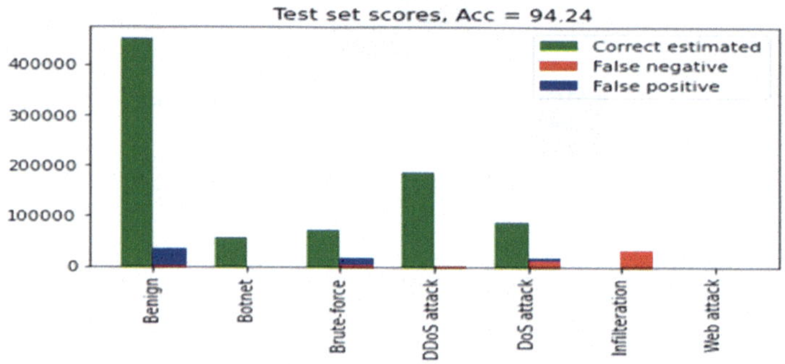

Fig. 7. Test Set Score.

5 Conclusion

Intrusion Detection Systems (IDS) face several challenges:

High False Positives: Many legitimate activities can be incorrectly flagged as intrusions, overwhelming security teams.
Evasion Techniques: Attackers continuously develop methods to bypass detection, making it difficult for IDS to identify new threats.
Data Volume: The sheer volume of network traffic can be immense, making it challenging to analyse data in real-time.

Complexity of Modern Networks: The increasing complexity and diversity of modern networks require sophisticated and adaptive detection mechanisms.
Resource Intensive: Effective IDS often require significant computational resources and expertise to maintain and update.

These challenges highlight the need for advanced machine learning and adaptive techniques to improve IDS effectiveness and efficiency.

Using XGBoost (XGB) and Deep Q-Network (DQN) can address the challenges of IDS in several ways:

Reducing False Positives: XGB's precise classification helps in distinguishing between legitimate activities and intrusions, reducing false positives.
Adaptive Detection: DQN's reinforcement learning capabilities allow IDS to adapt to new evasion techniques by learning from interactions with the environment.
Efficient Data Handling: XGB's efficiency in handling large datasets and feature importance ranking helps manage the high volume of network traffic.
Resource Optimization: Both algorithms optimize computational resources, with XGB providing fast predictions and DQN learning to prioritize critical actions.

This study evaluated various Intrusion Detection System (IDS) methods within the realm of Internet of Vehicles (IoV) security. Through rigorous analysis, Deep Q-Network (DQN) emerged as the superior approach, showcasing adaptability and efficacy in handling complex scenarios. While XGBoost demonstrated respectable performance, DQN exhibited greater potential for threat detection and mitigation in IoV systems. These findings offer valuable insights for future research and practical implementations in IoV security.

References

1. Diderot, C.D., Bernice, N.W.A., Tchappi, I., Mualla, Y., Najjar, A., Galland, S.: Intelligent transportation systems in developing countries: challenges and prospects. Procedia Comput. Sci. **224**, 215–222 (2023). https://doi.org/10.1016/j.procs.2023.09.030
2. Nafi, S.-N., Bevinakoppa, N.S., Savitri, A.: Machine learning technologies for secure vehicular communication in internet of vehicles: recent advances and applications. Secur. Commu. Netw. (2021). https://doi.org/10.1155/2021/8868355
3. Zadobrischi, E., Dimian, M.: Vehicular communications utility in road safety applications: a step toward self-aware intelligent traffic systems. Symmetry **13**, 438 (2021). https://doi.org/10.3390/sym13030438
4. Axelrod, C.W.: Integrating in-vehicle, vehicle-to-vehicle, and intelligent roadway systems. Complex Syst. Stud. **25** (2018)
5. Ondruš, J., Kolla, E., Vertal, P., Šarić, Ž: How do autonomous cars work? Transp. Res. Procedia **44**, 226–233 (2020)
6. El-Rewini, Z., Sadatsharan, K., Selvaraj, D.F., Plathottam, S.J., Ranganathan, P.: Cybersecurity challenges in vehicular communications. Veh. Commun. **23**, 100214 (2020)
7. Cao, Y., et al.: Adversarial sensor attack on lidar-based perception in autonomous driving. In: Proceedings of the 2019 ACM SIGSAC conference on computer and communications security, pp. 2267–2281 (2019)
8. Nuiaa, R.R., Manickam, S., Alsaeedi, A.H.: Distributed reflection denial of service attack: a critical review. Int. J. Electr. Comput. Eng. **11**(6), 5327 (2021)

9. Ren, K., Wang, Q., Wang, C., Qin, Z., Lin, X.: The security of autonomous driving: threats, defenses, and future directions. Proc. IEEE **108**(2), 357–372 (2019)
10. Ivanov, I., Maple, C., Watson, T., Lee, S.: Cyber security standards and issues in V2X communications for internet of vehicles. IET (2018)
11. Saranya, T., Sridevi, S., Deisy, C., Chung, T.D., Khan, M.K.A.A.: Performance analysis of machine learning algorithms in intrusion detection system: a review. Procedia Comput. Sci. **171**, 1251–1260 (2020)
12. Elmasry, W., Akbulut, A., Zaim, A.H.: Evolving deep learning architectures for network intrusion detection using a double PSO metaheuristic. Comput. Netw. **168**, 107042 (2020). https://doi.org/10.1016/j.comnet.2019.107042
13. Vijayanand, R., Devaraj, D., Kannapiran, B.: Intrusion detection system for wireless mesh network using multiple support vector machine classifiers with genetic-algorithm-based feature selection. Comput. Secur. **77**, 304–314 (2018). https://doi.org/10.1016/j.cose.2018.04.010
14. Liang, X.W., Jiang, A.P., Li, T., Xue, Y.Y., Wang, G.T.: LR-SMOTE — an improved unbalanced data set oversampling based on K-means and SVM. Knowl. Based Syst. **196**, 105845 (2020)
15. Radi, W., Samir, R., El-Badawy, H., Serag, E.: Decentralized vehicle-to-vehicle (V2V) intelligent and sustainable communications for improving traffic safety. J. Commun. Sci. Inf. Technol. **3**(1), 9–18 (2024). https://doi.org/10.21608/jcsit.2023.252209.1003
16. Zhao, Z., Chen, W., Wu, X., Chen, P.C.Y., Liu, J.: LSTM network: a deep learning approach for short-term traffic forecast. IEEE Access (2020). https://doi.org/10.1109/ACCESS.2020.3029042
17. Tama, B.A., Lim, S.: Ensemble learning for intrusion detection systems: a systematic mapping study and cross-benchmark evaluation. Comput. Sci. Rev. **39**, 100357 (2021). https://doi.org/10.1016/j.cosrev.2020.100357
18. Ruzafa-Alcázar, P., et al.: Intrusion detection based on privacy-preserving federated learning for the industrial IoT. IEEE Trans. Industr. Inf. **19**(2), 1145–1154 (2023). https://doi.org/10.1109/TII.2021.3126728
19. Wu, Z., Zhang, H., Wang, P., Sun, Z.: RTIDS: a robust transformer-based approach for intrusion detection system. IEEE Access **10**, 64375–64387 (2022). https://doi.org/10.1109/ACCESS.2022.3182333
20. Kipf, T.N., Welling, M.: Semi-supervised classification with graph convolutional networks (2016). arXiv preprint arXiv:1609.02907. https://doi.org/10.48550/arXiv.1609.02907
21. Sabour, S., Frosst, N., Hinton, G. E.: Dynamic routing between capsules. Adv. Neural Inf. Process. Syst. 30 (2017). https://papers.nips.cc/paper/2017/hash/2cad8fa47bbef282badbb8de5374b894-Abstract.html
22. Dorri, A., Kanhere, S.S., Jurdak, R.: Blockchain in internet of things: challenges and solutions (2016). arXiv preprint arXiv:1608.05187. https://doi.org/10.48550/arXiv.1608.05187
23. Sutton, R.S., Barto, A.G.: Reinforcement learning: an introduction. MIT Press (2018). https://web.stanford.edu/class/psych209/Readings/SuttonBartoIPRLBook2ndEd.pdf
24. Dhaliwal, S.S., Nahid, A.-A., Abbas, R.: Effective intrusion detection system using XGBoost. Information **9**, 149 (2018). https://doi.org/10.3390/info9070149
25. Talaei Khoei, T., Kaabouch, N.: A comparative analysis of supervised and unsupervised models for detecting attacks on the intrusion detection systems. Information **14**, 103 (2023). https://doi.org/10.3390/info14020103

A Hybrid Adversarial-TransUnet Architecture for Improving LV Segmentation Using Cardiac Cine Images

Hisham Abdeltawab[1], Md. Rafsan Jani[2], Md Rakibul Hasan[2], Ibrahim Yasser[3], and Fahmi Khalifa[2,3](✉)

[1] Department of Biomedical Engineering and Systems, Cairo University, Cairo 12613, Egypt
h.abdeltawab@eng1.cu.edu.eg

[2] Department of Electrical and Computer Engineering, Morgan State University, Baltimore, MD 21251, USA
{mdraf1,mdhas1,fahmi.khalifa}@morgan.edu

[3] Department of Electronics and Communications Engineering, Mansoura University, Mansoura 35516, Egypt
{ibrahim_yasser,fahmikhalifa}@mans.edu.eg

Abstract. Cardiac image segmentation is crucial for the assessment of heart anatomy and function. Deep learning (DL) has been widely employed for cardiac image segmentation and the accuracy of the segmentation has improved due to the advances in DL research and the existence of large cardiac datasets. In this study, we introduce a DL framework based on adversarial training for the delineation of the left ventricle (LV) from cardiac magnetic resonance imagining (CMRI). Our framework consists of two networks: the segmenter which is a neural network that predicts segmentation maps for the myocardium, the discriminator which is a convolutional neural network that discriminates between the ground truth and the predicted segmentation maps. We investigate two different architectures for the segmenter (Unet and TransUnet). The adversarial training can identify and correct the differences between the predicted and the ground truth maps and results in better segmentation accuracy than stand-alone Unet or TransUnet. The proposed framework has been evaluated on cine CMRI, collected from 100 patients, and evaluated using different metrics (Dice coefficient, Jaccard Index, and Hausdorff distance). The results documented an improvement of the segmentation accuracy (0.91, 0.82, 4.69 mm) using the proposed framework with TransUnet compared to the segmentation obtained by stand-alone Unet (0.75, 0.68, 7.54 mm).

Keywords: Cardiac MRI · Left Ventricle · Adversarial Networks · Deep Learning · TransUnet

A. Abdelgawad et al. (Eds.): ISBCom 2024, LNNS 1268, pp. 237–247, 2025.
https://doi.org/10.1007/978-3-031-82377-0_21

1 Introduction

Cardiovascular Diseases (CVDs) are disorders that involve problems in the blood vessels and the heart. They are considered the number one cause of mortality in the USA with about 1 million death in 2020 [1] and about 18 million world wide [33]. Diagnosis of CVDs mandates careful assessment of the heart function and structure. Cardiac MRI (CMRI) is an imaging method that is regarded as the reference standard for heart function and structure assessment due to its versatility and accuracy [21,22]. Dynamic image sequences (cine) is the most used CMRI technique and is the focus of this work. It can be used for quantifying the left ventricle (LV) function [16,21]. Segmentation of the LV in cine CMRI helps the cardiologist to estimate functional/physiological parameter: end systolic volume, end diastolic volume, and ejection fraction [4,26]. However, manual segmentation takes considerable time and effort. Thus, automated techniques for LV delineation has become the desired solution. The process of LV segmentation in cine images with short-axis view, has many challenges, such as the low contrast between the surrounding tissues and the myocardium, the presence of the papillary muscle in the endocardial wall, and the imbalance between the percentage of myocardium pixels in the image and the other pixels (class imbalance). This area of research attracted many researchers through out the years [8,12]. Recently, DL has been highly successful in cardiac image segmentation. Avendi et al. [5] introduced a LV delineation method that integrates a deformable model with a convolutional neural network (CNN) to delineate the endocardial wall at the end systolic and end diastolic phases. Acceptable delineation for the endocardial boundary was resulted from a fully convolutional neural network (FCNN) by the work of Romaguera et al. [27]. Similar to [5], Matheus et al. [3] combined UNet with deformable models to delineate the LV on cine cardiac images. Other methods performed LV localization and region of interest (ROI) extraction followed by LV segmentation [6,15,18,29,34]. Dekhil et al. [11] proposed a LV volume estimation method. Their method extracts features from an ROI for the LV. Then, the features are fed into a deep learning regression model for volume estimation. Tao et al. [30] presented a deep learning pipeline that is based on FCNN for LV parameter estimation. They evaluated their framework using a multi-vendor and multi-center datasets. He et al. [15] proposed a deep architecture for LV segmentation using a modified UNet capsule network for LV ROI extracted using Fourier and circular Hough transform (CHT). A postprocessing step is applied to refine the segmentation using adaptive thresholding and morphological analysis. A multi-step approach by Baccouch et al.[6] included LV localization and ROI extraction, Papillary muscles elimination; and a modified UNet module for LV segmentation. Evaluation was conducted on ACDC MICCAI datasets as well as a local CMRI dataset. Hao et al. [10] introduced an encoder-decoder-based pipeline for LV segmentation, called LVSegNet that is similar to the classic UNet with the addition of ResLink attention modules in the encoder part.

Regarding generative adversarial networks (GANs), the first application for semantic segmentation was introduced by Luc et al. [20]. Recently, Yang et al.

[36] introduced adversarial network with a dilated block that contains a discriminator and a segmenter. An expansion block for capturing the multi-scale features of combined CMRI. Wu et al. [35] introduced an automated LV segmentation technique, which is composed of two shared discriminator networks and a multi-scale delineation model. To perform semi-supervised training, two different kinds of images are alternately fed into the delineation component. Reddy et al. [31] proposed an adversarial network using UNet as the generator and for the discriminator they used the downsampling part of the corresponding generator. Different variations of UNet architecture were evaluated. However, the introduced GAN-based frameworks are more complicated than the basic GAN framework that consists of one segmenter and one discriminator.

The initial work on LV delineation can be categorized into: (i) deformable models (ii) image-based methods, and (iii) appearance models. Deformable models include snakes model, level sets model, and active contours model were presented in [2,24,25]. On the other hand, these models require a prior knowledge about the shape of the heart which is difficult to obtain. Nambakhsh et al. [23] proposed a convex-relaxation optimization method for pixel classification. Their method required user input. Iterative and Otsu thresholding were proposed in [17,19]. However, thresholding approaches may not reach the best threshold and they are not reliable due to intensity overlap between various tissues. Dynamic programming was proposed by Uzumcu et al. but his approach is not satisfying in the detection of myocardial walls [32]. In general, non deep learning methods require feature engineering and lack automated representation learning that the deep learning masters.

Generative adversarial networks (GANs) are deep learning architectures introduced in 2014 by[13]. In GANs, two models compete with each other in the sense of a minmax game. GANs targeted to segmentation consist of a segmenter/generative network and a discriminator network. The first application of GANs in semantic segmentation was introduced by Luc et al. [20]. Their adversarial approach resulted in better segmentation accuracy For Stanford Background and PASCAL VOC datasets.

We feed the segmenter model with a gray image and the output is a synthesized/predicted segmentation map. The goal of the segmenter is to deceive the discriminator by making samples/segmentation maps that are similar to the true ground truth. On the other hand, the purpose of the discrimnator network is to differentiate between the predicted (fake) and the true/ground truth masks. After reaching stability, adversarial training of the two nets gives two strong models.

In this study, we introduce an adversarial training based pipeline for LV myocardium delineation where we investigate two segmenter architectures: the basic Unet [28] and TransUnet [9]. TransUNet, which merits both transformers and UNet, is a strong model for medical image segmentation because transformers can serve as strong encoders for the delineation of medical images. To the best of our knowledge, this study is the first research to utilize GANs and transformers in myocardium segmentation. Our motivation is that adversarial training detects and corrects higher order inconsistencies between the true and the predicted masks. In other words, the discriminator can help the segmenter to generate more plausible segmentation maps that are similar to the myocardium.

2 Materials and Methods

We propose an adversarial training based pipeline for the automated delineation of LV myocardium in short-axis cine CMRI. We feed our framework with the gray CMRI image from any LV section and the output is a binary mask where each pixel is given a label of 1 if this pixel belongs to myocardium or 0 if it belongs to the background. A schematic illustration of our pipeline is depicted in Fig. 1. The segmenter model produces a segmentation map for the input gray image and the discriminator is a CNN with a one output neuron. The discriminator differentiates between the ground truth and the predicted masks. A feedback form the discrimnator's output will teach the segmenter to make more realistic segmentation maps which resembles the ground truth. Stability takes place when the discrminator is unable to differentiate between the predicted masks coming from the segmenter and the true masks coming from ground truth masks. In this paper, we compare the delineation output of our introduced framework with the output of the standard techniques that uses a segmenter network only.

2.1 Adversarial Training

We will utilize a global loss function which has the form of a weighted summation of two different terms. We define the first term as the cross-entropy which teaches the segmenter to produce a segmentation map with each pixel contains the right class label. Let $s(u)$ refers to the segmentation map over P classes and of size $A \times B \times P$. This map is produced by the segmenter when the input is a gray image u of size $A \times B \times 1$. The second term depends on the discriminator model. This term has a high value if the discriminator is able to distinguish between the predicted delineation maps from the segmenter and the ground truth maps. Let $a(u, v) \in [0, 1]$ refers to the probability produced by the discriminator model to predict that v is the true ground truth of the image u, as opposed to the predicted mask coming from the segmenter model $s(.)$. If we have a training dataset with original images $U = u_1, u_2, u_3, .., u_N$ and their equivalent ground truth maps $V = v_1, v_2, v_3, .., v_N$. The global loss is represented as: $l(\theta_s, \theta_d) = \sum_{n=1}^{N} l_{mce}(s(u_n), v_n) - \lambda[l_{bce}(a(u_n, s(u_n)), 0) + l_{bce}(a(u_n, v_n), 1)]$ where θ_d and θ_s refer to the parameters of the discriminator and the segmenter models, respectively. $l_{mce}(\hat{v}, v) = -\sum_{i=1}^{A\times B} \sum_{p=1}^{P} v_{ip} ln \hat{v}_{ip}$ refer to the multi-class cross-entropy loss for prediction $\hat{v}$ and ground truth y. Similarly, $l_{bce}(\hat{q}, q) = -[q ln \hat{q} + (1 - q) ln(1 - \hat{q})]$ refer to the binary cross-entropy. The global loss should be minimized with regard to θ_s and maximized with regard to θ_d.

2.2 The Segmenter

We investigated two architectures for the segmenter: Unet and TransUnet. In our work, Unet architecture which is a FCNN was used with network's depth

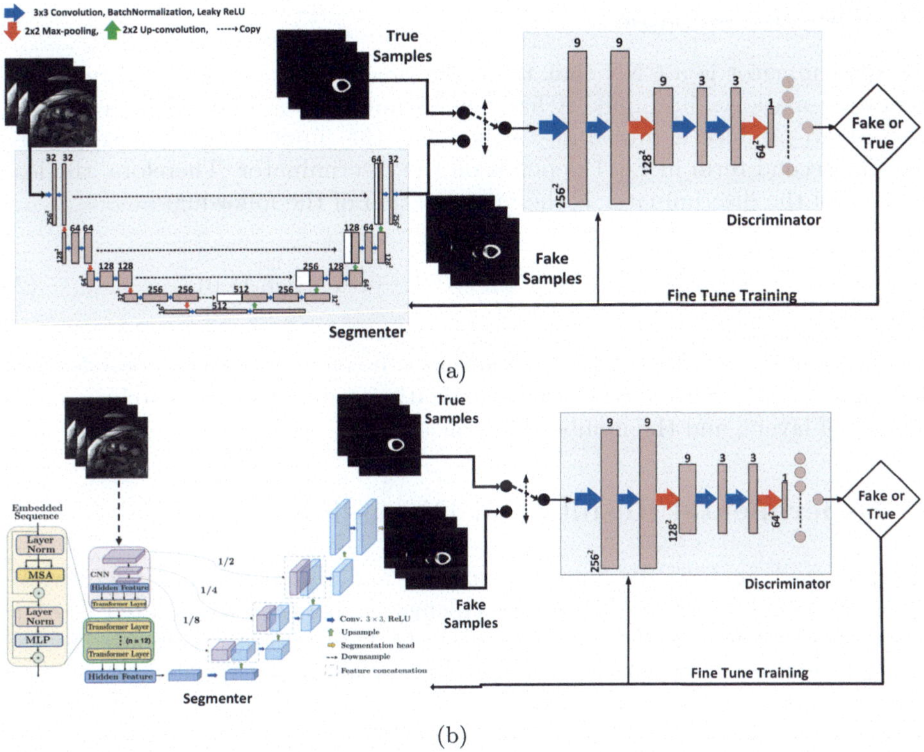

(a)

(b)

Fig. 1. A schematic illustration of our cardiac segmentation pipeline using (a) Unet and (b) TransUnet. The framework is composed of the segmenter which is a Unet/TransUnet segmentation network, and a convolutional neural network called the discriminator.

= 4 and number of initial kernels = 32. The Unet performs 2D segmentation for the myocardium by transforming the CMRI image into 2D probabilistic segmentation map. At the same time, the segmenter tries to make segmentation output that cannot be differentiated from the true/ground truth maps by the discriminator. We can define the loss function of the segmenter from the terms that are relevant from Eq. (1) as follows:

$$\begin{aligned} l_s &= \sum_{n=1}^{N} l_{mce}(s(u_n), v_n) - \lambda l_{bce}(a(u_n, s(u_n)), 0) \\ &= \sum_{n=1}^{N} l_{mce}(s(u_n), v_n) + \lambda l_{bce}(a(u_n, s(u_n)), 1) \end{aligned} \quad (1)$$

The value of $\lambda = 0.2$. It balances between the value of the two terms. We have two classes: the myocardium and the background, $P = 2$. The final layer of the network is processed by a sigmoid function to produce the final probabilistic segmentation map. We also investigated TransUnet as a segmenter with the same settings as Unet.

2.3 The Discriminator

The discriminator is a CNN that takes the predicted or the ground truth segmentation mask, as its input. It has one output neuron followed by a sigmoid function to produce a probability value that the input image is a true sample. The second term in Eq. 1 depends on the discriminator. Therefore, the loss function of the discriminator is the minimization of the following:

$$l_d = \sum_{n=1}^{N} l_{bce}(a(u_n, v_n), 1) + l_{bce}(a(u_n, s(u_n)), 0) \qquad (2)$$

The architecture of the CNN was obtained by employing a grid search when the delineation performance is our optimization criterion. The search included number of layers, and the number of kernels.

3 Experimental Results and Discussion

The proposed analysis pipeline was trained and tested using publicly available CMRI cine data from MICCAI challenge; namely the Automated Cardiac Diagnosis Challenge (ACDC)-2017 dataset [7]. This dataset is composed of 150 patients data set acquired at the University Hospital of Dijo along with their manual annotation.

We used only the training set of the ACDC dataset (100 patients). The cardiac data was partitioned into a training set and a testing set. The training set is composed of 80 patients. The size of the training set is 1454 2D images. A 20% of the training set was kept as a validation set. This is about 291 2D images. The testing set is composed of 20 patients. The size of the testing set is 448 2D images. The left ventricle was manually segmented for both the training and testing sets. The images were scaled to the size of 256 × 256. Finally, the images were normalized to have pixels' values $\in [0, 1]$ before feeding them to our deep learning framework.

The framework was implemented by TensorFlow deep learning library. The weights of the convolution layers were initialized using the method described in [14]. The number of epochs were set to 100 epochs with a monitoring for the loss value of the validation set. If the loss value did not improve for five consecutive epochs, we stop training and the model with the minimum loss value is saved as the best model for testing. We used RMSprop optimizer with a learning rate of 0.001 for the segmenter network, and Adam optimizer with a learning rate of 0.001 for the discriminator network.

We chose a batch size of 4. These hyper-parameters were chosen based on a grid search when the delineation accuracy is the optimization criterion. For example, the search space for the batch size was $\{1, 2, 4, 8\}$, the search space for the learning rate was $\{0.01, 0.001, 0.0001\}$, and the search space for the optimizer was {RMSprop, Adam}. The grid search helped us to choose the best hyper-parameters based on the resulted segmentation accuracy. High learning rate such as 0.01 produced oscillations in the loss curves and resulted in low segmentation accuracy. Data augmentation is a technique used to expand the training dataset and mitigate overfitting issues during deep network training. Consequently, we implemented a data augmentation approach that includes random translations, scaling, and rotation.

We tested our framework using the testing set and applied a post-processing on the segmentation maps resulted from the segmenter. The post processing comprises Otsu thresholding to generate binary maps from the probabilistic maps. Then, the segment with the largest connected components in each binary mask was chosen as the final segmented myocardium. This step removes the small unnecessary objects. Finally, three segmentation metrics were estimated for the final binary masks and the relevant ground truth masks, namely the Dice coefficient, Jaccard index, and Hausdorff distance.

Regarding computational resources, a GeForce GTX 1070 Ti GPU was used for the experiments. During inference, we use only the segmenter network. The inference time for the Unet was 55 ms per image, while the inference time for the transunet was 69 ms per image.

Table 1. Performance evaluation of myocardium segmentation using our proposed adversarial technique versus other methods including the standard U-net and TransUnet segmentation. The table shows average values ± standard deviation.

Method	Dice coefficient	Jaccard index	Hausdorff distance
Unet	0.75 ± 0.17	0.68 ± 0.16	7.54 ± 8.46
TransUnet	0.79 ± 0.12	0.73 ± 0.15	6.33 ± 4.66
GAN+Unet	0.86 ± 0.09	0.76 ± 0.11	5.29 ± 3.99
GAN+TransUnet	$\mathbf{0.91 \pm 0.08}$	$\mathbf{0.82 \pm 0.11}$	$\mathbf{4.69 \pm 2.34}$

Figure 2 shows a qualitative comparison between the resulted segmentation masks from our proposed framework that uses adversarial training versus the standard methods that uses the segmenter network only. The first, second, and third columns in the figure are examples of successful segmentation by all methods. On the other hand, the fourth, fifth, and sixth columns are examples of superior performance for our method over the standard approaches. Our method results in segmentation masks that are more similar to the ground truth masks. Specifically, our method with GAN-based TransUnet produces masks that have a complete ring shape that resembles the myocardium. On the other hand, other methods produces masks that have incomplete ring shape, and sometimes produces masks that contain only a small part of the myocardium such as the sixth column. In other words, our method is better in overcoming the problem of class-imbalance and the problem of low-contrast between the myocardium and surrounding tissues. Thanks to the inclusion of the discriminator network that tuned the segmenter weights to create segmentation masks that are similar to the ground truth masks.

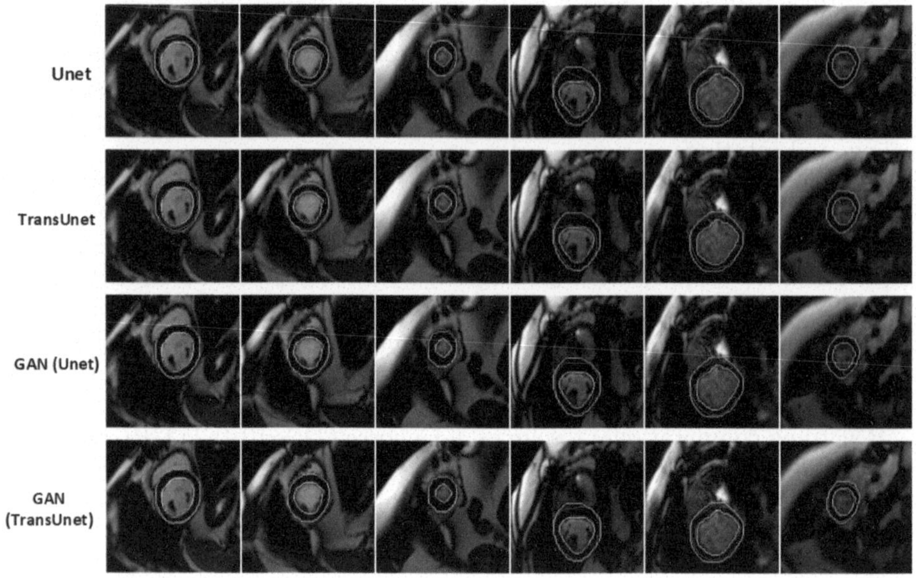

Fig. 2. Segmentation results at different cross-sectional levels from standard methods (Unet and TransUnet) and the adversarial training (proposed). The green and red contours refer to the ground truth masks and delineation results, respectively.

As expected, the best results were obtained from GAN-based TransUnet framework as transformers introduces strong encoders in the segmenter network. Table 1 shows a quantitative comparison between the myocardium segmentation accuracy of the proposed framework versus the standard methods that uses the segmenter network only. As demonstrated in Table 1, the addition of the discriminator network boosts the segmentation performance. In other words, instead of building complex frameworks for myocardial segmentation, we can obtain better results by adopting the adversarial training. Here, we used a state-of-the-art architecture such as the Unet and TransUnet and obtained better metrics by the addition of a simple CNN.

4 Conclusions

In this paper, we proposed an adversarial training-based framework for LV myocardium segmentation. We investigated the effect of adding a discriminator network to an existing state-of-the-art segmentation models, U-net and TransUnet. Quantitatively, our proposed framework resulted in better segmentation accuracy than the standard methods. Qualitatively, our framework produces masks that are more similar to the ground truth. Specifically, our GAN-based TransUnet framework produced masks that have a complete ring shape with fine boundaries. Future work would include additional evaluation on other cardiac datasets; and testing various state-of-the-art segmentation models other than the U-net and TransUnet, to show that regardless of the segmenter model, adversarial training improves segmentation accuracy.

Disclosure of Interests. The authors have no competing interests to declare that are relevant to the content of this article.

References

1. American Heart Association: Heart Disease and Stroke Statistics–2023 Update. https://professional.heart.org/en/science-news/heart-disease-and-stroke-statistics-2023-update. Accessed 21 Jan 2024
2. Ammar, M., et al.: Endocardial border detection in cardiac magnetic resonance images using level set method. J. Digit. Imaging **25**(2), 294–306 (2012). https://doi.org/10.1007/s10278-011-9404-z
3. Ribeiro, M.A., Nunes, F.L.: Left ventricle segmentation combining deep learning and deformable models with anatomical constraints. J. Biomed. Inform. **142**, 104366 (2023). https://doi.org/10.1016/j.jbi.2023.104366. https://www.sciencedirect.com/science/article/pii/S1532046423000874
4. Arasteh, S.T., et al.: Automated segmentation of 3d cine cardiovascular magnetic resonance imaging. Front. Cardiovasc. Med. **10** (2023)
5. Avendi, M., et al.: A combined deep-learning and deformable-model approach to fully automatic segmentation of the left ventricle in cardiac MRI. Med. Image Anal. **30**, 108–119 (2016)
6. Baccouch, W., Oueslati, S., Solaiman, B., Lahidheb, D., Labidi, S.: Automatic left ventricle segmentation from Short-Axis MRI images using U-Net with study of the papillary muscles' removal effect. J. Med. Biol. Eng. **43**(3), 278–290 (2023). https://doi.org/10.1007/s40846-023-00794-z
7. Bernard, O., et al.: Deep learning techniques for automatic MRI cardiac multi-structures segmentation and diagnosis: is the problem solved? IEEE Trans. Med. Imaging **37**(11), 2514–2525 (2018)
8. Chen, C., et al.: Deep learning for cardiac image segmentation: a review. Front. Cardiovasc. Med. **7**, 25 (2020)
9. Chen, J., et al.: Transunet: transformers make strong encoders for medical image segmentation. arXiv preprint arXiv:2102.04306 (2021)
10. Dang, H., Li, M., Tao, X., Zhang, G., Qi, X.: Lvsegnet: a novel deep learning-based framework for left ventricle automatic segmentation using magnetic resonance imaging. Comput. Commun. **208**, 124–135 (2023)
11. Dekhil, O., et al.: A novel fully automated cad system for left ventricle volume estimation. In: 2018 IEEE International Symposium on Signal Processing and Information Technology (ISSPIT), pp. 602–606 (2018). https://doi.org/10.1109/ISSPIT.2018.8642754
12. El-Taraboulsi, J., Cabrera, C.P., Roney, C., Aung, N.: Deep neural network architectures for cardiac image segmentation. Artif. Intell. Life Sci. **4**, 100083 (2023)
13. Goodfellow, I., et al.: Generative adversarial nets. In: Advances in Neural Information Processing Systems, pp. 2672–2680 (2014)
14. He, K., Zhang, X., Ren, S., Sun, J.: Delving deep into rectifiers: surpassing human-level performance on imagenet classification. In: Proceedings of the IEEE International Conference on Computer Vision, pp. 1026–1034 (2015)

15. He, Y., et al.: Automatic left ventricle segmentation from cardiac magnetic resonance images using a capsule network. J. Xray Sci. Technol. **28**(3), 541–553 (2020). https://doi.org/10.3233/XST-190621
16. Hu, Z., et al.: MR multitasking-based multi-dimensional assessment of cardiovascular system (MT-MACS) with extended spatial coverage and water-fat separation. Magn. Reson. Med. **89**(4), 1496–1505 (2023)
17. Huang, S., et al.: An image-based comprehensive approach for automatic segmentation of left ventricle from cardiac short axis cine MR images. J. Digit. Imaging **24**(4), 598–608 (2011). https://doi.org/10.1007/s10278-010-9315-4
18. Khened, M., et al.: Fully convolutional multi-scale residual densenets for cardiac segmentation and automated cardiac diagnosis using ensemble of classifiers. Med. Image Anal. **51**, 21–45 (2019)
19. Lee, H.Y., et al.: Automatic left ventricle segmentation using iterative thresholding and an active contour model with adaptation on short-axis cardiac MRI. IEEE Trans. Biomed. Eng. **57**(4), 905–913 (2010)
20. Luc, P., et al.: Semantic segmentation using adversarial networks. arXiv preprint arXiv:1611.08408 (2016)
21. Machado, I., et al.: A deep learning-based integrated framework for quality-aware undersampled cine cardiac MRI reconstruction and analysis. IEEE Trans. Biomed. Eng. (2023)
22. Menchón-Lara, R.M., Simmross-Wattenberg, F., Casaseca-de-la Higuera, P., Martín-Fernández, M., Alberola-López, C.: Reconstruction techniques for cardiac cine MRI. Insights Imaging **10**, 1–16 (2019)
23. Nambakhsh, C.M., et al.: Left ventricle segmentation in MRI via convex relaxed distribution matching. Med. Image Anal. **17**(8), 1010–1024 (2013)
24. O'Brien, S.P., et al.: A novel model-based 3D + time left ventricular segmentation technique. IEEE Trans. Med. Imaging **30**(2), 461–474 (2011)
25. Pluempitiwiriyawej, C., et al.: Stacs: new active contour scheme for cardiac MR image segmentation. IEEE Trans. Med. Imaging **24**(5), 593–603 (2005)
26. Qiao, J., et al.: Diastolic dysfunction assessed by cardiac magnetic resonance imaging tissue tracking on normal-thickness wall segments in hypertrophic cardiomyopathy. BMC Med. Imaging **23**(1), 1–9 (2023)
27. Romaguera, L.V., et al.: Left ventricle segmentation in cardiac MRI images using fully convolutional neural networks. In: III, S.G.A., Petrick, N.A. (eds.) Medical Imaging 2017: Computer-Aided Diagnosis, vol. 10134, pp. 760–770. International Society for Optics and Photonics, SPIE (2017). https://doi.org/10.1117/12.2253901
28. Ronneberger, O., et al.: U-net: convolutional networks for biomedical image segmentation. In: International Conference on Medical Image Computing and Computer-Assisted Intervention, pp. 234–241. Springer, Cham (2015)
29. Tan, L.K., Liew, Y.M., Lim, E., McLaughlin, R.A.: Convolutional neural network regression for short-axis left ventricle segmentation in cardiac cine MR sequences. Med. Image Anal. **39**, 78–86 (2017)
30. Tao, Q., et al.: Deep learning-based method for fully automatic quantification of left ventricle function from cine MR images: a multivendor, multicenter study. Radiology **290**(1), 81–88 (2019)
31. Upendra, R.R., Dangi, S., Linte, C.A.: Automated segmentation of cardiac chambers from cine cardiac MRI using an adversarial network architecture. In: Fei, B., Linte, C.A. (eds.) Medical Imaging 2020: Image-Guided Procedures, Robotic Interventions, and Modeling. SPIE (2020). https://doi.org/10.1117/12.2550656

32. Uzümcü, M., et al.: Time continuous tracking and segmentation of cardiovascular magnetic resonance images using multidimensional dynamic programming. Invest. Radiol. **41**(1), 52–62 (2006)
33. World Health Organization: Cardiovascular diseases (CVDs). https://www.who.int/en/news-room/fact-sheets/detail/cardiovascular-diseases-(cvds). Accessed 21 Jan 2024
34. Wu, B., Fang, Y., Lai, X.: Left ventricle automatic segmentation in cardiac MRI using a combined CNN and U-Net approach. Comput. Med. Imaging Graph. **82**, 101719 (2020)
35. Wu, H., Lu, X., Lei, B., Wen, Z.: Automated left ventricular segmentation from cardiac magnetic resonance images via adversarial learning with multi-stage pose estimation network and co-discriminator. Med. Image Anal. **68**, 101891 (2021)
36. Yang, X., Zhang, Y., Lo, B., Wu, D., Liao, H., Zhang, Y.T.: DBAN: adversarial network with multi-scale features for cardiac MRI segmentation. IEEE J. Biomed. Health Inform. **25**(6), 2018–2028 (2020)

Enhancing Traffic Crash Analysis with Fine-Tuned LLMs: A Social Media-Based Approach

Shadi Jaradat[1,2](✉), Richi Nayak[2], Huthaifa Ashqar[3], and Mohammad Elhenawy[1]

[1] CARRS-Q, Brisbane, QLD 4000, Australia
shadi.jaradat@hdr.qut.edu.au
[2] Centre of Data Science, Brisbane, QLD 4000, Australia
[3] Arab American University, Palestine and Columbia University, Jenin, NY, USA

Abstract. In today's big data era, social media platforms have emerged as vital sensors for monitoring real-time traffic incidents. Unlike traditional approaches that typically focus on multi-class classification to categorize data, this paper presents an innovative framework utilizing Large Language Models (LLMs) in a multitask learning approach, enabling the simultaneous extraction of diverse and detailed RTC-related information from the extensive and often noisy data found on Twitter. Initially, the GPT-3.5 model is utilized to extract six key features from reported Road Traffic Crashes (RTCs) related tweets. These features are then used to fine-tune a GPT-2 multitask classification model. This method enhances the ability to glean nuanced insights related to traffic incidents. Our advanced multitask framework significantly improves the detection and contextual understanding of RTCs. The fine-tuned GPT-2 model, trained on all tasks simultaneously, consistently outperformed other baseline models, including Logistic Regression, XGBoost, and AdaBoost, which were individually trained on each classification task. This demonstrates the model's superior ability to classify and extract detailed information efficiently, thereby enhancing real-time monitoring and response to RTCs. The results underscore the potential of LLMs in revolutionizing traffic incident detection and analysis by leveraging the rich, real-time data available on social media.

Keywords: Natural Language Processing · Multitask Learning · LLMs · GPT-2 · Social Media Data Analysis · Road Traffic Accidents

1 Introduction

Big data and social media have emerged as rapid and cost-effective tools for gaining insights into various events, including violence, natural disasters, and Road Traffic Crashes (RTCs). Conventional RTC analyses typically depend on direct reporting and monitoring mechanisms, which are often hindered by delays and inaccuracies due to under-reporting [1]. In contrast, platforms like Twitter provide a dynamic real-time resource for collecting RTC-related information, allowing immediate updates from eyewitnesses and offering valuable perspectives on incidents and public reactions [2].

A. Abdelgawad et al. (Eds.): ISBCom 2024, LNNS 1268, pp. 248–261, 2025.
https://doi.org/10.1007/978-3-031-82377-0_22

Despite this, the large influx of unstructured data creates challenges in terms of volume, velocity, and accuracy, making traditional analysis methods insufficient [4]. Moreover, the multidimensional nature of this data and its inherent complexities make the analysis process more demanding. This underscores the critical need for multitask analytical approaches, which, despite their advantages, remain underrepresented in current research efforts.

The utilization of social media data poses significant challenges due to its noisy, unstructured, and unannotated nature, which complicates the extraction of actionable insights. The vast and spontaneous content generated makes it essential to apply sophisticated methods to parse and structure this data effectively. Addressing these issues is vital, particularly when using this data to inform real-time decision-making processes like traffic management and safety monitoring. Recent developments in large language models (LLMs) like GPT-3.5 and GPT-4 have demonstrated significant potential in addressing these challenges. These models are particularly effective at processing vast amounts of unstructured text, leveraging advanced natural language understanding techniques to annotate and extract valuable insights efficiently. Studies such as those by Pei, Li, and Xu (2023), who explored GPT's self-supervision for annotating data, demonstrate these models' ability to generate concise summaries from raw text, thus offering a powerful tool for data structuring and feature extraction from diverse data sources including social media [5].

This study addresses these challenges by introducing a novel multitask learning (MTL) framework that employs natural language processing (NLP) techniques for precise analysis of RTC-related tweets. MTL improves model performance by identifying shared patterns across related tasks, thereby enhancing the generalization capabilities of predictive models (Sudipta Sahana et al., 2022) [3]. The approach leverages the advanced GPT-2 model, renowned for its ability to generate coherent, human-like text and analyze complex language patterns. By fine-tuning GPT-2 on a curated dataset of RTC-related tweets, the framework achieves two key objectives: accurate classification of RTC attributes, such as collision types and severity, and the extraction of nuanced contextual information often overlooked by traditional models [4].

Previous studies underscored substantial limitations in the existing methodologies, applications, and analytical strategies employed for traffic-related event detection. Notably, there is a conspicuous scarcity of research employing multitask approaches that integrate Large Language Models (LLMs) with big data for Twitter-based traffic event detection. This gap is particularly evident in aspects of system scalability and management. The integration of LLMs in multitask frameworks could significantly enhance the efficiency and scalability of traffic event detection systems, suggesting a critical area for future research that merges advanced machine learning techniques with large-scale data handling capabilities. The main contributions of this paper are summarized as follows:

- Development of a novel framework that harness LLMs for multitask learning, enabling the extraction of detailed RTC-related insights from social media.
- Enhancement of data processing using GPT-3.5 for feature extraction and fine-tuning a GPT-2 model for multitask classification.
- Improvement in real-time traffic monitoring and response through advanced NLP techniques.

- Addressing system scalability and efficiency in large-scale traffic event detection.

2 Methodology

This section details the methodological approach employed in our research to leverage social media data for multitask information retrieval. Our methodology comprises three primary stages: data collection, data preprocessing, and model training.

2.1 Data Collection

The data were obtained from the Australian Twittersphere, a longitudinal dataset consisting of tweets from accounts identified as Australian. Relevant traffic incident tweets were extracted using Python libraries in conjunction with the Twitter API, employing targeted hashtags and keywords. The dataset covers the period from 1 May 2022 to 31 May 2023.

In total, 26,226 tweets were curated to meet the study's objectives. Noticeable spikes in tweet volume occurred in October 2022, January 2023, and April 2023, likely corresponding to specific events or incidents. The tweets were stored systematically in CSV files with key details such as tweet ID, username, content, timestamp, and retweet metrics, providing a structured foundation for detailed analysis (Fig. 1).

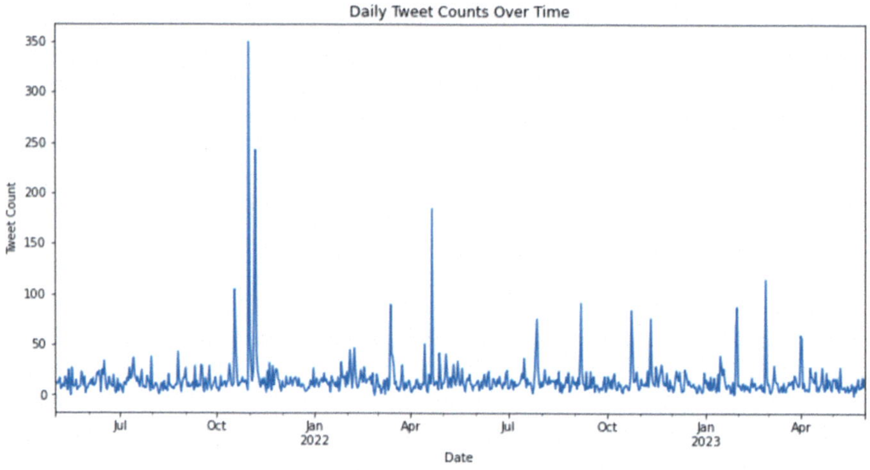

Fig. 1. Daily Tweet Counts Over Time.

2.2 Data Preprocessing

To ensure compatibility with contextual word embeddings, our preprocessing method emphasizes preserving the original context of the data, setting it apart from traditional approaches that often discard essential details. Using regular expressions, we filtered out common tweet-specific elements such as usernames (@username), hashtags (#hashtag), emojis, smileys, non-alphanumeric characters, and URLs. To maintain consistency, non-English tweets were removed, and duplicate entries were identified and excluded. This rigorous process refined the dataset to 19,384 unique tweets, delivering a high-quality dataset for analysis while retaining critical contextual information.

2.3 Data Labeling

To improve data labeling and feature extraction for traffic analysis, we employed GPT-3.5, an advanced large language model, to process 19,834 tweets with a custom-designed prompt tailored to extract six key features related to traffic crashes. The prompt, outlined in Table 1, facilitated automated labeling, generating a detailed dataset of 119,004 labeled entries (calculated as 19,834 tweets multiplied by six features per tweet), as shown in Table 2. This method significantly streamlined feature extraction and enriched the dataset for traffic crash analysis by minimizing the reliance on manual labeling.

The use of GPT-3.5 aligns with findings from recent studies highlighting the efficacy of GPT-based models in data annotation. For instance, Pei, Li, and Xu (2023) demonstrated that GPT models enhance accuracy and efficiency in data generation through self-supervision [5]. Similarly, studies on GPT-3 and GPT-4 have shown their capacity for handling complex, multilingual, and multi-feature data extraction tasks, validating their application in diverse domains [6]. A 2022 study, "Is GPT-3 a Good Data Annotator?", further underscored GPT-3's superiority over traditional annotation methods in NLP tasks, emphasizing its transformative potential in data processing [7].

To manage large-scale data and comply with API rate limits, the workflow processed tweets in batches of 128 with 120-s intervals between batches. Post-processing was performed to standardize and clean the output, addressing inconsistencies such as reconciling variations in spelling (e.g., "Hit-and-run" vs. "hit-and-run"). This step ensured consistency and reliability in the dataset, preparing it for effective use in traffic crash analysis models.

GPT-3.5 model was tasked to generate a JSON dictionary combining the cleaned tweets with their corresponding labels. This structured output is tailored for efficient model training and evaluation. Figure 2 shows the steps followed for dataset curation.

Table 1. GPT 3.5 GPT-3.5 Prompt Structure and Dataset Creation

"""

Given this tweet below:

<tweet>

{text}

</tweet>

I want you to extract information for the following fields:

- Road Accident: Yes or No

- Severity: mild, moderate, fatal

- Driver apprehended:

- collision type:

- sentiment: (positive, negative or neutral)

- emotions: (pick only one of these: fear, anger, sadness, happy, neutral, disgust, love, confusion, gratitude, sympathy or empathy)

Provide your answer in a JSON dictionary.

RULES

- You must provide a value to each field above.

- if the tweet does not contain enough information to answer any of the above field, you must set that field to the value 'None'.

Your response:"""

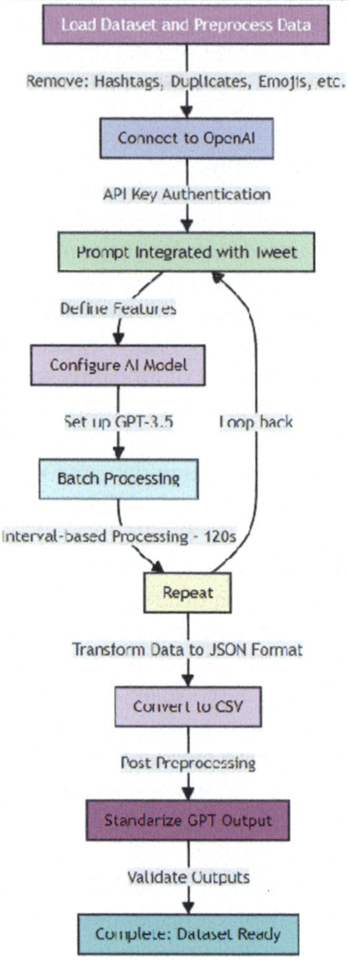

2.4 Data Splitting

The dataset comprised 19,004 samples. We divided this into training and testing sets to effectively evaluate the model's performance. The training set consisted of 112,104 samples, while the test set included 4,200 samples (700 tweets x 6 classes) derived from 700 records out of the 19,834 total samples, scaled to meet the requirement for each of the six targeted outputs. Table 2 shows the data distribution.

2.5 Model Architecture and Implementation

This section details the architecture and implementation of our framework. Table 3 illustrates the comprehensive workflow, encompassing data collection, preprocessing,

Table 2. Dataset Statistics

Total Tweets Collected	26,226
Tweets After Preprocessing	19,384
Features Extracted per Tweet	6
Total Labeled Data Points	19,384 * 6 = 116,304
Training Dataset Size	112,104
Testing Dataset Size	4,200

labeling, feature extraction, model fine-tuning, and evaluation stages, providing clarity on the methodology used for this study.

3 Experiment

3.1 Fine-Tuning the GPT-2 Model

The GPT-2 medium model was selected for its optimal balance between computational efficiency and performance, particularly for tasks requiring nuanced text generation and contextual understanding. Fine-tuning was conducted using a curated dataset generated by GPT-3.5, which extracted six key features from tweets related to Road Traffic Crashes (RTCs). Table 5 outlines these features. The process was executed in the Google Colab environment utilizing the Hugging Face Transformers library, ensuring efficient adaptation of tokenization and data loading to the dataset's unique requirements. Key hyperparameters used in training, including batch size, learning rate, and epochs, are provided in Table 3 for transparency and reproducibility.

The GPT-2 medium model architecture, consisting of 24 transformer layers and 12 attention heads, was pre-trained on a vast text corpus to generate coherent and contextually relevant text. For our specific application, fine-tuning involved the following steps:

Data Preparation: The dataset was cleaned and preprocessed to ensure high-quality input.

Tokenization: Text was tokenized using the GPT-2 tokenizer, converting it into a format suitable for training.

Training Configuration: Hyperparameters such as batch size and learning rate (2e-5) were selected for optimal performance.

Model Training: Fine-tuning was performed over 12 epochs using the Adam optimizer, allowing the model to refine its parameters for the specific tasks.

Table 3. Implementation Overview: Pseudocode and Flowchart

```
# Step 1: Data Collection
def collect_data(hashtags, keywords, start_date, end_date):
    tweets = []
    for hashtag in hashtags:
        tweets.extend(api.search_tweets(hashtag, start_date,
end_date))
    for keyword in keywords:
        tweets.extend(api.search_tweets(keyword, start_date,
end_date))
    return tweets

# Step 2: Data Preprocessing
def preprocess_data(tweets):
    preprocessed_tweets = []
    for tweet in tweets:
        tweet = remove_special_characters(tweet)
        tweet = remove_non_english(tweet)
        preprocessed_tweets.append(tweet)
    return preprocessed_tweets

# Step 3: Data Labeling
def label_data(tweets, gpt_model):
    labeled_data = []
    for tweet in tweets:
        prompt = generate_prompt(tweet)
        labels = gpt_model.generate_labels(prompt)
        labeled_data.append((tweet, labels))
    return labeled_data

# Step 4: Feature Extraction
def extract_features(labeled_data):
```

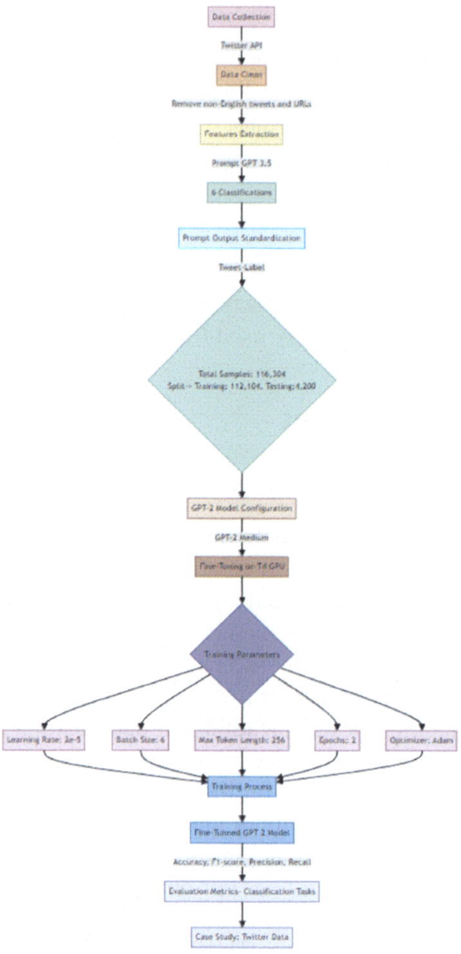

The fine-tuned model was evaluated on a separate test set to measure its ability to classify and extract RTC-related features effectively. This workflow aligns with best practices in natural language processing, drawing from foundational works by Vaswani et al. (2017) on transformers, Wolf et al. (2020) on the Hugging Face library, and Radford et al. (2019) on GPT-2's capabilities [9, 11–13]. These practices ensured the model's ability to process complex patterns and dependencies in RTC-related data from Twitter (Table 4).

Table 4. GPT-2 Hyper parameter values

Hyper-parameter	Value
Maximum Length	256 tokens
Batch Size	6
Learning Rate	2e−5
Number of Epochs	2
Optimizer	Adam
Number of Epochs	12
Model/Tokenizer Name	gpt2-medium
Training/Testing Split	287,010/10,500 samples

Table 5. Classification Labels Extracted from Twitter Data using GPT 3.5

No.	Feature
1	Did a road traffic accident occur?
2	What was the severity of the road traffic accident?
3	What type of collision was it?
4	What sentiment is expressed in the tweet?
5	What emotion does the tweet convey?
6	Is there any information about the person responsible?

3.2 Model Evaluation

To assess the performance of our models on classification tasks, we utilized a range of well established statistical metrics designed for these specific tasks. Metrics such as F-score, Recall, Accuracy, and Precision were applied to evaluate how effectively the model identified tweets related to traffic incidents. Tweets mentioning traffic crashes were classified as positive cases, while those unrelated were labeled as negative. These metrics provide valuable insights into the model's capability to distinguish between relevant and irrelevant tweets, as outlined by Powers (2011) [8]. Figure 2 presents confusion matrices for the six classification features, offering a visual summary of the model's predictive accuracy for each feature.

4 Results

4.1 Model Performance Metrics

This study provides valuable insights into the understanding and management of Road Traffic Crashes (RTCs) by employing innovative big data analysis techniques leveraging social media platforms. By leveraging a multitask learning (MTL) framework integrated with Large Language Models (LLMs) such as GPT-3.5 and GPT-2, we were able to extract and classify detailed, actionable insights from Twitter data, an approach that transcends traditional traffic safety analysis methods [14, 15].

Our findings illustrate that social media is a veritable goldmine of real-time data that, when harnessed with advanced NLP techniques, can greatly enhance traffic incident

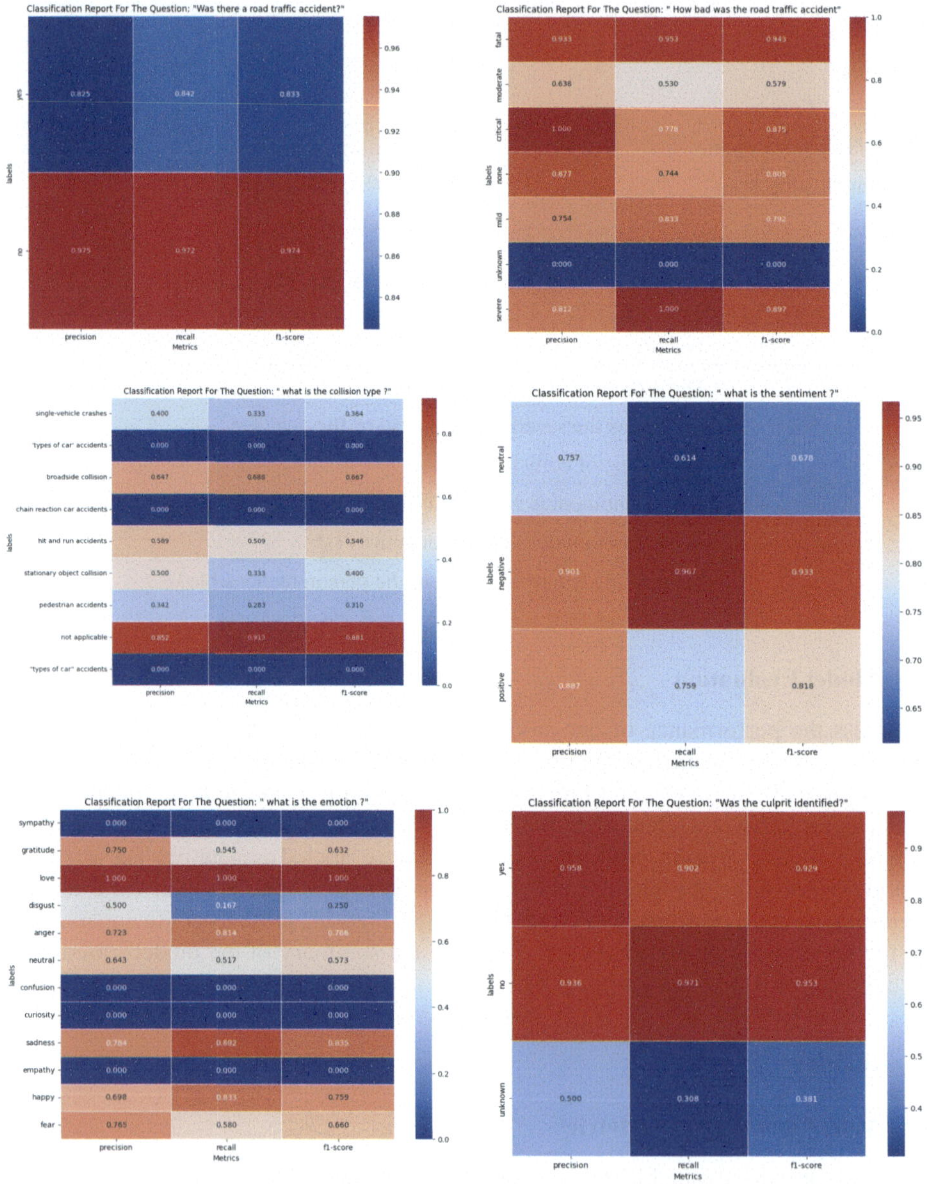

Fig. 2. Confusion matrices illustrating the performance across six classification features

analysis and response strategies. The application of MTL allowed for the simultaneous processing of multiple data types within the tweets, facilitating a more comprehensive analysis. The classification reports for multiple classification tasks are summarized in Fig. 2, with performance metrics detailed for each classification label:

- Did a road traffic accident occur?

  Accuracy: 97.5%
  F1 Score: Yes (94.3%), No (87.4%)

- What was the severity of the road traffic accident?

  Accuracy: 82.3%
  F1 Score: Fatal (82.9%), Mild (83.9%), Moderate (77.2%), None (80.0%), Severe (84.7%), Critical (71.4%), Unknown (0%)

- What type of collision was it?

  Accuracy: 76.0%
  F1 Score: Single-vehicle crashes (83.4%), Types of car crashes (85.2%), Multi-vehicle collision (84.8%), Chain reaction accidents (80.0%), Hit and run crashes (84.4%), fixed object collision (79.3%), Pedestrian crashes (83.3%), N/A(0%)

- What sentiment is expressed in the tweet?

  Accuracy: 90.1%
  F1 Score: Negative (84.5%), Positive (83.1%), Neutral (83.3%)

- What emotion does the tweet convey?

  Accuracy: 93.4%
  F1 Score: Sympathy (88.2%), Gratitude (100.0%), Anger (83.2%), Disgust (72.5%), Confusion (85.7%), Curiosity (50.0%), Sadness (85.7%), Neutral (83.9%), Fear (86.0%), Empathy (50%), Happiness (66.7%), Love (0%)

- Is there any information about the person responsible?

  Accuracy: 85.0%
  F1 Score: Yes (81.3%), No (87.7%), Unknown (73.9%), Reportedly (0%), Some (0%)

These results highlight the model's strong performance, evidenced by high accuracy and F1 scores across multiple classification tasks, demonstrating its effectiveness in analyzing and handling diverse aspects of Road Traffic Crashes (RTCs) using Twitter data. The successful application of our MTL framework not only supports the efficacy of using LLMs in real-world applications but also opens up new avenues for improving real-time monitoring and management of public safety issues. However, the study also identified challenges, such as the need for improved data collection and annotation methods. Future research could explore the integration of more dynamic data sources and extract additional information. This study not only enhances our ability to understand and react in real-time to traffic incidents but also sets a new benchmark for public safety by showcasing the potential of advanced NLP techniques to improve data extraction from unstructured sources like social media.

4.2 Comparison with Existing Models in the Literature

Given the novelty of our multitask learning framework leveraging large language models (LLMs) for traffic crash analysis using social media data, direct comparisons with existing literature are limited. However, related studies provide a baseline for evaluation. Al Mamun et al. demonstrated that optimizing kernel size and filter numbers allowed convolutional neural networks (CNNs) to outperform other machine learning models in detecting traffic-related information from social media, achieving higher F1 scores and generalization accuracy [16]. Similarly, Ramadhani et al. found that a CNN + Fast-Text model achieved the highest accuracy (86.33%) for detecting traffic congestion from Twitter data, surpassing CNN + Word2Vec and SVM models, the latter achieving only 67.62% accuracy [17].

Xintao et al. advanced event detection by integrating Graph Attention Networks (GAT) and Graph Convolutional Networks (GCN), achieving an F1 score of 75.2% on the ACE-2005 dataset, outperforming both BGCN and Cross-Event models [18]. Additionally, Dinesh et al. reviewed road accident detection methods and developed a tweet classification approach using models such as SVM, logistic regression, ANN, and LSTM combined with TF-IDF and BERT. Their LSTM + BERT model achieved the highest precision and recall scores of 0.88 and 0.87, respectively [19].

In contrast, our fine-tuned GPT-2 model achieved an overall accuracy of 95% and an F1 score of 90% for detecting road traffic crashes, significantly outperforming the models in these studies. The multitask learning approach allowed our model to excel across multiple classification tasks by leveraging shared information between tasks, resulting in nuanced and contextually accurate predictions. The advanced natural language understanding of GPT-2 enabled it to handle the complexities and noise inherent in social media data more effectively than traditional models, making it highly suited for real-time traffic crash analysis.

4.3 Baseline Model Comparison

To facilitate a direct comparison, we trained and evaluated several baseline models on our dataset, including Logistic Regression, XGBoost, AdaBoost, and GPT-2. Among these, our fine-tuned gpt-2 model consistently outperformed the baseline models, showcasing its superior ability to handle the complexity and noise inherent in social media data for traffic incident classification. The detailed performance metrics for each classification task are presented in Table 6, highlighting the significant advantage of the GPT-2 model over traditional approaches.

These results underscore the effectiveness of our approach in leveraging advanced NLP techniques and large language models to enhance the analysis and interpretation

Table 6. Comprehensive Performance Comparison

Classification Task	Model Name	Accuracy	Precision	Recall	F1-Score
Did a road traffic accident occur?	Logistic Regression	0.93	0.88	0.78	0.82
	XGBoost	0.95	0.92	0.85	0.88
	AdaBoost	0.92	0.83	0.80	0.81
	GPT-2	0.95	0.90	0.90	0.90
Is there any information about the person responsible?	Logistic Regression	0.92	0.89	0.67	0.73
	XGBoost	0.93	0.87	0.70	0.75
	AdaBoost	0.82	0.58	0.40	0.42
	GPT-2	0.92	0.80	0.73	0.75
What was the severity of the road traffic accident?	Logistic Regression	0.80	0.71	0.63	0.66
	XGBoost	0.82	0.78	0.73	0.74
	AdaBoost	0.45	0.28	0.22	0.18
	GPT-2	0.83	0.72	0.69	0.70
What emotion does the tweet convey?	Logistic Regression	0.65	0.38	0.27	0.28
	XGBoost	0.68	0.42	0.34	0.36
	AdaBoost	0.36	0.11	0.17	0.13
	GPT-2	0.74	0.49	0.45	0.46
What type of collision was it?	Logistic Regression	0.75	0.29	0.21	0.23
	XGBoost	0.77	0.53	0.38	0.43
	AdaBoost	0.72	0.09	0.12	0.10
	GPT-2	0.78	0.37	0.34	0.35
What sentiment is expressed in the tweet?	Logistic Regression	0.84	0.82	0.68	0.73
	XGBoost	0.86	0.83	0.72	0.77
	AdaBoost	0.82	0.79	0.64	0.68
	GPT-2	0.88	0.85	0.78	0.81

of social media data for road safety applications. The GPT-2 model, in particular, consistently provided higher accuracy, precision, recall, and F1-scores across various tasks compared to other models, demonstrating its superior capability in handling complex and noisy social media data for traffic incident classification.

5 Conclusion and Future Work

This research showcased the success of a multitask learning framework powered by the GPT-2 model in analyzing road traffic crash (RTC) data extracted from Twitter. The model achieved impressive performance metrics, with an accuracy of 95.43%, a precision of 97.51%, and a recall of 97.19%. These results highlight the model's capability to efficiently classify and extract detailed information from noisy social media data, contributing to enhanced real-time RTC monitoring and response. Building on these findings, future work will focus on expanding the framework to include additional information retrieval tasks and integrating diverse data sources, such as CCTV footage and emergency dispatch records, to enrich the model's contextual understanding. We also aim to explore the use of more advanced large language models (LLMs) for improved performance and to develop a real-time predictive system that can proactively alert authorities to high-risk incidents, reducing response times and potentially saving lives. This comprehensive approach seeks to set new benchmarks in traffic safety analytics and real-time incident management.

Acknowledgments. We extend our heartfelt thanks to the Digital Observatory at Queensland University of Technology (QUT) for their collaboration in data collection. We also express our sincere gratitude to QUT for their invaluable support in disseminating this work.

References

1. Ashqar, H.I., Shaheen, Q.H., Ashur, S.A., Rakha, H.A.: Impact of risk factors on work zone crashes using logistic models and random forest. In: 2021 IEEE International intelligent transportation systems conference (ITSC), pp. 1815–1820. IEEE (2021)
2. Elhenawy, M., Rakha, H.A., Ashqar, H.I.: Joint impact of rain and incidents on traffic stream speeds. J. Adv. Transp. **2021**, 1–12 (2021)
3. Gutierrez-Osorio, C., Gonzalez, F.A., Pedraza, C.: A deep learning ensemble model for predicting traffic accidents from social media data. Computers **11**(9), 126 (2022)
4. Pei, X., Li, Y., Xu, C.: GPT self-supervision for a better data annotator (2023). arXiv.org. https://doi.org/10.48550/arXiv.2306.04349
5. Rathje, S., Mirea, D.M., Sucholutsky, I., Marjieh, R., Robertson, C., Van Bavel, J.J.: GPT is an effective tool for multilingual psychological text analysis (2023)
6. Ding, B., Qin, C., Liu, L., Chia, Y. K., Joty, S., Li, B., Bing, L.: Is gpt-3 a good data annotator? (2022). arXiv preprint arXiv:2212.10450
7. Powers, D.M.W.: Evaluation: from precision, recall and F-measure to ROC, informedness, markedness & correlation. J. Mach. Learn. Technol. **2**(1), 37–63 (2011)
8. Radford, A., Wu, J., Child, R., Luan, D., Amodei, D., Sutskever, I.: Language models are unsupervised multitask learners. OpenAI Blog (2019). Available at: https://openai.com/blog/better-language-models/
9. Brown, T., et al.: Language models are few-shot learners (2020). arXiv preprint arXiv:2005.14165
10. Vaswani, A., et al.: Attention is all you need. In: Proceedings of the 31st international conference on neural information processing systems, pp. 6000–6010 (2017)
11. Jaradat, S., Nayak, R., Elhenawy, M.: explainable language models for the identification of factors influencing crash severity levels in imbalanced datasets. In: Proceedings of the 2024 3rd international conference on computing and machine intelligence (ICMI). IEEE, Michigan (2024)
12. Wolf, T., et al.: Transformers: state-of-the-art natural language processing. In: Proceedings of the 2020 conference on empirical methods in natural language processing: system demonstrations, pp. 38–45 (2020)
13. Jaradat, S., Hossain, A., Alhadidi, T.I., Elhenawy, M., Ashqar, H.I.: Exploring traffic crash narratives in Jordan using text mining analytics (In review)

14. Uniyal, D., Nayak, R.: Twitter's pulse on hydrogen energy in 280 characters: a data perspective. social network analysis and mining (2024)
15. Al Mamun, A., Rahman, M.M., Al Mamun, A.: A deep learning-based traffic event detection from social media. IEEE (2020)
16. Almassar, R.R.: Detection of traffic congestion based on Twitter using convolutional neural network model. IAES Int. J. Artif. Intell. **1**, 21–30 (2022)
17. Xintao, J., Zhishan, C., Guoyi, C.: Event detection model based on graph attention network and graph convolutional network. In: 2023 IEEE 6th information technology, networking, electronic and automation control conference (ITNEC), pp. 322–329 (2023)
18. Dinesh, L., Kuhaneswaran, B., Ravikumar, N.: A multifaceted machine learning approach to understand road accident dynamics using twitter data. In: Handbook of research on advancements of contactless technology and service innovation in library and information science, 1st ed., pp. 21–30 (2023)

Water Quality Classification Using an Artificial Neural Network

Saeed Mohsen[1,2], Wael M. F. Abdel-Rehim[2,3], and M. Abdel-Aziz[2,4](✉)

[1] Electronics and Communications Engineering Department, Al-Madinah Higher Institute for Engineering and Technology, Giza 12947, Egypt
g17082131@eng.asu.edu.eg

[2] Department of Artificial Intelligence Engineering, Faculty of Computer Science and Engineering, King Salman International University (KSIU), South Sinai 46511, Egypt
m.abdel-aziz@ksiu.edu.eg

[3] Department of Computer Science, Faculty of Computers and Information, Suez University, Suez 43512, Egypt

[4] Department of Basic Sciences, Faculty of Computer and Information Sciences, Ain Shams University, Cairo 11566, Egypt

Abstract. Water quality is vital for public health and ecosystem sustainability. However, approximately 2 billion people worldwide do not have access to drinking water that is appropriately managed, contributing to significant health risks and environmental issues. This paper proposes an efficient approach to classifying water quality as potable or impotable water using artificial neural networks (ANNs). This work highlights the perspective of ANNs in reducing water pollution and ensuring safe water access. The ANN is implemented with a dataset consisting of 2 water classes based on 3,276 water samples. The proposed ANN is tested via the Keras framework with a tuning technique to obtain the most significant water quality classification (WQC) precision. An error matrix is used to evaluate the ANN efficiency. The results show that the ANN is highly efficient in the classification of water quality. Moreover, The ANN provides a test accuracy of 68.38%, a mean precision of 69.64%, and the average area under the ROC curve for the implemented ANN is 51%.

Keywords: Water Quality · Deep Learning · Artificial Neural Network · Classification

1 Introduction

Water is a crucial thing that individuals utilize in drinking, washing, cooking, and other daily activities related to water. However, water pollution is escalating daily, because of several factors: the disposal of industrial waste products, migration of persons from rural to urban places, untreated sewage disposal, crowded population, wastewater, and other harmful things from the industries [1]. There are more than seven billion individuals on the planet who need clean/good water for domestic purposes, especially drinking. Therefore, it should present a scientific method to consequently assess water quality

A. Abdelgawad et al. (Eds.): ISBCom 2024, LNNS 1268, pp. 262–273, 2025.
https://doi.org/10.1007/978-3-031-82377-0_23

that can assist in overcoming this issue [2]. For example, artificial intelligence (AI) techniques can be applied in this field, such as ANNs. It is important to classify water quality for its utilization through various techniques. Furthermore, ANNs can learn a selected non-linear relation, and it is used to tackle several issues especially the prediction of water quality, and these ANNs are evaluated to classify the type of water [3]. For example, realizing different predictive algorithms based on applying ANN and ensemble algorithms, which is a fast process required these days. So, it is important to study the advancement of easy-to-learn neural network tools. [4].

High-quality water should exhibit the following characteristics [5].

- Devoid of harmful agents.
- Lower turbidity.
- Lack of saline.
- Lower levels of hazardous elements for instance lead which give negative or unsafe health effects either in the long or short term.

In the literature, various deep learning (DL) networks were introduced for the classification of water quality (WQ). Also, various works presented on the classification of water quality and assessed by different machine learning (ML) algorithms. Thus, many research papers included various scientific approaches and areas that were applied for a classification of WQ.

In [6], Wechmongkhonkon, et al., implemented a multilayer perceptron (MLP) neural network based on Levenberg-Marquardt (LM) algorithms to a collection of the water nature of Bangkok, Thailand. The results showed that the MLP reached a high accuracy of 96.52%. Sheppard, et al., [7] presented how the utilization of geographic information systems, remote sensing (RS), and global positioning system (GPS) device were used to improve the quality of water in Northern Alabama. The findings enhanced their ability to assess and foresee contamination sources [7].

In [8], Xiang and Jiang developed two techniques: particle swarm optimization (PSO) and least squares support vector machine (LS-SVM) to classify the water quality and overcome the disadvantage of the backpropagation approach. These two techniques applied to a simulation testing dataset, the LS SVM illustrated high efficiency in predicting the WQ of the LiuxiRiver. Chi, et al., [9] introduced a PSO to prepare an ANN to forecast water quality. The results showed that the contamination is more serious on the Yangtze River. Generally, all the mentioned works discussed how to predict and classify water quality using various approaches to obtain diverse results.

Afaq et al. [10] presented a prediction of WQ using MLP and *k*-nearest neighbor (KNN) algorithms. The work efficiently considered the dropped attributes and yielded the best classification efficiency. The authors utilized nine layers for the MLP algorithm and KNN. This work used traditional ML algorithms. The findings proposed that the MLP with KNN achieved a total prediction accuracy of 95% in the classification of WQ. On the other hand, the grouping of the KNN and MLP was assessed using different metrics to ensure results were reached. In [11], Nasir et al. implemented an efficient technique to predict WQ. The authors applied traditional ML techniques i.e. decision tree (DT), MLP, random forest (RF), SVM, and CATBoost algorithms. The results of the CATBoost reached a high precision of 94.5% to predict WQ. Also, stacking ensemble (SE) techniques were implemented, and the mentioned algorithms gave convincing results. Radhakrishnan et al. [12] utilized ML algorithms for WQC. The algorithms were Naïve Bayes (NB), DT, and SVM, they were trained on several datasets. Eventually, the ML algorithms compared their efficiency and achieved the highest prediction accuracy of 96.5% utilizing the DT algorithm. The method of tuning hyperparameters for the DT contributed to the experimental results reached by the ML algorithms.

In [13], Theyazn et al. proposed two DL models of nonlinear auto-regressive neural networks (NARNETs) and long short-term memory (LSTM) for WQC. The authors tested and evaluated SVM, KNN, and NB to compare experimental results. The LSTM and NARNET models overlapped with dissimilar regression results of NARNETs and RLSTM at 96.17% and 94.21%. Further, the SVM gave an overall efficiency of 97% precision. According to this work, a kernel function was used with the SVM which achieved experimental results more convincing than other techniques. For water distribution systems, DL-based structures were presented by Shahra et al. [14]. This work attained better accuracy with few computations. The authors used ANN and SVM models, where the ANN performed higher than the SVM model. So, ANN and SVM yielded a maximum accuracy of 94% and 89%. The experimental results were obtained for SVM because of the employment of the correct selection of filter functions.

In this paper, an ANN is developed to classify WQ. The efficiency of this network is assessed utilizing some measures, for instance, sensitivity, precision, accuracy, and confusion matrix. The water is categorized into two classes: Impotable and Potable. Impotable water indicates harmful while Potable water refers to beneficial. The testing accuracy of the ANN achieved 68.38% while the sensitivity and f1-score are 62.17% and 61.41%, respectively. A hyperparameter tweaking approach is utilized to obtain the best efficiency of the proposed ANN.

The paper has several sections: Sect. 2 lists an overview about the MLP network. In Sect. 3, the proposed ANN is described. In Sect. 4, the results are analyzed. The paper presents the conclusion in Sect. 5.

2 Overview of MLP Neural Network

MLP is one of the feed-forward networks, it is the most widely used ANN structure, learned with a backpropagation loss, and created via Rumehart et al. [15]. Figure 1 demonstrates a particular MLP network design. it is implemented with three layers: input, hidden, and output [16]. The purpose of the input layer (IL) is to give a number of data features from an external environment, these features are sent to the hidden layer (HL) and then it feds to the output layer (OL) to produce a particular target [17]. The prediction of the network depends on the HL, which includes a certain number of neurons. For an ANN, a few features provided to the IL and multiplied by weights are added to a bias and after that, a specific transfer function is used as illustrated in Eq. (1).

$$a_k = ReLU\left(\sum_{i=1}^{n} W_{xh}.x_i + b_k\right) \tag{1}$$

W_{xh} is the weight between the IL and HL, a_k indicates the HL output, b_k is the bias for the present layer, and a ReLU is the transfer function to activate the layers, used in the learning phase to reduce the loss. Moreover, ReLU has a better speed than other transfer functions [18]. It is defined in Eq. (2) as.

$$ReLU(x) = max(0, x) \tag{2}$$

The OL is used for the MLP structure of the ANN. The OL is determined from Eq. (3). The MLP network process is repeated until the last input and weight as illustrated in Fig. 2.

$$y_k = \sigma\left(\sum_{i=1}^{n} W_{hy}.a_{k-1} + b_k\right) \tag{3}$$

W_{hy} indicates the weight between OL and the HL, y_k is the predicted target, a_{k-1} is the output from the previous layer, and σ is the sigmoid function. It is more convenient for applications of binary classification. This sigmoid is obtained from Eq. (4).

$$\sigma(x) = \frac{1}{1 + e^{-x}} \tag{4}$$

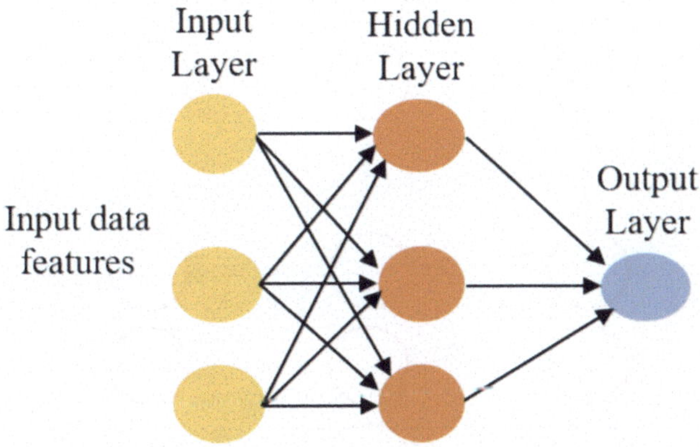

Fig. 1. A particular MLP neural network architecture.

3 The Proposed ANN

Figure 2 shows the proposed ANN architecture. It has five layers: input, three hidden, and output. The IL contains 288 neurons with 12 different features. In the first, a HL is employed for the features of data, this layer is activated by a ReLU function to extract features. This layer is implemented by a dense function, followed by a second HL which contains twenty-four neurons, and uses a ReLU. Then, a third HL is employed with a setting of twelve neurons and a ReLU function. The third HL is placed to enable the ANN to detect more features, which may be lost in the two previous HLs. A dropout technique is employed after the first, second, and third HLs with a 30% drop percentage from the neurons to solve the overfitting issue [19]. Eventually, the OL includes only one neuron, with a sigmoidal function to classify the output based on two classes [20]. This ANN is implemented through the Keras library. There is an advantage of the implemented ANN: the simplicity of the structure, also this ANN consumed little time during the learning stage, and it has good efficiency in predicting water quality.

In this ANN, the error is optimized using Adam algorithm. It obtains the best weights of the ANN, reduces the losses, and maximizes the learning precision. The error cost function, called cross-entropy for binary classes, is employed for measuring the error between the true and prediction values [21]. This ANN is learned with a batch size of 100 and learning epochs of 60. This ANN has overall parameters of 31,197. Table 1 demonstrates a summary of the designed layers for the ANN.

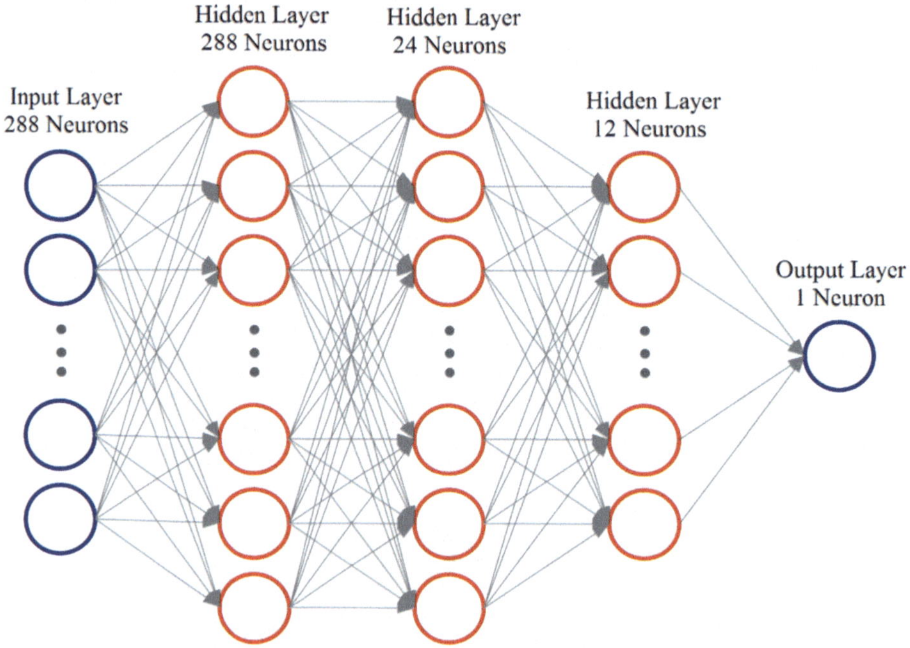

Fig. 2. The proposed ANN architecture.

The dataset utilized for this work is obtained from the Kaggle website [22], with 3,276 water samples of two classes Impotable and Potable, and the sample numbers for each class are 1,998 and 1,278, respectively. The dataset is gathered from India for ten features: PH, hardness (HD), chloramines level (CL), solids, sulfate percentage (SP), conductivity, Trihalomethanes, turbidity, organic carbon (OC), and potability. The dataset has two parts: a test part of 20% and a train group of 80%, the test set employed to assess the ANN. Figure 3 shows the number of samples for the two classes: Impotable "0" and Potable "1". Thus, the percentage of samples for the Potable sample is 39.02%, while the Impotable has 60.98% samples.

Table 1. Layers summary.

Layer (Type)	Output Shape	Parameters
Dense	(None, 150)	1500
Dropout	(None, 150)	0
Dense	(None, 128)	19328
Dropout	(None, 128)	0
Dense	(None, 64)	8256
Dropout	(None, 64)	0
Dense	(None, 32)	2080
Dense	(None, 1)	33

Total Parameters: 31,197.

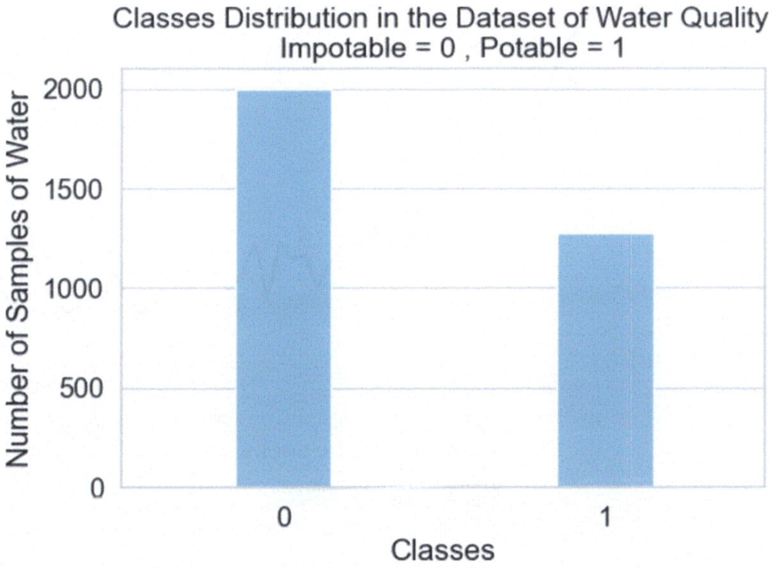

Fig. 3. Classes distribution in the dataset of water quality.

Equations (5), (6), and (7) illustrate the precision, accuracy, and sensitivity as metrics in terms of the true positives (TP) and true negatives (TN) [23, 24].

$$Precision = \frac{TP}{TP + FP} \tag{5}$$

$$Accuracy = \frac{TP + TN}{TP + TN + FP + FN} \tag{6}$$

$$Sensitivity = \frac{TP}{TP + FN} \tag{7}$$

4 Results Analysis

The ANN is created via Python based on a Spyder IDE. This ANN is implemented using a workstation that has Windows 11, Intel-Core i7, 8 CPU_S, and 16 GB RAM memory. Figure 4 illustrates the loss rates for the ANN. In this figure, the minimum testing loss and training loss are 0.6034 and 0.5824, respectively.

Figure 5 shows the two curves for the accuracy of the ANN. The orange curve is training accuracy, which continuously changes and achieves a peak value of 68.99% at 60 epochs. The blue curve refers to the test accuracy, which begins at 60.56% and increases to 68.38% at 60 epochs. Figure 6 shows the ROC curve. The average area under this curve reaches 0.51 (51%) based on both classes "Impotable and Potable".

Fig. 4. Loss for ANN.

Figure 7 shows the confusion matrix (CM) determined from the ANN based on the testing set. In this CM, the diagonal values refer to the accuracy of the prediction, while the values above and below the diagonal show losses that occurred.

The CM reveals data of 454 and 106 as TP and TN, respectively for the two classes: Impotable and Potable. The normalized CM for the ANN is shown in Fig. 8. It has an accuracy of 0.92 and 0.33 for the two classes – Impotable and Potable, respectively, with errors of 0.085 and 0.67 for Impotable and Potable, respectively.

Figure 9 shows the correlation matrix for the used dataset. In this figure, the correlated features between ten classes are: PH, hardness (HD), chloramines level (CL), solids, sulfate percentage (SP), conductivity, Trihalomethanes, turbidity, organic carbon (OC), and potability. In the red diagonal of this matrix, we see that all classes achieved 1 or 100% between the predicted and true data. Figure 10 illustrates the distribution of the sulfate class in a water sample. The maximum value is roughly 0.015 for 350 samples.

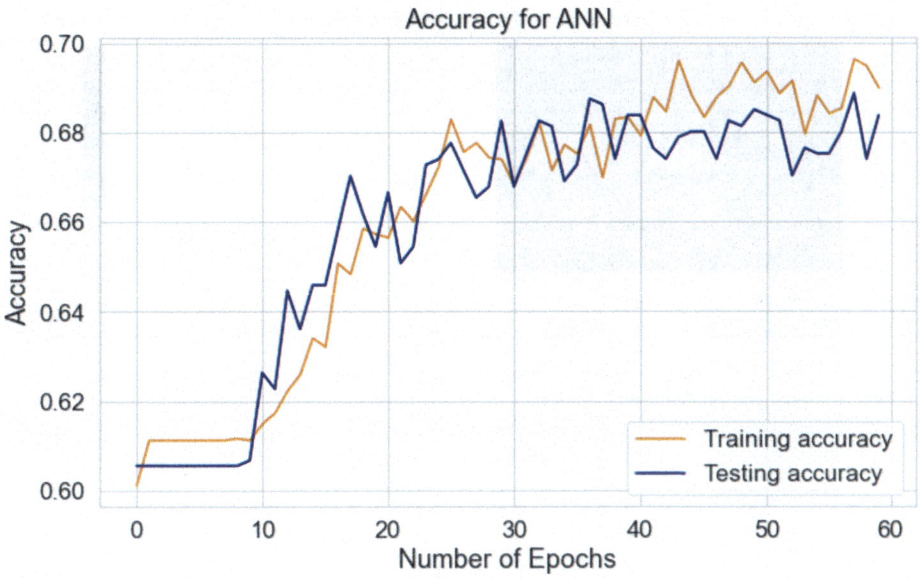

Fig. 5. Accuracy curves for ANN

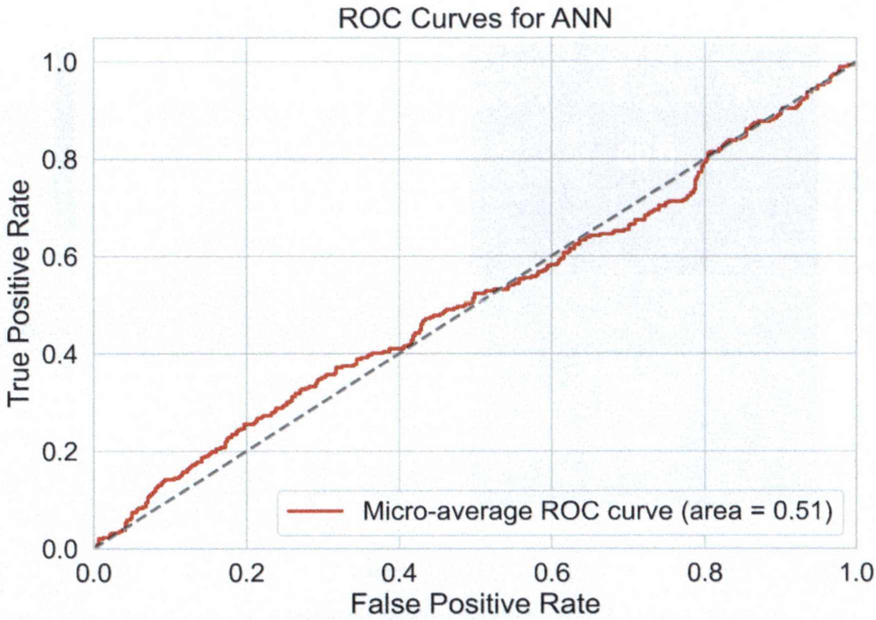

Fig. 6. ROC curves for ANN

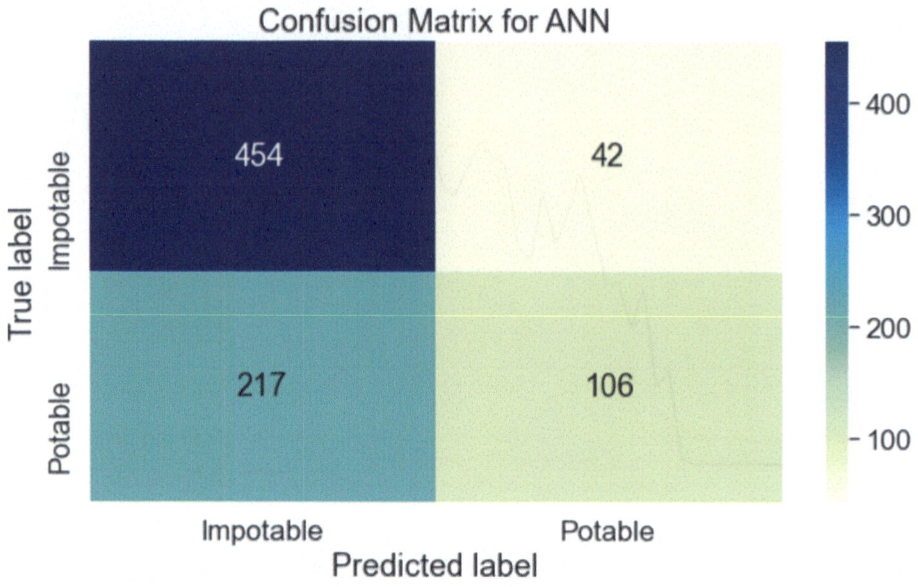

Fig. 7. CM for ANN

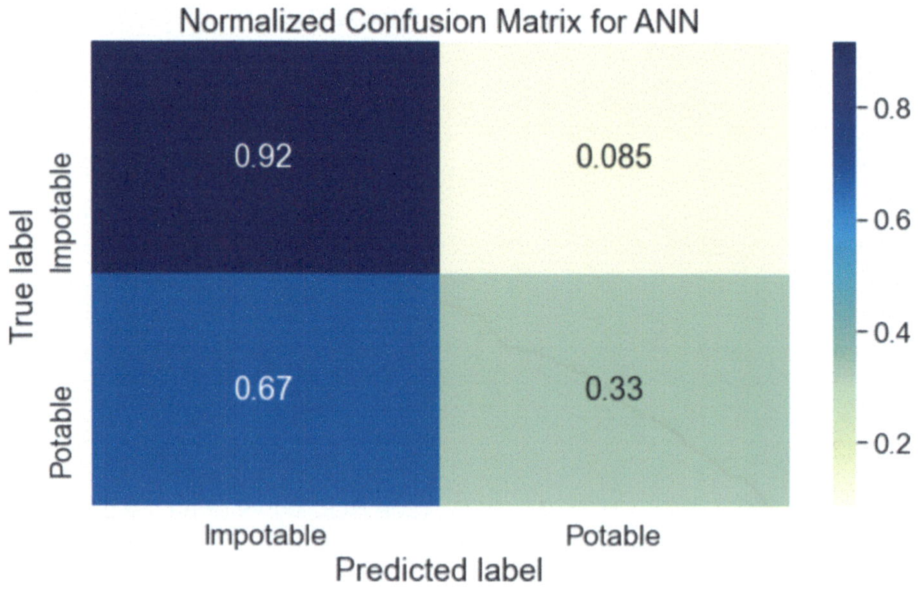

Fig. 8. Normalized CM for ANN

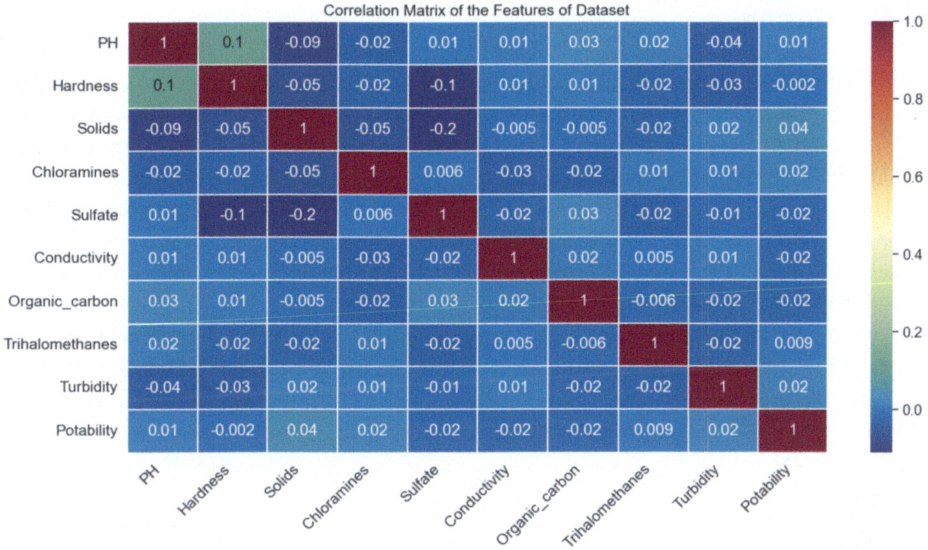

Fig. 9. Correlation matrix

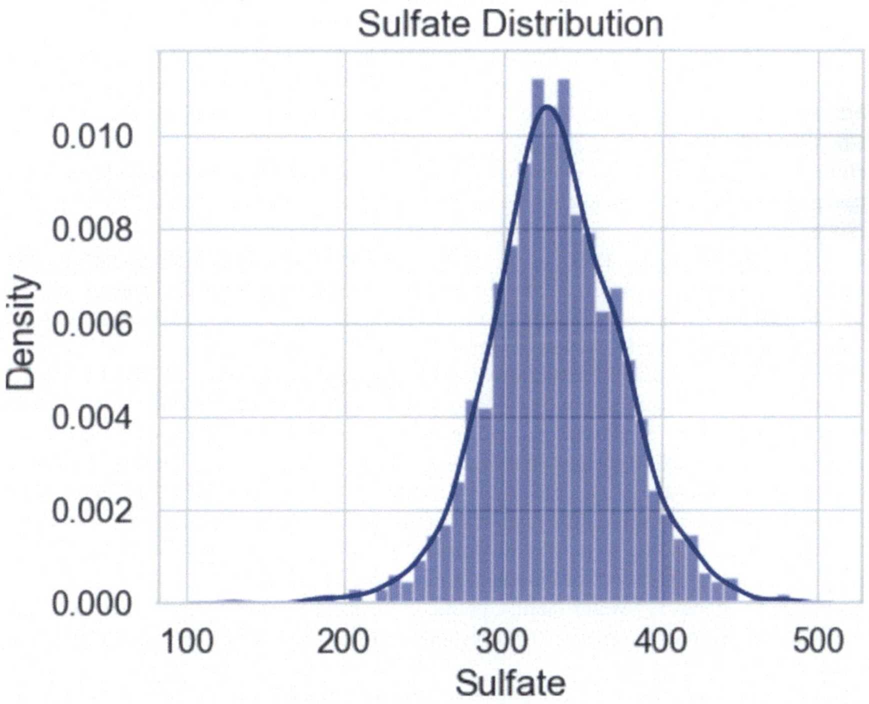

Fig. 10. Sulfate distribution.

5 Conclusion

This article proposes an efficient technique for the classification of two types of water: potable and impotable, which are obtained from a water source in India. This technique has presented an effective design of an ANN to classify WQ. This ANN is implemented with data containing 1,278 potable water samples, and 1,998 impotable water samples.

The experiment prepared to assess the accuracy of the ANN technique gives a testing accuracy of 68.38%. The error matrix, recall, precision, and f1-score are presented to allow assessment of the ANN's efficiency. The ANN achieved a precision of 69.64%, a recall of 62.17%, and f1-score of 61.41%.

Eventually, the learning curves of accuracy and loss rate provide high performance for the WQC. The developed technique is still in the proof-of-concept phase to classify the water quality. In the future, one may train other techniques i.e. convolutional neural network (CNN). Also, one may use huge datasets to increase the performance of the proposed ANN.

References

1. Zhartybayeva, M., Muntayev, N., Tulegenova, S., Oralbekova, Z., Lamasheva, Z., Iskakov, K.: Monitoring and forecasting of water pollution by heavy metals. IEEE Access **11**, 1593–1602 (2023)
2. Tejaswi, T., Manoj, C., Venkata Daivakeshwar Naidu, P., Santhosh, T., Venkata Sai Akhil, P., Ganesan, V.: nexus of water quality prediction by ANN. In: 2022 international conference on innovative computing, intelligent communication and smart electrical systems (ICSES), Chennai, India, pp. 1–5 (2022)
3. Huiqun, M., Ling, L.: Water quality assessment using artificial neural network. In: International conference on computer science and software engineering, Wuhan, China, pp. 13–15 (2008)
4. Ladjal, M., Bouamar, M., Djerioui, M., Brik, Y.: Performance evaluation of ANN and SVM multiclass models for intelligent water quality classification using Dempster-Shafer Theory. In: 2016 international conference on electrical and information technologies (ICEIT), Tangiers, Morocco, pp. 191–196 (2016)
5. Chakravarthy, S.R.S., et al.: Prediction of water quality using SoftMax-ELM optimized using adaptive crow-search algorithm. IEEE Access **11**, 140900–140913 (2023)
6. Wechmongkhonkon, S., Poomtong, N., Areerachakul, S.: Application of artificial neural network to classification surface water quality. World Acad. Sci. Eng. Technol. **6** (2012)
7. Sheppard, D., Tsegaye, T.D., Tadesse, W., Mckay, D., Coleman, T.L.: The application of remote sensing, geographic information systems, and global positioning system technology to improve water quality in Northern Alabama. Centre for hydrology, soil climatology, and Remote sensing Department of Plant and soil science (2001)
8. Yunrong, X., Jiang, L.: Water quality prediction using LS-SVM with particle swarm optimization. In: Second International Workshop On Knowledge Discovery And Data Mining. IEEE (2009)
9. Zhou, C., Liang, G., Haibing, G., Chuanyong, P.: Pattern classification and prediction of water quality by neural network with particle swarm optimization. In: Proceedings of the 6th world congress on intelligent control and automation, June 21–23, Dalian, China (2006)
10. Juna, A., Umer, M., Sadiq, S., Karamti, H., Eshmawi, A.A., Mohamed, A., et al.: Water quality prediction using KNN imputer and multilayer perceptron. Water **14**(17), 2592 (2022)

11. Nasir, N., et al.: Water quality classification using machine learning algorithms. J. Water Process Eng. **48**, 102920 (2022)
12. Radhakrishnan, N., Pillai, A.S.: Comparison of water quality classification models using machine learning. In: 2020 5th international conference on communication and electronics systems (ICCES), pp. 1183–1188. IEEE (2020)
13. Aldhyani, T.H., Al-Yaari, M., Alkahtani, H. and Maashi, M.: Water quality prediction using artificial intelligence algorithms. Appl. Bionics Biomech. **2020** (2020)
14. Shahra, E.Q., Wu, W., Basurra, S., Rizou, S.: Deep learning for water quality classification in water distribution networks. In: International conference on engineering applications of neural networks, pp. 153–164 (2021). Springer, Cham
15. Agatonovic-Kustrin, S., Beresford, R.: Basic concepts of artificial neural network (ANN) modeling and its application in pharmaceutical research. J. Pharm. Biomed. Anal. **22**, 717–727 (2000)
16. McClelland, D.R.J.L.: Explorations in parallel distributed processing. MIT Press, Cambridge (1998)
17. Bourquin, J., Schmidli, H., Van Hoogevest, P., Leuenberger, H.: Basic concepts of artificial neural networks (ANN) modeling in the application to pharmaceutical development. Pharm. Dev. Technol. **2**, 95–109 (1997)
18. Vinayakumar, R., Alazab, M., Soman, K.P., Poornachandran, P., Al-Nemrat, A., Venkatraman, S.: Deep learning approach for intelligent intrusion detection system. IEEE Access **7**, 41525–41550 (2019)
19. Xiong, J., Zhang, K., Zhang, H.: A vibrating mechanism to prevent neural networks from overfitting. In: Proceedings of 15th international wireless communications & mobile computing conference (IWCMC), pp. 1929 (2019)
20. Agostinelli, F., Hoffman, M., Sadowski, P., Baldi, P.: Learning activation functions to improve deep neural networks, pp. 1–9 (2015)
21. Ho, Y., Wookey, S.: The real-world-weight cross-entropy loss function: modeling the costs of mislabeling. IEEE Access **8**, 4806–4813 (2019)
22. Water Quality Classification (WQC), Kaggle Repository (2024). https://www.kaggle.com/datasets/drsaeedmohsen/water-quality-classification-wqc
23. Mohsen, S., Alharbi, A.G.: EEG-based human emotion prediction using an LSTM model. In: 2021 IEEE international midwest symposium on circuits and systems (MWSCAS), Lansing, MI, USA, pp. 458-461 (2021)
24. Mohsen, S., Scholz, S.G., Elkaseer, A.: Detection of COVID-19 in chest x-ray images using CNN model toward medical applications. Wirel. Person. Commun. Lansing (2024). https://doi.org/10.1007/s11277-024-11309-7

Automated Question Generation for Science Tests in Arabic Language Using NLP Techniques

Mohammad Tami[1], Huthaifa I. Ashqar[2,3], and Mohammed Elhenawy[4](✉)

[1] Arab American University, Jenin, Palestine
[2] Arab American University Jenin, Jenin, Palestine
Huthaifa.ashqar@aaup.edu
[3] Columbia University, New York, NY, USA
[4] CARRS-Q, Queensland University of Technology, Brisbane, Australia
mohammed.elhenawy@qut.edu.au

Abstract. Question generation for educational assessments is an expanding field within artificial intelligence, particularly in educational technology domains such as intelligent tutoring systems and dialogue-based platforms. These tools are crucial for creating assessment questions with clear-cut answers, typically relying on syntactical and semantic cues within declarative sentences to transform them into questions. Recent efforts have focused on generating educational questions in Arabic, but performance has been hindered by issues like sentence parsing inaccuracies, named entity recognition problems, and errors from rule-based question transformation, exacerbated by the complexity of lengthy Arabic sentences. This research introduces an innovative Arabic question-generation system built on a three-stage process: extraction of keywords and key phrases, question generation, and subsequent ranking. The proposed framework leverages an existing pre-trained model without fine-tuning, achieving results comparable to fine-tuned models. Notably, it achieved impressive metrics with a BLEU score of 17.31, ROUGE-L of 51.6, and METEOR of 32.74. These results underscore the framework's effectiveness and potential in generating high-quality assessment questions in Arabic. Human evaluation further validated the model's performance, highlighting its significant contribution to the field of automatic question generation for educational assessments.

Keywords: Natural Language Processing (NLP) · Question generation (QG) · Arabic Language Processing (ANLP)

1 Introduction

Automated Question Generation (AQG) has become a significant area of research in Natural Language Processing (NLP), with the goal of generating assessment questions from textual content automatically [1]. This technology holds promise for educational applications, particularly in creating assessments and practice questions for students. Although AQG has been extensively explored for English and other widely spoken languages, its development for the Arabic language is relatively nascent.

A. Abdelgawad et al. (Eds.): ISBCom 2024, LNNS 1268, pp. 274–285, 2025.
https://doi.org/10.1007/978-3-031-82377-0_24

Early approaches to AQG often relied on rule-based systems, which utilized predefined templates and syntactic rules to generate questions and extract information [2, 3]. These systems were limited by their inflexibility and the extensive manual effort required to create rules. As machine learning techniques evolved, they began to be applied to AQG, allowing systems to learn from large datasets of question-answer pairs. These machine learning approaches improved the efficiency and adaptability of AQG systems but still faced challenges in generating high-quality questions consistently [4, 5].

The advent of deep learning models, particularly those based on transformer architectures such as Bidirectional Encoder Representations from Transformers (BERT) [6] and Text-to-Text Transfer Transformer (T5) [7], marked a significant advancement in AQG. These models leverage large-scale datasets and sophisticated training techniques to produce high-quality, contextually relevant questions. For instance, BERT and T5 have been successfully fine-tuned for various AQG tasks, demonstrating superior performance in generating nuanced questions compared to earlier models [8].

In the context of Arabic NLP, AQG presents unique challenges due to the language's morphological richness and syntactic complexity. The development of AraT5 [9], a transformer-based model tailored specifically for Arabic, has been a crucial step forward. AraT5 has shown effectiveness in generating coherent and contextually appropriate questions by leveraging extensive Arabic corpora and fine-tuning on specific tasks [10].

Nwafor and Onyenwe [11] highlighted the application of NLP techniques in automating the generation of multiple-choice questions. Their approach involved fine-tuning pre-existing NLP models to generate both correct answers and plausible distractors, significantly reducing the manual effort involved in question creation. This study underscored the potential of NLP to streamline the development of educational assessments in Arabic.

Sharma et al. [12] further explored AQG from educational content such as textbooks and lecture notes. Their model employed a sequence-to-sequence architecture with attention mechanisms, capable of generating questions that align with the educational objectives of the source material. This method demonstrating the versatility and applicability of advanced NLP techniques across different languages.

Neural Question Generation (NQG) has advanced significantly with deep neural networks, addressing issues in traditional methods that either focus too much on entities or produce unanswerable questions. A novel approach integrates question generation and answer selection, achieving state-of-the-art results on the SQuAD dataset[13]. Meanwhile, the Unified pre-trained Language Model (UniLM) [14] excels in both understanding and generating natural language by leveraging a shared Transformer network and specific self-attention masks, outperforming models like BERT on various benchmarks.

To further tackle pre-training challenges such as exposure bias, ERNIE-GEN [15] introduces a multi-flow sequence-to-sequence framework that aligns training with human writing patterns, achieving impressive results with less data. In educational assessments, a new Arabic Automatic Question Generation (AAQG) [16] model based on the Transformer architecture addresses the complexity of generating questions from extensive texts, enhancing linguistic diversity and adaptability. These innovations collectively push the boundaries of natural language generation and its practical applications.

Despite these advancements, several challenges remain. Integrating AQG systems with existing educational technologies is a promising future direction. This integration can create adaptive learning environments, allowing educators to customize and refine generated questions to better fit their curriculum needs [17]. Moreover, exploring multimodal AQG, which involves generating questions from a combination of text, images, and video content, could further enhance the learning experience by catering to different learning styles [18].

Building on these advancements, this paper introduces a novel contribution to the field by developing an Arabic Question Generation (QG) framework leveraging an existing pre-trained model (specifically AraT5 and AraBert) without the need for fine-tuning. This approach capitalizes on the strengths of pre-trained models while bypassing the resource-intensive fine-tuning process. Remarkably, the proposed framework achieves results comparable to those of fine-tuned models, demonstrating its efficacy and potential in generating high-quality questions in Arabic. This work not only addresses the gap in Arabic QG but also showcases the power of pre-trained models in achieving significant performance gains across languages and tasks.

2 System Methodology

The proposed system as illustrated in Fig. 1 initiates with an Arabic text or book as the primary input. This text undergoes a segmentation process, split into overlapping sentence chunks. This overlapping technique ensures that contextual continuity is preserved between adjacent chunks, enhancing the semantic understanding of the text. These chunks are subsequently processed using the AraBERT model, a transformer-based model specifically fine-tuned for Arabic language tasks, to generate high-dimensional embeddings that capture the contextual nuances of the Arabic language.

These embeddings, paired with their corresponding text chunks, are stored in a vector database optimized for handling and querying high-dimensional data. When a query is introduced to the system, it computes the cosine similarity between the query's embedding and the embeddings stored in the database. This similarity measurement facilitates the retrieval of the top-10 most semantically similar sentences from the database.

Following retrieval, these sentences are input into KeyBERT, a minimalistic approach for keyword extraction using BERT embeddings. KeyBERT identifies the most relevant keywords within these sentences, capturing the core topics and terms. Lastly, these keywords, along with the retrieved sentences, are used as input for the AraT5 model, an Arabic adaptation of the T5 (Text-to-Text Transfer Transformer) model, which is tasked with generating candidate questions. This end-to-end system leverages deep learning models and techniques specifically tailored for Arabic text processing, from initial text segmentation to sophisticated question generation, embodying a comprehensive approach to semantic search and content analysis.

2.1 Dataset

This research employed a dataset derived from Palestinian biology textbooks for grades eleven and twelve. The text was automatically extracted and processed using python.

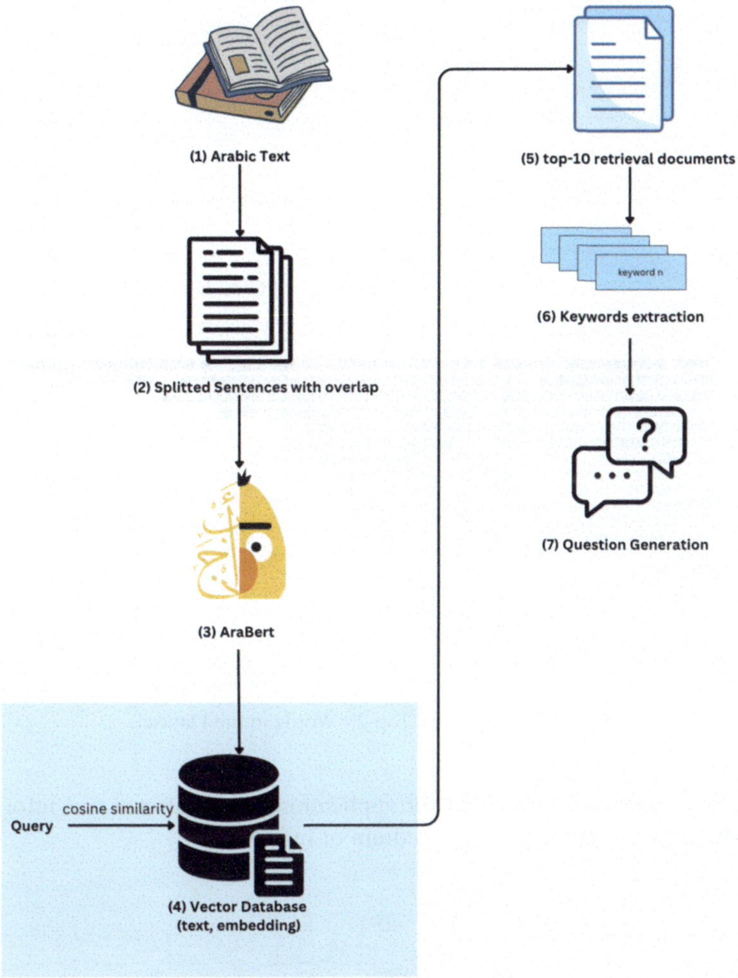

Fig. 1. Semantic Processing Pipeline for Arabic Text Analysis and Question Generation.

Analysis of the dataset's content was conducted using a word cloud, as depicted in Fig. 2. Comprising 111 entries, the dataset spans various biology subjects. A basic bag-of-words method revealed that the most common words pertain to blood, cells, DNA, and neurons. The dataset's top 20 words are presented in Fig. 3.

In addition to utilizing traditional educational materials, this research also integrated the ArabicaQA dataset [19] to establish a robust comparative evaluation framework with prevailing studies in the Question Generation (QG) field. ArabicaQA, with its rich variety of question types and comprehensive scope, provided a pivotal benchmarking tool as shown in Figs. 4 and 5 respectively. This allowed for a nuanced assessment of our model's performance relative to well-established models in the domain. Employing ArabicaQA enabled an in-depth exploration of the model's capabilities in handling complex QA tasks

Fig. 2. Arabic Dataset Word Cloud.

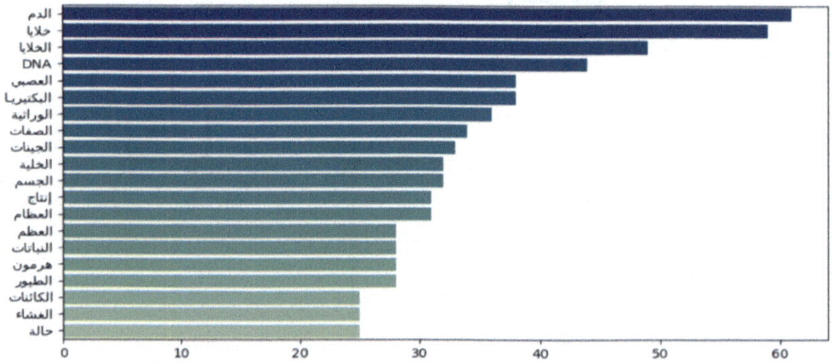

Fig. 3. Frequency of Top 20 Words in the Dataset.

in Arabic, underscoring its potential for application in educational and informational contexts where Arabic is the primary medium of instruction.

Fig. 4. ArabicaQA Dataset Word Cloud.

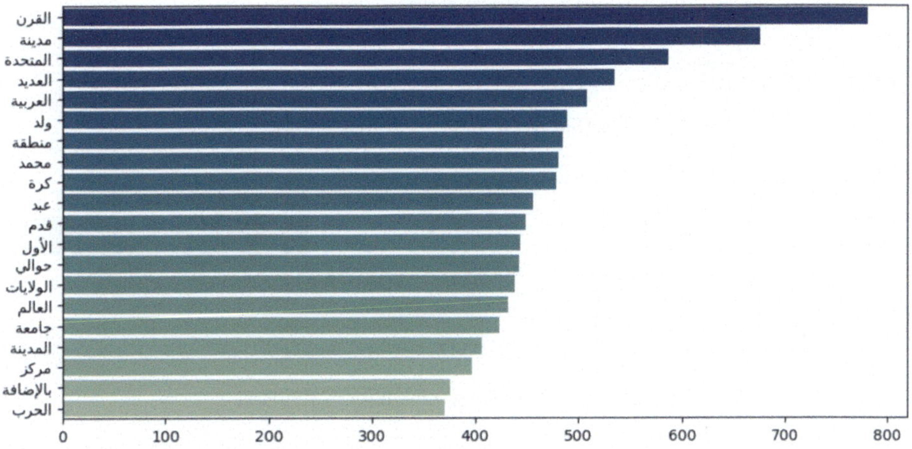

Fig. 5. Frequency of Top 20 Words in the ArabicaQA Dataset.

2.2 Key Phrase Extraction

In our proposed framework, we integrate KeyBert [20], a method that stands out for its simplicity and effectiveness in keyword extraction. KeyBert leverages BERT's sophisticated language models, focusing on the text's context and semantics. Unlike conventional keyword extraction techniques based on frequency, KeyBert embeds the entire text in a high-dimensional vector space to discern the most representative words or phrases of the text's overall theme.

KeyBert functions by first generating document-level representations using BERT embeddings as shown in Fig. 6. It then extracts embeddings for N-gram words or phrases as illustrated in Fig. 7. Then it applies cosine similarity to identify the words or phrases that are closest to the document's overall vector representation as in Fig. 8. Those words or phrases that show the highest similarity are then considered the most descriptive of the document. This method ensures that the keywords extracted are not only contextually right but also encapsulate the essential meaning and theme of the text.

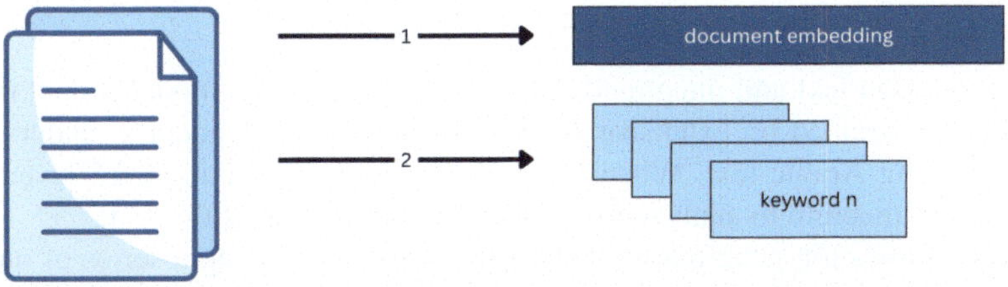

Fig. 6. Document-Level Representation Generation with KeyBert.

A primary drawback of the KeyBert-generated phrases is that they can sometimes deviate from the main topic. This occurs because KeyBert relies on basic n-grams to select phrases initially, before applying embedding techniques. Although experimenting

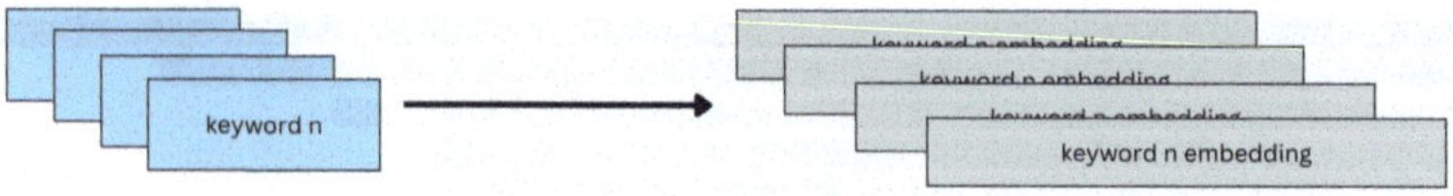

Fig. 7. Extracting N-Gram Embeddings with KeyBert.

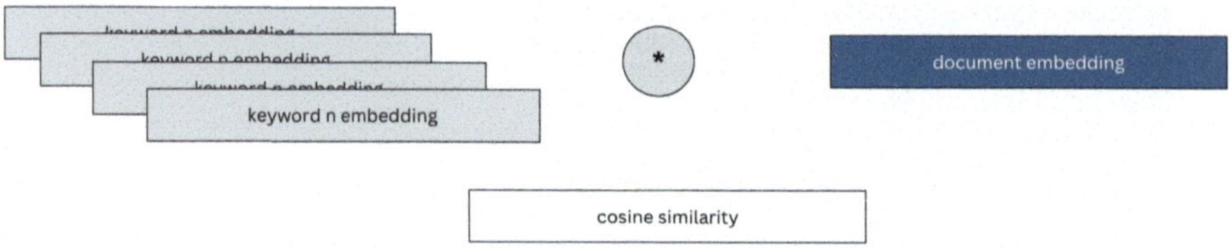

Fig. 8. Cosine Similarity Matching in KeyBert.

with various n-gram sizes might optimize results for a specific dataset, this method is not scalable to other datasets. To address this issue, a more advanced approach, such as incorporating part-of-speech (PoS) tagging, as mentioned in [21, 22], has been integrated into our framework. This technique involves tagging the original text with PoS tags before calculating embeddings. The KeyphraseVectorizer is comprised of a suite of vectorizers designed to identify keyphrases in text documents using specific part-of-speech patterns. These key phrases are then transformed into a document-keyphrase matrix. This matrix is a structured representation where each row corresponds to a distinct text document and each column represents unique keyphrases. The primary function of this matrix is to track the occurrence frequencies of these keyphrases across the document collection. Advantages of using the KeyphraseVectorizer include the ability to accurately extract keyphrases based on their part-of-speech tags, eliminating the need for predefining n-gram ranges. It also offers the capability to produce document-keyphrase matrices efficiently. The vectorizer supports multiple languages, enhancing its versatility. The outcome of this process is a set of key phrases that effectively summarize the original texts as shown in Table 1. These phrases are then utilized in the subsequent question-generation step.

2.3 ARAT5 Invocation

With the processed text and integrated key phrases, the ARAT5 model is then invoked. This model, pre-trained on extensive Arabic language data, is adept at understanding and manipulating Arabic text. When fed with the prepared input, araT5 leverages its transformer architecture to analyze the context and semantics of the text.

The actual question generation involves the model producing a series of questions based on the input text. The transformer model's capability to understand context allows it to generate questions that are not only syntactically correct but also semantically relevant to the text. The model uses the extracted key phrases from the previous step as potential question worthy phrases or concepts within the text and formulates questions accordingly as shown in Table 1.

Table 1. Arabic Extracted KeyPhrases and Generated Questions Samples.

Document	Extracted Key Phrase	Generated Question (ARAT5)	Generated Question (GPT4)
يوجد في الطبيعة مجموعة من العناصر يبلغ عددها حوالي 92 عنصرًا، (20 - 25 %) منها عناصر أساسية للكائن الحي ...	عناصر أساسية للكائن الحي	ما هي العناصر الأساسية في الطبيعة؟	ما هي العناصر الأساسية التي يحتاجها الكائن الحي ليعيش بصحة جيدة؟
يُعد (غريغور مندل) واضع حجر الأساس لعلم الوراثة، وهو أول من توصل إلى نتائج ذاتِ ...	الصفات المندلية لدى الإنسان	ما هي الصفات الوراثية عند الإنسان؟	ما هي الصفة الوراثية التي كانت تظهر في جميع فراد الجيلين الأول والثاني في تجارب مندل؟

2.4 Evaluation

For a comprehensive evaluation of the questions produced by our proposed framework, we contrast these questions with benchmark questions derived from the same corpus using the state-of-the-art GPT-4. Our assessments indicated that our proposed framework attained a precision of 83.50%, a recall of 78.68%, and an Fl score of 80.95% relative to GPT- 4. Furthermore, a human expert assessed the quality of the generated questions on a scale from 1 to 5, where 1 signifies a very poor question format and 5 represents an outstanding question format. This human evaluation process revealed additional insights into the qualitative aspects of our model's performance. The expert's ratings indicated that most questions scored above 3, suggesting a generally high standard in question formulation. Specifically, the average rating was found to be 4.2 (84%), reflecting the model's ability to formulate coherent and relevant questions. This humancentric evaluation complements the quantitative metrics, offering a more nuanced understanding of the model's capabilities in generating contextually appropriate and linguistically sound questions.

3 Results and Discussion

For a comparison evaluation of our proposed model, this study utilizes the ArabicaQA dataset [19] to demonstrates competitive performance in comparison with other established models, particularly in the METEOR metric where it significantly outperforms its counterparts. Specifically, our model achieved a METEOR score of 32.75, surpassing models like Alhashedi et al. [16] (23.00), ERNIE-GENBASE Xiao et al. [15] (25.13), and UNILMLARGE Dong et al. [14] (25.06), where [16] was tested on Arabic while the other two studies in [14, 15] were tested on English. This suggests a strong ability of our model to generate semantically relevant responses that align closely with human judgments, a key strength in METEOR's evaluation approach.

In terms of the BLEU score, our model posted a 17.31, which is lower compared to the models evaluated on the English SQuAD dataset but slightly below the Alhashedi model evaluated on the Arabic mMARCO dataset. This indicates a modest capability in matching the exact word sequences expected in the reference texts, which could be attributed to linguistic and structural differences in the datasets, as BLEU tends to be more sensitive to syntactic variations.

The ROUGE-L score of 51.60 for our model is commendably high, illustrating robustness in capturing the longest common subsequences and thus maintaining the salient points of the references. This score is competitive with the Alhashedi model (51.99) and surpasses that of the ERNIE-GENBASE model (50.58), highlighting our model's proficiency in retaining critical information in generated responses.

Overall, the proposed model exhibits a strong performance across key metrics as shown in Fig. 9, particularly excelling in semantic accuracy as evidenced by the METEOR scores. The results suggest that our approach is particularly effective for the Arabic language, where it manages to maintain a balance between semantic relevance and critical information retention. This capability marks a significant step forward in developing models tailored to Arabic, a language that poses unique challenges due to its rich morphology and syntax. The variations in performance across different metrics underscore the importance of a multi-faceted evaluation to fully capture a model's capability in handling the intricacies of language processing in different linguistic and data contexts (Table 2).

Table 2. Model Evaluation Metrics.

Model	Dataset	BLEU	METEOR	ROUGE-L	Language
Wang et al. [13]	SQuAD	14.78	18.61	41.87	English
UNILMLARGE Dong et al.[14]	SQuAD	22.12	25.06	51.17	English
ERNIE-GENBASE Xiao et al. [15]	SQuAD	22.28	25.13	50.58	English
Alhashedi et al. [16]	mMARCO	19.12	23.00	51.99	Arabic
The Proposed model	ArabicaQA	17.31	32.75	51.60	Arabic

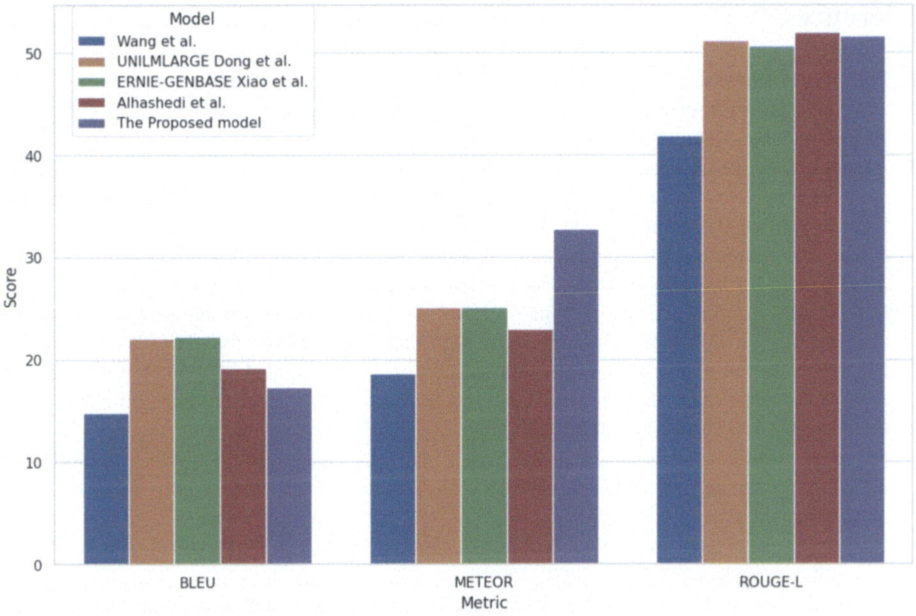

Fig. 9. Comparison of Model Performance on Evaluation Metrics.

4 Conclusion and Future Work

This study introduced an innovative Arabic Question Generation (QG) framework that utilizes an existing pre-trained model without the need for fine-tuning, achieving results comparable to those of fine-tuned models. Our approach integrated KeyBert for keyword extraction and the araT5 transformer model for question generation, achieving high accuracy and efficiency. The model's impressive performance, with a precision of 83.50%, recall of 78.68%, and Fl score of 80.95%, coupled with a human expert rating of 4.2 out of 5, underscores its effectiveness in generating contextually relevant and educationally valuable questions. In addition, the proposed framework achieved a comparable results for many metrics like Blue (17.31), ROUGE-L (51.6) and METEOR (32.74).

This contribution significantly addresses the challenges in the Arabic language processing domain, providing an efficient and effective solution for generating high-quality educational questions from extensive texts. Our approach demonstrates the power and versatility of pre-trained models, showcasing their potential to perform well across different languages and tasks without extensive resource investment in fine-tuning. The success of this project underscores the role of NLP in creating dynamic, adaptive educational tools, and sets the stage for future exploration in NLP and education. The findings suggest promising directions for expanding these methods to other subjects and languages, potentially transforming how educational content is delivered and assessed. This research not only contributes a novel approach for Arabic question generation but also establishes a foundation for future advancements in automated question generation for diverse linguistic contexts.

References

1. Patil, P.M., Bhavsar, R.P. and Pawar, B.V.: A review on natural language processing based automatic question generation. In 2022 international conference on augmented intelligence and sustainable systems (ICAISS), pp. 01–06. IEEE (2022)
2. Alazani, S.A., Mahender, C.N.: Rule based question generation for Arabic text: question answering system. In: Proceedings of the international conference on data science, machine learning and artificial intelligence, pp. 7–12 (2021)
3. Jaradat, S., Alhadidi, T.I., Ashqar, H.I., Hossain, A., Elhenawy, M.: Exploring traffic crash narratives in Jordan using text mining analytics (2024). arXiv preprint arXiv:2406.09438
4. Blšták, M., Rozinajová, V.: Automatic question generation based on sentence structure analysis using machine learning approach. Nat. Lang. Eng. **28**(4), 487–517 (2022)
5. Bachiri, Y.A., Mouncif, H.: Applicable strategy to choose and deploy a MOOC platform with multilingual AQG feature. In: 2020 21st International Arab conference on information technology (ACIT), pp. 1–6. IEEE (2020)
6. Masri, S., Raddad, Y., Khandaqji, F., Ashqar, H.I., Elhenawy, M.: Transformer models in education: summarizing science textbooks with AraBART, MT5, AraT5, and mBART (2024). arXiv preprint arXiv:2406.07692
7. Raffel, C., et al.: Exploring the limits of transfer learning with a unified text-to-text transformer. J. Mach. Learn. Res. **21**(140), 1–67 (2020)
8. Wang, Z., Xu, X., Li, X., Wei, H., Huang, D.: An improved nested named-entity recognition model for subject recognition task under knowledge base question answering. Appl. Sci. **13**(20), 11249 (2023)
9. Elmadany, A., Abdul-Mageed, M.: Arat5: text-to-text transformers for Arabic language generation. In: Proceedings of the 60th annual meeting of the association for computational linguistics, vol. 1: Long papers, pp. 628–647 (2022)
10. Einieh, Y., Almansour, A., Jamal, A.: Fine tuning an AraT5 transformer for Arabic abstractive summarization. In: 2022 14th international conference on computational intelligence and communication networks (CICN), pp. 194–198. IEEE (2022)
11. Nwafor, C.A., Onyenwe, I.E.: An automated multiple-choice question generation using natural language processing techniques. Int. J/ Natl. Lang. Comput. **10**(2), 26–39 (2021)
12. Sharma, H., Mathur, R., Chintala, T., Dhanalakshmi, S., Senthil, R.: An effective deep learning pipeline for improved question classification into bloom's taxonomy's domains. Educ. Inf. Technol. **28**(5), 5105 (2023)
13. Wang, B., et al.: Neural question generation with answer pivot. In: Proceedings of the AAAI conference on artificial intelligence (2020)
14. Dong, L., et al., Unified language model pre-training for natural language understanding and generation (2019). arXiv pre-print server
15. Xiao, D., et al.: ERNIE-GEN: an enhanced multi-flow pre-training and fine-tuning framework for natural language generation (2020). arXiv pre-print server
16. Alhashedi, S.S., Suaib, N.M., Bakri, A.: Arabic automatic question generation using transformer model. In: AIP conference proceedings, vol. 2991, no. 1. AIP Publishing (2024)
17. Radwan, A., Amarneh, M., Alawneh, H., Ashqar, H.I., AlSobeh, A., Magableh, A.A.A.R.: Predictive analytics in mental health leveraging LLM embeddings and machine learning models for social media analysis. Int. J. Web Serv. Res. (IJWSR) **21**(1), 1–22 (2024)
18. Al-Maroof, R.S., et al.: Students' perception towards behavioral intention of audio and video teaching styles: an acceptance study. Int. J. Data Netw. Sci. **6**(2), 603 (2022)
19. Abdallah, A., et al.: Arabicaqa: A comprehensive dataset for Arabic question answering (2024). arXiv preprint arXiv:2403.17848

20. Sharma, P., Li, Y.: Self-supervised contextual keyword and keyphrase retrieval with self-labelling (2019)
21. Schopf, T., Klimek, S., Matthes, F.: Pattemrank: leveraging pretrained language models and part of speech for unsupervised keyphrase extraction. In: Proceedings of the 14th international joint conference on knowledge discovery, knowledge engineering and knowledge management - KDIR, INSTICC. SciTePress, pp.243–248 (2022)
22. Sammoudi, M., Habaybeh, A., Ashqar, H.I., Elhenawy, M.: Question-answering (QA) model for a personalized learning assistant for Arabic Language (2024). arXiv preprint arXiv:2406.08519

Transformer Models in Education: Summarizing Science Textbooks with AraBART, MT5, AraT5, and mBART

Sari Masri[1], Yaqeen Raddad[1], Fidaa Khandaqji[1], Huthaifa I. Ashqar[2,3], and Mohammed Elhenawy[4,5](✉)

[1] Department of Natural, Engineering and Technology Sciences, Arab American University, P.O. Box 240, Ramallah, Palestine
{s.masri3,y.raddad,f.khandaqji}@student.aaup.edu
[2] Civil Engineering Department, Arab American University, P.O. Box 240, Jenin, Palestine
Huthaifa.ashqar@aaup.edu
[3] Columbia University, New York, NY, USA
[4] CARRS-Q, Queensland University of Technology, Brisbane, Australia
mohammed.elhenawy@qut.edu.au
[5] Centre for Data Science, Queensland University of Technology, Brisbane, Australia

Abstract. In an era of rapidly expanding digital archives, effective text summarization tools are crucial for educators and learners. This study presents an advanced text summarization system specifically designed for Arabic textbooks, leveraging state-of-the-art natural language processing (NLP) models. We trained and evaluated four models - AraBART, mBART50, AraT5, and MT5 - on biology textbook content for 11th and 12th grades in the Palestinian curriculum using the ROUGE metric to compare their performance. Our results show that AraBART and mBART50 outperformed the other models in terms of ROUGE-1 (score: 0.24), ROUGE-2 (score: 0.118–0.139) and ROUGE-L (score: 0.24), indicating their effectiveness in summarizing complex biological information while preserving essential details. To further validate the performance of these models, we conducted a subjective evaluation by an educational expert who assessed the summaries for their accuracy, relevance, and overall quality, with AraBART ranking highest due to its ability to capture key concepts and reduce noise in the original text. The study's findings provide valuable insights into the performance of these models and contribute to the development of effective tools for understanding and generating Arabic text.

Keywords: Text Summarization · Arabic Textbooks, · Natural Language Processing · Transformer · AraBART · Rouge Metric

1 Introduction

Text summarization technology is fundamental and has many uses such as information retrieval and language analysis. Summarization can be divided into two types: Abstractive and Extractive. Extractive summarization creates summaries using exact

S. Masri, Y. Raddad and F. Khandaqji—Authors contributed equally to this work.

A. Abdelgawad et al. (Eds.): ISBCom 2024, LNNS 1268, pp. 286–300, 2025.
https://doi.org/10.1007/978-3-031-82377-0_25

words from the original text, while abstractive summarization generates new sentences or paraphrases, often using new words not found in the source [1]. As the volume of textual information on the internet grows, it becomes increasingly challenging for readers to search through and find valuable content. This challenge has led researchers to develop automated text summarization technologies that generate concise summaries capturing the key information of texts. Although this field is rich in research, Arabic text studies face significant challenges due to the language's complexity and limited NLP (Natural Language Processing) resources. The intricate word structure in Arabic adds to the Morphological Complexity, complicating text analysis. Script Issues further challenge text segmentation due to the cursive and variable letter shapes in Arabic writing. Additionally, the lack of Annotated Corpora restricts the availability of essential data for developing robust NLP models, particularly for complex tasks like text summarization that demand a thorough linguistic and contextual grasp.

The field of Natural Language Processing (NLP) endeavors to develop techniques capable of handling various language tasks like translation, summarization, and generation. These tasks all require extracting meaning from text [2]. However, the current single-task training approach produces limitations, as it requires creating specific models for each task and domain, leading to inefficiencies. Recurrent Neural Networks (RNNs) have been widely used in NLP but face challenges with long sentences and parallelization due to their inherent structure. To address these limitations, the Transformer model was introduced. Unlike RNNs, Transformers rely only on attention mechanisms, eliminating issues like vanishing gradients and enabling easier parallelization. This advancement facilitates the training of more flexible networks for NLP tasks [3].

Pre-training language models have proven to be beneficial in enhancing the performance of various natural language processing (NLP) tasks; this includes tasks at the sentence level, such as natural language inference and paraphrasing, which focus on identifying relational ships between sentences through holistic analysis. It also covers token-level tasks like named entity recognition and question answering, which provide detailed outputs for each token [4].

The underlying problem in this paper is how to create efficient and effective Arabic text summarization specifically for education, the challenge stems from the complexity of the morphology of Arabic, the works based on scripts, and the limitation of annotated corpora. Therefore, we proposed to use state-of-the-art language models as a potential solution to these challenges, operating to enhance the summarization of Arabic text by reducing the need to a high-level of user specified details. We looked to creating dataset and fine-tuning pre-trained models in a diverse way to increase their understanding and generation-performance for educational content in Arabic. By overcoming challenges related to the organization and linguistic standardization of educational content, we have contributed innovative solutions that enhance the models' ability to accurately understand and generate educational texts.

Our contribution to Arabic text processing and summarization can be summarized as follows:

1. Collection and revision of a specialized dataset: A dataset has been created from the biology textbooks for the 11th and 12th grades of the Palestinian curriculum, focused on the unique addition provided by the paper to the research field, this selection of

the textbook was based on its comprehensive and accurate content, and the texts were carefully collected to ensure diversity of the curriculum. The dataset is diverse and rich in covering complex scientific concepts, providing a strong foundation for training deep language models.

2. Training MT5 [5], AraBART [6], AraT5 [7] and mBART50 [8] models to enhance understanding of educational content and efficient text generation and summarization using the specialized datasets collected from the biology textbooks for the 11th and 12th grades of the Palestinian curriculum. The training process was designed to improve the models understanding of the educational content presented in the textbooks and to increase their ability to generate and summarize texts efficiently. During the training process, the training parameters and performance metrics were fine-tuned based on the unique characteristics of the educational content, this contributed to improving the quality and accuracy of the models in understanding and generating summaries for educational materials.
3. Manual evaluation by an expert: this method involves evaluating summaries generated by pre-trained models where model expert assigns a rating from 1 to 10 based on specific criteria related to the quality of the summary, its accuracy, and its capture and of the key elements from the original text. We employed this approach to assess the summaries generated by the MT5, AraBART, AraT5 and mBART50 models. An expert with domain knowledge in the subject matter was tasked with evaluating the generated summaries based on criteria such as coherence, informativeness and relevance to the original text.

This research paper stood out for its application of advanced training methodology focused on enhancing the model's performance in understanding educational content and generating comprehensive and accurate summaries. This paper begins with a careful review of previous studies on language representation, especially Arabic. The second section highlights the achievements and reveals the research gaps that our research intends to fill. The third section details the methodology used to develop the models, including the data collection and processing steps and the strategies used to train the models. In particular, we highlight how MT5, AraBART, AraT5 and mBART50 models can be modified. In addition to the subsequent tasks and datasets used to evaluate the models with a framework for understanding their performance. In Section IV, we discussed the experimental setup and achieved results. We finally conclude our work in Section V, presenting our conclusions and suggestions for future research, and emphasizing the importance of continued development and improvement along this path.

2 Literature Review

Most of the related work in text summarization has utilized extractive and abstractive approaches, which involve identifying important sentences in the original document to obtain the final summary. Al-Maleh et al. [9] on the utilization of deep learning techniques, employing extractive and abstractive summarization methods to summaries Arabic texts. El-shishrawy et al. [10] rephrased sentences to align with the research context and they used machine learning for text summarization. Sobh et al. [11] followed an extractive approach in extracting Arabic texts by dividing the problem into two subproblems, identifying extracting sentences using machine learning techniques. On the other hand, utilized deep learning in natural language processing for text summarization, employing sequence-to-sequence models by Bahdabau et al. [12], and Sutskever et al. [13] neural networks were employed for text summarization, and the LSTM model was used for translating phrases from English to Arabic. Rush et al. [14] the researchers utilized an attention-based approach for creative sentence summarization by employing deep learning techniques. In this method, each word in the summary is generated based on the original sentences in the original text. Reda et al. [15] the researchers employed both extractive and abstractive summarization approaches to summarize articles using AraBERT, they introduced a sequence-based model that relies on machine transformer techniques and the pre-trained MT5 model. In addition, the researchers in this study introduced pre-trained transfer learning models, such as BERT, which can perform various tasks including text summarization, question answering, and language inference [16, 17]. These models leverage deep learning techniques and have shown significant effectiveness in handling diverse natural language processing tasks by Devlin et al. [18] adding an important contribution to previous studies is the summarization of Arabic texts based on a general engineering approach for Natural Language Understanding (NLU) and Natural Language Generation (NLG). This involves utilizing pre-trained transfer models such as AraBERT, BERT, XLNet, and XLM. The goal is to extract the most important sentences from the original text, enhancing the process of producing and comprehending natural language through NLU and NLG techniques by Abu Nada et al. [19], Wisam et al. [20], researchers also trained transfer models (BERT) for processing Arabic texts, and the results showed advanced performance in most natural language processing tasks. Kieuvongngam et al. [21] the researchers utilized modern models such as BERT and GPT-2 to summarize medical articles related to COVID-19. ALmarjeh et al. [22] researchers compared the model architectures based on RNN and how they were utilized in summarizing Arabic texts, such as BERT, AraGPT-2, AraT5. Kahla et al. [23] researchers developed models for summarizing extractive texts in Arabicusing the AraSum dataset, which contains 50,000 articles they trained and improved BART and MT5 models, and the results showed that these models outperformed previous models in this field dataset.

All previous studies must improve the understanding and processing of natural language, specifically addressing Arabic challenges. All studies employed extractive and abstractive approaches in text summarization, with some combined both Methods. They used transformer-based models such as BERT, GPT and AraT5, trained on substantial data. Previous studies focused on utilizing word embedding techniques [24], such as Word2Vec, to understand Arabic language models. In Arabic text generation, RNN models were first used, but BERT models surpassed them in quality, readability, and linguistic sequence. All studies used evaluation tools, such as ROUGE and F1-SCORE, to assess the performance of models. Some studies also relied on manual evaluation with the assistance of language experts, translating to English. The studies provided focus on the development of an extractive Arabic text summarizer using a general-purpose architecture for Natural Language Generation (NLG) and Natural Language Understanding (NLU) models, such as AraBART, BERT, XLNet, XLM, etc. The goal is to summarize Arabic documents by evaluating and extracting the most important sentences. The researchers propose an extractive summarizer based on AraBART and compare its efficiency using the Rouge measure and human evaluation.

3 Materials and Methods

The main aim of this study is to present and evaluate a new method for summarization of Arabic school textbooks by modifying and extending existing methods to include the use of mBART50, AraBART, MT5 and AraT5 models while applying improvement to these models to improve the summarization process, this improvement aim compares the results with expert human summarization to demonstrate the effectiveness and improved achieved. Figure 1 presents the structured methodology employed in our study to ensure systematic handling and analysis of the data collected. This methodology is segmented into four principal stages: Data Collection, Data Preparation, Modeling, and Summarization and Evaluation.

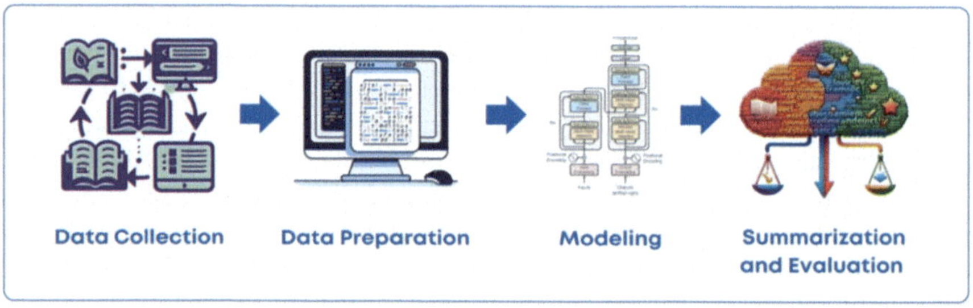

Fig. 1. Biology schoolbook summarization methodology.

3.1 Data Collection and Preparation

In the framework of our current project, which aims to develop a model for the synthesis of scientific texts, we have taken a strategic step by focusing on data collection from a rich and unique source: Biology school textbooks for 11th and 12th grades.

This approach to data collection is qualitative in the development of automated summary models. The text extraction process has been carried out with great care while maintaining the context and meaning of the original content. This process covered not only the basics but also graphics and interactive schemes that were possible and useful, the dataset we built consists of 6 features and 111 rows, the dataset scheme is as shown below (Table 1).

Data cleaning and preparation are crucial steps in enhancing the quality of the dataset and ensuring the effectiveness of training and accuracy of summarization for language models. In our research, a series of methodical techniques were applied to process the texts and summaries extracted from the dataset of school textbooks. Initially, we removed unnecessary marks that could distort the text. Secondly, we eliminated leading

Table 1 Dataset Details and Column Descriptions [26]

Features	Description
id	A unique identifier for each section of the book
unit_title	The title of the educational unit within the book
lesson_title	The title of the specific lesson within the unit
section_content	The title of the section within the lesson
questions	Questions related to the section content which will enhance understanding
expert_summary	The expert's summary to be used in the evaluation process

and trailing spaces and unified multiple spaces between words into a single space to achieve consistency in text formatting. Thirdly, to avoid complexities associated with varying case sensitivities, all English characters in texts and summaries were converted to lowercase. After these processes, any empty records were discarded to ensure incomplete data did not affect the quality and effectiveness of training. These methodical steps for data cleaning and preparation guarantee the creation of a unified and clean dataset, which contributes to enhancing the training capability and overall performance of the language model in the task of text summarization.

The figure below indicates that within the dataset, a significant number of entries have a word count exceeding 1000 in the section content feature. This results in a maximum input tokenizer length of 1024 for the section content having most of the observations into consideration, and 256 for the expert summary. The TokenizerFast library from Hugging Face is employed for tokenizing sentences. The dataset was divided into 70% for training, with the remaining 30% equally split between validation and testing (Fig. 2).

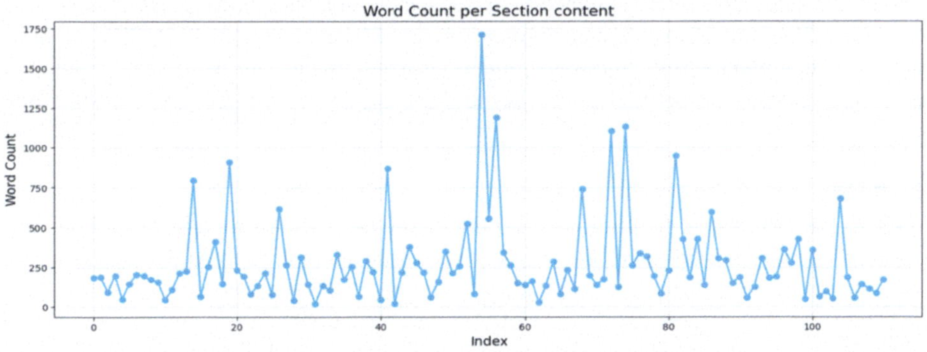

Fig. 2. The word count in section_content feature.

3.2 Modeling

After optimizing the mBART50, AraBART, MT5 and AraT5 models, they were directly used to generate summaries, this approach benefits from the advantages in text understanding and generation capabilities of the models to create concise and useful summaries, providing a valuable tool for students and teachers for quick access to essential information (Table 2).

Table 2. Hyperparameters of the transformer model [26]

Hyperparameter	Value
Epoch	10
Learning rate	5×10^{-5}
Weight decay	1%
Batch size	4
Optimizer	AdamW

In our research, we focused on exploring a range of advanced neural network models for Arabic text processing tasks, including translation and summarization. Each model was trained under a consistent set of hyperparameters: we employed a learning rate of 5×10^{-5}, conducted training over 10 epochs, and set a weight decay of 1%. The models were trained with a batch size of 4 using the AdamW optimizer, aiming to optimize performance while maintaining computational efficiency.

MBART50

mBART50, a multilingual sequence-to-sequence model, was utilized for Arabic text translation. Pre-trained in a diverse corpus covering 50 languages, mBART50 is particularly adept at understanding and translating contextual nuances in Arabic. We evaluated its performance by assessing the accuracy and fluency of the translations, ensuring they retained the semantic integrity of the original texts [8].

AraBART

AraBART, tailored specifically for the Arabic language, extends the BART architecture to support generative summarization tasks. Its unique adaptation to Arabic morphology and script nuances enhances its capability to produce coherent and concise summaries of Arabic text. The effectiveness of AraBART was measured using standard metrics like ROUGE scores, which helped in evaluating the relevance and information retention in the summaries [6].

AraT5

The Summarization-arabic-english-news model specializes in bilingual summarization, converting Arabic news articles into English summaries. This model bridges language barriers and facilitates understanding for non-Arabic speakers. We focused on the quality of the English summaries, evaluating them based on accuracy, readability, and the preservation of essential information [25].

MT5

The Multilingual T5 (MT5) model was instrumental in summarizing Arabic texts, showcasing its versatility across multiple languages. We focused on MT5's summarization capabilities, evaluating the quality and coherence of its outputs. The performance assessments were based on ROUGE scores, highlighting the model's efficiency in producing relevant and high-quality summaries [5].

3.3 Evaluation

1. Manual evaluation by experts: This method relies on evaluating summaries generated by pre-training models, where the expert assigns a rating from 1 to 10 based on specific criteria related to the quality of the summary, its accuracy and its capture of the key elements from the original text.
2. Comparing expert summaries with pre-training models' summaries and the original text: this method involves comparing the summaries prepared by experts with those generated by pre-training models, as well as considering the original text.
3. Using automated evaluation metrics:

$$ROUGE - 1 = \frac{\sum_{unigramsinreference} Count_{match}(unigram)}{\sum_{unigramsinreference} Count(unigram)} \quad (1)$$

$$ROUGE - 2 = \frac{\sum_{bigramsinreference} Count_{match}(bigram)}{\sum_{bigramsinreference} Count(bigram)} \quad (2)$$

$$ROUGE - L = \frac{LengthofLongestCommonSubsequence}{LengthofReferenceSummary} \quad (3)$$

ROUGE-1 (Eq. (1)) assesses the degree of overlap between unigrams present in the generated summary and those found in the reference summaries. It quantifies the number of individual words from the generated summary that appear in the reference summary, thereby providing a straightforward metric for content overlap at the word level. ROUGE-2 (Eq. (2)) assesses the degree of overlap between bigrams present in the generated summary and those found in the reference summaries. ROUGE-2 scores consecutive word pairs, providing insight into the coherence and ordering of information in the summaries and reflecting the extent to which phrases and more subtle content structures are captured by the summary system. ROUGE-L (Eq. (3)) focuses on the longest common subsequence between the generated summary and the reference summaries. Unlike ROUGE-2, it does not require consecutive matches, but instead scores the longest sequence of words that occur in the same order in both texts. This measure is particularly useful for assessing the structural similarity of the summary to the reference. It provides insight into the extent to which the key thematic elements are retained, even if the exact wording differs.

Table 3. Summarization Model Evaluation Metrics (Rouge score)

Model	rouge1	rouge2	rougeL
AraBART	0.2462	0.1184	0.2462
mBART50	0.2431	0.1392	0.2431
AraT5 (summarization-arabic-english-news)	0.2065	0.1153	0.2065
MT5	0.0566	0	0.0566

Table 4. Summarization Model Evaluation by Expert

Model	Rating (0–10)
AraBART	8.409
mBART50	8.18
AraT5 (summarization-arabic-english-news)	7.727
MT5	7.17

4 Results and Discussion

Table 3 provides a comprehensive overview of the performance of the different summary models evaluated using the ROUGE metric, which measures the overlap of n-grams between the generated summaries and the reference summaries. The metrics included in this analysis are ROUGE-1 (unigram overlap), ROUGE-2 (bigram overlap), ROUGE-L (longest common subsequence).

The models' performance as reflected in Table 4 likely indicates their effectiveness in handling specialized academic content. AraBART, with the highest rating of 8.409, demonstrates a strong capability in accurately summarizing complex biological information, which is critical for educational tools and resources. The slightly lower scores for the other models, including mBART50 and AraT5, suggest that while they are also competent, they might not manage specific subtleties or specialized terms as precisely as AraBART. The ratings also imply that these models, including the MT5 with the lowest score of 7.17, could be useful in educational settings, aiding in the creation of concise study materials and summaries of dense textbook content.

The AraBART algorithm demonstrates robust performance across all ROUGE metrics, with scores of 0.1184 for ROUGE-1, ROUGE-L and a ROUGE-2 score of 0.2462. This indicates a notable degree of unigram and bigram overlap with the reference summaries. The degree of overlap with the reference summaries indicates that AraBART is capable of effectively capturing both the surface form and some finer details of the source texts. mBART50 also performs well, particularly in capturing bigram overlap, as evidenced by its ROUGE-2 score of 0.1392, the highest among the models tested. This indicates that mBART50 is particularly adept at comprehending and replicating intricate linguistic structures. Nevertheless, its overall performance in terms of unigrams and longest common subsequence is slightly inferior to that of AraBART.

mBART50 also performs well, particularly in capturing bigram overlap, as evidenced by its ROUGE-2 score of 0.1392, the highest among the models tested. This indicates that mBART50 is particularly proficient in comprehending and replicating intricate textual structures, although its overall performance in terms of unigrams and longest common subsequence is slightly inferior to that of AraBART.

The AraT5 model exhibits moderate effectiveness, with scores of 0.2065 for ROUGE-1, ROUGE-L and a ROUGE-2 score of 0.1153. Although the performance is satisfactory, it is not as robust as that of AraBART or mBART50, indicating that while essential content can be captured, there is room for improvement in capturing detailed content.

MT5 model shows limited performance in these evaluations, with low scores in all metrics. The scores for ROUGE-1 and ROUGE-L are only 0.06 and for ROUGE-2 0, meaning that there is no bigram overlap. This indicates that these models are not able to summarize the given Arabic texts in a way that matches well with the reference summaries. Their poor performance could be due to several factors, such as insufficient training data, suboptimal model tuning, or inherent limitations in handling the complex Arabic language.

The analysis of the model-generated summaries presented in Table 5 provides a comprehensive assessment of the capabilities of different AI models in synthesizing specialized biological content related to sponges. Each model displays specific strengths and demonstrates varying abilities in addressing complex scientific summarization tasks. The expert summary serves as a standard for the efficient summarization of key anatomical and reproductive information about sponges. This summary serves as a standard against which the accuracy and conciseness of the summaries produced by the AI can be assessed. The MT5 model provides a comprehensive summary of the anatomical structure and taxonomic classification of sponges. The approach of this model ensures thorough coverage of the relevant details, although it could benefit from refinement to improve conciseness and maintain the focus observed in the expert summary.

The AraBART algorithm performs commendably, producing a summary that is both comprehensive and closely aligned with the expert's summary. The summary deftly incorporates essential structural and reproductive details, and the inclusion of specific anatomical features -such as inhalation and exhalation openings- adds depth to the summary, even if it goes beyond the core elements prioritized in the expert summary. The summary provided by Summarization-Arabic-English-News model is coherent and accurately reflects the anatomical structure and classification within the animal kingdom. While the summary effectively captures primary structure, expanding its scope to address reproductive strategies more comprehensively could bring it more closely in line with the expert summary and thus improve its completeness.

In conclusion, the performance of these models demonstrates the potential and existing capabilities of AI in the automated summarization of specialized scientific content. Models such as AraBART and MT5 exhibit robust summarization capabilities, suggesting that minor adjustments could enhance their alignment with expert-generated summaries. The summarization models, and namely Summarization-Arabic-English-News, demonstrate potential for improvement, particularly in ensuring comprehensive coverage of all critical aspects of the content. This evaluation highlights the necessity

Table 5. Comparison between summary of expert and model's summary

Model	Summary
Original text	قبيلة الإسفنجيات (المساميات) تعد الإسفنجيات من أبسط الكائنات التي تنتمي إلى المملكة الحيوانية، وهي في مقدّمة سُلّم تصنيف المملكة. نستخدم الإسفنج الصناعي في حياتنا اليومية للاغتسال، وتنظيف الأواني، وأسطح المكاتب، تُعد الإسفنجيات من الكائنات الحية متعدّدة الخلايا، وتُسمّى أيضاً المساميات؛ نظراً لكثرة المسامات في جسمها. تركيب الإسفنجيات: تتركب الإسفنجيات من طبقتين يفصل بينهما هلام متوسط, الطبقة الخارجيّة أشبه ما تكون بالنسيج الطلائي الذي يغطي الجسم، أما الطبقة الداخلية (السطح الداخلي) فهي تتكوّن من خلايا مطوّقة (سوطية) لها عدة وظائف، بينما يحتوي الهلام المتوسّط على مجموعة من الخلايا الأميبية. ويتخلل الطبقات عدداً كبيراً من الفتحات تُسمّى ثقوباً شهيقية، وفتحة زفيرية واحدة أو أكثر موجودة في أعلى الكائن. تكاثر الإسفنجيات: تتكاثر الإسفنجيات لا جنسيا وجنسيا. أولاً: التكاثر اللاجنسي 1- التجزؤ (التشظي): يحدث عند تعرض الإسفنج لقوة مؤثرة كبيرة، مثل مرور تيارات مائية؛ ما يؤدي إلى انفصال جزء منها ما يلبث أن يكوّن كائناً جديداً في منطقة أخرى. 2- التبرعم: ويحدث في جميع الظروف. 3- البريعمات (الدرائر): وتلجأ إليها إسفنجيات المياه العذبة خلال الظروف غير المناسبة، مثل الجفاف، أو تجمد المياه. ثانياً: التكاثر الجنسي: تتكاثر الإسفنجيات جنسياً من خلال إنتاج حيوانات منوية وبويضات من الخلايا الأميبية، وتنتقل الحيوانات المنوية من إسفنج إلى إسفنج آخر، حيث يتم الإخصاب داخلياً في الهلام المتوسط
Expert	قبيلة الاسفنجيات (المساميات): وهي من أبسط الكائنات الحية كالإسفنج نظرا لكثرة المسامات في جسمها. تتركب الإسفنجيات من طبقتين يفصل بينهما هلام متوسط, الطبقة الخارجيّة أشبه ما تكون بالنسيج الطلائي الذي يغطي الجسم، أما الطبقة الداخلية فهي تتكوّن من خلايا سوطية، بينما يحتوي الهلام المتوسّط على مجموعة من الخلايا الأميبية. وتتكاثر الإسفنجيات لا جنسيا من خلال التجزؤ أو التبرعم أو البريعمات, وجنسيا من خلال إنتاج حيوانات منوية وبويضات من الخلايا الأميبية
MT5	قبيلة الإسفنجيات (المساميات): وهي من أبسط الكائنات الحية متعدّدة الخلايا، وتُسمّى أيضاً المساميات؛ نظراً لكثرة المسامات في جسمها. تصنيف المملكة الحيوانية: وهي في مقدّمة سُلّم تصنيفها الحيواني. تركيب الإسفنجيات: تتركب الإسفنجيات من طبقتين يفصل بينهما هلام متوسط، الطبقة الخارجيّة أشبه ما تكون بالنسيج الطلائي الذي يغطي الجسم، أما الطبقة الداخلية فهي تتكوّن من خلايا مطوّقة (سوطية) لها عدة وظائف, بينما يحتوي الهلام المتوسّط على مجموعة من الخلايا الأميبية
AraBART	قبيلة الإسفنجيات (المساميات): وهي كائنات حية متعددة الخلايا, وتسمى أيضا المساميات؛ نظرا لكثرة المسامات في جسمها. مكوناتها: تتركب من طبقتين يفصل بينهما هلام متوسط, الطبقة الخارجية أشبه ما تكون بالنسيج الطلائي الذي يغطي الجسم, أما الطبقة الداخلية (السطح الداخلي) فهي تتكون من خلايا مطوقة (سوطية) لها عدة وظائف بينما يحتوي الهلام المتوسط على مجموعة من الخلايا الأميبية. ويتخلل الطبقات عددا كبيرا من الفتحات تسمى ثقوبا شهيقية، وفتحة زفيرية واحدة أو أكثر موجودة في أعلى الكائن. ثانيا: التكاثر اللاجنسي: تتكاثر الإسفنج لا جنسيا وجنسيا من خلال إنتاج حيوانات منوية وبويضات من الخلايا الاميبية وتنتقل الحيوانات المنوية من إسفنج إلى إسلفنج آخر حيث يتم الإخصاب داخليا في الهلام المتوسط
AraT5 (summarization-arabic-english-news)	تعد الإسفنجيات من أبسط الكائنات التي تنتمي إلى المملكة الحيوانية، وهي في مقدّمة سُلّم تصنيف المملكة. تتركب الإسفنجيات من طبقتين يفصل بينهما هلام متوسط, الطبقة الخارجيّة أشبه ما تكون بالنسيج الطلائي الذي يغطي الجسم، أما الطبقة الداخلية فهي تتكوّن من خلايا مطوّقة (سوطية) لها عدة وظائف، بينما يحتوي الهلام المتوسّط على مجموعة من الخلايا الأميبية

of continuous refinement and domain-specific training to enhance the precision and relevance of AI-generated summaries in specialized fields such as biology. These findings contribute to the ongoing discourse on the further development and optimization of AI-driven summarization tools in the academic and research environment.

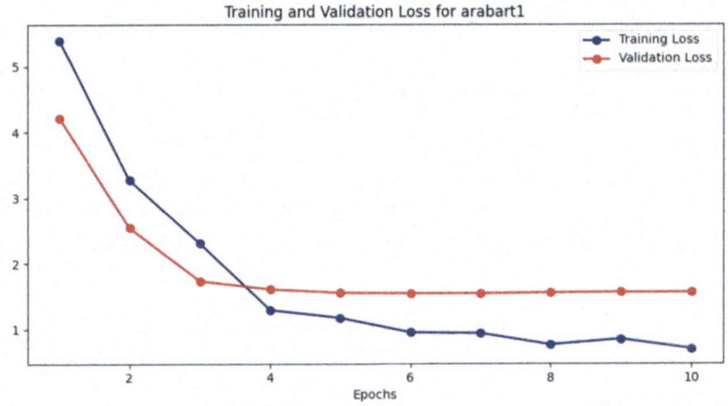

Fig. 3. Training and validation loss for AraBART model.

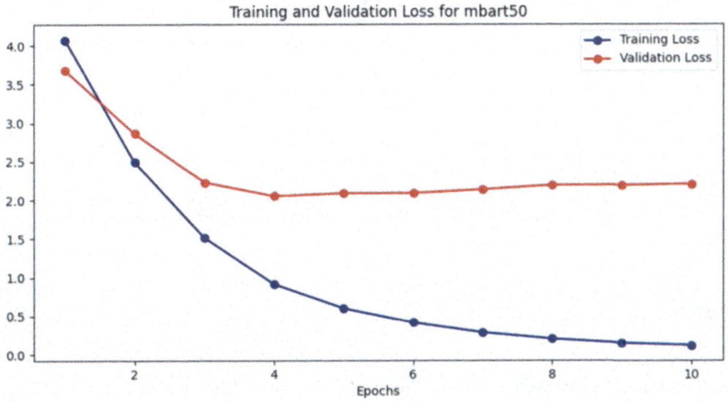

Fig. 4. Training and validation loss for mBART50 model.

In Figs. 3, 4, 5, and 6, the training and validation loss graphs of the AraBART, mBART50, AraT5 (summarization-arabic-english-news), and MT5 models reveal varied learning behaviors and model efficiencies. Figure 3 (AraBART) and Fig. 5 (summarization-arabic-english-news) show promising trends where both training and validation losses decrease sharply and consistently, indicating strong learning capabilities and effective generalization without significant signs of overfitting. In contrast, Fig. 4 (mBART50) and Fig. 6 (mt5) display effective initial learning but suggest potential overfitting in later stages, as evidenced by the validation losses that either flatten or increase while training losses continue to decrease. These observations underscore the need for precise model tuning and the adoption of strategies such as regularization and data augmentation to enhance the models' performance and ensure their ability to generalize to new, unseen data.

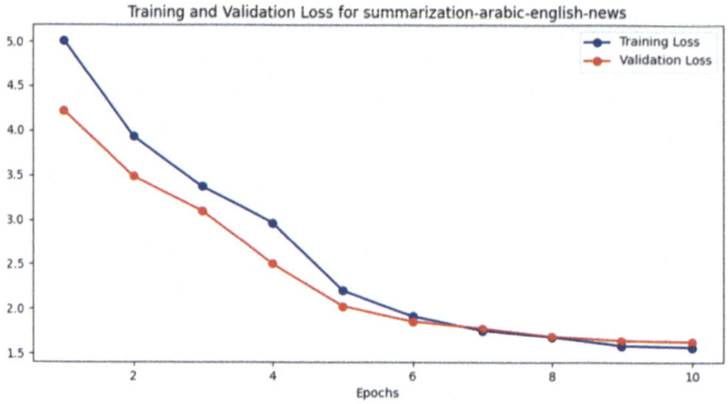

Fig. 5. Training and validation loss for AraT5 model.

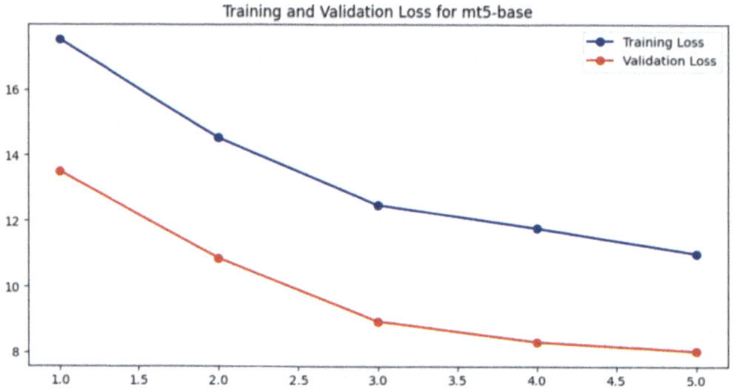

Fig. 6. Training and validation loss for MT5 model.

5 Conclusion

We created a specialized dataset of biology textbooks for 11th and 12th grade students to improve the understanding of advanced linguistic models such as BERT and MT5 for educational content. The study focuses on training NLP models, specifically Ara-BART, mBART50, araT5, and MT5, using this specialized suite to enhance their ability to understand and produce educational content efficiently. Comprehensive evaluation criteria were adopted to evaluate the performance of these models, including manual evaluation by experts in education and textbook writing, and automated evaluation metrics such as ROUGE.

Future work in NLP looks to push the boundaries toward achieving a deeper and more accurate understanding of complex scientific texts, focusing on improving the ability of machine models to summarize these texts efficiently. This includes developing new technologies to overcome the unique linguistic challenges presented by the Arabic language and expanding the scope of training databases to cover a more comprehensive range of scientific topics and concepts. This direction also includes improving the methods used in evaluations, whether automated or manual, to ensure the production of high-quality and effective summaries. In addition, advanced optimization techniques such as LoRA (Low-Rank Adaptation) are being explored to enhance the performance of trained models, promising significant advances in how scientific texts in Arabic are understood and processed in the future.

Acknowledgment. We are deeply grateful to Mr. Ayman Raddad - 16 years' experience in teaching Biology for 11th and 12th grades - and the Palestinian Ministry of Education for providing all the summaries and evaluations as experts to conduct this study.

References

1. Widyassari, A.P., et al.: Review of automatic text summarization techniques & methods. J. King Saud Univ. Comput. Inf. Sci. **34**(4), 1029–1046 (2022). https://doi.org/10.1016/J.JKSUCI.2020.05.006
2. Jaradat, S., Alhadidi, T.I., Ashqar, H.I., Hossain, A., Elhenawy, M.: Exploring traffic crash narratives in Jordan using text mining analytics (2024). arXiv preprint: arXiv:2406.09438
3. Gillioz, A., Casas, J., Mugellini, E., Khaled, O.A.: Overview of the transformer-based models for NLP tasks. In: Proceedings of the 2020 federated conference on computer science and information systems, FedCSIS 2020, pp. 179–183 (2020). https://doi.org/10.15439/2020F20
4. Devlin, J., Chang, M.-W., Lee, K., Toutanova, K.: BERT: pre-training of deep bidirectional transformers for language understanding (2018)
5. Xue, L., et al.: mT5: a massively multilingual pre-trained text-to-text transformer (2020). Available: https://arxiv.org/abs/2010.11934
6. Kamal Eddine, M., Tomeh, N., Habash, N., Le Roux, J., Vazirgiannis, M., Polytechnique, É.: AraBART: a pretrained Arabic sequence-to-sequence model for abstractive summarization, pp. 31–42 (2022). Available: https://huggingface.co/moussaKam/AraBART. Accessed: 05 May 2024
7. marefa-nlp/summarization-arabic-english-news Hugging Face. Available: https://huggingface.co/marefa-nlp/summarization-arabic-english-news. Accessed 05 May 2024
8. Tang, Y., et al.: Multilingual translation with extensible multilingual pretraining and finetuning (2020)
9. Al-Maleh, M., Desouki, S.: Arabic text summarization using deep learning approach. J Big Data **7**(1), 1–17 (2020). https://doi.org/10.1186/S40537-020-00386-7/TABLES/6
10. El-Shishtawy, T., El-Ghannam, F.: Keyphrase based Arabic summarizer (KPAS). In: 2012 8th international conference on informatics and systems (INFOS), Giza, Egypt, pp. NLP-7–NLP-14 (2012)
11. Sobh, I., Darwish, N., Fayek, M.: An optimized dual classification system for Arabic extractive generic text summarization. Giza, Egypt (2009)
12. Bahdanau, D., Cho, K.H., Bengio, Y.: Neural machine translation by jointly learning to align and translate. In: 3rd international conference on learning representations, ICLR 2015 - conference track proceedings (2014). Available: https://arxiv.org/abs/1409.0473v7. Accessed 16 June 2024
13. Sutskever, I., Vinyals, O., Le, Q.V.: Sequence to sequence learning with neural networks. In: Ghahramani, Z., Welling, M., Cortes, C., Lawrence, N., Weinberger, K.Q. (Eds.) Advances in neural information processing systems. Curran Associates, Inc. (2014). Available: https://proceedings.neurips.cc/paper_files/paper/2014/file/a14ac55a4f27472c5d894ec1c3c743d2-Paper.pdf
14. Rush, A.M., Chopra, S., Weston, J.: A neural attention model for abstractive sentence summarization (2015). Available: http://arxiv.org/abs/1509.00685

15. Reda, A., et al.: A hybrid Arabic text summarization approach based on transformers. In: 2022 2nd international mobile, intelligent, and ubiquitous computing conference (MIUCC), Cairo, Egypt, pp. 56–62 (2022). https://doi.org/10.1109/MIUCC55081.2022.9781694
16. Sammoudi, M., Habaybeh, A., Ashqar, H.I., Elhenawy, M.: Question-answering (QA) model for a personalized learning assistant for Arabic language (2024). arXiv preprint arXiv:2406.08519
17. Tami, M., Ashqar, H.I., Elhenawy, M.: Automated question generation for science tests in Arabic language using NLP techniques (2024). arXiv preprint: arXiv:2406.08520
18. Devlin, J., Chang, M.-W., Lee, K., Toutanova, K.: BERT: pre-training of deep bidirectional transformers for language understanding (2018). Available: http://arxiv.org/abs/1810.04805
19. Abu Nada, A., Alajrami, E., Alsaqqa, A., Abu-Naser, S.: Arabic text summarization using AraBERT model using extractive text summarization approach (2020)
20. Antoun, W., Baly, F., Hajj, H.: AraBERT: transformer based model for Arabic language understanding (2020)
21. Kieuvongngam, V., Tan, B., Niu, Y.: Automatic text summarization of COVID-19 medical research articles using BERT and GPT-2 (2020)
22. Bani Almarjeh, M., Kurdy, M.-B.: Arabic abstractive text summarization using RNN-based and transformer-based architectures. Inf. Process. Manag. **60**, 103227 (2023). https://doi.org/10.1016/j.ipm.2022.103227
23. Kahla, M., Novák, A., Yang, Z.: Fine-tuning and multilingual pre-training for abstractive summarization task for the Arabic language. Ann. Math. Inform. (2022). https://doi.org/10.33039/ami.2022.11.002
24. Radwan, A., Amarneh, M., Alawneh, H., Ashqar, H.I., AlSobeh, A., Magableh, A.A.A.R.: Predictive analytics in mental health leveraging LLM embeddings and machine learning models for social media analysis. Int. J. Web Serv. Res. (IJWSR) **21**(1), 1–22 (2024)
25. Bani-Almarjeh, M., Kurdy, M.B.: Arabic abstractive text summarization using RNN-based and transformer-based architectures. Inf. Process. Manag.Manag. **60**(2), 103227 (2023). https://doi.org/10.1016/J.IPM.2022.103227
26. Masri, S.: Biology Textbook for Grades 11 and 12 with Summary. Kaggle (2024). https://www.kaggle.com/datasets/sarimasri/biology-textbook-for-grades-11-and-12-with-summary

CosmiCode: Automated COSMIC Measurement Tool for Arduino Using RegEx and NLP

Mireille Bishay[1(✉)], Salma Salem[1], Milad Ghantous[1], and Hassan Soubra[2]

[1] German University in Cairo, New Cairo, Egypt
{mireille.bishay,salma.abdallahsalem}@student.guc.edu.eg,
milad.michel@guc.edu.eg

[2] ECE - École centrale d'électronique, Lyon, France
hsoubra@ece.fr

Abstract. Efficient software project management relies on sizing and measurement, facilitating control and performance comparison across different projects. COSMIC ISO 19761, presents a method to quantifying software functional size, catering to specific information requirements regardless of the underlying technology. Concurrently, Arduino's accessibility, characterized by its user-friendly interface, open-source nature, and affordability, holds promise for a myriad of applications, including IoT devices and smart cities. The identification of common patterns in Arduino code within the COSMIC framework has spurred the development of CosmiCode. This tool, designed to measure software functional size in Cosmic Function Points (CFPs) from source code, yields an accuracy rate of 98.28% in initial tests. Moreover, an extension utilizing machine learning has been integrated into CosmiCode, showcasing promising capabilities in CFP measurement, with an average accuracy of 88.943%.

Keywords: Functional Size Measurement · IoT · RegEx · NLP · Arduino · COSMIC · ISO 19761

1 Introduction

Like other engineering disciplines, software engineering practitioners and industry are concerned with understanding and assessing software development. By using efficient measuring techniques, organisations can more efficiently plan, control, estimate, and execute software work [1]. "You cannot control what you cannot measure", as Tom DeMarco stated in [2]. A key engineering idea that aids managers in developing practical plans for enhancing software management is measurement.

Abran *et al.* [3] state that software size measurement serves two main purposes: enabling effective project management by providing key performance indicators like project delivery within set timelines and budgets, and facilitating the estimation of effort and duration for future projects using past performance data.

A. Abdelgawad et al. (Eds.): ISBCom 2024, LNNS 1268, pp. 301–315, 2025.
https://doi.org/10.1007/978-3-031-82377-0_26

COSMIC (Common Software Measurement International Consortium) was established in 1998 to develop a new, advanced FSM method based on solid software engineering principles. The COSMIC method represents a significant evolution in FSM, offering a structured and principled approach to software size measurement. Automation of COSMIC measurements is deemed essential for broadening its adoption and acceptance within the industry, improving time-reduction, effort required for size measurement, bench-marking, maintenance, and working with legacy software [4].

Arduino's widespread adoption is attributed to its user-friendly and versatile nature, facilitating a broad spectrum of projects from hobbyist endeavors to sophisticated industrial applications [5]. Measuring the size of software functions, particularly in platforms like Arduino, is acknowledged as challenging and time-consuming and current methods are not well-suited to the specific requirements of Arduino programming.

Since automation is crucial for enhancing planning and decision-making in Arduino-based projects, the aim of this project is to develop an automated tool for measuring the functional size of Arduino codes using the COSMIC method. The tool, CosmiCode, will utilize both RegEx and NLP techniques.

CosmiCode improves upon existing tools in the following ways:

1. Unlike other tools, CosmiCode, along with the tool developed by Soubra and Salem [15], focuses on measuring the size of Arduino codes using the COSMIC method.
2. CosmiCode enhances accuracy results compared to previous tools.
3. To the best of our knowledge, CosmiCode is the only tool that outputs the total number of COSMIC Functional Points (CFPs) per data movement for Arduino codes.
4. This paper introduces a pattern for Arduino functions in each data movement, which the CosmiCode was built upon. This is a contribution not previously associated with other tools.

The rest of this paper is organized as follows: Sect. 2 explores previous studies related to the COSMIC method and its automation and Sect. 3 introduces CosmiCode, the FSM tool. Section 4 discusses the tests done on CosmiCode and their outcomes and finally, Sect. 5 tackles the conclusion and future work.

2 Background

2.1 Overview of the COSMIC Method

The COSMIC method outlines guidelines, principles, and a procedure for quantifying a uniform functional size of software. It is designed based on core software engineering principles, independent of technology and development methodology, versatile and applicable to any type of software across all architectural layers. Additionally, it serves as a basis for metrics on resources, quality, and performance benchmarks and utilizes a ratio scale for measurement, ensuring mathematical validity in all operations on measured sizes, a feature lacking in most first-generation methods [6].

The COSMIC measurement process is a threefold which can be broken down into these phases [7]:

The Measurement Strategy Phase. The COSMIC functional size measurement process begins by determining the *Purpose* and *Scope* - the "why" and the "what" - of the measurement. As the scope of the software to be measured is defined, the process involves identifying functional users, software layers, and the boundaries. Following this initial setup, the focus shifts to the Functional User Requirements (FUR), which are central to measuring the software's functional size. The next stage of the process involves pinpointing the different layers and boundaries of the software, as well as identifying the users who interact with it - humans, hardware devices or other pieces of software that enter data to and/or receive data from the software being measured. Persistent storage also sends data to and/or receives data from the software.

The Mapping Phase. In the mapping phase, the emphasis is on connecting each functional process to specific FUR and dissecting the processes into distinct data movements. These movements are categorized into four types:

- **Entry (E):** moves data from a functional user (e.g. sensor) to a functional process, marking the start of the process.
- **Exit (X):** transfers data from a functional process back to a functional user (e.g. actuator), completing a transaction or interaction.
- **Read (R):** brings data from persistent storage (e.g. EEPROM) to a functional process, supplying necessary information.
- **Write (W):** saves data from a functional process to persistent storage (e.g. EEPROM), ensuring its availability for future use.

Processes must include at least two data movements: an Entry and either an Exit or a Write for measurement. This ensures they are confined to a single layer and fulfill their FURs and simplifies understanding their purposes and trigger points. Additionally, this phase identifies relevant objects - "object of interest" - and their data groups, facilitating the understanding of data movement within processes and across software boundaries.

The Measurement Phase. In the measurement phase, each data movement within a functional process is assigned a unit of measurement, 1 COSMIC Functional Point (CFP). To calculate the functional size of a process, the method involves counting each type of data movement, assigning the unit measurement to each, and then summing up these units to obtain a singular value representing the functional size of that process. For the overall functional size of software, or specific changes to the software's FUR, the method aggregates the sizes of all functional processes within the FUR for each piece of software.

2.2 Automation of the COSMIC Method

Software projects continue to be difficult to size, despite the abundance of research on the subject [8]. Manually applying FSM procedures takes a lot of time, which is problematic for organisations that need to measure a lot of projects quickly for productivity studies or project estimating [9]. Furthermore, when the source code has been developed in many languages, the human application of FSM to a very broad collection of inputs demands specialized knowledge [1]. Numerous automatable COSMIC FSM processes have been presented in the literature for various areas, such as automotive [10,11], quantum computing [12], aviation [13,14], and the Internet of Things (IoT) [5].

Using regular expressions, Moulla *et al.* [1] presented a prototype automated tool for measuring CFP from C program source code. The automated tool's ability to reliably compute functional size and provide COSMIC CFP for C programs was demonstrated through validation against manual measurements made by trained professionals. The validation process involved using a test C program; however, it was not tailored specifically to Arduino codes.

Mnkandla *et al.* [8] introduced an automated COSMIC FSM tool for X86 assembly programs. By applying COSMIC-ISO 19761 rules, it effectively sizes X86 assembly programs through identifying functional processes, data groups, and data movements. The tool was validated through case studies on Fibonacci and Consecutive sum programs and showed a measurement accuracy of 80.74% and 100%, respectively.

Salem and Soubra [15] explored the integration of NLP with the COSMIC method for FSM in the context of IoT devices. They presented a new approach proposing a generalized tool that utilizes NLP to automatically measure the functional size of any source code. Additionally, they detailed modifications to a previously presented tool, incorporating machine learning and regular expressions to detect data group movements in source code. A real-life test using an IoT code snippet demonstrates the tool's accuracy, achieving an 82.3% accuracy with a neural network model and 53% with a naive bias approach, while the regular expression method accurately detects data group movements as either CFP or non-CFP.

Inspired by previous work and specifically the studies by Moulla *et al.* [1] and Salem and Soubra [15], this paper proposes CosmiCode, a COSMIC-based FSM tool that utilizes NLP and regular expressions to automatically measure the size of Arduino codes.

3 Methodology

3.1 Finding the RegEx Patterns

To successfully automate the functional size measurement process using the COSMIC method, it was necessary to first identify and categorise the relevant functions within the Arduino programming environment.

3.2 Data Collection

The first step was to visit the official Arduino website and look up the language reference documentation [16]. By reading through the documentation, a complete list of functions was constructed, covering various elements of hardware and software interactions within the Arduino environment. Additionally, more functions were discovered by examining codes which Arduino hobbyists have uploaded on the Arduino Project Hub website [17].

3.3 Further Categorization

The gathered list of functions was then categorized based on their potential impact on the software module's functionality. Functions were classified into several categories:

- **Entries (E):** Functions that signify the initiation or entry points of certain operations or tasks within the software module.
- **Exits (X):** Functions that denote the conclusion or exit points of operations or tasks within the software module.
- **Reads (R):** Functions primarily involved in reading data from external sources or devices.
- **Writes (W):** Functions primarily involved in writing or sending data to external destinations or devices.

3.4 Pattern Identification

After categorising the functions, the next step was to find common patterns within each category. This involved examining the behaviours and usage contexts of functions to identify repeating patterns that may be used for automation. Table 1 provides a comprehensive overview of the patterns associated with categorized functions. It should be noted that in order to ensure the quality of work, categorization of the functions found in Table 1 was validated by COSMIC professionals.

It is important to clarify that functions written in italics (such as *.beginTransaction()*) imply that any class invoking a function with that name will be involved in that particular data movement process, i.e.: *SPI.beginTransaction()*, *Wire.beginTransaction()*, etc. are Exits.

Additionally, functions that were deemed irrelevant to the functional size measurement process, such as object instantiations like *LiquidCrystal lcd()*, were excluded from consideration.

Table 1. Patterns in Function Categories

Entry	Exit	Read	Write
digitalRead()	digitalWrite()	EEPROM.read()	EEPROM.write()
analogRead()	analogWrite()	EEPROM.get()	EEPROM.update()
pulseIn()	noTone()		EEPROM.put()
pulseInLong()	tone()		EEPROM.commit()
shiftIn()	shiftOut()		
micros()	delay()		
millis()	delayMicroseconds()		
SPI.transfer()	shiftOut()		
Wire.requestFrom()	delayMicroseconds()		
Wire.onReceive()	Wire.onRequest()		
Mouse.isPressed()	WiFi.disconnect()		
WiFi.SSID()	WiFi.config()		
WiFi.BSSID()	WiFi.setDNS()		
WiFi.RSSI()	client.connected()		
WiFi.encryptionType()	client.connect()		
WiFi.macAddress()	client.stop()		
WiFi.getSocket()	WiFiUDP.beginPacket()		
WiFi.status()	WiFiUDP.endPacket()		
IPAddress.localIP()	lcd.setCursor()		
IPAddress.subnetMask()	display.clearDisplay()		
IPAddress.gatewayIP()	display.display()		
WiFiUDP.parsePacket()	*.print()*		
WiFiUDP.remoteIP()	*.println()*		
WiFiUDP.remotePort()	*.begin()*		
readBit()	*.flush()*		
bitRead()	*.write()*		
.requestTemperatures()	*.end()*		
.getTempCByIndex()	*.beginTransaction()*		
.parseFloat()	*.endTransaction()*		
.parseInt()	*.beginTransmission()*		
.peek()	*.endTransmission()*		
.read()	*.press()*		
.readBytes()	*.release()*		
.readBytesUntil()	*.releaseAll()*		
.readString()	*.click()*		
.readStringUntil()	*.move()*		
.get()	*.stop()*		
.find()	*.update()*		
.findUntil()	*.set()*		
.available()			

3.5 Extracting Regular Expressions

By recognising patterns within each category, it was feasible to create regular expressions that could identify function calls in source code. They were extracted from the Patterns in Function Categories Table (Table 1) and validated by both COSMIC professionals and university professors alike.

Entry Pattern:

```
r'\b('
r'digitalRead|analogRead|pulseIn(?:Long)?|shiftIn|micros|millis|'
r'SPI\.transfer|Wire\.requestFrom|Wire\.onReceive|Mouse\.isPressed|'
r'WiFi\.(SSID|BSSID|RSSI|encryptionType|macAddress|getSocket|status)|'
r'IPAddress\.(localIP|subnetMask|gatewayIP)|client\.available|'
r'WiFiUDP\.(available|parsePacket|remoteIP|remotePort)|'
r'(?!EEPROM\.(read|get))[a-zA-Z_]+\.read[a-zA-Z_]*|'
r'.*\.find|.*\.findUntil|.*\.parseFloat|.*\.parseInt|.*\.peek|.*\.get.*|'
r'.*\.readBytes|.*\.readBytesUntil|.*\.readString|
r'.*\.readStringUntil||.*\.available|'
r'sensors\.requestTemperatures|sensors\.getTempCByIndex|readBit|bitRead'
r')\s*\('
```

Exit Pattern:

```
r'\b('
r'digitalWrite|analogWrite|noTone|tone|pinMode|shiftOut|delay(?:Microseconds)?|'
r'(Serial|Stream|Wire|WiFi|client|WiFiUDP|lcd|display)\.flush|'
r'Wire\.onRequest|WiFi\.(disconnect|config|setDNS)|'
r'client\.(connected|connect|stop|flush)|'
r'WiFiUDP\.(beginPacket|endPacket|flush)|'
r'lcd\.setCursor|'
r'display\.(clearDisplay|display)|'
r'(?!EEPROM\.write)[a-zA-Z_]+\.write[a-zA-Z_]*|'
r'(?:.*\.)?(.*print.*|.*update.*|begin|end|beginTransaction|endTransaction|'
r'beginTransmission|endTransmission|press|release(?:All)|'
r'click|move|flush|stop|\.set.*)'
r')\s*\('
```

Read Pattern:

```
r'\b(EEPROM\.(read|get))\s*\('
```

Write Pattern:

```
r'\b(EEPROM\.(write|commit|update|put))\s*\('
```

Function Definition Pattern:

```
r'\b(?:void|int|char|short|long|float|bool|double)\s+(\w+)\s*\('
```

Function Call Pattern:

```
r'\b(\w+)\s*\((?:[^()]|\([^()]*\))*\);'
```

3.6 CosmiCode

CosmiCode is a Python-based tool designed to automate the COSMIC method. It accepts .txt or .c files as input, which are then preprocessed to remove unnecessary spaces, resulting in a single string of code. CosmiCode uses RegEx to analyze Arduino code and calculate CFPs. It applies specific patterns to find function definitions, calls, and key function types relevant to CFPs. First, it identifies function definitions by matching return types and function names. Then, it detects function calls to see which functions are used. The **entry_pattern** finds functions that start actions or read data, while the **exit_pattern** finds those that finish actions or write data. Additional patterns focus on EEPROM read and write operations in a similar fashion. By processing each line of code with these patterns, CosmiCode classifies lines as entries, exits, reads, or writes and computes the total CFPs. Additionally, it features a user-friendly graphical user interface (GUI) that visually represents the code with color-coded categories. Furthermore, the GUI provides a summary of the number of entries, exits, reads, writes, and the total CFPs calculated from the code. CosmiCode went through multiple stages of development.

Version 1. In the first version of CosmiCode, the tool employed a set of predefined regular expressions to categorize each line of code as an entry, exit, read, write, or non-CFP. The tool iterated through each line of the Arduino code, applying the regular expressions to determine the category of each function call. Additionally, it counted the occurrences of each category to calculate the total number of CFPs present in the code. Alongside this functionality, the tool featured a simple GUI that displayed the counts of entries, exits, reads, writes, and the total number of CFPs.

To test the tool, three distinct approaches were employed to ensure its accuracy and reliability. The initial approach involved testing with properly formatted code wherein the functions, excluding setup and loop, were called. Subsequently, testing proceeded with code that was not properly formatted but still included calls to functions, excluding setup and loop. Finally, the tool was tested using properly formatted code where functions, excluding setup and loop, were not called. These approaches aimed to assess the tool's performance under various scenarios.

This version exhibited several limitations. Firstly, it required each meaningful line of code to be written on a separate line for accurate calculation of CFPs.

For instance, multiple statements on the same line would be counted as a single CFP, even if they represented distinct function calls. To illustrate this point more clearly, consider the following example:

```
System.out.println("hello"); System.out.println("world");
```

In this form, the output would indicate only one exit (i.e., 1 CFP), which is inaccurate. To obtain the correct result, the code should be formatted as follows:

```
System.out.println("hello");
System.out.println("world");
```

By adopting this format, the expected outcome of 2 exits (i.e., 2 CFPs) would be accurately reflected. Furthermore, function calls (e.g.: *OpenDoor()*) were not included in the CFP exit count. Additionally, all functions were included in the CFP calculation process, regardless of whether they were actually invoked in the code. This approach overlooked the fact that functions not called during runtime effectively served no functional purpose. Moreover, the user interface lacked user-friendliness, as it did not provide a display of the processed code or categorize each statement based on its data movement. The latter three shortcomings were addressed in the second version of the tool.

Refer to Fig. 1 for a snapshot of the first version of CosmiCode when ran on a Door Lock project.

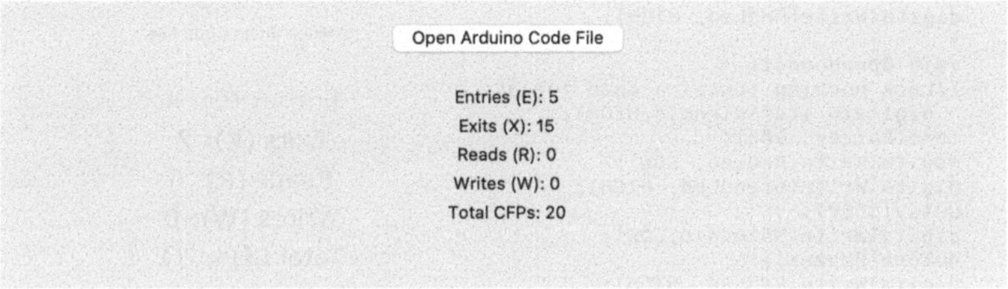

Fig. 1. A snapshot of the first version of CosmiCode.

Version 2 A more friendly GUI was implemented, which presented the code back to the user with color-coded statements indicating their respective categories: green for entries, blue for exits, red for reads, purple for writes, pink for comments, and black for non-CFPs. The label of each data movement is displayed in its respective color, acting as a visual key. Additionally, two RegEx expressions, namely the Function Definition and Function Call Patterns, were incorporated to analyze the code before calculating the number of CFPs. This analysis aimed to identify any functions present in the code other than *setup()*

and *loop()*, and to determine whether they had been called. If a function was found and called, its CFPs were calculated accordingly; if not, the tool skipped the function. Finally, the Function Call pattern was used to match function calls as Exits.

Although this version represented a notable improvement over version 1, like yielding expected distinct results for codes with and without function calls, it still had its limitations. Notably, it did not resolve the formatting issue.

Version 3. The third version of CosmiCode introduced a preprocessing step to address the formatting issue. This preprocessing function utilizes regular expressions to identify and trim segments of the code, including strings, comments, block delimiters, and statements ending in a semicolon. Finally, improperly formatted codes are now returned to the user in a proper format.

Refer to Fig. 2 for a snapshot of the result of the 3rd version of CosmiCode when ran on a Door Lock project without a function call.

Fig. 2. A snapshot of the outcome of the third version of CosmiCode.

3.7 CosmiCode in ML

To enhance the functionality of CosmiCode, an extension named CosmiCode in ML was developed, utilizing machine learning (ML) techniques, namely NLP. The objective was to improve the efficiency of CFP estimation by employing supervised learning on a dataset comprised of Arduino code snippets annotated with CFP labels.

Dataset Preparation. Initially, Salem *et al.*'s [15] dataset served as the foundation, containing Arduino code segments annotated with CFP or non-CFP labels. To build on the dataset, additional samples were incorporated, resulting in a final dataset comprising 1865 lines of code, each labelled as either a CFP or non-CFP instance.

Training and Validation The aforementioned dataset is sourced from an Excel file. Following that, text preprocessing techniques are used to clean the code lines, removing parameters, spaces, and punctuation. Once preprocessed, the data is divided into training and validation sets. The text data is tokenized and padded to ensure that the model's input dimensions are uniform. A sequential model is then built with TensorFlow Keras API. The model consists of an embedding layer for word representation, followed by a flattening layer. Dense layers with ReLU activation functions are used, along with a layer employing sigmoid activation for binary classification. Additionally, two dropout layers with a dropout rate of 0.5 are included to prevent overfitting after the flattening layer and after the dense layer with ReLU activation. The Adam optimizer and binary cross-entropy loss are utilized in constructing the model. It then goes through 30 epochs of training with the training and validation data in order to track performance. In the final epoch, the model had an accuracy of 77.47% and a loss of 0.4110 on the training data, but a somewhat lower accuracy of 75.54% and a loss of 0.3979 on the validation data. Lastly, utility functions are provided to plot accuracy and loss graphs over epochs to visualise the training process. Refer to Figs. 3 and 4 for the accuracy and loss graphs respectively.

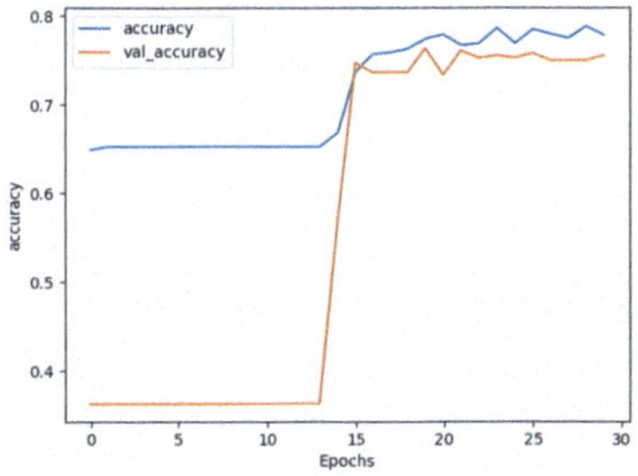

Fig. 3. Training and Validation accuracy results.

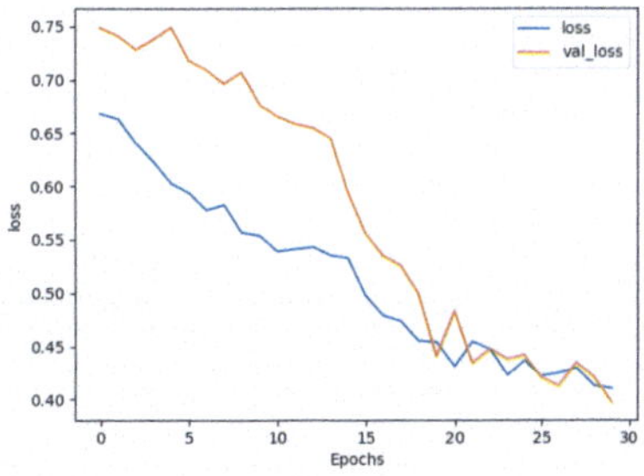

Fig. 4. Training and Validation loss results.

Testing. The user is prompted to enter the path to a .txt or .c file containing code to be assessed. The file's code is then read and preprocessed in two different stages: initially to remove unnecessary spaces and extract meaningful segments to be able to asses improperly written codes, similar to the method used in the RegEx implementation, and then to standardize function calls by removing parameters and extra spaces in order to match the training data. Next, the preprocessed code is tokenized and padded to conform to the model's required input format. Each line of code undergoes predictions based on the loaded model. The prediction for each line is printed beside the code line, indicating whether or not it is a CFP. Additionally, a counter tracks the overall number of CFP estimates made. Finally, the total number of CFP predictions is printed to provide an overall summary of the model's classification results on the provided code file. Figure 5 showcases the result of running the ML extension on a Line Following Robot Arduino project.

```
Line      | Prediction
-----------------
int mr18           | Not CFP
int mr29           | Not CFP
int ml110          | Not CFP
int ml211          | Not CFP
int sr6 | Not CFP
int sl7 | Not CFP
int svr0           | Not CFP
int svl0           | Not CFP
int led13          | Not CFP
int enr3           | Not CFP
int enl5           | Not CFP
int vspeed100      | Not CFP
int tspeed255      | Not CFP
int tdelay20       | Not CFP
void setup         | CFP
pinmode | CFP
pinmode | CFP
pinmode | CFP
pinmode | CFP
pinmode | CFP
pinmode | CFP
pinmode | CFP
delay   | CFP
        | Not CFP
```

(a)

```
void loop       | CFP
svrdigitalread  | Not CFP
svldigitalread  | Not CFP
if      | Not CFP
forward | Not CFP
//maju  | Not CFP
        | Not CFP
if      | Not CFP
left    | CFP
//belok kiri    | Not CFP
        | Not CFP
if      | Not CFP
right   | CFP
//belok kanan   | Not CFP
        | Not CFP
if      | Not CFP
stop    | CFP
//berhenti      | Not CFP
        | Not CFP
        | Not CFP
void forward    | CFP
digitalwrite    | CFP
digitalwrite    | CFP
digitalwrite    | CFP
digitalwrite    | CFP
analogwrite     | CFP
analogwrite     | CFP
        | Not CFP
```

(b)

```
void backward    | CFP
digitalwrite     | CFP
digitalwrite     | CFP
digitalwrite     | CFP
digitalwrite     | CFP
analogwrite      | CFP
analogwrite      | CFP
         | Not CFP
void right       | CFP
digitalwrite     | CFP
digitalwrite     | CFP
digitalwrite     | CFP
digitalwrite     | CFP
analogwrite      | CFP
analogwrite      | CFP
delay    | CFP
         | Not CFP
void left        | CFP
digitalwrite     | CFP
digitalwrite     | CFP
digitalwrite     | CFP
digitalwrite     | CFP
analogwrite      | CFP
analogwrite      | CFP
delay    | CFP
         | Not CFP
void stop        | CFP
analogwrite      | CFP
analogwrite      | CFP
         | Not CFP
-----------------
Total CFP lines: 46
```

(c)

Fig. 5. A snapshot of the outcome of the ML extension of CosmiCode

4 Results

A dataset comprising 15 distinct codes has been compiled, sourced from the Arduino Project Hub [17] website. Additionally, a manual computation of CFPs has been performed for each code file. These computed CFPs were subsequently submitted to COSMIC professionals for thorough review and validation.

4.1 RegEx Implementation Results

To compute the results of the RegEx implementation of CosmiCode, a .txt or .c file of each code obtained from the website was utilized as input. As illustrated in Fig. 2, the code is presented with color codes, indicating whether each statement corresponds to an entry, exit, read, write, comment, or non-CFP statement. The total number of CFPs for each data movement is then exhibited alongside the final aggregate count. These outcomes are subsequently juxtaposed with the manually computed CFPs for validation and comparative analysis.

Out of the 15 codes assessed, 13 codes achieved a perfect accuracy of 100% when compared to the manual computation. The remaining two codes demonstrated accuracies of 97.14% and 77.14% respectively. This results in an average accuracy score of 98.28%.

It should be noted that conditional statements within "if", "while", or "'switch" structures are not correctly recognized. For example, expressions such as "if(Serial. available() == 0)" are incorrectly detected as non-CFP statements because the RegEx patterns were not designed to match with functions within conditional statements. Additionally, some statements that serve as both entry and exit points are wrongly labeled only as entries. For instance, the statement "Serial.println(digitalRead(startled))"; is incorrectly classified as an entry. This happens because the algorithm assumes a CFP has been found when it detects a data movement and doesn't check for its dual nature.

4.2 ML Extension Implementation Results

The 15 aforementioned projects were used to augment Salem *et al.*'s [15] dataset, increasing its size from 404 to 1865 lines. After this expansion, the model underwent testing with an additional 4 projects obtained from the Arduino Project Hub website. Initially, each of these new projects underwent manual analysis, following a similar process to the older projects. Subsequently, the results obtained from manual assessment were compared with those generated by the machine learning extension. The results were promising, with accuracies of 83.95%, 83.33%, 93.18%, and 95.312% achieved across the four projects, respectively. This resulted in an average accuracy of 88.943%.

Unlike CosmiCode's RegEx implementation, the machine learning extension does not provide the total number of CFPs per data movement; instead, it offers the overall count of CFPs for the entire code. Another challenge is the inconsistency in predictions, where the same sentence might be identified as a CFP at times but not consistently, despite efforts to refine the dataset for inconsistencies. Moreover, the extension lacks the ability to discern whether a function has been invoked, disregarding the fact that unused functions serve no practical purpose. Hence, the body of such unused functions should be excluded from the CFP counting process. It remains unclear whether the model accurately predicts statements within conditional statements (“if”, “switch” or “while”) or not.

5 Conclusion and Future Work

The research focused on improving the accuracy and efficiency of measuring the functional size of Arduino programs in C using the COSMIC approach, complemented by RegEx and NLP techniques. Automating this process addressed challenges in manual measurement and aimed to optimize resource allocation and project planning for Arduino applications. CosmiCode marks a significant advancement in automating software function measurement, though further refinement is needed.

In conclusion, both the RegEx and ML methods showed promising results. The RegEx method achieved 98.28% accuracy in identifying code elements and unused functions but struggled with conditional statements. The ML extension, with an average accuracy of 88.943%, provided a broader analysis but faced challenges with function invocations. These findings highlight the potential for enhanced automation in software development, improving efficiency and accuracy.

Future research should focus on refining RegEx patterns for complex code structures and improving ML algorithms to better handle function invocations. Exploring hybrid approaches and applying these techniques to other programming languages could further enhance functional size measurement and validate their broader applicability.

We hope that future improvements for CosmiCode would include:

1. Enhanced Recognition of Conditional Statements: Develop algorithms to accurately identify and classify conditional statements within code structures like “if,” “while”, or “switch”, which are currently overlooked.
2. Refinement of Entry and Exit Point Identification: Implement a more robust method to precisely label statements serving as both entry and exit points, or any other two (or more) data movements.
3. Incorporation of Function Invocation Detection in the ML Extension: Integrate functionality to determine whether functions are invoked in the code, enabling more precise identification of unused functions and their exclusion from function point calculations.
4. Addressing Prediction Inconsistencies: Investigate and fix issues related to prediction inconsistencies, where the same statement may be inconsistently classified as a function point, despite efforts to refine the dataset.
5. Improving ML Extension Accuracy: Further refine the machine learning extension to enhance its accuracy in predicting function points.

References

1. Moulla, D.K., Oumate, Mnkandla, E., Soubra, H., Abran, A.: Automated COSMIC function points measurement for C program using regular expressions. In: Proceedings of IWSM-Mensura 2022, İzmir, Turkey (2022)
2. DeMarco, T.: Controlling software projects (1982)
3. Abran, A., Symons, C., Ebert, C., Vogelezang, F., Soubra, H.: Measurement of software size: contributions of COSMIC to estimation improvements. In: International Training Symposium (2016)
4. Hrechko, A., Melashchenko, A.: Application of the functional size measurement method on the example of the national register of Ukraine. Cybern. Syst. Anal. **55**, 1–8 (2019)
5. Soubra, H., Abran, A.: Functional size measurement for the internet of things (IoT): an example using cosmic and the arduino open-source platform. In: Proceedings of IWSM Mensura 2017, Göteborg, Sweden (2017)
6. COSMIC Sizing. Benefits of the COSMIC Method (2023). https://cosmic-sizing.org/cosmic-sizing/intro/benefits-of-cosmic/. Accessed 09 Mar 2024
7. Abran, A., Fagg, P., Lestherhuis, A.: Measurement manual v5.0 part 2: guidelines. COSMIC Group, version 5.0 (2021). https://cosmic-sizing.org/measurement-manual-v5-part-2/
8. Moulla, D.K., Kitikil, A.A., Mnkandla, E., Soubra, H., Abran, A.: Functional size measurement for X86 assembly programs. In: Proceedings of IWSM-Mensura 2023, Rome, Italy (2023)
9. Soubra, H., Abufrikha, Y., Abran, A.: Towards universal COSMIC size measurement automation. In: Proceedings of IWSM-Mensura 2020, Mexico City, Mexico (2020)
10. Soubra, H., Azib, T.: Functional size measurement for energy needs early estimation in autonomous drones. In: Proceedings of IWSM-Mensura 2018, Beijin, China (2018)
11. Soubra, H., Chaaban, K.: Functional size measurement of electronic control units software designed following the AUTOSAR standard: a measurement guideline based on the COSMIC ISO 19761 standard. In: 2012 Joint Conference of the 22nd International Workshop on Software Measurement and the 2012 Seventh International Conference on Software Process and Product Measurement, pp. 78–84 (2012)
12. Khattab, K., Elsayed, H., Soubra, H.: Functional size measurement of quantum computers software. In: Proceedings of IWSM-Mensura 2022, İzmir, Turkey (2022)
13. Soubra, H., Jacot, L., Lemaire, S.: Manual and automated functional size measurement of an aerospace real-time embedded system: a case study based on SCADE and on COSMIC ISO 19761. Int. J. Eng. Res. Sci. Technol. **4**, 79–100 (2015)
14. Soubra, H., Abran, A., Sehit, M.: Functional size measurement for processor load estimation in AUTOSAR. In: Software Measurement, pp. 114–129. Springer, Cracow (2015)
15. Salem, S., Soubra, H.: Using NLP for functional size measurement of IoT devices. In: Proceedings of 2023 Eleventh International Conference on Intelligent Computing and Information Systems (ICICIS), Cairo, Egypt (2023)
16. Arduino. Arduino Language Reference (2024). https://www.arduino.cc/reference/en/. Accessed 18 Apr 2024
17. Arduino Project Hub (2024). https://projecthub.arduino.cc. Accessed 18 Apr 2024

An NLP-Based WhatsApp Group Chat Summarizer

Kareem Fayed(✉) and Milad Ghantous

German University in Cairo, New Cairo, Egypt
kareem.ashraf@student.guc.edu.eg, milad.michel@guc.edu.eg

Abstract. The advancement in Natural Language Processing (NLP), particularly in text summarization, alongside the growing volume of chat data on platforms like WhatsApp, necessitates tools to process this data efficiently. This paper presents a detailed study on analyzing and summarizing WhatsApp group chats using NLP techniques. It explores message classification, chat clustering, and summarization to extract significant information from conversations. Four machine learning models (Logistic Regression, Naive Bayes, SVM, and Random Forest) were compared for message classification, with Random Forest emerging as the best model after optimization. The classification process included manual message categorization and essential preprocessing. For chat clustering, a combined approach based on message category, author, and timestamp was developed, yielding better results than individual methods. Both extractive and abstractive summarization methods were investigated, with BERT and Gensim used for extractive summarization, albeit with limitations in capturing context. Abstractive summarization using BART and PEGASUS models from Hugging Face showed superior results, especially with the SAMSum-trained model. The findings underscore the effectiveness of the combined clustering approach and the importance of advanced abstractive summarization models for capturing context in WhatsApp group chats.

Keywords: NLP · WhatsApp · Chat Summarization

1 Introduction

The widespread adoption of WhatsApp has transformed global communication, with over 2 billion users in 180 countries utilizing its text messaging, multimedia sharing, and group conversation features according to [1]. This ubiquity has made WhatsApp essential for both personal and professional interactions. However, the high volume of messages exchanged within group chats often leads to information overload, making it challenging for users to stay updated, especially when they miss discussions due to scheduling conflicts or limited access.

This paper addresses the need for innovative solutions to manage this information overload by developing an NLP-based model to summarize lengthy WhatsApp group chat conversations. Our model aims to go beyond simple summarization by identifying and extracting key details, such as important points and

A. Abdelgawad et al. (Eds.): ISBCom 2024, LNNS 1268, pp. 316–329, 2025.
https://doi.org/10.1007/978-3-031-82377-0_27

decisions made within the chat. The objective is to create an automated tool that not only summarizes long group conversations but also highlights crucial information, processing both textual and voice messages for comprehensive information management.

The paper is structured as follows: Sect. 2 provides background and a literature review, exploring NLP techniques and relevant research on chat summarization, highlighting gaps specifically for WhatsApp group chats. Section 3 details the methodology, including the chosen approach, data collection, and preprocessing techniques, focusing on the model's architecture and training process. Section 4 presents the results, demonstrating the model's performance in summarization and key detail extraction. Section 5 discusses the conclusions, summarizing the paper's findings. Finally, Sect. 6 outlines limitations and future work, discussing the challenges faced and suggesting directions for future research based on the findings.

2 Background and Literature Review

2.1 Background

Machine Learning
According to Mitchell, Machine Learning (ML) is a core component of AI, allowing computers to learn from data without explicit programming [2]. Supervised learning involves training algorithms on labeled data as defined by James et al. [3], while unsupervised learning deals with unlabeled data to uncover hidden patterns. ML has revolutionized various fields, including NLP, by continuously improving its performance through experience with data.

Deep Learning
As stated by Schmidhuber in [4], Deep Learning (DL) is a subset of ML inspired by the human brain's structure, utilizing Artificial Neural Networks (ANNs) with multiple interconnected layers. ANNs process information and adjust weights during training to learn complex patterns from data. DL is particularly powerful for NLP tasks, as it can automatically extract features from text data through its layered architecture.

Natural Language Processing
NLP focuses on enabling computers to understand, process, and generate human language. It bridges the gap between rigid programming languages and the complex nature of human language, unlocking a wide range of applications across various domains. NLP techniques are crucial for tasks like machine translation, speech recognition, information extraction, sentiment analysis, and chatbots.

Text Summarization

Text summarization is a critical task in NLP, aiming to generate concise summaries of lengthy text documents while preserving key points and essential information. There are two main approaches to automatic text summarization: extractive and abstractive. Extractive summarization selects important sentences from the original text, while abstractive summarization generates a new summary using different words and phrasing.

Transformers

Transformers are a recent advancement in deep learning architectures designed for sequence-to-sequence tasks in NLP. Unlike traditional architectures like LSTMs, Transformers utilize an attention mechanism to focus on specific parts of the input sequence, enabling them to efficiently capture long-range dependencies within the data. Transformers, such as BERT, have revolutionized NLP, pushing the boundaries of what machines can achieve with language understanding and generation.

2.2 Literature Review

Chat Summarization

Recent advancements in chat summarization have explored both extractive and abstractive methods. Extractive techniques select key sentences directly from the text but often fail to capture the full context. Abstractive methods, on the other hand, generate new sentences that encapsulate the main ideas, providing more coherent summaries. Tepper et al. developed an abstractive-based system to tailor summaries to user needs, demonstrating the effectiveness of integrating machine learning and user preferences in enhancing information management within group chats in [5].

Our work builds upon these foundations, focusing specifically on WhatsApp group chats. We explore both extractive and abstractive methods, integrating clustering and classification techniques to improve summary relevance and organization.

WhatsApp Chat Analysis

Several studies analyze WhatsApp conversations, but they often focus on different objectives than this paper. For example, In [6], Saqib et al. presented a tool for chat activity analysis, which calculates metrics like message count and word frequency but does not summarize conversations. In [7], Ramana et al. explored sentiment analysis in WhatsApp groups to understand emotional tone but did not address key point extraction.

3 Methodology

This section outlines the approach and implementation details of the WhatsApp Chat Summarizer. Figure 1 illustrates the architecture of the summarizer.

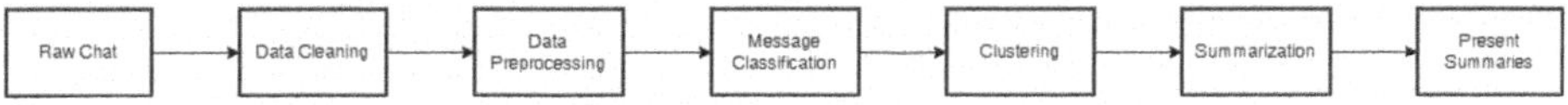

Fig. 1. Architecture of the WhatsApp Chat Summarizer

3.1 Dataset

Data Collection
A single WhatsApp chat with 22 participants and over 500 messages, used for communication between foreign med students and a local team in a student exchange progch1ram, was obtained with participants' consent. To protect privacy, all names were replaced with random names, and sensitive information was omitted.

Data Cleaning
Data cleaning involved removing system messages such as "This message was deleted", member addition/removal notifications, changes in group icon or description, and metadata of attachments (except audio files).

3.2 Data Preprocessing and Representation

The dataset, exported in a .txt file, was read line by line and organized into a pandas dataframe with four columns:

1. **Date:** Contains the date in yyyy-mm-dd format.
2. **Time:** Contains the time of the message.
3. **Author:** Contains the author of the message.
4. **Message:** Contains the message body.

Voice Messages
Voice messages were detected if they were attachments with a .opus extension. These were transcribed into text using AssemblyAI's API, chosen for its accuracy and ease of integration. The transcribed text replaced the original message in the dataframe.

3.3 Chat Clustering

Group chat conversations are complex, often involving multiple participants discussing various topics simultaneously. This complexity arises from interleaving topics, simultaneous discussions, grammar and punctuation errors, and the use of slang and informal language, which complicate clustering and summarization. Two initial approaches were considered for clustering.

Author-Based Clustering
Messages were grouped by author using a python dictionary, but this approach lacked conversational context and coherence in summaries.

Timestamp-Based Clustering
Messages were clustered within a 60-min time window from the first message's timestamp. This approach preserved the context by prefacing messages with the author's name but failed to capture conversational threads accurately, leading to unrelated messages being clustered together or single conversations being split across clusters.

Given the limitations of these methods, a combined approach using message category, author, and timestamp was developed to create more contextually relevant clusters that better capture distinct conversations within the group chat.

3.4 Message Classification

Message classification is essential for analyzing chat conversations, particularly in group chats where messages vary in content and intent. Despite challenges such as unstructured data, informal language, and varying message lengths, effective classification leads to more accurate clustering and better summaries.

This work aimed to classify each message into one of five categories:

- **Information:** Factual information or details.
- **Suggestion:** Ideas or recommendations.
- **Question:** Requests for information or clarification.
- **Statement:** Thoughts, opinions, or assertions.
- **Reply:** Responses to previous messages.

Messages were manually labeled with one of the five categories, adding a "Category" column to the dataframe. Preprocessing steps included:

- **Lowercasing:** Ensured consistency in word representation.
- **Removing Punctuation, Special Characters, and Emojis:** Reduced noise.
- **Stopword Removal:** Focused on meaningful words using NLTK's stopwords.

Four models were initially considered for this task:

1. **Logistic Regression:** A simple, linear model for baseline comparison.
2. **Multinomial Naive Bayes:** Efficient and effective for text classification.
3. **Support Vector Machine (SVM):** Captures complex relationships in high-dimensional spaces.
4. **Random Forest:** Robust against noise, suitable for text classification.

The best-performing model was selected based on accuracy, precision, recall, and F1-score.

The dataset was split into training (80%) and testing (20%) sets. TF-IDF vectorization (max 5000 features) transformed text into numerical features. Models were trained and evaluated on accuracy, precision, recall, and F1-score, with the Random Forest model performing best.

In order to improve the model's performance, several techniques were applied to achieve better results in all four evaluation metrics.

Grid Search for Hyperparameter Tuning
Optimal hyperparameters for the Random Forest model were determined using GridSearchCV:

1. **n_estimators:** 100
2. **max_depth:** None
3. **min_samples_split:** 5
4. **min_samples_leaf:** 1

Feature Engineering

- **Stemming:** Used SnowballStemmer to reduce words to their root forms.
- **Addition of Bigrams:** Enhanced TF-IDF vectorization to capture more context.

Feature Selection
SelectKBest with chi-squared test were used to select the top 500 features, improving generalization and focusing on the most relevant aspects of the text.

Handling Class Imbalance with SMOTE
SMOTE generated synthetic samples for minority classes, balancing the dataset and improving model accuracy for underrepresented categories.

The message classification process included manual categorization, preprocessing, model selection, and optimization. Feature engineering and selection, along with handling class imbalance, significantly improved model performance. The robust classification model laid the groundwork for effective chat clustering.

3.5 Combined Approach for Clustering

Clustering messages in a WhatsApp group chat presents challenges due to the dynamic nature of conversations and the presence of multiple participants. Previous attempts at clustering have shown limitations in capturing the complex relationships within group chats. To address these challenges, A combined approach that considers the category of each message, the author, and the timestamp of the message to create more related and meaningful clusters was proposed as followed:

- **Message Classification:** The first step and the core of this clustering approach, as discussed thoroughly in the previous section.
- **Clustering Algorithm:** The main challenge in this algorithm is to decide whether the current message should be added to the current cluster or initiate a new cluster. A set of priority-based rules that govern the process of when to initialize a new cluster is provided:

- **Questions:** Questions are likely to indicate a topic change. The number of messages since the last question is tracked, and if the current message is a question and the count exceeds a threshold (e.g., 5 messages), a new cluster is initiated.
- **Suggestions:** Treated similarly to questions, with a threshold for initiating a new cluster.
- **Replies:** Replies indicate ongoing discussions and are added to the current cluster regardless of the previous message's category.
- **Author:** Consecutive messages from the same author are clustered together unless a predefined time window (e.g., 30 min) is exceeded.
- **Different Consecutive Categories:** If two consecutive messages are of different categories, the time difference exceeds a larger predefined time window (e.g., 90 min) is checked to decide on initiating a new cluster.
- **Similar Consecutive Categories:** Consecutive messages of the same category are grouped together unless the time difference exceeds the predefined time window.
- **Otherwise:** A new cluster is initiated with the current message.

To implement the clustering algorithm, a function cluster_messages was declared that takes three input parameters:

- **max_messages_without_question:** Set to 5.
- **author_time_window:** Set to 30 min.
- **category_time_window:** Set to 90 min.

Python lists clusters and current_cluster were initialized, and the dataframe was traversed row by row. The current_cluster is added to clusters and reinitialized with the current message if necessary.

The function belongs_to_cluster checks all the cases and returns True if the message belongs to the current cluster and False otherwise.

A Python list paragraphs was initialized, and the clusters list was traversed. Each cluster's text was prefaced with the author's name to maintain context.

In conclusion, our combined approach for clustering WhatsApp messages based on message category, author, and timestamp has shown promising results on the available dataset. The algorithm's ability to capture the dynamic nature of conversations and group related messages together has outperformed other clustering approaches. In the next chapter, the results of our clustering approach will be presented and its implications for message organization and summarization will be discussed.

3.6 Chat Summarization

Representing the primary objective of this paper, chat summarization plays a crucial role in condensing lengthy conversations into concise summaries, particularly for users catching up on missed messages. An original chat is fed into the system and a summary is generated. Figure 2 demonstrates the process of the

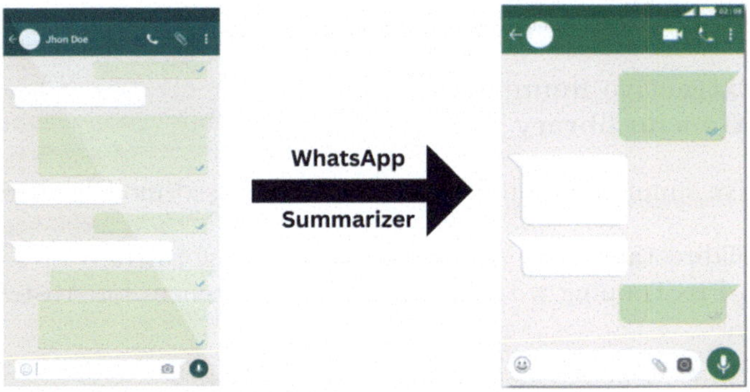

Fig. 2. Explanation of the Tool's Objective; A Long WhatsApp Chat is Summarized into a Shorter and More Concise Form

WhatsApp Summarizer. In this section, the methodologies and techniques used to tackle the complexity of WhatsApp group chats will be discussed, aiming to provide users with informative summaries that capture the core ideas and topics discussed. Our journey from evaluating extractive summarization libraries to leveraging advanced abstractive models, highlighting the challenges faced is also discussed.

As previously discussed in all three clustering techniques, a list paragraphs is returned from each function containing several paragraphs, where each paragraph represents a cluster. Paragraphs are constructed in a suitable format based on whether there is one or more authors in the cluster as mentioned. The data is now in a suitable format to be used as input to the summarizer.

In deciding the approach for chat summarization, Both extractive and abstractive methods were considered for the task of creating summaries for users who missed parts of the conversation. Extractive summarization, which involves selecting important sentences from the text, may not be the most effective approach based on logical thinking. The rationale behind this is that WhatsApp chats can be informal, unstructured, and complex in nature, and extractive summarization may lead to summaries that lack context and contain random messages that do not hold significant value.

Abstractive summarization, on the other hand, seems like a better approach. The ability of abstractive summarization models to analyze the conversations and understand what is being said allows them to identify key points more effectively. Furthermore, abstractive models can paraphrase the text, condensing the conversations while preserving the main points being discussed.

Although abstractive models require more computational power than extractive models, their ability to understand context and generate summaries that capture the essence of the conversation makes them a better choice for WhatsApp group chats.

For extractive summarization, two models were considered:

- **BERT Extractive Summarizer**
- **Python Gensim library**

For abstractive summarization, Two transformer-based models were explored:

- **BART:** Bidirectional and Auto-Regressive Transformers
- **Pegasus:** Pre-Training with Extracted Gap-Sentences for Abstractive Summarization

The BART model trained on the CNN/DailyMail dataset was tried at first. However, it did not perform satisfactorily for our task. The SAMsum dataset, which contains messenger-like conversations with summaries, was considered next. Using both BART and Pegasus models trained on this dataset yielded better results, providing summaries that were more coherent and relevant to the original conversations.

Summarizing the available dataset using all three clustering techniques mentioned is attempted:

- **Author-Based Clustering**
- **Timestamp-Based Clustering**
- **Category-Based Clustering**

For the abstractive summarization, The Transformers library from Hugging Face, a powerful tool for natural language processing tasks, was utilized. Specifically, The pipeline module, which provides a simple API for various NLP tasks, including text summarization, was employed.

Two advanced abstractive summarization models available in the Transformers library were considered: BART and Pegasus. These models are pretrained on large-scale datasets and fine-tuned on specific tasks, making them suitable for our chat summarization task.

To summarize a given chat cluster, The cluster's text represented as a paragraph was passed through the summarization pipeline, specifying the desired model (*facebook/bart-large-cnn* for BART pretrained on the CNN/DailyMail dataset, *philschmid/bart-large-cnn-samsum* for BART fine-tuned using the SAMSum dataset, and *transformersbook/pegasus-samsum* for PEGASUS fine-tuned also using the SAMSum dataset) and other relevant parameters. The pipeline internally handled the tokenization, model inference, and decoding, returning a summary of the input text.

The choice of using the Transformers library and these specific models was driven by their state-of-the-art performance in abstractive summarization tasks and their ease of use through the pipeline interface, allowing us to focus more on the overall summarization process.

4 Results

4.1 Introduction

In this chapter, The results of our study on using NLP techniques to analyze and summarize WhatsApp group chat conversations are presented. The primary focus was on message classification and chat summarization, with a brief overview of chat clustering. The performance metrics of different models for message classification, the clustering techniques used, and the effectiveness of the summarization methods are discussed.

4.2 Message Classification

Table 1. Evaluation Metrics Comparison of the Four Machine Learning Models Considered for Message Classification

Model	Accuracy	Precision	Recall	F1-Score
Logistic Regression	0.4950	0.4631	0.4950	0.4599
Naive Bayes	0.5149	0.4631	0.5149	0.4657
Support Vector Machine	0.5149	0.4855	0.5149	0.4958
Random Forest	0.5248	0.4906	0.5248	0.4995

Messages were classified into five categories: Information, Question, Statement, Reply, and Suggestion. Initially, four models were evaluated: Logistic Regression, Naive Bayes, SVM, and Random Forest. Table 1 presents the evaluation metrics for each of the four models considered. The Random Forest model performed the best and was further optimized through parameter tuning, feature engineering, and handling class imbalance.

The performance of the classification models was evaluated using precision, recall, accuracy, and F1-score. These metrics helped assess the effectiveness of the models in predicting message categories.

Four baseline models were trained and evaluated, with the Random Forest model achieving the highest scores across all metrics.

Model Enhancements

Several enhancements were applied to improve the Random Forest model:

- **Parameter Tuning**
- **Feature Engineering**
- **Feature Extraction**
- **Handling Class Imbalance**

After these enhancements, the model's performance saw significant improvements, particularly with the application of SMOTE for handling class imbalance. Table 2 shows the enhancement achieved by the application of parameter tuning, feature engineering, feature selection and handling class imbalance using SMOTE.

Table 2. Accuracy, Precision, Recall, and F1-Score of the Random Forest Model Before and After Enhancements

Model	Accuracy	Precision	Recall	F1-Score
Random Forest (Before)	0.6138	0.6302	0.6138	0.5900
Random Forest (After)	0.8268	0.8481	0.8268	0.8225

4.3 Chat Clustering

Three chat clustering techniques were employed: author-based, timestamp-based, and a category-based combined approach.

Author-Based Approach
This approach grouped messages by author but often included unrelated messages due to neglecting the context of conversations.

Timestamp-Based Approach
Messages were clustered based on their sequential timestamps. While better at capturing context, this method sometimes resulted in unnecessary clusters.

Category-Based Combined Approach
This approach integrated message category, author, and timestamp, creating more coherent and contextually relevant clusters. It showed better results than the other two approaches. Figure 3 shows a cluster.

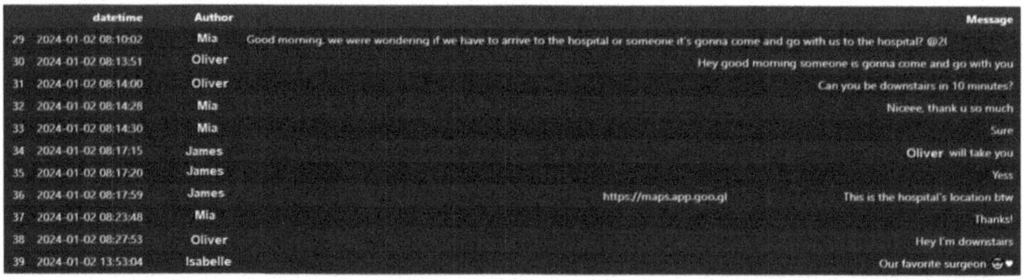

	datetime	Author	Message
29	2024-01-02 08:10:02	Mia	Good morning, we were wondering if we have to arrive to the hospital or someone it's gonna come and go with us to the hospital? @2l
30	2024-01-02 08:13:51	Oliver	Hey good morning someone is gonna come and go with you
31	2024-01-02 08:14:00	Oliver	Can you be downstairs in 10 minutes?
32	2024-01-02 08:14:28	Mia	Niceee, thank u so much
33	2024-01-02 08:14:30	Mia	Sure
34	2024-01-02 08:17:15	James	Oliver will take you
35	2024-01-02 08:17:20	James	Yess
36	2024-01-02 08:17:59	James	https://maps.app.goo.gl This is the hospital's location btw
37	2024-01-02 08:23:48	Mia	Thanks!
38	2024-01-02 08:27:53	Oliver	Hey I'm downstairs
39	2024-01-02 13:53:04	Isabelle	Our favorite surgeon ♥

Fig. 3. A Category-Based Cluster, Containing the Timestamp, Author Name, and Message Body for Each Message Within The Cluster

4.4 Chat Summarization

Both extractive and abstractive summarization methods were explored.

Extractive Summarization
Extractive summarization was briefly explored but found unreliable for chat summarization, leading us to focus on abstractive methods.

Abstractive Summarization

The BART and Pegasus models were used, both trained on the SAMSum dataset. The BART model performed best overall, generating more accurate summaries. Figure 4 shows a summary generated for the category-based cluster in Fig. 3 by the BART model trained on the SAMSum dataset.

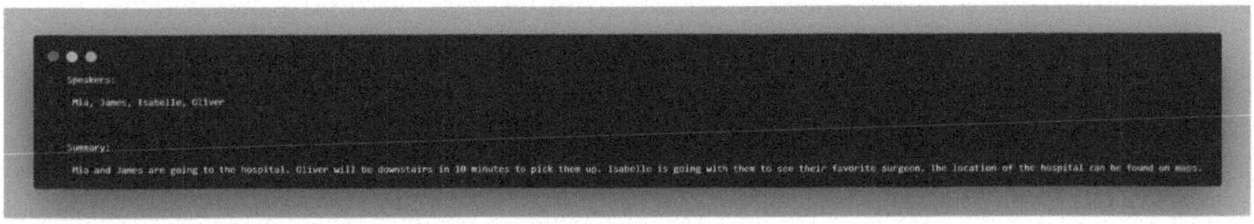

Fig. 4. A Summary of a Category-Based Cluster Generated by the BART SAMSum Model

To evaluate the summaries, a survey was conducted with 25 participants who rated the summaries based on accuracy and coherence. The BART model's summaries received an average score of 3.49 out of 5, suggesting acceptable quality despite the lack of objective metrics for a thorough evaluation.

5 Conclusion

The main aim of this paper was to build a tool that can summarize and extract key points from long WhatsApp group chats, alleviating the need for participants to read through a large number of messages to catch up. To achieve this, a comprehensive approach involving message classification, chat clustering, and summarization was developed.

For message classification, a Random Forest (RF) model was trained, achieving an accuracy of 80.51%. Enhancements such as parameter tuning, feature engineering, and handling class imbalance significantly improved the model's performance. The RF model outperformed other models considered in this study.

In chat clustering, a combined approach based on message category, author, and timestamp was utilized, leveraging the classification model. This method produced more coherent clusters compared to author-based and timestamp-based clustering alone.

For summarization, abstractive models like BART and Pegasus were more effective than extractive methods. The BART model trained on the SAMSum dataset was particularly successful in capturing conversational context and generating meaningful summaries.

6 Limitations and Future Work

6.1 Limitations

A primary limitation of this paper was the lack of available WhatsApp chat datasets. Despite extensive search efforts, the study relied on a single English dataset containing only 500 messages. This limited the effectiveness and generalizability of the machine learning models used for classification, clustering, and summarization.

Another limitation was the format of exported WhatsApp data. The inability to detect message replies and map mentions within messages hindered accurate identification of conversation threads. Additionally, the omission of photo captions in exported data reduced accuracy in clustering and summarization.

The absence of quantitative metrics for evaluating clustering and summarization processes posed a significant limitation. Reliance on qualitative assessments limited the precision and depth of the evaluation, affecting the ability to objectively measure the performance of the developed models.

6.2 Future Work

Future work should focus on acquiring larger and more diverse WhatsApp chat datasets to improve the performance and generalizability of the developed models. This would enable more comprehensive training and testing, leading to more robust models for message classification, clustering, and summarization.

Providing human-written summaries for larger datasets, with the help of professional linguists, would enhance summary evaluation through ROUGE scores. Continued research into abstractive summarization models trained on large conversational datasets could further improve the quality of chat summaries.

Acquiring datasets in other languages and focusing on summarizing multilingual WhatsApp chats, especially Arabic, would be beneficial. Developing models capable of handling multilingual input would lead to more inclusive and effective summarization solutions, particularly in regions with diverse linguistic populations.

In conclusion, while this paper faced several limitations, it demonstrated the potential of using NLP techniques to analyze and summarize WhatsApp group chats. Future research and development in this area could overcome these challenges and significantly enhance the accuracy and effectiveness of chat summarization tools.

References

1. WhatsApp (2024). https://www.whatsapp.com/about. Accessed 5 Mar 2024
2. Mitchell, T.M.: Machine Learning. McGraw-Hill (1997)
3. James, G., Witten, D., Hastie, T., Tibshirani, R.: An Introduction to Statistical Learning, vol. 112. Springer (2013)
4. Schmidhuber, J.: Deep learning in neural networks: an overview. Neural Netw. **61**, 850–863 (2015)
5. Tepper, N., Hashavit, A., Barnea, M., Ronen, I., Leiba, L.: Collabot: personalized group chat summarization. In: WSDM 2018: Proceedings of the Eleventh ACM International Conference on Web Search and Data Mining, pp. 771–774 (2018). https://doi.org/10.1145/3159652.3160588

6. Saqib, S.M., Bhosle, S.: WhatsApp chat analyzer. Int. Res. J. Mod. Eng. Technol. Sci. **4**(5) (2022)
7. Ramana, D.V.S., Anusha, K.M., Bhagya, K., Keerthana, M., Lakshmi, G.: WhatsApp group chat analysis. Int. J. Adv. Res. Sci. Commun. Technol. (IJARSCT) **3**(5) (2023)

ELT: Easy Label Trainer

David Nyarko, Kelechi Nwachukwu, Anjolie Anthony, and Kofi Nyarko(✉)

Morgan State University, Baltimore, MD 21251, USA
kofi.nyarko@morgan.edu

Abstract. In this paper, we present ELT-Easy Label Trainer (ELT), a comprehensive application designed to streamline the process of labeling and training object detection models while tracking objects for analysis. Leveraging advanced models such as YOLOv8, YOLOv9, and YOLO-World, ELT provides an intuitive interface built with Gradio for users to load video streams and perform object detection or segmentation. YOLO-World, known for its open-vocabulary detection using descriptive text prompts, is enhanced within ELT by custom prompts and additional parameters to improve detection confidence. Despite its robustness, YOLO-World's performance decreases for objects outside the COCO128 dataset. To address this, ELT employs pseudo-labeling to capture low-confidence labels and retrains models with YOLOv8 or YOLOv9, enhancing detection accuracy. ELT also automates object tracking and labeling using the Segment Anything Model (SAM) and Long-Term Video Object Segmentation with an Atkinson-Shiffrin Memory Model (XMem), inspired by Track Anything, to facilitate precise and efficient labeling. We demonstrate the effectiveness of ELT on GMOT-40: A Benchmark for Generic Multiple Object Tracking datasets, showing robust object detection in real-time. The results indicate a significant reduction in manual labeling effort and enhanced model performance, proving the viability of ELT in diverse practical applications.

Keywords: Object Tracking · YOLO · SAM · XMem · Labeling · Training · Pseudo-Labeling · Computer Vision

1 Introduction

Object detection and tracking are critical tasks in computer vision, with applications ranging from surveillance and autonomous driving to medical imaging and industrial automation. Traditional models like YOLO (You Only Look Once) have demonstrated impressive performance within predefined classes due to their balance of high accuracy and real-time processing capabilities [1]. However, these models struggle significantly with objects outside their training datasets, leading to reduced detection accuracy and requiring extensive manual effort for labeling and training new datasets [2].

To address these challenges, we introduce the ELT-Easy Label Trainer (ELT), a comprehensive solution designed to simplify the labeling and training process by automating object tracking and segmentation. The core of ELT is built on advanced models such as YOLOv8, YOLOv9, and YOLO-World, which are integrated into a user-friendly

A. Abdelgawad et al. (Eds.): ISBCom 2024, LNNS 1268, pp. 330–340, 2025.
https://doi.org/10.1007/978-3-031-82377-0_28

web application developed using Gradio [3]. This application allows users to load video streams and perform object detection or segmentation with minimal effort.

One of the key innovations of ELT is the incorporation of pseud-labeling with YOLO-World, an advanced variant of YOLOv8 designed for open-vocabulary detection. YOLO-World excels in identifying objects using descriptive text prompts, making it highly adaptable for various vision-based applications [1]. However, its performance declines when dealing with objects outside the COCO128 dataset [2]. To mitigate this, ELT employs pseudo-labeling techniques to identify low-confidence labels and retrain the models using YOLOv8 or YOLOv9, thereby enhancing detection accuracy [4].

Furthermore, ELT leverages the Segment Anything Model (SAM) and Long-Term Video Object Segmentation with an Atkinson-Shiffrin Memory Model (XMem) for automating object tracking and labeling. Inspired by the Track Anything methodology, ELT utilizes SAM to create segmentation masks [5] and XMem to track objects across video frames [6]. This combination allows for precise and reliable object tracking, reducing the manual effort required for labeling data [7].

The effectiveness of ELT is demonstrated using GMOT-40 datasets, including Stock-3 and Fish-3, which serve as ground truth for evaluation [8]. Our experimental results indicate a significant reduction in manual labeling effort [2] and enhanced model performance, proving the viability of ELT in diverse practical applications [9]. This paper provides a detailed overview of the ELT system, its implementation, and its performance evaluation, highlighting its potential to transform the field of object detection and tracking.

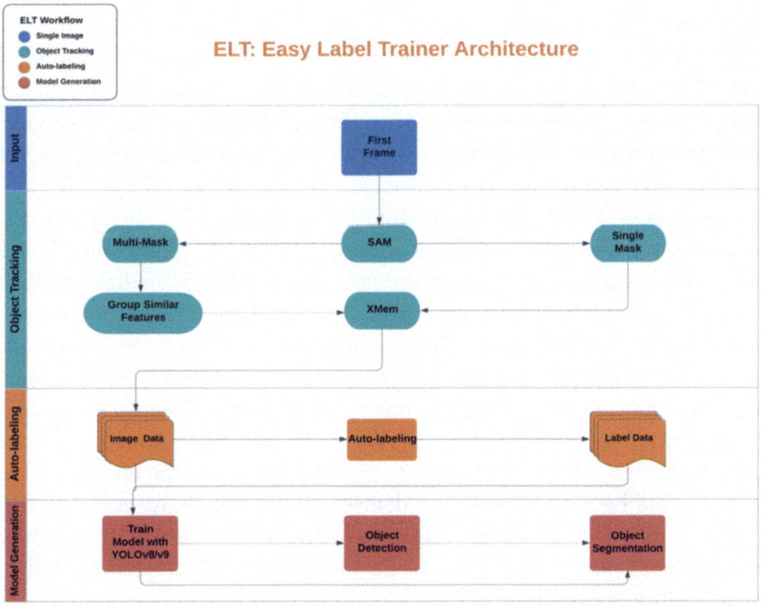

Fig. 1. ELT system architecture showing integration of YOLO models and workflow from input to output.

2 Objectives and Contributions

2.1 Objectives

The primary objective of ELT-Easy Label Trainer is to enhance the efficiency of labeling and training object detection models, particularly in environments with dynamically changing object contexts. This is achieved through the automation of object tracking and pseudo-labeling, significantly reducing manual effort and improving model accuracy.

2.2 Contributions

This paper introduces a novel framework integrating YOLOv8, YOLOv9, and YOLO-World within a unified architecture to deliver robust object detection capabilities. By leveraging Gradio for a user-friendly interface and implementing advanced pseudo-labeling techniques, ELT effectively addresses the challenges of open-vocabulary detection and training model precision on previously unseen objects. Importantly, ELT also eliminates the traditional need for manual labeling of data, further enhancing the efficiency and scalability of deploying object detection models in varied application settings.

3 Background

The field of object detection has seen significant advancements with the development of various deep-learning models. One of the most influential models is YOLO (You Only Look Once), which revolutionizes real-time object detection by balancing speed and accuracy [10]. YOLO's foundational architecture, which is based on convolutional neural networks (CNNs), and is present in all current iterations, enables it to detect multiple objects within an image using a single forward pass. This efficiency has made YOLO a popular choice for applications requiring real-time processing, such as autonomous driving, surveillance, and robotics.

Despite its success, YOLO faces challenges when dealing with objects not included in its training datasets. This limitation arises because YOLO, like many other deep learning models, relies heavily on the diversity and representativeness of the training data. When encountering objects outside the scope of its training data, YOLO's performance can degrade significantly, leading to false detections or missed objects [9]. Researchers have developed various techniques to address this issue, including transfer learning, domain adaptation, and pseudo-labeling [11], to improve the model's generalization capabilities.

YOLO-World, an advanced variant of YOLOv8, has been developed to tackle the open-vocabulary detection problem. By using descriptive text prompts, YOLO-World can identify objects beyond its training dataset, making it more versatile for a wide range of applications [1]. However, YOLO-World's performance still declines when dealing with objects entirely unfamiliar to the model, necessitating additional strategies to enhance detection accuracy.

ELT (Easy Label Trainer) leverages these advancements by integrating YOLOv8, YOLOv9, and YOLO-World into a single architecture. This architecture has been implemented into a user-friendly application that is built using Gradio, which provides an intuitive interface for users to load video streams and perform object detection or segmentation tasks effortlessly. ELT also incorporates the Segment Anything Model (SAM) and XMem for automating the object tracking and labeling process, further reducing the manual effort required for creating training datasets [12].

4 Methodology

4.1 Preliminaries

ELT integrates YOLOv8, YOLOv9, and YOLO-World within a single architecture to perform object detection and segmentation. YOLO-World's open-vocabulary detection is particularly leveraged for its ability to identify objects using descriptive text prompts. However, ELT incorporates pseudo-labeling and retraining strategies to address its limitations with objects outside the COCO128 dataset [13].

4.2 Implementation

ELT supports fine-tuning the YOLO-World model with custom classes, enabling precise and acceptable detections. For objects not detected by YOLO-World or detected but at low confidence, users can utilize ELT to train YOLOv8 or YOLOv9 models by selecting and tracking objects while labeling data in the YOLO format. This process is facilitated by SAM and XMem when a segmentation model is needed [14]. ELT integrates advanced object detection and tracking models within a streamlined user interface. Using YOLO-World's open-vocabulary capabilities, ELT can detect objects described by text prompts. The pseudo-labeling technique further refines the detection process by identifying and retraining the model on low-confidence detections using additional data. This method ensures higher accuracy and adaptability across varied datasets.

4.3 Automated Object Tracking and Labeling

ELT automates the object tracking process by applying binary masks to each segmentation mask created by SAM. The application analyzes the image's color, texture, and keypoint characteristics within the masked region. It uses normalized histograms for color in HSV space, Local Binary Patterns (LBP) for texture, and ORB features for key points and descriptors. One or a combination of these features can be selected for analysis. The Bhattacharyya distance between feature sets for color and texture is computed, and these distances are averaged to measure overall dissimilarity. [15].

Tracked objects are labeled, and data is saved in the YOLO format. Subsequently, a YOLOv8 or YOLOv9 model is trained for robust object detection on similar datasets (Fig. 6).

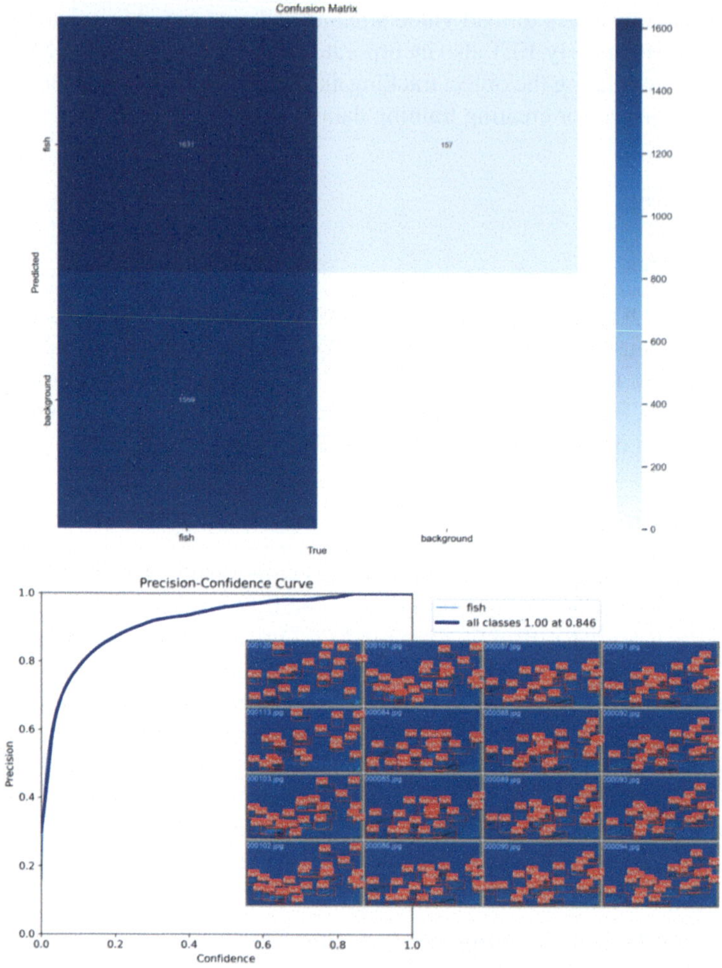

Fig. 2. Inference Results from the Fine-Tuned YOLO-World Model on GMOT Fish-3 Data. This figure presents the detection accuracy improvements post fine-tuning on YOLO-World, showcasing the effectiveness of pseudo-labeling techniques in recognizing low-confidence objects and enhancing overall model precision.

5 Experiments

5.1 Quantitative Results

To evaluate the performance of ELT, we conducted extensive experiments using the GMOT-40 datasets (Stock-3 and Fish-3) as ground truth. These datasets include a diverse range of objects and scenarios, providing a comprehensive benchmark for assessing the effectiveness of ELT. Our experimental setup involved comparing the detection accuracy, processing time, and manual labeling effort required when using ELT versus traditional manual labeling methods.

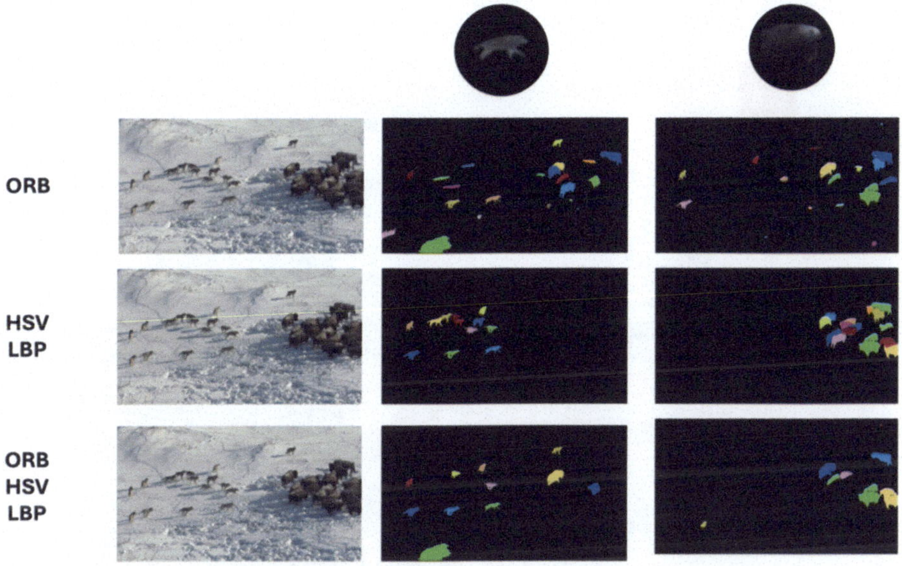

Fig. 3. Comparison of Mask Features Using HSV and LBP Techniques. This figure compares the object detection capabilities when using Hue, Saturation, Value (HSV) color space and Local Binary Patterns (LBP) for texture analysis. The side-by-side results demonstrate the superior performance of combining these techniques for accurate object segmentation.

Fig. 4. Comparison of Labeled Objects with Ground Truth Data. This figure displays the precision of the automated labeling process of ELT, highlighting the correlation between the manually annotated labels and those generated by ELT. It underscores the system's ability to label objects in complex scenes accurately.

Our results indicate that ELT significantly reduces the manual effort required for labeling (Fig. 7) and training while achieving high accuracy and efficiency in object detection and segmentation tasks. Specifically, this ELT provides a reduction in manual labeling time by 100%, which translates to a substantial increase in productivity for researchers and practitioners when training models with ELT. Additionally, the automated labeling and training process enhances the model's detection capabilities, particularly for objects outside the initial training dataset [17].

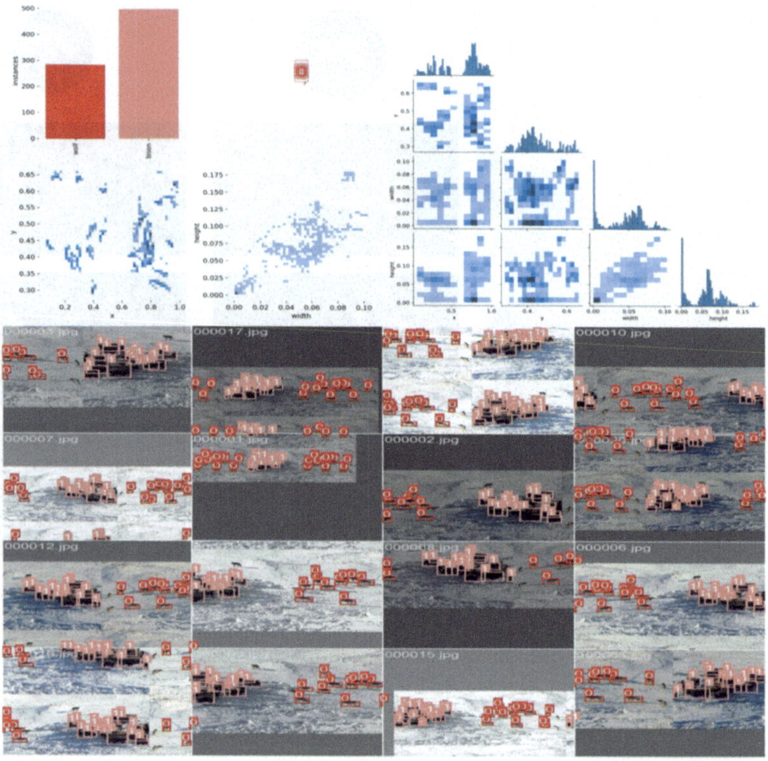

Fig. 5. Preview of Automated Labels Generated by ELT. This visual representation shows the labels generated by ELT for a set of detected objects, illustrating the clarity and detail of labeling that the system achieves. This figure emphasizes the practical utility of ELT in reducing manual labeling workload.

5.2 Qualitative Results

The qualitative analysis of ELT showcases its precision and reliability in real-time object detection tasks. Figures 1, 2, 3, 4 and 5 provide a visual representation of the various stages in the ELT process, from initial object selection to final training outputs. These results highlight ELT's effectiveness in automating the labeling and training process, leading to improved detection accuracy [18].

6 Applications

The ELT-Easy Label Trainer (ELT) is a versatile tool with wide-ranging applications across various industries. By simplifying the labeling and training process for object detection models, ELT can be utilized in several impactful ways:

6.1 Industrial Automation

In industrial settings, object detection models are used for quality control, monitoring production lines, and managing inventory. ELT can streamline the training of these

Fig. 6. Output from Single-Step Training by ELT. Displaying class IDs for Wolf and Bison, this figure highlights the results of ELT's auto-training capability. Despite Wolves being the primary subject in the ground truth, the inclusion of Bison underscores ELT's broad detection and labeling capacity."

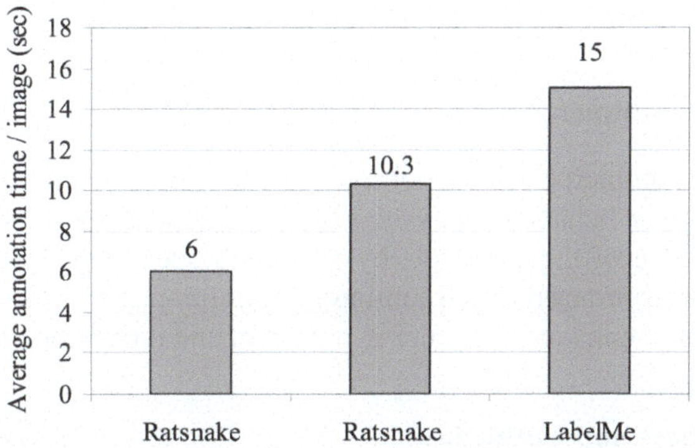

Fig. 7. Average time per image (in seconds) required with Ratsnake and LabelMe for the annotation of the image sequences used in the experiments [16].

Table 1. Compares validation results for Yolo-World and ELT models on GMOT-4: Stock3

Model	Box(P)	R	mAP50	mA P50-95
YOLO-World	0.412	0.588	0.404	0.237
ELT (Single Step Train)	0.86	0.65	0.76	0.41

models, ensuring they can accurately identify defects, track products, and manage resources effectively. The ability to automate labeling and improve model accuracy through pseudo-labeling makes ELT an invaluable tool for maintaining high standards in manufacturing and logistics.

6.2 Agriculture

In agriculture, object detection is used for tasks such as monitoring crop health, detecting pests, and managing livestock. ELT can be employed to train models that analyze images and videos from drones or stationary cameras to monitor large agricultural areas efficiently. By automating the labeling process and enhancing model accuracy, ELT helps farmers and agricultural professionals make informed decisions, improve yield, and manage resources more effectively.

6.3 Environmental Monitoring

ELT can be applied in environmental monitoring, where object detection models are used to track wildlife, monitor deforestation, and observe changes in natural habitats. Additionally, ELT can be used for tracking the evolution of events like floods, providing real-time data for disaster management and environmental protection. The automated labeling and tracking features of ELT allow for the efficient training of models that can analyze large volumes of environmental data, providing valuable insights for conservation efforts and environmental protection.

6.4 Retail and E-Commerce

In the retail and e-commerce sectors, object detection models are used for inventory management, customer behavior analysis, and automated checkout systems. ELT can streamline the training of these models by providing accurately labeled datasets and improving detection accuracy through continuous retraining. This helps retailers manage stock more efficiently, enhance customer experience, and reduce operational costs.

6.5 For Researchers and Enthusiasts

Researchers and individuals interested in utilizing detection and segmentation technologies but lacking extensive labeling and training knowledge can greatly benefit from ELT. With just a few clicks in the ELT app, users can achieve customized use cases for their specific needs. ELT provides an intuitive interface and robust backend, enabling users to focus on their research and applications without the need for deep expertise in model training.

7 Limitations and Future Research

7.1 Limitations

The ELT-Easy Label Trainer (ELT) marks a significant advancement in reducing manual effort and enhancing the accuracy of object detection models. However, it is notably constrained by its high dependency on computational resources, specifically when utilizing the Segment Anything Model (SAM) and XMem, which are intensive in their GPU requirements (in our studies, the NVIDIA RTX TITAN was utilized). Such intensive computational demands, necessary for processing advanced algorithms and maintaining real-time capabilities, may restrict its deployment in environments with limited computational capacity, such as remote or mobile settings. This limitation is particularly impactful in regions where such advanced technology is most needed, potentially hindering broader adoption and effectiveness.

7.2 Future Research

Future research should focus on several key areas to enhance the capabilities of the ELT system and address its current limitations. Firstly, the integration of lightweight model architectures that demand less computational power could significantly expand ELT's usability in resource-constrained environments. These architectures could benefit from recent advancements in model pruning and knowledge distillation, achieving high accuracy while reducing the processing load.

Secondly, there is substantial merit in exploring adaptive learning techniques that adjust the model's complexity based on the available computational resources. Such adaptive techniques could provide a dynamic solution tailored to various deployment scenarios, allowing ELT to offer scalable solutions that adjust in real-time to the hardware capabilities of the deployment environment. This direction not only aligns with the need to overcome the system's current limitations but also enhances its applicability and efficiency in diverse operational contexts.

8 Conclusion

The ELT-Easy Label Trainer offers a comprehensive solution for simplifying the labeling and training of object detection models. By integrating advanced detection models with automated tracking and pseudo-labeling techniques, ELT enhances the performance and usability of object detection systems in various applications. Our experimental results demonstrate the effectiveness of ELT in improving detection accuracy and reducing manual effort, making it a valuable tool for the computer vision community.

The results presented in Table 1 illustrate the enhanced performance of the ELT model over the YOLO-World across various metrics. Notably, ELT shows a significant improvement in Box Precision (Box(P)) and mean Average Precision (mAP50), which are critical indicators of model accuracy and reliability in real-world scenarios. The increase in Box(P) from 0.412 to 0.86 reflects a more precise bounding box detection by ELT, suggesting a substantial reduction in false positives and more accurate object localization. Similarly, the improvement in mAP50 from 0.404 to 0.76 indicates that ELT effectively increases the true positive rate across all detection thresholds. These quantitative enhancements underscore ELT's capability to not only reduce manual labeling efforts significantly but also to improve the overall efficiency and accuracy of object detection systems. This advancement is pivotal for applications requiring high precision and robustness, further proving ELT's utility in practical, operational settings.

References

1. Cheng, T., et al.: YOLO-World: Real-Time Open-Vocabulary Object Detection. arXiv (2024)

2. Lin, T.-Y., et al.: Microsoft COCO: Common Objects in Context. arXiv (2015)
3. Abid, A., et al.: Gradio: Hassle-Free Sharing and Testing of ML Models in the Wild. arXiv (2019)
4. Reis, D., Kupec, J., Hong, J., Daoudi, A.: YOLOv8: Next Generation Real-Time Object Detection. arXiv (2023)
5. Kirillov, A., et al.: Segment Anything. Meta AI Research (2023)
6. Cheng, H.K., Schwing, A.G.: XMem: Long-Term Video Object Segmentation with an Atkinson-Shiffrin Memory Model. ECCV (2022)
7. Yang, J., et al.: Track Anything: Segment Anything Meets Videos. SUSTech VIP Lab (2023)
8. Bai, H., et al.: GMOT-40: A Benchmark for Generic Multiple Object Tracking. arXiv (2021)
9. Reis, D., Kupec, J., Hong, J., Daoudi, A.: Real-Time Flying Object Detection with YOLOv8. Georgia Institute of Technology (2023)
10. Wang, C.Y., Bochkovskiy, A., Liao, H.Y.M.: YOLOv4: Optimal Speed and Accuracy of Object Detection. arXiv (2020)
11. Chen, Z., Zhang, W., Wang, X., Chen, K., Wang, Z.: Mixed Pseudo Labels for Semi-Supervised Object Detection. arXiv (2023)
12. Gao, M., Zheng, F., Yu, J.J.Q., Shan, C., Ding, G., Han, J.: Deep learning for video object segmentation: a review. Artif. Intell. Rev. **56**(1), 457–531 (2023)
13. Oh, S.W., Lee, J.Y., Xu, N., Kim, S.J.: Video Object Segmentation Using Space-Time Memory Networks. ICCV (2019)
14. Yang, Z., Wei, Y., Yang, Y.: Associating Objects with Transformers for Video Object Segmentation. NeurIPS (2021)
15. Kailath, T.: The divergence and bhattacharyya distance measures in signal selection. IEEE Trans. Commun. Technol. **15**(1), 52–60 (1967)
16. Iakovidis, D., Smailis, C.: Efficient Semantically-Aware Annotation of Images. In: Proceedings of the 2011 IEEE International Conference, pp. 146–149 (2011)
17. Wang, C.Y., Bochkovskiy, A., Liao, H.Y.M.: Scaled-YOLOv4: Scaling Cross Stage Partial Network. CVPR (2020)
18. Kristan, M., et al.: The Tenth Visual Object Tracking VOT2022 Challenge Results. ECCV Workshops (2022)

Eyeballing Combinatorial Problems: A Case Study of Using Multimodal Large Language Models to Solve Traveling Salesman Problems

Mohammed Elhenawy[1,2](✉), Ahmed Abdelhay[3], Taqwa I. Alhadidi[4], Huthaifa I. Ashqar[5], Shadi Jaradat[1], Ahmed Jaber[6], Sebastien Glaser[1], and Andry Rakotonirainy[1]

[1] Accident Research and Road Safety Queensland, Queensland University of Technology, 130 Victoria Park Rd, Kelvin Grove, Brisbane, QLD 4059, Australia
mohammed.elhenawy@qut.edu.au
[2] Centre for Data Science, Queensland University of Technology, 130 Victoria Park Rd, Kelvin Grove, Brisbane, QLD 4059, Australia
[3] Computer and Systems Engineering Department, Faculty of Engineering, Minia University, Minia, Egypt
[4] Civil Engineering Department, Al-Ahliyya Amman University Al-Saro Al-Salt, Amman, Jordan
[5] Arab American University, Palestine and Columbia University, Jenin, NY, USA
Huthaifa.ashqar@aaup.edu
[6] Department of Transport Technology and Economics, Faculty of Transportation Engineering and Vehicle Engineering, Budapest University of Technology and Economics, Műegyetem Rkp. 3., Budapest 1111, Hungary

Abstract. Multimodal Large Language Models (MLLMs) have demonstrated proficiency in processing diverse modalities, including text, images, and audio. These models leverage extensive pre-existing knowledge, enabling them to address complex problems with minimal to no specific training examples, as evidenced in few-shot and zero-shot in-context learning scenarios. This paper investigates the use of MLLMs' visual capabilities to 'eyeball' solutions for the Traveling Salesman Problem (TSP) by analyzing images of point distributions on a two-dimensional plane. Our experiments aimed to validate the hypothesis that MLLMs can effectively 'eyeball' viable TSP routes. The results from zero-shot, few-shot, self-ensemble, and self-refine zero-shot evaluations show promising outcomes. We anticipate that these findings will inspire further exploration into MLLMs' visual reasoning abilities to tackle other combinatorial problems.

Keywords: Multimodal Large Language Models · Traveling Salesman Problem · Combinatorial Optimization · Zero-shot learning

1 Introduction

Currently, researchers are investigating the potential of using Large Language Models (LLMs) to solve combinatorial problems such as the traveling salesman problem (TSP). Liu et al. [1] discussed the first study on solving combinatorial problems by using

A. Abdelgawad et al. (Eds.): ISBCom 2024, LNNS 1268, pp. 341–355, 2025.
https://doi.org/10.1007/978-3-031-82377-0_29

LLMs with evolutionary algorithms. They developed an approach named LLM-driven evolutionary algorithms (LMEA). Yang et al. [2], proposed a different approach to Optimization by PROmpting (OPRO). For each instance, describe the optimization task in natural language. Next, the LLM generates a new solution from the prompt that contains previously generated solutions with their values. Then, the new solutions are evaluated and added to the prompt for the next optimization step. Also, combining ensemble learning techniques with LLMs shows a promising increase in solution optimality such as methods suggested in Mingjian et al. and Silviu et al. works [3, 4].

Meanwhile, combining LLMs with other optimization methods showed promising capabilities for solving TSP; for each group of methods, the in-context prompting techniques such as (zero-shot, few-shot, Chain-of-thoughts CoT, etc.) can increase the correctness of LLMs responses. Researchers investigate the effect of using each of these techniques individually and the effect of combining multiple in-context techniques on increasing LLMs' response correctness such as papers [5–11].

In Huang et al. paper [14], researchers investigate the potential of using LLMs to solve vehicle routing problems (VRPs). From many prompt paradigms for LLMs, they found directly feeding natural language into LLMs provides the best performance. Also, they propose a self-refinement framework allowing LLMs to iteratively improve their solutions. Additionally, they studied the importance of details in explaining tasks and proposed a mechanism to help LLMs get thorough task descriptions.

While LLMs excel in text-based tasks, they often struggle to understand and process other data types [12]. MLLM addresses this limitation by combining various modalities, enabling a more comprehensive understanding of diverse data. MLLMs combine multiple data types, overcoming the limitation of pure text models and opening possibilities for handling diverse data types such as (text, image, video, audio, etc.).

In the meantime, few research discusses the potential of using MLLM to solve optimization problems. To the best of our knowledge, Huang et al. research [13] is the first to develop a multimodal LLM-based optimization framework; whereas they emulate the workflow of human beings in solving optimization problems. Their proposed method incorporates both textual and visual prompts simultaneously to facilitate a comprehensive understanding of optimization problems and further improve optimization performance. They evaluate the proposed method over the capacitated vehicle routing problem.

The main contributions of this paper are as follows:

1. We investigate the visual reasoning capabilities of the MLLM by instructing the model to estimate solutions for the Traveling Salesman Problem (TSP) through visual analysis.
2. We explore the performance of the proposed MLLM in estimating TSP solutions using various in-context learning techniques.
3. We adopt a self-ensemble strategy to enhance the quality of the MLLM's estimated solutions.
4. We propose two self-refinement strategies that leverage the visual capabilities of the MLLM to solve the TSP more effectively.

The remainder of this paper is organized as follows: In Sect. 2, we describe our approach and methods for in-context prompting, self-ensemble, and self-refining. Section 3 presents the results of experimented methodologies. Section 4 concludes.

2 Methodology

In this section, we outline our proposed methodology to "eyeball" solutions to the Traveling Salesman Problem (TSP) using visual prompts within a multimodal large language model (MLLM). Our approach is designed to incrementally increase complexity throughout the series of experiments. Initially, we conduct a baseline assessment to determine whether zero-shot or few-shot methods yield superior outcomes without delving into the optimization of demonstration example quantities or their selection strategies. Subsequently, the insights gained from these initial experiments inform our decisions for more advanced techniques in subsequent stages. Specifically, we then explore the integration of self-ensemble and self-refinement methods, selecting the most effective based on earlier results. This structured progression ensures that each phase builds upon the learnings of the previous, enhancing our methodology's effectiveness in addressing the TSP through visual analysis.

2.1 Dataset Generation

In this research, we used a dataset of journeys, where each journey is a number of unique 2-dimensional points. Whereas we want to test MLLM response against journeys with different numbers of points, we created the data as follows:

- To simplify the analysis, we selected this journey size {5, 10, 15, 20}
- For each journey size, we generate 30 journeys with several points equal to the journey size.
- We calculate the optimal route, and minimum distance for each journey using the Miller-Tucker-Zemline formulation.

2.2 Zero-Shot Model

In our zero-shot evaluation, we utilized the following Python prompt in Table 1 to instruct the MLLM to generate solutions for the TSP based solely on the visual representation of the nodes. The prompt is designed to direct the model to analyze a visual depiction of the nodes and derive an improved route sequence:

The prompt comprises two instructions:

1- The model is asked to suggest a trip sequence based solely on its visual assessment of the provided visualization of the nodes. The response must be delineated using <<start>> and <<end>> markers to encapsulate the sequence.
2- It ensures that the model's response adheres to the constraints of a Hamiltonian circuit, passing through each of the num_node points exactly once, thus maintaining the integrity of the TSP's requirements. This structured query facilitates the model's focus on generating a feasible and optimized route based on visual cues alone.

Table 1. Zero-shot Prompt

```
prompt = (f"Inspect the above visualization of the TSP nodes and do the following:\n"
        f"-1- Return a trip sequence based on your visual inspection only. "
        f"The sequence should be enclosed within <<start>> and <<end>>markers.\n"
      f"-2- Make sure to return the Hamiltonian circuit that passes by all {num_nodes} points with IDs from 1 to {num_nodes}.")
```

2.3 Few-Shot Model First Version

In the few-shot approach to solving TSP, the prompt given to the model is meticulously designed to encapsulate the specifics required for generating a valid solution. The prompt reads as follows:

Table 2. XXX

```
prompt = (f"This task requires a solution for exactly {num_nodes} points that has IDs from 1 to {num_nodes}. Your output must precisely follow this node count and IDs.\n"
        f"-1- Strictly identify a Hamiltonian circuit for exactly {num_nodes} points. Ensure the path visits each point once and returns to the starting point.\n"
        f"-2- Accurately sequence the circuit: List the IDs of these {num_nodes} points in the exact order they are visited, starting and ending at the same point.\n"
        f"The sequence must contain only and exactly these {num_nodes} points, formatted as: <<start>> 1 , 2 -> ... -> 1 <<end>>. Include no additional points or IDs.")
```

This prompt in Table 2 clearly defines the task for the model: to generate a Hamiltonian circuit for a predefined number of nodes, each uniquely identified by IDs from 1 to {num_nodes}. The model is instructed to provide a sequence in which these nodes are visited, starting and ending at the same point, without introducing any extraneous nodes or IDs.

The format for the output is specified to ensure clarity and correctness in the response, which is essential for validating the model's ability to adhere to the strict requirements of the TSP.

Table 3 shows a Python line that shows how we organize the input to operationalize the prompt in a real-world scenario using a multimodal large language model, integrating both visual and textual data. Using this organization to generate responses based on a combination of provided images and textual descriptions. Here's how the function orchestrates the input and output:

In this setup, the model is exposed to multiple pairs of images representing TSP scenarios ('input') and their optimal solutions ('output'), arranged sequentially as 'Example

1', 'Example 2', and 'Example 3'. This structured data feed is designed to teach the model the pattern of analyzing TSP routes from visual data and generating corresponding optimal paths. The training examples are carefully ordered to progressively build the model's understanding, culminating in its application to a new, unseen problem ('The New Problem') at the end. This methodology leverages the few-shot learning capabilities of the model, where it learns from a few examples before being tested on novel data, demonstrating the practical application of transfer learning in complex optimization tasks.

Table 3. Few-shot prompt's output section

```
responses = model.generate_content(
    ["""Training Examples:
    Example 1 Input:""", image1, """Example 1 output:""", image2,
    """Example2 input:""", image3, """Example2 output:""", image4,
    """Example3 input:""", image5, """Example3 output:""", image6,
    """The New Problem:""", imag7, prompt],
        stream=True )
```

2.4 Few-Shot Model Second Version

The Few Shots 2 methodology uses the Few Shots 1 prompt and builds upon the Few Shots 1 approach by modifying how optimal solutions are presented to the MLLM. Instead of using images to show optimal solutions, Few Shot 2 incorporates these solutions as textual data. This alteration aims to refine the model's text-processing capabilities in generating precise route sequences from visual cues.

This function initializes the model and dynamically generates responses based on a series of input images and corresponding textual outputs, detailing the optimal route sequences. Each set, termed as 'Example 1', 'Example 2', and 'Example 3', provides the model with a visual input followed by a textual description of the optimal solution, encapsulated within <<start>> and <<end>> tags as shown in Table 4. This method reinforces learning from textual representations of solutions, which can be crucial for tasks where precise text output is required. The function concludes with applying the learned patterns to a new problem, aiming to evaluate how well the model transfers its learned knowledge to unseen scenarios. This iterative learning and testing are vital for enhancing the model's accuracy in practical applications.

2.5 Self-Ensemble Model

To experiment with self-ensemble, we prompted MLLM 13 times and collected 13 responses. Then, we assessed the effect of self-ensemble using varying ensemble sizes: {3, 5, 7, 9, 11, 13}. As shown in Fig. 1, determining the solution of an ensemble of size S starts by selecting the first S responses from the collected 13 responses. Next, we

Table 4. Few-shot 2 prompt example

```
responses = model.generate_content(
    ["""Training Examples:
    Example 1 Input:""", image1, """Example 1 output:""", f'<<start>> {load_trip_data(FB, n, 0)} <<end>>',
    """Example2 input:""", image3, """Example2 output:""", f'<<start>> {load_trip_data(FB, n, 1)} <<end>>',
    """Example3 input:""", image5, """Example3 output:""", f'<<start>> {load_trip_data(FB, n, 2)} <<end>>',
    """The New Problem:""", imag11, prompt],
    generation_config=generation_config,
    safety_settings=safety_settings,
    stream=True,
)
```

filtered out any responses that were deemed to be hallucinations. Then, we calculated the traveling distance of each solution and selected the one with the lowest traveling distance as the definitive solution for that instance.

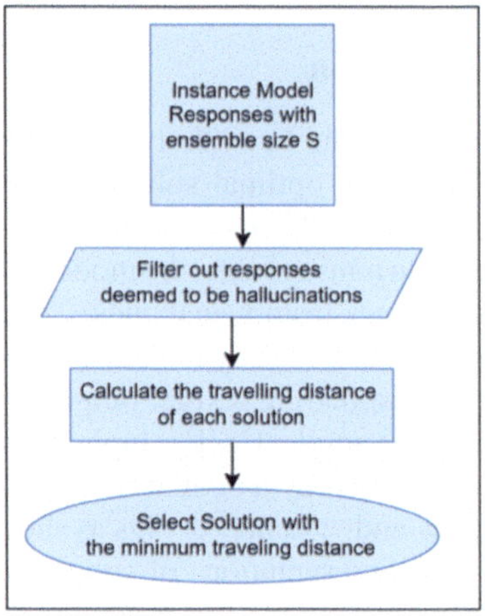

Fig. 1. Self-Ensemble method diagram

2.6 Self-Refine Model

We Experiment with two self-refinement configurations. Self-refine outlines an iterative process entirely reliant on visual reasoning, tailored explicitly for optimizing TSP solutions using LLMs. However, it is quickly adopted to solve other problems that can be visualized by changing the prompt. This visual-based approach purposely avoids mathematical equations, leveraging the model's capacity for visual analysis to propose enhancements to TSP routes. Although the example demonstrates a zero-shot in-context learning scenario, the framework is designed to be adaptable, accommodating various in-context learning strategies. The models employed in this methodology belong to the "Gemini" family of LLMs, although it is constructed to be compatible with other LLM architectures. The stopping criterion for this iterative process is initially set to a fixed number of iterations, which is suggested to correlate with the problem's complexity. However, alternative stopping criteria can be developed, such as using prompts to allow the model to assess visual cues to determine whether the current solution is optimal or if further iterations are necessary. For this research, the prompt is intentionally kept simple to explore the concept of "eyeballing" solutions, focusing on the fundamental idea of visually driven optimization without the complexities of additional computational checks. Self-refine 1 shown in Fig. 2 uses a text-based prompt to generate the first instance; Meanwhile, self-refine 2 shown in Fig. 3 uses a visualization of a non-connected points graph.

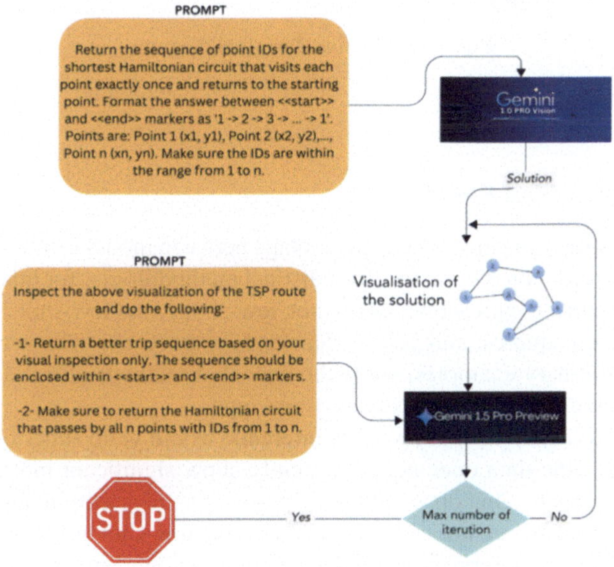

Fig. 2. Self-refine1 method diagram

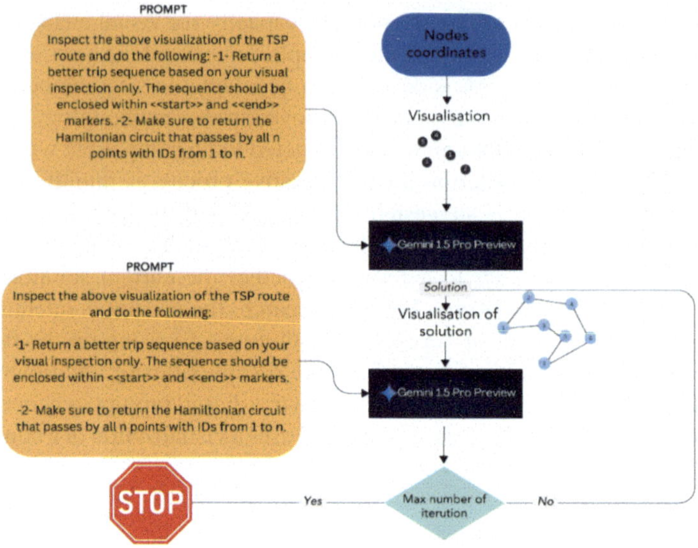

Fig. 3. Self-refine 2 method diagram

3 Results

3.1 Zero and Few-Shot Results

Figure 4 (A) illustrates the median gap percentage between model solutions as a function of the problem size, denoted by n. The three distinct methodologies for inferring solutions to a routing problem are represented: zero-shot, few-shot 1, and few-shot 2, differentiated by blue circles, red squares, and yellow diamonds respectively. The zero-shot method employs no preliminary examples, while the Few-shot methods utilize a select number of examples, potentially varying in quantity and type.

Contrary to expectations that few-shot inference would yield superior results due to tailored examples, the data does not consistently show significant improvements over the zero-shot approach. This observation suggests that the selection and formatting of demonstration examples are not trivial tasks. Effective few-shot inference requires optimization in terms of the number of examples, their sequencing, and the variability among these examples, which influences how the model learns to solve the given problem.

Furthermore, the median gap increases with the problem size across all inference methodologies, indicating that larger problems are inherently more difficult to solve accurately. This trend is likely due to the increased complexity and solution space. Additionally, practical difficulties such as node crowding or overlapping identifiers in graphical representations exacerbate these challenges, complicating the model's ability to infer correct solutions, particularly at larger scales. This graph underscores the necessity for careful consideration in the design of training and inference methodologies, especially in the face of escalating problem complexities.

Figure 4 (B) depicts the Interquartile Range (IQR) for the three in-context techniques: Zero Shot, Few Shot 1, and Few Shot 2. The x-axis represents the problem size (n) and the y-axis represents the IQR gap, highlighting how performance varies across different problem sizes. The three techniques show an increase in IQR as the problem size grows. Initially, Few Shot 2 closely followed Few Shots 1, but around a problem size of 13, it starts fluctuating dramatically. Overall, the Few Shots 1 consistent increase in IQR gap indicates stable performance. Meanwhile, Zero Shot exhibits lower IQR gaps at multiple instances, suggesting optimized solutions, but its fluctuation implies less consistency.

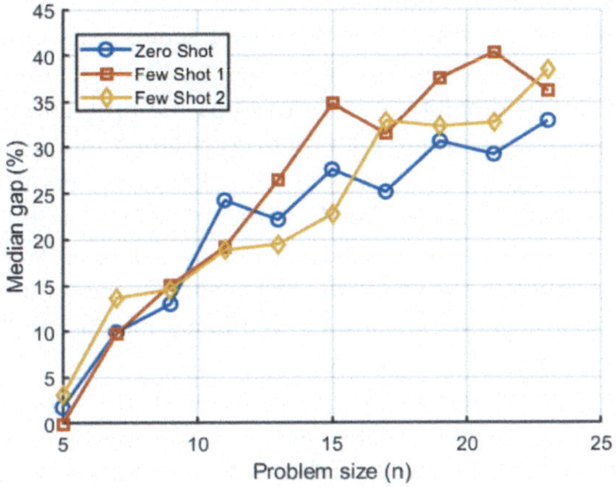

(A)

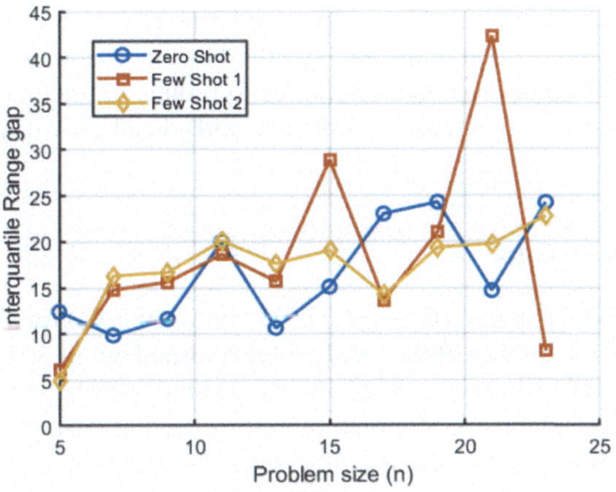

(B)

Fig. 4. Illustrates the median and interquartile range (IQR) for the three tested in-context learning techniques zero-shot, few-shot 1, and few-shot 2. Chart (A) shows the effect of increasing problem size on the different in-context learning techniques. Chart (B) shows the interquartile values of increasing problem size on different in-context learning techniques.

The bar charts in Fig. 5 depict two types of hallucinations observed while eyeballing TSP (Traveling Salesman Problem) solutions across three methodologies: Zero Shot, Few Shot 1, and Few Shot 2. The first type, labeled as "Incorrect nodes ID," likely refers to errors in identifying or labeling nodes correctly during the problem-solving process.

The second type, "Incomplete route," suggests issues with not fully connecting the tour, possibly missing some nodes or connections.

From the charts, a clear trend emerges: the percentage of hallucinations generally increases with the problem size (number of nodes). This is understandable as larger problem sizes lead to more complex visualizations, which can become crowded and more challenging to interpret accurately. As the visualization gets crowded, it's easier to mislabel nodes (Incorrect node ID) or overlook parts of the route (Incomplete route), leading to these types of hallucinations.

In terms of methodology differences, zero-shot seems to exhibit a relatively consistent level of hallucinations across problem sizes for "Incorrect nodes ID" but shows an increasing trend in "Incomplete route" errors as the problem size increases. This suggests that Zero Shot may struggle with route completion in larger visual fields.

Few Shot 1 and Few Shot 2 methodologies display varied performances. Few Shot 1 shows an increase in hallucinations for "Incorrect nodes ID" with increasing problem sizes, perhaps indicating that this methodology does not scale well with complexity in terms of node identification. Meanwhile, Few Shot 2 maintains a more consistent error rate for node identification but shows a significant increase in "Incomplete route" hallucinations as the problem size increases.

The differences between methodologies in terms of the percentage of hallucinations highlight the strengths and weaknesses of each approach. Zero Shot appears more robust in node identification but weaker in route completion, whereas Few Shot 1 struggles with node identification in larger sizes. Few Shot 2, while consistent in node identification, shows vulnerabilities in completing routes as complexity increases. These observations suggest that while each method has merits, there may be a need for refined approaches or combined methodologies to reduce hallucinations, especially in larger problem visualizations where complexity and visual crowding pose significant challenges (Fig. 6).

3.2 Self-Ensemble Results

In Fig. 7, the graphs display gap statistics (i.e. Median and IQR) for TSP solutions from MLLM, showcasing the quality of the valid routes returned out of 30 instances evaluated at each size. Figure 7 (A) illustrates the median gap percentage between model solutions as a function of the problem size, denoted by n for different ensemble sizes {1, 3, 5, 7, 9, 11, 13}. The median gap for all ensemble sizes shows an increasing trend with the problem's complexity growth. The chart shows a clear relation between median gap and ensemble size, as increasing ensemble size reduces the median gap, at the same problem complexity level. Figure 7 (B) depicts the Interquartile Range (IQR) for the gap similar to the median chart. Unlike the median gap, the IQR does not follow a discernible pattern of increase with the instance size. The fluctuation in the IQR suggests variability in the model's performance consistency.

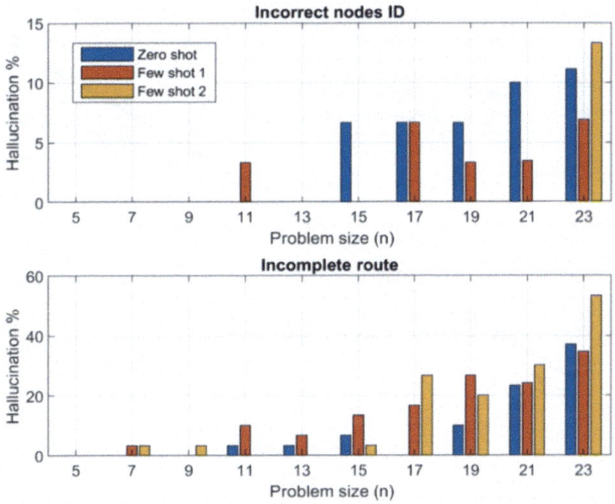

Fig. 5. Observed hallucination across different methodologies. The first chart describes the effect of increasing problem complexity on the number of solutions with incorrect nodes. The second chart presents the number of solutions with an incomplete route.

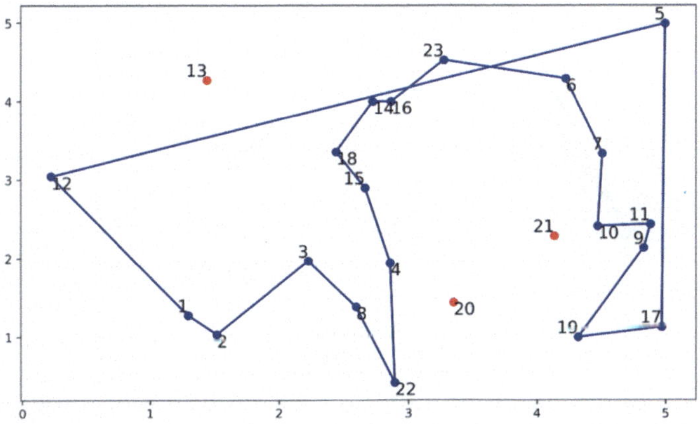

Fig. 6. Visualization of incomplete route hallucination example. Red points (13, 20, 31) are miss-visited points

3.3 Self-Refine Results

Figure 8 visually contrasts the median gaps across various problem sizes for two distinct methodologies: self-refine 1 and self-refine 2.

In the self-refine method, an initial Hamiltonian circuit is generated by Gemini 1.0 Pro based on textual descriptions of the points, which is subsequently visualized and refined through ten iterative feedback loops with Gemini 1.5, focusing on visual enhancements and route optimization. Each iteration aims to refine the route further, culminating in the selection of the best solution based on distance. Conversely, the

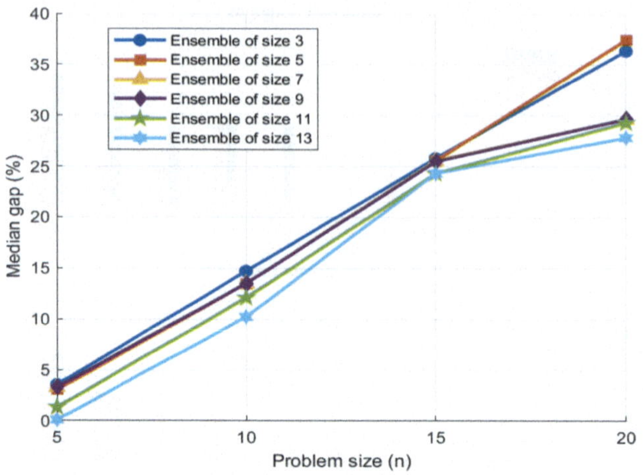

(A)

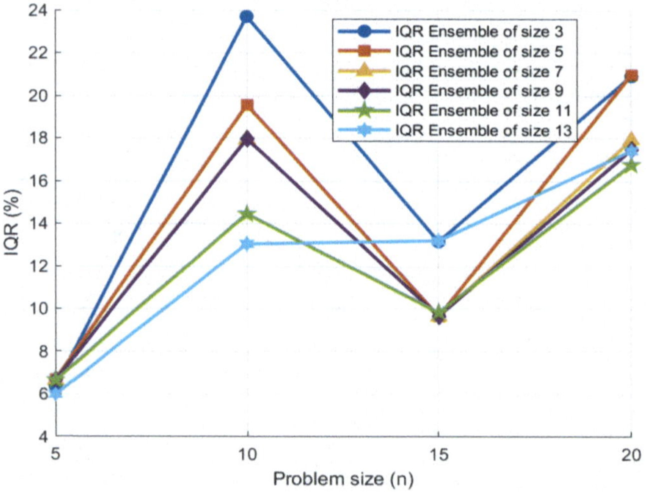

(B)

Fig. 7. Illustrates the median and interquartile range (IQR) for a different ensemble of sizes. Chart (A) shows the effect of increasing problem size on different ensemble sizes. Chart (B) shows the interquartile values of increasing problem size on different ensemble sizes.

Refine Two approach involves a straightforward process where the nodes of the problem are visualized, and a Hamiltonian circuit is directly solicited from Gemini 1.5. This proposed solution undergoes 10 iterative visualizations and enhancements to refine the route, with the final selection predicated on identifying the most optimal solution from these iterations (Figs. 9 and 10).

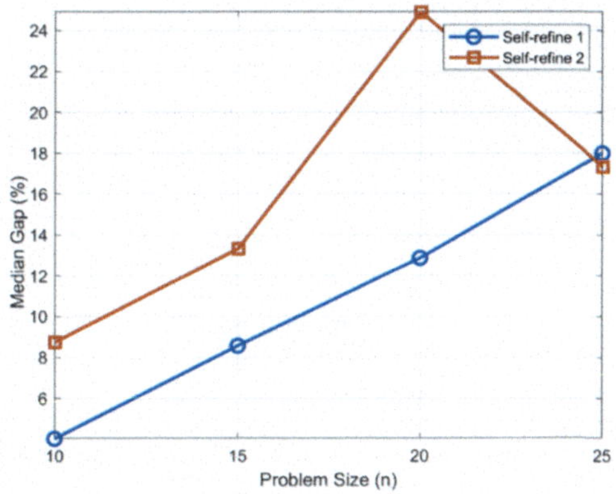

Fig. 8. Contrasts the median gaps across various problem sizes for two distinct methodologies: Self-refine 1 and Self-refine 2.

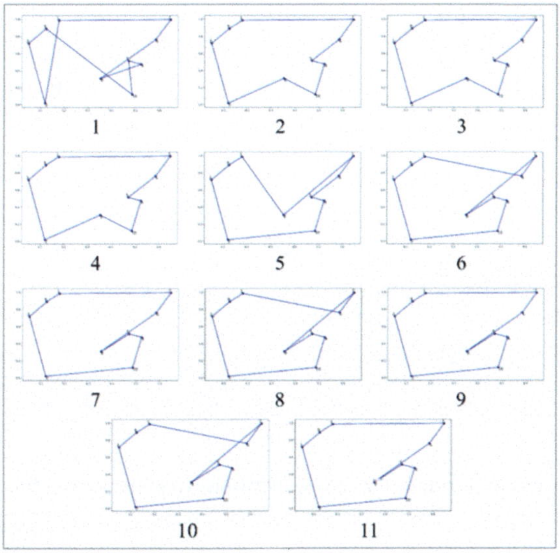

Fig. 9. Example of self-refine 1 full iteration to generate the final solution

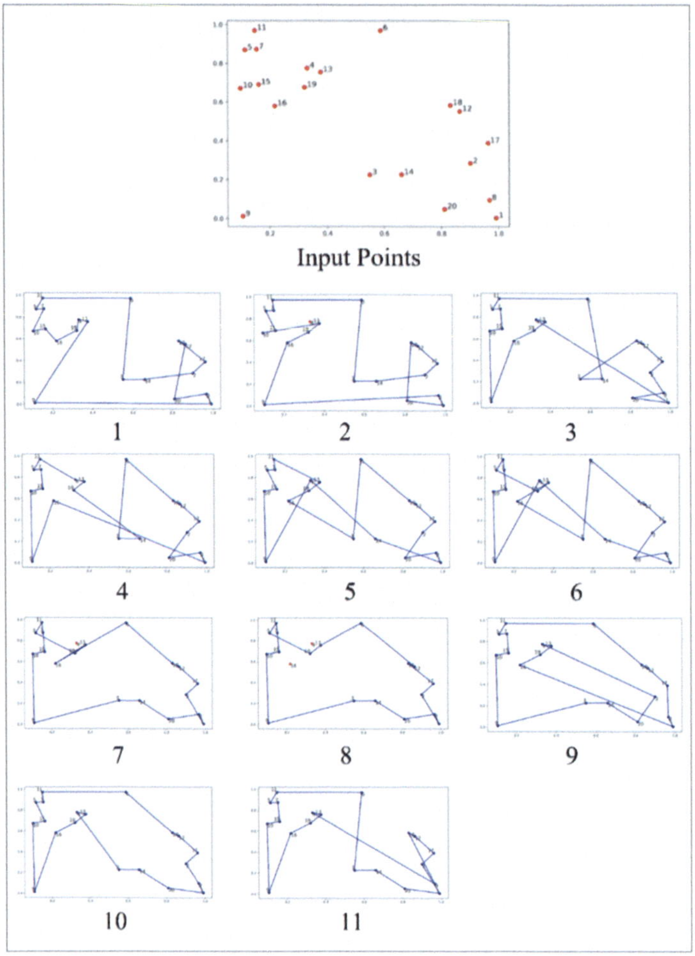

Fig. 10. Example of self-refine 2 full iterations to generate the final solution

4 Conclusion

This paper has explored the application of Multimodal Large Language Models, particularly the "Gemini" family of LLMs to combinatorial problems, focusing on the TSP. Our findings indicate that MLLMs are indeed capable of effectively solving TSP instances, utilizing prior knowledge in text and visual parts of the input prompts. We investigate zero-shot scenarios and enhance this with few-shot approaches; specifically, one-shot and two-shot solved examples. Contrary to expectation, zero-shot in-context prompting shows a more stable median gap analysis than few-shot techniques. Next, the implementation of self-ensemble methods has further enhanced the model's performance, delivering improved solution quality. At last, we adapted self-refinement methodologies allowing LLMs to iteratively refine solutions to complex optimization problems, highlighting how iterative visual feedback can influence the quality of the solutions and improve their solutions. From the two self-refinement techniques we experimented with, we found visualizing the input points is more effective than feeding the points as in text format.

Future research includes exploring other open-source MLLMs. Another promising avenue is experimenting our methodology on 3-dimensional data points, as well as using other valuation metrics such as the combination of time and distance cost.

References

1. Liu, S., Chen, C., Qu, X., Tang, K., Ong, Y.-S.: Large Language Models as Evolutionary Optimizers. arXiv:2310.19046 (2023)
2. Yang, C., et al.: Large Language Models as Optimizers. arXiv preprint arXiv:2309.03409 (2023)
3. Mingjian, J., et al.: Calibrating Language Models via Augmented Prompt Ensembles. openreview (2023)
4. Silviu, P., Zhang, M.R., Andrew, W., Ba, J.: Boosted Prompt Ensembles for Large Language Models. arXiv preprint arXiv:2304.05970 (2023)
5. Takeshi, K., Gu, S.S., Machel, R., Yutaka, M., et al.: Large Language Models are Zero-Shot Reasoners. arXiv preprint arXiv:2205.11916 (2023)
6. Jason, W., et al.: Finetuned Language Models Are Zero-Shot Learners. arXiv preprint arXiv: 2109.01652 (2022)
7. Sewon, M., et al.: Rethinking the Role of Demonstrations: What Makes In-Context Learning Work? arXiv preprint arXiv:2202.12837 (2022)
8. Brown, T.B., et al.: Language Models are Few-Shot Learners. arXiv preprint arXiv:2005. 14165 (2020)
9. Zhuosheng, Z., Aston, Z., Mu, L., et al.: Automatic Chain of Thought Prompting in Large Language Models. arXiv preprint arXiv:2210.03493 (2022)
10. Wei, J., et al.: Chain-of-Thought Prompting Elicits Reasoning in Large Language Models. arXiv preprint arXiv:2201.11903 (2023)
11. Xuezhi, W., el al.: Self-Consistency Improves Chain of Thought Reasoning in Language Models. arXiv preprint arXiv:2203.11171 (2022)
12. Wu, J., Gan, W., Chen, Z., Wan, S., Yu, P.S.: Multimodal Large Language Models: A Survey. arXiv arXiv:2311.13165 (2023)
13. Huang, Y., Zhang, W., Feng, L., Wu, X., Tan, K.C.: How Multimodal Integration Boost the Performance of LLM for Optimization: Case Study on Capacitated Vehicle Routing Problems. arXiv arXiv:2403.01757 (2024)
14. Huang, Z., Shi, G., Sukhatme, G.S.: From Words to Routes: Applying Large Language Models to Vehicle Routing. arXiv arXiv:2403.10795v1 (2024)

Question-Answering (QA) Model for a Personalized Learning Assistant for Arabic Language

Mohammad Sammoudi[1], Ahmad Habaybeh[1], Huthaifa I. Ashqar[2], and Mohammed Elhenawy[3,4](✉)

[1] Department of Natural, Engineering and Technology Sciences, Arab American University, Ramallah 240, Palestine
[2] Civil Engineering Department, Arab American University, Palestine and Columbia University, Jenin, NY 240, USA
Huthaifa.ashqar@aaup.edu
[3] CARRS-Q, Queensland University of Technology, Brisbane, Australia
mohammed.elhenawy@qut.edu.au
[4] Centre for Data Science, Queensland University of Technology, Brisbane, Australia

Abstract. This paper describes the creation, optimization, and assessment of a question-answering (QA) model for a personalized learning assistant that uses BERT transformers customized for the Arabic language. The model was particularly fine-tuned on science textbooks in Palestinian curriculum. Our approach uses BERT's brilliant capabilities to automatically produce correct answers to questions in the field of science education. The model's ability to understand and extract pertinent information is improved by fine-tuning it using 11th and 12th grade biology book in Palestinian curriculum. This increases the model's efficacy in producing enlightening responses. Exact match (EM) and F1 score metrics are used to assess the model's performance; the results show an EM score of 20% and an F1 score of 51%. These findings show that the model can comprehend and react to questions in the context of Palestinian science book. The results demonstrate the potential of BERT-based QA models to support learning and understanding Arabic students' questions.

Keywords: Learning Assistant · NLP · BERT · Transformers · Question-Answer

1 Introduction

Wars, strikes and pandemics cause and create a gap between the students and their teachers. The students are having their book (contexts) and can self-study these books. However, when they are studying, they face questions and challenges that require some help. The use of technology in the field of education has become essential to improve learning experiences in the constantly changing education methods. The developing of learning assistant systems, especially those suited for classroom use, is one potential direction. These systems use Question-Answer (QA) and Natural Language Processing (NLP) to build intelligent and interactive platforms that meet each student's unique demands [1].

A. Abdelgawad et al. (Eds.): ISBCom 2024, LNNS 1268, pp. 356–367, 2025.
https://doi.org/10.1007/978-3-031-82377-0_30

A field of artificial intelligence called natural language processing gives computers the ability to comprehend, interpret, and produce language that is similar to that of humans. When NLP is included into educational systems, it makes it easier for students and educators to communicate smoothly, which promotes a more customized and interesting learning environment [2].

The Question-Answer model, which is the core of these systems, is made to efficiently understand and react to user inquiries. Because of the large amount of data that this model has been trained on, it can recognize linguistic nuances and offer precise, contextually appropriate answers. Learning Assistant Systems can serve as virtual tutors, helping students understand difficult ideas, work through issues, and reinforce what they learn in the classroom by putting QA models into action [3].

A system like this is important because it can adjust to the different learning styles and speeds of each learner. Learning Assistant Systems, fueled by NLP and QA models, provide a dynamic and customized learning experience. These tools may adjust to meet the particular requirements of every learner, offering tailored assistance and direction.

In order to explain how NLP and QA models work together to create an intelligent and engaging teaching aid, this paper examines the design and implementation of a learning assistant system for students in schools.

2 Literature Review

Many studies have studied the idea of personalized learning assistant (PLA). These studies aim to build models that can help students in their study and examination [4]. Most of these studies used Natural Language Processing (NLP) techniques to extract information from students and support them with suitable answers, exercises, and materials [5].

The process of extract information from users to understand their query and give them an answer is the main challenge. A lot of models are used for this purpose. In [6] TF-IDF was used to match questions with similar questions that are found on search engines, this was a step in the right direction but a problem is faced when some question is not present on Internet, so we will not get an answer.

NLP tasks can be used for any language, but you need to be careful to use the version of model which support the desired language. For example, a study by Yilmaz and Toklu use NLP with Turkish language for analysis on question classification task. They use Word2Vec method with skip gram on a large corpus consists of user questions. They also use Convolutional Neural Networks (CNN) for classification task [7].

The study [8] explored the usage of chatbots in the pedagogical setting between 2017 and 2021. It evaluated the literature concerning the implementation of chatbots in education. The findings encompass the advantages and disadvantages of employing chatbots in e-learning, strategies to enhance and address the drawbacks, and the role of personalization in student advancement.

A paper [9] introduces the idea of interactive agents, such as Siri and Alexa, and their usage in everyday life, specifically in the context of higher education. It highlights the potential use of these agents as digital tutors. The paper concludes by analyzing the creation of conversational agents and their effect on learning outcomes.

A project explained in [10] suggest the use of AI and NLP techniques to generate personalized feedback automatically, including hints and explanations based on Wikipedia, without the need for expert intervention. This approach aims to improve student performance outcomes and is demonstrated through the implementation of the Korbit ITS, which serves approximately 20,000 students. The strategy employed entails utilizing ML and NLP techniques to create tailored feedback.

Chatbots as PLAs was used a lot in this field. For example, A study [11] explores the fusion of AI and NLP in the design of an ITS for school education, particularly focusing on computer science subjects. In this study a chatbot was developed and they conclude that chatbots have significant potential in the realm of school education, particularly if enhancements are made to the knowledge base and language models.

Another chatbot called LANA, proposed in [12], is a Conversational Intelligent Tutoring System (CITS) specifically designed for Arabic-speaking children with Autism Spectrum Disorder (ASD) who exhibit high-functioning autism or Asperger's Syndrome. LANA aims to provide a personalized learning experience in Arabic for individuals aged 10 to 16 by leveraging the structured nature of technology, which is often preferred by individuals with ASD.

An AI learning assistant called HAnS is the outcome of a collaboration between German universities and research institutes. It incorporates Natural Language Processing (NLP), speech recognition, and AI-based indexing to promote self-directed and personalized learning [13].

All the above studies use unidirectional techniques, which have many constraints. The inherent recurrent structure of these models makes it difficult to handle parallel processing. Additionally, they struggle with long sentences due to the vanishing gradient problem [14].

In 2017, Vaswani et al. proposed a new neural network model named Transformer [15]. This model introduced a modern architecture that inspired new developments like BERT and GPT, which rely heavily on the Transformer architecture. GPT is an autoregressive model and is considered a unidirectional language model [16]. In contrast, BERT provides a bidirectional representation, allowing for a better understanding of the text [17].

The implications of these advancements are significant. LANA and HAnS, while innovative, are limited by their use of unidirectional models, which face challenges in parallel processing and managing long sentences. The Transformer model, with its ability to handle parallel processes and mitigate the vanishing gradient problem, represents a major leap forward. This has led to the development of BERT and GPT models, which offer improved performance. BERT, with its bidirectional representation, enhances contextual understanding, making it particularly powerful for tasks requiring nuanced text comprehension. GPT, being autoregressive, excels in text generation tasks, but its unidirectional nature poses certain limitations in understanding context compared to BERT. These insights highlight the importance of selecting appropriate model architectures to address specific challenges in AI and NLP applications [18, 19].

In our study, we focus on using a pre-trained model called BERT and fine-tuning it with a dataset containing Arabic texts from the Palestinian curriculum biology books for

11th and 12th grades. This approach differentiates our work from other models described in the literature review in several significant ways.

Firstly, while existing systems like LANA and HAnS have made strides in providing personalized learning experiences, they typically use unidirectional models or focus on different user demographics and educational content. For instance, LANA is designed for Arabic-speaking children with Autism Spectrum Disorder, leveraging the structured nature of technology to support individuals aged 10 to 16. However, LANA does not specifically utilize a pre-trained BERT model fine-tuned with educational content from a specific curriculum.

Secondly, the HAnS assistant integrates NLP, speech recognition, and AI-based indexing to facilitate self-directed learning but does not target Arabic texts from a specific educational curriculum. Additionally, HAnS is a product of collaboration between German institutions and is primarily designed with their context in mind, rather than focusing on Arabic educational content.

2.1 Study Objective

Our study's uniqueness lies in the use of BERT, a bidirectional transformer model that provides a deeper contextual understanding of the text. By fine-tuning BERT with Arabic texts from the Palestinian curriculum's biology books, we ensure that the model is tailored to the specific language nuances and educational needs of Arabic-speaking students in Palestine. This curriculum-specific fine-tuning enhances the model's relevance and accuracy in understanding and responding to educational queries related to biology for grades 11 and 12.

Moreover, our model is designed to predict suitable responses based on the student's level, input context, and query, providing a highly personalized learning experience. This contrasts with the generic or less specialized models discussed previously, which may not offer the same level of contextual understanding and specificity in their responses.

By focusing on fine-tuning BERT with a targeted Arabic dataset from the Palestinian curriculum, our work aims to fill a critical gap in educational AI tools, offering a solution that is not only linguistically and contextually relevant but also highly specialized for the intended educational content. This approach leverages the strengths of BERT's bidirectional representation to improve the precision and relevance of the responses, ultimately supporting more effective and personalized learning experiences for Arabic-speaking students. By tailoring the model specifically to the Palestinian educational context, this study contributes to the development of AI-driven educational resources that are culturally and linguistically attuned to the needs of Arabic-speaking learners, fostering better engagement and comprehension.

3 Materials and Methods

For our research we follow the main NLP methodology which followed for almost every NLP Project. However, to make it suitable for our project, we modify it in some way to be suitable for our main goal. Our main project goal is to create a Personalized Learning Assistant (PLA) that will help school students. The first step was to choose a subject from the Palestinian curriculum and extract useful information from it. We choose the science subject for the school students for the secondary schools.

3.1 Data Collection and Preparation

To achieve the research goals two datasets were used, the first dataset was collected by a colleague team as a part of an NLP related project in a master's Natural Language Processing class held in Arab American University /Palestine in fall/2023. This Arabic dataset was manually collected from the 11th and 12th grade's biology books which are taught in Palestine, the dataset contains 111 observations and 7 attributes, the attributes are: id, unit_title, lesson_title, section_title, section_content, questions, avaliable_summary.

The second dataset which contains the student information is a publicly available dataset. The origin of the dataset is The University of Jordan, Amman, Jordan. It was collected by Elaf Abu Amrieh, Thair Hamtini, and Ibrahim Aljarah from learning management system (LMS) called Kalboard 360. The dataset consists of 480 student records and 16 features, Table 1 shows features descriptions and distinct values of each feature. The students in the dataset are classified into three categories: Low-Level 0 to 69, Middle-Level 70 to 89, High-Level 90 to 100.

3.2 Problem Statement and Proposed Framework

During In recent years, Palestine has faced significant challenges, including wars and financial difficulties, leading to frequent teachers' strikes. These disruptions have created a gap between students and their educators, making it difficult for students to receive the necessary guidance and support. Despite having access to their science books from the Palestinian curriculum, students often struggle to find the answers to their questions without teacher assistance.

To address this issue, we propose a model designed to retrieve answers for students' questions based on the context provided in their science books. Our model is fine-tuned specifically for the Arabic language using texts from the Palestinian curriculum's science books. This fine-tuning ensures that the model understands the specific linguistic and educational context of Palestinian students.

Our model serves as a learning assistant, leveraging advanced NLP techniques to provide accurate and contextually relevant answers. By incorporating personalization, the model tailors its responses based on the student's educational level, ensuring that the answers are appropriate for their understanding and needs. This personalized approach helps bridge the gap left by the absence of teachers, offering students a reliable resource to continue their education independently.

Using a pre-trained BERT model, fine-tuned with domain-specific content, we enhance the model's ability to comprehend and respond to queries effectively. This method not only improves the accuracy of the answers but also ensures that they align closely with the curriculum, providing a seamless learning experience. The deployment of this model supports continuous learning despite external challenges, empowering students to progress in their education autonomously [20].

Table 1. Students Profiles Dataset.

Id	Feature name	Description	Data Type	Available values
ID	Feature Name	Description	Data Type	Available Values
1	Gender	Student's Gender	Nominal	Male, Female
2	Nationality	Student's Nationality	Nominal	Kuwait, Lebanon, Egypt, Saudi Arabia, USA, Jordan, Venezuela, Iran, Tunis, Morocco, Syria, Palestine, Iraq, Libya
3	Place of birth	Student's Place of Birth	Nominal	Kuwait, Lebanon, Egypt, Saudi Arabia, USA, Jordan, Venezuela, Iran, Tunis, Morocco, Syria, Palestine, Iraq, Libya
4	Educational Stages	Educational Level Student Belongs	Nominal	Lower Level, Middle School, High School
5	Grade Levels	Grade Student Belongs	Nominal	G-01, G-02, G-03, G-04, G-05, G-06, G-07, G-08, G-09, G-10, G-11, G-12
6	Section ID	Classroom Student Belongs	Nominal	A, B, C
7	Topic	Course Topic	Nominal	English, Spanish, French, Arabic, IT, Math, Chemistry, Biology, Science, History, Quran, Geology
8	Semester	School Year Semester	Nominal	First, Second
9	Parent responsible for student	Parent responsible for student	Nominal	Mom, Father

Our solution also alleviates the pressure on the educational system by providing an alternative means of instruction during times when teachers are unavailable. By ensuring students have access to accurate and relevant information, we contribute to maintaining educational standards and continuity. This initiative highlights the potential of AI-driven solutions in overcoming educational barriers and enhancing the learning experience in challenging environments.

The above flowchart in Fig. 1 explains the methodology followed in our project. The project is consisting of two stages: the first stage is student's personalization and the second is the model building and fine tuning. Figure 1 illustrates a comprehensive workflow for a personalized learning assistant (PLA) system designed to cater to the unique needs of students by leveraging NLP and machine learning techniques. The system begins with the collection of various datasets, including student profiles, science textbooks, and

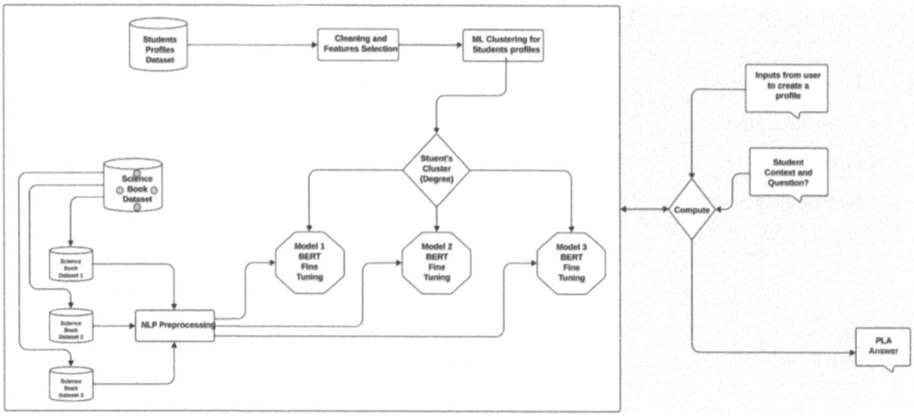

Fig. 1. Project Implementation Flowchart.

additional reference materials. These datasets undergo NLP preprocessing to ensure that the text data is appropriately formatted and ready for further processing. The student profiles dataset is subjected to cleaning and feature selection to extract relevant characteristics of each student. These features are then used in a machine learning clustering algorithm to group students into clusters based on their profiles. This clustering helps in identifying students with similar learning needs and preferences, facilitating targeted personalization.

Simultaneously, the science textbooks and additional resources are processed and fine-tuned using BERT (Bidirectional Encoder Representations from Transformers). The fine-tuning process is tailored for each student cluster, resulting in three distinct BERT models, each optimized for a specific student cluster. This ensures that the generated educational content and responses are highly relevant to the students' contextual and educational needs.

When a student interacts with the PLA system, they provide inputs to create or update their profile. Based on this profile, the system determines the appropriate student cluster and selects the corresponding fine-tuned BERT model. The student then submits their content and questions, which are processed using the chosen BERT model to generate precise and relevant answers. The PLA system ultimately delivers personalized answers to the students, enhancing their learning experience by providing contextually accurate and tailored educational support. This approach ensures that the learning assistant is not only responsive to individual queries but also aligned with the specific educational backgrounds and needs of the students it serves.

Students Personalization. The student's profiles dataset is used for this part of project. We get this dataset form Kaggle. Next, we clean the dataset by deleting any NA values and ensure the dataset is cleaned and all features' data types are suitable to be used by the machine learning model. Then, we did a classification task using K-Means clustering unsupervised ML model. We used the elbow method to find the optimal number of clusters, we found using elbow method that the students are distributed among three clusters, we will have three students' profiles. The three clusters that we have are classified based on their behavior (Excellent students, Good students and weak students).

Question-Answer Bert Fine Tuning Model. The second part of our project will be fine tuning BERT model to get a suitable answer for the student inquires. We will use a dataset collected from the Palestinian science book which is collected by our colleagues.

The main dataset is subdivided into three datasets. Each one will be different from the other by the question part which will be suitable for the cluster students from the first part of the model. We then fine-tuned the BERT model for Question-Answering task one for each cluster.

We used a pretrained BERT model for question answering (base-multilingual-cased). This model was created by Google and fine-tuned on XQuAD like data for 11 different languages including Arabic. Number of parameters in Bert model is 100 M [12].

User Interaction with the model. When the user interacts and use the model, he will be asked to input some information for the purpose to create his profile and for user classification later.

Then user will put context from the Palestinian science book. After that, he will ask a question to the model.

The model will use BERT fine-tuned model with the suitable dataset for the cluster that the student belongs to and gives an answer for the student.

We will use the BERT-base-multilingual-cased and fine-tuned it with our datasets.

4 Results and Discussion

The proposed framework illustrates a comprehensive process for tailoring educational content to students based on their profiles using NLP techniques. The process begins with a dataset of students' profiles, which undergoes cleaning and feature selection to prepare the data for clustering. This step ensures that the data is accurate and that only the most relevant features are used. The cleaned data is then processed using ML algorithms to cluster students into groups based on their profiles, categorized by degree. Concurrently, datasets from science books (Dataset 1, Dataset 2, and Dataset 3) are subjected to NLP preprocessing to extract relevant information and prepare the text data for model training. Each student cluster is mapped to a specific BERT model that has been fine-tuned on these preprocessed datasets. This fine-tuning allows the models to provide answers tailored to the context of each student cluster. New inputs from users (i.e., students) are collected to create or update their profiles, ensuring the system has current and accurate data. The system considers the student's context and specific questions, and based on this information, selects the appropriate BERT model to compute the most relevant answer. The final output is a PLA answer that is specifically tailored to the student's needs, leveraging advanced ML and NLP techniques to enhance the learning experience. This framework ensures that educational content is personalized and relevant, thereby optimizing the learning process for each individual student.

We used a QA system using multilingual BERT model. This model is trained on SQuAD v1 dataset. We fine-tuned this model using our dataset in Arabic language for the science book in Palestinian curriculum. Quantifying the success of question answering model is not straight and can be tricky. This is because when you ask a question to multiple people, each of them will answer differently from others. The answers may be the same in context but mostly they differ in synonyms or the order of words.

There are two primary evaluation metrics for question answering models. The first one is Exact Match (EM), and the other is F1-Score. The overall EM and F1 scores are computed by taking the average over the individual example scores.

EM is simple and very clear. It compares the predicted answer obtained from the model with the true answer found in the dataset. Its value will be either 1 if an exact match is found between the two answers, and 0 if no match is found. The model we fine-tuned uses the Arabic language, and as we know, Arabic words are very diverse, leading to a lower expectation for the EM metric. We used the Arabic SQuAD dataset to evaluate our model. Specifically, we took 693 question-answer pairs from the dataset, then used only the questions to predict the answers. The average EM for our model predictions is 20%, which means that our model is able to provide the exact answer 20% of the time, as shown in Table 2.

The implications of a 20% EM score highlights the inherent challenges in processing and understanding Arabic text, given the language's morphological richness and syntactic complexity. This relatively low EM score suggests that while the model has some capability to generate accurate answers, there is significant room for improvement. This could be due to various factors, such as the quality and size of the training data, the model architecture, or the preprocessing steps used. Furthermore, a 20% EM indicates that the model might be generating plausible but not exact answers for a large portion of the questions, which can still be useful in many practical applications where an approximate answer suffices. However, for applications requiring high precision and exact matches, additional efforts in model refinement, such as incorporating more diverse training data, enhancing the model's understanding of context, or using more advanced NLP techniques, are necessary. This evaluation underscores the importance of continuous iteration and enhancement in developing robust NLP models for complex languages like Arabic.

F1 score is also a common metric for all classification problems. It depends on recall and precision. F1 score can be found using the following formula in Eq. (1):

$$F1 = 2 \times \frac{\text{Precision} \times \text{Recall}}{\text{Precision} + \text{Recall}} \tag{1}$$

In our case, the F1 score is computed by finding the probabilities of individual words in the prediction against words in the true answer. The number of shared words between the predicted answer and the true answer is used to calculate precision, recall, and the F1-score. Precision is the ratio of the number of shared words to the total number of words in the prediction, while recall is the ratio of the number of shared words to the total number of words in the true answer. The average F1 score for our model predictions, based on 693 question-answer pairs, is 51%, which is considered good for a model using the Arabic language, as shown in Table 2.

The implications of a 51% F1 score are significant. This metric, which balances precision and recall, provides a more nuanced evaluation of the model's performance than the EM score alone. An F1 score of 51% indicates that the model is fairly effective in capturing the essence of the true answers, even if it does not always produce exact matches. This is particularly impressive given the complexities and variations in the Arabic language, including its rich morphology, diverse dialects, and complex syntactic structures.

A 51% F1 score suggests that the model can generate answers that are partially correct and contextually relevant, even if they do not match the true answers word-for-word. This level of performance is useful in real-world applications where an exact match is not always necessary, and approximate correctness is acceptable. For example, in information retrieval and question-answering systems, providing a relevant and contextually accurate answer can be more valuable than an exact match.

Table 2. Results of the proposed method.

Metrics	Results
EM	20%
F1-score	51%

However, this also indicates that there is room for improvement. To further enhance the model's performance, additional training data, especially more diverse and representative datasets, could be used. Advanced preprocessing techniques and model architectures might also be explored to better capture the nuances of the Arabic language. Moreover, continuous model evaluation and refinement are essential to gradually increase both the EM and F1 scores, making the model more robust and accurate.

5 Conclusion

The building of personalized learning assistant and use of a question-answering model for school science textbooks and Arabic language that makes use of BERT transformers offers encouraging developments in natural language processing in educational environments. We have shown through this study that it is feasible and efficient to use pre-trained language models such as BERT to automatically produce correct answers to questions in Arabic, especially in the field of science education in Palestinian schools. Moreover, incorporating this kind of technology has a great deal of promise to help Arabic-speaking students to learn and understand. This concept can be a useful tool for teachers and students, providing students with immediate access to precise explanations and clarifications that are customized to the content of school science books and personalized to their performance level.

The effective creation and assessment of a learning assistant BERT-based question answering model for Arabic language and school science textbooks highlights the new advantages of NLP in learning environments. We may improve Arabic-speaking students' ability to access to high-quality educational resources and allow them to benefit from AI and NLP techniques. This will ultimately contribute to the growth of education in this digital age.

The practical implications of this study are substantial, particularly in the realm of education for Arabic-speaking students. By fine-tuning BERT with a dataset specifically curated from the Palestinian curriculum, the study offers a highly contextualized AI tool that can provide accurate and relevant educational support. This can revolutionize how students interact with educational content, enabling more effective and personalized learning experiences. For example, students will benefit from AI-generated summaries and explanations of complex subjects that are not only linguistically appropriate but also aligned with their cultural and educational context. This targeted approach ensures that the AI tool addresses specific learning needs and challenges faced by students within this demographic, potentially leading to improved educational outcomes.

Moreover, educators and curriculum developers can leverage this fine-tuned model to create more efficient and effective teaching strategies. Automated question generation tailored to the Palestinian curriculum can save teachers significant time and effort, allowing them to focus more on personalized instruction and student engagement. The tool can also assist in identifying gaps in students' understanding by providing detailed, contextually relevant feedback. This level of personalization and relevance in educational tools can help bridge the gap between traditional teaching methods and the growing demand for adaptive, technology-driven learning environments.

Additionally, the success of this study can serve as a model for other regions and languages, demonstrating the feasibility and benefits of developing AI tools that are fine-tuned to specific linguistic and cultural contexts. It highlights the importance of contextualized AI in education and could encourage further research and investment into similar projects for different curricula and languages. This not only advances the field of educational technology but also promotes more inclusive and accessible learning opportunities worldwide.

References

1. Farea, A., Yang, Z., Duong, K., Perera, N., Emmert-Streib, F.: Evaluation of question answering systems: complexity of judging a natural language. arXiv preprint arXiv:2209.12617 (2022)
2. Tami, M., Ashqar, H.I., Elhenawy, M.: Automated Question Generation for Science Tests in Arabic Language Using NLP Techniques. arXiv preprint arXiv:2406.08520 (2024)
3. Sharma, Y., Gupta, S.: Deep learning approaches for question answering system. Procedia Computer Science **132**, 785–794 (2018)
4. Tang, Y., Liang, J., Hare, R., Wang, F.-Y.: A personalized learning system for parallel intelligent education. IEEE Trans. Computat. Soc. Sys. **7**, 352–361 (2020)
5. McHugh, D., et al.: ncovering themes in personalized learning: Using natural language processing to analyze school interviews. J. Res. Technol. Educ. **52**, 391–402 (2020)
6. Qaiser, S., Ali, R.: Text mining: use of TF-IDF to examine the relevance of words to documents. Int. J. Comp. Appl. **181**, 25–29 (2018)
7. Masri, S., Raddad, Y., Khandaqji, F., Ashqar, H.I., Elhenawy, M.: Transformer Models in Education: Summarizing Science Textbooks with AraBART, MT5, AraT5, and mBART. arXiv preprint arXiv:2406.07692 (2024)
8. Ashok, M., Ramasamy, K., Snehitha, G., Keerthi, S.R.: A systematic survey of cognitive chatbots in personalized learning framework. In: 2021 Sixth International Conference on Wireless Communications, Signal Processing and Networking (WiSPNET), pp. 241–245. IEEE (2021)

9. Wellnhammer, N., Dolata, M., Steigler, S., Schwabe, G.: Studying with the help of digital tutors: design aspects of conversational agents that influence the learning process (2020)
10. Kochmar, E., Vu, D.D., Belfer, R., Gupta, V., Serban, I.V., Pineau, J.: Automated data-driven generation of personalized pedagogical interventions in intelligent tutoring systems. Int. J. Artif. Intell. Educ. **32**, 323–349 (2022)
11. Mathew, A.N., Rohini, V., Paulose, J.: NLP-based personal learning assistant for school education. Int. J. Electr. Comput. Eng. **11**, 4522–4530 (2021)
12. Aljameel, S.S., et al.: Development of an Arabic conversational intelligent tutoring system for education of children with ASD. In: 2017 IEEE international conference on computational intelligence and virtual environments for measurement systems and applications (CIVEMSA), pp. 24–29. IEEE (2017)
13. Schmohl, T., et al.: Combining NLP, speech recognition, and indexing: an AI-based learning assistant for higher education. In: Conference Proceedings. The Future of Education 2023 (2023)
14. Gillioz, A., Casas, J., Mugellini, E., Abou Khaled, O.: Overview of the Transformer-based Models for NLP Tasks. In: 2020 15th Conference on Computer Science and Information Systems (FedCSIS), pp. 179–183. IEEE (2020)
15. Vaswani, A., et al.: Attention is all you need. Adv. Neural Info. Proc. Sys. **30** (2017)
16. Radford, A., Narasimhan, K., Salimans, T., Sutskever, I., et al.: Improving language understanding by generative pre-training (2018)
17. Devlin, J.A.C.M.-W.A.L.K.A.T.K.: Bert: pre-training of deep bidirectional transformers for language understanding. arXiv preprint arXiv:1810.04805 (2018)
18. Radwan, A., Amarneh, M., Alawneh, H., Ashqar, H.I., AlSobeh, A., Magableh, A.A.A.R.: Predictive analytics in mental health leveraging LLM embeddings and machine learning models for social media analysis. Int. J. Web Ser. Res. (IJWSR) **21**(1), 1–22 (2024)
19. Jaradat, S., Alhadidi, T.I., Ashqar, H.I., Hossain, A., Elhenawy, M.: Exploring Traffic Crash Narratives in Jordan Using Text Mining Analytics. arXiv preprint arXiv:2406.09438 (2024)
20. Yilmaz, S., Toklu, S.: A deep learning analysis on question classification task using Word2vec representations. Neural Comput. Appl. **32**, 2909–2928 (2020)

Role of ERP as an Integrative Intelligent System for Business Enterprises

Ibrahim Yasser[1(✉)] and Fahmi Khalifa[1,2]

[1] Electronics and Communications Engineering Department, Mansoura University, Mansoura 35516, Egypt
ibrahim_yasser@mans.edu.eg

[2] Department of Electrical and Computer Engineering, Morgan State University, Baltimore, MD 21251, USA
fahmik.khalifa@morgan.edu

Abstract. Enterprise Resource Planning (ERP) systems are now vital for organizations to integrate and manage their core business processes. This paper provides an overview of ERP systems, their functionalities, benefits, and challenges. ERP systems are comprehensive software solutions that enable organizations to streamline and automate various functions, such as finance, human resources, supply chain management, manufacturing, and customer relationship management. By integrating these functions into a centralized system, ERP promotes data consistency, enhances operational efficiency, and facilitates informed decision-making. Successful ERP implementation requires a thorough understanding of organizational requirements, effective change management strategies, strong leadership, and ongoing support and training. Furthermore, we touch base on emerging trends in ERP systems, such as cloud-based ERP solutions, mobile access, and integration with emerging technologies like artificial intelligence and analytics. These developments offer organizations greater flexibility, scalability, and real-time insights.

Keywords: Enterprise resource planning · ERP implementation · Artificial Intelligence · Big Data · Business Intelligence System

1 Introduction

In the last few decades, companies have invested in various aspects focusing on cost reduction, maximizing profits, and accuracy in decision making [1,2]. Many companies invest in implementing Enterprise Resource Planning (ERP) systems. An ERP system is a business software package that allows organizations to integrate business functions, such as Customer Relationship Management (CRM), finance, manufacturing, logistics, administrative reporting, management project, human resources [3]. Also, it facilitates purchasing management, data sharing, communication of information throughout the enterprise, automated business processes, creation and access of real real-time information in a database. ERP

A. Abdelgawad et al. (Eds.): ISBCom 2024, LNNS 1268, pp. 368–383, 2025.
https://doi.org/10.1007/978-3-031-82377-0_31

can be analogous to the backbone of the information system in a company [4]. Implementing an ERP system can result in higher quality, reduced time to market, improved communication, support in decision-making, shorter lead times, higher productivity, and lower costs [5]. ERP systems have been widely used in large-scale companies but are still limited to Small and Medium Enterprises (SMEs). Recently, many open-source ERP systems have been quite powerful and can be used by SMEs. Open-source ERP generally has a free license but limited modules and customization [6].

There are many different types of ERP solutions on the market today that may be tailored to meet the demands and structures of different organisations. Because of its scalability, flexibility, and decreased requirement for internal IT resources, cloud-based ERP solutions are becoming more and more popular. Businesses may use the internet to access their ERP system because these systems are housed on external servers. Conversely, servers housed within the company are used to deploy and operate on-premises ERP systems. They may need a larger initial outlay, but they provide more customisation and control over data. Furthermore, there are ERP systems designed specifically for certain industries, such as healthcare, retail, or manufacturing, that are adapted to meet their particular needs. The features and functions of these customised systems are preset to meet the unique problems faced by each sector [7].

The reminder of this paper is organized as follows: In Sect. 2, key components of ERP systems are discussed followed by top ERP modules in Sect. 3. Types and benfits of ERP systems are given in Sect. 4. In Sect. 5, a summary of the implementation procedure is presented, while challenges and considerations of ERP Systems are given in Sect. 6. The emerging trends in ERP systems, how ERP systems are employed in different sectors, and benefits realization are presented in Sects. 7, 8, 9, and 10, respectively. Finally, Sect. 11 details concluding remarks and future works.

2 Key Components of ERP Systems

ERP systems are often made up of a number of essential parts that cooperate to offer an all-inclusive solution for organising and integrating different business activities inside a company. Organisations thinking about using an ERP system must comprehend these elements [8]. The key components of ERP systems are:

Modules and Functionality: ERP systems are made up of several modules, each of which serves a particular department or company function. These programs include HR, supply chain management, finance and accounting, and CRM, among other topics. Every module provides a range of features designed to meet certain requirements.

Customization and Configuration: ERP systems are flexible enough to be configured and customised to match the unique requirements of a company. The system's features, workflows, and user interfaces can all be customised by administrators to meet the particular needs and procedures of the company.

Centralized Database: A centralised database, which acts as a storehouse for all pertinent data from various functional areas of the company, is the primary component of an ERP system. A centralised database gives the company a single source of truth, guarantees data consistency, and removes data redundancy.

Integration and Workflow Management: The ability of ERP systems to connect and optimise operations across many departments and functions is one of its main benefits. Data is effortlessly transferred across modules via integration, which removes the need for human data entry and offers real-time operational insight.

Security and Access Controls: Security is an essential component of ERP systems since they handle private and sensitive company data. To guard against illegal access and data breaches, ERP systems include security features including user authentication, access limitations, and data encryption.

Reporting and Analytics: ERP systems give businesses powerful analytics and reporting features that let them extract meaningful insights from their data. These systems let users to measure progress, keep an eye on important performance indicators, and make data-driven decisions by providing pre-defined reports and configurable dashboards.

3 Top ERP Modules

Most ERP solutions begin with financial and then add extra capabilities, depending on your business priorities, your industry, and changing conditions. Heres a list of the 10 top ERP modules [9], how they can help your organization, and why companies decide to use them; see Fig. 1.

Financial management is the cornerstone of an ERP system. By tracking accounts payable and receivable and managing the general ledger, this module provides a clear financial picture and outlook. It generates and stores balance sheets, payment receipts, and other essential financial documents, with automation playing a crucial role.

Procurement improves your ability to purchase the products or materials you need to make or sell goods. It automates, tracks, and analyzes quotes while helping to prepare and send purchase orders. Using a list of approved vendors, the module can automate purchases of specific items, which speeds up the process, enforces compliant spending, improves supplier relationships, and ultimately helps increase profitability.

Risk management and compliance module uses AI and ML to enhance financial controls, enabling finance teams to prevent cash leaks, enforce audits, and protect against emerging risks-all while saving hours of manual work.

Supply Chain Management (SCM) monitors the flow of supplies and goods from suppliers to consumers, enabling efficient planning of demand, supply, and

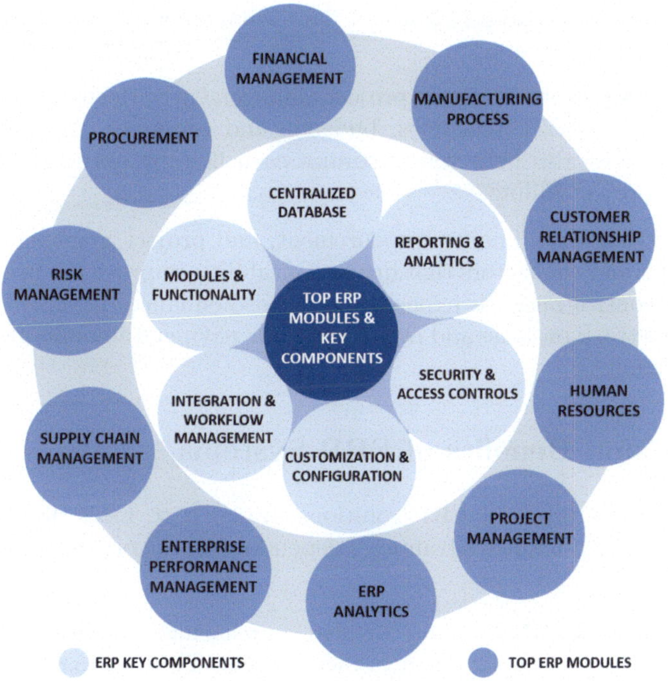

Fig. 1. Top ERP Modules & key components of ERP systems.

production. This reduces disruptions, minimizes costs, and ensures the right inventory is available at the right time and place.

Enterprise Performance Management (EPM) enhances your business's planning, budgeting, forecasting, and performance reporting. These tools empower comprehensive and connected planning through a unified approach.

Manufacturing Process: makes it easier for your business to plan and manage production runs. That includes making sure your facility has the raw materials on hand to meet demand, can efficiently run a shop floor, and keep costs down-all while ensuring the quality of finished goods.

Customer Relationship Management (CRM) modules enhance customer management by optimizing sales and marketing workflows, particularly those related to sales leads. They store comprehensive information on customers and prospects, including interactions and purchase histories, and provide segmentation data along with targeted cross-sell recommendations.

Human Capital Management (HCM) sometimes is referred to as human resources management (HRM)-contains detailed records on all employees, including performance reviews, job descriptions, benefits selections, and attendance/time off. Like other ERP modules, it supplants annual spreadsheets with automation that continuously gathers data companywide, reducing duplicate data and increasing accuracy.

Project Management improves project management with shared visualization tools that help to track schedules, budgets, and resources. Advanced project management tools also help you maximize cash flow by automating customer invoices and project billing.

ERP Analytics helps finance, procurement, and project management professionals understand the factors that drive profitability, improve the use of working capital, and control business expenditures.It also enhances payables by helping finance teams track on-time and overdue vendor payment amounts to determine payment urgency.

4 Types and Benefits of ERP Systems

There are various types of ERP solutions available to meet diverse business needs, each offering distinct advantages depending on the organization's requirements [10].

- **On-Premise ERP** systems involve purchasing software licenses and installing them on the company's own servers. This approach provides full control over data and customization options but requires dedicated IT resources for maintenance and security.
- **Cloud ERP** systems operate on remote servers accessed via the internet, offering scalability without significant upfront hardware costs. Managed by the provider, these systems use a subscription model and are ideal for businesses needing flexibility and rapid deployment.
- **Industry-Specific ERP** systems cater to particular sectors with tailored modules and functionalities. For instance, manufacturing ERP solutions focus on inventory management and supply chain logistics, differing greatly from those designed for financial institutions.
- **Open-Source ERP** allows access to source code for customization, making it suitable for companies with technical expertise seeking tailored solutions.
- **Small Business ERP** systems are designed for SMBs, balancing affordability with essential functionalities. They are easier to implement and require less customization compared to enterprise-level ERP systems.
- **Tiered ERP** systems offer scalable solutions with varying functionalities. Businesses can select tiers based on current needs and expand or customize modules as operations grow, such as adding international currency modules for global expansion.

Organisations in a variety of sectors and industries can profit greatly from ERP systems. These advantages result from ERP systems' ability to integrate, automate, and streamline corporate activities [11]. ERP solutions offer a range of benefits crucial for modern businesses:

- **Improved Data Integrity and Accuracy:** ERP consolidates data from various departments into a single database, eliminating silos and reducing data duplication. This ensures reliable, up-to-date information for informed decision-making and agile responses to market changes.
- **Enhanced Efficiency:** Automation of repetitive tasks such as data entry and inventory management streamlines operations, reduces errors, and optimizes resource utilization. This frees up staff to focus on strategic initiatives, boosting overall productivity.
- **Enhanced Customer Service:** By integrating customer data, ERP enables personalized customer care by providing a comprehensive view of preferences and interactions. This fosters customer loyalty and satisfaction, leading to improved retention rates.
- **Streamlined Collaboration and Communication:** Centralized databases and standardized procedures in ERP systems facilitate seamless collaboration across departments. Enhanced teamwork and communication improve efficiency and coordination within the organization.
- **Real-time Reporting and Analytics:** ERP systems offer robust reporting tools and customizable dashboards for real-time insights into business performance. This enables proactive decision-making and rapid responses to market dynamics based on accurate data analysis.
- **Regulatory Compliance:** ERP systems include built-in controls and audit trails to help businesses adhere to industry regulations. Automated compliance processes ensure accurate reporting and mitigate the risk of non-compliance penalties.
- **Scalability and Flexibility:** ERP solutions are scalable to accommodate growing user bases, transactions, and complex business processes. Customization options allow organizations to tailor the system to their specific operational needs, enhancing flexibility and adaptability over time.

5 Implementation of ERP Systems

A lot of meticulous planning, coordinating, and administration go into the complicated and important process of implementing an ERP system. There are several crucial phases and considerations for a successful ERP deployment. An summary of the implementation procedure may be seen below [12]:

Vendor Selection: Businesses must assess and choose an ERP provider who meets both their general demands and those unique to their sector. This entails evaluating elements including the reputation of the vendor, the features of the product, its scalability, its support services, and its total cost of ownership.

Needs Assessment and Planning: Evaluating the organization's needs in detail to ascertain its goals and requirements is the first stage in implementing ERP. This include figuring out what the ERP system's shortcomings are, setting objectives, and outlining precise requirements.

Data Migration: Transferring data from old systems to the new ERP system is known as data migration. To guarantee accuracy and integrity, data must be validated and cleaned. A thorough data migration strategy that includes data mapping, data cleaning, and conversion techniques should be developed by organisations.

Business Process Mapping and Customization: Mapping current company processes and workflows is essential before putting the ERP system into place. This aids in pinpointing areas in need of modification and enhancement. Businesses could have to adjust their procedures to fit the ERP system's built-in features or modify the system to suit unique needs.

System Configuration and Testing: The ERP system is configured in this phase in accordance with the needs and procedures of the organisation. Configuring modules, processes, user roles, security controls, and system integration are all included in this. Thorough testing is done to make sure the system works as planned and satisfies the requirements of the company.

Go-Live and Post-Implementation Support: Moving from the old systems to the new ERP system is part of the go-live process. To reduce interruptions to corporate activities, the cutover procedure must be properly planned. Support following implementation is necessary to handle any problems or difficulties that crop up after the system goes live.

Training and Change Management: To guarantee that staff members are able to operate the ERP system efficiently, proper training is essential. The unique demands of various user groups should be taken into consideration while designing and delivering training programmes. Change management techniques should also be used to handle any opposition to change and get staff members ready for the shift.

6 Challenges and Considerations of ERP Systems

Organisations must handle a number of issues and concerns while implementing an ERP system in order to guarantee a successful deployment and optimise the system's advantages. Here are some key challenges and considerations [13]:

Customization vs. Standardization: To guarantee system stability and simplicity of maintenance, organisations need to find a balance between utilising conventional functions and tailoring the ERP system to their unique demands. Excessive personalisation may result in more work, longer implementation schedules, and more expensive bills.

Project Management: The adoption of ERPs is a complicated undertaking that calls for skilled project management. Establishing specific project goals and timetables, allocating specialised resources, and regularly monitoring progress are all necessary for organisations.

Change Management: Implementing ERP sometimes necessitates considerable adjustments to job descriptions, roles, and company procedures. Employee resistance to change may impede the process of implementation. Establishing a change management plan is crucial for raising awareness, involving staff, and resolving issues.

Data Migration and Integration: A difficult and time-consuming process is data migration from old systems to the new ERP system. Ensuring accurate and dependable data in the new system requires careful attention to data quality, completeness, and mapping.

Vendor Selection and Support: Choosing the appropriate ERP supplier is essential to the implementation's success. Based on their knowledge of the industry, product functionality, installation process, support services, and long-term sustainability, organisations should carefully assess their providers.

Resource Allocation: A project team, subject matter experts, and technical staff are among the specialised resources needed for ERP deployment. For the implementation, organisations must set aside enough money, time, and qualified staff.

Organizational Readiness: Establishing preparedness for ERP adoption is a crucial step before starting a project. This entails assessing the culture of the company, its openness to change, and the availability of qualified employees.

Ongoing Maintenance and Upgrades: As technology advances and company demands change, ERP systems need to be updated, maintained, and upgraded on a regular basis. For the purpose of utilising the ERP vendor's new features and bug patches, organisations should budget for routine system maintenance, user assistance, and updates.

7 Emerging Trends in ERP Systems

In a technology environment where advancements are occurring at a rapid pace, ERP systems are evolving and adapting to suit the changing demands of organisations. The future of ERP systems is being shaped by a number of new developments. Here are some of the key trends [14]:

AI and Machine Learning (ML): ERP systems are progressively using AI and ML technologies to provide predictive analytics, automated routine processes, and advanced analytics. AI-enabled ERP systems can analyze massive amounts of data, identify trends, and provide insightful information to help with decision-making.

Internet of Things (IoT) Integration: The integration of ERP systems with IoT devices is becoming more and more common. IoT sensors and devices may send real-time data from a range of sources, like as inventory, supply networks, and manufacturing equipment, directly into the ERP system.

Blockchain Integration: Blockchain technology is becoming more and more popular because of its promise to improve security, traceability, and transparency across a range of businesses. The combination of blockchain technology and ERP systems enables secure, immutable transaction records, simplified supply chain processes, and verifiable provenance of goods.

Cloud-Based ERP: Recent years have seen a major increase in the use of cloud computing, and cloud-based ERP systems are growing in acceptance. Cloud-based ERP offers several advantages, including scalability, flexibility, affordability, and accessibility from any place with an internet connection.

Mobile ERP: With the increasing prevalence of mobile devices, mobile ERP apps are growing in popularity. Users may access and use ERP capabilities on smartphones and tablets using mobile ERP, allowing them to make decisions while on the road and access real-time data.

Enhanced User Experience: ERP providers are emphasising the use of intuitive and user-friendly interfaces in order to enhance the user experience. More aesthetically pleasing, easily navigable, and customisable modern ERP systems are built into their architecture.

Focus on Industry-Specific Solutions: For different sectors, ERP providers are developing industry-specific systems that meet the unique needs and requirements. These solutions facilitate the faster and more efficient deployment of ERP systems within businesses by making use of pre-configured procedures and industry best practises.

Integration with Third-Party Applications: ERP systems are becoming more and more connected with various external applications and systems, including productivity suites, e-commerce platforms, business intelligence tools, and CRM.

8 ERP Systems in Different Industries

ERP systems are utilized across various industries to streamline business processes, enhance efficiency, and improve overall organizational performance, see Fig. 2. Here are some examples of how ERP systems are employed in different sectors [15]:

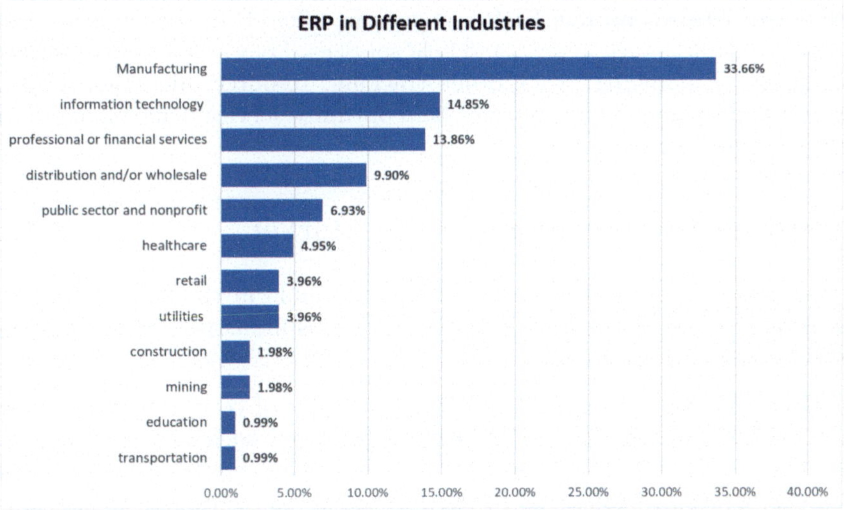

Fig. 2. ERP in different industries [16].

Manufacturing Industries: ERP systems are quite helpful in manufacturing throughout the working day, handling logistics, billing, and data organisation with efficiency. Beyond these features, they also let staff members keep their personal data accessible.

Healthcare Industries: Healthcare operations such as patient scheduling, inventory control, and billing are greatly streamlined by ERPs. Enhanced operational efficacy and adherence to healthcare data privacy regulations, such as the Health Insurance Portability and Accountability Act (HIPAA), are among the primary advantages.

Education Sector: ERPs are essential to education because they make a variety of jobs easier, including grading, student registration, and resource organisation. They offer seamless assistance for efficient operations to institutions and schools.

Financial Sector: In finance, ERPs are indispensable for managing financial data, reducing data loss risks, and ensuring compliance with industry laws, emphasizing the importance of rapid and secure access to accounting information.

Hospitality and Tourism: ERP systems assist hotels, resorts, and travel agencies in managing reservations, front-office operations, housekeeping, inventory control, point-of-sale (POS) systems, and customer relationship management. They help streamline guest services, optimize room occupancy, track customer preferences, and enhance overall guest satisfaction.

Service and Maintenance: ERP systems are valuable in service-based industries such as maintenance, repair, and field services. They support service request management, dispatching, work order tracking, resource allocation, and customer communication. These systems help optimize service delivery, scheduling, and resource utilization.

9 Existing ERP Systems in the Market

There are several established ERP systems available in the market that cater to the diverse needs of organizations across various industries. Here are some of the well-known ERP systems:

SAP ERP: Developed by SAP, a leading provider of corporate software solutions, SAP ERP automates critical business functions such as accounting, sales, and human resources. It includes SAP ECC for managing transactions, reporting, and integration, while SAP S/4HANA enhances operations with real-time data processing and advanced analytics. Known for its intuitive interface, SAP ERP is favored by industry giants like PayPal and NBA. Recently integrated with cloud capabilities, SAP ERP offers customizable solutions for global business models, promoting revenue growth [17].

Oracle ERP: Oracle ERP, developed by Oracle Corporation, optimizes finance, supply chain management, and human resources. It is known for reliability and scalability, serving both small enterprises and large corporations. Highly customizable to meet specific industry needs, Oracle ERP offers robust capabilities in data analytics, reporting, and system integration. With extensive support services and a strong market presence spanning over a decade, Oracle ERP remains a leader in enterprise solutions [18].

Microsoft Dynamics 365: Part of Microsoft's business applications suite, Microsoft Dynamics 365 is widely adopted for accounting, supply chain management, sales, and customer support. It integrates seamlessly with Microsoft products like Office 365, providing a unified user experience. Offering advanced analytics, AI-driven insights, and flexible cloud deployment options, Microsoft Dynamics 365 enhances operational efficiency and automation [19].

Odoo ERP: Odoo ERP is a versatile open-source solution known for rapid adoption and growth. It offers comprehensive tools for accounting, manufacturing, and sales management, featuring a modular architecture that allows organizations to integrate specific modules as needed. With a user-friendly interface, community-driven ecosystem, and cloud-based deployment options, Odoo provides an affordable and scalable ERP solution suitable for SMEs and businesses of all sizes and sectors [20] (Table 1).

Table 1. Showing the comparison of the ERP providers.

ERP Provider	SAP	Oracle	Microsoft 365	Odoo
Deployment	On-premise, Cloud	On-premises, Cloud	On-premises, Cloud	On-premises, Cloud
Open source	No	No	No	Yes
Customization	Moderate	Moderate	Moderate	Highly
Scalability	Yes	Yes	Yes	Yes
User-Friendly	Yes	Yes	Yes	Yes
Cost-effective	Varies	Varies	Varies	Yes
Popularity	Popular	Popular	Popular	Growing
Community	Active	Active	Strong	Open

10 Benefits Realization and Return on Investment (ROI)

Measuring ERP success is critical because distinguishing between implementation success and true ERP success is often overlooked, leading to failure in many organizations post-implementation. In the 1980s, ERP success was primarily gauged by improved performance and user satisfaction. Today, success is multifaceted due to fierce competition, where even minor factors like reduced lead times, free shipping, and product quality can determine a company's competitive edge. Delone and McLean's framework is widely respected for assessing implementation success, see Fig. 3.

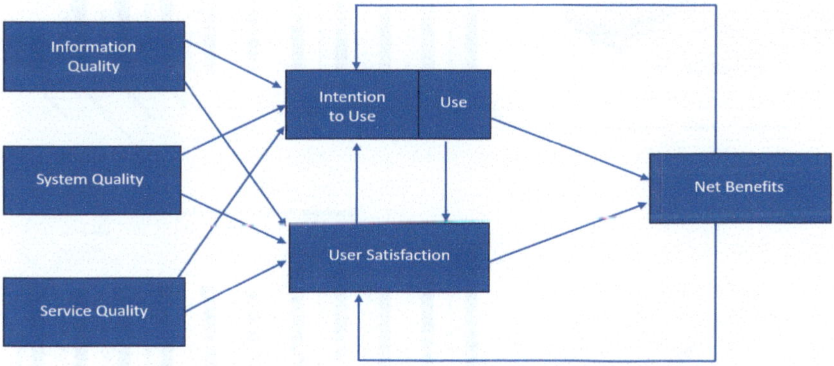

Fig. 3. Updated DeLone and McLeans success model [21].

Information Quality: The accuracy with which the system generates its data System quality includes both the system's hardware and overall build quality. Intention to use: "Intention to use" is a useful substitute metric in some situations. "Intention to use" refers to a mindset, whereas "use" denotes an action. Use: Making timely and effective use of the information produced by the system. User satisfaction: The response that is returned as a result of using ERP data. Overall Gains: Prior to the modified approach, individual and organisational impacts made up the net benefits. Many researchers use diverse effect metrics, including workgroup, industry, and interorganizational impacts (and consumer impacts), however as time goes on, a number of factors add up to [21].

Many researchers critique the model for its broad approach to assessing net benefits, arguing that success criteria should vary by industry and stakeholder perspectives. Despite these criticisms, the framework remains adaptable across industries and phases of the ERP lifecycle. Organizations can use these metrics to stay aligned with their technology improvement goals. Ultimately, achieving benefits relies on achieving Return on Investment (ROI). Last year's ERP reports from Panorama Consulting evaluated outcomes like success, failure, and uncertainty, illustrated in a pie chart Fig. 4(a). Key metrics considered include Key Performance Indicators, User Adoption rates, and adherence to Project Budget and Timeline.

Benefits can be financial or through innovation and improved employee retention. In Fig. 4, among those who conducted pre-implementation ROI analyses and were operational for over a last two years (2022 and 2023), 81% met their ROI expectations. Over 70% of organizations that quantified benefits reported

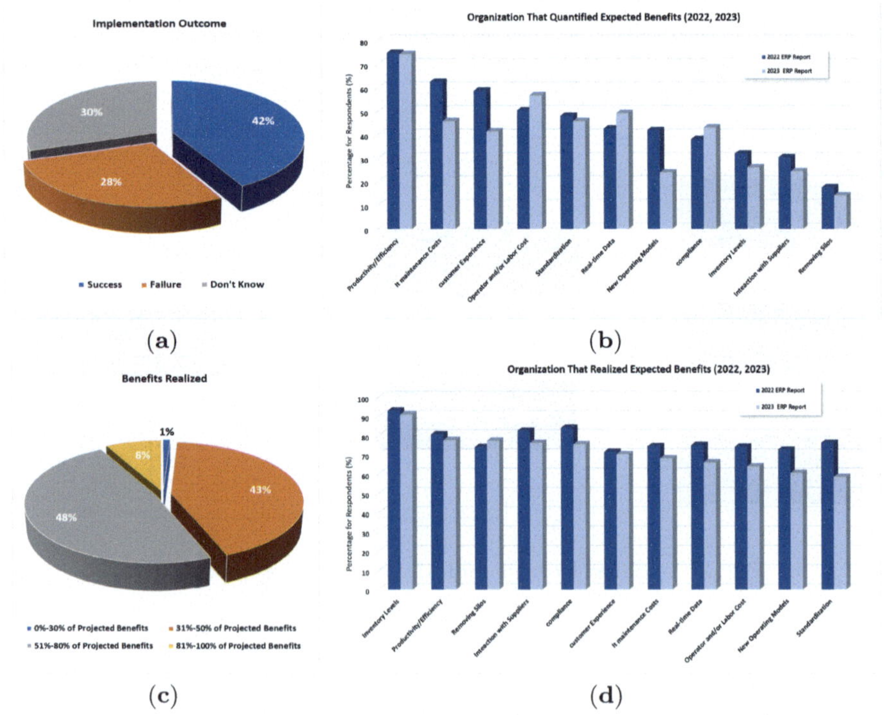

Fig. 4. Source adapted form ERP reports [22–24]: (a) implementation outcome; (b) Organizations that realized expected benefits; (c) Benefits realized; and (d) Organizations that realized quantified benefits

achieving them across specific categories, indicating widespread success. Despite not always outlining anticipated benefits beforehand, recognized benefits like enhanced efficiency and improved customer experience stand out. Establishing a benefits realization plan is critical for measuring ROI and demonstrating tangible benefits to executives. It's recommended to form a dedicated team involving stakeholders from across the organization to ensure all expected benefits are quantified.

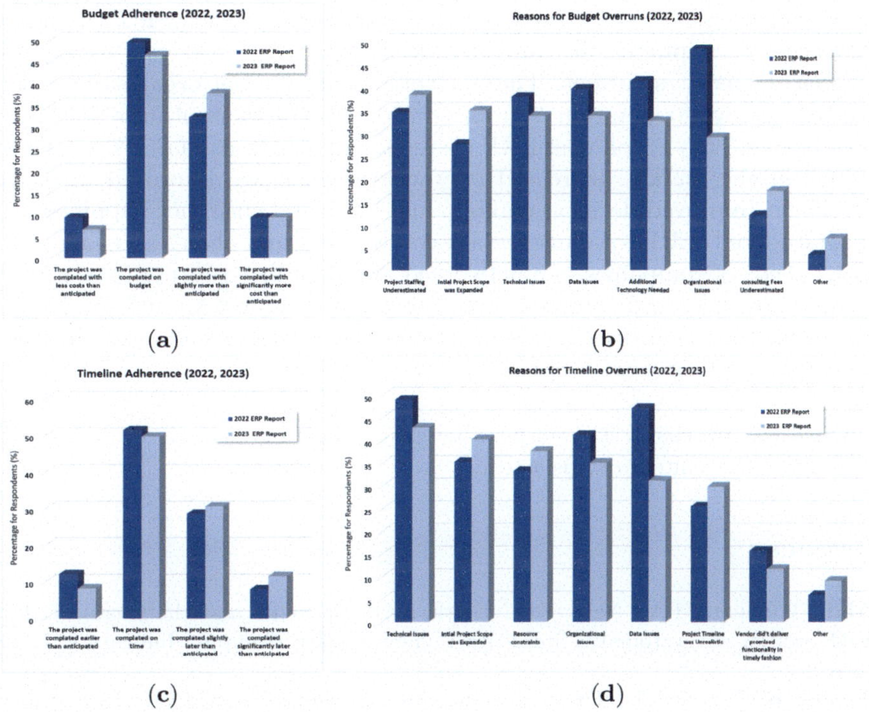

Fig. 5. Source adapted form ERP report [23,24]: (a) Budget Adherence; (b) Reasons for Overruns; (c) Timeline Adherence; and (d) Reasons for Timeline overruns

As shown in Fig. 5, estimating project costs for enterprise software can be challenging due to overlooked activities, leading to unexpected costs and overruns. Encouragingly, more than half (59%) of organizations stayed within budget, up from 40% previously. Moreover, a majority (64%) completed projects on time, reflecting a strong understanding of project attributes and business readiness.

11 Conclusions and Future Work

ERP systems have emerged as critical tools for organizations to integrate and manage their core business processes. These comprehensive software solutions provide a centralized platform for streamlining operations, optimizing resources, and facilitating informed decision-making. ERP systems have become indispensable for organizations seeking to optimize their operations, improve efficiency, and gain a competitive advantage. Successful ERP implementation requires a holistic approach, considering organizational needs, effective change management, ongoing support, and a focus on continuous improvement. By harnessing the power of ERP systems, organizations can drive growth, improve customer satisfaction, and achieve operational excellence in today's dynamic business environment. **The future of ERP systems** is impacted by new developments including cloud-based deployment, mobile accessibility, IoT connectivity, blockchain acceptance, AI and ML integration, enhanced user experience, and industry-specific solutions. These developments present organisations with chances to improve cooperation, spur innovation, and further optimise their operations. By effectively employing ERP systems, organisations may get a competitive edge in the market, increased operational efficiency, improved inventory management, improved decision-making, and increased customer happiness. All things considered, ERP systems are always evolving and changing to meet the changing needs of companies in a networked and digitalized corporate world. As technology advances and companies seek to become more lean and productive, ERP systems will remain a vital tool for overseeing complex corporate processes and encouraging long-term growth.

Disclosure of Interests. The authors have no competing interests to declare that are relevant to the content of this article.

References

1. Abd Elmonem, M.A., Nasr, E.S., Geith, M.H.: Benefits and challenges of cloud ERP systems-a systematic literature review. Future Comput. Inform. J. **1**, 1–9 (2016)
2. Adams, R., Ferreira, D.: One share-one vote: the empirical evidence. Rev. Financ. **12**, 51–91 (2008)
3. Yulianto, N., Anon, M., Prabowo, H., Hidayanto, A.N.: ERP system selection for small medium enterprises (SMEs): a systematic literature review. Int. J. Mech. Eng. Technol. (IJMET) **11**, 12 (2020). https://doi.org/10.34218/IJMET.11.12.2020.001
4. Ibrahim, M.F., Dharmawan, Y.S., Ngatini, N., Mustofa, M.I., Utama, D.M.: A literature review on ERP implementation: Methodologies, module, software, and policy. In: AIP Conference Proceedings, vol. 2927. AIP Publishing (2024)
5. Weerasekara, U., Gooneratne, T.: Enterprise resource planning (ERP) system implementation in a manufacturing firm: rationales, benefits, challenges and management accounting ramifications. Account. Manage. Inf. Syst. **22**(1), 86–110 (2023)
6. Alaskari, O., Pinedo-Cuenca, R., Ahmad, M.M.: Framework for implementation of enterprise resource planning (ERP) systems in small and medium enterprises (SMEs): a case study. Procedia Manuf. **55**, 424–430 (2021)
7. Wahab, N.A.B.A., Nor, R.B.M.: The role of enterprise resource planning (ERP) systems in facilitating sustainable business practices. AI IoT Fourth Industr. Revolution Rev. **13**(9), 29–39 (2023)

8. medium. What is ERP (enterprise resource planning) software? ERP systems for businesses big and small (2023). https://medium.com/1erp/what-is-erp-enterprise-resource-planning-software-87d9c3b5a07f. Accessed 1 May 2024
9. Jackley, M.: 10 Top ERP Modules and Their Features (2023). https://www.oracle.com/eg/erp/erp-modules/. Accessed 1 May 2024
10. Investopedia team (2024). https://www.investopedia.com/terms/e/erp.asp. Accessed 1 May 2024
11. Aptean (2024). https://www.aptean.com/en-US/insights/blog/15-benefits-implementing-erp-software. Accessed 1 May 2024
12. Mahar, F., Ali, S.I., Jumani, A.K., Khan, M.O.: ERP system implementation: planning, management, and administrative issues. Indian J. Sci. Technol. **13**(01), 1–22 (2020)
13. Mahmood, F., Khan, A.Z., Bokhari, R.H.: ERP issues and challenges: a research synthesis. Kybernetes **49**(3), 629–659 (2020)
14. Kenge, R., Khan, Z.: A research study on the ERP system implementation and current trends in ERP. Shanlax Int. J. Manage. **8**(2), 34–39 (2020)
15. Aggarwal, I., Anirudh, A., Buddala, R.: Literature review: ERP implementation in various industries. In: 2021 Innovations in Power and Advanced Computing Technologies (i-PACT), pp. 1–6 (2021)
16. Financesonline (2024). https://financesonline.com/erp-statistics/. Accessed 1 May 2024
17. SAP (2024). https://www.sap.com/mena/products/erp.html. Accessed 1 May 2024
18. Oracle (2024). https://www.oracle.com/eg/erp/. Accessed 1 May 2024
19. Microsoft (2024). https://www.microsoft.com/en-us/dynamics-365. Accessed 1 May 2024
20. Odoo (2024). https://www.odoo.com/. Accessed 1 May 2024
21. DeLone, W.H., McLean, E.R.: The Delone and Mclean model of information systems success: a ten-year update. J. Manage. Inf. Syst. **19**(4), 9–30 (2003)
22. Panorama Consulting Solutions (2018). https://www.panorama-consulting.com/resource-center/erp-industry-reports/panoramas-2018-erp-report/. Accessed 27 June 2024
23. Panorama Consulting Solutions (2022). https://www.panorama-consulting.com/resource-center/erp-report-archives/. Accessed 27 June 2024
24. Panorama Consulting Solutions (2023). https://4439340.fs1.hubspotusercontent-na1.net/hubfs/4439340/Reports/ERP20Report/2023-ERP-Report-Panorama-Consulting.pdf. Accessed 27 June 2024

COSMIC-REC: A Recursive COSMIC Functional Size Measurement Tool for Arduino IoT Edge Devices

Sana Abdullah[1](✉), Salma Salem[1], Milad Ghantous[1], and Hassan Soubra[2]

[1] German University in Cairo, New Cairo, Egypt
{sana.hamadallah,salma.abdallahsalem}@student.guc.edu.eg,
milad.michel@guc.edu.eg
[2] ECE-Ecole centrale d'electronique, Lyon, France
hsoubra@ece.fr

Abstract. In the expanding realm of the Internet of Things (IoT), the connectivity of Edge devices is crucial for seamless data exchange. These devices, encompass embedded software operating in real-time, are integral to IoT systems. Functional Size Measurement (FSM), particularly the COSMIC-ISO 19761 method, has become vital for measuring Edge devices' software functionality in different software development phases. This paper builds on previous research by proposing a novel approach to recursively measure COSMIC Function Points (CFPs) for function calls within Arduino libraries and nested libraries, using regular expressions and caching techniques. This paper explores Arduino libraries, focusing on the implementation of function calls within them. Implemented in Python, this approach aims to provide a comprehensive framework for accurately measuring the functional size of IoT Edge Device code. Through rigorous experimentation, our results demonstrate the effectiveness of our approach.

Keywords: Functional Size Measurement-FSM · IoT · COSMIC ISO 19761 · Regex · Edge Computing

1 Introduction

The Internet of Things (IoT) refers to a network of Edge Devices that transmit and receive data, requiring processing and analysis. These devices are often equipped with sensors and actuators, and can vary from everyday household objects, such as smart vaccum cleaners, all the way to sophisticated industrial machines. This paper specifically sheds a light on Arduino. Arduino is an open-source hardware and software platform that is used popularly in the development of IoT projects.

Software size measurement plays a vital role in software engineering, as it serves various purposes in the software development life-cycle (SDLC). During the design phase, it helps in giving early technical indicators, such as network and

A. Abdelgawad et al. (Eds.): ISBCom 2024, LNNS 1268, pp. 384–397, 2025.
https://doi.org/10.1007/978-3-031-82377-0_32

processor loads. As well as, facilitating the estimation of development effort and measuring productivity [2]. Therefore, it is only logical, that Functional Size Measurement (FSM) holds major significance in the Internet of Things (IoT) realm, specifically as it aids in using resources efficiently, managing memory effectively, and writing maintainable code that will eventually run on limited-capacity Arduino boards.

The COSMIC method stands out as a second generation sizing method among Functional Size Measurement (FSM) methods [3]. It offers a standardized approach to evaluate the functional size of software, accommodating to both Management Information Systems viewpoints and real-time requirements. Furthermore, in 2023, [4] presented a tool to measure COSMIC functional size of any IoT Edge Device code automatically. This paper, however, will be mainly addressing the gap in [4] where the tool only covers a limited number of function calls within a few of the Arduino libraries, and does not take into consideration the CFP size of their source code. This oversight may lead to misleading size measurements, as modern software systems rely heavily on libraries to supply fundamental functionality. Also, such omissions may generate underestimation of a project's complexity, potentially resulting in insufficient resource allocation, project planning, and maintainability. Thus, extending the tool's capabilities to encompass a wider range of function calls across a greater variety of libraries is vital for gaining a comprehensive understanding of the software's functional size and it's associated development requirements.

2 Background

The COSMIC FSM, an International Standard: ISO/IEC 19761 Software Engineering, was developed with several key objectives in mind. Firstly, it aimed to abide to the fundamental principles of software engineering. In addition to being applicable to various types of domains including, quantum computing [9], aeronautical fields [7], automotive branch [8], and computer hardware [4,10,11]. Importantly, this method focuses on producing measurements that are independent of technical requirements and project methods, thus ensuring consistency across multiple various contexts. Also, it aims to be flexible, by enabling measurement during any stage of the software development life-cycle (SDLC). This is clearly highlighted through the fact that COSMIC can be used to measure software size using the Functional User Requirements (FUR), as well as measuring the actual source code during the development phase. Lastly, it was intended to be available freely for use, which in turn emphasizes openness and accessibility [1].

In the COSMIC method, software functionality is measured using the functional flows of data groups. These data flows exhibit 4 types of data movements that represent a fundamental component of the Functional User Requirement (FUR) for the measured software, and they can be categorized as follows:

1. Entry (E): this data movement moves the data group from a functional user to the functional process where it is needed across the boundary

2. Exit (X): this data movement moves the data group from a functional process to the functional user that necessitates it across the boundary
3. Read (R): this data movement retrieves the data group from persistent storage to the functional process that requires it.
4. Write (W): this data movement transfers the data group from within the functional process to the persistent storage [2,5].

COSMIC function points (CFP) is the unit of measurement used to represent the functional size of the software measured. CFP unit is given for each identified data movement, therefore the functional size of a software is equivalent to the sum of all of its functional processes, independent of the technology used and free from bias. The mapping of these COSMIC elements to Arduino elements is clearly highlighted in Fig. 1 as introduced in [2].

COSMIC	Arduino	Examples
Functional users	sensors, actuators, computers or other systems	LED, switch, push button, LCD display, etc.
Functional processes	setup() loop()	-

Entry data group movements	Functions via INPUT pin(s)	digitalRead();
		analogRead();
		Serial. readString();
		WiFi.getSocket();

Exit data group movements	Functions via OUTPUT pin(s)	digitalWrite();
		analogWrite();
		Serial.println();
		lcd.print();
		server.write();

Read data group movements	EEPROM Input functions	EEPROM.read(); EEPROM.get();
Write data group movements	EEPROM Output functions	EEPROM.write(); EEPROM.put();
Persistent storage	EEPROM	-

Fig. 1. COSMIC elements mapped to Arduino elements

3 Methodology

3.1 Arduino Libraries

Arduino is an open-source electronics platform, that is composed of a microcontroller board, and an environment that gives users the chance to implement electronic projects. The versatility in applications such as internet of things (IOT) has seen Arduino boards become quite popular due their low cost compared with other similar devices such as Raspberry PI or BeagleBone Black machines which serve similar functions. Arduino has several different types of boards with various functionalities. It is possible to write programs for these boards by means of an easy-to-understand programming language ('Arduino') originating from simple modification of C/C++ codes. It, also provides a wide range of libraries, making it convenient for beginners and advanced users, as well.

Arduino libraries are pre-written code sets offering more functionality and simplicity to the process of programming the boards. They are groups of methods and classes which wrap up hard actions allowing developers to handle different hardware parts easily or to perform certain jobs more easily.

Below are the different types of Arduino libraries:

1. Core Libraries
2. Standard Libraries
3. Third-Party Libraries
4. User-Created Libraries

Generally, including Arduino libraries in the project can be done by adding `# include` directive to the top of the sketch. Consequently, that action allows users to deploy functions and classes from the library into your program. Users can search, install, update or manage them through the Library Manager that is one of the tools within the Arduino IDE.

By using Arduino libraries, instead of repeating simple and routine tasks from scratch, it is possible to implement pre-existing code for common tasks to spare some time and energy, concentrate on the special functions needed, and create difficult systems.

Through vigorous research, a collection of the most widely used libraries was created, and each library's corresponding source was gathered, mainly from [12], and saved according to signature of source codes in a Firebase database.

3.2 Firebase Database of Function Calls

Firebase offers a set of powerful utilities for web and mobile application developers that are quite robust. Firebase has a real-time database feature to store data so that it can be used on different computers at the same time. Real-time databases possess many benefits over keeping information in the application.

By leveraging a real-time database like Firebase, the source code of frequently used Arduino libraries can be saved in a centralized location. This way, the code can be accessed and pulled from any device or client with the permission to interact with the database. Moreover, since it is real-time, any changes or updates done on the code will reflect on all connected clients immediately making it feel like one smooth operation.

A hierarchical structure was used to organize the Arduino libraries' source code in the Firebase database. By doing this, it is possible to simplify the process for retrieving and managing the codebase. The format employed was as shown in Fig. 2.

```
"Library Name": {
        "Function Name":{
            "Has a return value or not":{
                "Number of parameters":"source code"
            }
        }
  }
```

Fig. 2. Structure used for saving libraries and function calls in the database

Each Arduino library is saved as a separate node, with the library name as a key, its value is a series of nodes each representing function calls in the library. The key of every function call is the function name, and its value is one or two nodes, where the key of these node(s) is 'Y' or 'N'. 'Y' is to represent that this function call has a return value, while 'N' is used to indicate this function is a void one. The value of these node(s) is, also, a group of one or more nodes, where the key of each of those is the number of parameters of the function call, and their values initially is the equivalent source code. By following this particular format, we can ensure the accuracy when returning source code of a particular function call by the function's signature, not just by function name, since there may exist several functions with the same function name, but with different signatures.

3.3 Regular Expressions

Regular expressions, usually shortened to regex, are powerful programming and text-processing tools that help in searching, matching or modifying string patterns in a text. In a large body of text, they provide a concise and flexible way of identifying specific patterns.

A search pattern is formed by regular expressions, which are a sequence of characters. These can be inclusive simple letters or figures and exclusive special symbols. For instance, the dot (.) represents any single character, and the asterisk (*) denotes zero or more occurrences of the preceding character or group.

It was noticed that every function call from an external library in Arduino had this format: libraryName.functionName(parameters); (e.g. Servo.detach();).

Therefore, the regular expression in Fig. 3 was used to detect all function calls from external libraries.

```
'(?:(\w+)\s=\s)?(\w+(?:.\w+)+)((.*?));'
```

Fig. 3. Regular Expression for function calls from libraries

To summarize, this regular expression is designed to match this specific pattern in text:

- "variable =" is an optional assignment statement which can be captured within the first group.
- In the second group, there is a word followed by one or several dots and another word.
- The third capturing group captures the rest of the text (including the brackets) until a semicolon is encountered.

Moreover, another regex pattern, shown in Fig. 4, was defined to handle the case where there is a function call to a function in the same library, with the following format: word(parameters);

```
'\b(\w+)(((?:\w+(?:,\s)?)));'
```

Fig. 4. Regular Expression for function calls in same library

To sum up, this regex is employed to match this pattern:

- A word representing the first identifier
- An open bracket "(" is supposed to mark the beginning of words separated by a list of commas (such as function parameters).
- Each word can consist of one or more word characters (letters, digits, or underscores), and an optional white-space is allowed after each comma.
- A close bracket ")" is supposed to mark the end of the words.
- The line terminates with a semicolon ";".

4 Tool Development and Cache

Building on the tool proposed in [6], a sequence of steps was implemented to analyze each line during the code parsing process. The first action done was checking whether the line being parsed satisfied the function call's regular expression pattern defined in Fig. 3. If it did, then important details like library name, function name, whether the function has a return value or not, and the number of parameters are extracted from this line of code.

Once this information is obtained, the next thing to be done is to check whether the function call exists in the database or not. In case it does, it's corresponding value is examined. By checking the data type of the corresponding value, it was easily identified whether it had been cached before as a number or if it contained the actual source code of the function call as a text. In the case where the value was a number, it served as an indication that the function call had been cached before, serving as a reference to it's calculated CFP (COSMIC Function Points). If it's value was a string then that means that it holds the source code of the function call.

Subsequently, the same function was called recursively on each line of code within the array to ensure that all nested function calls are comprehensively examined along with their details. The recursive nature of the tool is clearly showcased in Fig. 5. The total CFP calculated for the initial function call is then saved in the value for the function call in the database to be used later directly instead of recalculating the same value, thus saving processing time and improving performance of the proposed tool.

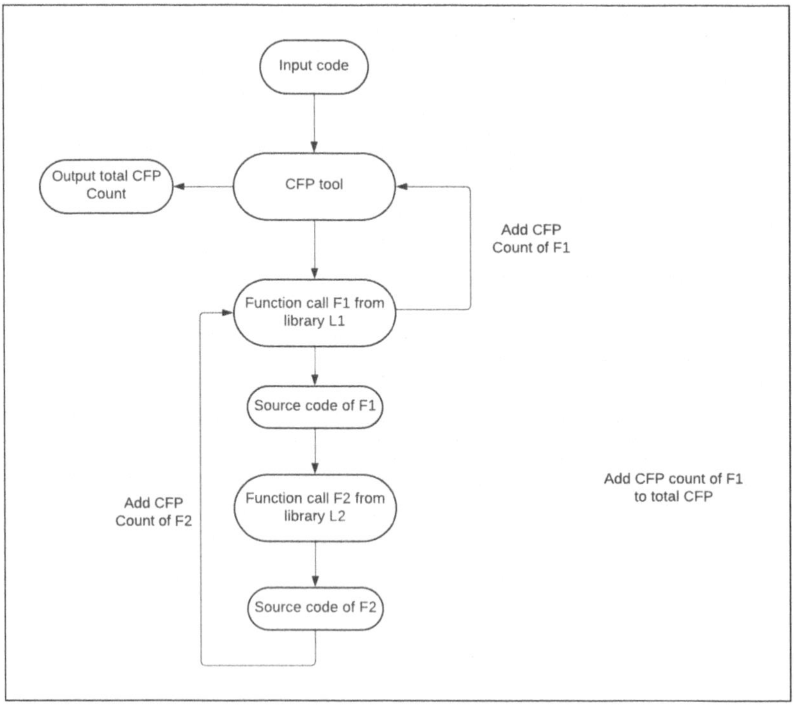

Fig. 5. Recursive architecture of tool

Another aspect to be addressed is that in some cases there is a function call to a function in the same library, with this format (word(parameters);). This was handled by first checking when parsing a new line of code, if this is a recursive call or not, by checking a flag value, and if it matches the function call regex pattern in Fig. 3.2. If it doesn't, then this line is checked again if it matches the regex pattern in Fig. 3.3, if it does then the length of the library attribute that is extracted must be zero, this indicates that this is a function call in the same library. Then, the `get_lines_of_code` function is called with the library value the same as the library value of the parent function, and the rest of the steps are followed the same.

5 Results

This chapter focuses on meticulously testing the tool as well as demonstrating the outcomes that have been generated. Testing is essential for authenticating the efficiency, precision and general performance of the tool in actual scenarios. Through an all-encompassing testing process, various aspects are meticulously examined, including the tool's functionality, robustness, and adherence to desired specifications.

To determine if the automated functional size measurement is accurate, the following line of code, in Fig. 6, was used to test the tool. This line of code that was picked for testing was carefully chosen because its source code covers every possible scenario that we wanted to test. This line specifically contains scenarios that test function calls from both separate and within the same library. Through the integration of these varied scenarios, the selected line guarantees a thorough and resilient evaluation of the tool's skills, enabling us to confirm its efficiency in managing different kinds of function calls and interactions across unique libraries. This deliberate choice makes it easier to analyze the tool's functionality in a variety of scenarios.

```
lines_of_code = [
    "Ethernet.begin(mac,ip,dns,gateway);",
  ]
print(CFPS(lines_of_code,False,"",False))
```

Fig. 6. Code example used for testing

The source code of the function call in Fig. 6 is the code displayed in Fig. 7.

```
void EthernetClass::begin(uint8_t *mac, IPAddress ip, IPAddress dns, IPAddress gateway)
{
        IPAddress subnet(255, 255, 255, 0);
        begin(mac, ip, dns, gateway, subnet);
}
```

Fig. 7. Source Code of Test function

When parsing the source code in Fig. 7, another function call to a method in the same library is discovered (begin(mac,ip,dns,gateway,subnet)), therefore this method's source code is retrieved and parsed as well, as shown in Fig. 8.

```
void EthernetClass::begin(uint8_t *mac, IPAddress ip, IPAddress dns, IPAddress gateway, IPAddress subnet)
{
        if (W5100.init() == 0) return;
        SPI.beginTransaction(SPI_ETHERNET_SETTINGS);
        W5100.setMACAddress(mac);
        W5100.setIPAddress(ip.raw_address());
        W5100.setGatewayIp(gateway.raw_address());
        W5100.setSubnetMask(subnet.raw_address());
        SPI.endTransaction();
        _dnsServerAddress = dns;
}
```

Fig. 8. Source code of function in the same library.

Moreover, when parsing the source code in 8, two function calls from the SPI library were discovered, and their source code, was the code snippets shown in Fig. 9.

The expected and actual CFP calculation of every line of code is shown in Table 1 and 2.

```
inline static void beginTransaction(SPISettings settings) {
  if (interruptMode > 0) {
    uint8_t sreg = SREG;
    noInterrupts();

    #ifdef SPI_AVR_EIMSK
    if (interruptMode == 1) {
      interruptSave = SPI_AVR_EIMSK;
      SPI_AVR_EIMSK &= ~interruptMask;
      SREG = sreg;
    } else
    #endif
    {
      interruptSave = sreg;
    }
  }

  #ifdef SPI_TRANSACTION_MISMATCH_LED
  if (inTransactionFlag) {
    pinMode(SPI_TRANSACTION_MISMATCH_LED, OUTPUT);
    digitalWrite(SPI_TRANSACTION_MISMATCH_LED, HIGH);
  }
  inTransactionFlag = 1;
  #endif

  SPCR = settings.spcr;
  SPSR = settings.spsr;
}
```

(a) Source Code of first function in SPI library

```
inline static void endTransaction(void) {
  #ifdef SPI_TRANSACTION_MISMATCH_LED
  if (!inTransactionFlag) {
    pinMode(SPI_TRANSACTION_MISMATCH_LED, OUTPUT);
    digitalWrite(SPI_TRANSACTION_MISMATCH_LED, HIGH);
  }
  inTransactionFlag = 0;
  #endif

  if (interruptMode > 0) {
    #ifdef SPI_AVR_EIMSK
    uint8_t sreg = SREG;
    #endif
    noInterrupts();
    #ifdef SPI_AVR_EIMSK
    if (interruptMode == 1) {
      SPI_AVR_EIMSK = interruptSave;
      SREG = sreg;
    } else
    #endif
    {
      SREG = interruptSave;
    }
  }
}
```

(b) Source Code of second function in SPI library

Fig. 9. Code Snippets from SPI library

Table 1. Expected and Actual (as per tool) CFP Calculation per line of code

Line of Code	Expected	Rec	Actual
Ethernet.begin(mac,ip,dns,gateway);	9	Yes	12
IPAddress subnet(255, 255, 255, 0);	0	No	1
begin(mac, ip, dns, gateway, subnet);	9	Yes	11
if (W5100.init() == 0) return;	1	No	1
`SPI.beginTransaction(SPI_ETHERNET_SETTINGS);`	2	Yes	3
W5100.setMACAddress(mac);	1	No	1
`W5100.setIPAddress(ip.raw_address());`	1	No	1
`W5100.setGatewayIp(gateway.raw_address());`	1	No	1
`W5100.setSubnetMask(subnet.raw_address());`	1	No	1
SPI.endTransaction();	2	Yes	3

(continued)

Table 1. *(continued)*

Line of Code	Expected	Rec	Actual
_dnsServerAddress = dns;	0	No	0
if (interruptMode >0) {}	0	No	0
uint8_t sreg = SREG;	0	No	0
noInterrupts();	0	No	1
#ifdef SPI_AVR_EIMSK	0	No	0
if (interruptMode == 1) {	0	No	0
interruptSave = SPI_AVR_EIMSK;	0	No	0
SPI_AVR_EIMSK &= ~interruptMask;	0	No	0
SREG = sreg;	0	No	0
} else	0	No	0
#endif	0	No	0
{	0	No	0
interruptSave = sreg;	0	No	0
}	0	No	0
}	0	No	0
#ifdef SPI_TRANSACTION_MISMATCH_LED	0	No	0
if (inTransactionFlag) {	0	No	0
pinMode(SPI_TRANSACTION_MISMATCH_LED, OUTPUT);	1	No	1
digitalWrite(SPI_TRANSACTION_MISMATCH_LED, HIGH);	1	No	1
}	0	No	0
inTransactionFlag = 1;	0	No	0
#endif	0	No	0
SPCR = settings.spcr;	0	No	0
SPSR = settings.spsr;	0	No	0
#ifdef SPI_TRANSACTION_MISMATCH_LED	0	No	0
if (!inTransactionFlag) {	0	No	0
pinMode(SPI_TRANSACTION_MISMATCH_LED, OUTPUT);	1	No	1
digitalWrite(SPI_TRANSACTION_MISMATCH_LED, HIGH);	1	No	1
}	0	No	0
inTransactionFlag = 0;	0	No	0
#endif	0	No	0
if (interruptMode >0) {	0	No	0
#ifdef SPI_AVR_EIMSK	0	No	0
uint8_t sreg = SREG;	0	No	0
#endif	0	No	0
noInterrupts();	0	No	1
#ifdef SPI_AVR_EIMSK	0	No	0

Table 2. Expected and Actual (as per tool) CFP Calculation per line of code

Line of Code	Expected	Rec	Actual
if (interruptMode == 1) {	0	No	0
SPI_AVR_EIMSK = interruptSave;	0	No	0
SREG = sreg;	0	No	0
} else	0	No	0
#endif	0	No	0
{	0	No	0
SREG = interruptSave;	0	No	0
}	0	No	0
}	0	No	0

5.1 Analysis of Results

After comparing expected and actual output of the tool, the accuracy of the tool was almost perfect except for a few miscalculations. This was due to the tool misinterpreting the following three lines as CFPs when they are not, and this was due to the fact that they were matched in the regex pattern, and not analyzed logically. These lines of code were:

- IPAddress subnet(255, 255, 255, 0);
- noInterrupts();
- noInterrupts();

A summary of the results obtained, is also displayed in Table 3.

Table 3. Summary of results obtained

Function call	CFP size manually	CFP size automatically	Accuracy
Ethernet.begin(5 parameters)	9	12	66.67%
Ethernet.begin(4 parameters)	9	11	81.81%
SPI.beginTransaction(1 parameter)	2	3	50%
SPI.endTransaction()	2	3	50%

6 Limitations and Future Work

One of the limitations of the implemented tool is its lack of logical reasoning capabilities. These instances occur when Non-CFP statements, although matching the regex pattern, are not actually CFPs, but are counted as ones. A potential solution for this limitation could be the integration of Natural Language

Processing (NLP) techniques or other advanced methods to enhance logical analysis and reduce the occurrence of such false positives. The tool's database, which only contains the most popular and extensively used Arduino libraries, is another drawback. For this database to stay comprehensive, constant updates are required. One potential solution to this problem could be to use web scripting techniques to automatically download and incorporate source codes of libraries that are found during code parsing. This would guarantee that the tool's library coverage is current and comprehensive.

7 Conclusion

In conclusion, this document emphasizes the importance of measuring software size in the Internet of Things (IoT) and concentrates particularly on the Arduino platform. This paper aims to address a gap in existing tools. Specifically, this tool seeks to broaden the functionality of a COSMIC-functional size measurement tool to accommodate a broader variety of function calls spread out in different Arduino libraries. The proposed tool developed using Python and regular expressions demonstrates its efficacy in measuring the functional size of IoT Edge Device code. Overall, this research contributes to advancing the understanding and measurement of software size in the IoT domain, enabling better resource allocation, project planning, and maintainability.

References

1. Abran, A., Symons, C., Ebert, C., Vogelzang, F., Soubra, H.: Measurement of software size: contributions of COSMIC to estimation improvements. In: ICEAA (2016)
2. Soubra, H., Abran, A.: Functional size measurement for the internet of things (IoT): an example using COSMIC and the Arduino open-source platform. In: IWSM-Mensura (2017)
3. Ecar, M., Kepler, F.N., da Silva, J.P.S.: AutoCosmic: COSMIC automated estimation and management tool. In: SBSI (2018)
4. Salem, S., Soubra, H.: Functional size measurement automation for IoT edge devices. In: IWSM-Mensura (2023)
5. Moulla, D. K., Mnkandla, E., Soubra, H., Abran, A.: Automated COSMIC function points measurement for C program using regular expressions. In: IWSM-Mensura (2022)
6. Salem, S., Soubra, H.: Using NLP for functional size measurement of IoT devices. In: Proceedings of 2023 Eleventh International Conference on Intelligent Computing and Information Systems (ICICIS), Cairo, Egypt (2023)
7. Soubra, H., Jacot, L., Lemaire, S.: Manual and automated functional size measurement of an aerospace realtime embedded system: a case study based on SCADE and on COSMIC ISO 19761. Int. J. Eng. Res. Sci. Technol. **4**(2) (2015)
8. Soubra, H., Abran, A., Sehit, M.: Functional size measurement for processor load estimation in AUTOSAR. In: IWSM-Mensura (2015)
9. Khattab, K., Elsayed, H., Soubra, H.: Functional size measurement of quantum computers software. In: IWSM-Mensura (2022)

10. Soubra, H., Abufrikha, Y., Abran, A.: Towards universal COSMIC size measurement automation. In: IWSM-Mensura (2020)
11. Soubra, H., Darwish, A.: COSMIC functional size of ARM assembly programs. In: IWSM-Mensura (2020)
12. Arduino Official Website. https://www.arduino.cc/reference/en/libraries/

Using Blockchain and Smart Contracts to Secure Medical Data Management System

Omar Khaled[1] and Abeer F. Ali[2](✉)

[1] British University in Egypt (BUE), Cairo, Egypt
[2] Electronic Research Institute (ERI), Cairo, Egypt
abeerfarouk@eri.sci.eg

Abstract. Recently, blockchain applications have been in the spotlight, and their entry into the medical field has shown great promise in many aspects. For example, Blockchain allows for decentralisation, transparency, and anonymization, which can provide fail-proof data protection, identity and data verification, supply chain management, and interoperability between platforms in healthcare. The new data protection laws govern how researchers handle medical data. The General Data Protection Regulation (GDPR) was chosen as the basis for the data protection regulations. This paper aims to integrate GDPR into an Ethereum Blockchain network. The proposed application involves auditing and tracking medical researchers' access to medical records in a medical data management system. The methodology is to utilise Smart Contracts to manage access to medical records in compliance with GDPR laws. The paper illustrates how to set up the environment of the Blockchain and build the Smart Contract to allow medical facilities that provide the data for medical researchers to track research details, requesters' credentials, verifications of selected GDPR attributes, statuses of request acceptance, and their relevant metadata. The paper examines the Solidity code deployment on the Sepolia Ethereum test network. The implementation shows different scenarios for medical facilities and requesters.

Keywords: Blockchain · Medical Record Auditing · Ethereum · dApps · Smart Contract · Solidity

1 Background

1.1 Blockchain in Healthcare

Bitcoin (BTC) was first introduced in a 2008 paper published under the pseudonym Nakamoto Satoshi [8]. Bitcoin introduced the concepts that paved the way for all cryptocurrencies. Nakamoto proposes a peer-to-peer distributed timestamp server to remove the need for central banks and the proof-of-work

A. Abdelgawad et al. (Eds.): ISBCom 2024, LNNS 1268, pp. 398–412, 2025.
https://doi.org/10.1007/978-3-031-82377-0_33

consensus algorithm that solves the problem of double-spending. He developed the first blockchain as a public, immutable ledger for storing Bitcoin transactions. It brought decentralization, transparency, and anonymization. Since then, many cryptocurrencies have emerged, such as Ethereum (ETH), Tether (USDT), USD Coin (USDT), etc. The success of the Bitcoin blockchain sparked discussions that led to the potential applications of blockchain technology in other data-driven industries, such as healthcare [7].

The healthcare industry is a people- and data-intensive, problem-driven field. It currently faces a higher demand for real-world data from industry and research organizations [1]. Health education and research institutions maintain strong connections with the healthcare industry. It provides them with access to test subjects' medical records and samples, which is necessary for the operation of these institutes. The activities necessitate the efficient exchange of consent from the patients, which efficiently entails the exchange of information between institutions. Additionally, healthcare facilities are required to safeguard the exceedingly sensitive information that patients elect to share-ensuring the patient's confidentiality while facilitating the sharing of information with other healthcare organizations. The most up-to-date tools and techniques are essential to achieving the desired outcomes.

Simultaneously, unauthorized information sharing, well-reported data breaches, and theft significantly undermine the public's confidence in healthcare establishments. Another issue arises from unethical practices inside the healthcare system that take advantage of the trust placed in it. Blockchain technology possesses essential features such as decentralisation, distribution, and data integrity. It does not require any third-party involvement. These properties offer numerous advantages that can be leveraged to enhance information sharing, access control, provenance, and data integrity among stakeholders. Consequently, it enables the development of a new infrastructure that supports establishing confidence in healthcare institutions. Blockchain technology can provide healthcare procedures and services that secure patient data management, interoperability among healthcare systems, and patient consent management. However, many challenges are associated with applying the new technology. We explore the considerations and potential barriers to adoption that healthcare organizations face when implementing blockchain solutions.

1.2 Smart Contracts

Ethereum is a Blockchain-powered application with its own cryptocurrency, Ether. In addition to cryptocurrency functionality, Ethereum also provides network programmability features. It enables the creation of decentralized network applications. Such applications run, and action is triggered by written programs called Smart Contracts. In [15], Xu et al. define Smart Contracts as programs in the form of data appended to the Blockchain ledger whose transactions are its form of execution. These decentralized apps are commonly called "dApps". Smart Contracts' decentralization property offers several advantages in the case of data exchange. One of them is the removal of third-party intermediaries. Smart

Contracts are "binding agreements" between two or more participants, with no chance for reputation, and hence lawfully imposed. The absence of intermediate parties, as opposed to regular contract service providers, reduces the transactional costs that would typically occur and be charged to the parties involved in the transaction [14]. Smart Contracts are written in Solidity, an object-oriented programming language developed by Ethereum exclusively for writing Smart Contracts. Using Solidity, the application's business logic and functions can be written to trigger transactions and manipulate data accordingly.

1.3 General Data Protection Regulation (GDPR)

The General Data Protection Regulation is a European law that came into effect in May 2018. It is issued to mandate how organizations should maintain integrity when collecting and processing personal data. According to the GDPR definition, personal data is any data that could make a living person identifiable. Those include the person's direct identification, such as his name, address, and mobile number, and indirect identification, such as his political opinions, interests, health, and medical data. The GDPR categories data processing as collecting, organizing, using, sharing, and erasing [5]. Therefore, collecting and processing patient medical data for research and study purposes must be governed by data protection regulations. Access to these records must be monitored and tracked when shared. Furthermore, it is imperative to conduct audits of these requests and access sessions in order to uphold accountability and prevent legal responsibility in the case of data breaches. Blockchain technology offers many of these functionalities, and its Smart Contracts utility provides a decentralized, secure resolution to the aforementioned issues.

This research paper aims to examine the Blockchain solution and develop a Blockchain-based application for securely managing and monitoring access permissions to patients' information. The paper proposes a solution to facilitate monitoring and auditing of patient medical record access in accordance with GDPR policies. Further analysis of the GDPR and its implications for Blockchain applications is discussed in Sect. 3.

The remainder of this paper is structured as follows: Sect. 2 discusses related work, and Sect. 3 introduces the proposed system. Section 4 shows the implementation and results, in Sect. 5, we discuss the conclusion. Finally, Sect. 6 the future work.

2 Related Work

We provide a comprehensive examination of the current state of blockchain research and applications in healthcare and health sciences, offering insights into both the opportunities and challenges associated with this emerging technology. Healthcare educational institutions and clinical facilities engage in numerous activities involving the institutional exchange of patient-related information. These institutions are obligated by law to protect such data during storage and exchange procedures [4]. [9] explores the use of blockchain technology for cross-domain image sharing in healthcare systems. It proposes a blockchain-based framework that eliminates third-party access to health information, meets interoperable health system criteria, and can be generalized to other fields. However, the framework faces challenges like privacy and security complexity and regulatory uncertainty. It introduces a method of sharing a patient's medical image. The scheme first establishes an agreement between the patient and the image source, the hospital, with a specific key. The patient grants access authorization. The image beneficiary node-the requesting doctor-requests such patient scans from an image source. The requester goes through authentication procedures and exchanges of public and private keys through the Blockchain. Consequently, the image source node verifies the requester's identity and validates his access key before transmitting the "imaging study".

Another method in [10] detailed utilizes the Fast Healthcare Interoperability Resources (FHIR). The FHIR is a standard for presenting medical data. It also provides open source Application Program Interfaces (APIs) for communicating Electronic Health Records. The Blockchain transactions, in this case, exchange the URL that provides access to FHIR resources. Security remains an important aspect of the approach, despite the choice of exchanging URLs instead of actual patient records. Therefore, employing a Smart Contracts framework is a crucial step, as was highlighted in [11].

In [13] a Blockchain-based secure storage scheme for medical information was proposed. The solution comprises three components: the Attribute Based Access Control (ABAC) architecture, IPFS, and the Hyperledger Fabric Blockchain network. The ABAC architecture provides access to medical records based on key attributes, while IPFS stores records in a distributed peer-to-peer file system. The Hyperledger Fabric Blockchain network is used to create a workflow, with administrators, users, doctors, and patients agreeing on access policies. The records are formatted with patient information, doctor information, and identification numbers.

Kushch et al. [6] devised a scheme in their paper that stores patient's personal and medical information in a complexly secure yet simply structured private Blockchain. The network consists of 3 chains. The main Blockchain contains a block for each patient. These blocks comprise the name, surname, date of birth and other credentials. Each block is a root for two other branchings, "Subchains". Subchain 1 registers the medical history of the patient including his diagnosis, test results, and other relevant medical and clinical data. Subchain 2 is a log keeper that keeps track of the history of access times, dates, and requester info doctor's name and id for its corresponding block in Subchain 1. Subchain 2 hash stores the hashes of the respective main block, the respective Subchain 1 block, and the hash of its predecessor block in Subchain 2. Thus, it provides a complex, secure scheme that is more challenging for attackers to compromise.

In [12]an access control framework for electronic medical records (EMR) using Smart Contract technology was proposed. It includes four Smart Contracts: a "Validation Contract" that validates new participant entrances, a "GetAccess Contract" that evaluates data requester eligibility, and a "Revoke Contract" that denies access. The "activity" list contains attributes like EMR Requester and Owner IDs, permission type, and the timestamp of the requester's latest request. The framework ensures secure storage and access control for EMR data.

3 Proposed System

In this section, we describe the proposed system. First, we analyze the GDPR to extract the rules related to medical data processing and the basis by which they are processed. We analyze the lawfulness of the processing based on the GDPR guidelines. Next, we illustrate the design as a framework and a general template to provide guidance throughout the design.

3.1 GDPR Analysis

GDPR mandates legal restrictions on how different entities treat personal data [2]. In GDPR, three main entities interact with the data. Data controllers are entities that hold and process people's personal data. On the paper's premises, it could be the medical facilities that store the patients' files and medical records. Data processors are entities that process the data as well, but as a third party on behalf of the data controller. Data processors could be mapped to the medical researchers who request access to the needed data to conduct their research. The data subject is the third entity. Data subjects are the data owners from whom the data controllers and processors collect. Moreover, the GDPR laws obligate data controllers to determine how data processors use data and audit these interactions to ensure that these actions comply with the GDPR laws [5].

The GDPR is governed by principles that ensure data processing is lawful, transparent, and fair. Article 6 of the GDPR outlines conditions for processing data for public benefit or under authorized authority. In this context, data is being processed under the supervision of official medical facilities and search labs, making it lawful. The Information Commissioner's Office (ICO) regulates data protection laws and safeguards public information rights. The ICO provides guidelines and checklists for data processors and controllers to ensure their processing procedures are executed lawfully and within the mandates of data protection laws, including GDPR. The six foundations for processing data are consent given by the data subject, contractual obligation with the data subject, compliance with the law, for the safety of the data subject or another person, serving the public interest, and pursuing a legitimate interest. These guidelines and indicators can be incorporated into the Blockchain's smart contract of the medical research facility, ensuring it operates legitimately under GDPR laws.

The GDPR also states that some rules about collection and processing must have a purpose. Data cannot be processed in the event that it is needed later. Secondly, the processing of personal data should be supported by a lawful foundation. The law presents six lawful bases on which the data processing needs to be based. Thirdly, the GDPR defines several rights that the data subjects are entitled to with regard to the processing of data. Data processors must respect and be obliged to exercise these rights whenever they are invoked. An example of these rights is the data subject's right to know how the collecting organization will process their data.

Blockchain technology offers security and transparency benefits but conflicts with GDPR laws. The GDPR assumes a centralized data controller for every record, which contradicts the idea of an open blockchain. The GDPR articles 16 and 17 mandate that data subjects have the right to rectification and erasure. The right to rectification allows data subjects to correct false or incomplete data without delay, while the right to erasure grants data subjects the right to erase their data without delay. However, Blockchain maintains data integrity and immutability, making modifying or deleting records problematic.

3.2 Design

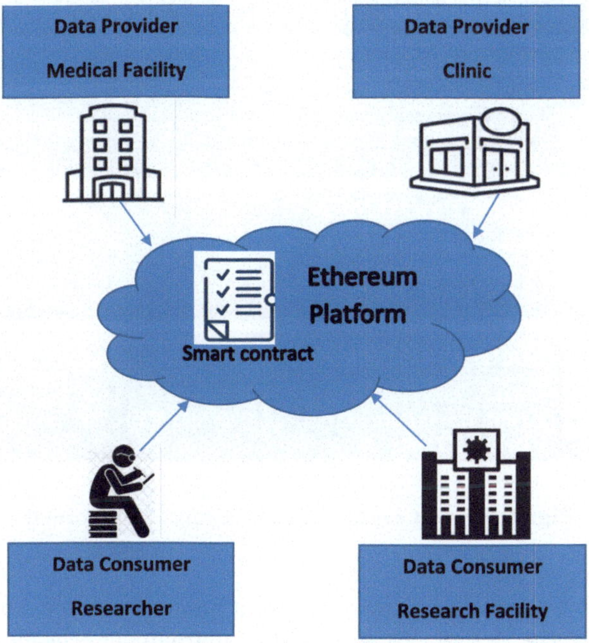

Fig. 1. System Diagram

The proposed solution aims to facilitate medical data access between medical researchers who request access to the data and the medical institutions that have the data through Blockchain technology. The design would exploit the technology of Blockchain's Smart Contracts. Smart Contracts contain rules that act as triggers to process data transactions. The Smart Contract rules will dictate the required attributes needed to authorize permissions for data requesters to access the medical data held by the medical facilities. Moreover, authorized permissions will be audited for accountability to trace back the requester's digital print in the event of data leakage. Additionally, the Smart Contract would validate requests to dispose of the data on the Blockchain. The rules and attributes written in the Smart Contract will be determined based on the analysis of GDPR rules. The GDPR has specified the legality of accessing, storing, securing, and disposing of data. Figure 1 shows the proposed systems over the Ethereum Blockchain.

Fig. 2. Smart Contract GDPR Transaction Format

We adopted A Blockchain Environment, a Smart Contract, and other tracking tools to test the design. To set up the environment, the first thing to do is set up an account on a cryptocurrency wallet to store some Ether for transactions. The cryptocurrency wallet is a client interface that sends and receives Ether or other cryptocurrencies. The most famous wallet is called Metamask. Metamask is a cryptocurrency wallet that stores Ethereum accounts with their respective Ether amounts. It also stores the address of the account and its private keys. Thus allowing decentralized applications to interact with them easily. The next component is the Blockchain environment where the smart contract will be deployed. Several open-source applications host both real Blockchains and test networks. Two that stand out the most are Truffle's Blockchain Suite Ganache and the other is the online Remix editor. Ganache's Blockchain test network is hosted locally on the PC running the application. Many accounts or nodes can be created, each having 100 ethers. Even though it has many features, interacting with Ganache and employing Smart Contracts requires a high knowledge of terminal commands to compile, deploy, and interact with. Although many tutorials are available online, using Ganache can be tiresome for beginners. Therefore, Remix editor is preferred for quick testing. Remix is an all-in-one tool that edits and deploys Smart Contract in its Blockchain network environment.

The general flow of the system starts by creating two accounts on the cryptocurrency wallet client interface, Metamask. The accounts represent the two different nodes that are parties to the Smart Contract: (1) The medical facility collecting and holding the data (data controller), (2) The data researcher processing the data (data processor). The two nodes will interact with the Smart Contract according to their respective roles. Upon creating the accounts, the next step is to buy some fake Ether to test from one of the Ether faucets provided for testing Blockchain networks. Next is writing the Smart Contract that satisfies the GDPR requirements. The transaction format will show different attributes based on the GDPR. Those attributes are audited and tracked along the Blockchain for each access request.Figure 2 shows the proposed system's intended transaction output format saved on the Ethereum Blockchain issued by the Smart Contract.

4 Implementation

4.1 Environment Setup

Metamask is a cryptocurrency wallet and Blockchain client used when performing transactions on the Ethereum Blockchain network and developing Smart Contract. Metamask can be easily integrated into the Google Chrome browser as an extension. Easy to install. After creating the account and validating the recovery phrase, the wallet configures and deals with Ethers. The account starts with zero Ethers. The Metamask can access two types of networks. One is the actual Ethereum network called Ethereum Mainnet. There are other real Blockchain networks as well. Ethereum Mainnet is Ethereum's main public Blockchain where actual transactions and monetary exchanges of Ether occur. The second type is test networks that utilize several forms of fake Ether. Those networks are used for application and Smart Contract transaction testing purposes and have no fiscal value in the real world. Several test networks can be used to deploy Smart Contracts and perform transactions with fake Ethers. Goerli, Sepolia, and Linea are among the commonly available test networks. The test network used here is called Goerli Test Network. After choosing the test network needed, the next step is to charge the wallet with some fake Ether, or Sepolia ETH in this case. For this purpose, we use an Ether faucet.

Ether Cryptocurrency Faucets: Sepolia Faucet. Ether faucets are tools that generate fake Ether test networks for testing and troubleshooting. Since

Sepolia Eth is the test network the account is on, the corresponding Sepolia Faucet is utilized. Sepolia Faucet supplies 0.5 Ether/day for the given account. To receive the fake Ether into the account, copy the account address from the Metamask crypto wallet, and paste it in the Sepolia Faucet website prompt. After the fake Ether is added to the account, any Blockchain application or Smart Contract transaction can spend the required Ether from that wallet. Keeping track of those transactions is important for analyzing them. Since a function in the Smart Contract triggers each transaction, which highly depends on the function itself, analyzing them can help further modify and optimize them to reduce the fee for those exchanges. The most common tracking tool is called Etherscan.

Etherscan: A Block Explorer. Etherscan is a block explorer. It shows the data and details of any Ethereum Blockchain. Examples of observed data are the transaction details, such as their hash, and the status of whether a transaction is successful, pending, or failed. The details also include the sender and receiver addresses. Moreover, they include transaction fees and gas price. Ethereum transactions on the Blockchain are not free of charge. The transactions require computational resources to complete. Therefore, each transaction is charged a fee that compensates for the mentioned computational resource of the node. Ethereum calls that fee Gas, representing the amount of computational effort required to execute specific operations on the Ethereum network" [3]. Gas is charged to the account regardless of whether the transaction succeeds or not. Gas is paid in Ether, but since 1 Ether has a large monetary value and the gas price is usually a significantly small fraction of Ether; another unit called gwei is used as a unit for gas. Gwei is about one billionth of an Ether or one nano Ether. This is important to understand as any change of data on the Blockchain or update is considered a transaction, thus efficiently optimizing the code and functions that change the Blockchain state to burn the least gas possible in any transaction.

Remix: Smart Contract Code Editor and Deploy Environment. Remix is an online integrated development environment (IDE) for writing Smart Contract Code using the Solidity programming language. It is also used to deploy those smart contracts on test Blockchain networks. Proving to be an almost all-in-one Ethereum Blockchain toolbox. Remix has a friendly user interface, a wide array of configuration options, and deployment settings.

4.2 Testing

The aim of this paper is to present a Smart Contract template for data controllers, the term coined by the GDPR laws, to track and audit the researchers' access to patients' medical information. Moreover, the code should allow the data controllers to enforce the appropriate GDPR laws that apply to the event of data processing at hand. This includes defining whether the patient gave consent for his data to be processed or defining the time for the record to be deleted to abide by the Right to be Forgotten rule dictated by the GDPR. Meanwhile, the Smart Contract will support the following:

- Check whether the Right to Forget is invoked, and if so define the deletion time
- Check and display the patient's consent for processing of his/her data
- Displays which researcher/data processor from an array of researchers is granted permission access
- to the required medical data for processing.
- Define and display the data processing purpose as specified by the GDPR.

Two types of nodes should be created before writing the solidity code for the Smart Contract. The first node is the medical facility/data controller node. The following figure shows the medical facility account. The account was initialized with 0.5 Sepolia Ethers at first. The second type of account to be created is the data researcher node, which requests the medical records to be studied. Two accounts of this type will be created. The first will be used in a scenario where the transaction format shows request acceptance and the rest of the attributes related to the GDPR. The second account will show an output where the request for the medical records was registered as denied, accompanied by the other GDPR attributes as well. After creating the required nodes to represent the system implementation, the next step is to write the Solidity Smart Contract code.

The Smart Contract code starts with specifying the license used. The most commonly used license identifier is the MIT license. It is an open-source license that allows room for freedom of general modification and reuse of the code. Then, we define which version of the Solidity programming language is being used. In this case, we used Solidity version 0.8.0. The contract has two main nodes that interact with each other: the medical facility and the medical researcher. The medical facility is the entity that holds the data and enforces the GDPR laws-in other words, it is the data controller. Then, the Smart Contract code declares variables relating to GDPR principles and user rights. The first is the isConsented Boolean variable. If it returns true, this variable indicates that the patient has consented for medical data to be processed by the agency that collected it. Thus, abiding by the GDPR law of acquiring consent and showing it. The second variable is the isRightToForget variable. This variable indicates that the holder of this data has invoked the Right to Be Forgotten Act and needs his data to be deleted. If true, a function will be triggered, and the data will be erased. The third Boolean value to explain is the isPublicTask variable. This variable indicates whether the research to be conducted on the patient's medical data is of a public service nature, as GDPR states there needs to be a lawful basis for processing personal data according to the GDPR. Obtaining consent from the data subject is one of them. Performing data processing to achieve a public service or task is another lawful basis. The fourth boolean variable is the isEncrypted variable. This variable will indicate whether the data is encrypted or not. The encryption of personal data is not explicitly compulsory but is one of the ways of protecting data. Protecting and securing data is one of the articles dictated by the GDPR laws. So, GDPR requires these identifications.

Another three variables have been declared to be used as follows: Firstly, there is the deletionTime variable. This variable can be initialized by the data controller to limit the time by which the data is accessible. One of the GDPR articles specifies that the data should be discarded upon the completion of data processing, or when the data is not necessary for the researcher. So when the deletion time is reached, the data holding this attribute will be deleted. The second variable is the string encryptiontype. The string indicates the type or name of encryption used to codify the data. Encryption algorithms such as AES or RSA are some of the examples that can be written in this string variable. The third string variable is researchPurpose. The research purpose is a crucial attribute that needs to be specified or stated when processing data. As mentioned before, one of the GDPR principles explicitly mandates that the data collection and processing purpose must be known and exposed to the data subjects. Then, we define a couple of metadata attributes and statuses. The status public string will define if the request was accepted or denied. The second public integer is the first access variable. The attribute records the first time the data researcher accessed the medical record. The attribute latestAccess shows the latest access date. A new datatype that will aid in the implementation. Mapping is a type of hash table that saves data as key-value pairs. Each hash key is associated with a corresponding value of a specified datatype. For example, the accessToResearcher mapping ties a key address to a Boolean value. This is used to set a certain researcher using the researcher's node hash address to a true or false value. Thus specifying whether that researcher has received permission to access the patient's medical records for processing.

The Solidity programming language allows users to create customized access modifiers. We created the onlyMedicalFacility access modifier. When used in a function definition, this modifier restricts the function to only the medicalFacility, the address variable that was explained earlier. The contract is to start with a constructor. The constructor initializes the medicalFacility address variable with the Metammask's created medical facility account address. If the Smart Contract is available to deploy at the medical facility only, then the variable can be initialized as msg.sender. The msg.sender is a built-in variable in the Smart Contract that holds the address of the node or block that called the Smart Contract. Since this Smart Contract is a template for the data controllers, they are initialized as the Smart Contract callers. We define the function initializeAttributes. This function takes input attributes as parameters and assigns them to the contract local attributes. The following snippet of code shows the function definition. The function body checks whether encryption will be applied to the stored data on the data processor's side. If the data is encrypted, the type of encryption algorithm is then recorded. Otherwise, it records the encryption type as no encryption. Depending on the organization and data controllers, lacking encryption may deny access to the data processor's request.

After compiling the Smart Contract and the compilation is successful, the next step is to deploy the Smart Contract on the Sepolia Ethereum network. Firstly, change the Remix environment to Injected Provider Metamask. This is done to connect our Metamask cryptocurrency wallet to the Remix editor and direct it to the attached Sepolia test network. Secondly, choose the account address of the medical facility, which can be checked from the Metamask client Chrome browser extension. Finally, the deployment of the contract is complete. When the deployment is successful, the Remix editor shows this in its terminal output. After waiting a few seconds, a notification from the Metamask wallet interface indicates that the transaction has succeeded. To further confirm the transaction's success and its various details, the Etherscan block explorer can track the addresses' activities. Whether it is transaction hashes, contract addresses, account addresses, or block numbers, Once you deploy the contract, you can use the Remix Editor to utilize the smart contract and manipulate the created variables on the Sepolia Ethereum Blockchain using the previously created functions.

4.3 Results

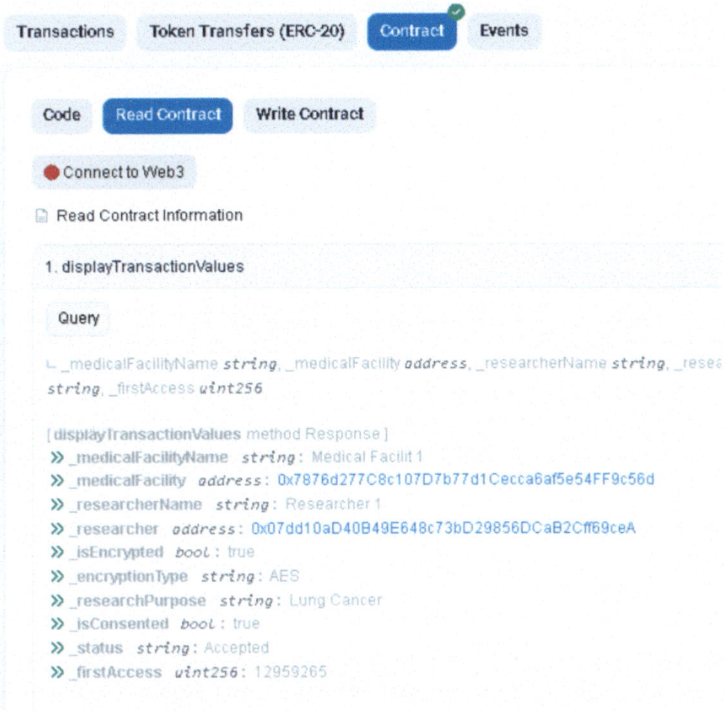

Fig. 3. Scenario 1: accepted request

Scenario 1. The first scenario is for the recording of an accepted medical data request. After initializing the variables and performing the transaction, Check with Etherscan by searching for the contract address and its details. The Smart Contract code is visible when clicking on Read Contract, and the state variables are loaded along with the Smart Contract code functions. Figure 3 illustrates the result.

Scenario 2. The second scenario is for the recording of a denied medical data request. After initializing the variables and performing the transaction with the appropriate variable values. Figure 4 illustrates this result:

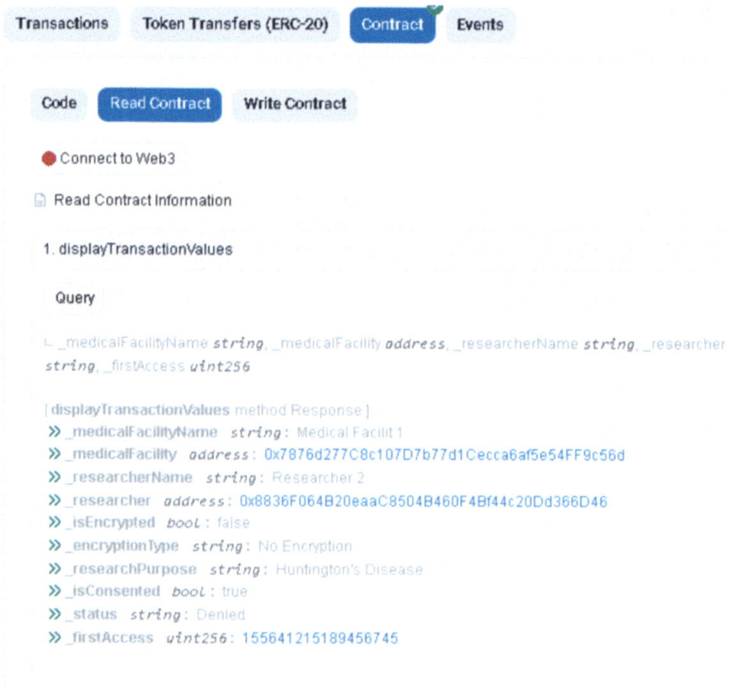

Fig. 4. Scenario 2: Denied request

The recorded transaction showed that the requested access to the medical data records in the second scenario was denied. Notice how some of the attributes were undesirable, which best explains why the request access was denied. The encryption, for example, needed to be more present. The Boolean value for it was false, hence the encryption type variable displayed the "No Encryption" string. Since there was no mention of other means for providing safety for the medical data records, a main requirement of the GDPR, the request to access was denied.

5 Conclusion

To conclude, the paper aims to analyze the European General Data Protection Regulation laws regarding data collection and processing. After analyzing the law, the next objective was to integrate these laws in the Blockchain Ethereum environment using Smart Contracts to address the logging and tracking of requested medical data accessing and to display the respectively supportive GDPR attributed based on which the data controllers provide such access to the requesters from the medical research community. The Smart Contract was implemented using the Remix Solidity Smart Contract editor and the Metamask cryptocurrency wallet and client.

6 Future Work

Upon experimenting with the Ethereum network and further analyzing the GDPR, the application has the potential to go further than just auditing and tracking law-abiding transactions. Providing access control through the Smart Contract would extend the implementation into a full-featured system application with access control, monitoring, and customized reporting. Moreover, the template can be even more generalized to fit any organizational data controller and its respective data processors. Additionally, a Javascript-based application can provide a user interface convenient for users to interact with and read. This may lead further to commercial usage later on for hospitals. The transactions can be extended later on to patients when Web3 applications become more mainstream and have a sufficient customer base to launch.

References

1. Coiera, E.: Guide to Health Informatics. CRC Press (2015)
2. European Commission: Data protection in the EU (2023). https://commission.europa.eu/law/law-topic/data-protection/data-protection-eu_en#national-data-protection-authorities
3. Ethereum.org: Gas and fees (2023). https://ethereum.org/en/developers/docs/gas/
4. Hasselgren, A., Kralevska, K., Gligoroski, D., Pedersen, S.A., Faxvaag, A.: Blockchain in healthcare and health sciences-a scoping review. Int. J. Med. Inform. **134**, 104040 (2020). https://doi.org/10.1016/j.ijmedinf.2019.104040, https://www.sciencedirect.com/science/article/pii/S138650561930526X
5. ICO, T.I.C.O.: UK GDPR guidance and resources (2022). https://ico.org.uk/for-organisations/uk-gdpr-guidance-and-resources/
6. Kushch, S., Ranise, S., Sciarretta, G.: Blockchain tree for eHealth. In: 2019 IEEE Global Conference on Internet of Things (GCIoT), pp. 1–5. IEEE (2019)
7. Mettler, M.: Blockchain technology in healthcare: the revolution starts here. In: 2016 IEEE 18th International Conference on e-Health Networking, Applications and Services (Healthcom), pp. 1–3. IEEE (2016)
8. Nakamoto, S.: Bitcoin: a peer-to-peer electronic cash system (2008)

9. Patel, V.: A framework for secure and decentralized sharing of medical imaging data via blockchain consensus. Health Inform. J. **25**(4), 1398–1411 (2019)
10. Peterson, K., Deeduvanu, R., Kanjamala, P., Boles, K.: A blockchain-based approach to health information exchange networks (2016). https://www.healthit.gov/sites/default/files/12-55-blockchain-based-approach-final.pdf. Accessed 05 Sept 2019
11. Sabu, S., Ramalingam, H., Vishaka, M., Swapna, H., Hegde, S.: Implementation of a secure and privacy-aware e-health record and IoT data sharing using blockchain. Glob. Transit. Proc. **2**(2), 429–433 (2021)
12. Saini, A., Zhu, Q., Singh, N., Xiang, Y., Gao, L., Zhang, Y.: A smart-contract-based access control framework for cloud smart healthcare system. IEEE Internet Things J. **8**(7), 5914–5925 (2020)
13. Sun, Z., Han, D., Li, D., Wang, X., Chang, C.C., Wu, Z.: A blockchain-based secure storage scheme for medical information. EURASIP J. Wirel. Commun. Netw. **2022**(1), 40 (2022)
14. Taherdoost, H.: Smart contracts in blockchain technology: a critical review. Information **14**(2), 117 (2023)
15. Xu, X., Weber, I., Staples, M.: Architecture for Blockchain Applications. Springer (2019)

Advancing ECT Imaging: Residual UNet Deep Neural Network Approach

Wael Deabes[1], Fahmi Khalifa[2](✉), Alaa E. Abdel-Hakim[3], and Kheir Eddine Bouazza[4]

[1] Computational, Engineering, Mathematical Sciences Department, Texas A&M University SA, San Antonio, TX 78224, USA

[2] Electrical and Computer Engineering Department, Morgan State University, Baltimore, MD 21251, USA
fahmi.khalifa@morag.edu

[3] Computer Science Department in Jamoum, Umm Al-Qura University, Makkah, Saudi Arabia

[4] Computer and Information Science Division, Higher Colleges of Technology, Abu Dhabi, United Arab Emirates

Abstract. Recently, there has been significant progress in Electrical Capacitance Tomography (ECT) image reconstruction, yet there remains room for enhancing both speed and quality in these reconstructions. Deep Learning (DL) offers a promising avenue due to its capacity to handle intricate nonlinear functions. This study introduces DL-based ECT image reconstruction model based on Residual UNet and termed as ECT_ResUNet. A sizable dataset containing permittivity distribution and corresponding capacitance measurements was created to train and evaluate the proposed model. The ECT_ResUNet processes modulated capacitance vectors, representing them as 66×66 images. The model's effectiveness and practicality were assessed using unseen samples, demonstrating precise image reconstruction with an average Image Error (IE) of 0.0918 and Correlation Coefficient (CC) of 0.9527.

Keywords: Electrical Tomography · DL · UNet

1 Introduction

Electrical Tomography is used to image multiphase flow processes by creating slice images. Electrical Capacitance Tomography (ECT) is employed in various industries, including renewable energy, oil and gas, pharmaceuticals, and food and beverage sectors [6]. An ECT system generally includes a capacitance-sensor, an electronic data collection components, and a computer for post-processing images. This technique measures capacitance between electrodes surrounding the imaging area and creates a permittivity distribution based on these measurements [16].

A. Abdelgawad et al. (Eds.): ISBCom 2024, LNNS 1268, pp. 413–423, 2025.
https://doi.org/10.1007/978-3-031-82377-0_34

ECT is challenging because the relationship between measurements and permittivity is nonlinear. Approaches encompasses non-iterative and iterative algorithms using simplified linear models. Non-iterative methods like Linear Back Projection (LBP) [7] are fast but yield limited image quality, often containing artifacts. Iterative methods like Tikhonov regularization produces better quality but require more computational time. Nonlinear model-based algorithms offer better image quality but demand intensive calculations, typically suitable for offline imaging and analysis.

Hence, there's an urgent necessity to develop a novel approach to enhance image reconstruction outcomes and expedite the reconstruction process. Deep Learning (DL) has garnered significant interest for its ability to map complex nonlinear functions using multiple sets of Artificial Neural Networks (ANN) [10]. ANN provides robust models and finds applications in various domains [19]. DL achieves classification and pattern recognition by processing nonlinear data through multiple layers for supervised or unsupervised feature extraction [18], and recently. DL techniques help automatically extract features for image reconstruction, which traditional methods struggle with due to the complexity of image composition [15].

Deep Neural Networks (DNNs), known for their ability to learn complex nonlinear functions through multiple hidden layers, are commonly used for initial data post-processing across various fields [14]. Adaptations of DNN algorithms, such as those based on Convolutional Neural Networks (CNNs), multi-scale CNNs, Dbar methods, autoencoders, and UNet, have been applied to expedite industrial process parameter extraction and improve ECT image reconstruction [12]. Additionally, efforts like those by Li et al. utilizing BP and RBF neural networks, along with the implementation of LSTM-IR algorithms, further contribute to addressing ECT image reconstruction challenges [5].

Implementing DNN techniques accelerates the estimating of industrial process characteristics and improves the ECT image. These methods include adaptations like CNN-based image reconstruction methods, multi-scale CNNs [12], Dbar methods [1], and autoencoders [3]. To address ECT image reconstruction challenges, Li et al. employed BP and RBF neural networks [11]. Additionally, an LSTM-IR algorithm was used to accurately generate material distribution images [4]. Authors in [2] utilized a Conditional-Generative Adversarial Network (CGAN) model specifically designed for reconstructing ECT images.

Accordingly, we introduce an ECT Residual UNet (ECT_ResUNet) as a novel approach, utilizing DNN methodology for ECT image reconstruction. Leveraging its unique parameter extraction capability, the UNet improves edge information extraction, which is applied in this paper to enhance ECT image reconstruction. UNet architecture has found wide applications in various domains such as image-to-image translation, image denoising [17], medical image processing, and image Segmentation [8]. He et al. [9] introduced Residual Network (ResNet) concepts to counteract performance degradation with increased layer count, showcasing superior performance in object recognition and image classification compared to conventional CNNs. ResNets leverage residual learning to achieve deeper depth, yielding better results and extracting more critical features.

The structure of the paper is as follows: Sect. 2 provides a detailed explanation of the ECT model. Section 3 explores the application of the residual UNet model in the context of ECT. Section 4 focuses on presenting the experimental outcomes of applying the ECT_ResUNet model. Lastly, Sect. 5 summarizes the study's findings and conclusions.

2 ECT Problem

In practical terms, as illustrated in Fig. 1, the ECT sensor contains twelve (i.e., $n = 12$) electrodes evenly spaced around the region of interest. It produces $M = 66$ independent measured mutual capacitance.

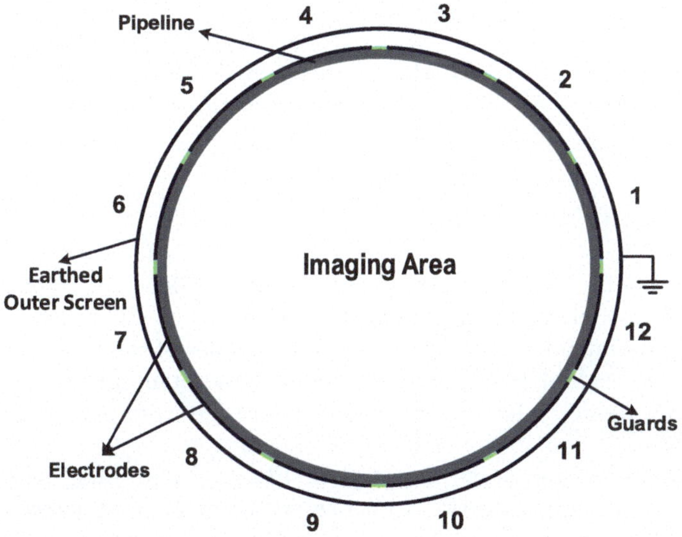

Fig. 1. Distribution of the ECT sensors around the imaging area

The ECT forward problem, aimed at determining the capacitance measurements for a given material distribution. the capacitance is computed numerically using the following equation:

$$C_{M\times 1} = S_{M\times N}(\varepsilon_0) \cdot G_{N\times 1} \tag{1}$$

here C represents the capacitance vector, M is the total number of measurements, S denotes the sensitivity matrix, N signifies the number of image pixels, and G indicates the permittivity distribution.

The inverse problem estimates G from C, a classic example of an ill-posed problem. Consequently, the generated ECT image is highly sensitive to measurement errors. Existing algorithms for this reverse problem often encounter

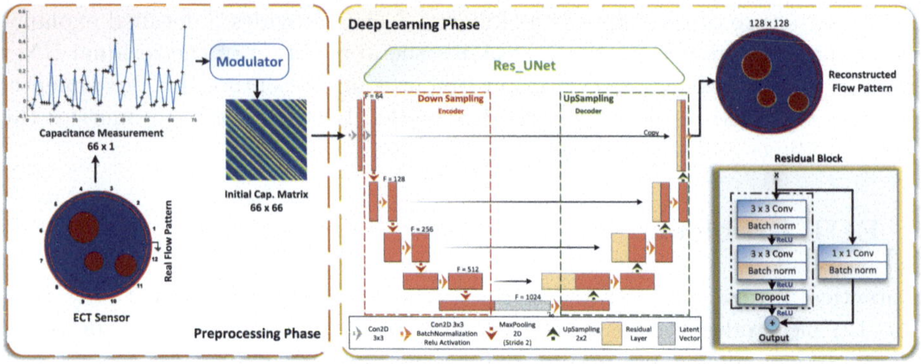

Fig. 2. Architecture of ECT_ResUNet model

quality or efficiency issues. To address these challenges in ECT, we introduce the ECT_ResUNet model. The subsequent subsections detail the primary components of our proposed ECT_ResUNet model.

3 ECT_ResUNetModel

In this section, we describe the configuration of the proposed ECT_ResUNetmodel, which is designed to address the ECT inverse problem. Additionally, we introduce the fundamental principles of the UNet network integrated with a residual block. Figure 2 depicts the structure of the proposed ECT_ResUNet model, which includes two mainstages: pre-processing and deep learning phases.

In the preprocessing stage depicted in Fig. 2, a new image is created by applying a rotate-right operation to a raw vector of $M = 66$ capacitance measurements. This operation simulates the counterclockwise rotation of the ECT sensor electrodes, repeated $(M-1)$ times to generate image of resolution 66×66. Simulating the rotation of the ECT sensor is vital to ensure the material distribution inside the imaging area remains invariant to rotation. This characteristic is especially important for asymmetric flow patterns, as it ensures a consistent input matrix regardless of how the pattern aligns spatially with the sensor's orientation.

The ECT_ResUNet model takes the modulated capacitance image and generates the ECT image. The UNet architecture is characterized by its encoder-decoder structure, aiming to accurately generate images by efficiently capturing both local and global context. In the encoder phase, UNet utilizes down-sampling methods such as max-pooling or convolutional layers with a stride to extract high-level features from the input, progressively condensing these features through the encoder layers. Conversely, in the decoder phase, up-sampling techniques like transposed convolutions or interpolation reconstruct the segmented image from these features. Skip connections between corresponding encoder and decoder layers preserve spatial information, enabling precise localization and segmentation. The UNet effectively balances feature extraction and spatial resolution preservation, ensuring high accuracy even for complex image structures.

The UNet architecture used in ECT_ResUNet is shown in Fig. 2, featuring four blocks in both the encoder and decoder sides, with a latent vector size of eight. During pre-processing, the low-resolution input image is concatenated with the final layer's generated image, and similar concatenation occurs between hidden layer inputs and outputs across encoder and decoder sides. The UNet consists of Down and Up sampling sections, with the Down sampling stage involving 3×3 convolutions, batch normalization, Relu activation, Residual Block (RB), and 2×2 maxpooling. This stage produces a downsized image of 64×64 size progressing to a latent vector of 8×8. The number of features are 128 and 1024 features for the downsampled and latent vector, respectively. In the Up sampling section, 2×2 upsampling layers post-convolutions are used, followed by residual blocks, reusing feature maps from the Down sampling part to reduce image distortion. The model is designed for a 12-electrode ECT sensor setup, requiring new dataset generation for changes in sensor count due to hardware configuration variations.

In contrast to traditional DNN methods, ResNets incorporated residual learning into the DNN framework. Deeper networks are typically constructed for better results, but increased depth can lead to gradient explosion and performance degradation. Literature suggests that simply adding pooling and convolution layers doesn't improve accuracy and may degrade performance [13]. To tackle this issue, residual learning is employed, aiming to minimize the discrepancy between the input and output. In the ECT_ResUNet model, a residual block is employed, as depicted in Fig. 2. This block enables direct connections between input and output, facilitating the learning of residuals. By integrating residual learning, the network effectively trains parameters, enhances performance, and learns deeper features without adding extra parameters or computations.

4 Experimental Results

A comprehensive ECT simulation dataset was generated using a custom Matlab GUI software package for training and testing. This dataset encompasses various two-phase flow patterns, with corresponding capacitance measurements accurately computed. Each dataset entry comprises a pair of vectors: one representing the permittivity distribution and the other the corresponding capacitance measurements.

Electrodes of the ECT sensor, illustrated in Fig. 1, are uniformly distributed around a PVC pipe with a relative permittivity of 2, a diameter of 126 mm and a thickness of 2.5 mm. Each electrode spans an angle of 26° with separation gaps of 4°. A big ECT dataset comprises $320k$ instances, encompassing five flow patterns such as ring, annular, stratified, and various bar configurations. To enhance dataset diversity, random variables are introduced: the width of the ring and the height of the stratified flow are randomly varied within defined ranges, and the number of circular and square bars is three. This approach ensures the dataset is comprehensive and applicable across a wide range of scenarios (Fig. 3).

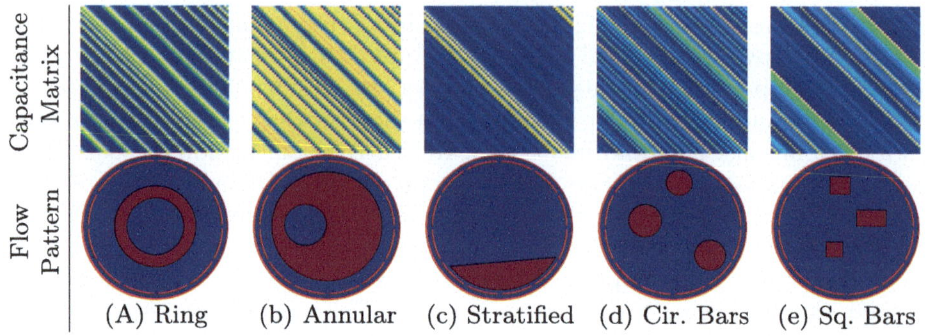

Fig. 3. Examples of various input modulated capacitance matrices alongside corresponding ground truth flow patterns.

The development of the ECT_ResUNet model commences with training the ResUNet using the assembled dataset. Following this, the model's efficacy is gauged through the reconstruction of the testing dataset. To mitigate overfitting, validation is conducted during training, where 10% of training samples are randomly selected each trial. The model's comprehensive performance and generalization ability are tested in various scenarios, including the original dataset, noisy test data, new data, and real experimental data. Generally, for evaluating the ECT problem, image quality and the effectiveness of reconstruction algorithms are measured using two standard metrics: Image Error (IE) and Correlation Coefficient (CC) [4].

To assess the reconstruction and generalization capabilities of the proposed ECT_ResUNet model, a testing dataset representing 30% of the ECT simulation dataset is randomly chosen. This testing set remains unseen during the training phase. The loss curves shown in Fig. 4 illustrate a gradual decrease over 250 epochs for both the training and validation sets.

Table 1 displays the minimum, maximum, and average relative values of IE and CC across each category of two-phase flow in the testing set. These results indicate that the ECT_ResUNet model effectively generates accurate images that closely match actual permittivity patterns. With average IE and CC values of 0.0918 and 0.9527 respectively, the overall performance of the proposed ECT_ResUNet model is exceptional. Importantly, most flow patterns demonstrate CC values exceeding 95%, and IE values are generally below 10%, except for the multiple circular and square bars flow pattern. Among the flow patterns analyzed, annular configurations show the highest relative IE and CC values.

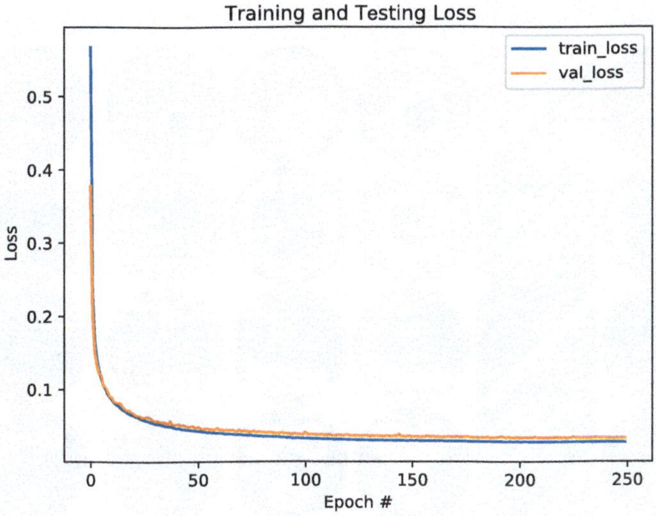

Fig. 4. Loss curves during Training and validation processes

Table 1. Min. and max. IE and CC values during the testing phase.

	Image Error (IE)			Correlation Coefficient		
	Min	Max.	Avg.	Min.	Max.	Avg
Annular	0.0447	0.1342	0.0560	0.9700	0.9981	0.9741
Ring	0.0435	0.1548	0.0640	0.9855	0.9533	0.9838
Stratified	0.0147	0.1281	0.0701	0.9882	0.9989	0.9911
Cir. Bars	0.0521	0.2571	0.1530	0.9267	0.9903	0.9351
Sq. Bars	0.0626	0.1404	0.1003	0.9423	0.9961	0.9540
Total Average			**0.0918**			**0.9527**

A study involving various flow patterns is devised to assess the effectiveness of the proposed ECT_ResUNet model against traditional and DNN-based image reconstruction algorithms. As shown in Fig. 5, the first column displays the actual phantom, followed by the modulated capacitance matrix in the second column. Subsequent columns present results from the LBP, ILM, LSTM-IR [4], and ECT_ResUNet algorithms, respectively. Notably, the ECT_ResUNet model produces more precise images compared to other algorithms, with clearer edges and reduced transition zones. Table 2 outlines the IE and CC values for all reconstruction algorithms, highlighting the exceptional ECT_ResUNet model performance.

To evaluate the scalability of the proposed model, measurements from new two-phase flow models, not included in the training dataset, are utilized. These models, labeled from 1 to 3, are depicted in the first column of Fig. 6. The generated images by the ECT_ResUNet model are shown in the third column

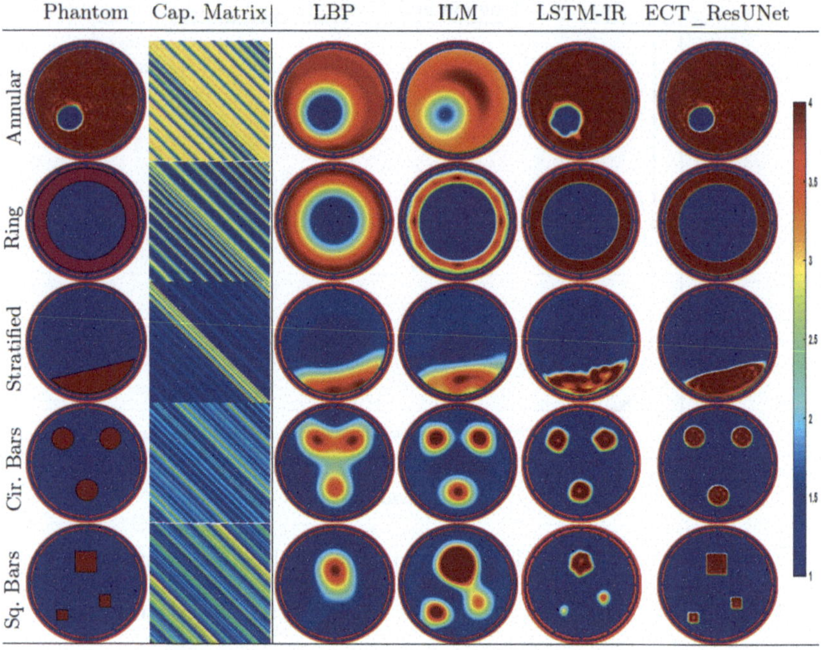

Fig. 5. Results of established algorithms and the proposed technique

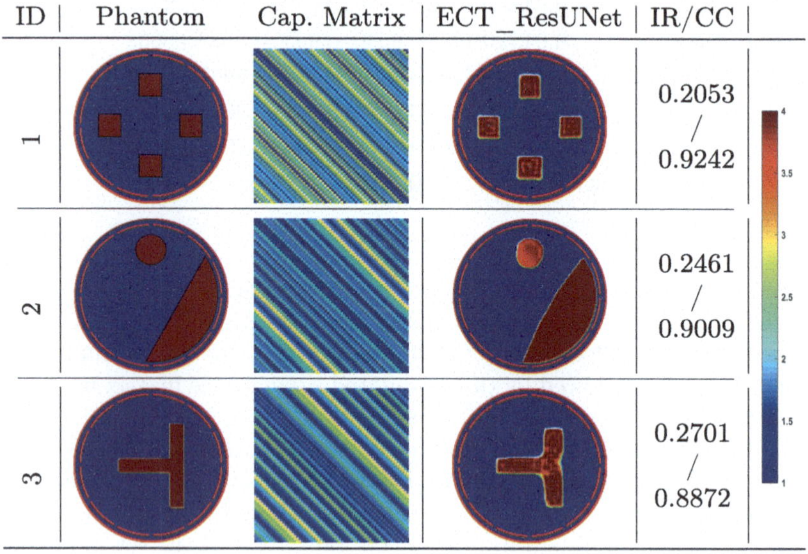

Fig. 6. Examples of reconstructed image of patterns absent from the training cohort.

Table 2. Comparative results of different ECT algorithms.

	Flow	LBP	ILM	LSTM-IR	ECT_ResUNet
IE	Annular	0.2412	0.3351	0.0561	0.0541
	Ring	0.3776	0.2984	0.0989	0.0941
	Stratified	0.2590	0.3203	0.2032	0.1994
	Cir. Bars	0.6083	0.4275	0.2043	0.1523
	Sq. Bars	0.5112	0.6909	0.2558	0.1283
CC	Annular	0.8701	0.9084	0.9913	0.9894
	Ring	0.8110	0.9576	0.9857	0.9870
	Stratified	0.9126	0.9100	0.9401	0.9587
	Cir. Bars	0.5498	0.7625	0.9363	0.9643
	Sq. Bars	0.5099	0.5688	0.8707	0.9803

of Fig. 6, along with their respective relative IE and CC values in the fourth column. Despite the absence of mixed flow patterns or configurations featuring four square bars, the ECT_ResUNet model demonstrates high-quality results in accurately identifying the actual patterns of these phantoms.

5 Conclusion

A novel ResUNet DNN for ECT image reconstruction is introduced. Utilizing a massive ECT simulation dataset, we train, validate, and test the proposed UNet-based model. Through various experiments involving noisy data, and unseen flow patterns, we evaluate the model's performance and feasibility. Results across different experiments demonstrate that our proposed UNet model outperforms traditional and DNN-based algorithms visually and quantitatively. Notably, the UNet model exhibits rapid reconstruction speeds comparable to the LBP algorithm for online imaging tasks. Moreover, the model demonstrates high generalization capabilities, achieving an average CC exceeding 95% and an average IE around 10%. Future work will focus on expanding the training dataset to include diverse flow patterns with gradually changing permittivity values, aiming to further enhance the performance of our DL model.

Disclosure of Interests. The authors have no competing interests to declare that are relevant to the content of this article.

References

1. Cao, Z., Xu, L., Fan, W., Wang, H.: Electrical capacitance tomography with a non-circular sensor using the dbar method. Meas. Sci. Technol. **21** (2009)
2. Deabes, W., Abdel-Hakim, A.E.: CGAN-ECT: reconstruction of electrical capacitance tomography images from capacitance measurements using conditional generative adversarial networks. Flow Meas. Instrum. **96**, 102566 (2024)

3. Deabes, W., Bouazza, K.E.: Residual autoencoder deep neural network for electrical capacitance tomography. Comput. Mater. Continua **73**(3), 6307–6326 (2022). https://doi.org/10.32604/cmc.2022.030420, http://www.techscience.com/cmc/v73n3/49055
4. Deabes, W., Khayyat, K.M.J.: Image reconstruction in electrical capacitance tomography based on deep neural networks. IEEE Sens. J. (2021)
5. Deabes, W., Sheta, A., Braik, M.: ECT-LSTM-RNN: an electrical capacitance tomography model-based long short-term memory recurrent neural networks for conductive materials. IEEE Access **9**, 76325–76339 (2021). https://doi.org/10.1109/ACCESS.2021.3079447
6. Deng, H., Yang, K., Liu, Y., Zhang, S., Yao, Y.: Actively exploring informative data for smart modeling of industrial multiphase flow processes. IEEE Trans. Industr. Inform. **17**(12), 8357–8366 (2021). https://doi.org/10.1109/TII.2020.3046013
7. Gamio, J.C., Ortiz-Aleman, C.: An interpretation of the linear back-projection algorithm used in capacitance tomography. In: Proceedings of 3rd WCIPT, Banff, Canada, pp. 427–432 (2003)
8. Gui, Y., Juming, D., Yihang, W., Xinglin, Z.: RUC-Net: a residual-UNet-based convolutional neural network for pixel-level pavement crack segmentation. Sensors **23**(1) (2023). https://doi.org/10.3390/s23010053, https://www.mdpi.com/1424-8220/23/1/53
9. He, K., Zhang, X., Ren, S., Sun, J.: Deep residual learning for image recognition. In: Proceedings of the IEEE Conference on Computer Vision and Pattern Recognition, pp. 770–778 (2016)
10. LeCun, Y., Bengio, Y., Hinton, G.: Deep learning. Nature **521**(7553), 436 (2015)
11. Li, J., Yang, X., Wang, Y., Pan, R.: An image reconstruction algorithm based on RBF neural network for electrical capacitance tomography. In: 2012 Sixth International Conference on Electromagnetic Field Problems and Applications, pp. 1–4. IEEE (6 2012). https://doi.org/10.1109/ICEF.2012.6310416, http://ieeexplore.ieee.org/document/6310416/
12. Lili, W., Xiao, L., Deyun, C., Hailu, Y., Wang, C.: ECT image reconstruction algorithm based on multiscale dual-channel convolutional neural network. Complexity **2020** (2020). https://doi.org/10.1155/2020/4918058
13. Man, J., Sun, G.: A residual learning-based network intrusion detection system. Secur. Commun. Netw. (2021)
14. Tan, C., Lv, S., Dong, F., Takei, M.: Image reconstruction based on convolutional neural network for electrical resistance tomography. IEEE Sens. J. **19**(1), 196–204 (2019). https://doi.org/10.1109/JSEN.2018.2876411, https://ieeexplore.ieee.org/document/8493596/
15. Xu, Z., Wu, F., Zhu, L., Li, Y.: LSTM model based on multi-feature extractor to detect flow pattern change characteristics and parameter measurement. IEEE Sens. J. **21**(3), 3713–3721 (2021). https://doi.org/10.1109/JSEN.2020.3024294
16. Yang, W.Q., Beck, M.S., Byars, M.: Electrical capacitance tomography-from design to applications. Meas. Control **28**, 261–266 (1995)
17. Zhang, H., Lian, Q., Zhao, J., Wang, Y., Yang, Y., Feng, S.: RatUNet: residual U-net based on attention mechanism for image denoising. PeerJ Comput. Sci. **23**(10) (2022). https://doi.org/10.7717/peerj-cs.970

18. Zheng, J., Peng, L.: A deep learning compensated back projection for image reconstruction of electrical capacitance tomography. IEEE Sens. J. 1 (2020). https://doi.org/10.1109/jsen.2020.2965731
19. Zhu, H., Sun, J., Long, J., Tian, W., Sun, S.: Deep image refinement method by hybrid training with images of varied quality in electrical capacitance tomography. IEEE Sens. J. **21**(5), 6342–6355 (2021). https://doi.org/10.1109/JSEN.2020.3036659

Analyzing the Effect of Feature Selection Algorithms on ML Classifiers

Aya Ahmed(✉), Minar El-Aasser, and Milad Ghantous

German University in Cairo, New Cairo, Egypt
aya.ibrahim@student.guc.edu.eg, {minar.elaasser,milad.michel}@guc.edu.eg

Abstract. High-dimensional datasets pose one of the main challenges that degrade the performance of IDS models, increase the classification models' complexity and deteriorate the detection speed. Dimensionality reduction techniques aim to reduce the dimensionality of the dataset by transforming the original feature space to a representable low-dimensional feature space. Dimensionality reduction techniques are classified into feature selection and feature extraction. This study aims to find the smallest set of features that best represents the original feature space while achieving high classification performance. We also aim to find the feature selection technique that selects features that enable the fast detection of attacks and reduce the model complexity. The experiment evaluates the effect of using 4 feature selection techniques Chi-squared, ANOVA, Recursive Feature Elimination (RFE), and Extra-trees on Decision Tree (DT), Random Forest (RF), XGBoost (XGB), Naive Bayes (NB) in terms of performance metrics, training and testing times in iterative fashion. The experiment was conducted using the benchmark dataset CICIDS2017. This analysis shows that the combination of RFE and DT achieves the highest accuracy of 97.9% accuracy using 20 features, requiring the least training and testing time, reflecting the model's simplicity and fast detection speed.

Keywords: IDS · Dimensionality Reduction · Feature Selection · Machine Learning

1 Introduction

According to recent statistics [4], almost 328.77 million terabytes of traffic is generated daily over the internet. This surge in the volume of traffic produced per day is due to the increased number of users on the Internet and the increased number of services that depend on the Internet such as Internet of Things (IoT) devices. This growth in Internet usage has resulted in the generation of more frequent and sophisticated network attacks. Hence, having efficient and effective Intrusion Detection Systems (IDS) is a must to safeguard the networks. An IDS function is based on monitoring the network traffic to identify malicious behavior Another aspect that is also affected by the increased internet utilization is the

A. Abdelgawad et al. (Eds.): ISBCom 2024, LNNS 1268, pp. 424–438, 2025.
https://doi.org/10.1007/978-3-031-82377-0_35

dimensionality of the data. For network traffic to be analyzed it is structured in datasets. The dimensionality of the dataset corresponds to the number of columns or features in the dataset. This means that the dataset's volume and dimensionality affect the IDS's performance in terms of accuracy, precision and other performance metrics and computational complexity to detect attacks.

The increased frequency of attacks makes it necessary to quickly and accurately detect attacks while monitoring vast amounts of data. Satisfying such a need is challenging when the datasets have huge volumes and dimensions. To address the curse of dimensionality reduction techniques are employed to transform higher dimensional datasets into lower dimensional datasets while preserving relevant features and discarding irrelevant and redundant information. Dimensionality reduction methods do not only aid fast detection but also reduce the complexity of the IDS models.

Given the critical role of IDS in protecting networks and the importance of real-time attack detection, it is mandatory to explore and evaluate the impact of dimensionality reduction techniques on the performance of IDS classifiers. In this study, we compare the impact of different dimensionality reduction methods such as ANOVA scores and recursive feature elimination on IDS performance so we can identify the optimal method for enhancing the scalability, accuracy, and efficiency of IDS.

2 Background and Related Work

Dimensionality reduction is the process of transforming the high-dimensional feature space into low-dimensional feature space or in other words reducing the number of features or columns in the dataset. Dimensionality reduction methods aim at minimizing the number of features by selecting a smaller set of features that best represent the original full set of features or creating a new set of features from the original set. Dimensionality reduction techniques are employed before training the classifiers to address the curse of dimensionality problem of datasets. The advantages of Low dimensional datasets include reduced computational complexity of the ML classifiers and elimination of redundant and irrelevant features hence reducing over-fitting and enhancing data quality. Dimensionality reduction techniques can be grouped into two main types: feature selection (FS) and feature extraction (FE) [15].

2.1 Feature Selection (FS)

Feature selection techniques reduce the dimension of the dataset by selecting a subset of features depending on their relevance that represents the original dataset the best. The relevance of the feature is evaluated based on scores calculated according to the FS technique used [11,15]. Feature selection can be classified into 3 categories: Filter, Wrapper, and Embedded.

- **Filter** methods calculate a score for each feature relying on the characteristics of the features, independent of other features, and independent of the

machine learning algorithms that the data will be trained on. The score measurement criteria can be based on information, dependency, consistency and distance. Information-based methods calculate scores depending on information theory concepts such as Information Gain. Dependency-based methods calculate scores based on the relevance between the feature and the labels such as chi-squared and ANOVA. Consistency-based methods aim to select a subset of features that ensure the consistency of the dataset regarding labels. Distance-based scores use distance metrics to evaluate how well features separate instances of different classes. Features are ranked by these scores to be able to select the best subset of features from them [11,15].
- **Wrapper** methods rely on the performance of the machine learning (ML) classifier used in the training process to select the best subset of features. Wrapper algorithm selects a subset of features, evaluates the accuracy after training the model with the subset of features and repeats the process of selecting subsets and training the model until the best accuracy is achieved [15]. There are different methods for selecting the subset of features including forward selection, backward elimination and recursive feature elimination (RFE).
- **Embedded** methods from the name they embed the feature selection process within the ML classifier used. Depending on the classifier, feature importance is evaluated. For example, in decision trees, the most important feature is the feature that results in the least impurity [15].

Filter-based methods are the fastest however they do not consider the dependence between the features. Wrapper-based methods create the smallest feature subset however, they are computationally expensive. Embedded methods combines the merits of both filter-based and wrapper-based.

2.2 Feature Extraction

Feature extraction methods create new features from the original dataset. The original feature space is transformed into a lower dimensional features space by combining features from the high dimensional feature space. The new feature set created is more descriptive and informative than the original dataset. However, the extracted features lose the semantic meaning due to the combination process, hence features lose their interpretability. The most used feature extraction technique is principal component analysis (PCA) [11].

Both Feature Selection and Feature extraction reduce the high-dimensional feature space into a lower dimension. FS achieves this by selecting the most relevant features subset, while FS creates a lower dimensional representation of the dataset.

2.3 Related Work

There are multiple dimensionality reduction techniques used in IDS to transform high-dimensional data into lower dimensions. This section reviews previous studies that compared different feature reduction techniques across several classifiers and the effect on the performance of the classifiers after performing feature reduction. The authors in [12] compared 3 dimensionality reduction techniques Chi-squared, RFE, and IG with ML-based classifiers SVM, DT, NB, RF, KNN, and LR and one DL-based classifier ANN. The experiments were performed using the benchmark NSL-KDD dataset. The performance of classifiers in terms of accuracy, precision, recall, and f-score is measured and compared. The analysis of the results shows that the use of RFE feature reduction and SVM yields the highest accuracies. In [7] authors compared the performance of ML-based classifiers RF, KNN, and SVM with feature reduction performed using ANOVA and RFE to select the most relevant features. The comparison also included training the classifiers with all features. Both feature reduction techniques were used to select the 13 most important features out of 41 features from the NSL-KDD dataset. The results of this study show that RF yields the highest accuracy compared to KNN and SVM. An iterative approach was implemented by authors in [5] to select the least number of features required to achieve the highest accuracy. The comparative analysis includes 3 feature reduction algorithms chi-squared and ANOVA are filter-based algorithms and extra-tree is an embedded-based algorithm and 4 ML-based classifiers RF, DT, KNN, and XGBoost. The results of this analysis show that the highest accuracy was achieved by ANOVA and XGBoost with 15 features only. Authors in [6] compared using 2 embedded-based feature selection algorithms extra-tree and random forest with 5 ML-based classifiers: KNN, SVM, DT, NB, and LR and 8 ensemble classifier: Bagging, AdaBoost, Gradient Boosting, Extreme Gradient Boosting Classifier, Light Gradient Boosting Classifier, Voting, Extra tree, and random forest. The analysis shows almost similar performance is achieved using either extra-tree or random-forest. The best performance achieved was by KNN in ML-based classifiers and RF in ensemble classifiers. In [10] authors evaluated the effect of using the filter-based feature selection method IG to rank and group features according to their weights on the 4 classifiers RF, Bayesian Network, Random Tree, NB, and J48. The experiment was conducted using the CICIDS2017 dataset. The experiment results that RF algorithm has the highest accuracy of 99.86 relevant selected features of 22, whereas the J48 classifier algorithm provides an accuracy of 99.87 with 52 relevant selected features with a longer execution time. In [8], a comparison between the performance of SVM and RF was performed before and after reducing the number of features to 13 features using RFE. Prior reduction RF outperformed SVM, however after reduction, SVM performed better.

3 Methodology

3.1 Data Preprocessing

Data preprocessing is a fundamental step performed on the dataset before the training phase to improve the quality of the dataset to enhance the performance of the classifiers. CICIDS2017 is a benchmark dataset that is widely used in recent research in the field of IDS. CICIDS2017 dataset consists of 8 separate

CSV files. Each file contains network flow data for one type of attack. These files are concatenated together to create 1 dataset. CICIDS2017 dataset is generated by the Canadian Institute for Cybersecurity to be used in studies related to network security and IDS. The dataset contains labeled network flow traffic and contains 6 attack profiles including DoS and DDoS, infiltration, web attacks, brute force, heartbleed and botnet. The network traffic was captured in 4 days two periods each day. Each period contains normal traffic and generated synthetic attacks using different attack-generating tools. CICFlowMeter is then used to extract more than 80 network traffic features [9]. The list below describes the preprocessing steps performed on the dataset in this study:

1. Removing duplicated rows, rows with NaN values and rows with "inf" values.
2. Removing duplicated features.
3. Removing labels that rarely occur in the dataset which are "Infiltration", "SQL injection", and "Heartbleed".
4. Removing features that contain negative values which are "Init Win bytes forward" and "Init Win bytes backward".
5. Removing rows that contain erroneous values.
6. Removing all-zero features which are "Bwd PSH Flags", "Bwd URG Flags", "Fwd Avg Bytes/Bulk", "Fwd Avg Packets/Bulk", "Fwd Avg Bulk Rate", "Bwd Avg Bytes/Bulk", "Bwd Avg Packets/Bulk", "Bwd Avg Bulk Rate".
7. Removing features that result in the classifiers over-fitting which are "Source IP", "Source Port", "Destination IP", "Destination Port" and "Timestamp"
8. Label encoding the labels column
9. Using Synthetic Minority Over-sampling Technique (SMOTE) to balance the dataset. After this step, Random Under Sampling (RUS) is used to reduce the size of the dataset to 1200000 rows.
10. Min-max scaler is used to scale the data.

4 Feature Selection

4.1 Chi-Squared

Chi-Square feature selection is a filter-based feature selection method. This FS method measures the strength of the association between that feature and the target variable where a high chi score indicates a strong dependence between the feature and labels [12]. To calculate the chi score for each feature contingency table is created with the data points that summarize the frequency distribution of the dataset features. Then for each cell in the table, the expected value is calculated using this formula 1. Where row total is the total occurrences across all features per row, column total is the total occurrences across all feature categories per feature and grand total is the total number of instances in the dataset

$$\text{Expected Frequency} = \frac{\text{Row Total} \times \text{Column Total}}{\text{Grand Total}} \tag{1}$$

The chi score is calculated using this formula 2, where the observed frequency is the actual occurrence recorded in the contingency table.

$$chi^2(x) = \frac{(\text{Observed Frequency} - \text{Expected Frequency})^2}{\text{Expected Frequency}^2} \tag{2}$$

4.2 ANOVA

Analysis of variance (ANOVA) F-score assesses the variance in the label variable across different groups or categories per feature. For each feature, the F-score is calculated using this formula 3. Where the variance between groups evaluates the variability of labels between the different feature groups and variance within groups evaluates the variability within each feature group.

$$F = \frac{variance_{\text{between}}}{variance_{\text{within}}} \tag{3}$$

4.3 Extra Tree

Extra Trees stands for Extremely Randomized Trees. It is an ensemble learning technique that builds multiple decision trees during training. This technique is categorized as an embedded feature selection method. It randomly selects subsets of features and splitting points. The importance of each feature is correlated with its impact on decreasing the impurity. The higher the importance of the feature the lower the impurity it results in. Since extra tree creates multiple trees, the importance of the feature is calculated by averaging the reduction in impurity that the feature provides across all trees [5].

4.4 Recursive Feature Elimination

It is a wrapper-based feature selection method. RFE trains the model with all features and calculates the feature importance depending on how the classifier measures the feature importance score. The features are ranked then the model is trained with the ranked features and the accuracy is measured. The least important features are eliminated and the process is repeated until a set of K-selected features is reached [12].

5 ML-Based Classifiers

5.1 Decision Tree (DT)

DT is a supervised ML classifier. The classifier constructs a tree with nodes, edges, and leaves. Each node represents a feature, the branch represents a decision and each leaf represents a class label. To create a tree, a root node must be chosen. The feature used to create the root node is the feature that results in the least impurity. For each splitting node created after the initial split a new feature is selected for each subsequent node using the same principal which is the feature that gives the least impurity. This process continues until no further splits can be made [1].

5.2 Random Forest (RF)

RF is an ensemble-supervised ML classifier that combines bagging with DT [14]. RF constructs multiple DT and each DT created is trained using different subsets of the training dataset. This process is called bootstrapping or bagging. Each decision tree gives a prediction and the final prediction is the mean of the predictions of the separate trees [2].

5.3 XGBoost

XGBoost is an ensemble-supervised ML classifier based on gradient boosting and decision trees. Like Random forest, XGBoost combines the prediction of multiple trees to reach the final classification. Gradient boosting means that the classifier learns sequentially, this is applied in XGBoost by creating one tree at a time and the new tree learns from the mistakes of the previous tree [3].

5.4 Logistic Regression (LR)

LR classifier is a supervised ML statistical classifier. The model classifies data points by calculating the probability by which a single point belongs to every class using the logistic probability function 4 which is a soft-max function. Where P(Y = k|X) is the probability that the target label is label k given the input vector x, b_{0k} is the bias or intercept for class k, b_{nk} is the weight for each feature and X_n is the input feature [12,13].

$$P(Y = k|X) = \frac{e^{b_{0j}+b_{1j}X_1+b_{2j}X_2+...+b_{nj}X_n}}{\sum_{j=1}^{K} e^{b_{0j}+b_{1j}X_1+b_{2j}X_2+...+b_{nj}X_n}} \tag{4}$$

5.5 Naive Bayes

NB classifier is a supervised ML probabilistic classifier based on Bayes theorem. NB assumes that all features are conditionally independent of each other, hence the probabilities of all the features of the class independently contribute to classifying the data points. Data points are classified using Bayes theorem formula 5 [12,13]. Where P(x|L) is the likelihood probability which is the probability of observing the feature vector X given that the class is Y, P(L—x) is the posterior probability which is the probability of the class L given the feature vector X, P(L) is the probability of class L and P(x) is the probability of observing the feature vector across all classes.

$$P(x|L) = \frac{P(L)}{P(L|x) \times P(x)} \tag{5}$$

5.6 Performance Metrics

TP is true positive, TN is true negative, FP is false positive and FN is false negative.

- **Accuracy** measures the total number of correctly classified instances over the total number of instances.

$$\text{Accuracy} = \frac{TP + TN}{TP + TN + FP + FN} \tag{6}$$

- **Precision** measures the proportion of true positive instances that are correctly identified by the model out of all instances predicted as positive by the model.

$$\text{Precision} = \frac{TP}{TP + FP} \tag{7}$$

- **Recall** measures the proportion of true positive instances that are correctly identified by the model out of all actual positive.

$$\text{Recall} = \frac{TP}{TP + FN} \tag{8}$$

- **F1-score** calculates the harmonic mean of precision and recall. The value reflects the balance between correctly identifying positive instances and minimizing false positive predictions

$$F1 = 2 \times \frac{\text{Precision} \times \text{Recall}}{\text{Precision} + \text{Recall}} \tag{9}$$

6 Experiment

We performed 4 separate experiments one experiment for each feature selection method. In each experiment, the features were ranked based on the score calculated by the FS algorithm or importance score. Each feature selection method ranks the features differently, hence the sets of features used to train and test the ML models are different for each experiment. Then the 5 classifiers DT, RF, XGB, LR, and NB were trained and tested using the top k ranked features iteratively, where $k \in \{5, 10, 15, 20, 25, 30, 35, 40, 45, 50, 55, 60, 65\}$. This is a total of 13 iterations, where in each iteration the number of features increments by 5. The ML classifiers were implemented with their default parameter values. In our analysis, we refer to the number of features per iteration as N. The accuracy, precision, recall, F1 score, training time and testing time were calculated in each iteration for each classifier.

RFE experiment iterative approach is different from chi-squared, ANOVA and extra-tree methods. RFE selects the top K features recursively. RFE relies on the feature importance characteristic of the classifier to rank features. Then the least important features are eliminated and the classifier's accuracy is calculated. This process is repeated until the k best features that achieve the highest accuracy remain in the dataset. NB can not be used with RFE feature selection without using another method to rank features so for fair comparison NB classifier was not included in this experiment.

7 Results and Analysis

Table 1 shows the accuracy for each iteration per classifier. Analyzing the table results shows that DT achieves 97.1% using 35 features, RF achieves 96.6% using 25 features, XGB achieves 96.6% using 35 features, LR achieves 83.8% using 45 features and NB achieves 75.3% using 50 features. Increasing the number of features from 35 to 45, results in a slight increase of 0.8% in the accuracy for the DT classifier. Increasing the number of features from 35 to 60 increases the accuracy slightly by 0.2%. RF utilizes the least number of features to obtain the maximum accuracy. DT achieves the highest accuracy among all classifiers. The accuracy, precision, recall and F1-score calculated in this experiment are equal to each other in all iterations.

Table 1. Accuracy of classifiers using Chi-squared FS

N	5	10	15	20	25	30	35	40	45	50	55	60	65
DT	0.729	0.896	0.948	0.951	0.965	0.965	0.971	0.972	0.979	0.979	0.979	0.979	0.979
RF	0.729	0.898	0.951	0.954	0.966	0.966	0.962	0.965	0.967	0.965	0.964	0.963	0.966
XGB	0.721	0.889	0.933	0.938	0.957	0.958	0.968	0.968	0.969	0.9694	0.970	0.970	0.971
LR	0.456	0.628	0.726	0.779	0.818	0.824	0.827	0.827	0.838	0.843	0.843	0.843	0.843
NB	0.297	0.373	0.455	0.528	0.577	0.592	0.662	0.655	0.663	0.753	0.737	0.733	0.719

Comparing the training time required by each classifier per iteration displayed in Table 2, shows that NB took the least training time which indicates the model's simplicity. However, this advantage does not leverage the low performance of the NB classifier. DT using 35 features requires 91% less time than RF using 25 features. XGB using 35 features also requires 69% less time than RF using 25 features. From these results, we can infer that DT is computationally less complex than XGB and that XGB is computationally less complex than RF. The most computationally complex classifier used is LR.

Table 2. Training time of classifiers using Chi-squared feature selection

N	5	10	15	20	25	30	35	40	45	50	55	60	65
DT	1	7	10	13	16	20	26	30	31	36	38	38	39
RF	85	215	223	265	300	301	295	326	321	365	345	333	357
XGB	40	48	54	59	68	78	92	100	110	123	128	132	141
LR	33	69	97	147	175	208	250	272	306	342	396	408	983
NB	0.23	0.32	0.41	0.51	0.68	0.79	0.78	0.94	0.97	1.01	1.15	1.29	1.33

Table 3 displays the time required for testing by each classifier. Comparing the testing times of each classifier shows that LR using 50 features requires 39.2% less time than DT using 35 features. DT using 35 features requires 86.5% less time than RF using 25 features. RF using 25 features requires 52.3% less time than NB using 50 features and NB requires 33.4% less time than XGB. Low testing time indicates fast attack detection.

Table 3. Testing time of classifiers using Chi-squared feature selection

N	5	10	15	20	25	30	35	40	45	50	55	60	65
DT	0.083	0.106	0.110	0.111	0.115	0.125	0.130	0.135	0.139	0.149	0.154	0.163	0.163
RF	0.729	0.898	0.951	0.954	0.966	0.966	0.962	0.965	0.966	0.967	0.965	0.964	0.963
XGB	3.031	3.407	3.678	3.538	3.562	3.556	3.328	3.596	3.537	3.560	3.598	3.598	3.621
LR	0.053	0.051	0.056	0.059	0.066	0.063	0.070	0.078	0.079	0.079	0.085	0.112	0.097
NB	0.300	0.523	0.712	0.929	1.225	1.533	1.611	1.818	2.009	2.215	2.410	2.670	2.905

Table 4 shows the accuracy of the classifiers per iteration. In this experiment, the accuracy, precision, recall and F1-scores are equal to each other in each iteration. The table results show that DT achieves the highest accuracy of 97.2% using 40 features, RF achieves 96.1% using 25 features, XGB achieves 96.8% using 40 features, LR achieves 83.9% using 45 features and NB achieves 74.9% using 45 features. Increasing the number of features from 40 to 50 results in a slight increase of 0.6% in the accuracy of DT. Increasing the number of features from 25 to 45 results in a slight increase of 0.7% in accuracy. XGB and LR also experience a slight increase in accuracy when trained with more features.

Table 4. Accuracies of the classifiers using ANOVA feature selection

N	5	10	15	20	25	30	35	40	45	50	55	60	65
DT	0.804	0.950	0.955	0.955	0.956	0.959	0.959	0.972	0.978	0.978	0.979	0.979	0.979
RF	0.805	0.954	0.957	0.958	0.961	0.961	0.962	0.965	0.966	0.968	0.963	0.963	0.958
XGB	0.799	0.933	0.941	0.941	0.943	0.943	0.944	0.968	0.970	0.970	0.969	0.970	0.971
LR	0.391	0.652	0.678	0.677	0.812	0.819	0.824	0.827	0.839	0.843	0.843	0.843	0.843
NB	0.362	0.577	0.567	0.488	0.539	0.592	0.649	0.655	0.749	0.746	0.741	0.728	0.719

Table 5 required by each classifier per iteration. The table shows similar patterns to that of chi-squared. NB took the least training time in all iterations however, the classifier's performance does not leverage this advantage. DT using 40 features took 88% less time than RF using 25 features. XGB using 35 features also took 73% less time than RF using 25 features. This indicates that DT is the least complex, followed by XGB, followed by RF then LR.

Table 5. Training time of classifiers using ANOVA feature selection

N	5	10	15	20	25	30	35	40	45	50	55	60	65
DT	1	3	5	6	8	10	12	14	15	17	17	17	18
RF	47	94	110	125	145	138	141	151	141	165	157	149	157
XGB	16	17	21	24	26	30	31	34	36	36	41	42	45
LR	13	23	34	43	51	60	79	90	100	112	123	129	314
NB	0.2	0.3	0.4	0.5	0.7	0.8	0.9	1.1	1.2	1.3	1.4	1.5	1.7

DT, RF and XGB are the top 3 performing classifiers comparing their testing times displayed in Table 6 shows the following: DT using 40 features took 93% less time than RF using 25 features, XGB using 45 features took 9.5% less time than RF using 25 features. It can be inferred from the results that DT has the fastest detection rate. Even though, LR has the highest detection rate among all classifiers its low performance does not leverage this advantage.

Table 6. Testing time of classifiers using ANOVA feature selection

N	5	10	15	20	25	30	35	40	45	50	55	60	65
DT	0.034	0.049	0.053	0.062	0.054	0.068	0.071	0.066	0.071	0.074	0.080	0.084	0.085
RF	0.805	0.954	0.957	0.961	0.961	0.9262	0.965	0.966	0.968	0.962	0.963	0.963	0.9590
XGB	0.627	0.688	0.833	0.840	0.861	0.858	0.897	0.870	0.856	0.858	0.865	0.880	0.895
LR	0.024	0.027	0.027	0.031	0.036	0.032	0.039	0.040	0.041	0.044	0.045	0.047	0.043
NB	0.180	0.316	0.423	0.532	0.686	0.833	0.924	1.091	1.212	1.304	1.442	1.513	1.653

DT attains the highest accuracy of 97.2% using 40 features, which is more features than RF uses to reach 96.1%. Not only does DT achieve 0.9% higher accuracy, but it also provides 93% faster detection than RF.

Table 7 displays the accuracy calculated for each classifier per iteration. The accuracy, precision, recall and F1 scores are equal to each other for every iteration. The results show that DT achieves an accuracy of 97.1% using 20 features, RF achieves 96.1% using 20 features, XGB achieves 96.7% using also 20 features, LR achieves 83.4% using 40 features and NB achieves 71.7% using 40 features. Increasing the number of features from 20 to 50 for DT and XGB results in a slight increase in the accuracy of both classifiers by 0.8% and 0.4% respectively.

Table 7. Accuracies of classifiers using Extra-tree feature selection

N	5	10	15	20	25	30	35	40	45	50	55	60	65
DT	0.892	0.950	0.952	0.971	0.971	0.971	0.973	0.973	0.977	0.979	0.979	0.979	0.979
RF	0.895	0.953	0.954	0.969	0.968	0.966	0.963	0.964	0.962	0.968	0.962	0.961	0.960
XGB	0.918	0.951	0.953	0.967	0.968	0.967	0.968	0.969	0.970	0.970	0.971	0.970	0.971
LR	0.560	0.743	0.785	0.798	0.806	0.823	0.829	0.834	0.835	0.841	0.841	0.843	0.843
NB	0.435	0.496	0.544	0.572	0.605	0.650	0.704	0.717	0.716	0.686	0.733	0.728	0.719

Analyzing training time in Table 8 also shows similar patterns to the previous two experiments. Training time for NB is significantly lower than other classifiers however, the low accuracy of the classifier does not leverage the high speed of the training process. Even though DT, RF and XGB achieve their maximum performance using only 20 features, the training time required by DT is 93% less than RF and XGB is 80% less than RF.

Table 9 which displays the testing times calculated during the experiments shows also similar patterns to the previous experiments. DT detects attacks 98.7% faster than RF and XGB detects attacks 81% faster than RF.

Table 8. Training time of classifiers using Extra-tree feature selection

N	5	10	15	20	25	30	35	40	45	50	55	60	65
DT	3	5	7	8	9	10	11	12	14	15	16	18	18
RF	120	123	125	131	133	131	144	141	147	151	157	161	163
XGB	17	19	22	25	29	28	32	34	36	39	40	42	43
LR	14	25	38	46	65	78	90	101	160	248	284	295	313
NB	0.3	0.4	0.8	0.6	0.8	0.9	1.	1.1	1.3	1.4	1.5	1.7	1.8

Table 9. Testing time of classifiers using Extra-tree feature selection

N	5	10	15	20	25	30	35	40	45	50	55	60	65
DT	0.054	0.054	0.054	0.058	0.054	0.061	0.061	0.066	0.073	0.074	0.077	0.077	0.077
RF	5.3	4.8	4.7	4.5	4.4	4.2	4.3	4.1	4.3	4.4	4.4	4.5	4.5
XGB	0.724	0.779	0.858	0.851	0.863	0.843	0.851	0.856	0.871	0.872	0.875	0.892	0.884
LR	0.034	0.026	0.034	0.036	0.037	0.037	0.041	0.042	0.050	0.045	0.055	0.056	0.045
NB	0.141	0.231	0.257	0.290	0.354	0.401	0.434	0.461	0.508	0.540	0.618	0.624	0.684

Evaluating the accuracies achieved by classifiers using RFE features selection displayed in Table 10 shows that DT achieves 97.9% using 20 features. RF achieves 96.1% accuracy using 15 features, XGB achieves 97% using 45 features and LR achieves 83.2% using 40 feature. The values of accuracy, precision, recall and f1- score are equal to each other in this experiment.

Table 10. Accuracies of classifiers using RFE feature selection

N	5	10	15	20	25	30	35	40	45	50	55	60	65
DT	0.929	0.959	0.967	0.979	0.979	0.979	0.979	0.979	0.979	0.979	0.979	0.979	0.979
RF	0.953	0.958	0.961	0.961	0.967	0.966	0.961	0.965	0.958	0.962	0.961	0.960	0.960
XGB	0.848	0.858	0.905	0.909	0.926	0.942	0.945	0.947	0.970	0.970	0.971	0.971	0.971
LR	0.548	0.732	0.740	0.770	0.783	0.788	0.801	0.832	0.839	0.842	0.843	0.843	0.843

Analyzing the training time displayed in Table 11 shows that DT requires the least training time followed by XGB then RF and lastly LR. DT using 20 features took 93% less time than RF using 15 features and XGB using 20 features took the 80% less time than RF. Lower training time indicates the low computational complexity of the classifier.

Table 12 shows the time needed for testing the classifiers. A shorter testing time indicates faster attack detection. Evaluating the table results shows that LR took the least testing time however, the poor performance of the classifier does not leverage this advantage. DT using 20 features took 98.8% less time than RF using 15 features to achieve its maximum accuracy, while XGB using 20 features took 81.9% less time than RF using 15 features.

Evaluating the effect of each feature selection algorithm shows that DT achieves the highest accuracy with all feature selection techniques of 97.1% with 35 features using chi-squared, 97.2% with 40 features using ANOVA, and 97.1%

Table 11. Training time of classifiers using RFE feature selection

N	5	10	15	20	25	30	35	40	45	50	55	60	65
DT	2	4	5	8	10	12	14	15	15	16	17	18	18
RF	90	102	115	11	153	146	135	157	149	163	156	151	164
XGB	17	19	21	23	25	26	29	32	35	38	41	43	43
LR	7	16	29	38	49	60	74	84	94	109	119	132	324

Table 12. Testing time of classifiers using RFE feature selection

N	5	10	15	20	25	30	35	40	45	50	55	60	65
DT	0.041	0.017	0.049	0.056	0.059	0.066	0.066	0.069	0.072	0.074	0.078	0.081	0.087
RF	4.60	4.58	4.77	4.61	4.33	4.50	4.11	4.11	4.14	4.22	4.33	4.12	4.37
XGB	0.826	0.860	0.856	0.860	0.860	0.860	0.860	0.860	0.871	0.875	0.859	0.890	0.890
LR	0.0628	0.0312	0.0312	0.040	0.047	0.030	0.047	0.031	0.031	0.047	0.049	0.040	0.044

with 20 features using extra-tree and 97.9% with RFE using 20 features. Considering the complexity and the speed of detection, DT took the least training and testing times using features selected by RFE feature selection.

RF achieved 96.6% accuracy with 25 features using chi-squared, 96.1% with also 25 features using ANOVA, 96.9% using 20 features with extra-tree and 96.1% with 15 features using RFE. While RF does not achieve the highest accuracy, it achieves reasonably high accuracies using the least features compared to other classifiers. RF lowest testing time was achieved when using features ranked according to scores calculated by extra-tree feature selection algorithm.

XGB achieved 96.8% with 35 features using chi-squared feature selection, also 96.8% with 40 features using ANOVA, 96.7% with 20 features using extra tree and 97 % using RFE with 45 features. Using features ranked with extra-tree feature selection results in the least testing time and a low training time.

From the table results analyzed previously, we can infer that LR and NB are not suitable classifiers to be used with this type of data. LR can be a very fast classifier when it comes to classifying data points and NB is a very fast classifier to train. However, they are under-performing classifiers compared to the accuracies achieved by DT, RF and XGB.

RFE and extra-tree reduce the testing time of classifiers the most compared to chi-squared and ANOVA. This finding reinforces the advantages of wrapper-based and embedded-based FS methods over filter-based. RFE selects the smallest possible subset of features compared to RFE, chi-squared and ANOVA. The best combination of FS and classifier is RFE and DT which provides a small feature subset, high accuracy, low model complexity and high detection speed.

8 Conclusion

The main objective of employing dimensionality reduction techniques is to reduce the feature space of the dataset. This reduction reduces the complexity of the classification models and reduces the detection speed. In this study, we adopted an iterative approach to evaluate the effect using different feature selection algorithms to in model complexity, detection speed and accuracy. We aim to find a combination of an FS and a classifier, where the FS technique is used to find the smallest subset of important features to train the classifier and get the best performance. We also aim to find the FS algorithm that enhances the detection speed of the classifier. After analysing our results we concluded that both extra tree and RFE have similar effects on reducing testing time. RFE can select the smallest subset of features to train the classifiers and still achieve high accuracy. The best combination is using RFE with DT to select 20 features only and achieve 97.9% accuracy. The paper provides a thorough analysis of the effect of FS techniques however, several limitations need to be acknowledged. While CICIDS2017 is a benchmark dataset it only covers certain network scenarios. This means it may not be the best representative of real-network traffic. That's why the findings need to be tested in a real network to validate the findings. For future work, we aim to deploy the combination of RFE and DT in a network model to test the applicability of our findings. Also, the analysis will be conducted on newer datasets such as CICDDoS2019 and compared with the current findings to improve generalizability of our results.

References

1. Ahmad, Z., Shahid Khan, A., Wai Shiang, C., Abdullah, J., Ahmad, F.: Network intrusion detection system: a systematic study of machine learning and deep learning approaches. Trans. Emerg. Telecommun. Technol. **32**(1), e4150 (2021)
2. Alkasassbeh, M., Al-Haj Baddar, S.: Intrusion detection systems: a state-of-the-art taxonomy and survey. Arab. J. Sci. Eng. **48**(8), 10021–10064 (2023)
3. Dhaliwal, S.S., Nahid, A.A., Abbas, R.: Effective intrusion detection system using XGBoost. Information **9**(7), 149 (2018)
4. Duarte, F.: Amount of data created daily (2024). https://explodingtopics.com/blog/data-generated-per-day. Aaccessed 13 Dec 2023
5. Gaur, V., Kumar, R.: Analysis of machine learning classifiers for early detection of DDoS attacks on IoT devices. Arab. J. Sci. Eng. **47**(2), 1353–1374 (2022)
6. Gbenga, F., Adetunmbi, A., Elohor, O.: Towards optimization of malware detection using extra-tree and random forest feature selections on ensemble classifiers. Int. J. Recent Technol. Eng. **9**, 223–232 (2021)
7. Ibrahim, Z.K., Thanon, M.Y.: Performance comparison of intrusion detection system using three different machine learning algorithms. In: 2021 6th International Conference on Inventive Computation Technologies (ICICT), pp. 1116–1124. IEEE (2021)
8. Patgiri, R., Varshney, U., Akutota, T., Kunde, R.: An investigation on intrusion detection system using machine learning. In: 2018 IEEE Symposium Series on Computational Intelligence (SSCI), pp. 1684–1691 (2018). https://doi.org/10.1109/SSCI.2018.8628676
9. Sharafaldin, I., Lashkari, A.H., Ghorbani, A.A., et al.: Toward generating a new intrusion detection dataset and intrusion traffic characterization. ICISSp **1**, 108–116 (2018)

10. Stiawan, D., Idris, M.Y.B., Bamhdi, A.M., Budiarto, R., et al.: CICIDS-2017 dataset feature analysis with information gain for anomaly detection. IEEE Access **8**, 132911–132921 (2020)
11. Tang, J., Alelyani, S., Liu, H.: Feature selection for classification: a review. Data Classif.: Algorithms Appl. **37** (2014)
12. Thakkar, A., Lohiya, R.: Attack classification using feature selection techniques: a comparative study. J. Ambient Intell. Humaniz. Comput. **12**(1), 1249–1266 (2021)
13. Thampi, M., Bose: Applied Soft Computing and Communication Networks. Springer (2021)
14. Yang, Z., et al.: A systematic literature review of methods and datasets for anomaly-based network intrusion detection. Comput. Secur. **116**, 102675 (2022)
15. Zebari, R., Abdulazeez, A., Zeebaree, D., Zebari, D., Saeed, J.: A comprehensive review of dimensionality reduction techniques for feature selection and feature extraction. J. Appl. Sci. Technol. Trends **1**(1), 56–70 (2020)

GuardedLearn: Safeguarding Federated Learning with Robust Defenses and Privacy Preserving Mechanisms

Malak Fathi(✉), Milad Ghantous, and Minar El Aasser

German University in Cairo, New Cairo, Egypt
malak.fathi@student.guc.edu.eg, {milad.michel,minar.elaasser}@guc.edu.eg

Abstract. Federated learning (FL), a distributed machine learning paradigm, has attained significant attention in recent years. FL addresses privacy concerns by enabling multiple parties to collaboratively train a model through gradient exchange instead of sharing raw data. However, inherent privacy risks persist in FL, such as the potential inference of user training samples from gradients. Additionally, the emergence of poisoning attacks poses a significant security threat to FL, where malicious users may tamper their local data or submit crafted gradients to compromise model integrity. The integration of privacy-preserving federated learning solutions with defenses against data and model poisoning attacks will significantly improve the overall security and privacy protection in FL. To address this dilemma of simultaneously mitigating data and model poisoning while ensuring the privacy of transmitted weights, this paper develops and introduces GuardedLearn (GL) framework. GL is a novel privacy-enhanced FL framework engineered to safeguard against poisoning attacks. It employs techniques, including local masking and gradient analysis, to empower the server in penalizing malicious behavior. Through meticulous feature extraction, resilient aggregation strategies, and streamlined clustering algorithms, GL sets a new standard in mitigating both data and model poisoning in FL settings. We conducted extensive experiments and comprehensive analyses, which clearly demonstrate the efficiency of GL and its superiority over existing models.

Keywords: Federated Learning · Privacy preservation · Poisoning attacks

1 Introduction

The proliferation of internet-connected mobile phones and vehicles has led to the collection of vast amounts of user activity data. Organizations increasingly leverage this data through machine learning (ML) applications to enhance services and develop intelligent mobile apps like personalized recommendations and mobile healthcare solutions. However, centralizing this data for analysis raises significant privacy concerns and high communication network costs.

A. Abdelgawad et al. (Eds.): ISBCom 2024, LNNS 1268, pp. 439–453, 2025.
https://doi.org/10.1007/978-3-031-82377-0_36

In 2016, Google introduced Federated Learning (FL) [13], a decentralized machine learning approach that trains algorithms collaboratively across multiple devices without requiring them to share raw data. Instead, only model output gradients are exchanged between clients and central server. This approach upholds data privacy, reduces communication costs, and enhances model training efficiency in decentralized environments. Despite its transformative potential, recent studies highlight challenges in privacy and security within FL frameworks.

Bhowmick et al. [2] confirmed that FL systems are vulnerable to inference attacks, where private data can be deduced from transferred weights. Solutions like differential privacy [6], homomorphic encryption [17], and secure multi-party computation (SMC) [21] address privacy issues but often increase communication, computational overhead or reduce overall accuracy.

Security is also a concern, as verifying the trustworthiness of all devices involved in training is challenging; some may submit malicious data, leading to incorrect model updates. Among security threats, poisoning attacks, including data poisoning [20] and model poisoning [10], pose significant threats aiming to disrupt model performance.

In response to these challenges, researchers have proposed some defense strategies, such as robust FL methods [3], but these struggle to defend against both data and model poisoning simultaneously, especially with encrypted gradients.

Our study introduces GuardedLearn (GL), a federated learning framework designed to tackle privacy breaches and malicious attacks. GL enhances privacy preservation and combats both data and model poisoning threats through advanced aggregation techniques and clustering methods. Additionally, it employs removable weights' masks to ensure privacy without compromising communication efficiency. Our main contributions are:

- We develop a novel federated learning framework featuring an advanced aggregation mechanism specifically designed to maintain model integrity and performance.
- We incorporate a sophisticated clustering technique to bolster our defenses against both data and model poisoning attacks, enhancing the robustness of the system even in adversarial scenarios with up to 70% malicious clients.
- Our method includes the use of detachable masks on clients' gradients. These masks not only preserve privacy but also facilitate the extraction of core features.
- Through extensive performance analysis, we demonstrate that our framework significantly outperforms existing methods in resilience against various poisoning attacks. This analysis underscores the superior effectiveness of our approach across diverse adversarial conditions.

The subsequent sections of this paper follow a structured outline. Section 2 delves into the background and related works, offering insights into the existing literature. Section 3 outlines the privacy and security threat models.Section 4 unveils our novel approach, GuardedLearn (GL), providing a detailed explanation of its architecture. Subsequently, Sect. 5 discusses the experiments and assessment of our framework. Finally, Sect. 6 summarizes the key findings and contributions of the paper.

2 Background and Related Work

2.1 Federated Learning

In federated learning (FL), a central server coordinates with multiple clients. Each client j has a local dataset. At each round t, a fraction m_t of clients are selected to participate. The server distributes the initial global model W_0 to the selected clients, who then train their local models based on W_0 and send back the updates W_m^j. The server commonly aggregates these updates using FedAvg, which averages the model updates weighted by the size of the clients' local datasets [16].

2.2 Poisoning Attacks and Their Defense Strategies

Federated learning is prone to numerous poisoning attacks [1,10] that can be classified to:

Data Poisoning Attacks. Data poisoning [1,19] is one of the types of adversarial attacks that comprise the corruption of the training data, so that the outcome of the ML models satisfy the attacker's target. Label Flipping attack [19] is a common data poisoning attack which modifies real labels of the training data and replaces them with incorrect ones. For instance, An attacker labels X-ray images of healthy lungs as having pneumonia, causing the model to misdiagnose healthy patients. Backdoor attacks [1] stands as another prevalent method of data poisoning. This strategy entails embedding a detrimental pattern or trigger within the training data. This hidden element can be leveraged to induce erroneous predictions from the model when presented with certain inputs containing the trigger. For example, an attacker embeds a specific accessory (e.g., a particular pair of glasses) in images of unauthorized users, causing a facial recognition system to recognize them as authorized, causing unauthorized access.

Model Poisoning Attacks. In model poisoning attacks [10], the adversary gets the chance to directly amend the gradients on the affected devices before forwarding them to the server. Gaussian attack [14], for instance, involves introducing Gaussian noise, which entails adding random noise to the updates before they are uploaded. Because of the direct alteration of gradients, model poisoning attacks exert greater impact on a FL environment[4].

Defense Strategies Against Poisoning Attacks. To defend against poisoning attacks in FL, various similarity-based methods have been proposed. These methods aim to assess the reliability of client submissions before aggregating the global model. One approach involves spectral anomaly detection [15], which distinguishes between benign and malicious updates. Cosine similarity [4]is utilized to assess the similarity between updates and those obtained from a clean dataset. However, these methods lack robustness, especially with large numbers of malicious clients. Regarding aggregation methods for the global model, conventional FedAvg, explained above in Sect. 2.1, is sensitive to malicious submissions. Novel statistical aggregation algorithms focus on adjusting weights based on trust. These methods include TMean, FedMedian abd RFA. TMean [22] selects gradients independently, removes extreme values, and computes the mean. FedMedian [22] selects the median value among received weights. Furthermore, [18] introduces RFA, robust federated aggregation algorithm that relies on the geometric median to safeguard the system against clients' compromised weights.

2.3 Privacy Preservation Coupled with Poisoning Attacks Defenses

Several studies focus on detecting malicious activities within encrypted data in the context of FL. Liu et al. [15] introduced a Privacy-Preserved FL framework (PEFL), utilizing Homomorphic Encryption (HE) to spot abnormal data within ciphertext. PEFL works by assessing the pearson correlation coefficient between gradients. [11] proposed, DPFLA, which enhances privacy preservation and defends poisoning attacks in FL. DPFLA performs a local masking technique, kmeans clustering and aggregates using FedAvg. However, it lacks defenses against model poisoning and is only effective in mitigating data poisoning when the proportion of malicious clients does not surpass 40% of the total clients.

3 Problem Formulation

In federated learning, multiple clients collaboratively train a global model on a server. Each client uses its local dataset, and the global model is derived by minimizing the combined loss function over clients' datasets as shown in 1:

$$w^* = \arg\min_{w} \sum_{j=1}^{n} f(D_j, w) \tag{1}$$

where $f(D_j, w)$ represents the loss for client j's local data. The process starts with an initial global model distributed to clients at each iteration. Clients compute and send back gradients of their local loss functions to the server. For simplicity, we assume clients perform one step of gradient descent.

3.1 Threat Model

In this section, we undertake a comprehensive examination of potential threats within our FL environment. **Privacy Threats:** Inference attacks pose significant risks to client data privacy during transmission. In FL, even though raw data isn't shared, attackers can still infer sensitive information from the communicated gradients. For example, if a health application uses FL, an attacker could deduce private health conditions of users by analyzing the gradients exchanged between them and the server.

Security Threats. We consider a scenario with m malicious clients among genuine participants, with the server being honest but curious attempting to deduce client data from the received updates. Benign clients compute and send authentic updates, but we simulate data and model poisoning attacks, including label flipping, backdoor, and Gaussian attacks. We also introduce combined strategies: ⟨**Label-flip**⟩ +⟨**Gaussian**⟩ and ⟨**Backdoor**⟩ + ⟨**Gaussian**⟩.

To protect FL protocols from data and model poisoning attacks while ensuring client privacy, we present GuardedLearn (GL). This section details the architecture, structure, and workflow of our framework.

4 GuardedLearn

4.1 Overview

We introduce a novel FL system designed to enhance privacy and security through a series of orchestrated stages. Our system includes the components of the typical FL setup:

1. **Central Server (CS):** Responsible for receiving updates and iteratively aggregating them to collaboratively update the global model.
2. **Clients:** A collection of devices that transmit their outputs from local training to CS.

Additionally, a **Reliable Entity** (RE) is utilised to generate two random seeds and distribute one seed for the clients and another for CS. These seeds play a pivotal role in locally masking updates, as detailed below in Sect. 4.5.

Here's the general process proposed in our system:

- **Stage 1:** RE initializes and distributes seeds for orthogonal matrices M and N to CS and clients, respectively.
- **Stage 2:** CS partitions matrix M into k submatrices, where k represents the number of clients in the system. Each client receives a submatrix, M_j, along with the initial global model.
- **Stage 3:** Clients conduct local training using their individual datasets and the received global model, producing output weights.
- **Stage 4:** Following training, clients apply a masking technique (detailed in Sect. 4.4) using matrices M_j and N to the generated weights.Securely, clients transmit these masked weights to CS.
- **Stage 5/6:** Subsequently, CS concatenates the received gradients and then employs the Singular Value Decomposition (SVD) technique (explained in Sect. 4.6) for feature extraction of the aggregated masked weights.
- **Stage 7:** CS removes the mask from the SVD output using matrix M_j, yielding the most dominant features of the weights unmasked.
- **Stage 8:** Next, the server applies DBSCAN clustering to the de-masked outputs to identify malicious clients.

- **Stage 9:** CS utilizes RFA technique to aggregate benign indices and update the global model accordingly.
- **Stage 10:** Finally, the updated global model is transmitted back to clients, initiating a new iteration. This iterative process continues for a fixed number of global rounds to ensure convergence.

Figure 1 depicts an illustrative figure of the system infrastructure.

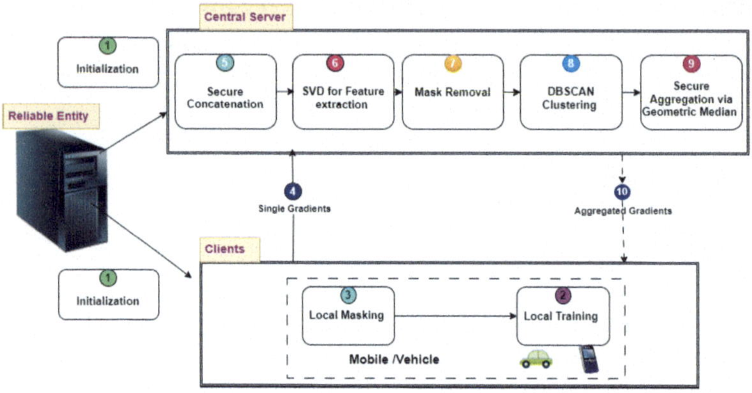

Fig. 1. Total System Architecture

4.2 Reliable Entity (RE)

To ensure secure communication and data masking of updates, the RE generates two random integer seeds: s_m and s_n. s_m is sent to CS, and s_n is sent to clients. Using s_m, CS constructs a random matrix M of dimensions $k \times k$ employing the Gram-Schmidt scheme [8]. Simultaneously, each client u_j utilizes s_n to generate S unique masking matrices $N_1, N_2, ..., N_r, ..., N_S$, where S is the number of layers of local model. Each N_r matrix has size $v \times v$, where v represents the number of neurons at each layer. Additionally, CS segments M into k components $(M_1, M_2, ..., M_j, ..., M_k)$ and transmits M_j, a submatrix of size $k \times 1$, to client u_j. Transmitting only the seeds, rather than the matrices, ensures communication overhead is not compromised.

4.3 Local Training

In our study, we utilized a simple CNN as the local model for client nodes. In each global round of training, every client (u_j) leverages their local dataset and received model to iteratively train a new gradient. The process unfolds in a sequence: firstly, u_j accesses the weight vectors, here the term "weight vector" encompasses the weights of all neurons in all hidden and output layers. These weight vectors of u_j are denoted as $X_j^{(i)}$, with 'i' spanning from 1 to c, where c is the number of neurons per layer.

4.4 Weights Masking

Next, u_j applies masking to these weight vectors using matrices M_j and N_r. The masking operation can be computed using the provided formula: $X_j'^{(i)} = M_j X_j^{(i)} N_r$, where $X_j'^{(i)}$ represents the resulting masked weight vector for the i-th neuron in layer r. Finally, u_j transmits the masked weight vectors of every layer r, $X_j' = (X_j'^{(1)}, X_j'^{(2)}, ..., X_k'^{(C)})$ to CS.

4.5 Secure Concatenation

The Central Server (CS) gathers all masked data of the same neuron from clients and combines it to form $Y'^{(i)}$. Denote the clients' raw data array as

$$Y^{(i)} = [X_1^{(i)}, \ldots, X_j^{(i)}, \ldots, X_k^{(i)}]^T.$$

Thus, the connection between Y_j and Y_j' is expressed as

$$\begin{aligned} Y'^{(i)} &= M_1 X_1^{(i)} N_r + \cdots + M_j X_j^{(i)} N_r + \cdots + M_k X_k^{(i)} N_r \\ &= [M_1, \ldots, M_j, \ldots, M_k] Y_j N_r, \quad j = 1, 2, \ldots, c \end{aligned}$$

4.6 Feature Extraction

Singular Value Decomposition (SVD) is a powerful method for breaking down matrices into three components, allowing for essential insights while reducing data dimensions. It works seamlessly with principal component analysis (PCA), a common tool for dimensionality reduction. In PCA, SVD is applied to the covariance matrix after centering the data, yielding principal components aligned with high variance directions. In our scenario, where matrix $Y_{c \times v}$ contains vectors $x_1, ..., x_i, ..., x_v$, the goal is to distill primary features while minimizing dimensionality. SVD on Y' yields $U \sum V^T$, enabling depiction of $\hat{Y}_{c \times v}$. This representation effectively captures fundamental features from the dataset, validated for integration into the server aggregation process.

$$Y_j'^{(i)} = U_j'^{(i)} \Sigma_j'^{(i)} V_j'^{T(i)} \tag{2}$$

Moreover, by applying SVD to Y_i, it can be decomposed into $U_i \Sigma_i V_i^T$. Based on this analysis, we can deduce the following relationship :

$$U_j'^{(i)} \Sigma_j'^{(i)} V_j'^{T(i)} = M_j U_j^{(i)} \Sigma_j^{(i)} V_j^{T(i)} N_r \tag{3}$$

4.7 Mask Removal

From the above equations and the detailed analysis in [5,11], Y_i has the same singular matrix with Y_i', i.e., $\Sigma_i = \Sigma_i'$. Therefore, we have $U_i' = M_i U_i$ and

$V_i'^T = V_i^T N_r$. $U_i = M_i^{-1} U_i'$. Because the CS holds the matrix M, U_i can be unmasked from U_i' using M_i by orthogonal transformations.

$$\hat{Y^{(i)}}_j = U_j^{(i)} \Sigma_j^{(i)} \tag{4}$$

Eventually, the CS obtains the low dimension of $X_j^{(i)}$ that contains primary feature information so can be used to represent each weight received by the client. The computational overhead imposed is fully explained in [5].

4.8 Clustering Using DBSCAN

This paper introduces a clustering algorithm designed to identify malicious clients or tampered local model output weights. The goal is to detect instances of data poisoning that compromise the integrity of output weights. Our method analyzes the distribution of features extracted from final-layer neuron weights, reducing computational overhead. We employ DBSCAN [9] (Density-Based Spatial Clustering of Applications with Noise), which defines clusters as dense regions of data points separated from regions of lower density.

DBSCAN operates based on two parameters: Eps (ϵ), the radius within which neighboring points are considered part of the same cluster, and MinPts, the minimum number of points required to form a dense cluster. It examines each data point to determine whether it is a core point (ϵ), a border point (within ϵ distance of a core point but with fewer than MinPts neighbors), or a noise point (neither core nor border).

DBSCAN effectively handles noise and outliers, identifying points that do not belong to any cluster as noise. This robustness is crucial when dealing with tampered weights or noisy data. DBSCAN's flexibility allows it to detect anomalies even when more than 50% of the clients are attackers. By fine-tuning Eps and MinPts, we achieve optimal clustering results, distinguishing between benign and malicious clients. This method overcomes a common assumption in previous research that the largest cluster always represents benign clients, which may not hold true in real-world scenarios.

4.9 Aggregation

Upon identifying the benign and malicious clients within our FL system, we meticulously filter out the malicious contributions, retaining only the benign clients. In our GL framework, the Central Server then undertakes the crucial task of aggregating the weight vectors of the neurons from all layers on benign clients only, employing a sophisticated technique known as RFA (Robust Federated Aggregation) [18]. RFA is mainly based on calculating the geometric median of the received clients' updates. The geometric median [7], also referred to as the Fermat-Weber point, stands as a pivotal method for discerning central tendencies within a dataset distributed across multiple dimensions. Unlike traditional averaging methods susceptible to outliers, the geometric median prioritizes minimizing the sum of distances from itself to all other points in the dataset. This attribute renders it particularly resilient in scenarios featuring malicious data points. In practical terms, the geometric median serves to identify the point that achieves the lowest total distance to all other data points in a multidimensional space as shown in Eq. 5.

$$W_{\text{aggregated}} = \arg\min_{w} \sum_{i=1}^{N} d(w, \hat{X^{(i)}}_j) \tag{5}$$

where $W_{\text{aggregated}}$ is the aggregated weight, arg min finds the argument that minimizes the function, w is the aggregated weight, and $d(w, \hat{X^{(i)}}_j)$ is the distance between the aggregated weight and the weight received of neuron i from client j. We aggregate all corresponding weights of neuron i at Layer r for all the clients. This process is repeated for all the neurons in all layers.

By focusing on minimizing distances rather than averaging values (as in FedAvg, it inherently guards against the excessive influence of outliers or malicious contributions. This quality makes it an indispensable tool, especially in FL settings where data integrity and security are paramount concerns.

5 Experiments

5.1 Experimental Setup

Within our experimental framework, we employed Python 3.12.2 and Pytorch 1.8.1 to orchestrate the training of a cnn via FL. We opt for a batch size of 64 during training, coupled with an SGD optimizer set at a learning rate of 0.01. Over the course of our investigations, we orchestrated 100 rounds of FL iterations. To align with previous research and ensure comparability, we simulated 20 clients.

Regarding our dataset selection, we selected **MNIST** dataset, which features 70,000 grayscale images showcasing handwritten digits spanning from 0 to 9. We conducted experiments on IID (Independent and Identically Distributed data) data from dataset. This dataset, renowned within the machine learning domain, is characterized by its allocation of 60,000 images for training and 10,000 for testing, with each image comprising a 28×28 pixel grid and pixel values ranging from 0 (white) to 255 (black).

In evaluating the resilience of our proposed methodology against adversarial intrusions, we simulated five distinct attack scenarios (explained in full details above in Sect. 3.1: label-flipping **(LF)**, backdoor **(BD)**, **Gaussian**, **LF combined with Gaussian**, and **BD combined with Gaussian**. LF attacks alter instances from a source class to a target class, while BD attacks tamper with samples using a 5×5 trigger pattern to induce misclassification. For the MNIST dataset, class 9 was the source, and class 6 was the target. We varied the ratio of malicious participants from 0% to 70% to evaluate resilience.

To evaluate the impact of attacks on our trained model, we employed several metrics on the test set. We measured **Test Accuracy (TACC)** to determine the percentage of correctly classified examples. We calculated the **Clustering F1-Score** to assess the classification of participants into benign and malicious categories. The **F1-score**, a harmonic mean of precision and recall, offers a balanced measure of model performance by combining both the proportion of true positive predictions out of all positive predictions (**precision**) and the proportion of true positives correctly identified (**recall**). Each experiment was conducted 10 times for each attack type, with the average results presented as the final outcome.

5.2 Experimental Results and Analysis

In this study, we conduct a comprehensive evaluation of our framework, **GuardedLearn**, against various attacks in FL by comparing it with other existing approaches.

Firstly, we analyze the performance of GL in terms of the clustering F1 score across different attack rates and types. In Fig. 2, we compare the performance of Kmeans clustering used by DPFLA [11] with DBscan clustering technique in GL. Remarkably, from Fig. 2, we observe that under all attack rates up to **70%** and across the 5 attack types, the clustering F1 score achieved using **DBSCAN** clustering remains consistently high at **1**. This robust performance is attributed to the inherent nature of DBSCAN, which excels in detecting clusters of varying shapes and sizes, thus effectively separating benign and malicious updates.

On the other hand, we investigate the behavior of the clustering F1 score using **Kmeans** [12] at **LF** and **BD** attacks. Notably, we find that the clustering technique performs well initially, with high scores even at moderate attack rates. However, as the attack rates exceed **40%**, the performance of Kmeans clustering deteriorates significantly as shown in Fig. 2. This decline is attributed to the increasing complexity and variability introduced by the higher proportion of malicious clients, making it challenging to accurately cluster the updates. In addition, from the analysis of the clustering F1 score results against attack rates of up to **70%** under the influence of model poisoning attacks-specifically **Gaussian** attacks, either in **isolation** or **combined** with **LF** or **BD**-we observe a consistently lower clustering F1 score compared to that observed under data poisoning. This observation underscores the heightened impact of model poisoning attacks relative to data poisoning attacks. It validates the notion that model poisoning attacks exert a more pronounced effect on the clustering performance, suggesting a greater threat posed by such attacks in FL environments. We primarily focused on model poisoning attacks due to their significant impact on test accuracy. We evaluated the test accuracy resulting from three aggregation techniques (FedAVG, Simple Median, and Geometric Median) combined with Kmeans clustering against Gaussian, Label flipping + Gaussian, and Backdoor + Gaussian attacks. This evaluation was repeated using DBSCAN clustering, as shown in Fig. 3. Figures 3a, 3b, and 3c indicate that Kmeans + Mean had the lowest accuracy, especially at attack rates above 30%. Switching to Simple Median improved accuracy but remained low above 50% attack rates. Geometric Median achieved the highest accuracy across all attack rates. Using DBSCAN

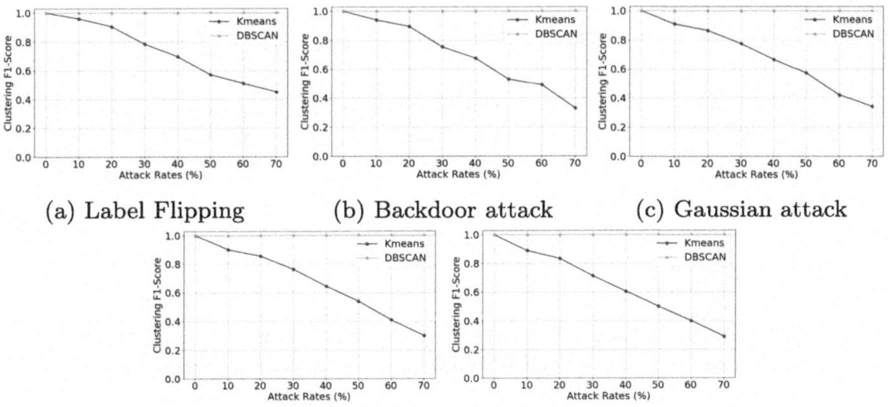

(a) Label Flipping (b) Backdoor attack (c) Gaussian attack

(d) LF + Gaussian (e) BD + Gaussian

Fig. 2. Clustering F1-Score Results on every attack type

clustering with the three aggregation techniques, as shown in Figs. 3d, 3e, and 3f, resulted in higher test accuracy under all attack types and rates. This improvement is due to DBSCAN's ability to identify clusters of arbitrary shapes and sizes. Additionally, test accuracy for BD + Gaussian attacks was lower than for LF + Gaussian, which was lower than for pure Gaussian attacks, highlighting the varying effectiveness of different poisoning attacks and the necessity of robust aggregation techniques in FL.

Furthermore, we have compared the performance of GL to existing models, DPFLA [11] and TMean [22]. As thoroughly explained previously in Sect. 2.2, DPFLA utilizes Kmeans for clustering the received updates at the server and FedAvg[16] for aggregation. TMean works on aggregating the updates by removing a known number of extreme values. In scenarios without malicious clients, both GL and DPFLA achieve a global test accuracy of 97.20% with a negligible deviation of 0.01%, while TMean results in 97.17% accuracy. FedAvg (baseline method) also achieves 97.20%, matching GL's performance. This demonstrates GL's effectiveness even without attackers, highlighting that GL does not require prior knowledge of the number of attackers in the system. In Table 1, we investigate the effect of LF and BD attacks at attack rates from 10 up to 40% on GL in comparison to DPFLA and TMean, where A represents the attack rate in the system. GL proves to achieve global test accuracy as high as DPFLA/TMean or higher in case of data poisoning with minor variations as the attack rate increases proving the robustness of GL.

In Table 2, as the attack rates reach or exceed half of the clients in the system, GL is the only technique that maintains a minor decrease in the test accuracy, remaining around 97%, in comparison to the others. In contrast, DPFLA and TMean experience a drop in accuracy by 13% as the attack rate increases from 10 to 70%. Consequently, GL demonstrates its robustness and effectiveness even under high attack rates, which makes it superior to the other techniques, even if

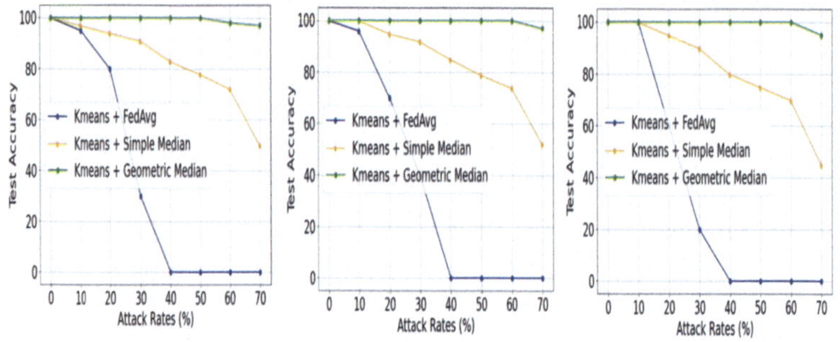

(a) Gaussian attacks with Kmeans (b) LF + Gaussian with Kmeans (c) BD + Gaussian with Kmeans

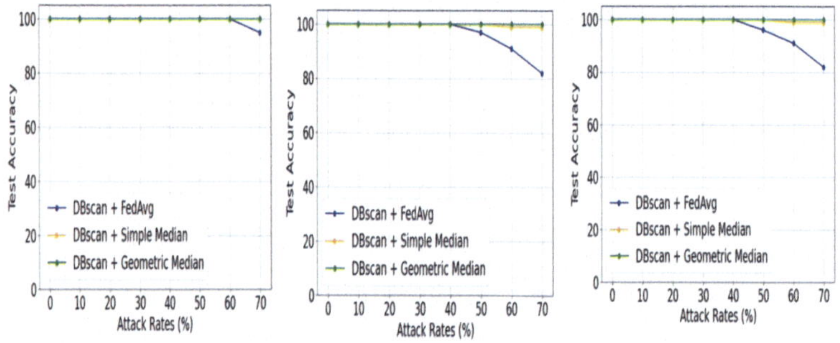

(d) Gaussian with DBSCAN (e) LF + Gaussian with DBSCAN (f) BD + Gaussian with DBSCAN

Fig. 3. Evaluating Aggregation techniques effect on Test Accuracy using Kmeans/DBSCAN clustering

Table 1. Test Accuracy Results of DPFLA Vs TrimmedMean Vs GL under Data poisoning attacks

Attack Types	A = 10%			A = 20%			A = 30%			A = 40%		
	DPFLA	TMean	GL	DPFLA	TMean	GL	DPFLA	TMean	GL	DPFLA	TMean	GL
LF	97.17	97.17	**97.18**	97.14	97.15	**97.15**	97.13	97.12	**97.13**	97.12	97.12	**97.12**
BD	97.13	97.12	**97.16**	97.12	97.10	**97.12**	97.10	97.10	**97.11**	97.08	97.08	**97.09**

they perform relatively similarly under lower attack rates. Additionally Tables 1, 2 prove that backdoor attacks impose stronger threat than label-flipping attacks as the overall accuracies are lower under backdoor than label-flipping attacks.

Table 3 illustrates the impact of model poisoning attacks across different attack rates. This table notably highlights the effectiveness of GuardedLearn (GL), as it consistently achieves high accuracy with only a slight decrease from 97.15 to 97.08 as the attack rate escalates from 10 to 40%, even under gaussian attacks. In contrast, DPFLA and TMean experience significant drops in test accuracy under model poisoning. For instance, with just 10% of clients acting as gaussian attackers, DPFLA's accuracy decreases from 97.17 to 91.36, and

Table 2. Test Accuracy Results of DPFLA Vs TrimmedMean Vs GL under Data Poisoning at Attack rates 50% and higher

Attack Types	A = 50%			A = 60%			A = 70%		
	DPFLA	TMean	GL	DPFLA	TMean	GL	DPFLA	TMean	GL
LF	96.62	96.32	**97.08**	85.80	85.78	**97.04**	84.30	84.23	**97.03**
BD	96.41	96.33	**97.06**	85.19	85.07	**97.03**	84.12	84.02	**97.02**

TMean experiences a 10% drop. As the percentage of attackers increases from by 10%, the test accuracy falls by 34–45%, severely compromising the global model. Moreover, when simulating LF or BD concurrently with Gaussian attacks, the test accuracy further decreases, underscoring the detrimental impact of combining both data and model poisoning in the same environment.

Table 3. Test Accuracy Results of DPFLA Vs TrimmedMean Vs GL under Model Poisoning up to 40% attack rates

Attack Types	A = 10%			A = 20%			A = 30%			A = 40%		
	DPFLA	TMean	GL	DPFLA	TMean	GL	DPFLA	TMean	GL	DPFLA	TMean	GL
Gaussian	91.36	89.82	**97.15**	48.88	46.32	**97.12**	14.72	12.02	**97.10**	10.42	10.23	**97.08**
LF + Gaussian	89.37	86.20	**97.13**	31.99	30.72	**97.13**	11.48	11.41	**97.10**	10.38	10.21	**97.06**
BD+ Gaussian	88.68	85.12	**97.13**	31.52	29.92	**97.10**	10.29	10.12	**97.08**	10.25	10.11	**97.04**

In Table 4, at high attack rates, where the majority of clients are compromised by model poisoning, the remarkable performance of GL stands out significantly compared to other techniques. GL demonstrates higher accuracy than other methods by approximately 74%, highlighting its unique and powerful capabilities. This substantial difference underscores the strength of GL in effectively mitigating the impact of model poisoning attacks and maintaining robustness in challenging environments.

Table 4. Test Accuracy Results of DPFLA Vs TrimmedMean Vs GL under Model Poisoning at Attack rates 50% and higher

Attack Types	A = 50%			A = 60%			A = 70%		
	DPFLA	TMean	GL	DPFLA	TMean	GL	DPFLA	TMean	GL
Gaussian	10.30	10.21	**97.06**	10.27	10.13	**97.02**	9.80	9.56	**97.02**
LF + Gaussian	10.22	10.10	**97.05**	10.08	10.06	**97.02**	9.57	9.52	**97.01**
BD+ Gaussian	10.09	10.06	**96.95**	9.84	9.72	**96.90**	9.26	9.22	**96.83**

6 Conclusion

In this study, we presented GuardedLearn, a federated learning framework designed to enhance privacy preservation and effectively combat data and model poisoning attacks. Through rigorous evaluations, GuardedLearn demonstrates superior resilience, outperforming conventional methods even under high rates of malicious activity up to 70%. The integration of DBSCAN clustering provides robust detection capabilities, crucial for maintaining model integrity, and consistently achieves high clustering F1 scores across various attack scenarios. Additionally, the adoption of RFA for aggregation significantly boosts test accuracy, even in severe attack conditions. Future work will focus on refining the aggregation mechanisms to enhance scalability and reduce computational overhead, aiming for broader adoption in diverse and larger-scale federated environments. Furthermore, we aim to make our framework fully functional in extreme non-iid environments.

References

1. Bagdasaryan, E., Veit, A., Hua, Y., Estrin, D., Shmatikov, V.: How to backdoor federated learning. In: International Conference on Artificial Intelligence and Statistics, pp. 2938–2948. PMLR (2020)
2. Bhowmick, A., Duchi, J., Freudiger, J., Kapoor, G., Rogers, R.: Protection against reconstruction and its applications in private federated learning. arXiv preprint arXiv:1812.00984 (2018)
3. Blanchard, P., El Mhamdi, E.M., Guerraoui, R., Stainer, J.: Machine learning with adversaries: byzantine tolerant gradient descent. In: Advances in Neural Information Processing Systems, vol. 30 (2017)
4. Cao, X., Fang, M., Liu, J., Gong, N.Z.: FLTrust: byzantine-robust federated learning via trust bootstrapping. arXiv preprint arXiv:2012.13995 (2020)
5. Chai, D., et al.: Practical lossless federated singular vector decomposition over billion-scale data. In: Proceedings of the 28th ACM SIGKDD Conference on Knowledge Discovery and Data Mining, pp. 46–55 (2022)
6. Chen, X., Yu, H., Jia, X., Yu, X.: APFed: anti-poisoning attacks in privacy-preserving heterogeneous federated learning. IEEE Trans. Inf. Forensics Secur. (2023)
7. Cohen, M.B., Lee, Y.T., Miller, G., Pachocki, J., Sidford, A.: Geometric median in nearly linear time. In: Proceedings of the Forty-Eighth Annual ACM Symposium on Theory of Computing, pp. 9–21 (2016)
8. Daniel, J.W., Gragg, W.B., Kaufman, L., Stewart, G.W.: Reorthogonalization and stable algorithms for updating the gram-Schmidt factorization. Math. Comput. **30**(136), 772–795 (1976)
9. Ester, M., Kriegel, H.P., Sander, J., Xu, X., et al.: A density-based algorithm for discovering clusters in large spatial databases with noise. In: kdd, vol. 96, pp. 226–231 (1996)
10. Fang, M., Cao, X., Jia, J., Gong, N.: Local model poisoning attacks to {Byzantine-Robust} federated learning. In: 29th USENIX Security Symposium (USENIX Security 2200), pp. 1605–1622 (2020)

11. Feng, X., Cheng, W., Cao, C., Wang, L., Sheng, V.S.: DPFLA: defending private federated learning against poisoning attacks. IEEE Trans. Serv. Comput. (2024)
12. Hartigan, J.A., Wong, M.A.: Algorithm as 136: a k-means clustering algorithm. J. Roy. Stat. Soc. Ser. C (Appl. Stat.) **28**(1), 100–108 (1979)
13. Kairouz, P., et al.: Advances and open problems in federated learning. Found. Trends® Mach. Learn. **14**(1–2), 1–210 (2021)
14. Li, S., Ngai, E., Voigt, T.: Byzantine-robust aggregation in federated learning empowered industrial IoT. IEEE Trans. Industr. Inf. **19**(2), 1165–1175 (2021)
15. Li, S., Cheng, Y., Wang, W., Liu, Y., Chen, T.: Learning to detect malicious clients for robust federated learning. arXiv preprint arXiv:2002.00211 (2020)
16. McMahan, B., Moore, E., Ramage, D., Hampson, S., y Arcas, B.A.: Communication-efficient learning of deep networks from decentralized data. In: Artificial Intelligence and Statistics, pp. 1273–1282. PMLR (2017)
17. Naseri, M., Hayes, J., De Cristofaro, E.: Local and central differential privacy for robustness and privacy in federated learning. arXiv preprint arXiv:2009.03561 (2020)
18. Pillutla, K., Kakade, S.M., Harchaoui, Z.: Robust aggregation for federated learning. IEEE Trans. Signal Process. **70**, 1142–1154 (2022)
19. Rosenfeld, E., Winston, E., Ravikumar, P., Kolter, Z.: Certified robustness to label-flipping attacks via randomized smoothing. In: International Conference on Machine Learning, pp. 8230–8241. PMLR (2020)
20. Sun, G., Cong, Y., Dong, J., Wang, Q., Lyu, L., Liu, J.: Data poisoning attacks on federated machine learning. IEEE Internet Things J. **9**(13), 11365–11375 (2021)
21. Xu, R., Baracaldo, N., Zhou, Y., Anwar, A., Ludwig, H.: HybridAlpha: an efficient approach for privacy-preserving federated learning. In: Proceedings of the 12th ACM Workshop on Artificial Intelligence and Security, pp. 13–23 (2019)
22. Yin, D., Chen, Y., Kannan, R., Bartlett, P.: Byzantine-robust distributed learning: towards optimal statistical rates. In: International Conference on Machine Learning, pp. 5650–5659. PMLR (2018)

Use of AI and Intelligent Algorithms to Detect and Prevent Network-Based Attacks

Sama ElHarras(✉), Milad Ghantous, and Minar ElAasser

German University in Cairo, Cairo, Egypt
sama.elharras@student.guc.edu.eg,
{milad.michel,minar.elaasser}@guc.edu.eg

Abstract. In today's interconnected digital environment, the rise of network-based attacks presents significant risks to individuals and organizations. Conventional security protocols have struggled to keep pace with the increasing complexity of cyber threats. In this study, we delve into the domain of cybersecurity with the aim of leveraging Artificial Intelligence (AI) methodologies such as Machine Learning (ML) and anomaly detection to fortify network defense mechanisms. We make use of different datasets, feature selection techniques, and ML algorithms, in order to compare them and decide which ML algorithm and feature selection technique help reach the best Network Intrusion Detection System (NIDS). Utilizing two distinct datasets, CICIDS2017 and CICIoT2023, the study investigates the efficacy of different ML algorithms, namely Decision Tree (DT), Naïve Bayes (NB), K-Nearest Neighbor (KNN), Logistic Regression (LR), Adaboost and Xgboost and feature selection techniques, namely Random Forest (RF) and Principal Component Analysis (PCA), in enhancing the performance of Intrusion Detection Systems (IDSs). Results show that DT and Xgboost outperform other algorithms in binary classification across both datasets and all feature selection techniques, and that DT with RF is the optimal option for multi-class classification.

Keywords: Cybersecurity · AI · ML · IDS · DT · NB · KNN · LR · Adaboost · Xgboost · RF · PCA

1 Introduction

The rapid evolution of technology and networking techniques and the increased use of the internet have lead to the increase of attackers' attempts to intrude networking systems. Although cybersecurity techniques have been widely investigated over the last decades, new types of attacks are emerging every day, and traditional detection and prevention methods such as firewalls, antivirus software and message encryption have proven inadequate against the sophistication and diversity of modern cyber threats especially with denial-of-service (DoS)

A. Abdelgawad et al. (Eds.): ISBCom 2024, LNNS 1268, pp. 454–467, 2025.
https://doi.org/10.1007/978-3-031-82377-0_37

and distributed denial-of-service (DDoS) attacks [5]. Recently, Intrusion Detection Systems (IDSs), both signature-based and anomaly-based, have become a necessary addition to the security infrastructure of most organizations. Anomaly-based IDSs offer distinct advantages over signature-based IDSs by focusing on detecting deviations from normal behavior rather than relying solely on known attack patterns, leveraging Artificial Intelligence (AI), Machine Learning (ML), and Deep Learning (DL) algorithms. AI encompasses technologies aiming to automate tasks traditionally performed by humans, such as learning, reasoning, and problem-solving [1]. Within AI, ML allows machines to learn from data and improve without explicit programming [2], while DL employs neural networks to extract complex patterns from raw data [3]. These intelligent algorithms are able to analyze network traffic and system behavior and identify unusual patterns that may indicate potential intrusions or security breaches. Many different AI, ML, and DL techniques, that can either rely on a single algorithm or a combination of different algorithms, have been investigated and used in the past years to enable anomaly-based IDS to autonomously learn and adapt to changing network conditions, improving accuracy and reducing false positives. The aim of this study is to explore and evaluate the efficacy of advanced AI and intelligent algorithms in enhancing the capabilities of intrusion detection and prevention systems. This study is conducted on two different datasets: CICIDS2017 and CICIoT2023, comprising diverse and evolving cyber threats, including DoS and DDoS attacks. These attacks aim to temporarily prevent legitimate users from accessing network resources by flooding computer systems or devices with malicious activities. The study employed DT, NB, KNN, LR, Adaboost, and XgBoost for both binary and multi-class classification machine learning techniques. Additionally, RF and PCA were utilized for feature selection. All algorithms and feature selection techniques are conducted on both datasets. Models evaluation is based on 7 metrics: accuracy, precision, recall, f1-score, confusion matrix, training time and prediction time. The proposed models are then compared in order to achieve the optimal choice of algorithm, feature selection technique and dataset, leading to the best results. To the best of the authors knowledge, this comprehensive comparative analysis of various machine learning algorithms and feature selection techniques across multiple datasets has not been previously reported in the literature. The study begins with an exploration of AI and network-based attacks, illustrating how AI-driven IDSs detect and prevent such attacks. Following this, a literature review discusses related work and various techniques used to address the research problem. The methodology section provides a detailed explanation of the study's approach and steps. Subsequently, the results are presented, discussed, and illustrated. Finally, a concise conclusion of the study's findings and suggestions of potential future research directions are offered.

2 Related Work

The issue of designing an effective Intrusion Detection System has been widely researched in the past years. Most researchers tend to use ML and DL for the development of anomaly-based IDSs since they offer the advantage of detecting new unknown types of attacks. In [8] and [9], hybrid models are developed using both signature-based and anomaly-based IDSs in order to benefit from the best of both worlds. Both models use ML algorithms that are trained and tested on the DARPA 1999 and DARPA 2000 datasets. Most of the studies are carried out using KDD Cup'99 and NSL-KDD datasets targeting mainly DoS, R2L, U2R and Probe attacks. However, some studies [6,16,19,21] used the CICIDS 2017 dataset detecting different types of attacks such as DDoS, botnet and Brute Force.

Many Machine Learning techniques have been adopted to implement efficient IDSs throughout the last decades. [6,16] and [7] use DT and RF to identify and mitigate potential security threats. In [10], the study focuses on a two-step approach of feature selection based on RF. SVMs offer several advantages for NIDS as illustrated in [11] and [12]. However, careful selection of features and optimization of SVM parameters are crucial for building accurate and reliable NIDS using SVMs. That is why in [12], PCA is used for feature selection. Furthermore, Bayesian networks serve as effective tools for representing domains characterized by inherent uncertainty, that is why [13] uses NB. In [14] and [15], the AdaBoost, short for Adaptive Boosting, is applied to intrusion detection, using decision stumps as weak classifiers, with the aim of achieving low computational complexity, high detection rates, and low false-alarm rates.

Many researchers tend to aggregate different ML techniques to benefit from the strength points of each individual algorithm and construct a more efficient IDS with enhanced performance. [16] proposes a multi-layer IDS using ensemble learning composed of DT, RF and Artificial Neural Networks (ANN). Similarly, [17] develops a new hybrid approach using NB, DT, RF and k-Nearest Neighbor (KNN). K-Centroid clustering and Genetic Algorithm (GA) are combined in [18] to achieve better detection and false positives rates. Alternatively, some studies tend to use different ML algorithms separately in order to compare their performance and results such as [19] where 7 algorithms are compared: DT, RF, NB, Adaboost, Multi-Layer Perceptron (MLP), Quadratic Discriminant Analysis (QDA) and KNN. The study shows that KNN outperforms the other algorithms in NIDS.

DL algorithms have been widely used in IDSs. This is usually done using ANNs [20]. ANNs are computational models consisting of interconnected nodes, or neurons, organized in layers. ANNs have proven to be effective in various tasks such as classification, regression, and pattern recognition. In the context of network intrusion detection, Convolutional Neural Networks (CNNs) can analyze network traffic data. In [21], CNNs are used to detect DDos attacks in Software defined networks (SDN). By learning hierarchical representations of features directly from the raw data, CNNs can automatically extract relevant features without the need for manual feature engineering [22]. Different approaches have been used to implement NIDSs using AI in the past years. However, Every implemented model lacks one or more of the following: recent dataset (not older than 2017), comparing algorithms to obtain best results, multi-class classification, maintaining both high accuracy and low prediction time. Our goal is to implement an IDS combining all of the above.

3 Methodology

3.1 Datasets

CICIDS2017. CICIDS2017 is a Dataset that encompasses both benign network traffic alongside various cyber attacks, drawn from real-world data captured in PCAPs. This Dataset includes 15 different network-based attacks categorized into 8 classes: DDoS, DoS, Web, Brute Force, Botnet, Heartblead, Infiltration and Port Scan.

CICIoT2023. The CICIoT 2023 dataset is a comprehensive collection of network traffic data specifically designed for research and analysis in the field of Internet of Things (IoT) security. It encompasses a wide range of network activities, including benign traffic as well as various types of cyber attacks. The CICIoT 2023 dataset likely addressed several shortcomings present in previous datasets related to Network security. These improvements include: enhanced diversity, increased scalability, updated threat landscape and improved realism.

As shown in Fig. 1, the CICIoT2023 dataset is composed of 34 network-based attacks categorized into 7 classes: DDoS, DoS, Web-Based, Brute Force, Recon, Spoofing and Mirai [4].

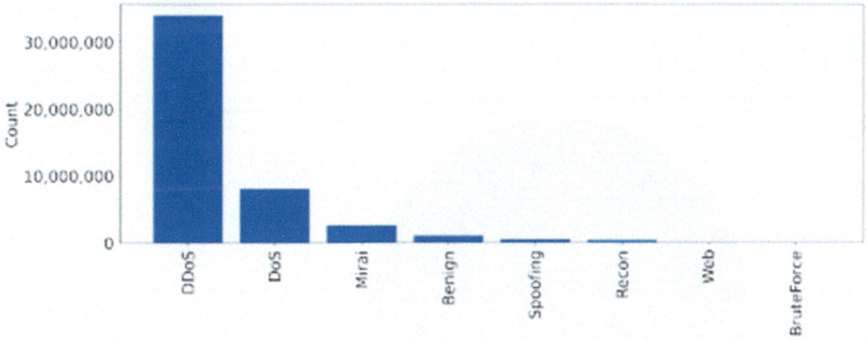

Fig. 1. CICIoT2023 attack categories

3.2 IDS Model

In addition to improving the performance of already existing IDSs that use CICIDS2017 Dataset, we make use of a more recent dataset, namely CICIoT2023, and carry out a complete study on it. Our aim is to compare different algorithms and feature selection techniques, in order to reach a new IDS that is more up to date with the new cyber attacks that exist nowadays. Each of the following steps was performed on both datasets.

Data Cleaning and Preprocessing. The CICIDS2017 dataset consists of 79 features (columns) and 2830743 records (rows), and the CICIoT2023 dataset consists of 47 features (columns) and 2366956 records (rows). Several data preprocessing steps were performed to prepare the dataset for the training and testing steps. First of all, we start by removing all records that have either "NaN" or "Infinity" in any of the columns. After this step, the dimensions of the CICIDS2017 dataset become 28276 rows × 79 columns while those of the CICIoT2023 do not change. Then, the dataset is randomly split into 70% training data and 30% testing data. Standard scaling is applied to the numerical features in both the training and testing sets, ensuring that they have a mean of 0 and a standard deviation of 1. One-hot encoding is then performed on the ('Label') column to convert it into a format suitable for machine learning algorithms. The label '0' represents benign records while '1' represents attack records. The last step addresses class imbalance in the dataset. However, this step is performed after feature selection, when applied. As seen in Fig. 1, the amount of benign traffic is very little compared to the attack traffic. Hence, oversampling is applied in order to have equal partitioning of benign and malicious traffics. The minority class is oversampled in the training data to avoid overfitting and ensure better model training. Finally, the preprocessed features and labels are stored in separate variables for both the training and testing sets, ready for use in our ML models (Fig. 2).

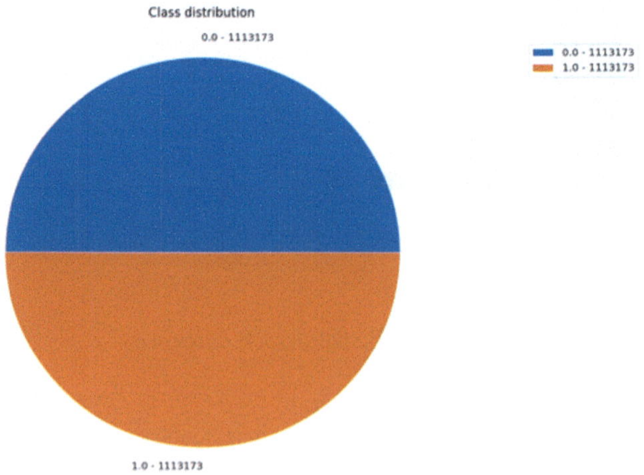

Fig. 2. Class distribution after oversampling

Feature Selection. All algorithms were trained and tested on the dataset 3 times. The first time all features of the dataset were included in the training and testing phases. The second time the 10 most important features were selected using RF algorithm. Finally, PCA was applied for feature selection before training and testing.

RF for feature selection: Random Forest is a ML technique that employs DTs. The algorithm calculates feature importance scores based on how much each feature contributes to decreasing the impurity in the tree nodes. RF can significantly impact model performance by improving accuracy, reducing overfitting and leading to faster training. However it has some limitations regarding computational complexity and bias towards features with unique values.

PCA for feature selection: PCA achieves feature selection by transforming the data into a lower-dimensional space, prioritizing components that capture the most significant variations. These components, acting as new features, are ranked by their importance. Using PCA helps achieve better performance and faster training through dimensionality reduction, variance retention and mitigation of multicollinearity. However, the new components created by PCA are linear combinations of the original features, which can be difficult to interpret. Furthermore, PCA might not always lead to performance improvement, especially if the original features are already well-optimized. We ran PCA several times using different number of components, and, using a binary search approach, choosing 40 components gave the best results.

Algorithms. 6 different ML algorithms were used for training and testing after each feature selection technique in order to compare their performance. The used algorithms are: DT, NB, KNN, Logistic Regression (LR), XGBoost, AdaBoost. In each case, hyperparameter tuning is applied using different techniques such as grid search, binary search and random search, in order to find the optimal hyperparameters combination that ensures better performance and faster training.

Decision Tree: A Decision Tree stands as one of the most powerful instruments in classification and prediction tasks [25]. Through iterative division of a dataset into smaller subsets, it progressively constructs an associated decision tree.

Naïve Bayes: Naive Bayes represents a group of probabilistic classification methods grounded in probability theory and Bayes' Theorem for predictive modeling [26].

K-Nearest Neighbors (KNN): KNN works by finding the K nearest data points to a query point based on a distance metric (e.g., Euclidean distance) in the feature space. For classification, the majority class among the K nearest neighbors is assigned to the query point [28].

Logistic Regression (LR): LR, often used for binary classification, predicts the odds of an instance belonging to a class (e.g., spam email) [29].

XGBoost: eXtreme Gradient Boosting operates through sequential learning, starting with a basic model and progressively integrating more trees into the ensemble. Each tree endeavors to correct errors from preceding iterations [30].

AdaBoost: AdaBoost is part of a family of ensemble methods called boosting. Boosting creates a strong classifier by combining multiple weak classifiers, such as DTs with limited depth [27].

3.3 Evaluation Metrics

In order to assess and compare machine learning models performance, several evaluation metrics were used which are: accuracy, precision, recall, F1 score, confusion matrix, training time and prediction time. In order to understand these evaluation metrics, the following explanations are needed:

- True Positives (TP): Describes instances where the model correctly predicts the positive class.
- True Negatives (TN): Describes instances where the model correctly predicts the negative class.
- False Positives (FP): Describes instances where the model incorrectly predicts the positive class.
- False Negatives (FN): Describes instances where the model incorrectly predicts the negative class.

$$\text{Accuracy} = \frac{\text{True Positives} + \text{True Negatives}}{\text{Total Predictions}} \tag{1}$$

$$\text{Precision} = \frac{\text{True Positives}}{\text{True Positives} + \text{False Positives}} \tag{2}$$

$$\text{Recall} = \frac{\text{True Positives}}{\text{True Positives} + \text{False Negatives}} \tag{3}$$

$$\text{F1-score} = 2 \times \frac{\text{Precision} \times \text{Recall}}{\text{Precision} + \text{Recall}} \tag{4}$$

$$\text{Confusion Matrix} = \begin{bmatrix} TN & FP \\ FN & TP \end{bmatrix} \tag{5}$$

3.4 Network-Based Attacks Analysis

All of the previous algorithms were used as binary classifiers to detect malicious traffic, which is useful for real-time IDSs. However, in the case of network-based attacks analysis, multi-class classifiers are more efficient than binary classifiers since they are able to classify attacks according to their type, hence, they give an insight about each existing type of attack and it's frequency of occurrence.

Since Cybersecurity is a very dynamic field, traffic analysis should be done on recent datasets that include the newest attack techniques. Consequently, in this part of the study we only use the CICIoT2023 dataset. We train and test the 2 algorithms that gave the highest accuracy (DT) and lowest prediction time (NB) in the binary classification model. However this time both algoirthms were used as multi-class classifiers. CICIoT2023 comprises 34 different types of attacks, so, the multi-classifiers work with 34 classes. The same steps for data cleaning and preprocessing were performed, except for the one-hot encoding. Instead, Label encoding was done to convert each categorical label into a numerical label (from 0 to 33) to represent each type of attack. The same 3 feature selection approaches were used in this model. Finally, data oversampling was applied to ensure equal partitioning of the different attack categories and avoid overfitting.

4 Results

4.1 CICIDS2017 Dataset Results

Results of binary classifiers using CICIDS2017 dataset are summarized in Tables 1, 2 and 3:

Table 1. CICIDS2017 Binary classifiers results using all features

	DT	KNN	NB	LGR	Xgboost	Adaboost
Accuracy	**1.00**	0.96	0.81	0.91	**1.00**	0.88
Precision	**1.00**	0.97	0.81	0.93	**1.00**	0.93
Recall	**1.00**	0.96	0.81	0.91	**1.00**	0.93
F1-Score	**1.00**	0.96	0.81	0.92	**1.00**	0.93
Training time (s)	359.05	1564.20	**74.02**	328.18	191.34	852.94
Prediction time (s)	196.60	1236.48	**194.86**	198.15	200.06	732.69

Using all features and no feature selection on the CICIDS2017 datasets, DT and Xgboost give the best accuracy, precision, recall and F1-Score. KNN and Adaboost come second in terms of the first 4 evaluation metrics but they give worst training and prediction times. NB gives the worst results but the best training and prediction times. LR gives slightly better results than NB but worse training and prediction times.

Table 2. CICIDS17 Binary classifiers results using 10 most important features

	DT	KNN	NB	LGR	Xgboost	Adaboost
Accuracy	**0.98**	0.98	0.75	0.86	**0.98**	0.96
Precision	0.98	**1.00**	0.76	0.91	0.98	0.97
Recall	**0.98**	0.97	0.75	0.86	**0.98**	0.96
F1-Score	0.98	**0.99**	0.76	0.87	0.98	0.96
Training time (s)	61.34	953.25	**14.42**	123.74	68.45	887.71
Prediction time (s)	71.63	986.45	**70.10**	73.10	75.04	897.32

In the case of using RF for feature selection, training and prediction times of all algorithms are less than those obtained when using all features, except for Adaboost. But results are consistent with the previous ones.

Table 3. CICIDS17 Binary classifiers results using PCA for feature selection

	DT	KNN	NB	LGR	Xgboost	Adaboost
Accuracy	0.94	**0.98**	0.74	0.91	**0.98**	0.96
Precision	0.94	**0.99**	0.84	0.93	0.98	0.97
Recall	0.94	0.98	0.74	0.91	0.98	**0.99**
F1-Score	0.94	**0.99**	0.76	0.91	0.98	0.98
Training time (s)	2145.01	3256.36	**44.91**	227.45	211.66	1925.35
Prediction time (s)	218.84	3156.49	**216.47**	221.10	223.87	1515.47

Using PCA for feature selection, results are similar to the previous feature selection techniques. However, as opposed to the previous cases, DT does not give the best results in terms of all 6 evaluation metrics.

As seen in the previous tables, the best results for the CICIDS2017 dataset are given when applying DT and XGboost and either using all features or RF for feature selection. NB always gives least training and prediction times.

4.2 CICIoT2023 Dataset Results

Results of binary classifiers using CICIoT2023 dataset are summarized in Tables 4, 5 and 6:

Table 4. CICIoT2023 Binary classifiers results using all features

	DT	KNN	NB	LGR	Xgboost	Adaboost
Accuracy	**1.00**	**1.00**	0.92	0.97	**1.00**	**1.00**
Precision	**1.00**	**1.00**	**1.00**	**1.00**	**1.00**	**1.00**
Recall	**1.00**	**1.00**	0.92	0.97	**1.00**	**1.00**
F1-Score	**1.00**	**1.00**	0.96	0.99	**1.00**	**1.00**
Training time (s)	103.63	3658.10	**36.78**	240.59	118.59	3289.17
Prediction time (s)	121.47	3854.62	**120.30**	122.75	124.18	3314.54

Using all features for training and testing, DT, KNN, Xgboost and Adaboost give the best results in terms of accuracy, precision, recall and F1-Score (1.00 for all). However, DT and Xgboost take significantly less time than KNN and Adaboost for training and prediction. NB and LR give lower accuracy, precision, recall and F1-Score even though NB gives the best training and testing times. DT and Xgboost can be considered to give the best results.

Table 5. CICIoT2023 Binary classifiers results using 10 most important features

	DT	KNN	NB	LGR	Xgboost	Adaboost
Accuracy	**1.00**	**1.00**	0.92	0.95	**1.00**	**1.00**
Precision	**1.00**	**1.00**	**1.00**	**1.00**	**1.00**	**1.00**
Recall	**1.00**	**1.00**	0.92	0.95	**1.00**	**1.00**
F1-Score	**1.00**	**1.00**	0.96	0.98	**1.00**	**1.00**
Training time (s)	42.67	2899.55	**14.52**	85.06	64.54	3715.22
Prediction time (s)	3717.49	3960.96	**3716.47**	3961.97	3963.12	3976.29

When using RF for feature selection, we obtain almost the same results as the previous model, however, DT and Xgboost take significantly more time for prediction than when using all features.

Table 6. CICIoT2023 Binary classifiers results using PCA for feature selection

	DT	KNN	NB	LGR	Xgboost	Adaboost
Accuracy	**1.00**	0.97	0.92	0.97	**1.00**	0.98
Precision	**1.00**	0.96	**1.00**	**1.00**	**1.00**	0.98
Recall	**1.00**	0.97	0.92	0.97	**1.00**	0.98
F1-Score	**1.00**	0.97	0.96	0.99	**1.00**	0.98
Training time (s)	236.18	3815.65	**20.85**	261.79	199.22	3658.25
Prediction time (s)	72.96	3974.58	**15.86**	198.59	200.49	3816.26

Using PCA for feature selection, results are consistent with the previous cases, but only DT and Xgboost give 1.00 in the first 4 metrics.

Overall, the optimal solution would be to use DT either when considering all features or when performing PCA for feature selection. Both approaches give similar and optimal results.

4.3 Multi-class Classifiers Results

CICIoT2023 dataset multi-class classifiers results are shown in Table 7:

Table 7. CICIoT2023 multi-class classifiers results

	All features		RF feature selection		PCA feature selection	
	DT	NB	DT	NB	DT	NB
Accuracy	0.56	0.80	**0.99**	0.52	0.98	0.83
Precision	0.61	0.82	**0.99**	0.53	0.98	0.91
Recall	0.56	0.80	**0.99**	0.52	0.98	0.83
F1-Score	0.53	0.79	**0.99**	0.43	0.98	0.86
Training time (s)	748.59	224.61	196.70	**80.39**	2957.80	193.15
Prediction time (s)	750.94	227.89	199.78	198.38	2960.06	**186.44**

As seen in the previous table, the best results are achieved when using DT and RF feature selection technique. DT with PCA for feature selection comes second with larger training and prediction times. Although NB gives the least training and prediction times with RF and PCA, it gives the worst results in terms of accuracy, precision, recall and f1-score. Using all features is considered the worst option for both DT and NB as it gives relatively bad results when compared to the cases where feature selection is applied. Hence, multi-class classification works best on the CICIoT2023 dataset when DT is used with RF feature selection technique.

4.4 Models Comparison

The comparative analysis underscores the superior performance of DT and Xgboost across both datasets, making them the preferred choices for binary classification tasks. In the case of using CICIoT2023, an accuracy, precision, recall and f1-score of 1.00 was obtained from both DT and Xgboost with all feature selection techniques, while these results were only obtained when considering all features from the CICIDS2017 dataset. Also, using NB on the CICIoT2023 dataset remains a competitive option for time-sensitive applications, owing to its efficient training and prediction times. Regarding the proposed multi-class classifier, the optimal solution will be to use DT with RF for feature selection, while DT with PCA comes second. In addition to the novelty and up-to-date attack types of CICIoT2023, this dataset achieves overall better results than CICIDS2017. Hence, developing a model using DT on CICIoT2023 is considered the best choice to be used in NIDSs nowadays, as well as in cyber attacks analysis.

5 Conclusion and Future Work

In conclusion, this study delved into the realm of AI with a focus on ML and its application in NIDSs. The study employed various datasets, feature selection techniques, and ML algorithms to compare their efficacy in NIDS.

Results revealed that DT and XGBoost consistently outperformed other algorithms across both CICIDS2017 and CICIoT2023 datasets and all feature selection methods, making them prime candidates for binary classification tasks in NIDS. Furthermore, the study extended its analysis to multi-class classification, where DT with RF feature selection emerged as the optimal solution for detecting diverse types of attacks, which is useful not only in IDSs but also in cyber attacks analysis and prevention of potential future attacks. The study not only enhanced our understanding of AI's role in cybersecurity but also provided valuable insights for developing robust and efficient NIDS capable of mitigating modern cyber threats.

Building on the findings of this study, the following ideas for future research emerge: First, exploring advanced ML techniques such as DL could offer deeper insights into complex attack patterns and enhance the detection capabilities of NIDS. Examples of DL algorithms include Recurrent Neural Networks (RNNs) for sequence-based data and Autoencoders for anomaly detection. Furthermore, incorporating real-time data streams and dynamic adaptation mechanisms could enhance the relevance and effectiveness of NIDS in practical scenarios. This can be done through stream processing frameworks, online learning algorithms and dynamic feature selection. Finally, developing hybrid approaches that combine the strengths of different ML algorithms to create more robust and adaptive NIDS solutions. Hybrid approaches may include ensemble methods, multi-stage detection systems and collaborative learning. By addressing these research directions, we can advance the state-of-the-art in NIDS and bolster our defenses against evolving cyber threats.

Disclosure of Interests. The authors have no competing interests to declare that are relevant to the content of this article.

References

1. Fathima Anjila, P.K.: What is artificial intelligence?. In: Success is No Accident. It is Hard Work, Perseverance, Learning, Studying, Sacrifice and Most of All, Love of What You are Doing or Learning to do, p. 65 (1984)
2. Choi, R.Y., Coyner, A.S., Kalpathy-Cramer, J., Chiang, M.F., Campbell, J.P.: Introduction to machine learning, neural networks, and deep learning. Transl. Vision Sci. Technol. **9**(2), 14–14 (2020)
3. Goodfellow, I., Bengio, Y., Courville, A.: Deep Learning. MIT press, Cambridge (2016)
4. Neto, E.C.P., Dadkhah, S., Ferreira, R., Zohourian, A., Lu, R., Ghorbani, A.A.: CICIoT2023: a real-time dataset and benchmark for large-scale attacks in IoT environment. Sensors **23**(13), 5941 (2023)
5. Nautiyal, N.: What is DDoS Attack and Top 5 Ways to Prevent DDoS Attack. Indies Education- Online Learning Library (2020). https://indieseducation.com/ddos-attack-and-ways-to-prevent-it/
6. Jamadar, R.A.: Network intrusion detection system using machine learning. Indian J. Sci. Technol. **7**(48), 1–6 (2018)

7. Kruegel, C., Toth, T.: Using decision trees to improve signature-based intrusion detection. In: Vigna, G., Kruegel, C., Jonsson, E. (eds.) RAID 2003. LNCS, vol. 2820, pp. 173–191. Springer, Heidelberg (2003). https://doi.org/10.1007/978-3-540-45248-5_10
8. Weon, I.-Y., Song, D., Lee, C.-H.: Effective intrusion detection model through the combination of a signature-based intrusion detection system and a machine learning-based intrusion detection system. J. Inf. Sci. Eng. **22**, 1447–1464 (2006)
9. Cepheli, Ö., Büyükçorak, S., Kurt, G.K.: Hybrid intrusion detection system for DDoS attacks. JECE **2016**, 4 (2016). https://doi.org/10.1155/2016/1075648
10. Hasan, M.A.M., Nasser, M., Ahmad, S., Molla, K.I.: Feature selection for intrusion detection using random forest. J. Inf. Secur. **7**(3), 129–140 (2016)
11. Hussain, J., Lalmuanawma, S., Chhakchhuak, L.: A two-stage hybrid classification technique for network intrusion detection system. Int. J. Comput. Intell. Syst. **9**(5), 863–875 (2016)
12. Ikram, S.T., Cherukuri, A.K.: Improving accuracy of intrusion detection model using PCA and optimized SVM. J. Comput. Inf. Technol. **24**(2), 133–148 (2016)
13. Panda, M., Patra, M.R.: Network intrusion detection using naive bayes. Int. J. Comput. Sci. Netw. Secur. **7**(12), 258–263 (2007)
14. Hu, W., Hu, W., Maybank, S.: Adaboost-based algorithm for network intrusion detection. IEEE Trans. Syst. Man Cybern. Part B (Cybern.) **38**(2), 577–583 (2008)
15. Hu, W., Hu, W.: Network-based intrusion detection using Adaboost algorithm. In: The 2005 IEEE/WIC/ACM International Conference on Web Intelligence (WI'05), pp. 712–717. IEEE (2005)
16. Agate, V., et al.: A behavior-based intrusion detection system using ensemble learning techniques. In: ITASEC, pp. 207–218 (2022)
17. Çavuşoğlu, Ü.: A new hybrid approach for intrusion detection using machine learning methods. Appl. Intell. **49**(7), 2735–2761 (2019). https://doi.org/10.1007/s10489-018-01408-x
18. Chakrabarty, B., Chanda, O., Saiful, M.: Anomaly based intrusion detection system using genetic algorithm and K-centroid clustering. Int. J. Comput. Appl. **163**(11), 13–17 (2017)
19. Alrowaily, M., Alenezi, F., Lu, Z.: Effectiveness of machine learning based intrusion detection systems. In: Wang, G., Feng, J., Bhuiyan, M.Z.A., Lu, R. (eds.) SpaCCS 2019. LNCS, vol. 11611, pp. 277–288. Springer, Cham (2019). https://doi.org/10.1007/978-3-030-24907-6_21
20. Karatas, G., Sahingoz, O.K.: Neural network based intrusion detection systems with different training functions. In: 2018 6th International Symposium on Digital Forensic and Security (ISDFS), pp. 1–6. IEEE (2018)
21. Haider, S., et al.: A deep CNN ensemble framework for efficient DDoS attack detection in software defined networks. IEEE Access **8**, 53972–53983 (2020)
22. Kumar, P., Kumar, A.A., Sahayakingsly, C., Udayakumar, A.: Analysis of intrusion detection in cyber attacks using DEEP learning neural networks. Peer-to-Peer Network. Appl. **14**(4), 2565–2584 (2021)
23. Fernández, G.C., Xu, S.: A case study on using deep learning for network intrusion detection. In: MILCOM 2019-2019 IEEE Military Communications Conference (MILCOM), pp. 1–6. IEEE (2019)
24. Goodfellow, I., Bengio, Y., Courville, A.: Deep Learning. MIT Press, Cambridge (2016). http://www.deeplearningbook.org
25. Song, Y.Y., Ying, L.U.: Decision tree methods: applications for classification and prediction. Shanghai Arch. Psychiatry **27**(2), 130 (2015)

26. Rish, I.: An empirical study of the naive Bayes classifier. In: IJCAI 2001 Workshop on Empirical Methods in Artificial Intelligence, vol. 3, no. 22, pp. 41–46 (2001)
27. Hastie, T., Rosset, S., Zhu, J., Zou, H.: Multi-class adaboost. Stat. Interface **2**(3), 349–360 (2009)
28. Guo, G., Wang, H., Bell, D., Bi, Y., Greer, K.: KNN model-based approach in classification. In: Meersman, R., Tari, Z., Schmidt, D.C. (eds.) OTM 2003. LNCS, vol. 2888, pp. 986–996. Springer, Heidelberg (2003). https://doi.org/10.1007/978-3-540-39964-3_62
29. LaValley, M.P.: Logistic regression. Circulation **117**(18), 2395–2399 (2008)
30. Chen, T., Guestrin, C.: Xgboost: a scalable tree boosting system. In: Proceedings of the 22nd ACM SIGKDD International Conference on Knowledge Discovery and Data Mining, pp. 785–794 (2016)

BCDM: A Novel AI-Based Model for Detection of Breast Cancer

Warda M. Shaban[1,2(✉)], Hossam El-Din Moustafa[3], and Mervat M. El-Seddek[4]

[1] Department of Communication and Electronics Engineering, Nile Higher Institute for Engineering and Technology, Mansoura, Egypt
warda_mohammed@nilehi.edu.eg
[2] Faculty of Artificial Intelligence and Informatics, Horus University, New Damietta, Egypt
[3] Electronics and Communications Engineering Department, Faculty of Engineering, Mansoura University, Mansoura, Egypt
Hossam_moustafa@mans.edu.eg
[4] Department of Electronics and Communications Engineering, Faculty of Engineering, Horus University, New Damietta, Egypt
mervat.elseddek@ieee.org

Abstract. Breast cancer is a prevalent and significant form of cancer that has a profound impact on women worldwide, and it is the leading cause of mortality. Timely identification of breast cancer is a crucial procedure that can enable suitable therapy, hinder the advancement of cancerous cells, and decrease rates of illness and death. Machine Learning (ML) is highly favored techniques for precise detection and classification of breast cancer. This paper introduces a unique method for categorizing breast cancer by utilizing a novel technique for selecting relevant attributes and employing ML algorithms. This method is referred to as the Breast Cancer Detection Model (BCDM). The BCDM system comprises three distinct phases, namely data preprocessing, feature selection, and patient detection. The primary objective of the data preprocessing phase is to extract distinctive characteristics from mammogram images and eliminate any anomalous items. This is accomplished by employing an innovative feature selection technique known as Enhanced Bat Algorithm (EBA) to identify the most influential and meaningful attributes from the extracted ones. Afterwards, these attributes are entered into the patient detection phase to determine whether the patient is unaffected or has either benign or malignant breast cancer. Based on the obtained results, the proposed method surpasses rivals in performance in terms of accuracy, precision, recall, and F-score.

Keywords: Breast Cancer · Machine Learning · Feature Selection · Bat Algorithm

1 Introduction

According to the American Cancer Society (ACS), breast cancer is the most perilous form of cancer that impacts women [1]. Early detection and classification of breast cancer are vital in preventing the spread of cancer cells and facilitating timely treatment,

A. Abdelgawad et al. (Eds.): ISBCom 2024, LNNS 1268, pp. 468–479, 2025.
https://doi.org/10.1007/978-3-031-82377-0_38

ultimately leading to the preservation of numerous lives. Medical imaging is one of the methods used to detect breast cancer [2, 3]. These methods procedure, advantages, and disadvantages are mentioned in Table 1.

Mammography is a diagnostic imaging procedure that takes precise pictures of the breasts by means of X-rays. Despite its usefulness, mammography has a few drawbacks that make it less than ideal for detecting breast cancer. For example, (i) a false-negative mammogram can make the cancer look normal, (ii) a false-positive can make the cancer look abnormal, and (iii) further exams can take longer to schedule than needed. Therefore, to improve sensitivity and specificity, it is advised that different radiologists revisit breast cancer imaging techniques on a multiple basis [4]. Double reading is more expensive for patients and requires radiologists with more experience [4]. An alternative approach is to utilize Artificial Intelligence (AI) for precise identification of breast cancer.

Table 1. A summary of the best imaging methods

Imaging Techniques	Procedure	Advantages	Disadvantages
Magnetic Resonance (MRI) Imaging	MRI imaging is conducted by utilizing radio waves and magnetic fields	It is appropriate for women who have a high susceptibility to breast cancer	People who have medical devices made of metal should not use it. Additionally, some people feel uneasy around MRI machines
Positron Emission Tomography (PET)	During a PET scan, Fluoro Deoxy Glucose is injected into a vein in the arm to distinguish between different tissues	Capable of identifying cancerous conditions	Expensive, sluggish, and lacking in accuracy
Thermography (Thermal Imaging)	Thermography is a technique that uses infrared scanning to create an image that shows the differences in temperature across the breasts	Secure, rapid, pain-free, and non-intrusive	Its efficacy remains unproven
Ultrasound imaging (Sonography)	The images in ultra-Sound imaging were generated through the utilization of sound waves	The lack of radiation makes it safe and painless	Doesn't cover the breasts entirely
Mammography	Mammography is a diagnostic procedure that uses X-rays to detect and diagnose breast cancer.	While it does not act as a preventative measure for breast cancer, early detection through these methods can be life-saving.	There are false positives and false negatives that it encounters.

AI has proven useful in healthcare. AI is getting faster, bigger, and more resilient. These advances are thought to be the reason AI can solve complex problems in many fields, including medicine [5]. Because of this, AI should be utilized by pathologists and radiologists to detect diseases. ML or AI employs programming techniques to acquire the ability to operate autonomously. Following this principle gradually improves efficiency and precision. All ML methods work similarly. Initially, the input training data is obtained in order to construct a mathematical model to solve the current problem [6]. Several AI and ML methods can identify and classify breast cancer [7–11]. They are more complicated and cannot achieve the required precision.

This paper presents a novel AI-based model called Breast Cancer Detection Model (BCDM) for identifying breast cancer patients. The proposed BCDM comprises three primary components, namely: (i) data preprocessing, (ii) feature selection, and (iii) patient detection. BCDM being proposed is founded on medical imaging. In the initial stage of data preprocessing, the input images undergo various stages, including feature extraction and outlier rejection. During the initial stage, characteristics are derived from input images through the utilization of Gray Level Co-occurrence Matrix (GLCM). Subsequently, Genetic Algorithm (GA) is applied to eliminate the outlier items that exhibit anomalous characteristics compared to the rest. The next step involves the selection of the most crucial and meaningful attributes through the utilization of the Enhanced Bat Algorithm (EBA). EBA is an upgraded version of the conventional BA. EBA integrates both BA and a crossover strategy. Utilizing a crossover strategy can effectively resolve the problem of a limited search space in the BA algorithm. Ultimately, these characteristics are inputted into the patient detection stage, where multiple ML algorithms are employed. The algorithms used are Logistic Regression (LR), Naïve Bayes (NB), K-Nearest Neighbor (KNN), Decision Tree (DT), and Support Vector Machine (SVM). Therefore, the decision is made based on the highest number of votes.

The remaining sections of the paper are structured in the following manner: Sect. 2 presents a comprehensive overview of the previous research and advancements made in the field of breast cancer detection techniques. Section 3 is dedicated to the proposed Breast Cancer Detection Model (BCDM). Section 4 introduces experimental results. Conclusions are in Sect. 5.

2 Literature Review

In this part, we reviewed the prior work that attempted to categorize breast cancer patients. A new Deep Learning Framework (DLF) was created to identify and categorize early-stage breast cancer, as shown in [12]. DLF is founded on the principle of knowledge transfer. There were two stages to implementing the suggested DLF. The initial stage involved extracting image features using various pre-trained Convolutional Neural Network (CNN) architectures, including ResNet, GoogLeNet, and Visual Geometry Group Network (VGGNet). Afterwards, a fully connected layer was trained to distinguish between benign and malignant cells by utilizing average pooling classification and the features that had been extracted. According to the results obtained, DLF performs well.

In [13], a novel methodology was presented that combines a pre-trained Deep Residual Network (DRN) model with SVM. DRN-SVM algorithm was utilized to classify and categorize both malignant as well as benign breast tumors using ultrasound images. Thus, DRN-SVM was created to reduce the burden on physicians and improve the precision of classification. The trained network's convolutional layer was initially trained using DRN to extract image features. The collected features were subsequently classified using SVM coupled with a sequential minimal optimization solver. A total of 2099 2D breast ultrasound images without labels were analyzed using the DRN-SVM method. Findings indicate that the proposed approach may help distinguish benign from malignant breast cancer.

As introduced in [14], a new approach to breast cancer diagnosis was developed using Random Forest (RF)-based Rule Extraction (RFRE). The proposed RFRE technique was employed to generate precise and easily understandable classification rules from a collection of decision trees. Initially, RF was employed to construct numerous decision tree models in order to generate a substantial a number of decision rules. After that, a methodology for extracting rules was devised to distinguish decision rules from the trained trees. Finally, an enhanced multi-objective evolutionary algorithm was employed to seek out an optimal base predictor in which the constituent rule set achieves the optimal balance between accuracy and interpretability.

As demonstrated in [15], an innovative Patient Detection Strategy (PDS) was implemented to promptly identify individuals with the disease in its initial stages. The proposed strategy comprises two components; a data preprocessing phase and a Patient Detection Phase (PDP). Initially, the most crucial characteristics are chosen utilizing a novel Hybrid Feature Selection Method (NHFSM). The NHFSM is based on a combination of filter and wrapper techniques. Subsequently, the preprocessed data is utilized in the PDP process to efficiently and precisely identify patients. Results indicate that the proposed NHFSM enhances the efficiency of patient classification compared to existing methods of feature selection.

As presented in [16], a Yolo-Based Model for Breast Cancer Detection in mammograms has been developed. The proposed model was based on CBIS-DDSM dataset. Moreover, transfer learning techniques were applied to enhance the models' ability to generalize when fine-tuning Yolo. A comparison was conducted among various Yolo architectures, namely YoloV3, YoloV5, and YoloV5-Transformer. Furthermore, Eigen-CAM was utilized to examine the model and provide explanations by emphasizing all the questionable areas of interest in the mammogram. The compact YoloV5 model yielded the most optimal solution achieved.

3 The Proposed Breast Cancer Detection Model (BCDM)

Among women, breast cancer is the most widespread tumor globally [16]. Hence, timely identification of breast cancer can enable physicians to preserve the lives of numerous individuals. This paper introduces a novel AI-based model. Figure 1 depicts the proposed model, referred to as the Breast Cancer Detection Model (BCDM). As depicted in Fig. 1, BCDM comprises three components: data preprocessing, feature selection, and patient detection. Subsequent subsections go into detail about each component.

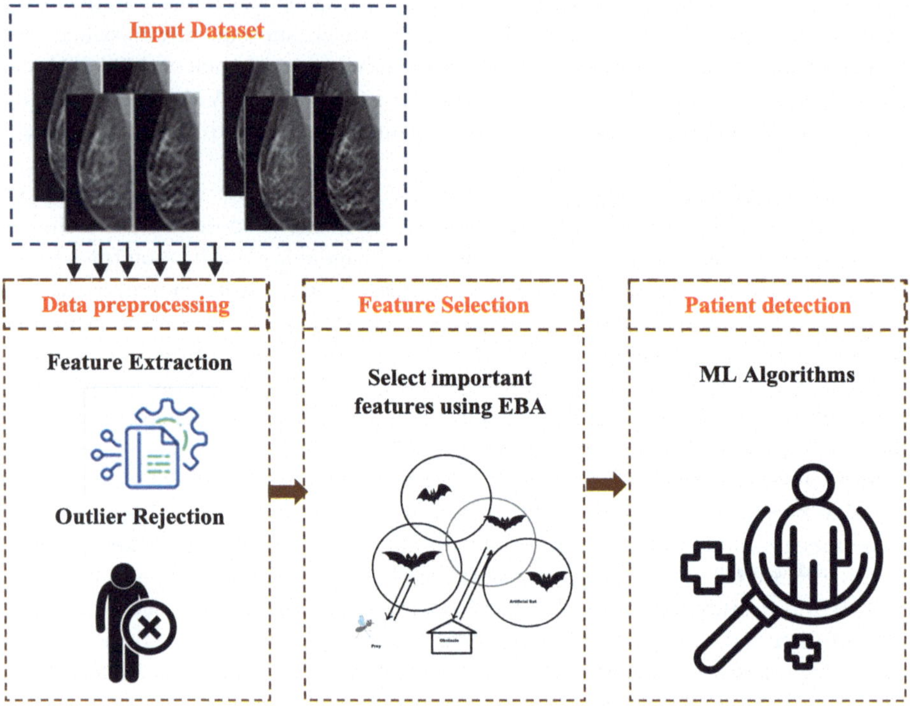

Fig. 1. The proposed Breast Cancer Detection Model (BCDM).

3.1 Data Preprocessing

In the initial phase of the proposed BCDM, the dataset is preprocessed and transformed into a suitable format for the subsequent phase. Hence, data preprocessing is a crucial component of the proposed BCDM. Next, a series of procedures are carried out, including feature extraction and outlier rejection. Feature Extraction (FE) is the process of creating a collection of characteristics from the input images. FE aims to decrease the number of elements in a dataset by creating new attributes based on existing ones and then removing the original features [17]. Utilizing FE from mammograms improves the classification model's effectiveness, allowing it to make more precise decisions [18, 19].

Recently, several methods have been used for FE. These methods include, co-occurrence matrix, wavelet transform-based features, etc. [17]. Gray level co-occurrence matrix (GLCM) is a highly used and reliable method for converting the used images into a set of features [18, 19]. The features include angular Second Moment, autocorrelation, maximum probability, contrast, homogeneity, correlation, entropy, energy, variance, and others [18, 19].

Then, after extracting all features from the images, outlier items are removed. Actually, outliers are items far away from the correct item. Thus, the existence of these items leads to wrong decision.

3.2 Feature Selection

The second part of the proposed BCDM is the feature selection process. The primary objective of the feature selection process is to identify and choose the most significant and valuable features from a given set of features. Actually, redundant features can detrimentally affect the performance of the classifier, lengthen the training process, and introduce complexity into the system. This paper presents a new feature selection methodology, known as Enhanced Bat Algorithm (EBA) to decide which features are most informative. EBA is based on combining conventional Bat Algorithm (BA) with crossover strategy.

Xin-She Yang created BA, a bio-inspired algorithm, in 2010, and it has been very effective [20]. This method replicates certain aspects of the micro-bat's echolocation traits in a straightforward manner [15, 20]. The fundamental framework of BA is constructed based on three key attributes of the micro-bat: (i) its echolocation behavior, (ii) the signal frequency it emits, characterized by a frequency f and a wavelength λ, and (iii) the loudness l_o of the emitted sound, which is utilized for prey localization [15, 20].

Bats can first use echolocation to gauge distance and, in a more enigmatic fashion, to determine how close they are to prey. The next thing they do is fly towards a position b_i at an unpredictable speed v_i in search of prey. After that, they produce noise with a constant minimum frequency (f_{min}) and variable wavelengths l_o. Afterwards, they can automatically modify the pulses' wavelength (or frequency) and rate of emission depending on how close their target is. In conclusion, the loudness can fluctuate in various ways, but it's fine if it drops from a high level l_o to a consistently low level l_{min}. The mathematical model for BA is calculated as follows:

I. For i^{th} bat the updated velocity, and location for the subsequent cycle $(n + 1)$ are determined by employing Eqs. (1) & (2) [20]:

$$VB_i(n+1) = VB_i(n) + (B(n) - B_G(n))f_i \tag{1}$$

$$B_i(n+1) = B_i(n) + VB_i(n+1) \tag{2}$$

II. The updated frequency at each cycle is computed using (3):

$$f_i = f_{min} + (f_{max} - f_{min})\beta \tag{3}$$

where β is a number chosen at random, $\beta\varepsilon[0, 1]$. Actually, f_i is utilized to regulate the extent and speed of bats' motion within the search area by modifying the velocity's rate of change.

Each bat generates a new solution locally through random walk. This is accomplished by employing Eq. (4).

$$B_i(n+1) = n_i(n) + \varepsilon\bar{l}_n \tag{4}$$

where ε is a random number, and ε belongs to the interval $[-1,1]$. $\bar{l}_n$ denotes the mean value of l for all bats in the subsequent cycle n, which bats utilize for exploration rather than exploitation as it progressively rises. The frequency value in the BA governs the spatial extent and direction of bats' movement. As bats approach their prey, the amplitude of their sound decreases and the rate at which they emit pulses increases. The revised values of loudness l and emission pulse rate r are determined by employing Eqs. (5) & (6) [20]:

$$l_i(n+1) = \alpha l_i(n) \tag{5}$$

$$r_i(n+1) = r_i(0)\left(1 - e^{-\gamma n}\right) \tag{6}$$

In this paper, a new methodology is introduced that combines conventional BA algorithm with crossover strategy. The proposed methodology called Enhanced Bat Algorithm (EBA). Figure 2 illustrated the steps of EBA. At first, the starting group and parameters of BA is created randomly. Then, calculate the fitness value of each bat to determine the strength of each solution using the following equation:

$$Fit(B_i) = 0.01*(1-\eta(B_i)) + 0.99*\left(\frac{number\ of\ selected\ features}{total\ number\ of\ features}\right) \tag{7}$$

After that, assign the best bat $B_{best}(n)$ and lowest bat solution $B_{lowest}(n)$ according to the fitness value. Then, perform crossover strategy between them according to the following equation:

$$B_c(n) = crossover(B_{lowest}(n), B_{best}(n)) \tag{8}$$

The *crossover*() function executes the crossover-based process on the bat that lost. Therefore, this technique involves further investigation and enhancements on the least performing bat by generating comparable random possibilities. These possibilities can be either selecting a specific number of features from the bat with the highest performance or keeping the current features based on Eq. (9).

$$B_G(n) = \begin{cases} B_{best}(n)\ if\ Fit(B_{best}(n)) > Fit(B_c(n)) \\ B_c(n) \quad otherwise \end{cases} \tag{9}$$

$B_G(n)$ refers to the global solution, which signifies the most optimal solution discovered by the group during cycle n. $B_{best}(n)$ represents the superior (best) bat at cycle n. Furthermore, $B_c(n)$ represents the newly created bat derived from the lowest and most optimal bats. The fitness value of the optimal bat at cycle n is denoted as $Fit(B_{best}(n))$, whereas the fitness value of the newly generated bat at cycle n is denoted as $Fit(B_c(n))$.

Due to its original purpose of addressing continuous search space problems, BA will undergo modifications to become binary to tackle discrete optimization problems, specifically feature selection [21]. The transformation of BA into binary BA was achieved by employing sigmoid transfer function that transforms the solutions from a continuous space to a discrete search space using the following equations [6].

$$sig\left(VB_i^j\right) = \frac{1}{1+e^{-VB_i^j}} \tag{10}$$

$$B_i^j(n+1) = \begin{cases} 0\ if\ rand(0,1) \geq sig\left(VB_i^j\right) \\ 1\ otherwise \end{cases} \tag{11}$$

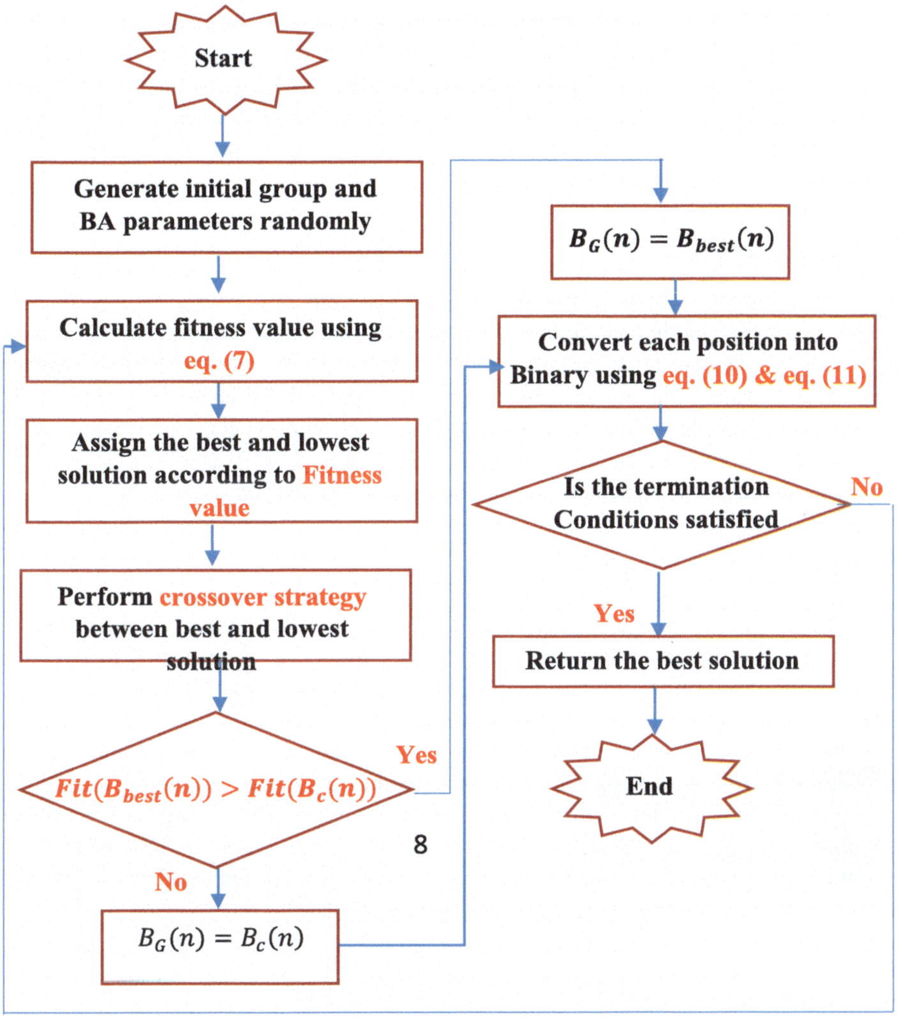

Fig. 2. Illustrated the steps of EBA

where e represents the base of the natural logarithm. Each bat in the group is assessed using the fitness function in Eq. (7), is used to evaluate each bat in the group, taking into account their new location $B_i^j(n+1)$. After updating the velocities and positions of each bat, a new solution is generated centered around the best solution that has been chosen. The new solution is generated only if the generated numerical value exceeds the pulse emission rate r_i, which can be calculated by employing Eq. (6). In addition, if the random number (rand) is less than the product of l_i and the difference between the fitness of B_i and the fitness of $B_G(n)$, then the loudness and emission rate are modified using Eqs. (5) & (6), and the new solution is deemed valid. These calculations are subsequently reiterated until the desired number of generations is attained.

Finally, the algorithm selects the most exceptional bat from the entire group of $B_G(n)$ as the output, and concludes. The features offered by 1 are the utmost crucial features for classifying breast cancer. Upon implementing the EBA algorithm on the breast's datasets, a total of seven features will be identified as the most relevant. The features include energy, angular second moment, homogeneity, autocorrelation, entropy, cluster prominence, contrast, and shade.

3.3 Patient Detection

The last component of the proposed model is patient detection, where the final decision is made regarding whether the input case is benign, malignant, or of normal condition. The primary goal of the detection procedure is to ascertain the category to which the new case will be allocated. Various ML algorithms are employed in this section to accomplish this objective. The algorithms used are Logistic Regression (LR), Naïve Bayes (NB), K-Nearest Neighbor (KNN), Decision Tree (DT), and Support Vector Machine (SVM). Therefore, the decision is made based on the highest number of votes.

4 Experimental Results

This section will evaluate the suggested BCDM. The MIAS mammogram dataset serves as the foundation for the implementation that we used [22]. The results presented in this paper are based on an online data compilation called the MIAS dataset. The MIAS dataset was created using Kaggle and is available to the public. There are 322 cases in the dataset; 209 are classified as normal, 61 as benign, and 52 as malignant breast cancer. The training set and the testing set are the two distinct categories into which the data has been separated. The model is trained using the training set. The testing set is employed to evaluate the suggested model's effectiveness. Thus, 226 cases, or 70% of the total, are in the training dataset, while 96 cases, or 30% of the total, are in the testing dataset. The distribution of the used dataset is displayed in Table 2.

Table 2. MIAS dataset distribution.

	Total	Malignant	Benign	Normal
Training	140	46	40	226
Testing	69	15	12	96
Total	209	61	52	322

4.1 Testing the Proposed BCDM

The proposed BCDM is now ready to be tested in this subsection. In order to demonstrate that our proposed strategy is effective, it is compared to several commonly used classification methods, as shown in Table 1. These methods are DLF [12], DNR-SVM [13], NHFSM [15], YoloV5 [16], and BCCNN [23]. The findings are displayed in Table 3.

Table 3. Comparison between BCDM and recently detection models

Used Method	Accuracy	Precision	Recall	F-score
DFL [12]	88%	89.2%	90.03%	89.61%
DNR-SVM [13]	90.2%	90.3%	87.09%	88.67%
NHFSM [15]	93%	91.87%	92.01%	91.93%
Yolo V5 [16]	95.5%	93.47%	88.5%	90.9%
BCCNN [23]	96.95%	93.5%	90.2%	91.8%
BCDM	98.9%	94.6%	91.06%	92.79%

As illustrated in Table 3, the proposed BCDM performs in comparison to the most recently used method. It introduces 98.9%, 94.6%, 91.06%, and 92.79% accuracy, precision, recall, and F-score in the same order. The reason is that, BCDM is based on an effective feature selection methodology (i.e., EBA) that helps the proposed BCDM reach the correct decision based on the most informative features. Then, BCDM is followed by YoloV5 as it introduces 95.5% accuracy, 93.47% precision, 88.5% recall, and 90.9% F-score.

5 Conclusions

Currently, breast cancer is a prevalent form of cancer that specifically impacts women. It ranks as the second most deadly. Both WHO and ACS concur that early detection of breast cancer enables physicians to make informed decisions regarding the most effective treatment options, ultimately leading to a higher likelihood of survival. This paper introduces a novel Breast Cancer Detection Model (BCDM) that utilizes AI. BCDM is comprised of three distinct phases: data preprocessing, feature selection, and patient detection. Initially, the mammogram images that have been previously used undergo preprocessing, which involves extracting relevant features and removing any outliers. The proposed Enhanced Bat Algorithm (EBA) is employed to select the most beneficial and efficient features. EBA is an enhanced iteration that integrates both traditional BA and crossover strategies. The crossover strategy produces random alternatives that are similar in quality to enhance the least optimal bat. Subsequently, these specifically chosen characteristics are inputted into the final stage, known as the patient detection phase. Based on the obtained results, BCDM demonstrates exceptional performance in terms of accuracy, precision, recall, and F-score. The percentages introduced are 98.9%, 94.6%, 91.06%, and 92.79%, respectively.

References

1. American Cancer Society. https://www.cancer.org/. Last accessed 5 February 2024
2. Moloney, B., O'Loughlin, D. Abd Elwahab, S., et al.: Breast Cancer Detection A Synopsis of Conventional Modalities and the Potential Role of Microwave Imaging. diagnostics **10**(2), 1–13 (2020)
3. Shaban, W., Abdullah, A., Ashraf, E.: An in-depth review of AI-based techniques for early diagnosis of breast cancer: Evaluation of CAD system design and classification methodologies. In: 2023 International Telecommunications Conference (ITC-Egypt), pp. 322–329. IEEE (2023)
4. Ramadan, S.: Methods Used in Computer-Aided Diagnosis for Breast Cancer Detection Using Mammograms: A Review. Journal of healthcare engineering, 1–21(2020)
5. Ahmed, Z., Mohamed, Kh., Zeeshan, S., et al.: Artificial intelligence with multi-functional machine learning platform development for better healthcare and precision medicine. Database, 1–35 (2020)
6. Shaban, W.: Early diagnosis of liver disease using improved binary butterfly optimization and machine learning algorithms. Multimedia Tools and Applications **83**(10), 30867–30895
7. Mushtaq, Z., Yaqub, A., Sani, Sh., et al.: Effective K-nearest neighbor classifications for Wisconsin breast cancer data sets. Journal of the Chinese Institute of Engineers **43**(1), 1–14 (2020)
8. Zakaria, N., Shah, Z., Kasim, S.: Protein structure prediction using robust principal component analysis and support vector machine. International Journal of Data Science **1**(1), 14–17 (2020)
9. Vijayarajeswari, R., Parthasarathy, P., Vivekanandan, S., et al.: Classification of mammogram for early detection of breast cancer using SVM classifier and Hough transform. Measurement **146**, 800–805 (2019)
10. Assiri, A., Nazir, S., Velastin, S.: Breast tumor classification using an ensemble machine learning method. Journal of Imaging **6**(6), 1–13 (2020)
11. Quist, J., Taylor, L., Staaf, J., et al.: Random Forest Modelling of High-Dimensional Mixed-Type Data for Breast Cancer Classification. Cancers **13**(5), 1–15 (2021)
12. Khan, S., Islam, N., Jan, Z., et al.: A novel deep learning-based framework for the detection and classification of breast cancer using transfer learning. Pattern Recognition Letters **125**, 1–9 (2019)
13. Shia, W., Chen, D.: Classification of malignant tumors in breast ultrasound using a pretrained deep residual network model and support vector machine. Comput. Med. Imaging Graph. **87**, 1–7 (2021)
14. Wang, S., Wang, Y., Wang, D., et al.: An improved random forest-based rule extraction method for breast cancer diagnosis. Applied Soft Computing **86**, 1–35 (2020)
15. Shaban, W.: Insight into breast cancer detection: new hybrid feature selection method. Neural Comput. Appl. **35**(9), 6831–6853 (2023)
16. Prinzi, F., Insalaco, M., Orlando, A., et al.: A yolo-based model for breast cancer detection in mammograms. Cogn. Comput. **16**, 107–120 (2024)
17. Xiao, Y., Wu, J., Lin, Z., Zhao, X.: Breast cancer diagnosis using an unsupervised feature extraction algorithm based on deep learning. In: 37th Proceedings on Chinese Control Conference (CCC), pp. 9428–9433. Wuhan, China (2018)
18. Kanagaraj, G., Suresh Kumar, P.: Pulmonary Tumor Detection by virtue of GLCM. Journal of Scientific & Industrial Research **79**(2), 132–134 (2022)
19. Zotin, A., Hamad, Y., Simonov, K., Kurako, M.: Lung boundary detection for chest X-ray images classification based on GLCM and probabilistic neural networks. Procedia Computer Science **159**, 1439–1448 (2019)

20. Alsalibi, B., Abualigah, L., Khader, A.T.: A novel bat algorithm with dynamic membrane structure for optimization problems. Appl. Intell. **51**, 1992–2017 (2021)
21. Shaban, W.: Detection and classification of photovoltaic module defects based on artificial intelligence. Neural Computing and Applications, 1–28 (2024)
22. Kaggle: https://www.kaggle.com/kmader/mias-mammography?select=Info.txt. Last accessed 10 February 2024
23. Abunasser, B., Al-Hiealy, M., Zaqout, I., Abu-Naser, S.: Convolution neural network for breast cancer detection and classification using deep learning. Asian Pac. J. Cancer Prev. **24**(2), 531–544 (2023)

UAV Control in Multi-access Edge Computing Using Markov Decision Process

Fadi Elghitani(✉) and Ibraheem M. Fayed

National Telecommunication Institute, Cairo, Egypt
{fadi.ghitani,ibrahim.fayed}@nti.sci.eg

Abstract. Cloud computing is the pradigim where service providers carry-on computational tasks for their customers. Multi-access Edge Computing (MEC) is a recent form of service where the computational tasks are carried in a distributed manner at the edge of the network to be as close to customer as possible. This new pradigim have the advantage of lowering the latency, making the pradigim suitable for real-time applications. One promising way to deploy MEC is to mount the servers over a UAV which can be moved dynamically to become closer to users. In this paper, we address the challenges associated with UAV-mounted servers like energy consumption and path routing. We propose a UAV control mechanism based on continuous-time Markov Decision Process (CT-MDP). We demonstrated the advantage of the proposed mechanism in terms of task-satisfaction rate and policy computational time.

Keywords: UAV · MEC · Path routing · Markov Decision Process

1 Introduction

Unmanned Aerial Vehicles (UAVs) have the potential to enhance mobile communication by providing temporary coverage in emergency situations, extending coverage to remote areas, improving network capacity in congested urban areas, and facilitating the deployment of next-generation technologies such as 6G. Multi-access Edge Computing (MEC) refers to a network architecture that brings computation and data storage closer to the location where it is needed, typically at the edge of the network [2]. This approach enhances the performance of applications by reducing latency and improving response times, as data processing occurs closer to the end users or devices. Key benefits of MEC includes 1) reduced latency by processing data locally rather than sending it to a central cloud and 2) bandwidth efficiency by offloading data processing and storage to the edge can help alleviate bandwidth bottlenecks, especially for applications that require real-time data access. MEC is becoming increasingly relevant in various industries, including telecommunications, smart cities, healthcare, and transportation, as the demand for real-time data processing continues to grow [16].

A. Abdelgawad et al. (Eds.): ISBCom 2024, LNNS 1268, pp. 480–491, 2025.
https://doi.org/10.1007/978-3-031-82377-0_39

There are two scenarios for UAV deployment: the first scenario is to assume the UAV is moving in a pre-defined path and the problem is to deduce the UAV optimal path prior to UAV dispatch. Such problem is usually referred to in literature as "UAV motion path planning". The second scenario is to assume the UAV is routed given the random information of the system which is revealed in real-time. This scenario is more challenging than the first due to the stochastic nature of its associated problems. Nevertheless, there are few studies which provided a general framework for real-time path control like in [9], which used Markov Decision Process (MDP) as a modeling approach and reinforcement learning as a solution method. The study in [21] had followed the same path, but it assumes obstacles exist in drone's path. Machine learning (ML) techniques are one of the most popular solution approaches needed for such systems. However, machine learning algorithms needs a huge amount of data to perform accurately. This attributed to testing these algorithms under various scenarios, which also requires discretizing any continuous variable, which will result in increasing computational complexity. Als, ML algorithms usually assume that the UAV is assigned to a private group of consumers instead of being publicly shared, making their scenarios less attractive economically [15,31].

2 Related Works

We begin by outlining the most prevalent presumptions found in the current research. Next, we explain how each study differs from the others and departs from these presumptions. Generally, one UAV is thought to be able to roam freely in the horizontal plane. The UAV does not change its height during its flight [7]. The drone's maximum speed limits its travelling range. The drone's energy consumption is based on the drone's kinetic energy which is equal to half the drone's mass multiplied by the square of the drone's speed [12]. The the CPU frequency ultimately determine the power consumption of the CPU [13]. Ground users are presumed to be mobile-less such that they do not change their locations from the time they send their requests to the time they receive the outcome [19]. The popular line-of-sight (LoS) path loss model is used to model the communication channel between the UAV and the consumers [11]. The Shannon capacity is used to compute the data rates from the users to the UAV and vice-versa [17]. There are main decision variables: Task scheduling and the UAV's location. While arranging the drone's position is known as drone's motion path planning, task scheduling is typically referred to as resource managment. The obhective function is to minimize the total energy consumption, which is divided into 1) communication energy, 2) flying energy, and 3) computational energy. Given all the previous assumptions, the research problem is formulated as a general, deterministic optimization problem that are usually solved using heurestic techniques [14].

Since we discussed the most common assumptions, here we discuss how the studies vary from each other. The objective functions can vary from minimizing data transfer rate [5] to maximizing total number of transmitted bits [26]. Some

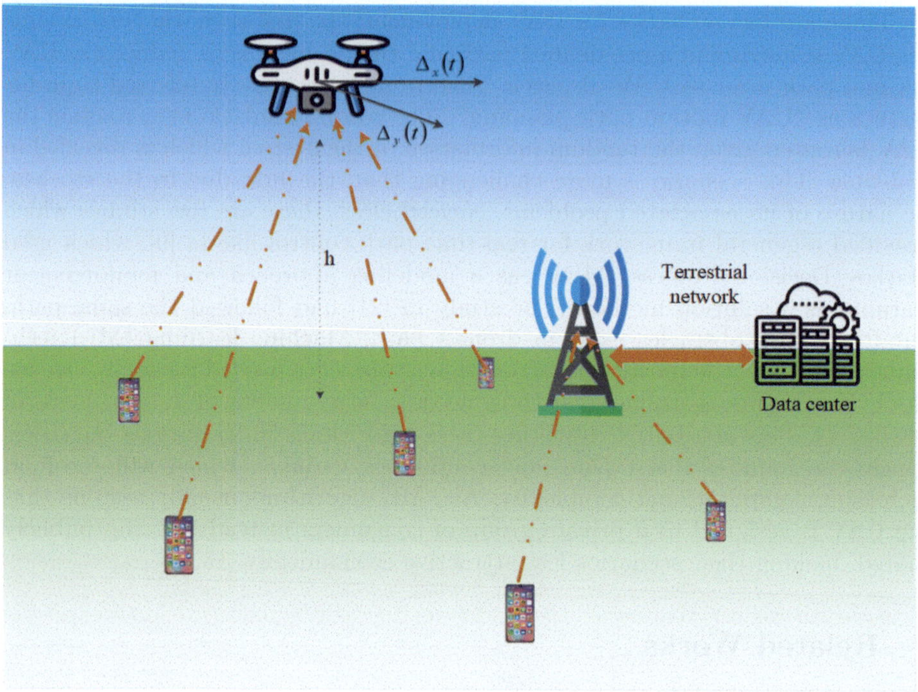

Fig. 1. System model under consideration which consists of the drone, mobile users, and terrestrial network.

studies employ a more thorough model for the UAV such that the model becomes more realistic [3,4,6,8]. Minimizing task processing time is considered in [3, 27,29]. As an alternative to the centralized optimization method, Study [23] suggested a decentralized optimization strategy which offers lower computational complexity. However, the drone is not allowed to change its position. The study in [32] cares about minimizing the energy consumption of the ground users which are basically sensors. Not only existing studies vary in their objective functions, they also vary in their service models: some studies assume minimum number of bits required [10,25,28]. Others set a minimum limit on the data rate [18,27,29]. Not every study uses SCA to solve the optimization problem. For instance, [22] provides a dual decomposition approach. Multi-UAV environment is also considered in some studies. For example, [20,30] assume resource allocation only to fixed-positions UAVs.

3 System Model

We assume the same system model used in [1]. Typically, we assume a single UAV that serves in a multi-access edge computing system (see Fig. 1). The servers inside the UAV attempts to execute these tasks in a greedy fashion. The computational task in its simplest for is assumed to consist of a number of bits,

ω. The task is then assigned to one of the virtual machines of the server, which has at most K virtual machines (VMs). The server also has a buffer that can store L computational tasks at most. Also, It is assumed that users are impatient, such that any tasks that was not executed by the UAV will be assigned to the data center. The UAV does not change its height during movement which is denoted by the constant h. The maximum distance a UAV can move given a specific time interval is defined by the maximum speed of the UAV, denoted by $\hat{v}$. We do not worry about if the drone runs out of charge, because before this happens a fully-charged UAV will be dispatched, which ensures the sustainability of our system. However, such replacement will not be done instantaneously; therefore, the flight time of the UAV must be long enough to allow more time for UAV replacement.

The rest of the model is defined as follows: 1) users' requests are generated in time and in space according to some random process, typically the Poisson Point Process (PPP), 2) the number of tasks that are needed to be computed per each request follows a general, finite-average distribution, and 3) the popular Line-of-Site (LoS) propagation model is assumed, whose gain is defined by:

$$g(m) = \frac{d_0}{(x^R(m))^2 + (y^R(m))^2 + h^2}, \tag{1}$$

where $\{x^R(m), y^R(m)\}$ represents the location of user m with respect to the current drone's location. Users' orthogonality is assumed such that power transmitted or received by a user does not affect the data rates of other users. Accordingly, data rate achieved by a user can be defined according to the following logarithmic function:

$$R(m) = R_0 \log_2 \left\{ \frac{P^r(m)}{N} + 1 \right\}, \tag{2}$$

where N is the channel interference power and $P^r(m)$ is the transmitted power of the requesting user. User's data rate is constant thanks to user power control, therefore, $R(m) = R$, under any circumstance. Accordingly, the received power is maintained to an acceptable level:

$$P^r(m) = N \left(2^{R/R_0} - 1 \right). \tag{3}$$

From which, one can deduce the following formula for power transmission level:

$$P^t(m) = \frac{P^r(m)}{g(m)} = P_0 \left[(x^R(m))^2 + (y^R(m))^2 + h^2 \right], \tag{4}$$

where $P_0 = \frac{N}{d_0} \left(2^{R/R_0} - 1 \right)$. The communication power need by the requesting user to transmit a single task is given by:

$$W^u(\alpha(t)) = T_u P_0 \left[(x^R(m) - \Delta_x(t))^2 + (y^R(m) - \Delta_y(t))^2 + h^2 \right]. \tag{5}$$

where T_u represents the upload duration. The UAV will start processing the task as soon as it is received, which will take several processor intervals and produce a lot of output bits. The required number of processor intervals is defined by $\omega\kappa(m)$, such that $\kappa(m)$ can take any finite-average general distribution. User application will generate a number of outputs which is exactly the same as the number of input bits. Accordingly, the energy required for task execution is determined by:

$$\begin{aligned} W^c(\alpha(t)) &= P^{\text{CPU}} T_u \frac{\text{Total cycles}}{\text{Maximum cycles}} \\ &= P^{\text{CPU}} T_u \frac{\kappa(m)\omega}{N_c f_c T_u} = P^{\text{CPU}} \frac{\kappa(m)\omega}{N_c f_c}, \end{aligned} \tag{6}$$

where P^{CPU} is the maximum power of the CPU. The maximum number of virtual machines (VMs) is chosen so that the data rate of transmission does not cause congestion. Accordingly, the computing delay of the task should be less than the transmission delay, i.e.,

$$\frac{\omega \hat{\kappa} K}{N_p N_c f_c} < \frac{\omega}{R}, \tag{7}$$

and therefore, calculating the required number of logical machines can be deduced by:

$$K < \frac{N_p N_c f_c}{R \hat{\kappa}}. \tag{8}$$

On the other hand, the drone consumes energy for movement. Such energy is independent of drone's location, and is given by:

$$W^f(\alpha(t)) = \frac{1}{2}\text{mass } v^2(t) = \frac{1}{2}\text{mass}\frac{\Delta_x^2(t) + \Delta_y^2(t)}{T_s^2}. \tag{9}$$

4 Problem Formulation

Our problem is similar to any ordinary optimization problem. We need to define objective function and decision variables. One can use the following for the former:

$$C(\alpha(t)) = \beta s(\alpha(t)) - W^f(\alpha(t)), \tag{10}$$

where $s(\alpha(t))$ represents the number of tasks accepted as a result of decision α and β represents a parameter that balances between task satisfaction and power consumption. For instance, if we choose β to be too high then the UAV tends to execute large number of tasks even if the power consumption is high. On the other hand, if β is chosen to be too low then the controller become more conservative and reduces the drone's movement. As mentioned earlier, the drone can always be replaced, therefore we can assume infinite time horizon, i.e.,

$$\mathbf{P1}: \max_{\psi \in \Psi} \bar{c} \triangleq \lim_{T \to \infty} \frac{1}{T} \sum_{t=0}^{T-1} \mathbb{E}\left\{C(\alpha(t))\right\} \tag{11}$$

Problem **P1** is a traditional stochastic optimization problem, which can be solved using Markov Decision Process (MDP) method. A challenging aspect for using MDP method is to define the system state such that the problem is accurately represented. Meanwhile, the resulting state space must be small enough to make the solution computationally feasible. If we choose to use discrete-time MDP method, then decisions are made every fixed period of time, T_s. System state can be defined by the number of users requesting to be served during T_s, their locations, and their demands of tasks. This definition is accurate, yet it can result in a huge state-space, since there are many variables included in that definition. Alternatively, we choose to use continuous-time MDP method, where decisions are event-driven. In other words, no decision is made unless there is a change in the system state. By applying CT-MDP to our problem, decisions will be made whenever a user requests to be served. Since the probability that more than one user request to be served at the exact moment is zero, the system state can be much simplified and defined as: $\phi(t) = \{r(t), d(t), q(t)\}$, where $r(t)$ represents the distance between the user and the UAV, $d(t)$ represents the user's demand, and $q(t)$ represents the number of tasks being currently held by the UAV. The state space will become three-dimensional as shown in Fig. 2. Also, the decision $\alpha(t)$ will be defined as the number of new tasks admitted to the UAV.

In order to complete the definition of the CT-MDP problem, we need to define the transition probability matrix the reward(cost) function. The latter can be easily deduced from (10). Let $\mathbb{P}\{s_2|s_1, \alpha_1\}$ denotes the probability that the system transits to state s_2 given the system was residing in state s_1 and decision α_1 has been made. Notice that request's location and its number of tasks are independent of each other and both are independent of the decision and number tasks being served by the UAV. Therefore, the transition probability can be written as:

$$\mathbb{P}\{s_2|s_1, \alpha_1\} = \mathbb{P}\{r_2\}\,\mathbb{P}\{d_2\}\,\mathbb{P}\{q_2|q_1, \alpha_1\} \tag{12}$$

Given the system model described in Sect. 3, both $\mathbb{P}\{r_2\}$ and $\mathbb{P}\{d_2\}$ are uniformly distributed. The last term can be written as:

$$\mathbb{P}\{q_2|q_1, \alpha_1\} = \mathbb{P}\{q_2 - \alpha_1|q_1\} \tag{13}$$

which is the probability that $z = q_1 - (q_2 - \alpha_1)$ tasks were served before the arrival of the next request. The probability that a task is served before the arrival of the next request can be given by:

$$p = \frac{1/T_u}{(1/T_u) + \lambda} = \frac{1}{1 + \lambda T_u} \tag{14}$$

Accordingly, the probability in (13) is given by the binomial distribution:

$$\mathbb{P}\{q_2|q_1, \alpha_1\} = {}^{q_1}\mathbb{C}_z \left(\frac{1}{1 + \lambda T_u}\right)^z \left(\frac{\lambda T_u}{1 + \lambda T_u}\right)^{q_1 - z} \tag{15}$$

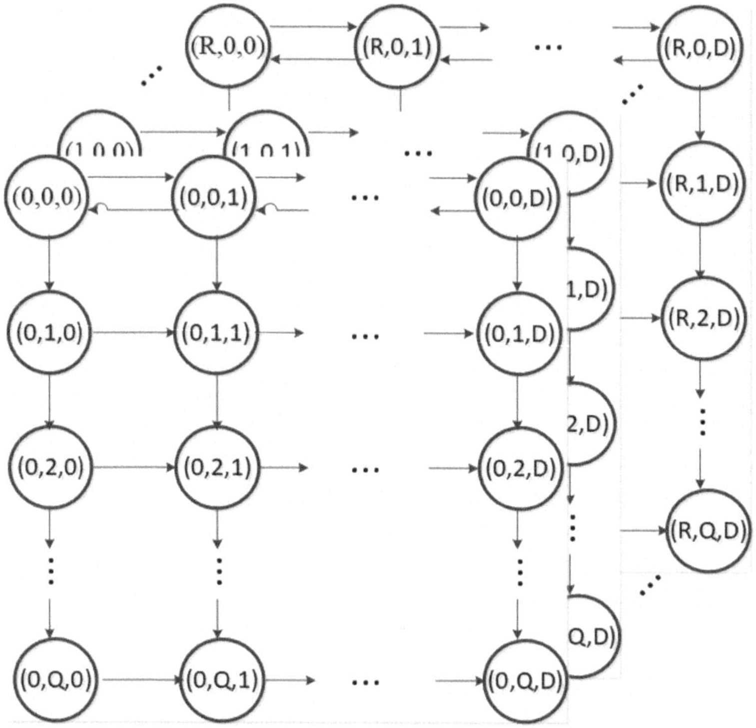

Fig. 2. A three-dimensional state-space resulting from our state definition. The state of the system is defined by three variables: relative distance of the newly arrived request to the UAV, number of tasks possessed by that request, and number of tasks currently served by the UAV.

5 Numerical Results

For the numerical results, we assume the following [1] task density is 8×10^{-4}, maximum number of bits per task is 10^5, maximum drone speed is 19 m/s, drone's mass is 0.9 kg, average data rate is 1 Mbps, maximum uplink power is 3 W, and drone's battery is 77 kWh. Each request takes a value between 1 and 10 tasks to be performed, while the number of CPU cycles needed to execute one bit ranges from 1000 to 2000. The parameters for UAV model can be found in [1] which is assumed to have one core i5 processor [24]. Two key performance metrics are considered: task satisfaction rate (Fig. 3) and UAV flight time (Fig. 4). The benefit of our methodology is depicted in Fig. 5, where the computational time needed to deduce the routing policy is almost fixed, regardless of the requests' density, given the fact that our simulation was performed on an Intel ®Core™i3-10100F CPU @ 3.60 GHz and 16 GB RAM machine.

We then investigate how other parameters can affect system's performance. Such parameters can be, for example, drone's height or mobile transmitting power. This is depicted in figures (Fig. 6) and (Fig. 7). Performance can be

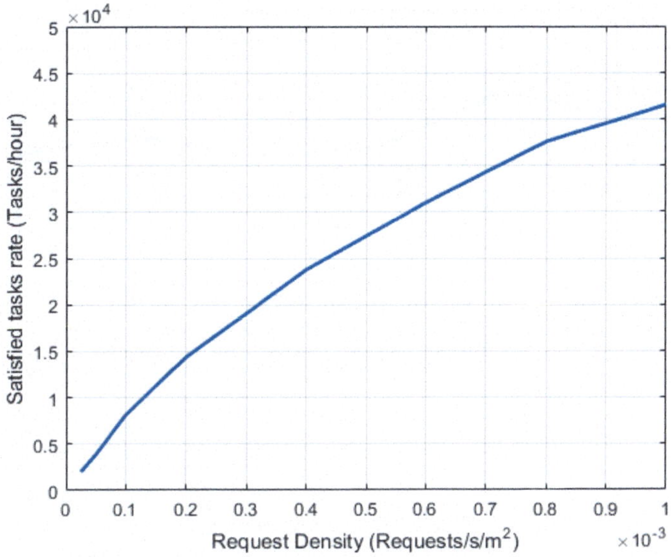

Fig. 3. Tasks rate increases as users' requests increase.

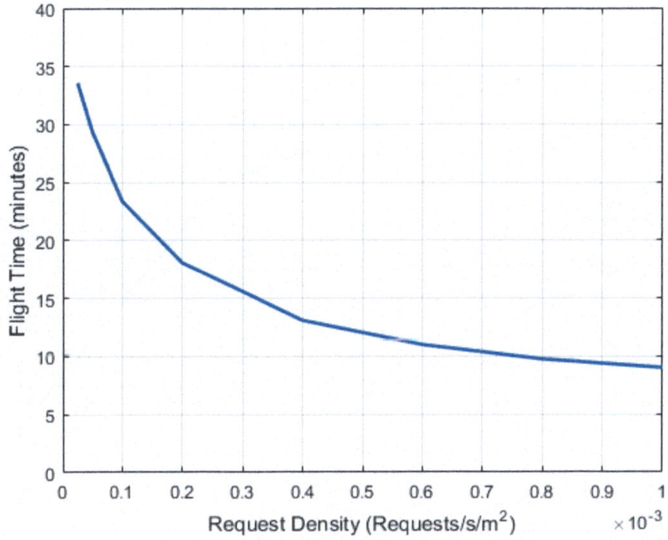

Fig. 4. Drone stays in air for a shorter time as users' requests increase.

improved by either increasing the power or decreasing the drone's height. Unfortunately, we cannot control these two parameters. The drone's height is defined by safety regulations also we cannot force mobile users to use elaborate equipment.

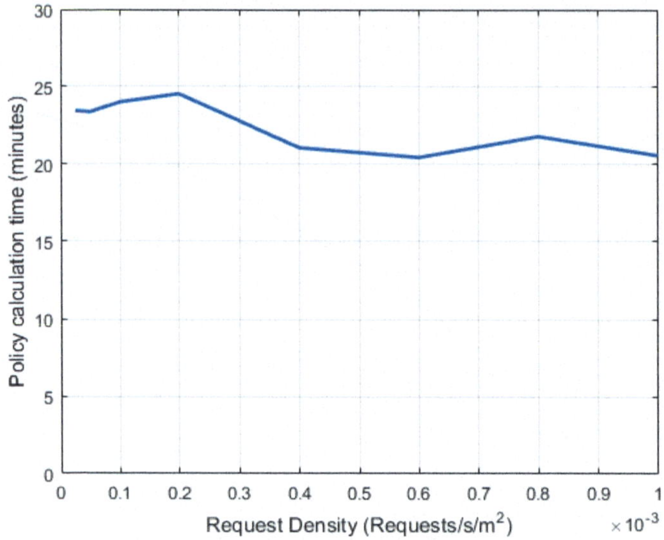

Fig. 5. The time (in minutes) required by our machine to deduce the policy.

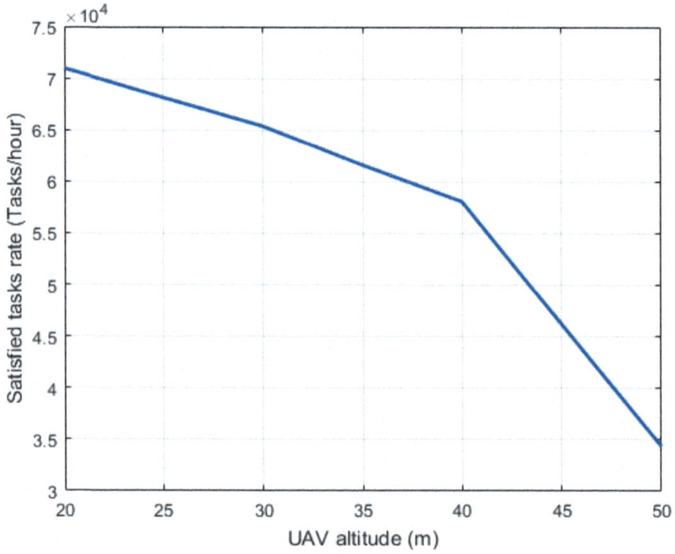

Fig. 6. As the UAV height increase, the performance decreases

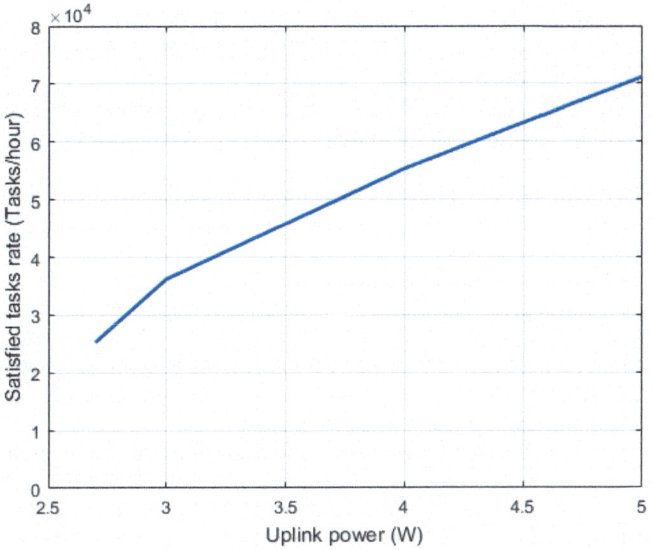

Fig. 7. As the uplink power increase, the performance increases

6 Conclusions

In this paper, we have discussed how to use a UAV to provide computational services to ground users, under the umbrella of multi-access edge computing. We developed a routing policy based on continuous-time Markov Decision Process to maximize tasks' satisfaction rate while taking drone's energy consumption into consideration. Our methodology offers a large task satisfaction rate and yet requires an adequate computational complexity. Numerical results shows that the computational time needed to deduce the routing policy is almost fixed, regardless of the requests' density.

References

1. Elghitani, F.: Dynamic UAV routing for multi-access edge computing. IEEE Trans. Veh. Technol. (2024)
2. Filali, A., Abouaomar, A., Cherkaoui, S., Kobbane, A., Guizani, M.: Multi-access edge computing: a survey. IEEE Access **8**, 197017–197046 (2020)
3. Gao, X., Zhu, X., Zhai, L.: Minimization of aerial cost and mission completion time in multi-UAV-enabled IoT networks. IEEE Trans. Commun. **71**(9), 5335–5347 (2023)
4. Goudarzi, S., Soleymani, S.A., Wang, W., Xiao, P.: UAV-enabled mobile edge computing for resource allocation using cooperative evolutionary computation. IEEE Trans. Aerosp. Electron. Syst. **59**(5), 5134–5147 (2023)
5. Javaid, S., et al.: Communication and control in collaborative UAVs: recent advances and future trends, vol. 24, pp. 5719–5739. IEEE (2023)

6. Ji, B., Li, Y., Zhou, B., Li, C., Song, K., Wen, H.: Performance analysis of UAV relay assisted IoT communication network enhanced with energy harvesting. IEEE Access **7**, 38738–38747 (2019)
7. Jiang, Y., Xu, X.X., Zheng, M.Y., Zhan, Z.H.: Evolutionary computation for unmanned aerial vehicle path planning: a survey. Artif. Intell. Rev. **57**(10), 267 (2024)
8. Jingming, X., Yufeng, L., Ling, T.: Research on multi-UAV energy consumption optimization algorithm for cellular-connected network. J. Commun./Tongxin Xuebao **44**(2) (2023)
9. Jones, M., Djahel, S., Welsh, K.: Path-planning for unmanned aerial vehicles with environment complexity considerations: a survey. ACM Comput. Surv. **55**(11), 1–39 (2023)
10. Jung, S., Jeong, S., Kang, J., Kang, J.: Marine IoT systems with space-air-sea integrated networks: hybrid LEO and UAV edge computing. IEEE Internet Things J. **10**(23), 20498–20510 (2023)
11. Kumar, K., Kumar, N.: Region coverage-aware path planning for unmanned aerial vehicles: a systematic review. Phys. Commun. **59**, 102073 (2023)
12. Liu, S., Jin, Z., Lin, H., Lu, H.: An improve crested porcupine algorithm for UAV delivery path planning in challenging environments. Sci. Rep. **14**(1), 20445 (2024)
13. Luo, J., Tian, Y., Wang, Z.: Research on unmanned aerial vehicle path planning. Drones **8**(2), 51 (2024)
14. Luo, J., Wang, Z., Xia, M., Wu, L., Tian, Y., Chen, Y.: Path planning for UAV communication networks: related technologies, solutions, and opportunities. ACM Comput. Surv. **55**(9), 1–37 (2023)
15. Ouamri, M.A., Barb, G., Singh, D., Adam, A.B., Muthanna, M.S.A., Li, X.: Nonlinear energy-harvesting for D2D networks underlaying UAV with SWIPT using MADQN. IEEE Commun. Lett. **27**(7), 1804–1808 (2023)
16. Porambage, P., Okwuibe, J., Liyanage, M., Ylianttila, M., Taleb, T.: Survey on multi-access edge computing for internet of things realization. IEEE Commun. Surv. Tutor. **20**(4), 2961–2991 (2018)
17. Poudel, S., Arafat, M.Y., Moh, S.: Bio-inspired optimization-based path planning algorithms in unmanned aerial vehicles: a survey. Sensors **23**(6), 3051 (2023)
18. Sharma, J., Mehra, P.S.: Secure communication in IoT-based UAV networks: a systematic survey. Internet Things, 100883 (2023)
19. Shen, Q., Zhang, D., He, Q., Ban, Y., Zuo, F.: A novel multi-objective dung beetle optimizer for Multi-UAV cooperative path planning. Heliyon **10**(17) (2024)
20. Shen, Y., Li, B., Zhang, R., Cheng, X., Yang, L.: A flexible load balancing scheme in multi-uav-enabled wireless networks. IEEE Trans. Veh. Technol. (2024)
21. Song, Z., Ma, C., Ding, M., Yang, H.H., Qian, Y., Zhou, X.: Personalized federated deep reinforcement learning-based trajectory optimization for multi-UAV assisted edge computing, pp. 1–6 (2023)
22. Tang, X., Zhang, H., Zhang, R., Zhou, D., Zhang, Y., Han, Z.: Robust trajectory and offloading for energy-efficient UAV edge computing in Industrial Internet of Things. IEEE Trans. Ind. Inf. **20**(1), 38–49 (2023)
23. Wang, M., Zhang, L., Gao, P., Yang, X., Wang, K., Yang, K.: Stackelberg-game-based intelligent offloading incentive mechanism for a multi-UAV-assisted mobile-edge computing system, vol. 10, pp. 15679–15689. IEEE (2023)
24. Website of Intel: (2022)
25. Wu, G., Liu, Z., Fan, M., Wu, K.: Joint task offloading and resource allocation in multi-uav multi-server systems: an attention-based deep reinforcement learning approach. IEEE Trans. Veh. Technol. (2024)

26. Wu, X., et al.: Rate maximization of UAV-assisted ambient backscatter communications. IEEE Wirel. Commun. Lett. (2023)
27. Xie, J., Fu, Q., Jia, R., Lin, F., Li, M., Zheng, Z.: Optimal energy and delay tradeoff in UAV-enabled wireless sensor networks. Drones **7**(6), 368 (2023)
28. Yan, M., Xiong, R., Wang, Y., Li, C.: Edge computing task offloading optimization for a uav-assisted internet of vehicles via deep reinforcement learning. IEEE (2023)
29. Zhan, C., Hu, H., Wang, J., Liu, Z., Mao, S.: Tradeoff between age of information and operation time for uav sensing over multi-cell cellular networks, vol. 23, pp. 2976–2991. IEEE (2023)
30. Zhang, Y., Gong, Y., Guo, Y.: Energy-efficient resource management for multi-UAV-enabled mobile edge computing. IEEE Trans. Veh. Technol. (2024)
31. Zhao, L., Qu, S., Xu, H., Wei, Z., Zhang, C.: Energy-efficient trajectory design for secure SWIPT systems assisted by UAV-IRS. Veh. Commun. **45**, 100725 (2024)
32. Zhu, Y., Wang, S.: Flying path optimization of rechargeable UAV for data collection in wireless sensor networks. IEEE Sens. Lett. **7**(2), 1–4 (2023)

Classification of Brain-MRI Images Using a Stacked-Deep-Network Ensemble Model into Multiple Region-Based Classes

Deepika Kumar[1](✉), Varun Srivastava[2], Shilpa Gupta[3], and Akhtar Jamil[4]

[1] Department of Computer Science and Engineering, Bharati Vidyapeeth's College of Engineering, New Delhi, India
deepika.kumar@bharatividyapeeth.edu
[2] Department of Computer Science and Information Technology, Jaypee Institute of Information Technology, Noida, Uttar Pradesh, India
[3] Department of Computer Science and Engineering, JIMS Engineering Management and Technical Campus, Greater Noida, Uttar Pradesh, India
[4] Department of Computer Science, National University of Computer and Emerging Sciences, Islamabad, Pakistan
akhtar.jamil@nu.edu.pk

Abstract. Brain tumour besides being lethal can also affect other human organs on a long-term basis if not detected at an early stage. Based on the region of its presence and speed of growth, it can be classified as Glioma, Meningioma. or Pituitary. In the given work, a Stacked Neural Net approach for the classification of brain tumour is presented. The images are first pre-processed using Histogram equalization and One Hot Encoding. Thereby, the improved images are fed into 3 popular pre-trained deep networks viz. Res-Net-50, VGG-16, and Xception as level-0 models in parallel. The features extracted are then fed into a neural network as level 1 learner for the final classification. A publicly available Figshare dataset has been used for testing the performance. The suggested approach has been thoroughly evaluated against its base models using accuracy, precision, and recall metrics and also the performance has been analysed using accuracy and loss curves along with epochs. The suggested model achieved a validation and training accuracy of 97.3% and 98.12% respectively. It has been observed that the suggested model referred here as Stacked-Neural-Net outperforms when compared to the conventional deep learning methods and state-of-the-art methods. An improvement of 3.35%, 1.3% and 1.7% is achieved over Inception ResNetV1, a basic CNN and Alex-Net CNN models.

Keywords: Brain MRI · Ensemble · Classification · Stacked deep network

1 Introduction

Human Brain is the most crucial and delicateorgan in the body. Even a minor disorder in the brain can lead to significant disruptions in other organs. A brain tumor, characterized by the abnormal growth of cells inside the brain, exemplifies this phenomenon [1]. It can

A. Abdelgawad et al. (Eds.): ISBCom 2024, LNNS 1268, pp. 492–505, 2025.
https://doi.org/10.1007/978-3-031-82377-0_40

be either primary or secondary metastatic based on the type of cells responsible for their origin [2] and Since can induce life-threatening conditions, therefore early diagnosis is crucial in many cases. There are many forms of brain tumor. Glioma (Gli.) is a violent one that forms inside the brain's supporting tissue and proliferates. Glioblastomais grade IV brain malignancy. Meningioma (MeG.) is a kind of malignant growth that starts inside the cells that line the boundary of brain and encompasses till spine [3]. The majority of growths are in fact benign and can be precisely eliminated [4].

Despite ongoing advances to some degree and completely computerized approaches for brain tumor segmentation and identification, there are as yet various critical issues with this work, primarily owing to variations in the form, dimensions, size, and consistency of the diverse appearance of the brain tumor [5]. The pre-processing phase of the tumor detection procedure on images obtained through magnetic resonance imaging (MReI), offers extensive information on the pathophysiology of brain tumors. These pre-processed images are then used for extraction of features for designing a Machine Learning (ML) or Deep Learning (DL) model for final MR-image classification.

In recent times, ensemble models have gained popularity, wherein various algorithms or classification models are combined to solve a computational problem. An ensemble learning method called stacking uses meta-learning to fuse various machine and deep learning models [6]. It works by training base-level algorithms on a whole training data set and then training a meta-model on the final results of all base-level models as a feature [7]. This technique is largely used to enhance precision and accuracy (classification, forecasting, statistical models, etc.) or to lessen the risk of unintentional poor modelling [8]. Transfer learning is an emerging and effective deep learning technique that involves applying an already trained model to a new problem using fresh and unique data, which may or may not be new to the model [9].

The organization of the paper in the form of sections is explained as follow. Section 2 explains the allied studies already done in this area. The approaches used in the image pre-processing phases are demonstrated in Sect. 3. In addition, a detailed description of classification using ensemble learning, its structure, and a flow diagram of the overall process and its architecture is explained. Section 4 explains the experimental outcomes as well as comparisons between approaches which is followed by the conclusion section.

2 Literature Review

2.1 Machine Learning-Based Approaches

Researchers have used many ML algorithms have to extract features and categorizeM-ReI images of brain tumour. MReI is considered the most preferable option for image acquisition due to its skills to detain soft tissue features [10]. The use of atlases improved the quality of segmentation, leading to a 1.5% rise in mean dice scores. An 85.5% dice score has been achieved by authors in [11], which starts by understanding batches of data using SVMs in an ensemble fashion, then dismisses the accurately classified input and a fresh SVM is trained on the incorrectly classified sample spot, and finally uses the weighted sum of all the accurately trained SVMs and of those which were based on the incorrectly classified points to produce the ultimate classification. The comparison indicates that the suggested technique has potential. In [12] authors proposed study utilizes CNN and SVM models for brain MRI image classification. The SVM classifier, employing HOG characteristics, categorizes the images into four classes: no tumor, glioma, meningioma, and pituitary. The CNN model, trained with three convolutional layers and utilizing a softmax classifier, achieves an accuracy estimate of 97%. In comparison, the SVM model estimates an accuracy of 92%. H. Mohsen et al. in [13] suggested a CAD approach to find brain cancers. To train a Support Vector Machine and categorize sarcoma, glioblastoma, or metastatic bronchogenic carcinoma as malignant tumors, this technique employs sequential minimal optimization. An automatic and supervised MRel Brain abnormalities identification system was created by S. T. Kebir et al. in [14]. The suggested approach uses a CNN deep learning network, segmenting data using the K-Means algorithm, and discriminating between typical and anomalous brain instances.

2.2 Ensemble Approach

In [15] Authors constructed a classification method that involves four key steps: brain tumor segmentation, ROI selection, feature extraction, and classification. The main challenge is a shortage of annotated training samples, addressed by randomizing scale, orientation, and flip during training. The method achieves impressive classification rates (96.6% for meningioma, 98.1% for glioma, 95.3% for pituitary tumor) on a public database, demonstrating its efficiency and robustness. In [16] features have been uprooted using the histogram of oriented gradients and Haar-discrete wavelet transform. The number of features is subsequently reduced using the Whale optimization procedure. The proposed hybrid scheme based on bagging ensemble techniques has the highest attained accuracy of 96.9% and a mean accuracy of 96.4%. Authors in [16] demonstrated and explained a neural net ensemble model using a novel technique called dynamic ensemble learning (DEL). Individual Neural Network learning is defined by the DEL technique, which specifies the ensemble dimension, the amount of constructive independent Neural Networks, the count of hidden nodes of constructive–pruning individual Neural Networks, and distinct training samples. Authors in [17] make use of a cascade of trained ensemble modules. A binary decision tree is used in the initial ensemble, whereas a random forest with neighbour-pixel correlation has been used as the second. Shafi et al. [18] presented atechnique to categorize lesions associated with auto-immune illness or neoplasms. The procedure entails feature-selection, feature-extraction, and classification. During the preprocessing phase, the lesion and tumour regions of interest (ROI), Collewet-normalization, and Lloyd-max-quantization techniques are utilized. The base learner consists of a Support-Vector-Machine (SVM) classifier and a majority voting prediction model. This model achieves an accuracy of 97.95% and 97.74% for training and testing respectively. Kang et al. [19] proposed an ensemble approach that employs deep features extracted from multiple pre-trained CNN models, along with ML classifiers. The processed images are fed into ML-trained models and the resulting features are evaluated using ML classification pedagogies.

2.3 Deep Learning- Based Approaches

Rehman et al. proffered a CNN approach [20] for classifying MR images of brain tumor. Here a separate CNN has been used to extract the affected area due to tumour and then obtained features are fed into a correlation selection model to fine most optimal features. Finally, a feed-forward neural network(NN)has been used for classification. Three BraTS datasets have been used and an accuracy of 98.32% has been achieved. A deep CNN model with six layers [21] is trained for classifying brain tumor images. The model was better in terms of time complexity when compared with sota algorithms and achieved an accuracy of up to 99% for a live clinical dataset. Authors in [22] classified brain tumor images using a nature-inspired Res-Net 152 hybrid model. Here the images are first pre-processed using Otsu binarization and then features like correlation, energy, entropy, contrast, and homogeneity are computed using grey-level occurrence. Thereby the images are classified when the batch-normalization part is removed in ResNet-152. The accuracy has been achieved up to 99.7%. Rai and Chatterjee [23] proposed U-Net Model with fewer layers and used minimum time complexity which is referred to as the LeU-Net model. It gives an accuracy of up to 98% on cropped images and 94% on uncropped ones. The model outperformed CNN architectures like LeNet, UNet and vgg-16. In [24], authors suggested a hybrid for classification. The images undergo pre-processing by removing the skull part, and then median filteringand segmentation were implemented. The features are then extracted using a CNN model which is finally classified using SVM. The model yielded an accuracy of 97.5%. Authors proposed a modified GoogleNet [25] for extracting features from MR images. The features are then fed into a classifier for final classification. The classifier used is either SVM or k-NN (Nearest Neighbor). Three classes of brain tumorsviz. Gli., Pituitary(Pit.) and MeG. have been classified using the model. The overall classification accuracy has been found to be 98%. Ozyurt et al. presented a Neutrosophy-based CNN to classify brain tumors as either benign or malignant. Brain MR-Images(Brain-MREI) are segmented using a neutrosophic-set-expert max-fuzzy-sure entropy method. The features extracted are then classified using SVM or KNN with 5-fold cross-validation. An accuracy of 95.62% has been obtained for the proposed model [26]. Authors [27] presented an architecture with two CNNs, an estimator and a fully-connected network. A neural autoregressive distribution estimation is used for the density estimation stage. An accuracy of 95.5% has been obtained for the proposed architecture.

3 Materials and Methods

The following section emphasizes and outlines the proposed methodology of Stacked Neural Net for the classification of Brain Tumor. This section is broadly divided into different subsections. The first subsection explains the dataset used for the experimentation which is followed by the data pre-processing and model architecture. Figure 1 depicts the flow diagram of the proposed approach:

3.1 Materials

Dataset This experimentation utilised the Figsharedataset. The dataset has 3154 Brain-MREIeI slices gathered from 233 people with MeG., Pit., and Gli. Brain tumours. The dataset is provided in MATLAB ". Mat" format and is open to the public. The mat file contains one patient ID, a distinct label specifying the Brain-Tumor (BT) type, 512 MReI

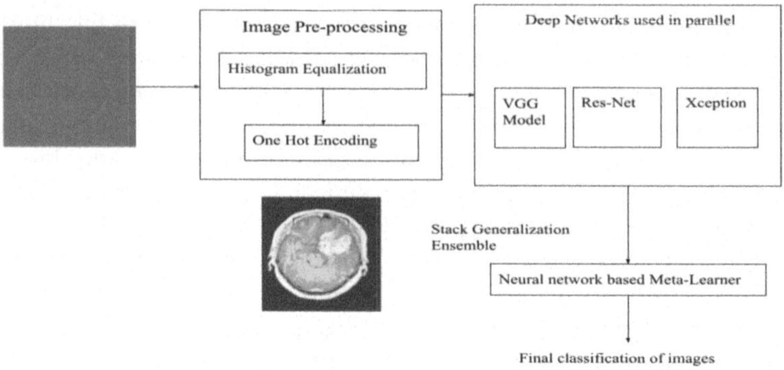

Fig. 1. The flow diagram of the proposed ensemble Stacked Neural Net approach.

image data in uint16 format, and a vector holding tumor border along with discrete point coordinates. For classification, a 512 X 512-dimension image dataset has been employed in uint16 format (MReI pictures) with labels. Then mat images are converted to image file format (.jpg). 113 images with the negative label have been included i.e. images with no tumor [28].

Data Preprocessing

In this step, the finer details of the MReI pictures are enhanced and noise is removed from the photos. For image improvement, the following method have been used.

Histogram Equalization is one of the most advanced approaches for increasing an image's contrast is to change the image's intensity histogram to a suitable shape. Depending on which modification function is used, it can be split into two branches. As a result, the regions with low contrast got a lot more contrast. Using this technique to adjust image intensity levels in order to improve visual contrast, the most common intensity values were efficiently spread apart.

One Hot Encoding

One-hot encoding [29] is then used for pre-processing the images and generate a new binary feature vector. The encodings used are given in Table 1.

Table 1. One Hot Encoding

S. No.	Encoding	Type of cancer
1	[1, 0, 0, 0]	Healthy image
2	[0, 1, 0, 0]	MeG. Brain Tumor (MBT)
3	[0, 0, 1, 0]	Gli. Brain Tumor (GBT)
4	[0, 0 0, 1]	Pit. Brain Tumor (PBT)

3.2 Methods

Convolutional Neural Network
CNNs are scale, shift and distortion invariants and this property makes them a very robust architecture to be used for classification [30].

Basic Architecture of a CNN:

1. Input layer: it feeds the input into the CNN. The exact no. of neurons in this layer are same as the number of inputs.
2. Convolutional layer: Here the feature selection is done. A kernel is selected to convolve over the image. Here hyper-parameters like stride (number of pixels that can be skipped while convolution), kernel size, and number of kernels etc. can be tuned to improve the performance of CNN. It also applies a non-linear activation function on the feature maps. It fires or deactivates neurons of the network depending on their activation values.
3. ReLU (Rectified Linear Unit): It is used for a threshold operation on its input by setting any value less than zero as zero. A transfer function can be used for it. It also performs local averaging for a group of pixels and reduces the filter size for the next layer. This reduction in the size of the filter is also known as down sampling.
4. Pooling layer: It minimizes the feature maps dimensions retrieved from the previous convolutional layer and also avoids overfitting. Here a 2-Dimensional (2D) filter slides over the received feature map to generate a single value. The size of the filter is termed as pool size and if max-filter is applied to reduce the feature map then it is termed as max pooling.
5. Dense layer: Every single neuron present in the dense layer are connected to the next layer of CNN and known as FCN or fully connected layer. This layer is used to adjust the input parameters by flattening them and then feeding into the final classification layer.
6. Classification layer: Number of output classes in the data determines the no. of neurons in this layer. Softmax classifiers give probabilities for each class label and the predicted target class is the one which has the maximum probability among the others.

3.2.1 Working of a CNN

For a given input image, I with C channels, K filters with dimension k1 x k2 and bias b for each filter, the convolution operation carried out on the first layer is given in Eq. (1) [31]

$$(\boldsymbol{I} * \boldsymbol{K})_{(i,j)} = \sum\nolimits_{p=0}^{k1-1} \sum\nolimits_{\mathrm{q}=0}^{\mathrm{k}2-1} \sum\nolimits_{\mathrm{c}=0}^{\mathrm{C}} \boldsymbol{K}(p,q,c) * \boldsymbol{I}(i+p,j+q,c) + \boldsymbol{b} \quad (1)$$

If the input image is grayscale, then C = 1, and corresponding convolution operation is shown in Eq. (2)

$$(\boldsymbol{I} * \boldsymbol{K})_{(i,j)} = \sum\nolimits_{p=0}^{k1-1} \sum\nolimits_{q=0}^{k2-1} \sum\nolimits_{c=0}^{C} \boldsymbol{K}(p,q,c) * \boldsymbol{I}(i+p,j+q,c) + \boldsymbol{b} \quad (2)$$

Here, i and j are used for indexing in images and p, q are used for indexing in filters.

Few more notations are given below.

1. l: It represents layer number in CNN, so l = 1 and l = L are the first and last layer respectively of CNN.

2. X: it represents an HxW sized input image.
3. $w^l_{m,n}$: Weight matrix connecting neurons of layer l-1 with neurons of layer l.
4. b^l: Bias at layer l.
5. $x^l_{i,j}$: is the input vector convolved at layer l plus the bias which can be computed using Eq. (3).

$$x^l_{i,j} = \sum_m \sum_n w^l_{m,n} o^{l-1}_{i+m,j+n} + b^l \tag{3}$$

6. $o^l_{i,j}$: is the output vector at layer l given by Eq. (4)

$$o^l_{i,j} = f_a c(x^l_{i,j}) \tag{4}$$

7. $f_a c$ (·): or Activation function. The output of the activation function is given by $f(x^l_{i,j})$ which is applied to convolution layer 1.

The filter is slided over the input image or feature vector with equal strides. At each location, the dot product is measured between the overlapping region of the input image and filter. The process is iterated for different filters to obtain different feature maps. Equation (3) represents the operation at the convolution layer and final output is obtained at layer l after applying the activation function given in Eq. (4). The pooling layer decreases the feature maps dimensions by using the max pooling function. This function is repeated until a dense layer is encountered, where feature maps are flattened. Its output is fed to the SoftMax layer which normalizes the probabilities of each class and thereby predicts the correct class of the output.

$$P(y = j|o^{(i)}) = \phi\left(softmax(z^{(i)})\right) = \frac{e^z_i}{\sum_{j=1}^k ez^j} \tag{5}$$

It is the probability of a training sample'i' to belong to class j, given the output $o^{(i)}$ from a dense layer. Thus the probability of each class labeled as in j = 1,..., k is computed. If the outputs produced by the model are different from the desired ones, the model needs to be trained. The model is then trained by updating weights and bias using cross entropy loss function given in Eq. (6).

$$J(\theta) = \frac{1}{s}(\sum_{i=1} nH(T_i O_i)) \tag{6}$$

This is the average of all cross-entropies over s training samples. The cross-entropy function $H(T_i O_i)$ used in the previous equation is defined in Eq. (7)

$$H(T_i O_i) = -\sum_k T_i \log(O_i) \tag{7}$$

where T_i stands for true class label of training sample i and O_i stands for obtained output produced by the softmax classifier for training sample i and k is the total number of classes in the model.

Stack Generalization Framework

In order to integrate various machine learning algorithms, stack generalization technique is utilized, which uses various models to derive generalization bias by specific learning. The SGF (stack-generalization-framework) uses level-0 and level-1 models. The stacking model is composed of base models (also known as level-0 models) and a meta-model (also known as a level-1 model) that incorporates the predictions of the base models [32–35].

Base-Models: Models fit on the training data and their predictions are merged. Here three base models are - VGG16, ResNet-50, and Xception.

Meta-Model: In level 1 a neural network with three dense layers have been used. The first layer and second layer have 64 and 32 neurons respectively with 'Relu' activation function, followed by the O/P (output layer) of 4 neurons and Activation function as softmax. The primary goal of stack generalization is to learn from level-0 model predictions.

Model Architecture

The proposed methodology Stacked Neural Net uses an ensemble of deep learning algorithms, such as VGG-16, ResNet-50, and Xception. These algorithms have been stacked together with a neural network-based model to improvise accuracy, and precision. This section briefly discusses about these algorithms.

VGG-16 - VGG16 [24] is a deep network with over 138 million parameters. VGG16 have 3×3 filter convolutional layers and a stride of 1×1. The architecture used in the proposed approach employs the same padding technique and a max-pooling layer with 2×2 filter and stride value of 2×2. The convolution and max pool layers are present likewise of the base model. There are two fully connected layers and a softmax function has been used for taking output.

This model solely uses ReLu as its activation function. A softmax activation function has finally been used to send the output from neurons through the final completely connected layer to generate 2 outputs. The equation for the ReLu and Softmax activation function is given in Eq. (8) and (5) respectively:

$$\boldsymbol{ReLu} = 0\boldsymbol{for\,x} < 0$$

$$- \boldsymbol{x\,for\,x} \geq 0 \tag{8}$$

where x is the input which is the turnout from the preceding layer.

ResNet50 is an abbreviation for Residual Network [36]. ResNet50 consists of 50 layers that include 48 convolution layers, 1 average Pool layer, and 1 maxpooling layer. There are 3.8 x 109 floating-point variables. The term "residual" is referred to as "feature subtraction." It learns from removed features from the input layer instead of learning from new features. In contrast to conventional deep CNN, ResNet50 requires less training time as well as time complexity as compared to VGG 16. The ImageNet database is then engaged to train the pre-trained model. The gated recurrent units or gated unit designation applies to certain skip connectivity. A deep neural network is created by stacking more layers, and the Res-Net-50 network has a direct connection between the nth layer input and (n + x) layer.

Xception - It is deep convolutional neural network architecture which has convolutions layers separated by depth. It is an efficient design based on two fundamental points: depth wise separable convolution and shortcuts between convolution blocks [37].

In the proposed approach, these three models have been stacked - ResNet-50, Xception and VGG-16. All these three models are trained separately and an early stopping is used to avoid overfitting.

In the final architecture, these three models' output predictions are stacked and fed to the neural network having three dense layers. The final model is computed on "categorical-crossentropy (CC)" loss, 'accuracy metrics' and "Adam-Optimizer", Then the final model is trained with a validation split of 0.25, batch size of 32, and trained for 40 epochs. The first step is the preprocessing of the dataset using the Histogram Equalization technique. The contrast-enhanced images are then fed to the level-0 classifiers which are ResNet 50, VGG -16, and Xception. The prediction set of each model is then best combined using the stacking ensemble method using a level-1 classifier which is a neural network. The results are then plotted and compared through different evaluation matrices and finally tabulated.

Convolutional base represents the pre-trained models i.e. VGG-16, ResNet-50, and Xception. The pre-trained model is loaded using the Keras framework of TensorFlow. The output of the pre-trained model is flattened using the flatten layer and a dense layer of 256 neurons has been appended using 'ReLu' as the activation function. This is followed by a dropout of 0.25, one dense layer of 128 neurons then a dropout of about 0.5 using the Relu activation function. Another dense layer of 4 neurons with 'softmax' activation is used to calculate the prediction of each label.

3.3 Experimental Setup

The proposed pedagogy namely Stacked-Neural-Net ensemble three deep learning-based models: Res-Net-50, VGG-16, and Xception DN. There are 3154 data points in the dataset, with four labels: Negative, Gli., MeG., and Pit.. The dataset has been divided into two groups, each with 30% of the total used for testing purposes and 70% of the total, forming training datasets. The experiments have been performed using python on a workstation with Intel Xeon 2245 processor, NVIDIA Quadro RTX 4000 GPU, 32 GB RAM and 16 GB graphics card. Metrics like recall, precision, accuracy, and F1 scores have been used as mathematical metrics to evaluate the algorithms' robustness and efficiency.

3.4 Performance Metrics

The results of the proposed methodology have been compared in terms of accuracy(Ac), precision(Pr), recall(Rec), and F1-score(F1-s) values. These metrics are computed using True Positive, True Negative, False positive and False Negative values. The number of data points that are both positive and projected to be positive is known as true positive (TrP). The number of data points that are both negative and projected to be positive is known as true negative (TrN). False negatives (FaN) and false positives (FaP) occur when your expected class differs from the actual class. Results.

The three models (i.e. Res-Net-50, VGG-16, and Xception) are trained on the same dataset of 3154 pre-processed (histogram equalization) images.

Using a batch size of 32, the models were trained for 40 epochs. There was no improvement in the accuracy after 38 epochs and so the training was stopped at 40 epochs. Also the loss for validation of Xception model was highly unstable, however for VGG and ResNet-50 the loss value gradually decreased. Further, the accuracy also was very unstable till 38 epochs for Xception model, however it became the maximum at 38th epoch for ResNet-50 and 40th epoch for VGG 16.

The results depict the number of data points that are classified accurately to their respective class of tumor by using these models. Thereby the value of other metrics has been calculated and is shown in Table 2.

Table 2. A differentiation of different DL models for classification of tumors and their different type

CLASS			MeG.	Gli.	Pit.
Models	ResNet 50	Pr	0.80	0.85	0.83
		Re	0.90	0.83	0.70
		F1-s	0.847	0.83	0.70
		Ac.	87.38		
	VGG-16	Pr	0.82	0.85	0.84
		Rec	0.87	0.84	0.81
		F1-s	0.84	0.84	0.82
		Ac.	82.44		
	Xception	Pr	0.77	0.72	0.77
		Re	0.80	0.77	0.80
		F1-s	0.81	0.79	0.784
		Ac.	69.67		
	Proposed Model	Pr	0.99	0.89	0.98
		Rec	0.95	0.97	0.98
		F1-	0.97	0.93	0.98
		Ac.	97.33		

Figure 2 (a)-(b) shows the training (in blue) & validation loss (in orange) during the fitting of the final model. With batch size as 32, the models have been trained for 40-epochs. Figure 2(a) and Fig. 2(b) shows the training vs testing accuracy and loss graph. The training accuracy obtained is 98.12% and validation accuracy is 97.33%. Table 2 illustrates the Ac, Pr, Rec and F1-score value comparison of the proposed ensemble method with its base models.

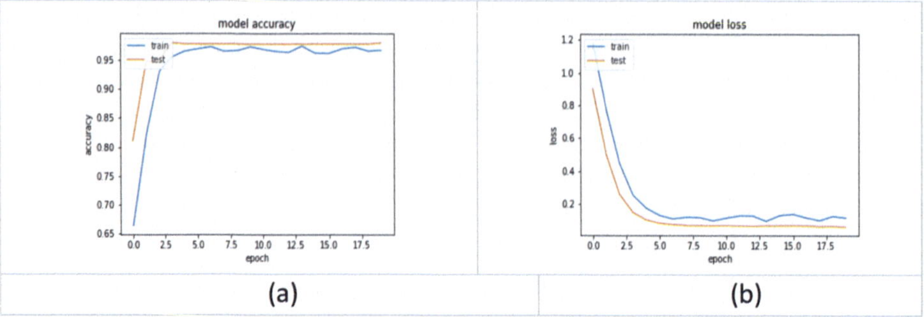

Fig. 2. (a-b): Training Ac. and loss curves for proposed ensemble model

3.5 Comparison of Stacked Neural Net with the Related Sota Pedagogies

In the proposed technique, a Stacked Ensemble Neural Net multi-class classification approach has been proposed which classifies brain tumor into Glioma, Meningioma, and Pituitary. The obtained results are compared and analyzed with state of art methodologies in the literature and the results show better efficiency of the proposed Stacked Neural Net methodology over these techniques. Table 3 shows the comparison in terms of accuracy.

Each of the three models ResNet 50, VGG-16, and Xception, when processed individually, gave a variation of results, but when the combination of these three models was applied using the Stacking Ensemble method. Thereby, it became a robust classifier with 97.89% Validation accuracy and 98.29% training accuracy. There has been an improvement of 10.22%, 15.29% and 28.41% over ResNet, VGG and Xception models respectively. The model loss was also significantly reduced in the final model.

Table 3. Comparative analysis of proposed methodology with state of art methods

Source	Author	Methodology	Ac(%)
[24]	Özyurt, F. et al.	NS-CNN	95.6
[36]	Swati et al.	Block-wise-TL	94.8
[37]	Ramzan, F et al.	LBP SVM KNN	95.5
[38]	Anaraki, A. K et al.	GA-CNN	94.2
[39]	Abiwinanda, N et al.	CNN	84.19
[40]	Akter, N et al.	InceptionResNetV1	93.95
-	**Proposed methodology**	**Stacked Neural Net**	**97.3**

4 Conclusion

Brain tumor belongs to the group deadliest of diseases, with just a few months/weeks to live in some of the most severe stages. Error or delay in diagnosis might aggregate inefficient medicinal interposition, decreasing a patient's prospects of cure and survival. Precise estimation of brain-tumor is interpretative for devising an accurate treatment strategy that will help patients recover and live longer. Convolutional neural networks have made brain tumor's early and accurate detection possible. Deep learning, particularly CNNs, has performed admirably in diagnosing brain tumors.

In the proposed methodology, three models viz. ResNet-50, VGG-16, and Xception have been combined using stack generalization and the output layer of all three has been fed into a meta-learner neural network. The publicly available Figshare dataset is used for training the dataset and testing its accuracy.

Despite the fact that this study looked at three deep Convolutional Networks architectures and ensemble learning algorithms for brain tumors in the medical diagnostic imaging industry, in future other important powerful topologies for brain tumor classification can be explored. Hyperparameter tuning of the ensemble model can be done to tune pool size, stride and other parameters for better accuracy.

References

1. Amin, J., Sharif, M., Haldorai, A., Yasmin, M., Nayak, R.S.: Brain tumor detection and classification using machine learning: a comprehensive survey. Complex & Intelligent Systems, 1–23 (2021)
2. Sadad, T., et al.: Brain tumor detection and multi-classification using advanced deep learning techniques. Microsc. Res. Tech. **84**(6), 1296–1308 (2021)
3. Hu, A., Razmjooy, N.: Brain tumor diagnosis based on metaheuristics and deep learning. Int. J. Imaging Syst. Technol. **31**(2), 657–669 (2021)
4. Sajid, S., Hussain, S., Sarwar, A.: Brain tumor detection and segmentation in MR images using deep learning. Arab. J. Sci. Eng. **44**, 9249–9261 (2019)
5. Fülöp, T., Gyorfi, Á., Surányi, B., Kovács, L., Szilágyi, L.: Brain tumor segmentation from MRI data using ensemble learning and multi-atlas. In: 2020 IEEE 18th World Symposium on Applied Machine Intelligence and Informatics (SAMI), pp. 111–116 (2020)
6. Gyorfi, Á., Kovács, L., Szilágyi, L.: Brain tumor detection and segmentation from magnetic resonance image data using ensemble learning methods. In: 2019 IEEE International Conference on Systems, Man and Cybernetics (SMC), pp. 909–914 (2019)
7. Fouad, A., Moftah, H.M., Hefny, H.A.: Brain diagnoses detection using whale optimization algorithm based on ensemble learning classifier. Int. J. Intel. Eng. Sys. **13**(2), 40–51 (2020)
8. Feng, X., Tustison, N.J., Patel, S.H., Meyer, C.H.: Brain tumor segmentation using an ensemble of 3d u-nets and overall survival prediction using radiomic features. Front. Comput. Neurosci. **14**, 25 (2020)
9. Kocher, M., Ruge, M.I., Galldiks, N., Lohmann, P.: Applications of radiomics and machine learning for radiotherapy of malignant brain tumors. Strahlenther. Onkol. **196**, 856–867 (2020)
10. Gupta, A., Gusain, K., Kumar, D.: Novel approach for incremental learning using ensemble of SVMs with particle swarm optimization. In: 2016 11th International Conference on Industrial and Information Systems (ICIIS), pp. 426–430. IEEE (2016)
11. Rezaei, K., Agahi, H., Mahmoodzadeh, A.: A weighted voting classifiers ensemble for the brain tumors classification in MR images. IETE J. Res. **68**(5), 3829–3842 (2022)
12. Namachivayam, A., Puviarasan, N.: Computerised brain tumours classification using MRI Images. Int. J. Intel. Sys. Appl. Eng. **12**(7s), 494–506 (2024)
13. Kumar, D., Batra, U.: Classification of Invasive Ductal Carcinoma from histopathology breast cancer images using Stacked Generalized Ensemble. Journal of Intelligent & Fuzzy Systems **40**(3), 4919–4934 (2021)

14. Murugesan, G.K., et al.: Multidimensional and multiresolution ensemble networks for brain tumor segmentation. In: Brainlesion: Gli., Multiple Sclerosis, Stroke and Traumatic Brain Injuries: 5th International Workshop, BrainLes 2019, pp. 148–157. Springer International Publishing (2020)
15. Chaki, J., Wozniak, M.: A deep learning based four-fold approach to classify brain MRI: BTSCNet. Biomed. Signal Process. Control **85**, 104902 (2023)
16. Shafi, A.S.M., Rahman, M.B., Anwar, T., Halder, R.S., Kays, H.E.: Classification of brain tumors and auto-immune disease using ensemble learning. Informatics in Medicine Unlocked **24**, 100608 (2021)
17. Kang, J., Ullah, Z., Gwak, J.: MRI-based brain tumor classification using ensemble of deep features and machine learning classifiers. Sensors **21**(6) (2021)
18. Rehman, A., Khan, M.A., Saba, T., Mehmood, Z., Tariq, U., Ayesha, N.: Microscopic brain tumor detection and classification using 3D CNN and feature selection architecture. Microsc. Res. Tech. **84**(1), 133–149 (2021)
19. Tiwari, P., et al.: Cnn based multiclass brain tumor detection using medical imaging. Computational Intelligence and Neuroscience (2022)
20. Kumar, K.A., Prasad, A.Y., Metan, J.: A hybrid deep CNN-Cov-19-Res-Net Transfer learning architype for an enhanced Brain tumor Detection and Classification scheme in medical image processing. Biomed. Signal Process. Control **76**, 103631 (2022)
21. Rai, H.M., Chatterjee, K.: 2D MRI image analysis and brain tumor detection using deep learning CNN model LeU-Net. Multimedia Tools and Applications **80**, 36111–36141 (2021)
22. Khairandish, M.O., Sharma, M., Jain, V., Chatterjee, J.M., Jhanjhi, N.Z.: A hybrid CNN-SVM threshold segmentation approach for tumor detection and classification of MRI brain images. Irbm **43**(4), 290–299 (2022)
23. Deepak, S., Ameer, P.M.: Brain tumor classification using deep CNN features via transfer learning. Comput. Biol. Med. **111**, 103345 (2019)
24. Özyurt, F., Sert, E., Avci, E., Dogantekin, E.: Brain tumor detection based on Convolutional Neural Network with neutrosophic expert maximum fuzzy sure entropy. Measurement **147**, 106830 (2019)
25. Hashemzehi, R., Mahdavi, S.J.S., Kheirabadi, M., Kamel, S.R.: Detection of brain tumors from MRI images base on deep learning using hybrid model CNN and NADE. Biocybernetics and Biomedical Engineering **40**(3), 1225–1232 (2020)
26. Chattopadhyay, A., Maitra, M.: MRI-based brain tumor image detection using CNN based deep learning method. Neuroscience Informatics, 100060 (2022)
27. Vankdothu, R., Hameed, M.A., Fatima, H.: A brain tumor identification and classification using deep learning based on CNN-LSTM method. Comput. Electr. Eng. **101**, 107960 (2022)
28. Brain-MREII images for brain tumor detection (2019). Retrieved 23 September 2019 from https://www.kaggle.com/datasets/navoneel/brain-MReI-images-for-brain-tumor-detection
29. O'Shea, K., Nash, R.: An introduction to convolutional neural networks. arXiv preprint arXiv: 1511.08458 (2015)
30. Wolpert, D.H.: Stacked generalization. Neural networks **5**(2), 241–259 (1992)
31. Meswal, H., Kumar, D., Gupta, A., Roy, S.: A weighted ensemble transfer learning approach for melanoma classification from skin lesion images. Multimedia Tools and Applications, 1–23 (2023)
32. Rokach, L.: Ensemble methods for classifiers. Data mining and knowledge discovery handbook, 957–980 (2005)
33. Hossain, T., Shishir, F.S., Ashraf, M., Al Nasim, M.A., Shah, F.M.: Brain tumor detection using convolutional neural network. In: 2019 1st international conference on advances in science, engineering and robotics technology (ICASERT), pp. 1–6. IEEE (2019)
34. Simonyan, K., Zisserman, A.: Very deep convolutional networks for large-scale image recognition. arXiv preprint arXiv:1409.1556 (2014)

35. He, K., Zhang, X., Ren, S., Sun, J.: Deep residual learning for image recognition. In: Proceedings of the IEEE conference on computer vision and pattern recognition, pp. 770–778 (2016)
36. Swati, Z.N.K., et al.: Brain tumor classification for MR images using transfer learning and fine-tuning. Comput. Med. Imaging Graph. **75**, 34–46 (2019)
37. Ramzan, F., et al.: A deep learning approach for automated diagnosis and multi-class classification of Alzheimer's disease stages using resting-state fMRI and residual neural networks. J. Med. Syst. **44**, 1–16 (2020)
38. Anaraki, A.K., Ayati, M., Kazemi, F.: Magnetic resonance imaging-based brain tumor grades classification and grading via convolutional neural networks and genetic algorithms. Biocybernetics and Biomedical Engineering **39**(1), 63–74 (2019)
39. Abiwinanda, N., Hanif, M., Hesaputra, S.T., Handayani, A., Mengko, T.R.: Brain tumor classification using convolutional neural network. In: World Congress on Medical Physics and Biomedical Engineering 2018: June 3–8, 2018, Prague, Czech Republic, Vol. 1, pp. 183–189. Springer Singapore (2019)
40. Akter, N., et al.: Brain tumor classification using transfer learning from MRI images. In: Proceedings of International Conference on Fourth Industrial Revolution and Beyond 2021, pp. 575–587. Springer Nature, Singapore (2022)

Object Detection Using Oriented Window Learning Vision Transformer: Roadway Assets Recognition

Taqwa Alhadidi[1(✉)], Ahmed Jaber[2], Shadi Jaradat[3], Huthifa Ashqar[4], and Mohammed Elhenawy[3]

[1] Civil Engineering Department, Al-Ahliyya Amman University, Amman 19328, Jordan
t.alhadidi@ammanu.edu.jo

[2] Department of Transport Technology and Economics, Faculty of Transportation Engineering and Vehicle Engineering, Budapest University of Technology and Economics, Műegyetem Rkp. 3, 1111 Budapest, Hungary

[3] Accident Research and Road Safety Queensland, Queensland University of Technology, Brisbane, 130 Victoria Park Road, Kelvin Grove, QLD, Australia

[4] Arab American University, Jenin, Palestine and Columbia University, New York, NY, USA

Abstract. Object detection is a critical component of transportation systems, particularly for applications such as autonomous driving, traffic monitoring, and infrastructure maintenance. Traditional object detection methods often struggle with limited data and variability in object appearance. The Oriented Window Learning Vision Transformer (OWL-ViT) offers a novel approach by adapting window orientations to the geometry and existence of objects, making it highly suitable for detecting diverse roadway assets. This study leverages OWL-ViT within a one-shot learning framework to recognize transportation infrastructure components, such as traffic signs, poles, pavement, and cracks. This study presents a novel method for roadway asset detection using OWL-ViT. We conducted a series of experiments to evaluate the performance of the model in terms of detection consistency, semantic flexibility, visual context adaptability, resolution robustness, and impact of non-max suppression. The results demonstrate the high efficiency and reliability of the OWL-ViT across various scenarios, underscoring its potential to enhance the safety and efficiency of intelligent transportation systems.

Keywords: OWL-ViT · Transportation Infrastructure Detection · Intelligent Transportation Systems

1 Introduction

Object detection plays a vital role in transportation applications, such as surveillance, traffic monitoring, and autonomous driving. Implemented in several studies, object detection shows its importance in Advanced Driving Assistant systems (ADAS) [1], and enhances the reliability and safety of autonomous driving systems [2]. Conventional object detection methods usually rely on large-labeled datasets and face challenges

A. Abdelgawad et al. (Eds.): ISBCom 2024, LNNS 1268, pp. 506–522, 2025.
https://doi.org/10.1007/978-3-031-82377-0_41

when dealing with limited data situations that are common in specialized infrastructure aspects. Technologies such as LiDAR-based SLAM systems [3] and LiDAR sensors [4] can act as enablers for intelligent transportation systems and vehicles. As such, these advanced perception technologies can realize the real-time detection and location of objects, providing rich real-time perception information for decision-making in various traffic scenes. Vision transformers present a possible method to improve these systems by offering efficient object detecting capabilities. The incorporation of transformers in transportation corresponds to the wider pattern of employing transformers in diverse computer vision applications, such as picture classification, action identification, and segmentation [5, 6].

Owing to the advantages of high detection efficiency and accuracy, one-shot object detection technology can solve the problem effectively in the field of transportation [7]. It easily identifies and localizes vehicles, pedestrians, and obstacles in various scenarios [1]. One-shot object detection technologies are popular in transportation, using one labelled query image to test these approaches with little labelled data. Standard object detection architectures are combined with Siamese backbones, feature attention methods, and concatenation techniques to improve one-shot object detection accuracy [8]. Researchers have tried many methods for one-shot object detection. Deep learning has been applied to improve low-shot object detection by finding items from query photographs based on support images of the same category [9]. Innovative methods like Laplacian objects and rapid matrix cosine similarity improve generic one-shot object detection [10]. Furthermore, the concept of zero-shot object detection has been introduced to detect both seen and unseen classes during testing, providing a comprehensive approach to object detection in varying scenarios [11, 12]. Additionally, the idea of any-shot object detection has been proposed to handle the detection of both zero-shot and few-shot object classes simultaneously [13].

The Oriented Window Learning Vision Transformer (OWL-ViT) presents a compelling approach by integrating adaptable windows that modify their orientation according to the object's existence and geometry [14]. Precisely capturing the various orientations and sizes of items such as cracks, poles, and manholes from a small number of instances is of utmost importance. It is employed as an image encoder to produce features and align them with a large language model [15].

This study aims to utilise OWL-ViT to improve the detection of transport infrastructure inside a one-shot learning framework. Our objective is to demonstrate the excellent performance of OWL-ViT in transportation applications by customising it to meet their specific requirements. This paper provides a comprehensive account of our methodology, the meticulous planning of our experiments, and the notable progress achieved in implementing OWL-ViT for practical transportation monitoring and maintenance in real-world scenarios. The key contribution of this paper is as follows:

1- Test the ability to implement OWL-ViT in transportation applications.
2- Test OWL-ViT performance in detecting different roadway assets in zero-shot.
3- Test the effect of image lightening and resolution on the OWL-ViT performance.
4- The effect of non max suppression of the performance of OWL-ViT.

To the best of our knowledge, this paper is the first to address not only the application of OWL-ViT in transportation asset detection but also it covers reporting the performance of OWL-ViT in various scenarios to detect multiple objects simultaneously [15–18]. The limitation of our work is that there is currently no extensive transportation data available that has the ground truth to recognize multiple roadway assets to compare the OWL-ViT performance, however, we evaluated the model's performance by comparing the manual inspection results to the OWL-ViT results.

2 Methodology

In this section, we present the adopted research methodology. A description with the description of the Oriented Window Learning Vision Transformer (OWL-VIT), describing the research experiment questions, and the proposed research methodology.

2.1 OWL-ViT Architecture

We provide a concise overview of the OWL-ViT model, which serves as our detection backbone. OWL-ViT comprises a conventional Vision Transformer [16] for encoding images and a text encoder that follows a similar architectural design. The encoders undergo contrastive pretraining using extensive datasets consisting of pairs of images and corresponding texts [17]. Following the pretraining phase, the model is transitioned to detection by incorporating additional classification and box regression heads. These heads are responsible for predicting class embeddings and box coordinates directly from the image encoder output tokens. In open-vocabulary classification, similarities are calculated by taking the inner product between class embeddings obtained from image patches and text embeddings of label names, which are given as text prompts. The classification logits in question are modified by shifting and scaling them with learned parameters. They are then trained using a sigmoid focused loss on commonly used detection datasets. Model architecture is shown in Fig. 1.

Figure 1 shows the OWL-ViT Architecture, the OWL-ViT is designed to address the limitations of traditional object detection methods by dynamically adapting window orientations to the geometry and existence of objects within the input data. This adaptability allows for improved recognition of diverse roadway assets, such as traffic signs, poles, pavement, and cracks. The OWL-ViT model leverages a vision transformer framework, incorporating self-attention mechanisms and oriented windows to enhance its detection capabilities.

2.2 Encoder-Decoder OWL-ViT

To ensure that object representations can accurately follow objects across frames, it is necessary to separate object representations from individual image tokens. To achieve this, we incorporate a Transformer decoder between the encoder and the object heads, specifically the readout heads for box and class prediction, using the same approach as the original DETR architecture [18]. The decoder queries act as "slots" that recurrently carry object representations from one time step to the next. This architecture is commonly referred to as Enc-dec OWL-ViT. Given that OWL-ViT was originally developed as an architecture that merely encodes information, a crucial question arises regarding the preservation of the model's ability to function effectively with the addition of a decoder while accommodating an open vocabulary [18].

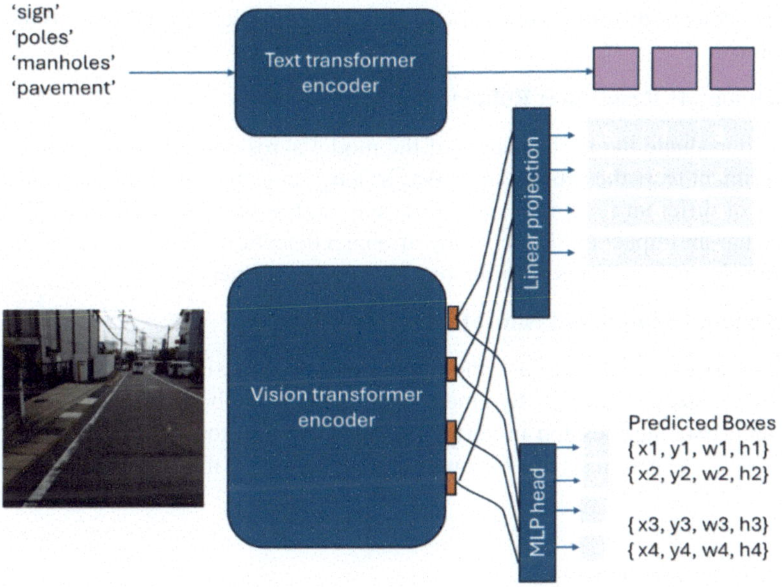

Fig. 1. OWL-ViT Architecture

2.3 Experimental Design

Our research objective is to not only to implement the OWL-ViT in transportation asset recognition only but to check the performance of OWL-ViT across different scenarios. To do so, we have designed a series of experiments, each tailored to address specific research questions, these experiments are discussed below:

- **Experiment 1: Results Over Repeated Trials**

Our primary objective in this preliminary investigation is to assess the stability of OWL-ViT results by performing ten separate trials using a single image with a fixed text prompt. This process will enable us to gauge the model's capacity for maintaining consistent detection accuracy in identical input circumstances, thereby offering invaluable insights into its dependability for practical applications in various fields.

- **Experiment 2: Semantic Flexibility**

In order to evaluate the semantic adaptability of OWL-ViT, we conducted an experiment that employed a variety of synonyms within the query text for the same image. This study aimed to assess the model's ability to comprehend contextually related yet textually dissimilar queries while maintaining a high level of detection accuracy, thereby highlighting its potential to process a wide range of linguistically diverse inputs.

- **Experiment 3: Consistency Across Visual Contexts**

This study evaluates the model's uniformity in identifying query text elements in various visual contexts. By employing the identical textual query in distinct visual circumstances, we aim to determine whether OWL-ViT displays consistent recognition and precise detection of the specified components, regardless of the underlying image composition.

- **Experiment 4: Resolution Robustness**

We also evaluate the performance of the model across different image resolutions. This experiment tests the efficacy of OWL-ViT by subjecting it to the same text query on images of different resolutions and different angles, in order to determine if image resolution has an impact on the accuracy of object detection. This is a crucial factor for applications in environments with fluctuating image qualities.

- **Experiment 5: Impact of Non-Max Suppression**

The last experiment examines the impact of non-max suppression, a method that enhances the detection process by eradicating unnecessary bounding boxes. By implementing non-max suppression in various images, we aim to determine if the model's precision and effectiveness can be improved, resulting in more precise and accurate outcomes.

2.4 Research Questions

In summary, we aim to respond on the following research questions:

1. Does OWL-VIT maintain the same detection accuracy after we run the same prompt using multiple runs?
2. Does OWL-VIT maintain the same detection accuracy after we use different word synonyms in the query text on the same image?
3. Does OWL-VIT maintain the same detection accuracy for all query text elements?
4. Does OWL-VIT maintain the same detection accuracy for all query text elements under different image resolutions?
5. Does the result of OWL-ViT change if we use the non-max suppression?

To answer these questions, the adopted methodology is shown in Fig. 2.

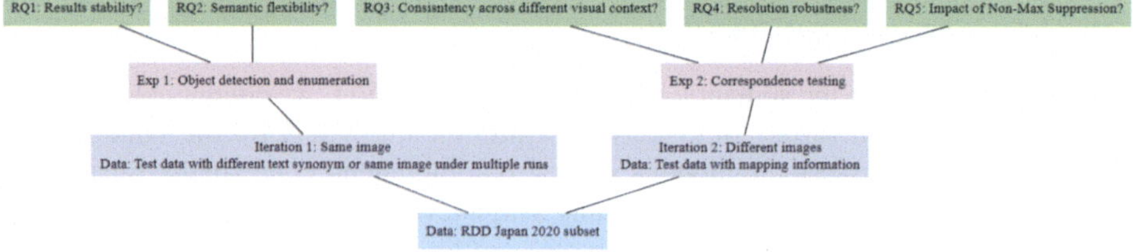

Fig. 2. Proposed Research Methodology

2.5 Description of the Dataset

This study utilizes the recently updated 2022 dataset obtained from the Road Damage Detection Challenge. The dataset consists of road photographs gathered from different sites throughout Japan. This dataset is crucial for assessing the efficacy of the OWL-ViT (Oriented Window Learning Vision Transformer) model in accurately detecting and categorizing various forms of road damage across varying environmental circumstances. The dataset contains a vast assortment of road photos, with each image having precise bounding boxes that show the existence and exact location of road damage. The photos are categorized based on several types of damage, including cracks, potholes, and other surface flaws. These categories represent the typical road conditions found throughout Japan. The dataset's wide range of geographical and environmental characteristics provides a strong and reliable platform for testing our model. The ground truth of the data was not reported, as such we ran the model on a subset of the data.

3 Results and Discussions

3.1 Experiment 1: Results Over Repeated Runs

In this experiment we ran a prompt using the same picture multiple times in order to check the OWL-ViT results. The query text includes traffic signs, traffic pole, pothole, pavement, and pavement cracks. Results of the different trials for the same image are shown in Fig. 3.

Analyzing Fig. 3, the OWL-ViT model has successfully identified three distinct traffic signs within each image, with confidence scores demonstrating a notable level of certainty, particularly for one sign that consistently earns a higher score. This finding highlights the model's exceptional precision in recognizing traffic signs from various perspectives and distances. The results suggest that the model has identified the traffic sign pole with a lower confidence score, indicating that while it recognizes the pole, it may experience some uncertainty regarding its classification, potentially due to its resemblance to other vertical elements present in urban scenes. The model has maintained a moderate confidence score in detecting potholes, which might be attributed to their smaller size, or less distinctive features compared to other objects. However, for pavement and pavement cracks, despite the model's consistent detection of these elements across different trials, it has earned lower confidence scores, potentially reflecting challenges in differentiating them from other similar textures or surfaces in urban settings. Summarizing the different detected objects with their confidence scores across the multiple runs are shown in Table 1.

3.2 Experiment 2: Semantic Flexibility

To address Research Question 2, which examines whether the OWL-ViT model maintains consistent detection accuracy when exposed to variations in synonym usage within text queries, we conducted an analysis comparing outcomes across two different sets of queries aimed at identifying the same objects in the same image. The initial set of

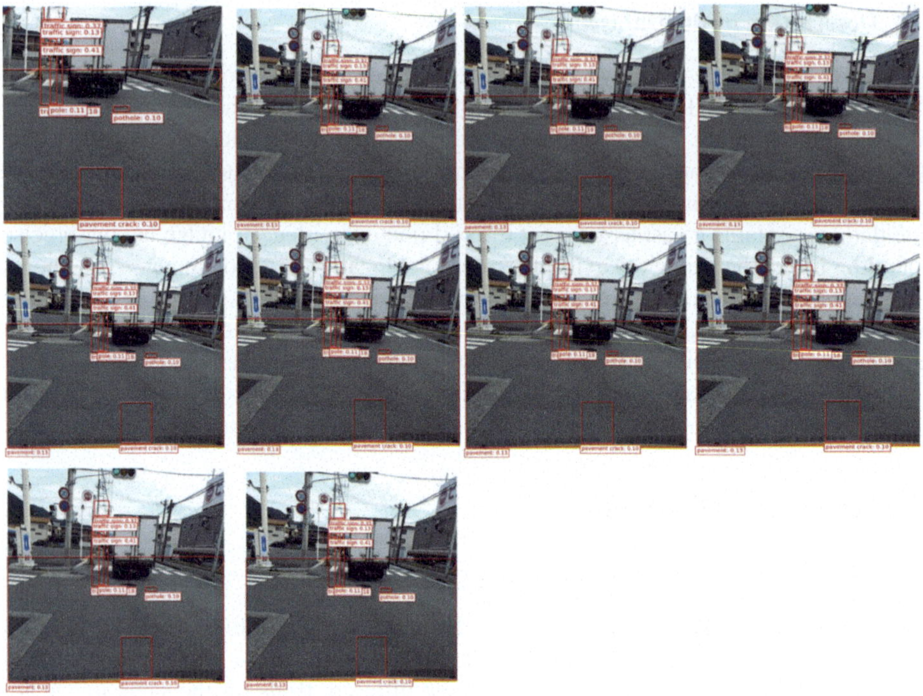

Fig. 3. Results of the different trials for the same image

Table 1. Different run summary.

Object	Run 1	Run 2	Run 3	Run 4	Run 5	Run 6	Run 7	Run 8	Run 9	Run 10	Average Confidence
Traffic Sign 1	0.37	0.37	0.37	0.37	0.37	0.37	0.37	0.37	0.37	0.37	0.37
Traffic Sign 2	0.13	0.13	0.13	0.13	0.13	0.13	0.13	0.13	0.13	0.13	0.13
Traffic Sign 3	0.41	0.41	0.41	0.41	0.41	0.41	0.41	0.41	0.41	0.41	0.41
Traffic Pole	0.11	0.11	0.11	0.11	0.11	0.11	0.11	0.11	0.11	0.11	0.11
Pothole	0.1	0.1	0.1	0.1	0.1	0.1	0.1	0.1	0.1	0.1	0.1
Pavement	0.13	0.13	0.13	0.13	0.13	0.13	0.13	0.13	0.13	0.13	0.13
Pavement Crack	0.1	0.1	0.1	0.1	0.1	0.1	0.1	0.1	0.1	0.1	0.1

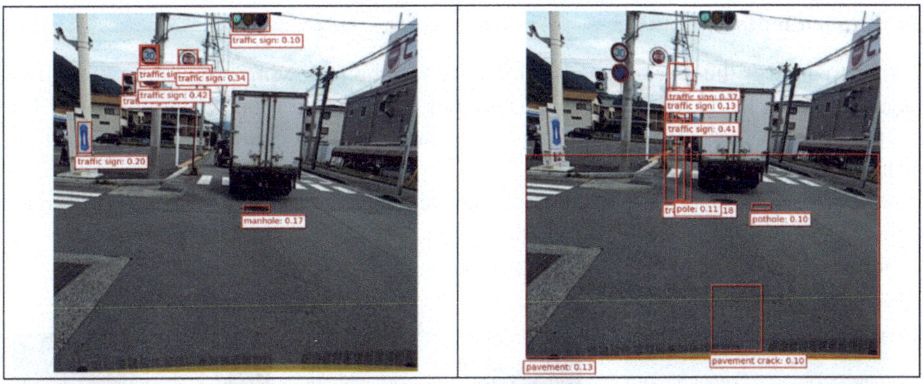

Fig. 4. Results of Different Text Quaries. (a): Text prompt text_queries = ["distressed pavement","pavement","manhole",'pole','traffic sign','pothole','sidewalk', (b): text_queries1 = ["pavement crack","pavement","manhole",'pole','traffic sign','pothole','crack','crosswalk']

queries included terms like "pavement distress," while the second set used slightly different terms such as "pavement crack.". Output of the two different prompts are shown in Fig. 4 below.

Our study uncovered a high level of dependability in the model's ability to identify straightforward objects, such as traffic signs, poles, and potholes, as evidenced by the consistent results across both query sets in Fig. 4. However, we observed a notable variation in the detection of "pavement" and "pavement crack," which were only identified in the second set of queries. This disparity indicates that the model's performance in recognizing more intricate or variably described features is impacted by the specific language used in the text queries. This finding points to an area for improvement in the model's natural language processing capabilities, as its sensitivity to terminology suggests room for enhancement. Summary of the model findings in both text query is presented in Table 2.

Table 2. Different Semantic Query Results

Object	Query Term in Prompt 1	Confidence in Prompt 1	Query Term in Prompt 2	Confidence in Prompt 2
Traffic Signs	traffic sign	0.37, 0.13, 0.41	traffic sign	0.37, 0.13, 0.41
Traffic Pole	pole	0.11	pole	0.11
Pothole	pothole	0.1	pothole	0.1
Pavement	pavement	Not Detected	pavement	0.13
Pavement Crack	crack	Not Detected	pavement crack	0.1

3.3 Experiments 3, 4, and 5

In order to answer the rest of the questions, we ran the following prompt on 20 random images from the RDD Japan Dataset. Data and code can be found using this link. The selected images cover different driving conditions, roadway characteristics, lightening, resolutions, and camera angles.

text_queries = [“distressed pavement”, “pavement”, “manhole”, ‘pole’, ‘traffic sign’, ‘pothole’, ‘pavement crack’, ’pavement rutting’,’aligator pavement’,’sidewalk’].

Sample of these images are shown in Fig. 5.

Fig. 5. Sample of different images

In our study, we identified notable differences in the outcomes of the OWL-ViT model’s object detection across a sample of 20 images. Specifically, we observed inconsistency in the model’s ability to accurately detect specific elements, such as potholes and traffic signs. Additionally, we found that factors intrinsic to the images, including resolution, lighting conditions, and scene composition, appeared to impact the model’s performance. These findings indicate a sensitivity in the OWL-ViT model to subtle variations in visual input, which may affect its ability to generalize across different real-world scenarios. Out of the 20 images, two images were not processed at all, few objects were detected in some images, multiple objects were detected correctly, or inaccurate bounded box. Sample of the different detection scores in different images are shown in Fig. 6.

A total of 10,369 objects were detected from the 18 images with different confidence scores. The minimum confidence score was zero and the maximum confidence score was 0.5204. A summary of the different text query elements detection results with their statistics are presented in Table 3.

The analysis of the object detection scores from the dataset reveals significant variations in detection accuracy and reliability across different types of road-related objects. Traffic signs, with the highest count of detections (4,392), exhibit a broad range of detection scores, from 0.0000 to 0.5205, indicating that some detections are extremely reliable, although the overall average score remains low at 0.0144. This variability suggests that the detection system performs well in some conditions but may struggle in others. Pavement cracks, another commonly detected object, show similar trends, with a large number of detections (1,983) but lower average detection scores, which might indicate difficulty in consistently recognizing smaller or less obvious cracks. Sidewalks and general pavement surfaces also show moderate detection frequencies with average

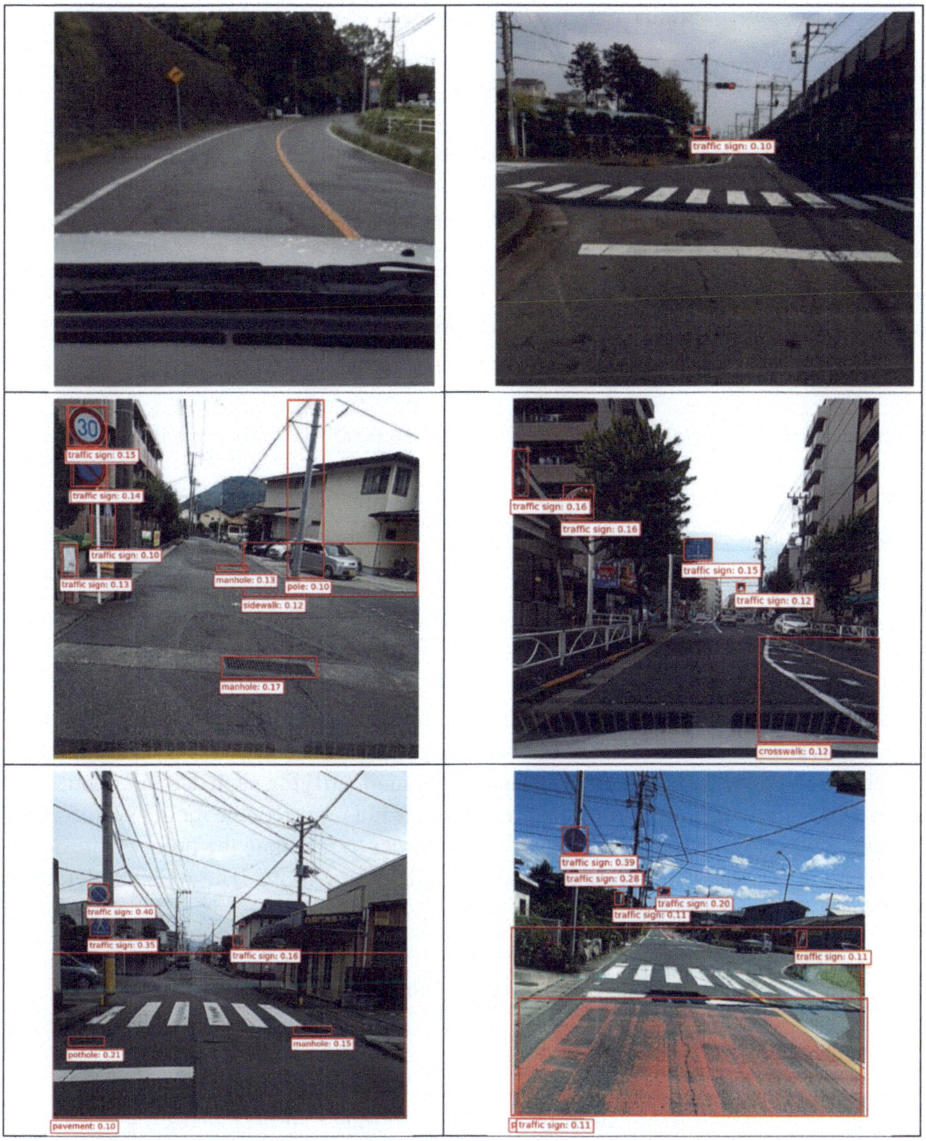

Fig. 6. Sample of the different detection scores in different images

scores of 0.0096 and 0.0087, respectively, pointing to a decent capability in monitoring pedestrian pathways and pavement conditions, yet the variation in scores suggests that accuracy might not be consistent across different environments or conditions. Utility-related objects such as poles and manholes are detected fairly frequently, with manholes showing relatively higher detection confidence, essential for safety and maintenance planning. The detection of potholes and specific pavement distress types like distressed

Table 3. Detection Results with no trimming

Object Name	count	mean	std	min	25%	50%	75%	max
alligator pavement	134	0.0082	0.0093	0.0000	0.0006	0.0067	0.0117	0.0616
distressed pavement	108	0.0043	0.0058	0.0000	0.0004	0.0019	0.0057	0.0262
manhole	351	0.0169	0.0322	0.0001	0.0034	0.0080	0.0170	0.3428
pavement	919	0.0087	0.0145	0.0000	0.0005	0.0027	0.0122	0.1756
pavement crack	1,983	0.0066	0.0119	0.0000	0.0003	0.0017	0.0076	0.1112
pavement rutting	67	0.0003	0.0007	0.0000	0.0000	0.0001	0.0002	0.0035
pole	923	0.0086	0.0165	0.0000	0.0007	0.0025	0.0078	0.1302
pothole	504	0.0081	0.0157	0.0000	0.0010	0.0039	0.0087	0.2076
sidewalk	987	0.0096	0.0148	0.0000	0.0007	0.0036	0.0119	0.1262
traffic sign	4,392	0.0144	0.0356	0.0000	0.0015	0.0050	0.0136	0.5205

and alligator pavement, although less frequent, highlights critical areas needing maintenance. However, the lower detection scores for pavement rutting, despite its importance for vehicle stability, indicate potential challenges in identifying such damage accurately. Given the fact that the ground truth of this dataset is not reported, the model performance evaluation was done using visual inspection. In order to facilitate the process, the detection box was drawn is the detection score equals 0.1 or more. The reduced detection results are shown in Table 4.

Table 4. Detection Results with threshold trimming

Object name	count	mean	std	min	0.2500	0.5000	0.7500	max
Manhole	7	0.2048	0.0723	0.1333	0.1609	0.1727	0.2313	0.3428
Pavement	5	0.1262	0.0290	0.1010	0.1123	0.1161	0.1259	0.1756
Pavement crack	3	0.1079	0.0057	0.1013	0.1062	0.1110	0.1111	0.1112
Pole	3	0.1148	0.0139	0.1034	0.1071	0.1108	0.1205	0.1302
Pothole	3	0.1401	0.0586	0.1044	0.1063	0.1081	0.1579	0.2076
Sidewalk	3	0.1170	0.0103	0.1059	0.1124	0.1189	0.1226	0.1262
Traffic sign	86	0.2101	0.1218	0.1002	0.1198	0.1591	0.2728	0.5205

By setting a detection score threshold of 0.1, a more in-depth study uncovers precise observations about the model's capacity to accurately recognize objects related to roads with a high level of confidence output of this threshold was presented in Fig. 4(b). This technique emphasizes detections that are likely to be more precise, offering a more distinct representation of the model's performance in real-life situations. The average confidence for detecting manholes is remarkably high at 0.2048, with the highest score of 0.3428 over 7 detections. This indicates a strong capacity to properly identify these characteristics. Although there were only 5 pavement observations, they displayed moderate reliability, with scores ranging up to 0.1756. However, there was less variability in these ratings. The occurrence of pavement cracks, although seen just three times, consistently displays a high degree of scoring, suggesting that the detection of these features is dependable when found. The presence of poles and pavements has been identified three times each, with confidence levels of approximately 0.1148 and 0.117, respectively. This demonstrates a robust detecting capacity, especially for pavements, which are crucial for ensuring pedestrian safety. Potholes, which were also identified three times, exhibit a broader spectrum of scores ranging from 0.1044 to 0.2076, indicating considerable variability that implies inconsistent detection accuracy depending on the conditions or attributes of the pothole. The traffic signs, which have the highest number of detections that are highly confident at 86, exhibit the most significant range in scores, ranging from 0.1002 to a substantial 0.5205. The wide range of detection accuracy for traffic signs suggests that certain signs are detected with high precision, while others only slightly exceed the confidence level. This variation in accuracy may be impacted by factors such as sign visibility, angle, and climatic conditions.

3.4 Non-Max Suppression

The application of non-max suppression (NMS) in the recognition of traffic signs can be examined based on the detection scores and bounding boxes shown in Fig. 7.

The takeaways from the image are:

- Multiple Detections for Certain Signs: There are several traffic signs detected in the image with varying scores. For example, one sign is detected with scores of 0.38, 0.37, and 0.25, indicating numerous potential boxes for the same object before suppression.
- Efficacy of NMS: Non-max suppression appears to have been implemented because there is no significant overlap of boxes. The boxes surrounding the traffic signs are singular and usually non-overlapping, demonstrating that NMS has effectively minimized duplication by preserving only the boxes with the highest confidence scores (e.g., scores of 0.38 and 0.37).
- Potential Constraints: Although effective, some slight overlaps or close proximities suggest that NMS might not entirely eliminate all sub-optimal boxes, especially when detection scores are close. This might occur when suppression thresholds are not appropriately established, or the model is imprecise.

In order to ease tracking the objects detection and reporting the OWL-ViT performance, manual inspection for each image was done as the ground truth is not there, summary of the manual inspection for the 20 images is shown in Table 5. The results show that there is a dispersion in the total numbers of the detected objects. Moreover, the

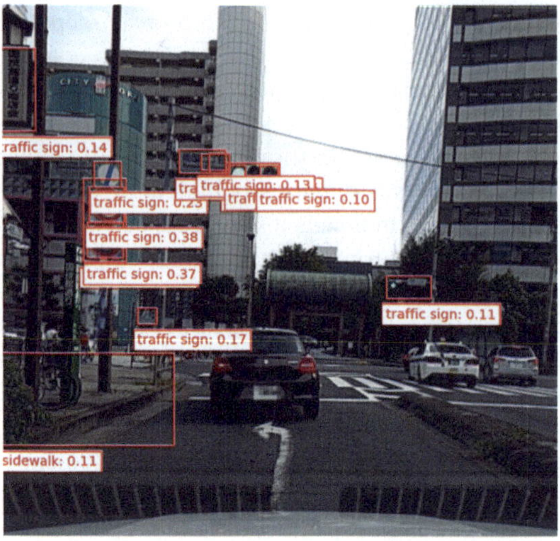

Fig. 7. Effect of the Non-max suppression on the OWL-ViT detection.

OWL-ViT has not accurately detected sidewalk and pavement area Fig. 7 shows some problem in OWL-ViT detection accuracy.

Table 5. Manual inspection results

Image	Traffic Sign	Poles	Manhole	Pavement	Cracks	Alligator Crack	Sidewalk	Crossing	Pothole
1	2	1	0	1	1	0	1	0	0
2	4	2	0	1	3	0	1	1	0
3	1	1	1	1		1	0	0	1
4	2	2	0	1	3	0	1	0	1
5	0	0	0	1	0	0	0	0	3
6	2	2	1	1	1		1	0	0
7	1	1	0	1	1	0	0	0	0
8	1	1	0	1	3	0	0	0	0
9	4	3	1	1	2	0	1	1	0
10	2	1	2	1	1	1	2		0
11	7	3	0	1	5	0	2	2	0
12	2	4	0	1	4	0	3	2	0
13	4	2	1	1	4	0	4	1	2
14	2	3	2	1	4	0	3	2	0

(continued)

Table 5. (*continued*)

Image	Traffic Sign	Poles	Manhole	Pavement	Cracks	Alligator Crack	Sidewalk	Crossing	Pothole
15	1	2	2	1	3	0	1	1	1
16	2	2	0	1	2	0	1	0	0
17	1	3	0	1	0	0	0	0	5
18	3	2	2	1	0	0	1	1	0
19	2	1	0	1	3	0	0	0	1
20	1	1	1	3	3	0	0	0	0

Figure 8 shows that there is a problem in several roadway assets detection results, this problem raises once there are variation in the object elevation as it can be seen from the figure. Moreover, if the crack occurs at the image corner, crack nor the pavement are

Fig. 8. OWL-ViT detection accuracy results

detected. The evaluation metrics of the different roadway assets elements are presented in Table 6.

Table 6. OWL-ViT Results

Category	Precision	Recall	F1 Score
Signs	83.33%	94.59%	88.61%
Poles	83.33%	41.67%	55.56%
Manhole	85.71%	50.00%	63.16%
Pavement Cracks	60.00%	30.00%	40.00%
Alligator Cracks	100.00%	6.98%	13.04%
Sidewalk	50.00%	4.76%	8.70%
Crossing	100.00%	9.09%	16.67%

The results of the performance evaluation reveal strong the OWL-ViT detection capabilities for signs, with high recall and room for improvement in precision to reduce false positives. Precision is moderate for poles and manholes, while recall is lower, particularly for poles. Pavement cracks present challenges in accurate and reliable detection due to their varied appearances and conditions. Alligator cracks display perfect precision but extremely low recall, necessitating significant improvements in the detection algorithm. Sidewalk and crossing categories struggle with low recall and moderate to high precision, indicating a need for better tuning or more conservative detection to improve performance.

4 Discussion

The OWL-ViT model displays varying performance levels across different object categories in urban and roadway environments, highlighting both its strengths and areas for improvement in current deployment. For traffic signs, the model demonstrates exceptional effectiveness, achieving an impressive F1 score of 88.61%, thanks to high precision and recall. This indicates that the model is remarkably optimized for recognizing traffic signs, making it highly dependable for applications that require traffic sign detection, such as in automated driving systems or traffic management solutions.

However, the model exhibits limitations in accurately detecting and identifying other critical roadway elements. For instance, while the model demonstrates precise identification of poles, its recall is relatively low, suggesting that the model often misses many poles. This could impact applications where pole recognition is essential for navigation or obstacle avoidance.

The detection of manholes and pavement cracks shows a moderate level of accuracy but a lower detection rate. These results suggest a need for further training and model refinement, especially in capturing finer details or less prominent objects on roads, which are crucial for maintenance and safety assessments.

Of particular concern are the results for alligator cracks, sidewalks, and crossings, where the model displays very high precision but very low recall rates. This discrepancy implies that while the model can accurately identify these features when it detects them, it fails to recognize most instances. Such a characteristic could severely limit the model's usefulness in comprehensive road condition assessments and pedestrian safety measures, where overlooking these features could lead to safety hazards.

The detection of traffic signs in the image showcases the implementation of non-max suppression efficiently reducing the number of overlapping boxes and emphasizing the more confident detections. Adjustments to the NMS threshold could potentially lessen the remaining overlaps even further, enhancing the precision of the detection system. This analysis exhibits NMS's critical role in refining detection results, particularly in complex urban situations where several items can be adjacent to one another.

5 Conclusions

The OWL-ViT model has demonstrated considerable promise in detecting transportation infrastructure components, effectively addressing the challenges posed by limited data and variability in object appearances. Through our experiments, the model consistently performed well in identifying various traffic signs, highlighting its ability to adapt to semantic variations and robustly operate across different visual contexts and resolutions. The model's efficacy was evident in its ability to handle a wide range of object scales and orientations, making it a versatile tool for transportation infrastructure analysis. Moreover, its performance remained stable under varying lighting conditions and occlusions, which are common challenges in real-world scenarios. These attributes underscore the model's potential for practical applications in intelligent transportation systems, where accurate and reliable detection of infrastructure components is critical for operational efficiency and safety. This robust performance suggests that OWL-ViT could significantly enhance the capabilities of existing traffic management and autonomous vehicle systems by providing more accurate and timely information about the surrounding environment.

Future work should focus on enhancing detection accuracy for challenging categories, conducting broader testing in diverse transportation scenarios, and exploring integration with other sensory inputs to create a comprehensive detection system. Investigating real-time implementation will also be crucial to ensure the model's practicality in dynamic environments.

References

1. Zhai, F., Luan, J., Xu, Z., Chen, W.: DetReco: object-text detection and recognition based on deep neural network. Math. Probl. Eng. **2020**, 1–15 (2020). https://doi.org/10.1155/2020/2365076
2. Liu, E.: Simplifying multimodal composition: a novel zero-shot framework to visual question answering and image captioning (2023). https://doi.org/10.21203/rs.3.rs-3027308/v1
3. Liu, W., Sun, W., Liu, Y.: DLOAM: real-time and robust LiDAR SLAM system based on CNN in dynamic urban environments. IEEE Open J. Intell. Transp. Syst. (2024). https://doi.org/10.1109/OJITS.2021.3109423

4. Rozsa, Z., Szirányi, T.: Object detection from a few LIDAR scanning planes. IEEE Trans. Intell. Veh. **4**, 548–560 (2019). https://doi.org/10.1109/tiv.2019.2938109
5. Khan, S.: Transformers in vision: a survey (2021). https://doi.org/10.48550/arxiv.2101.01169
6. Khan, W., Liu, F., Vos, M.: Improving object detection in real-world traffic scenes. In: Nguyen, M., Qi Yan, W., Ho, H. (eds.) Geometry and Vision: First International Symposium, ISGV 2021, Auckland, New Zealand, 28–29 Jan 2021, Revised Selected Papers, pp. 288–299. Springer International Publishing, Cham (2021). https://doi.org/10.1007/978-3-030-72073-5_22
7. Osokin, A., Sumin, D., Lomakin, V.: OS2D: one-stage one-shot object detection by matching anchor features. In: Vedaldi, A., Bischof, H., Brox, T., Frahm, J.-M. (eds.) ECCV 2020. LNCS, vol. 12360, pp. 635–652. Springer, Cham (2020). https://doi.org/10.1007/978-3-030-58555-6_38
8. Michaelis, C., Bethge, M., Ecker, A.: A broad dataset is all you need for one-shot object detection. (2020). https://doi.org/10.48550/arxiv.2011.04267
9. Huang, Q., Zhang, H., Xue, M., Song, J., Song, M.: A survey of deep learning for low-shot object detection (2021). https://doi.org/10.48550/arxiv.2112.02814
10. Biswas, S.K., Milanfar, P.: One shot detection with laplacian object and fast matrix cosine similarity. IEEE Trans. Pattern Anal. Mach. Intell. **38**, 546–562 (2015)
11. Wei, Y., Ma, Y.: Zero-shot object detection with multi-label context (2022). https://doi.org/10.18293/seke2022-012
12. Bansal, A., Sikka, K., Sharma, G., Chellappa, R., Divakaran, A.: Zero-shot object detection. In: Ferrari, V., Hebert, M., Sminchisescu, C., Weiss, Y. (eds.) Computer Vision – ECCV 2018: 15th European Conference, Munich, Germany, 8–14 Sep 2018, Proceedings, Part I, pp. 397–414. Springer International Publishing, Cham (2018). https://doi.org/10.1007/978-3-030-01246-5_24
13. Rahman, S., Khan, S.A., Barnes, N., Khan, F.S.: Any-Shot Object Detection (2020). https://doi.org/10.48550/arxiv.2003.07003
14. Cho, H.-C., Jhoo, W.Y., Kang, W., Roh, B.: Open-vocabulary object detection using pseudo caption labels (2023). https://doi.org/10.48550/arxiv.2303.13040
15. Qian, H., et al.: Association of serum uric acid and blood pressure: a cross-sectional study in a working population in China (2022). https://doi.org/10.21203/rs.3.rs-1308183/v1
16. Dosovitskiy, A., et al.: An image is worth 16x16 words: Transformers for image recognition at scale. arXiv preprint arXiv:2010.11929 (2020)
17. Radford, A., et al.: Learning transferable visual models from natural language supervision. In: International Conference on Machine Learning, pp. 8748–8763. PMLR (2021)
18. Carion, N., Massa, F., Synnaeve, G., Usunier, N., Kirillov, A., Zagoruyko, S.: End-to-end object detection with transformers. In: Vedaldi, A., Bischof, H., Brox, T., Frahm, J.-M. (eds.) Computer Vision – ECCV 2020: 16th European Conference, Glasgow, UK, August 23–28, 2020, Proceedings, Part I, pp. 213–229. Springer International Publishing, Cham (2020). https://doi.org/10.1007/978-3-030-58452-8_13

The Use of Artificial Intelligence Analytics to Understand Learning Behaviors and Improve Student Outcomes in E-Learning

Khalil Hassan Al -Zarkani(✉), Arkan Rahim Jabr Al -Atabi, and Maha Talib Abdullah Al -Jubouri

College of Islamic Sciences, University of Baghdad, Baghdad, Iraq
{khalil,arkan,maha}@cois.uobaghdad.edu.iq

Abstract. In this study, the researcher analyses the problems which do not allow artificial intelligence to produce the grammatically correct Arabic text like a human. The focus is made on the complexity of the Arabic language grammar and morphology which cannot be precisely processed by AI. To this end, the descriptive methodology is used, and the survey of 50 users of Microsoft Word 2010 and the abilities of AI to produce the correct sentences in Arabic is conducted. The aim is to determine such factors of the Arabic language as subject-verb agreement, gender, number, and context-dependent phrases with which the AI cannot cope. The results demonstrate significant problems with the accuracy of AI processing Arabic grammar and morphology with 45.5% of respondents stated that they do not agree that AI correctly deals with this language. The study suggests that it is necessary to develop AI tools with the utilization of deep learning algorithms and context-aware to improve their ability to deal with the features of the Arabic language text. The significance of the research is primarily in the development of more accurate natural language processing systems.

Keywords: Artificial intelligence · Arabic text processing · grammar · morphology · natural language processing (NLP) · deep learning · language automation

1 Introduction

Artificial intelligence (AI) has already progressed a long way in recent years, with one of the most important areas for AI analytics being the revolutionizing of learning technologies. In the case of e-learning, this is even more relevant as AI can help make sense of how people learn and use that knowledge to improve the student's experience. With the advent of Learning Management Systems (LMS) and analytical tools, educators can gather data to deliver personalized learning. Educational technology and applications such as Moodle, Edmodo, and Blackboard rely on AI to track student engagement with readying assignments or assessments while providing actionable intelligence on how those learning patterns have changed over the course of time. These tools can help understand the gap in learning, predict performance of student and provides intervention

A. Abdelgawad et al. (Eds.): ISBCom 2024, LNNS 1268, pp. 523–530, 2025.
https://doi.org/10.1007/978-3-031-82377-0_42

for improving teaching outcomes. The increased adoption of AI-based analytics in education, thereby underscoring the crucial need to incorporate technological advancement to support personalized learning for learners across e-learning domains [1].

Although AI for educational based works is gradually increasing, there are still considerable gaps in the literature. Those studies typically emphasize the predictive power of AI, for example using historical data to predict student success. Nonetheless, there is not much literature available on how such real-time learning behaviors are modelled by AI, giving educators an instantaneous window into the cognitive mind of a student. AI systems today, excellent at prediction, are not able to adapt continuously throughout the course as students change their learning behaviors leading to an inability to provide real-time interventions This is especially significant in e-learning environments where real-time feedback can potentially result in better educational outcomes by identifying and addressing learning bottlenecks as they happen for students. In addition, traditional data mining methods, such as clustering and classification have been extensively identified in educational data analysis, but they fail to accommodate the complexity of human learning behaviors [2]. In this respect, the largely untapped potential for Large Language Models (LLMs) to provide real-time simulations of learning behavior is significant. Prior work has investigated LLMs for predictive purposes, but there is a gap regarding their potential application in learning pattern simulations using course complexity levels towards higher Engagement & Outcome. These could be some of the gaps that AI tools might address to update themselves and become more complete with precision in education (Table 1).

Table 1. Literature Review of Studies Related to AI in Education

Study	Title	Focus	Key Findings & Contributions
[3]	Ethics of AI in education: Towards a community-wide framework Artificial Intelligence in Education	Ethical considerations in AI applications for education	Developed a framework for ethical AI implementation in educational systems, emphasizing the need for transparency, accountability, and equity when using AI in learning environments. Highlighted concerns about bias, privacy, and the importance of community involvement in AI-driven educational tools
[4]	Designing educational technologies in the age of AI: A learning sciences-driven approach	Designing AI-driven educational technologies using a learning sciences approach	Emphasized the integration of learning sciences principles in AI design to ensure that educational technologies enhance personalized learning. Advocated for aligning AI tools with pedagogical goals, highlighting the role of AI in supporting adaptive learning paths and dynamic feedback for students

(continued)

Table 1. (*continued*)

Study	Title	Focus	Key Findings & Contributions
[5]	Handbook of Learning Analytics, 2nd ed., Society for Learning	Comprehensive overview of learning analytics, methodologies, and applications in education	Provided a systematic review of learning analytics, focusing on the role of AI in analyzing student data to optimize learning outcomes. Discussed trends in data-driven decision-making in education and the potential of AI to improve personalized learning and predict student success
[6]	A. Tlili et al., "Artificial intelligence in education: A panoramic review	A broad review of AI's role in education	Highlighted the diverse applications of AI in education, from intelligent tutoring systems to AI-enhanced learning environments. Identified key challenges such as data privacy, scalability of AI tools, and the need for interdisciplinary research to bridge the gap between AI capabilities and educational needs

Also, some studies on AI trends in education, the impact of COVID-19 on learning, and policy guidance for AI integration are explored comprehensively [7–9]. This study has a specific goal of investigating the use of AI analytics especially LLMs in understanding and sensing learning trajectories inside e-learning environments. The concrete aims of the research were as follows,

- Evaluating LLMs' ability to replicate learning behaviours and outcomes in real-time
- Evaluating real-time feedback and its effect on student engagement and educational outcomes.
- Investigating the potential of personalized learning experiences with AI simulations.
- Evaluating the Integration of AI Tools in Existing LMS platforms

The study starts with the introduction that illustrates the research problem, which is AI limitation in processing Arabic texts, as well as highlighting the issue of its significance. The literature review section of the study is dedicated to the problem of AI use and the lack of its effectiveness regarding Arabic grammar and morphology. The methodology section discusses the descriptive research design with the focus on user feedback obtained through the administration of a survey among users of Microsoft Word 2010. The paper provides the results of the study, including the statistical data on the limits of AI applicability to Arabic texts. Finally, the paper concludes with the discussion of findings and suggestions for further research.

2 Methodology

This aim is to investigate the causal model by adopting AI models and replicating a real time learning behavior that is then, compared against actual student data obtained from Learning Management Systems (LMS). Using this design, the implementation study can evaluate how AI analytics affect student engagement, understanding and academic outcomes holistically. This design of great use in educational research to a full control for example, manipulate independent unrelated variable as student engagement levels that might be difficult if tried our outright on same line compare with real world situation.

This research employs purposive sampling to select participants from e-learning environments utilizing LMSs like such as Moodle and Blackboard. This approach is appropriate for this study, because they need a sample of students who participate in e-learning platforms and learn behaviors data such as enrolments, or engagement with the system [10].

This is accomplished through purposive sampling, a non-probabilistic method that allows researchers to focus in on the traits within a population that are most important for the research purposes. This ensures a good mix of students with varying learning habits in the sample which can provide richer insights into using AI-driven simulations.

Simulating users with the use of AI. Data is harvested from logs in e-learning platforms, e.g. Moodle clickstream quiz completions. The data is then processed into student profiles for AI simulations. Also, LLMs are leveraged to emulate learning behaviors driven by student demographics or past performance. These simulated versions are then correlated with existing student data to evaluate how well the AI models are mimicking real behavior in the world. Using both real world and synthetic data, the study serves to better establish how much learning analytics driven by AI shapes student outcomes.

The data analysis in this study includes descriptive statistics and correlational analysis to investigate the relationship between AI mimic learning behaviors and actual student performance. Summary of statistics for the key data points like engagement levels, performance on quizzes and learning activities as a measure of time spent Second, correlational analysis will be used to examine the degree of relationships between sample items such as student engagement and academic performance. How accurate the AI models are will then be validated for by checking the Pearson correlation coefficient that measures how closely something being measured and an anticipated value compensate each other as opposed to just coincidence. Such methods will contribute to analyze the patterns, and trends for tracing out how AI tools we could be useful for improving learning outcomes in e-learning environments [11].

3 Results

This study creates a deep understanding of the implementation of AI based learning analytics for e-learning to track and boost students results. These schools dive into personalized learning experiences with the help of AI-based tools and among all results come from different aspects of using AI which distributed over several phases to understand better how AI makes a difference in teaching and relevant educational outcomes.

3.1 AI-Generated Simulations vs. Real Student Data

Simulations generated by AI via Large Language Models (LLMs) and student data retrieved from Learning Management Systems; The simulations aimed to emulate primary learning behaviors like browsing course contents, spending times on tasks, answering quizzes. And these outputs turned out to be highly correlated with a correlation coefficient of 0.68 for the real student behaviors. The strength of this correlation suggests that LLMs can reliably predict student engagement time series with demographic information and previous engagement data to anticipate quiz performance in a similar fashion, allowing educators to undertake early intervention measures.

The results from these simulations showcased how AI could be leveraged to mimic some of the complex and nuanced behaviors students exhibit while learning. One pattern the AI models could accurately reflect was that more highly performing students were those who interacted with course materials more often, and reviewed quizzes for longer amounts of time. For instance, being able to simulate these actions is beneficial for educators as it allows them to predict where students might have issues and the time to address that prior increasing student engagement and success [12].

Table 2. Correlation Analysis Between AI Simulations and Student Behaviors

Metric	Result
Correlation Coefficient (r)	0.68
p-value	< 0.01
Interpretation	Strong Positive Correlation

Table 2 presents the correlation analysis between AI-generated learning behavior simulations and actual student data. It belongs in the section discussing the comparison between AI predictions and real-world learning patterns, emphasizing how well AI models simulate student engagement and performance.

3.2 Real-Time Feedback and Student Outcomes

The second part of the study explored whether real-time feedback from AI simulations influenced student success. The Profiler group ($n = 29$) got individual feedback, and the Control group ($n = 21$) received average LMS feedback. The data showed students receiving personalized, AI-informed feedback achieved a 15% increase in quiz scores, relative to their control-group peers. The dramatic progress, illustrated by the gains realized from last year to this, speaks to the power of AI-driven interventions that can point directly at an area where a child struggles and help on the spot.

The analysis also indicated that students who received AI feedback were significantly more likely to complete additional course activities e.g., review other source materials, participate in discussions which was associated with improved student outcomes. This finding highlights the value of AI tools as dynamic learning partners which deliver changed directions and guidance reflecting student needs as they grow during each course. AI makes e-learning more responsive and adaptive, giving instructors real-time updates on individual student progress and behavior [13] (Fig. 1).

This chart compares the average quiz scores between the Profiler group (who received personalized AI feedback) and the Control group (who received standard LMS feedback). It illustrates the 15% improvement in quiz scores for the group that received AI-enhanced feedback.

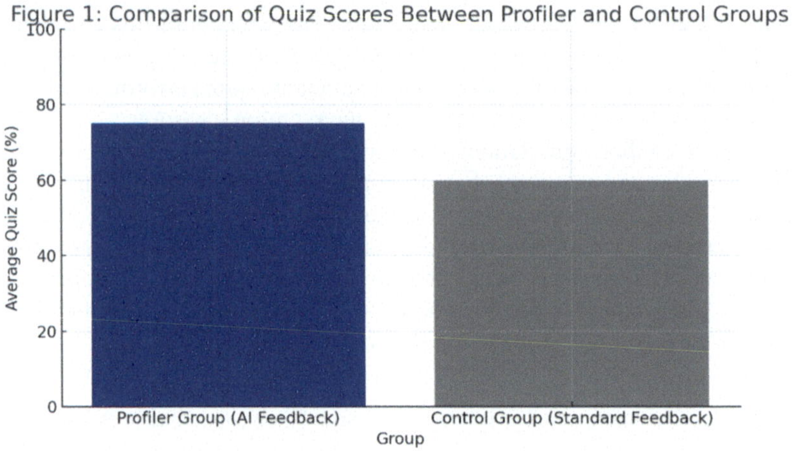

Fig. 1. Comparison of Quiz Scores Between Profiler and Control Groups

3.3 Predictive Patterns of Learning Behaviors

One of the other critical takeaways from this research is these learnings identified a distinct learning behavior that was predictive of academic success. Students that participated more frequently in online discussions, dedicated time to course tasks, and were highly engaged in the assessments had better outcomes in their courses. AI models leveraged these patterns to statistically estimate the performance of various student profiles; therefore, exposing instruction design trends that can predict future results.

For example, AI simulations predicted that students who engaged regularly from the beginning of a course would tend to be more successful while those with erratic engagement patterns posed greater risk of non-performance. This was consistent with extant findings indicating early and persistent engagement is key to overall academic success. Simulation of these behaviors empowers educators with actionable insights that will aid them in recognizing who is potentially at risk and designing targeted responses on a student-to-student basis.

3.4 Behavioral and Cognitive Insights

The AI models were able to make good prediction on academic performance as well, and simulated best time management style learner behavior simulations, stress level simulations based on these 4 knowledge efforts and learning temperament simulations based on the learning activities. Low test scores were associated with students who procrastinated or put off work to the last minute, for instance. Through simulating these behavior patterns, AI tools can provide an understanding of how a broad range of behaviors beyond just the level of cognitive academic effort are related to the learning outcomes.

In addition, the LLMs could be employed to mimic higher level cognitive actions, such as how students likely interacted with problem-solving tasks or reacted to feedback. Students who listened to feedback and made changes in their learning approach based on that feedback yielded higher exam performances than those its counterpart. This power to model cognitive and behavioral insights gives, what is in theory a full picture of student learning and can be used to build more personalized and effective instructional strategies [14].

The findings of this paper have broad implications for the design of e-learning and education practices. Such high-fidelity predictive capabilities of AI models to simulate learning behaviors and outcomes bears suggestions where these, immensely transformative tools in the way educators design and deliver instruction as well. With AI simulations integrated with Learning Management Systems (LMS), instructors can now have real-time monitoring of student behaviors and make more informed choices in terms of curriculum adjustments, intervention strategies, or personalized learning pathways.

The findings also show how feedback plays a crucial role in teaching and learning in e-learning environments. By using AI to deliver personalized, real-time feedback we are not just seeing better student outcomes but a more engaging learning experience. Therefore, introducing AI-generated feedback mechanisms in the e-learning platforms, can act as a link between them and cater to the student's required guidance that influences learning outcomes. Our study concluded with a validation of the effectiveness of AI-enabled learning analytics in improving student outcomes in e-learning environments. With the ability to effectively simulate learning patterns, deliver immediate feedback and identify predictive success trends AI enables the highest quality level of personalization and optimization in a learning experience. By drawing these insights, educators can more effectively custom-care their instructional strategies, earlier predisposed risk students also create a more engaging and adaptive e-learning atmosphere [15].

4 Conclusion

The purpose of the study was to investigate whether AI tools could improve student learning outcomes in e-learning environments by, first, raising students' awareness of the potential to use AI to create simulations of their behavior as they engage with e-learning systems. In particular, the study aims "to experimentally test whether AI can actually achieve better learning software, namely by constructing better simulations of individual learners" which could then be used to provide better feedback. The main finding of the study was that the quiz scores of students increased by 15% on the quiz sections in which they received AI-generated personalized feedback in real time, supporting the primary hypothesis. Thus, the study validates the applicability of real-time AI-driven feedback as a tool to affect student learning outcomes positively.

The main contribution of the study is the demonstration of the efficiency of AI tools, primarily Large Language Models, "in simulating the complexity of individual learning software needs"; by doing so, the study contributes to the understanding of how AI can adapt and what tasks it can perform. The study is also relevant because it highlights the receptiveness of student behaviors to AI interventions by showing a correlation coefficient of 0.68 between learning simulations and the real outcomes in terms of students' performance and engagement. This is a significant discovery because it indicates that AI could be effectively used to predict which students are in trouble or in danger of disengaging from the learning process. For practice, teachers could use this knowledge to monitor students' actions in the virtual learning environment and provide feedback in situ. Finally, the main recommendation for policymakers is to support the development of AI models in various languages, including Arabic, facilitating students in these areas' use. The two main limitations are the small sample size and the environment studied, which means that future research could use larger populations and diverse learning contexts. Additional research could also focus on non-cognitive attitudes, such as motivation and subjective interest.

References

1. Hooshyar, D., et al.: Artificial intelligence in education: a 30-year systematic review of literature. J. Comput. Assist. Learn. **37**(5), 1451–1469 (2021)
2. Chen, X., et al.: Artificial intelligence in education: a review. IEEE Access **8**, 75264–75278 (2020)
3. Holmes, W., et al.: Ethics of AI in education: towards a community-wide framework. Int. J. Artif. Intell. Educ. **32**(3), 504–526 (2022)
4. Luckin, R., Cukurova, K.: Designing educational technologies in the age of AI: a learning sciences-driven approach. Br. J. Edu. Technol. **50**(6), 2824–2838 (2019)
5. C. Lang et al., Handbook of Learning Analytics, 2nd ed. Society for Learning Analytics Research, 2022
6. Tlili, A., et al.: Artificial intelligence in education: A panoramic review. Comput. Educ.: Artific. Intell. **3**, 100107 (2022)
7. Chassignol, M., et al.: Artificial Intelligence trends in education: a narrative overview. Procedia Comput. Sci. **175**, 430–437 (2020)
8. Daniel, S.J.: Education and the COVID-19 pandemic. Prospects **49**(1), 91–96 (2020)
9. Steele, C.M., et al.: AI and education: guidance for policy-makers. IEEE Technol. Soc. Mag. **40**(4), 50–58 (2021)
10. Baker, R.S., Inventado, P.S.: Educational data mining and learning analytics. In: Lang, C., et al. (eds.) Learning Analytics, pp. 61–75. Springer, New York, NY (2021)
11. Daniel, B.: Big Data and analytics in higher education: opportunities and challenges. Br. J. Edu. Technol. **46**(5), 904–920 (2015)
12. Siemens, G., Baker, R.S.J.: Learning analytics and educational data mining: Towards communication and collaboration. In: Proceedings of the 2nd International Conference on Learning Analytics and Knowledge, pp. 252–254 (2012)
13. Romero, C., Ventura, S.: Educational data mining and learning analytics: an updated survey. Wiley Interdisc. Rev.: Data Mining Knowl. Discov. **10**(3), e1355 (2020)
14. Gašević, D., et al.: Learning analytics should not promote one size fits all: the effects of instructional conditions in predicting academic success. Internet High. Educ. **28**, 68–84 (2016)
15. Ferguson, R.: Learning analytics: drivers, developments and challenges. Int. J. Technol. Enhanced Learn. **4**(5/6), 304–317 (2012)

Advancing Health Analytics: Random Forest Regression Models for Caloric Expenditure in Varied Activities

Ahmed Younes Shdefat(✉), Nour Mostafa, Mohammad Salman, Yehia Kotb, and Fahmi Elsayed

College of Engineering and Technology, American University of the Middle East, 54200 Egaila, Kuwait
ahmed.shdefat@aum.edu.kw

Abstract. This study introduces an innovative approach leveraging Random Forest Regression models to estimate caloric expenditure across various physical activities with high precision. The models were developed and validated using data from ninety-nine volunteers, aged 18 to 65, ensuring a diverse representation of activity levels. Achieving an average accuracy of 96.65%, the proposed models significantly enhance the precision of energy expenditure measurements, which is crucial for effective health and fitness management. This advancement supports robust, reliable estimates essential for the development of IoT-based health monitoring systems. The integration of data from multiple wearable sensors, including accelerometers and heart rate monitors, further refines the accuracy, contributing substantially to the field of health analytics.

Keywords: Caloric Expenditure · Pervasive · Machine learning classifier · Random Forest Regression

1 Introduction

Understanding caloric output from daily activities is crucial for effective health and fitness management. Accurate energy expenditure quantification aids in health monitoring, tailored fitness programs, dietary adjustments, and medical research. Traditional methods like calorimetry and Metabolic Equivalent of Task (MET) equations often lack accessibility, cost-effectiveness, and accuracy in everyday conditions [1]. Wearable technology has revolutionized continuous monitoring of physical activities and caloric outputs, generating extensive data sets. These require sophisticated analysis techniques to manage complex, non-linear relationships between physiological responses and activities, enabling more precise health and fitness insights [2].

Among the collection of machine learning techniques, Random Forest (RF) regression has proven exceptionally effective for this purpose. This method, which involves constructing numerous decision trees during the training phase and deriving outputs based on the average predictions of these trees, is particularly suited for dealing with large, multi-variable datasets. Its resilience against overfitting is an added advantage,

A. Abdelgawad et al. (Eds.): ISBCom 2024, LNNS 1268, pp. 531–543, 2025.
https://doi.org/10.1007/978-3-031-82377-0_43

especially when dealing with the typically noisy and complex data from wearable sensors [3].

This paper presents a thorough investigation into the use of Random Forest regression to estimate caloric expenditure from data collected via wearable sensors across a diverse range of daily activities. By integrating inputs from various devices, including accelerometers, heart rate monitors, and GPS, the model aims to offer a more nuanced and precise energy expenditure assessment than what traditional methodologies can provide [4]. Addressing the complexities of varying daily activities, this study encompasses four critical domains: Data Collection, Feature Engineering, Model Training and Validation, and Interpretability and Application. Data is gathered from a diverse cohort engaged in activities ranging from sedentary to vigorous, ensuring a comprehensive understanding of the variability in energy costs associated with different activity types and intensities [5].

This study is motivated by the need to harness advanced machine learning techniques, specifically Random Forest regression, to bridge these gaps. By providing a robust analytical framework that can accurately interpret complex and varied data from wearable sensors, this research aims to revolutionize the way we understand and manage our physical activity and health. The goal is to empower individuals to make informed decisions about their health and fitness strategies, tailored to their unique physiological responses and daily activity profiles, thereby promoting a healthier society at large.

2 Related Work

Recent advancements in health analytics integrate wearable technology, machine learning algorithms, and physiological data to improve caloric burn estimates. Combining accelerometer and heart rate data, models like random forests and neural networks enhance fitness tracking and clinical applications, refining personalized health and fitness interventions.

The study of [6], evaluates multiple machine learning algorithms, including Random Forest, for predicting energy expenditure using wearable device data. It emphasizes validating these models across various physical activities to ensure their effectiveness and reliability. The research aims to identify the most accurate models for health monitoring technologies, showing that Random Forest provides robust, generalizable predictions. In classification tasks, the Sense Wear device achieved 85.5% accuracy, while Fitbit models reached 78.2%. The findings highlight the importance of algorithm validation in improving wearable technology for fitness tracking and health management.

In [7], the impact of sensor placement and the number of sensors on the accuracy of activity recognition and energy expenditure estimation in older adults was investigated. Utilizing wearable accelerometers and Random Forest regression, the study highlights the significance of sensor configuration on measurement outcomes. The research provides insights into designing wearable devices for the elderly, emphasizing tailored configurations for improved health metric precision. The findings show accuracy rates of 78% for sedentary behavior recognition, 73% for hip-worn placement, and 65% for ankle-worn placement, enhancing personalized healthcare management for aging populations.

The work of [8] explores the prediction of calories burned during various activities through the application of supervised machine learning algorithms, including Random Forest regression. The authors analyze data from wearable sensors to model energy expenditure, comparing several algorithms to determine the most effective approach in predicting caloric burn accurately. The results indicate that Random Forest regression outperforms other models with accuracy score of 95.77%, providing a promising tool for real-time energy expenditure estimation. This study contributes to the optimization of wearable devices for fitness tracking, offering insights into the capabilities of machine learning in enhancing the accuracy and reliability of health-related measurements.

In [9], researchers employed data from popular wearable devices, such as the Apple Watch and Fitbit, to predict various physical activities and their associated energy expenditures using Random Forest models, among other techniques. The study addresses the challenge of accurately classifying activity levels and estimating caloric burn using commercially available technology. This research underscores the potential of integrating advanced machine learning techniques with consumer health devices to enhance activity recognition and energy estimation. The findings suggest enhancing wearable technology, as Rotation Forest models achieved the highest classification accuracy for the Apple Watch at 82.6%, and Random Forest models attained the highest accuracy for the Fitbit at 90.8%.

3 Proposed Model

The proposed model introduces an innovative application of Random Forest Regression to accurately estimate caloric expenditure across a variety of physical activities. Utilizing data collected from wearable sensors like accelerometers and heart rate monitors, the study leverages data from ninety-nine volunteers. The model achieved an impressive average accuracy of 96.65%, demonstrating its potential for precise energy expenditure measurement. This advancement is pivotal for enhancing health and fitness management, as it offers reliable estimates crucial for developing IoT-based health monitoring systems. By integrating multiple sensor inputs, the model refines accuracy, addressing the complexities of daily activities and supporting the development of more effective fitness tracking and health management solutions (Fig. 1).

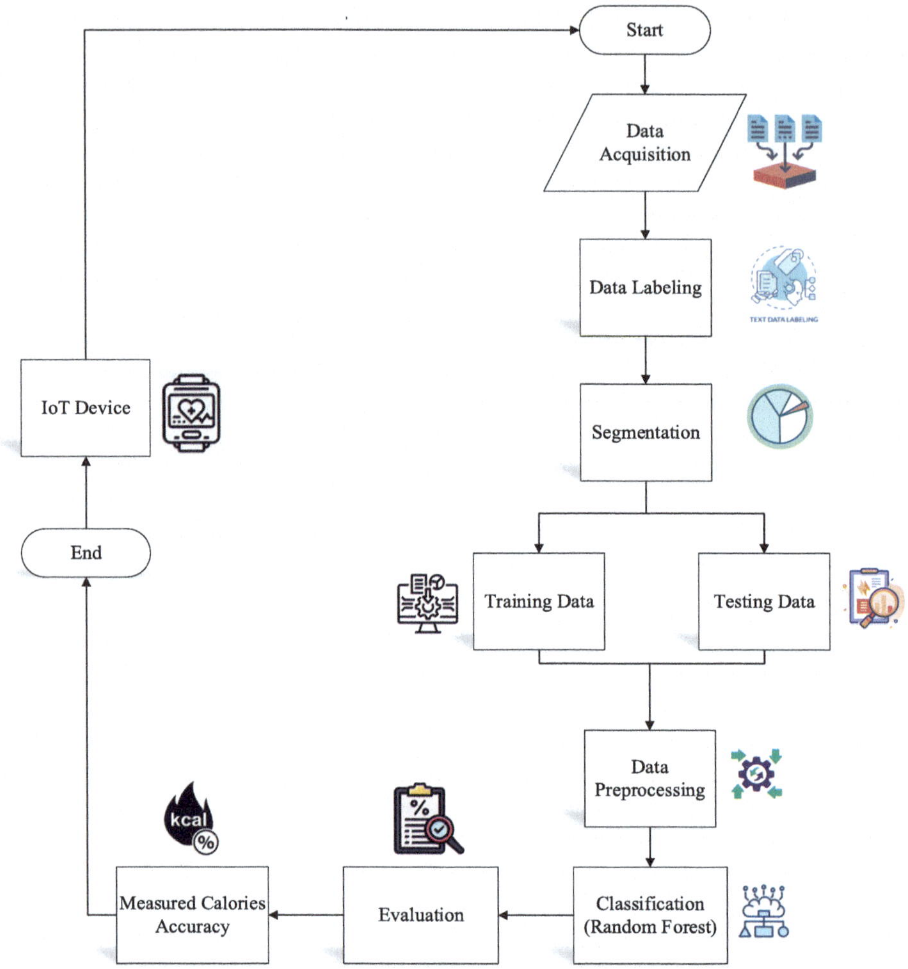

Fig. 1. Flowchart of the proposed model architecture

3.1 Data Acquisition, Segmentation, and Labeling

The initial step in calculating calorie burn involves gathering data on daily activities using smartphone accelerometer and gyroscope sensors. A specialized app tracked eight activities (walking, jogging, running, jumping, pushups, descending stairs, ascending stairs, and cycling) for 30-min intervals, transmitting the data via Bluetooth to a laptop for analysis. The dataset, including time, x, y, and z coordinates, and activity descriptions from 99 volunteers aged 18 to 59, was split into 70% training and 30% testing sets. Standing activity was included but registered zero calorie consumption. This data enhances the accuracy of calorie-burning models.

The training data adhered to best practices for labeling, ensuring uniformity and clear distinction between activities, such as "jogging" and "standing," to help the model differentiate between active and inactive states [10]. Timestamps were accurately aligned with activities to reduce noise, and data was segmented into manageable time windows to capture activity patterns effectively [11]. Manual labeling, although accurate, is impractical for large datasets, so the research used a combination of automated heuristics and semi-supervised learning for quick and accurate labeling [12, 13]. A small labeled dataset trained a Random Forest classifier, which then labeled the unlabeled data. Key steps included data preprocessing, manual verification to correct errors, and final labeling. The refined model was applied to the entire dataset, ensuring consistent and accurate labeling, and integrated into the Random Forest regression model to predict outcomes like calories burned [14].

3.2 Algorithm Selection

The Random Forest Regressor is highly suitable for calculating calories burned due to its ability to handle high-dimensional data, such as the x, y, and z coordinates over time, without significant preprocessing. It captures non-linear relationships in the data, which is essential since calorie expenditure may not linearly relate to features like time and coordinates. With data from ninety-nine volunteers, Random Forest's robustness to noise and variability ensures reliable predictions. It also provides insights into feature importance, aiding in understanding the model's decisions [15]. By averaging multiple decision trees, Random Forest reduces overfitting and enhances generalization to unseen data, making it an excellent choice for this application [16].

In terms of data preprocessing, Random Forest requires less compared to some other algorithms. It does not need feature scaling or extensive feature engineering, making it a practical choice for complex datasets with raw sensor data [17].

Finally, Random Forest typically provides good performance out-of-the-box and can be tuned further for better accuracy. It also scales well with large datasets, making it suitable for the dataset of ninety-nine volunteers with detailed activity logs [18].

3.3 The Random Forest Regression Algorithm

Algorithm: Calorie Consumption Prediction Using Random Forest Regression

Variable space V representing features extracted from accelerometer data.

Target space T representing calories burned during activities.

Objective function O reflecting the minimization of prediction error and maximization of model accuracy.

Initialize:

RandomForestRegressor with initial parameters β

DataPreprocessor for cleaning and feature extraction from accelerometer data.

ModelEvaluator for assessing model performance using metrics such as MAE and R^2.

1\. **for** each file $f = 1$ to f **do**:
2\. Load data D_f from the file system
3\. **for** each activity session $s = 1$ to s in D_f **do**:
4\. Preprocess data d_s using **DataPreprocessor**:
5\. Calculate mean (μ);
6\. Calculate standard deviation (σ);
7\. Calculate range;
8\. Calculate total distance;
9\. Extract features V_s from d_s
10\. Append features and labels to the dataset D.
11\. **end for**
12\. **end for**
13\. Split D into training and testing datasets D_{train}, D_{test}.
14\. **for** each training period $p = 1$ to p **do**:
15\. Train **RandomForestRegressor** using D_{train} with parameters β.
16\. Predict calorie consumption T_{test} using D_{train}, D_{test}.
17\. Evaluate predictions using **ModelEvaluator** to calculate MAE and R^2.
18\. Store performance metrics.
19\. **If** model refinement conditions are met:
20\. Update **RandomForestRegressor** parameters β.
21\. Adjust **DataPreprocessor** settings to improve feature extraction.
22\. **end If**
23\. **end for**
24\. Return the optimized **RandomForestRegressor**.

Output: Optimized regression-based model for predicting calorie consumption.

The algorithm for calorie consumption prediction using random forest regression involves several key parameters. The parameter V represents the features extracted from accelerometer data, while T denotes the calories burned during activities. The objective function O is designed to minimize prediction error and maximize model accuracy. The initial parameters for the **RandomForestRegressor** are denoted by β. The **DataPreprocessor** is a tool used for cleaning and extracting features from accelerometer data, and the **ModelEvaluator** assesses model performance using metrics such as Mean Absolute Error (MAE) and R^2 (coefficient of determination).

Each file in the dataset is represented by the parameter f, and D_f refers to the data loaded from file f. Within the loaded data D_f, each activity session is denoted by s, and d_s is the preprocessed data for each activity session. The mean of the preprocessed data for each activity session is represented by μ, and the standard deviation by σ. The dataset containing features and labels is denoted by D. This dataset is split into a training subset D_{train} and a testing subset D_{test}. The parameter p represents each training period. T_{test} refers to the predicted calorie consumption for the testing dataset. The model's performance is evaluated using the metrics MAE and R^2.

Objective Function.

The objective function O is to minimize the prediction error and maximize the model accuracy:

$$O = \min\left(MAE\left(T_{pred}, T_{true}\right)\right) \tag{1}$$

And

$$O = \max\left(R^2\left(T_{pred}, T_{true}\right)\right) \tag{2}$$

where:

T_{pred} is the predicted calorie consumption.
T_{true} is the true calorie consumption.

Algorithm modelling Steps:

1. Loading Data: For each file f in the dataset:

$$D_f = \text{Load data from file } f$$

2. Processing Activity Sessions: For each activity session s in D_f:

$$d_s = \text{Extract data for session } s \text{ in } D_f$$

- Preprocess data d_s using **DataPreprocessor** to calculate mean, standard deviation, range, total distance and energy expenditure in Eqs. (3,4,5,6 and 7) as follows:

$$\mu = \frac{1}{n}\sum_{i}^{n} d_{si} \tag{3}$$

$$\sigma = \sqrt{\frac{1}{n}\sum_{i=1}^{n}(d_{si} - \mu)^2} \tag{4}$$

$$Range = \max(d_s) - \min(d_s) \tag{5}$$

$$Total\ Distance = \sum_{i=2}^{n} \sqrt{(x_i - x_{i-1})^2 + (y_i - y_{i-1})^2 + (z_i - z_{i-1})^2} \tag{6}$$

$$E = \sum_{i=1}^{n} \left(\frac{1}{2} m v_i^2 \right) \tag{7}$$

where:

m is the mass and v_i is the velocity at step i.

- Right after that the feature extraction take place as extract features V_s from d_s:

$$V_s = \{\mu, \sigma, Range, Total\ Distance, E\}$$

- Append features V_s and labels T_s to the dataset D:

$$D = D \cup \{(V_s, T_s)\}$$

3. Splitting the Data:

$$D_{train}, D_{test} = Split(D)$$

4. Training the Model: For each training period p:

$$RandomForestRegressor.fit(D_{train}, T_{train})$$

Predict calorie consumption T_{pred} for D_{test}:

$$T_{pred} = RandomForestRegressor.predict(D_{test})$$

5. Evaluating Predictions: Evaluate predictions using **ModelEvaluator** for MAE and R^2 using the following equations:

$$MAE = \frac{1}{n} \sum_{i=1}^{n} \left| T_{pred,i} - T_{true,i} \right| \tag{8}$$

$$R^2 = 1 - \frac{\sum_{i=1}^{n} \left(T_{true,i} - T_{pred,i} \right)^2}{\sum_{i=1}^{n} \left(T_{true,i} - T_{true} \right)^2} \tag{9}$$

6. Storing the calculated MAE and R^2 in the Performance Metrics:

$$metrics = \left\{ MAE, R^2 \right\}$$

7. Model Refinement:

$$\beta = update(\beta)$$

8. Output: Return the optimized RandomForestRegressor.

4 Results and Discussion

In this study, sensor data was split into 70% training and 30% testing sets. A mobile app collected the testing data over 30-min intervals. A Python algorithm, trained on standard data, achieved an average calorie consumption accuracy of 96.65% across 99 samples per activity as illustrated in Table 1 and Fig. 2. However, jogging and running had the lowest accuracy, necessitating gyroscope and accelerometer data for better differentiation. Notably, standing registered zero calorie consumption. Data from 99 volunteers (aged 18–59) included time, x, y, z coordinates, and activity descriptions, covering eight activities to enhance model accuracy.

Table 1. Random Forest Regression classifier average calories consumption accuracy score per activity.

Activity	Accuracy score
Walking	0.96130
Jogging	0.94322
Running	0.95895
Jumping	0.98922
Pushups	0.96790
Stairs down	0.97882
Going upstairs	0.97225
Cycling	0.96017

The error rate calculation results, as presented in Table 2, provide a useful summary of the performance of various classification models. These results are expressed in terms of the error rate percentage calculated over all the test samples. Additionally, the error rate data for different classification models has been illustrated in the visualization output, as shown in Fig. 2.

Given an accuracy score A, the error rate E can be calculated using:

$$E = 1 - A \tag{10}$$

In Fig. 2, the bar chart shows the average calories predicted by the model and the actual calories recorded for each activity. It helps assess how well the model performs across different activities in terms of average calorie estimation.

Figure 3 shows activity counts per minute over 30 min, color-coded by activity. A significant spike at the 14th minute is dominated by jogging and running. Other minutes vary, with walking and standing most consistently recorded, effectively visualizing activity types and frequencies.

Figure 4 uses a heatmap to illustrate calories burned over 30-min intervals. Darker colors show higher expenditure, with significant burns during running and pushups, reaching up to 1080 cal. Standing shows zero burn. The heatmap highlights the variability in calorie consumption between intense and sedentary activities.

Table 2. Random Forest Regression classifier error rate score.

Activity	Error rate
Walking	0.038
Jogging	0.056
Running	0.041
Jumping	0.010
Pushups	0.032
Stairs down	0.021
Going upstairs	0.027
Cycling	0.039

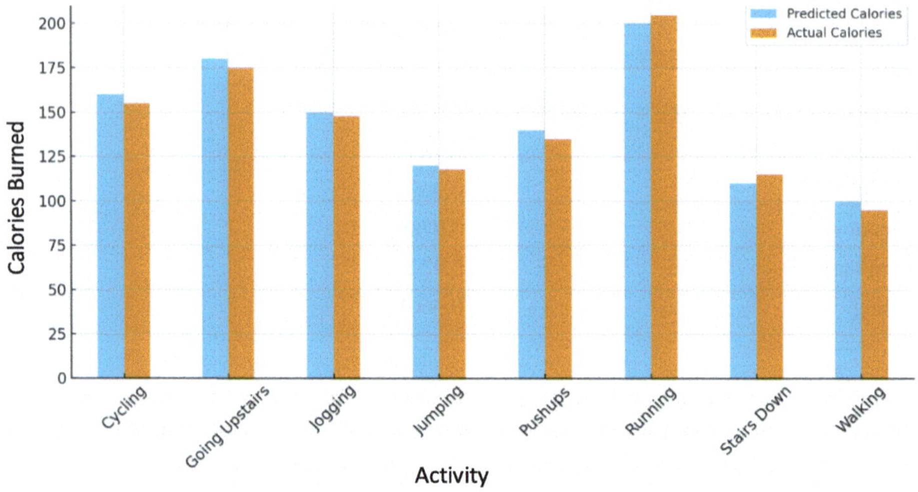

Fig. 2. Average Predicted vs. Actual Calories Burned Per Activity

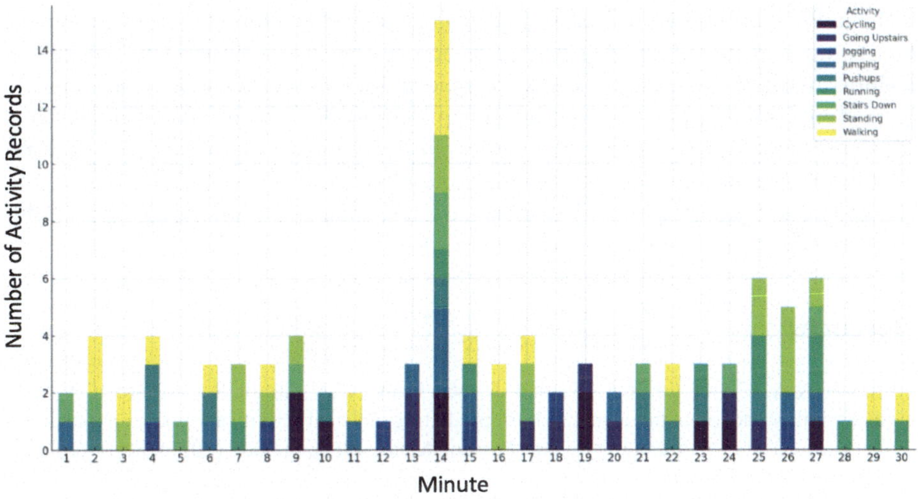

Fig. 3. Time Series Plot of Calories Burned Over Time

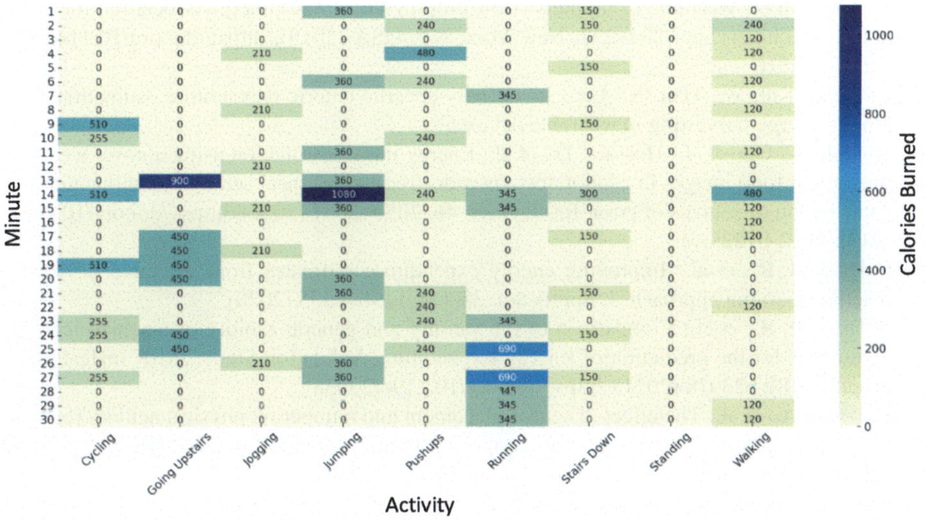

Fig. 4. Heatmap of Calories Burned by Activity

5 Conclusion

This study advances health analytics by developing Random Forest Regression models to accurately estimate caloric expenditure across various activities. Leveraging data from wearable sensors, including accelerometers and heart rate monitors, the models achieve an impressive 96.65% accuracy. This precision supports reliable energy expenditure estimates, crucial for effective health and fitness management.

The integration of diverse data sources and advanced machine learning techniques enhances the understanding of energy costs across activities. The models' robustness and ability to handle large datasets make them ideal for real-world applications, improving individual fitness tracking and clinical applications like obesity management. This study demonstrates significant improvements in the accuracy and reliability of caloric expenditure estimates, paving the way for more effective IoT-based health monitoring systems. The findings are expected to influence future wearable sensor designs and health monitoring algorithms, promoting a healthier society.

References

1. Pfeiffer, K.A., Clevenger, K.A., Kaplan, A., Van Camp, C.A., Strath, S.J., Montoye, A.H.K.: Accessibility and use of novel methods for predicting physical activity and energy expenditure using accelerometry: a scoping review. Physiol. Meas. **43**(8), 083001 (2022). https://doi.org/10.1088/1361-6579/ac89ca
2. Ling, X., Kai, W., Juan, L.: Caloric expenditure estimation from human kinetic energy in wearable devices. In: Adjunct Proceedings of the 2019 ACM International Joint Conference on Pervasive and Ubiquitous Computing and Proceedings of the 2019 ACM International Symposium on Wearable Computers (UbiComp/ISWC '19 Adjunct). Association for Computing Machinery, pp. 234–237. New York, NY, USA (2019). https://doi.org/10.1145/3341162.3343816
3. Ling, X., Kai, W., Tian, X., Luo, J.: Activity-specific caloric expenditure estimation from kinetic energy harvesting in wearable devices,
4. Doulah, A., Ghosh, T., Hossain, D., et al.: Energy intake estimation using a novel wearable sensor and food images in a laboratory (pseudo-free-living) meal setting: quantification and contribution of sources of error. Int. J. Obes. **46**, 2050–2057 (2022). https://doi.org/10.1038/s41366-022-01225-w
5. O'Driscoll, R., et al.: Improving energy expenditure estimates from wearable devices: a machine learning approach. J. Sports Sci. **38**(13), 1496–1505 (2020)
6. O'Driscoll, R., et al.: Comparison of the validity and generalizability of machine learning algorithms for the prediction of Energy Expenditure: Validation Study. JMIR mHealth and uHealth **9**(8), e23938 (2021). https://doi.org/10.2196/23938
7. Davoudi, A., et al.: The effect of sensor placement and number on physical activity recognition and energy expenditure estimation in older adults: validation study. JMIR mHealth and uHealth **9**(5), e23681 (2021). https://doi.org/10.2196/23681
8. Nipas, M., et al.: Burned calories prediction using supervised machine learning: regression algorithm. In: 2022 Second International Conference on Power, Control and Computing Technologies (ICPC2T), Raipur, India, pp. 1–4 (2022). https://doi.org/10.1109/ICPC2T53885.2022.9776710
9. Fuller, D., et al.: Predicting lying, sitting, walking and running using Apple Watch and Fitbit data. BMJ Open Sport Exerc. Med. **6**(1), e001004 (2020). https://doi.org/10.1136/bmjsem-2020-001004
10. Lopes, J.M., Figueiredo, J., Fonseca, P., Cerqueira, J.J., Vilas-Boas, J.P., Santos, C.P.: Deep learning-based energy expenditure estimation in assisted and non-assisted gait using inertial, EMG, and heart rate wearable sensors. Sensors **22**(20), 7913 (2022). https://doi.org/10.3390/s22207913
11. Kim, J., Lee, J., Park, M.: Identification of smartwatch-collected lifelog variables affecting body mass index in middle-aged people using regression machine learning algorithms and

SHapley additive explanations. Appl. Sci. **12**(8), 3819 (2022). https://doi.org/10.3390/app12083819
12. Wang, K., He, J., Zhang, L.: Sequential weakly labeled multiactivity localization and recognition on wearable sensors using recurrent attention networks. IEEE Trans. Human-Mach. Syst. **51**(4), 355–364 (2021). https://doi.org/10.1109/THMS.2021.3086008
13. Mishra, A., Bhattacharjee, V.: Applying semi-supervised learning on human activity recognition data. In 2022 International Conference on IoT and Blockchain Technology (ICIBT), pp. 1–6. IEEE (2022). https://doi.org/10.1109/ICIBT52874.2022.9807808
14. Zhang, Y., Shao, Y., Xiong, L., Zhang, J.: Human activity recognition based on on-body antennas and semi-supervised learning. In: 2023 17th European Conference on Antennas and Propagation (EuCAP), pp. 1–5 (2023). IEEE. https://doi.org/10.23919/EuCAP57121.2023.10133047
15. Radhika, V., Prasad, C.R., Chakradhar, A.: Smartphone-based human activities recognition system using random forest algorithm. In: 2022 International Conference for Advancement in Technology (ICONAT), pp. 1–4. IEEE (2022). https://doi.org/10.1109/ICONAT53423.2022.9726006
16. Feng, Z., Shi, Y., Zhou, D., Mo, L.: Research on human activity recognition based on random forest classifier. In: 2023 IEEE International Conference on Control, Electronics and Computer Technology (ICCECT), pp. 1507–1513. IEEE (2023). https://doi.org/10.1109/ICCECT57938.2023.10140545
17. Nia, G., Amiri, A., Nasab, A., Kaplanoglu, E., Liang, Y.: The power of ANN-random forest algorithm in human activities recognition using IMU data. In: 2023 IEEE EMBS International Conference on Biomedical and Health Informatics (BHI), pp. 1–7. IEEE (2023). https://doi.org/10.1109/BHI58575.2023.10313507
18. Wang, A., Chen, H., Zheng, C., Zhao, L., Liu, J., Wang, L.: Evaluation of random forest for complex human activity recognition using wearable sensors. In: 2020 International Conference on Networking and Network Applications (NaNA), pp. 310–315. IEEE (2020). https://doi.org/10.1109/NaNA51271.2020.00060

Enhanced Heart Attack Detection with Neural Networks

Kasem Khalil[1](✉), Md. Rahat Khan[1], Tamador Mohaidat[1], and Magdy Bayoumi[2]

[1] Department of Electrical and Computer Engineering, University of Mississippi, Oxford, MS, USA
kmkhalil@olemiss.edu, {mkaderk,tnmohaid}@go.olemiss.edu

[2] Department of Electrical and Computer Engineering, University of Louisiana at Lafayette, Lafayette, LA, USA
mab0778@louisiana.edu

Abstract. Heart attack detection is crucial for timely medical intervention and can significantly reduce mortality rates. Early and accurate detection enables prompt treatment, which is vital for improving patient outcomes and saving lives. This paper presents a novel method for heart attack detection, leveraging Fast Fourier Transform (FFT) for feature extraction and Artificial Neural Networks (ANN) for classification. The proposed method aims to provide accurate and real-time detection of heart attack events by transforming time-domain ECG signals into the frequency domain and utilizing a well-trained ANN to classify the patterns indicative of a heart attack. Implemented on a Xilinx Virtex-7 FPGA using VHDL, the system demonstrates efficient resource utilization and low power consumption, making it suitable for portable and embedded medical devices. Experimental results show that the proposed method achieves an accuracy of 92.41%, outperforming traditional methods such as Logistic Regression, k-Nearest Neighbors, Random Forest, and Support Vector Machine. This work underscores the potential of combining FFT and ANN for effective heart attack detection, offering a promising solution for real-time health monitoring systems.

Keywords: Artificial neural network · heart attack detection · low-power · FPGA

1 Introduction

Heart attacks, medically known as myocardial infractions, occur when the blood flow to part of the heart muscle is blocked, usually by a blood clot. This interruption deprives the heart muscle of oxygen, causing tissue damage or death. Risk factors such as high blood pressure, high cholesterol, smoking, obesity, and a sedentary lifestyle significantly contribute to the likelihood of experiencing a heart attack [6,8,26]. Symptoms often include chest pain, shortness of breath,

A. Abdelgawad et al. (Eds.): ISBCom 2024, LNNS 1268, pp. 544–553, 2025.
https://doi.org/10.1007/978-3-031-82377-0_44

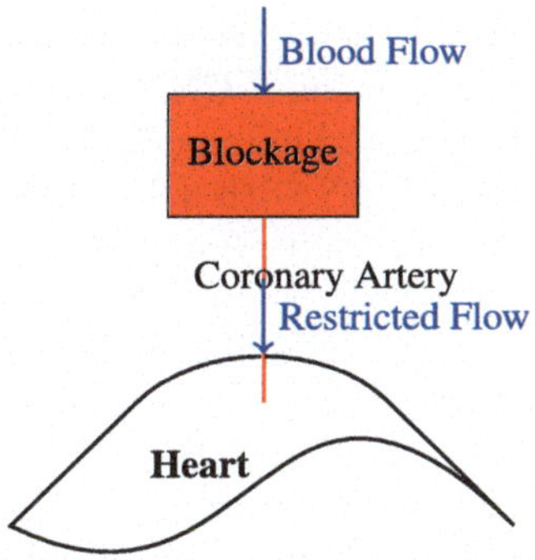

Fig. 1. Illustration of a heart attack caused by blockage in a coronary artery.

nausea, and lightheartedness, making early detection and prompt treatment critical. Early detection of heart attacks is crucial for timely medical intervention, which can substantially reduce heart muscle damage, improve survival rates, and enhance recovery outcomes. Advances in medical technology, such as Electrocardiograms (ECGs), cardiac biomarker tests, and imaging techniques, have improved our ability to diagnose heart attacks quickly and accurately [13,14,23]. Public awareness campaigns and routine health screenings are essential for identifying at-risk individuals before they suffer a heart attack. By recognizing early warning signs and risk factors, individuals and healthcare providers can take proactive measures to prevent heart attacks and manage cardiovascular health more effectively.

The mechanism of a heart attack caused by a blockage in a coronary artery is shown in Fig. 1. The coronary artery is depicted as a vital conduit responsible for supplying oxygen-rich blood to the myocardium. A blockage, shown as a conspicuous red obstruction, hinders blood flow, impeding the delivery of oxygen and essential nutrients to the heart muscle. This obstruction leads to restricted blood flow, as indicated by the blue arrows, ultimately causing a heart attack. This visual narrative encapsulates the pathophysiological cascade underlying myocardial infarction, serving as an educational tool for understanding the devastating consequences of coronary artery disease.

Neural networks are commonly used in several applications such as medical diagnostics [1,12,18,19,22], fault prediction [17], security [9,20,21], and self-healing [15,16,27]. Using neural networks for heart attack detection represents a significant advancement in medical diagnostics, harnessing the power of artificial intelligence to improve accuracy and speed. Neural networks, particularly deep learning models, can analyze vast amounts of data from sources [2,25] such as ECGs and patient history to identify patterns indicative of an impending heart attack [3,10,11,24]. These models are trained on extensive datasets, enabling them to recognize subtle and complex features that might be missed by human clinicians, leading to more accurate predictions and early detection.

Asmare *et al.* [4] introduced a Convolutional Neural Network (CNN) based deep learning algorithm for automatic auscultation and classification of heart sounds as normal or indicative of rheumatic heart disease. Their approach processes unsegmented data directly, using a CNN architecture with convolutional layers, batch normalization layers, and max pooling layers. This network demonstrates the application of a deep CNN with a Mel Spectro-temporal representation for heart sound classification. Again, Avanzato *et al.* [5] developed a system for the automatic diagnosis of heart disease, including five different pathology classes, utilizing ECG signals and CNNs. Their methodology employs advanced techniques for automatic detection of heart disease using ECG and phonocardiogram signals, using a dataset combining the MIT-BIH Arrhythmia Database and the MIT-BIH Atrial Fibrillation Database.

This paper focuses on heart attack detection based on FFT for feature extraction with ANN for classification. This hybrid approach leverages the FFT's ability to efficiently process and transform time-domain signals into frequency-domain data, which the ANN then analyzes to accurately detect heart attack patterns. By integrating these two techniques, the method aims to provide reliable and real-time detection of heart attacks. The rest of the paper is organized as follows. Section 2 introduces the proposed heart attack detection method while Sect. 3 presents the implementation and experimental results, and Sect. 4 concludes the work as well as the future plan.

2 The Proposed Method

The proposed heart attack detection method utilizes FFT and ANN. The methodology is divided into two primary stages: feature extraction through FFT and subsequent classification via ANN, as illustrated in the Fig. 2. The first step in our proposed method involves the extraction of features from ECG signals using the FFT. This technique is a powerful tool for transforming time-domain signals into their frequency-domain representations. This transformation allows us to capture the essential characteristics of the ECG signals that are indicative of heart abnormalities. Given an ECG signal, we apply the FFT to convert the signal from its time domain to the frequency domain. The resulting frequency components provide a comprehensive representation of the signal's behavior, capturing periodicities and other spectral features that are crucial for identifying patterns associated with heart attacks. The frequency-domain features obtained through FFT serve as inputs to the subsequent classification stage.

The FFT is an efficient algorithm for computing the Discrete Fourier Transform (DFT) and its inverse. The DFT transforms a sequence of complex numbers x_n in the time domain into another sequence of complex numbers X_k in the frequency domain. The mathematical formulation of the DFT is given by:

$$X_k = \sum_{n=0}^{N-1} x_n e^{-i2\pi kn/N} \quad \text{for} \quad k = 0, 1, \ldots, N-1$$

where x_n is the input time-domain signal. N is the number of points in the signal. k is the index of the output frequency component. i is the imaginary unit. The

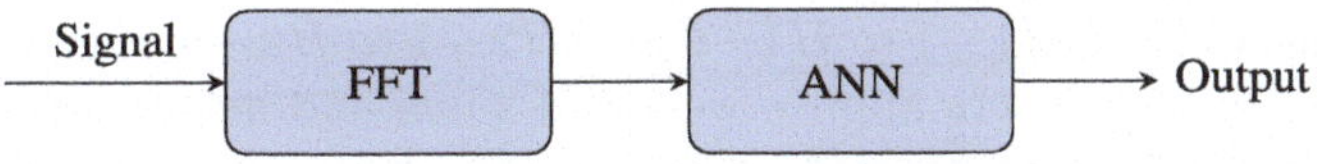

Fig. 2. Block diagram for heart attack detection using FFT and ANN.

inverse DFT, also called IDFT, which transforms the frequency-domain signal back to the time domain, is given by:

$$x_n = \frac{1}{N}\sum_{k=0}^{N-1} X_k e^{i2\pi kn/N} \quad \text{for} \quad n = 0, 1, \ldots, N-1$$

The FFT algorithm reduces the complexity of computing the DFT from $O(N^2)$ to $O(N \log N)$, making it practical for real-time signal processing. To apply FFT to an ECG signal, we follow these steps: for preprocessing, the ECG signal is preprocessed to remove noise and artifacts, ensuring a clean signal for analysis. For windowing, the signal may be segmented into smaller frames or windows, each of which is transformed separately to capture local frequency characteristics. For FFT computation, for each window, the FFT is computed using the Cooley-Tukey algorithm or another efficient FFT algorithm, resulting in a set of complex frequency components. For magnitude and phase extraction, the magnitude $|X_k|$ and phase $\angle X_k$ of the frequency components are extracted, where $|X_k| = \sqrt{(\Re(X_k))^2 + (\Im(X_k))^2}$ and $\angle X_k = \arctan\left(\frac{\Im(X_k)}{\Re(X_k)}\right)$. These frequency-domain features, particularly the magnitudes, provide important information about the energy distribution of the signal across different frequencies. By analyzing these features, we can identify characteristic patterns associated with heart attacks, which are then used as inputs to the classification stage involving an ANN.

After feature extraction, the next step is the classification of the extracted features to detect heart attacks. For this purpose, we employ an ANN, which is well-suited for recognizing complex patterns in data. The ANN used in our method consists of multiple layers, including an input layer, several hidden layers, and an output layer. The input layer receives the frequency-domain features derived from the FFT. These features are then processed through the hidden layers, where the network learns to map the input features to the output labels using a backpropagation algorithm. The output layer provides the final classification, indicating whether the input features correspond to a normal heart condition or a heart attack. Training the ANN involves using a labeled dataset of ECG signals, where each signal is pre-classified as either normal or indicative of a heart attack. The network learns from this dataset by adjusting its weights to minimize the classification error. Once trained, the ANN can accurately classify new ECG signals based on their frequency-domain features.

The ANN consists of input, hidden, and output layers, each consisting of neurons, as shown in Fig. 3. The input layer, comprised of five neurons labeled $I1$ to $I5$, receives input signals representing features or attributes of the data. These signals are then propagated through two hidden layers, each containing six neurons denoted as $H1-1$ to $H1-6$ and $H2-1$ to $H2-6$, respectively. Each neuron in the hidden layers applies weighted sums and activation functions to the incoming signals, thereby extracting and transforming information. Finally, the processed data flows to the output layer, represented by a single neuron labeled O, which computes the final output of the network. Arrows connecting neurons indicate the flow of information, with each neuron in the hidden layers connected to all neurons in the subsequent layer. The ANN comprises several key components: neurons, layers, activation functions, and weight matrices. The network consists of an input layer, multiple hidden layers, and an output layer. Each neuron in a given layer is connected to every neuron in the subsequent layer, forming a fully connected network.

Let $\mathbf{x} = [x_1, x_2, \ldots, x_n]$ be the input feature vector obtained from the FFT process. Each neuron j in the first hidden layer computes a weighted sum of the inputs, applies a bias term, and passes the result through an activation function σ. Mathematically, this can be expressed as:

$$z_j^{(1)} = \sigma\left(\sum_{i=1}^{n} w_{ij}^{(1)} x_i + b_j^{(1)}\right), \tag{1}$$

where $w_{ij}^{(1)}$ represents the weight connecting the i-th input neuron to the j-th neuron in the first hidden layer, and $b_j^{(1)}$ is the bias term for the j-th neuron.

In subsequent hidden layers l, each neuron k computes its activation as:

$$z_k^{(l)} = \sigma\left(\sum_{j=1}^{m_{l-1}} w_{jk}^{(l)} z_j^{(l-1)} + b_k^{(l)}\right), \tag{2}$$

where m_{l-1} is the number of neurons in the $(l-1)$-th layer, $w_{jk}^{(l)}$ is the weight connecting neuron j in layer $(l-1)$ to neuron k in layer l, and $b_k^{(l)}$ is the corresponding bias term.

The output layer produces the final classification result. If there are C classes, each output neuron o_c calculates:

$$o_c = \sigma\left(\sum_{k=1}^{m_{L-1}} w_{kc}^{(L)} z_k^{(L-1)} + b_c^{(L)}\right), \tag{3}$$

where L is the index of the output layer, $w_{kc}^{(L)}$ and $b_c^{(L)}$ are the weights and bias term for the c-th output neuron, respectively. For binary classification in heart attack detection, the output layer often uses the sigmoid activation function:

$$\sigma(z) = \frac{1}{1+e^{-z}}, \tag{4}$$

producing a probability value between 0 and 1, which indicates the likelihood of a heart attack.

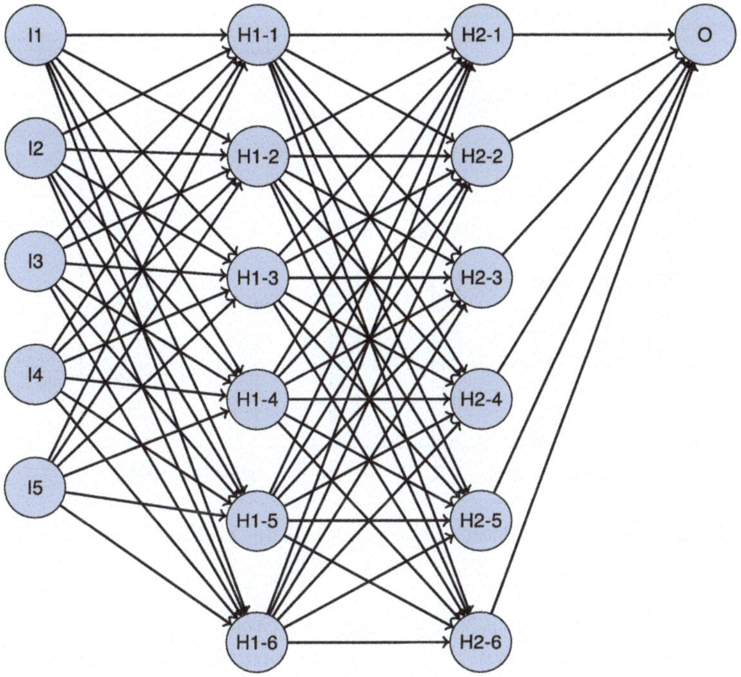

Fig. 3. Diagram of an ANN showing input, hidden, and output layers with neurons.

The ANN is trained using a backpropagation algorithm that minimizes a loss function, typically the cross-entropy loss for classification tasks, defined as:

$$\mathcal{L} = -\frac{1}{N}\sum_{i=1}^{N}\left[y_i \log(\hat{y}_i) + (1 - y_i)\log(1 - \hat{y}_i)\right], \tag{5}$$

where N is the number of training samples, y_i is the true label, and $\hat{y}_i$ is the predicted probability. The weights are updated using gradient descent to minimize this loss, improving the network's ability to accurately detect heart attacks.

3 Implementation and Experimental Results

The proposed method is tested using the heart attack dataset [7,28], which is an open-access resource designed to evaluate heart attack risks. This dataset includes records for 303 patients, with 138 patients (45.5%) classified as having a low risk of a heart attack and 165 patients (54.5%) categorized as having a high risk. This balanced dataset is essential for developing and testing predictive models for heart attack risk assessment.

The accuracy comparison of various methods for detecting heart attacks, including the proposed method and four other techniques: Logistic Regression

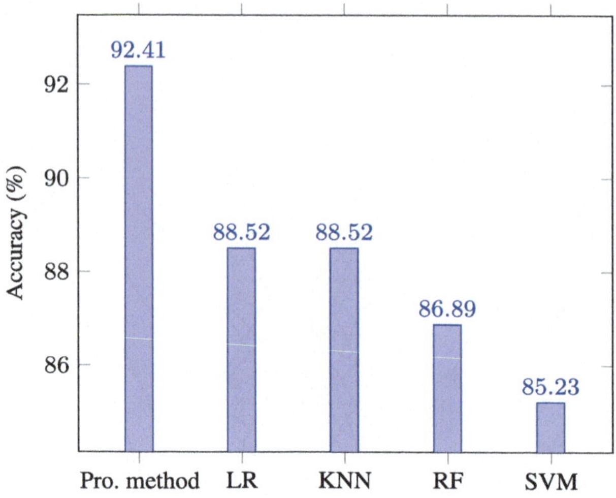

Fig. 4. Accuracy comparison of the proposed method and other methods

(LR), k-Nearest Neighbors (KNN), Random Forest (RF), and Support Vector Machine (SVM), is shown in Fig. 4. According to the results, the proposed method achieves the highest accuracy rate of 92.41%, outperforming LR, KNN, RF, and SVM, which achieved accuracies of 88.52%, 88.52%, 86.89%, and 85.23%, respectively. The significant margin by which the proposed method surpasses the others suggests its effectiveness in accurately detecting heart attacks. Moreover, the proposed method's superior performance indicates its potential for practical implementation in real-world scenarios, where accurate and timely detection of heart attacks is crucial for patient outcomes.

The proposed method for heart attack detection has been implemented using VHDL on a Xilinx Virtex-7 FPGA. This FPGA platform was chosen due to its high performance and efficient resource utilization capabilities. The implementation focuses on leveraging the computational power and flexibility of the Virtex-7 to achieve accurate and real-time heart attack detection. Resource utilization is studied for the proposed method. Resource utilization on an FPGA refers to the extent to which the available hardware resources, such as slice registers, slice LUTs (Look-Up Tables), and DSPs (Digital Signal Processors), are used during the implementation of a digital design. It provides an indication of how efficiently the FPGA's capacity is being leveraged for a specific application. High resource utilization means that a significant portion of the FPGA's resources are being employed, which can affect the performance and scalability of the design. Conversely, low resource utilization suggests that there is ample room for further optimization, additional features, or the inclusion of more complex algorithms without exceeding the FPGA's limits. This metric is crucial for understanding the balance between performance and resource consumption in FPGA-based designs.

Table 1. Logic Utilizing Statistics

Logic Utilizing	Available	Used	Utilization
Number of slices registers	20800	2684	12.90%
Number of slices LUTs	8150	1320	16.20%
Number of DSPs	90	25	27.78%
Power Consumption	28.4 mW		

The resource utilization and power consumption for implementing the proposed heart attack detection method on a Xilinx Virtex-7 FPGA using VHDL are shown in Table 1. The FPGA resources considered include the number of slice registers, slice LUTs, and DSPs. Out of the 20,800 available slice registers, the implementation uses 2,684, which corresponds to a utilization rate of 12.90%. Similarly, for the 8,150 available slice LUTs, 1,320 are used, resulting in a 16.20% utilization rate. Additionally, 25 out of the 90 available DSPs are employed, leading to a utilization of 27.78%. These figures indicate a balanced distribution of resource usage, ensuring that the FPGA can efficiently handle the computational demands of the heart attack detection algorithm without reaching its capacity limits. The power consumption for this implementation is measured at 28.4 mW, which is relatively low, suggesting that the design is not only resource-efficient but also power-efficient. This is a critical factor for portable or embedded medical devices where power efficiency is paramount. The low utilization percentages of slice registers and LUTs imply that there is considerable headroom for further optimization or for adding additional functionalities without requiring a more powerful FPGA. Overall, the table demonstrates that the proposed heart attack detection method can be effectively implemented on the Xilinx Virtex-7 FPGA with VHDL, making it a viable solution for real-time, efficient, and reliable heart health monitoring.

4 Conclusion

In this paper, we present a novel method for heart attack detection that combines FFT for feature extraction with ANN for classification. This method is implemented on a Xilinx Virtex-7 FPGA using VHDL, resulting in efficient resource utilization and low power consumption, which makes it highly suitable for real-time, portable medical applications. The experimental results underscore the superior accuracy of our method compared to traditional classification techniques, achieving an impressive accuracy rate of 92.41%. These findings confirm the effectiveness of our approach in accurately detecting heart attacks, providing a reliable and efficient solution for continuous heart health monitoring. Additionally, our work demonstrates significant potential for future improvements as we plan to further optimize the system and explore its application in broader healthcare contexts, thereby enhancing its utility and impact in the medical field.

References

1. Agarwal, D., Berbís, M.Á., Luna, A., Lipari, V., Ballester, J.B., de la Torre-Díez, I.: Automated medical diagnosis of alzheimer's disease using an efficient net convolutional neural network. J. Med. Syst. **47**(1), 57 (2023)
2. Aggarwal, C.C., et al.: Neural Networks and Deep Learning, vol. 10. Springer, Heidelberg (2018)
3. Arabasadi, Z., Alizadehsani, R., Roshanzamir, M., Moosaei, H., Yarifard, A.A.: Computer aided decision making for heart disease detection using hybrid neural network-genetic algorithm. Comput. Methods Programs Biomed. **141**, 19–26 (2017)
4. Asmare, M.H., Woldehanna, F., Janssens, L., Vanrumste, B.: Rheumatic heart disease detection using deep learning from spectro-temporal representation of unsegmented heart sounds. In: 2020 42nd Annual International Conference of the IEEE Engineering in Medicine & Biology Society (EMBC), pp. 168–171. IEEE (2020)
5. Avanzato, R., Beritelli, F.: Heart disease recognition based on extended ecg sequence database and deep learning techniques. In: 2022 IEEE International Conference on Internet of Things and Intelligence Systems (IoTaIS), pp. 117–121. IEEE (2022)
6. Bhardwaj, A., Singh, S., Joshi, D.: Explainable deep convolutional neural network for valvular heart diseases classification using pcg signals. IEEE Trans. Instrument. Meas. (2023)
7. Bhat, N.: Health care: Data set on heart attack possibility (2020). https://www.kaggle.com/nareshbhat/health-care-data-set-on-heart-attack-possibility. Accessed 30 May 2024
8. Bhushan, M., Pandit, A., Garg, A.: Machine learning and deep learning techniques for the analysis of heart disease: a systematic literature review, open challenges and future directions. Artif. Intell. Rev. **56**(12), 14035–14086 (2023)
9. Bian, L.: Design of computer network security defense system based on artificial intelligence and neural network. Wirel. Pers. Commun. 1–20 (2023)
10. Calderon-Vilca, H.D., Callupe, K.E.C., Aliaga, R.J.I., Cuba, J.B., Mariño-Cárdenas, F.C.: Early cardiac disease detection using neural networks. In: 2019 7th International Engineering, Sciences and Technology Conference (IESTEC), pp. 562–567. IEEE (2019)
11. Chang, V., Bhavani, V.R., Xu, A.Q., Hossain, M.: An artificial intelligence model for heart disease detection using machine learning algorithms. Healthc. Anal. **2**, 100016 (2022)
12. Elsersy, M., Sherif, A., Imam, A.A.A., Mamun, M.M.K., Khalil, K., Haitham, M.: Federated learning model for early detection of dementia using blood biosamples. In: 2023 IEEE International Conference on Artificial Intelligence, Blockchain, and Internet of Things (AIBThings), pp. 1–5. IEEE (2023)
13. Haimovich, J.S., et al.: Artificial intelligence-enabled classification of hypertrophic heart diseases using electrocardiograms. Cardiovasc. Dig. Health J. **4**(2), 48–59 (2023)
14. Jyothi, P., Pradeepini, G.: Heart disease detection system based on ecg and pcg signals with the aid of gkvdlnn classifier. Multimedia Tools Appl. **83**(10), 30587–30612 (2024)
15. Khalil, K., Eldash, O., Kumar, A., Bayoumi, M.: Self-healing approach for hardware neural network architecture. In: 2019 IEEE 62nd International Midwest Symposium on Circuits and Systems (MWSCAS), pp. 622–625. IEEE (2019)

16. Khalil, K., Eldash, O., Kumar, A., Bayoumi, M.: Intelligent fault-prediction assisted self-healing for embryonic hardware. IEEE Trans. Biomed. Circuits Syst. **14**(4), 852–866 (2020)
17. Khalil, K., Eldash, O., Kumar, A., Bayoumi, M.: Machine learning-based approach for hardware faults prediction. IEEE Trans. Circuits Syst. I Regul. Pap. **67**(11), 3880–3892 (2020)
18. Khalil, K., et al.: A federated learning model based on hardware acceleration for the early detection of alzheimer's disease. Sensors **23**(19), 8272 (2023)
19. Khalil, K., Mamun, M.M.K., Sherif, A., Elsersy, M., Imam, A.A.A., Abouzaid, K.: A dementia diagnosis technique based on ai and hardware acceleration. In: 2023 IEEE International Conference on Artificial Intelligence, Blockchain, and Internet of Things (AIBThings), pp. 1–5. IEEE (2023)
20. Khalil, K., et al.: Privacy-preserving and hardware acceleration-based authentication scheme for data collection in e-health applications. In: 2023 IEEE 66th International Midwest Symposium on Circuits and Systems (MWSCAS), pp. 703–707. IEEE (2023)
21. Khan, S., Khan, M.A., Alnazzawi, N.: Artificial neural network-based mechanism to detect security threats in wireless sensor networks. Sensors **24**(5), 1641 (2024)
22. Kshatri, S.S., Singh, D.: Convolutional neural network in medical image analysis: a review. Arch. Comput. Methods Eng. **30**(4), 2793–2810 (2023)
23. Malakouti, S.M.: Heart disease classification based on ECG using machine learning models. Biomed. Signal Process. Control **84**, 104796 (2023)
24. Mehrankia, A., Mollakhalili Meybodi, M.R., Mirzaie, K.: Prediction of heart attacks using biological signals based on recurrent GMDH neural network. Neural Process. Lett. **54**(2), 987–1008 (2022)
25. Mohaidat, T., Khalil, K.: A survey on neural network hardware accelerators. IEEE Trans. Artif. Intell. (2024)
26. Naeem, A.B., et al.: Heart disease detection using feature extraction and artificial neural networks: a sensor-based approach. IEEE Access (2024)
27. Sujatha, D., TF, M.R., Ramesh, G., Agoramoorthy, M., et al.: Neural networks-based predictive models for self-healing in cloud computing environments. In: 2024 International Conference on Intelligent and Innovative Technologies in Computing, Electrical and Electronics (IITCEE), pp. 1–4. IEEE (2024)
28. Yilmaz, R., Yağin, F.H.: A comparative study for the prediction of heart attack risk and associated factors using mlp and rbf neural networks. J. Cogn. Syst. **6**(2), 51–54 (2021)

Scholarly Article Classification Leveraging DistilBERT Transformer and Transfer Learning

Rasha S. Gargees(✉)

Central Michigan University, Mount Pleasant, MI 48859, USA
garge1r@cmich.edu

Abstract. Efficient organization and access to scholarly information are paramount in the digital age, where the volume of research papers is ever-growing. Research paper classification is a crucial task that aids in organizing and retrieving this vast amount of information. This study leverages the DistilBERT model, a distilled version of the BERT (Bidirectional Encoder Representations from Transformers) model, to classify research papers based on their titles and summaries. DistilBERT offers a computationally efficient solution for natural language processing tasks, making it ideal for large-scale document classification. The proposed method involves fine-tuning DistilBERT on a dataset of research papers to develop a classification model. Initially, the titles and summaries of research papers are preprocessed to convert them into vectors using DistilBERT's pre-trained language model. These vectors are then used as input features to train a classifier that predicts the category of the research paper. The proposed approach was evaluated on a dataset of research papers across various disciplines. The results demonstrate the method's effectiveness, with classification performance metrics such as F1 score, recall, precision, and accuracy all exceeding 81%. These results indicate that the proposed method can accurately classify research papers while significantly reducing the computational resources required compared to the original BERT model. Efficient classification of research papers can greatly benefit researchers, educators, and students by providing them with a more organized and accessible repository of scholarly information. The study highlights the utility of transformer models and transfer learning in scholarly article classification.

Keywords: DistilBERT Transformer · Classification · Transfer Learning · Scholarly Article

1 Introduction

With the increase of published research papers, an information explosion problem is formed in the process of paper retrieval, and to overcome the retrieval difficulties, papers can be classified by tagging them. Usually, manual classification implies a lot of time and effort and requires relevant workers in each segment

A. Abdelgawad et al. (Eds.): ISBCom 2024, LNNS 1268, pp. 554–566, 2025.
https://doi.org/10.1007/978-3-031-82377-0_45

to complete the classification task, which is a boring and inefficient process. In the era of deep learning, with the improvement of computing power and the increasing number of methods for processing text sequence data, people must use the tools of deep learning to combine the already labeled data for modeling and complete the task of document classification efficiently and accurately [1]. Consequently, innovative methods and techniques are urgently needed to streamline the management and categorization process of scientific documents [2].

Text classification is one of the numerous tasks being solved by Natural Language Processing (NLP) [3]. With the development of deep learning, more and more researchers use deep neural networks to classify short text. The main disadvantage of this kind of method is that it can not effectively alleviate the problem of feature sparsity and keyword ambiguity of short text. Directly inputting short text into the network for classification will make the classification effect very unsatisfactory. Recently, some researchers have extended the topic feature of the neural topic model. Although it can solve the problem of feature sparsity to some extent and can train the model end-to-end, it still can not effectively solve the problem of polysemy and noise. With the development of depth map learning network, researchers have begun to apply graph learning and graph neural networks to NLP or text classification in recent years. However, effectively applying graph neural network to short text classification remains challenging [4].

Deep learning based pre-trained language models (PLMs) like ELMo, GPT, BERT etc. have given rise to a technique called transfer learning (TL) in NLP. It is a process in which knowledge acquired while training a model on a particular task or domain can be reused to solve downstream task or domain. There are many NLP tasks and domains for which it is very difficult to find large scaled labelled data. PLMs are pre-trained on one or more tasks and can be reused to solve several other NLP tasks using the process of fine-tuning. Unlike machine learning, TL eliminates the need to build task specific language models. Transfer learning has demonstrated impressive outcomes across diverse domains of computer vision, including applications in medicine, remote sensing, and other areas. Recently, its adoption has been accelerating within the realm of natural language processing (NLP) [5–11].

The DistilBERT model is a distilled version of the BERT model, designed to be smaller and more efficient. By applying knowledge distillation, DistilBERT reduces the number of parameters by 60%, significantly speeding up the learning process. The distillation method minimizes the difference between the teacher model's and the student model's label distributions, known as soft target loss, alongside MLM loss. DistilBERT retains the structure of BERT but with fewer layers (six instead of twelve), allowing for efficient fine-tuning by initializing parameters from specific layers of the original BERT model. This approach enables domain-specific pre-training when using a target domain corpus. DistilBERT maintains approximately 97% of BERT's language understanding capabilities while being 60% faster and 40% smaller, addressing BERT's operational challenges related to size and computational requirements [5,12–15].

While existing research has made significant strides in automating the classification of textual data, there remains a gap in the application of advanced language models, such as DistilBERT, specifically for research paper classification using titles and summaries. Most studies focus on traditional machine learning algorithms or simpler neural network architectures, which may not fully capture the complex semantics and context present in research paper titles and summaries. This work aims to bridge this gap by leveraging the contextual understanding of DistilBERT for research paper classification. By fine-tuning DistilBERT on a dataset of research paper titles and summaries, the model can learn to extract and encode the most relevant features, leading to more accurate and nuanced classification results. This approach not only improves the accuracy of classification but also allows for a more comprehensive analysis of research papers, enabling researchers to efficiently navigate and explore vast amounts of scholarly literature.

The remainder of this paper is structured as follows: Sect. 2 provides a review of related work, discussing previous methodologies pertinent to our study. Section 3 explains our methodology, including dataset description, preprocessing techniques, and the design of our classification model. Section 4 outlines the experimental setup and presents and analyzes the results. Section 5 concludes with a summary of our findings and recommendations for future research.

2 Related Work

Several studies have addressed the classification of textual data in various domains, aiming to improve the accuracy and efficiency of these processes through different machine learning and natural language processing techniques.

Authors in [16] focus on classifying news articles by violation type under Indonesia's Electronic Information and Transaction (ITE) Law using a Support Vector Machine (SVM) algorithm with an RBF kernel to enhance public understanding of ITE Law violations. Another research [17] classifies newspaper articles into multiple categories using a multi-class SVM, transforming the task into several binary classification problems that surpass a Binary Classifier.

In the realm of ancient literature, Yihe et al. [18] utilize natural language processing and SPSS data analysis to classify and extract themes, organizing data to reduce redundancy and highlight research hotspots in ancient book research. Additionally, Hossain et al. in [19] studied improved academic information retrieval by classifying research papers based on their abstracts using GloVe word embeddings and deep learning models (LSTM, GRU, Bi-LSTM, Bi-GRU), achieving high precision and accuracy in categorizing papers across various scientific disciplines.

Another study [20] examines the complex discourse structures in scientific articles within the wastewater-based epidemiology domain, proposing a two-layer hierarchical discourse model and creating an annotated dataset to train an effective classifier for automatic annotation. Further, study [21] on scientific entity recognition addresses multi-label, low-resource, and nested scenarios, proposing an MRC framework and a Knowledge-Driven Prototype Guidance (KDPG) module to enhance few-shot learning and improve entity recognition accuracy. Feature selection's importance in handling large datasets is highlighted in a study [22] reviewing methods to reduce data dimensionality and select relevant features for accurate models. Similarly, a paper classification system [23] uses Word2Vec, network modeling, and community discovery facilitates efficient access to relevant literature for researchers.

Another study [1] addresses information overload in research paper retrieval, a joint embedding method for multi-label paper classification combines title embedding (TEM) and abstract embedding (AEM) models, outperforming baseline models on the CiteULike dataset. Lastly, a study explored [24] automates the assignment of conference papers to topics or reviewers using a classification model with TF-IDF vectorization, with SVM achieving the highest F-measure score among six tested algorithms.

Despite the advancements in classifying textual data, there is still a need for more efficient and accurate methods in research paper classification, particularly using titles and summaries. This work addresses this gap by leveraging pre-trained DistilBERT to classify research papers based on their titles and summaries. The use of DistilBERT, a smaller and faster variant of BERT, allows for effective transfer learning and fine-tuning on specific datasets, improving classification accuracy while maintaining computational efficiency. This approach not only builds on the strengths of existing methodologies but also introduces an application of DistilBERT in the domain of research paper classification, potentially setting a new standard for automated literature categorization.

3 Methodology

This section details the systematic approach taken to address the research questions and achieve the study objectives. It includes an explanation of the dataset description, the analytical techniques applied, and the rationale behind the choices made. Figure 1 illustrates the workflow for research papers classification. The process begins with text preprocessing, followed by label encoding and dataset splitting. The text is then tokenized before the model is built. After model construction, the model is trained using the tokenized data. Finally, the trained model is evaluated to assess its performance. Each step is crucial for preparing the data, constructing an effective model, and ensuring accurate classification results.

3.1 Dataset

The dataset utilized in this study, known as "arXiv-10," [25] comprises research papers sourced from the arXiv repository, spanning various disciplines such as computer science, physics, mathematics, and others. Renowned for its comprehensive array of papers, the dataset is well-suited for research endeavors across diverse fields. The "arXiv-10" dataset is a benchmark dataset consisting of 100,000 ArXiv scientific papers, meticulously balanced with exactly 10,000 papers per class across 10 categories. This dataset provides essential metadata,

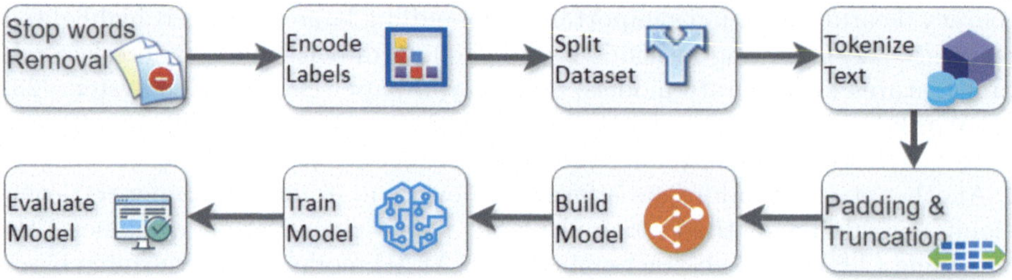

Fig. 1. Text Classification Workflow. The process flowchart illustrates the steps in the text classification pipeline.

including paper titles, abstracts, and categories, which play a crucial role in training the classification model. This enables the model to discern patterns and correlations within different research papers, thereby enhancing its effectiveness in categorizing papers accurately.

3.2 Preprocessing

In the preprocessing pipeline for text classification with DistilBERT, stop words common words like "the," "is," and "and" that add little meaning-are removed using NLTK's stop words list. This helps focus on more meaningful content. The text labels are then encoded into integers. This step is crucial as machine learning models require numeric labels for training. Next, the dataset is split into training and validation sets, with a split percentage of 70% training and 30% validation, ensuring both sets have a similar distribution of classes. This approach is important for training and evaluating the model effectively.

Finally, the text inputs are tokenized using DistilBERT's tokenizer. This process converts the text into numerical tokens that the model can understand. Additionally, padding is applied to ensure all sequences have the same length, truncation removes any excess tokens, and a maximum sequence length of tokens is set to manage computational resources efficiently. Overall, this preprocessing prepares the text data for training a DistilBERT model for sequence classification by cleaning the text, encoding labels, and formatting the inputs for the model.

3.3 Classification

The DistilBERT model is selected for its efficiency and effectiveness in natural language processing tasks. DistilBERT, a compact variant of the BERT model developed by Hugging Face, has gained popularity for its ability to efficiently process and understand natural language while maintaining impressive performance. By distilling the original BERT model into a smaller, faster version, DistilBERT retains much of the original's power while being more lightweight and easier to deploy in various NLP applications [26–29]. In the context of text classification, DistilBERT offers a compelling solution. Its tokenizer breaks down text inputs into numerical tokens, which are then processed by the model. This tokenization process enables the model to understand the semantic meaning of the text and make predictions based on that understanding [30,31].

The classification process with DistilBERT involves several key steps. Firstly, the text labels are converted into tensors, a format suitable for TensorFlow, ensuring compatibility with the model. The tokenized inputs are then converted into dictionaries of tensors, serving as the input format for the model. To enhance generalization and prevent overfitting, the DistilBERT model for sequence classification is customized with dropout and attention dropout rates. This configuration helps the model learn robust representations of the text data.

For training, an Adam optimizer is employed with a learning rate of 5e-6, and sparse categorical crossentropy is used as the loss function. These choices optimize the model's performance during training and enable it to make accurate predictions.

Mixed precision training is enabled to reduce memory usage and accelerate computation. Dataset pipelines are created to efficiently handle the training data, ensuring that the model receives data in the most optimized way possible. Overall, the classification process with DistilBERT is a comprehensive approach to text classification, encompassing data preprocessing, model setup, training, and evaluation.

4 Experimental Setup and Result Analysis

The experimental setup involved training a DistilBERT model for sequence classification using a dataset of research paper titles and abstracts. The dataset was split into training (70%) and validation (30%) sets. Text preprocessing included the removal of stop words and tokenization using DistilBertTokenizer. The model was compiled with the Adam optimizer and a SparseCategoricalCrossentropy loss function. Training was performed with a batch size of 32, incorporating mixed precision training and callbacks for detailed metric reporting. The final model utilized a dropout rate and achieved validation accuracy and loss monitoring through detailed plots of training progress.

Additionally, we present the performance metrics of our classification model evaluated on the test dataset. The metrics used to evaluate the model include accuracy, precision, recall, and F1-score. These metrics provide a comprehensive overview of the model's performance, both overall and for individual classes.

Figure 2 illustrates the per-class accuracy of our classification model across ten distinct categories. The red dashed line indicates the overall average accuracy of the model, calculated to be 81.76%. From the figure, it is evident that the model performs exceptionally well in certain categories, achieving accuracies of 94% for classes like astro-ph, hep-ph, and quant-ph. These high accuracies suggest that the model can effectively identify and classify instances within these categories, likely due to distinct and well-represented features.

Conversely, the model shows lower performance in classes such as cs (64%), cond-mat (69%), and physics (72%). The relatively lower accuracy in these categories might be attributed to several factors, including feature overlap with other

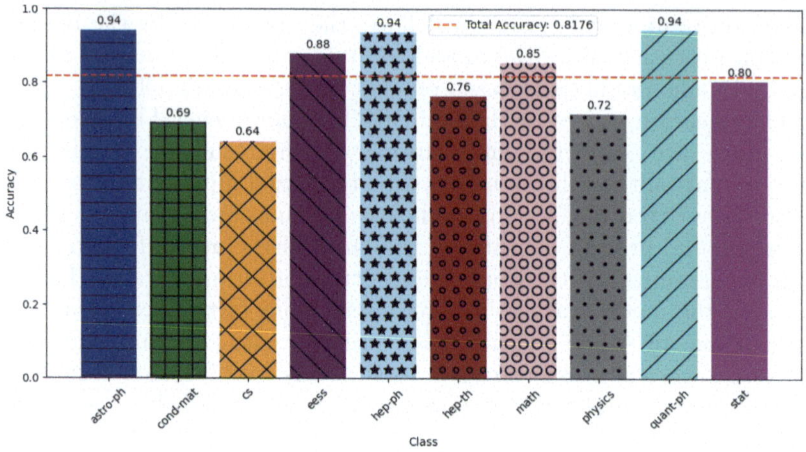

Fig. 2. Per-Class accuracy. The red dashed line showing the overall average accuracy of 81.76%. High accuracy is observed in astro-ph, hep-ph, and quant-ph, while lower accuracy is noted in cs and cond-mat.

classes, a higher degree of intra-class variability. For instance, the cs category's accuracy stands out as notably low, suggesting a need for targeted improvements, such as augmenting the dataset with more diverse examples or refining feature extraction methods specific to this class.

The red dashed line representing the overall accuracy provides a benchmark against which individual class performances can be compared. Classes with accuracies significantly below this benchmark, such as cs and cond-mat, highlight areas where the model's classification capabilities are less reliable, thus warranting further investigation and optimization. On the other hand, classes performing well above this benchmark underscore the strengths of the model and the effectiveness of the current classification strategy for those categories.

Figure 3 shows the overall precision, recall, and F1-score of the model across all classes. The precision of the model is 0.8264, indicating the proportion of true positive predictions among all positive predictions made by the model. The recall is 0.8176, reflecting the proportion of true positive predictions among all actual positives in the dataset. The F1-score, which is the harmonic mean of precision and recall, is 0.8168. These metrics indicate a well-balanced performance between precision and recall, suggesting that the model is effective in identifying relevant instances while maintaining a low rate of false positives.

Figure 4 breaks down these metrics by individual classes, providing a detailed view of the model's performance across different categories. The precision, recall, and F1-score are relatively high for most classes, such as 'astro-ph' (precision: 0.92, recall: 0.94, F1-score: 0.93) and 'hep-ph' (precision: 0.86, recall: 0.94, F1-score: 0.90), indicating that the model performs particularly well in these areas. However, some classes like 'cs' (precision:0.78, recall: 0.64, F1-score: 0.71) and 'physics' (precision: 0.66, recall: 0.72, F1-score:0.69) show more significant vari-

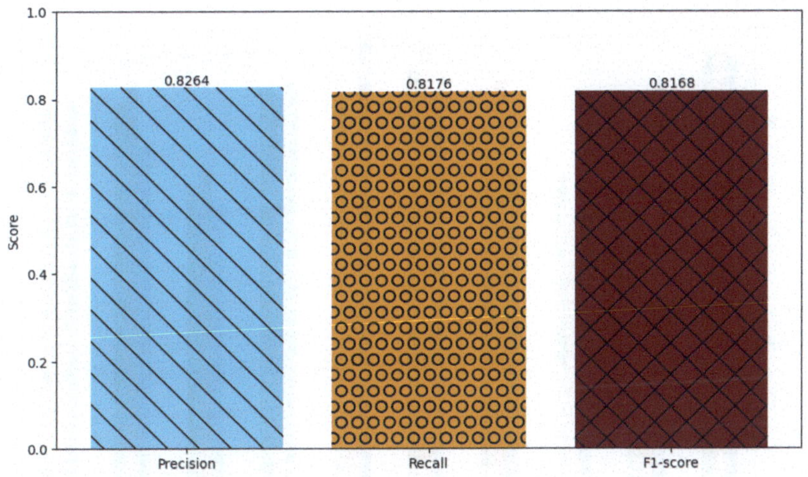

Fig. 3. Overall performance metrics of the classification model. The chart shows the precision, recall, and F1-score across all classes. These metrics indicate a balanced performance between precision and recall.

ability, highlighting areas where the model's performance could be improved. This class-wise analysis provides insights into the strengths and weaknesses of the model in different domains, guiding future enhancements and targeted improvements.

Figure 5 presents the confusion matrix, illustrating the model's performance in detail by showing the number of correct and incorrect predictions for each class. The confusion matrix reveals that the model correctly classified 2823 samples of 'astro-ph' but misclassified a few into other categories such as 'hep-ph' and 'quant-ph.' Misclassifications are more prominent in classes with overlapping features, such as 'cs' and 'eess,' where the model confused several instances. For example, the model misclassified 630 'cs' samples as 'eess,' and 522 'cond-mat' samples as 'physics.' The confusion matrix helps identify specific areas where the model struggles, providing a basis for future model refinements and feature engineering efforts.

The results indicate that while our model performs well overall, there are disparities in performance across different classes. The high precision and recall in certain classes suggest that the model effectively captures the defining features of these categories. However, the variability in other classes indicates a need for further optimization. The confusion matrix analysis highlights the importance of understanding the specific errors made by the model. By identifying patterns in misclassifications, we can target specific areas for improvement. For instance,

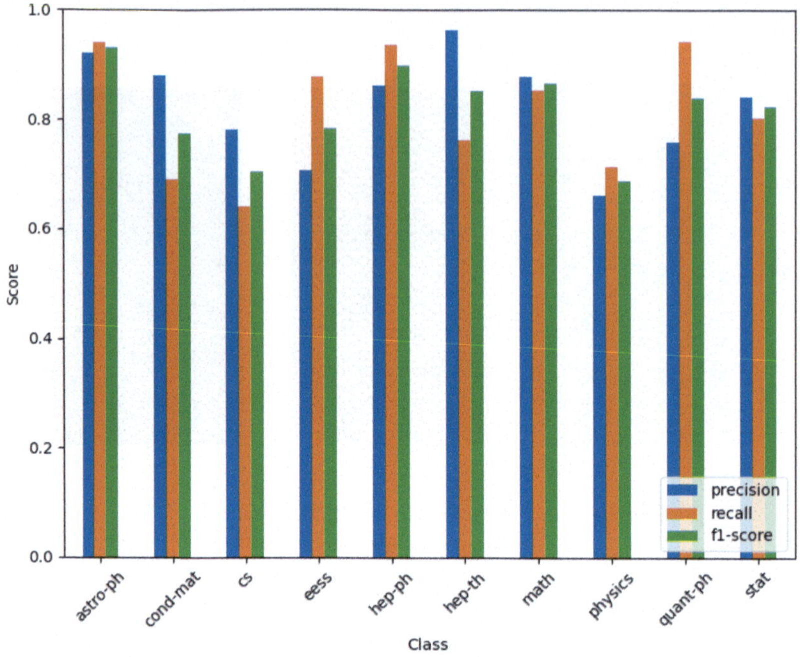

Fig. 4. Class-wise performance metrics of the classification model. The chart shows the precision, recall, and F1-score for each class, highlighting the variability in the model's performance across different categories. Some classes, such as 'astro-ph' and 'hep-ph,' have high scores, while others, like 'cs' and 'physics,' show more significant variability.

enhancing the model's ability to distinguish between closely related classes like 'cs' and 'eess' could significantly improve overall performance.

The presented metrics and analyses provide a comprehensive evaluation of our classification model, highlighting both its strengths and areas for potential improvement. The overall accuracy of 0.8176 demonstrates the model's effectiveness, while the detailed performance metrics and confusion matrix analysis offer valuable insights for future work.

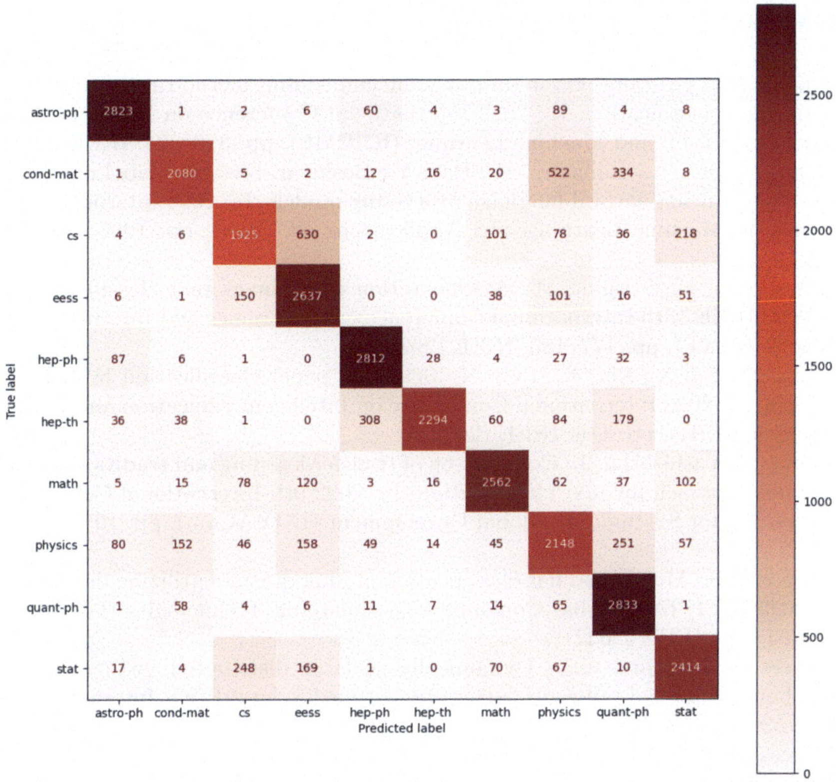

Fig. 5. Confusion matrix of the classification model. The matrix illustrates the number of true and false predictions for each class.

5 Conclusion

In the study, DistilBERT transformers were utilized for the classification of research papers, employing transfer learning techniques. The approach leveraged pre-trained language models, demonstrating exceptional performance compared to existing methods. A thorough analysis was conducted on a diverse dataset, with metrics such as accuracy, F1 score, recall, and precision meticulously evaluated on a per-class basis and collectively.

The findings highlighted the effectiveness of transformer models for research paper classification. By using transfer learning with DistilBERT, the study captured the semantic nuances inherent in scientific literature, enabling precise categorization. This approach not only streamlines the management of research documents but also enhances broader accessibility to scholarly knowledge. The research significantly advances methodologies for classifying research papers, showcasing the potential of transformer-based models. Future work could focus on refining the approach, including hyperparameter tuning and the incorporation of additional features, to further improve performance.

References

1. Wei, Y., He, Y., Yang, C.: Jetam: a joint embedding method for research paper mutil-label classification. In: 2022 International Conference on Image Processing, Computer Vision and Machine Learning (ICICML), pp. 390–394. IEEE (2022)
2. Zhang, Z., Das, A., Rahgouy, M., Bao, Y., Baskiyar, S.: Multi-label classification of cs papers using natural language processing models. In: 2023 International Conference on Machine Learning and Applications (ICMLA), pp. 1907–1912. IEEE (2023)
3. Pavlyshenko, B., Stasiuk, M.: Augmentation in a binary text classification task. In: 2023 IEEE 13th International Conference on Electronics and Information Technologies (ELIT), pp. 177–180. IEEE (2023)
4. Zhang, X., Yu, X., Liu, X., Lyu, X.: Scientific paper classification by fusing bert and gcn. In: 2023 International Conference on Intelligent Education and Intelligent Research (IEIR), pp. 1–6. IEEE (2023)
5. Taneja, K., Vashishtha, J.: Comparison of transfer learning and traditional machine learning approach for text classification. In: 2022 9th International Conference on Computing for Sustainable Global Development (INDIACom), pp. 195–200. IEEE (2022)
6. Gargees, R.S.: Multi-class flat classification of lung diseases utilizing deep learning. In: 2022 IEEE IAS Global Conference on Emerging Technologies (GlobConET), pp. 804–809. IEEE (2022)
7. Gargees, R.S., Scott, G.J.: Dynamically scalable distributed virtual framework based on agents and pub/sub pattern for iot media data. IEEE Internet Things J. **6**(1), 599–613 (2018)
8. Gargees, R.S., Scott, G.J.: Multi-stage distributed computing for big data: evaluating connective topologies. In: 2020 10th Annual Computing and Communication Workshop and Conference (CCWC), pp. 0626–0633. IEEE (2020)
9. Gargees, R.S., Scott, G.J.: Deep feature clustering for remote sensing imagery land cover analysis. IEEE Geosci. Remote Sens. Lett. **17**(8), 1386–1390 (2019)
10. Gargees, R.S.: Leveraging fog computing for geographically distributed smart cities. In: 2022 IEEE 2nd Conference on Information Technology and Data Science (CITDS), pp. 80–85. IEEE (2022)
11. Gargees, R.S., Scott, G.J.: Large-scale, multiple level-of-detail change detection from remote sensing imagery using deep visual feature clustering. Rem. Sens. **13**(9), 1661 (2021)
12. Bai, J., Cao, R., Ma, W., Shinnou, H.: Construction of domain-specific distilbert model by using fine-tuning. In: 2020 International Conference on Technologies and Applications of Artificial Intelligence (TAAI), pp. 237–241. IEEE (2020)
13. Mukherjee, A., et al.: Resume ranking and shortlisting with distilbert and xlm. In: 2024 IEEE International Conference for Women in Innovation, Technology & Entrepreneurship (ICWITE), pp. 301–304. IEEE (2024)
14. Jojoa, M., Eftekhar, P., Nowrouzi-Kia, B., Garcia-Zapirain, B.: Natural language processing analysis applied to covid-19 open-text opinions using a distilbert model for sentiment categorization. AI & Soc. 1–8 (2022)
15. Prema, V., Elavazhahan, V.: Sculpting distilbert: enhancing efficiency in resource-constrained scenarios. In: 2023 12th International Conference on System Modeling & Advancement in Research Trends (SMART), pp. 251–256. IEEE (2023)

16. Rahmat, R.F., Faza, S., Adnan, S., Situmorang, D.T.E., Gunawan, D., Lini, T.Z.: News articles classification for electronic information and transaction law in indonesia using support vector machine. In: 2021 International Conference on Data Science, Artificial Intelligence, and Business Analytics (DATABIA), pp. 106–110. IEEE (2021)
17. Manoj, V., Devi, T.: An effective approach for newspaper article classification using multi-class support vector machine in comparison with binary classifier to improve accuracy. In: 2023 Eighth International Conference on Science Technology Engineering and Mathematics (ICONSTEM), pp. 1–5. IEEE (2023)
18. Yihe, A., Shi, J., Yu, Z., Xie, Q.: Research on ancient text classification and topic extraction model based on natural language processing and spss analysis. In: 2023 International Conference on Data Science & Informatics (ICDSI), pp. 148–153. IEEE (2023)
19. Hossain, A., et al.: Utilizing glove embeddings for deep learning-based analysis of research paper abstracts. In: 2023 5th International Congress on Human-Computer Interaction, Optimization and Robotic Applications (HORA), pp. 1–6. IEEE (2023)
20. Gao, M., et al.: A novel hierarchical discourse model for scientific article and it's efficient top-k resampling-based text classification approach. In: 2022 IEEE International Conference on Systems, Man, and Cybernetics (SMC), pp. 774–781. IEEE (2022)
21. Hu, M., Qian, L., Chang, Z., Zhang, Z.: Kdpg-enhanced mrc framework for scientific entity recognition in survey papers. IEEE/ACM Trans. Audio Speech Lang. Process. (2024)
22. Adamov, A.Z.: Analysis of feature selection techniques for classification problems. In: 2021 IEEE 15th International Conference on Application of Information and Communication Technologies (AICT), pp. 1–6. IEEE (2021)
23. Gündoğan, E., Kaya, M.: Research paper classification based on word2vec and community discovery. In: 2020 International Conference on Decision and Sciences and Application (DASA), pp. 1032–1036. IEEE (2020)
24. Ngan, S.C.H., Lee, M.J., Khor, K.C.: Automating conference paper assignment using classification algorithms incorporated with tf-idf vectorisation. In: 2023 IEEE Symposium on Industrial Electronics & Applications (ISIEA), pp. 1–6. IEEE (2023)
25. Farhangi, A., Sui, N., Hua, N., Bai, H., Huang, A., Guo, Z.: Protoformer: embedding prototypes for transformers. In: Pacific-Asia Conference on Knowledge Discovery and Data Mining, pp. 447–458. Springer, Heidelberg (2022)
26. Ng, S.Y., Lim, K.M., Lee, C.P., Lim, J.Y.: Sentiment analysis using distilbert. In: 2023 IEEE 11th Conference on Systems, Process & Control (ICSPC), pp. 84–89. IEEE (2023)
27. Jingga, K., Hidayaturrahman, H.: The utilization of distilbert features space for topics suggestion in the context of learning content. In: 2023 International Workshop on Artificial Intelligence and Image Processing (IWAIIP), pp. 243–247. IEEE (2023)
28. Thiruthuvaraj, R., Jo, A.A., Raj, E.D.: Explainability to business: demystify transformer models with attention-based explanations. In: 2023 2nd International Conference on Applied Artificial Intelligence and Computing (ICAAIC), pp. 680–686. IEEE (2023)
29. Su, J., Wu, Y.: Optimizing pre-trained language models for efficient vulnerability detection in code snippets. In: 2023 9th International Conference on Computer and Communications (ICCC), pp. 2139–2143. IEEE (2023)

30. Joshy, A., Sundar, S.: Analyzing the performance of sentiment analysis using bert, distilbert, and roberta. In: 2022 IEEE International Power and Renewable Energy Conference (IPRECON), pp. 1–6. IEEE (2022)
31. Kongsumran, N., Phimoltares, S., Panthuwadeethorn, S.: Thai tokenizer invariant classification based on bi-lstm and distilbert encoders. In: 2022 19th International Conference on Electrical Engineering/Electronics, Computer, Telecommunications and Information Technology (ECTI-CON), pp. 1–6. IEEE (2022)

Portable Executable Analysis for Malware Detection for Reducing Risk

Avinash Kumar(✉) and Meena Chaudhary

Department of Computer Science Engineering, Mangalayatan University, Aligarh, UP, India
avinashkr338@gmail.com

Abstract. In the current era where data and information is the main fuel for business, education, banking, marketing and digital applications used by people. These applications consist of personal and sensitive data. These digital applications are using various technologies that makes it heterogeneous. This heterogeneity creates vulnerabilities in many cases and thus resulting in various threats like, virus, Trojan, spyware and other malicious programs. Malware is one of the deadliest threats in cyber security and it is very challenging to detect malware when entering a system because of its intelligent coding. The reverse engineering is used for detecting malware in classical way which is a time taking process and involves maximum human interventions. Therefore, there is also need of automated defensive mechanism to detect these malwares in order to reduce the duration of mitigation. The less time for accurate detection would result in stopping malware spreading in other part of the organization once it attacks the organization. Therefore, this paper focuses on the Machine Learning concept as one of the major solution for malware detection. This paper also addresses the classical reverse engineering concept that is used to analysis malware. The paper uses, Random Forest, XGBoosting Decision Tree, Gradient Boosting Classifier and SGD Classifier for analysis. Further the Machine Learning models that gives highest accuracy is XGBoosting that gives accuracy of 94.86% followed by random forest which 94.66%.

Keywords: Malware · Portable Executable · Random Forest and XGBoosting

1 Introduction

Malware is a type of malicious program which has the capability to steal, delete, modify or encrypt the data and information on the victim's system [1]. The malware depending upon its nature can either encrypt the file or lock the system through their malicious codes. There are mainly two types of malwares based upon previously described nature, it is termed as Crypto or Locky respectively. Malware is very powerful threat that can damage smart devices, laptops, servers, and other digital assets and has become major challenge to be tackled in the computation world [2]. The malware has become more dangerous because if its nature of hiding its code and hence malware detection is now more crucial [3].

A. Abdelgawad et al. (Eds.): ISBCom 2024, LNNS 1268, pp. 567–575, 2025.
https://doi.org/10.1007/978-3-031-82377-0_46

The classical way to analyze the malware sample is done by dissembling the malware or by debugging the malware sample and this process is termed as reverse engineering [4]. The malware analysis is powerful concept of reverse engineering where malware is analyzed based upon different techniques. The first is the Static Analysis (SA) which is done without execution of the malware. The next is called Dynamic Analysis (DA) which is done after the execution of the malware. Code Analysis (CA) is another technique used for malware analysis where inner working of malware is analyzed. Finally, the memory analysis also known as malware forensics is used to analyze the evasive technique of the malware.

Machine Learning (ML) is a concept where a software is trained in a way that it takes decision and produce suitable output. ML is data driven approach which is used to improve decision making. Moreover, the best part of ML is that it has the capability to automatically improve its decision. ML has three major important models that are known as Supervised, Unsupervised and Semi-Supervised ML. The supervised ML uses labelled data set in order to train the data set for producing output. Whereas, the Unsupervised ML uses unlabeled data in order to train itself to produce the results. Finally, the Semi-supervised ML combines the labeled and unlabeled data set to produce the result.

The detection of malware using ML would be very effective as it can detect the malware in less duration than required by human manual detection. Also, the Indicators of Compromise (IOC) that could be accidently ignored by human could be tackled by ML in more efficient manner. The rise of malware has gone high so much that banking sector has received 23.2 percent loss [5].

Many works have been done in the detection of the malware, but this work tries its best to improve the malware detection and also tries to produce subjective in-depth analysis to understand the need for automated detection. In summary, major contribution of this work is enumerated below.

- This work focuses the facts and research that demonstrate that severity of malware threat in current time which prepares the baseline of this work
- This work presents the classical reverse engineering concept used for malware analysis to understand the need for automated analysis or detection.
- This work proposes efficient Portable Executable based State-of-Art mechanism for malware detection to reduce malware risk attack.

Remainder of this work is organized as follows Sect. 2 includes Literature Review of various ML based malware detection. Section 3 deals with the classical Reverse Engineering approach for malware detection whereas Sect. 4 deals with the Methodology and proposed framework. Section 5 deals with the results and discussion followed by conclusion and future research directions in Sect. 6.

2 Literature Review

There is huge rise in use of ML for detecting malware because it reduces the human intervention and also reduce the time for detection [6]. The use of deep learning has also been considered by various scholars for malware detection [7]. However, this lacks the comparison from the classical reverse engineering detection efficiency. The use of ML has one major advantage in terms of duration for detection which would ultimately reduce the mitigation duration in a cyber-attack. The use of ML for detection is very powerful which has been also proposed but sometimes it lacks the performance parameters [8]. The performance parameters are very important aspect in order to identify malware more accurately so that even advanced and crafted malware can be detected. The forensics techniques have also been proposed by scholar for detection of malware [9]. But, it lacks to address the evolution strategies. The use of malware detection and prevention strategies has been address by various scholars [10, 11].

There are various works that has been carried out to detect malware and Advanced Persistent Threat (APT) that are very crucial as these can bring the target to complete stop [12]. Malware is very offensive in nature, they can either wipe out or encrypt the data. The Antivirus, Web application Firewall, and other defensive devices, needs continuous improvement with advancement in malware threat which are coming with time. The malware is having sophisticated deceptive techniques to evade the target before encrypting it or wiping it. Therefore, the behavior based detection of these malware is very important. The malware is also affecting Internet of Things (IoT) system at a greater extent and various research has been done to identify malware in an IoT system [13–15]. This can be noticed that most of the research that has been proposed to detect malware is done using ML concepts. There are two vital aspects for using ML that is minimizing the mitigation duration once malware attack takes place and also reducing any human ignorant done accidently. Table 1 present vital literature review done in this field.

Table 1. Literature Review

Reference	Year	Highlights
Gibert et al. [6]	2020	• Paper discussed the static, dynamic malware analysis as well as hybrid analysis • Paper provides comparative analysis among discussed malware analysis techniques
Taiyab et al. [7]	2022	• Paper discuses most of the reverse engineering approaches • Paper also cover some of the automated concepts like ML and Deep learning briefly
Aslan et al. [8]	2020	• The malware analysis is achieved through deep learning • IoT system also been considered for malware analysis
Zhang et al. [9]	2023	• Paper has used Convoluted Neural Network for malware analysis • The paper also covers the detection of encrypted files
Tsai et al. [10]	2023	• The paper used deep learning for analysis of malware • The paper also takes into account the PowerShell
Khalid et al. [11]	2023	• The paper covered RAM as important part for malware analysis • The paper worked on Fileless malware
Yang et al. [12]	2021	• The paper worked on APT detection • The paper proposed DBAR model for malware detection

3 Reverse Engineering for Malware Detection

Reverse Engineering is the concept where, the executable in any format is used to reach at the source code. This concept when used in malware analysis become very useful. The malware analysis helps to understand the main intension of the malware developer and also helps to develop most appropriate defensive mechanism for the same. There are various tools that are used to either dissemble or debug the malware sample.

Malware Analysis is used to know the details of the malware sample that could help to detect malware when it enters the system. Malware analysis is categorized mainly in four types, which are explained in below subsections.

3.1 Static Analysis

This is the most primitive method to analysis any sample in order to understand whether it is malware or not. In this concept the malware sample in not executed rather tools and scripts are used to figure out maximum nature of the Portable Executable (PE) file. The main advantage of Static Analysis is to identify the path for further analysis of the PE file.

3.2 Dynamic Analysis

This process is very useful to understand the behaviors of the malware and since it relies on behavioral concept of PE to identify as benign or malicious. It is also termed as Behavioral Analysis (BA). It is very important to understand the behavior of the malware because many advanced malware comes in packed format which could not be easily analyzed by Static Analysis. Also, signature based detection is carried out by many devices to identify malicious PE file which may not work when new malware or zero-day malware comes into picture whose signature has not yet been developed and stored by defensive mechanisms. Therefore, the attributes such as accessing the memory, retrieving process of the victim system could be triggered and identified only by the means of behavior based analysis or behavior based detection.

3.3 Code Analysis

It is very important to understand the inner working of the malware in order to develop more resilient cyber counter measures. The Code Analysis helps in deep diving the inner working of the PE file and identify whether it a malware or benign file. This analysis is also useful for understanding the threat level of the malware as code are deeply analyzed. The code analysis is further categories as Static Code Analysis and Dynamic Code Analysis.

3.4 Memory Analysis

This is another powerful malware analysis technique that could be used to understand if the PE file whose traces in RAM is found is a potential malware or a benign. The idea is taken from RAM forensics but when combined with malware analysis, it becomes very powerful to detect if the PE file is malware or not. This analysis is also very useful to detect advanced malware which consists of higher level of obfuscation (Table 2).

Table 2. Reverse Engineering Techniques

Reverse Engineering Techniques	Highlights	Limitations
Static Analysis [16]	Focuses on malware analysis without execution of Suspicious sample	It lacks efficiency to predict the behavior of the malware
Dynamic Analysis [1]	The main aim to determine the behavior of the suspicious sample	Inner working is tough to be analyses through this technique
Code Analysis [1]	It helps in understating the inner working of the malicious sample	It a lengthy process as the malware codes are having huge lines of code
Memory Analysis [17]	It deals with analysis of such suspicious sample that trigger from RAM	It lacks the speed of detection which is required in real time attack

4 Methodology

This section focuses on the generalized approach for multi-classification where Fig. 1 denotes different steps of ML algorithm. The source of the malware is taken from https://www.kaggle.com/datasets/joebeachcapital/windows-malwares. The malware here taken is in the form of PE files consisting of nine field values of ten different PE sections.

The first step is the Data Acquisition (DA) which is done from Kaggle. Data pre-processing is the next step where this work has handles null value and duplicate values. Then Data is splitted where 80% is used for training and 20% for testing. Then the classification model has been used to classify PE files into its categories. The algorithms used here are Decision Tree, Random forest, Logistic Regression, Gradient Boosting, XGBoosting and Stochastic Gradient Descent for detection of malware from PE file.

4.1 Data Acquisition

Data set is collected from the https://www.kaggle.com/datasets/joebeachcapital/windows-malwares which consisting benign samples. The consists of SHA256 signature values of the labelled data.

4.2 Data Pre-preprocessing

This is one of the most important step where it ensures that the data is clean, consistent as well as usable. Null value and duplicate values exiting in the collected data set were removed and were replaced by categorical mean.

4.3 Data Split

In this process the data is splitted into two parts which is 80% for the training and 20% for the testing. Then the next step goes with classification.

4.4 Classification Model

This phase deals with multi-classification of the data that goes with various supervised learning algorithms. The algorithms used in this case are Decision Tree (DT), Random Forest (RF), XGBoosting, Gradient Boosting Classifier (GBC) and Stochastic Gradient Descent (SGD).

The DT is used for predictive learning. The RF uses the concept of ensemble learning. GBC works by combining weaker and stronger learner in ML. XGBoosting is improved version of GBC and the SGD work upon the concept of maxima and minima in ML.

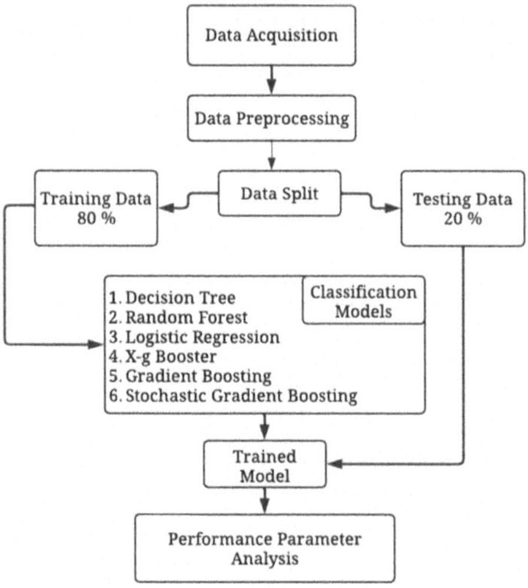

Fig. 1. Generalized approach for Multi-classification of Malware.

4.5 Performance Parameter Analysis

This helps to identify that which model work better. The comparative analysis is done here in this case. The accuracy, precision, F1 Score and recall has been used to identify the level of optimality of model achieved for detection of malware in multi-classification approach. These parameters could be calculated as mentioned below.

$$Accuracy = \frac{Number\ of\ Correct\ Predictions}{Total\ number\ of\ Predictions}$$

$$Precision = \frac{True\ Positive}{True\ Positive + False\ Positive}$$

$$Recall = \frac{True\ Positive}{True\ Positive + False\ Negative}$$

$$F1Score = 2 * \frac{Precision * Recall}{Precision + Recall}$$

The malware detection approach and its result is further explained in next section. The multi-classification consists of Redline Stealer, Downloader and Remote Access Trojan.

5 Results and Discussions

This work focuses on multi-classification of different types of malwares. A dataset of PE is taken which has various classes of malware samples. The classification task is performed mainly in 4 classes i.e., Class 0 as Benign, Class 1 as Redline Stealer, Class 2 as Downloader, Class 3 as Remote Access Trojan (RAT). The dataset is split into 4:1 ratio, where 80% of the samples are used for training the four classification models and 20% is used for testing the trained model. The classification models used for this purpose are Decision Tree, Random Forest, XGBoosting, Gradient Boosting Classifier, SGD Classifier. The performance metrics like accuracy, F1 score, recall, precision are calculated to find the efficiency of the models (Tables 3, 4, 5 and 6).

Table 3. Comparative analysis based on Accuracy.

Classes of Malware	Classification Models				
	Decision Tree	Random Forest	XGBoosting	Gradient Boosting Classifier	SGD Classifier
Benign	93%	94.66%	94.86%	92.14%	85.69%
Redline Stealer					
Downloader					
RAT					

Table 4. Comparative analysis based on Precision.

Classes of Malware	Classification Models				
	Decision Tree	Random Forest	XGBoosting	Gradient Boosting Classifier	SGD Classifier
Benign	0.81	0.80	0.85	0.62	0.45
Redline Stealer	0.97	0.97	0.98	0.96	0.93
Downloader	0.97	0.99	0.99	0.97	0.94
RAT	0.90	0.92	0.91	0.87	0.840

The tables above show the values of the performance metrics of the different classification models. Based on the values, it is observed that XGBoosting classifier and

Table 5. Comparative analysis based on Recall

Classes of Malware	Classification Models				
	Decision Tree	Random Forest	XGBoosting	Gradient Boosting Classifier	SGD Classifier
Benign	0.77	0.83	0.80	0.88	0.57
Redline Stealer	0.97	0.98	0.97	0.96	0.92
Downloader	0.97	0.99	0.99	0.97	0.91
RAT	0.92	0.91	0.93	0.85	0.91

Table 6. Comparative analysis based on F1 score

Classes of Malware	Classification Models				
	Decision Tree	Random Forest	XGBoosting	Gradient Boosting Classifier	SGD Classifier
Benign	0.79	0.81	0.82	0.72	0.57
Redline Stealer	0.97	0.98	0.98	0.96	0.92
Downloader	0.97	0.99	0.99	0.97	0.91
RAT	0.91	0.92	0.92	0.89	0.81

Random Forest outperform the best among rest algorithms and the least satisfactory accuracy was achieved by SGD classifier. The highest accuracy obtained by XGBoosting is 94.86% followed by random forest 94.66%. The lowest accuracy was achieved by SGD classifier which is 85.69%.

6 Conclusion and Future Research Directions

This work has tried it best to increase the efficiency of the malware detection. This work also explains the reverse engineering concept which is used for malware analysis or malware detection. The paper tries its best to conclude the need for automated malware detection mechanism with higher accuracy. This will help to reduce the malware from infecting other system of the victim organization. Random Forest and XGBoosting was able to achieve highest level of accuracy. Further this paper can provide a baseline for multi-classification for future research directions. This paper can also provide path for zero-day malware detection as it works on analyzing the behaviors rather than only relying on signature.

References

1. Monnappa, K.A.: Learning Malware Analysis: Explore the Concepts, Tools, and Techniques to Analyze and Investigate Windows Malware. Packt (n.d.)
2. Alodat, I.: Malware: Detection and defense (2023).https://doi.org/10.5772/intechopen.108434
3. JovanovicAslan, Ö.A., Samet, R.: A comprehensive review on malware detection approaches. IEEE Access **8**, 6249–6271 (2020)
4. Akram, B., Ogi, D.: The making of indicator of compromise using malware reverse engineering techniques. In: 2020 International Conference on ICT for Smart Society (ICISS), Bandung, Indonesia, pp. 1–6 (2020). https://doi.org/10.1109/ICISS50791.2020.9307581
5. 2023 Cyber Security Statistics Trends & Data. PurpleSec (22 February 2023). https://purplesec.us/resources/cyber-security-statistics/
6. Gibert, D., Mateu, C., Planes, J.: The rise of machine learning for detection and classification of malware: research developments, trends and challenges. J. Netw. Comput. Appl. **153**, 102526 (2020). https://doi.org/10.1016/j.jnca.2019.102526
7. Tayyab, U.-H., Khan, F.B., Durad, M.H., Khan, A., Lee, Y.S.: A survey of the recent trends in deep learning based malware detection. J. Cybersecurity Priv. **2**(4), 800–829 (2022). https://doi.org/10.3390/jcp2040041
8. Aslan, O., Samet, R.: A comprehensive review on malware detection approaches. IEEE Access **8**, 6249–6271 (2020). https://doi.org/10.1109/access.2019.2963724
9. Zhang, S., Hu, C., Wang, L., Mihaljevic, M., Xu, S., Lan, T.: A malware detection approach based on deep learning and memory forensics. Symmetry **15**(3), 758 (2023). https://doi.org/10.3390/sym15030758
10. Tsai, M.-H., Lin, C.-C., He, Z.-G., Yang, W.-C., Lei, C.-L.: PowerDP: De-obfuscating and profiling malicious PowerShell commands with multi-label classifiers. IEEE Access **11**, 256–270 (2023). https://doi.org/10.1109/access.2022.3232505
11. Khalid, O., et al.: An insight into the machine-learning-based fileless malware detection. Sensors **23**(2), 612 (2023). https://doi.org/10.3390/s23020612
12. Yang, L.-X., Huang, K., Yang, X., Zhang, Y., Xiang, Y., Tang, Y.Y.: Defense against advanced persistent threat through data backup and recovery. IEEE Trans. Netw. Sci. Eng. **8**(3), 2001–2013 (2021). https://doi.org/10.1109/tnse.2020.3040247
13. Malik, A., Bhushan, B., Bhatia Khan, S., Kashyap, R., Chaganti, R., Rakesh, N.: Security attacks and vulnerability analysis in mobile wireless networking. In: 5G and Beyond, pp. 81–110 (2023). https://doi.org/10.1007/978-981-99-3668-7_5
14. Malik, A., Parihar, V., Bhushan, B., Chaganti, R., Bhatia, S., Astya, P.N.: Security services for wireless 5G internet of things (IOT) systems. In: 5G and Beyond, pp. 169–195 (2023). https://doi.org/10.1007/978-981-99-3668-7_9
15. Parihar, V., Malik, A., Bhushan, B., Bhattacharya, P., Shankar, A.: Innovative smart grid solutions for fostering data security and effective privacy preservation. In: Data Analytics for Smart Grids Applications—A Key to Smart City Development, pp. 351–380 (2023). https://doi.org/10.1007/978-3-031-46092-0_19
16. Sikorski, M., Honig, A.: Practical malware analysis the hands-on guide to dissecting malicious software. No Starch Press (2012)
17. Ligh, M.H., Case, A., Levy, J., Walters, A.: The Art of Memory Forensics: Detecting Malware and Threats in Windows, Linux, and Mac Memory. Wiley, Hoboken (2014)

A Comprehensive Framework for Electronic Medical Records (EMRs) Leveraging Blockchain Technology

Shraiyash Pandey[1(✉)], Esraa Mohammed Alazzawi[2], Surendra Pandey[3], Bharat Bhushan[1], Akhtar Jamil[4], and Pashupati Baniya[1]

[1] Department of Computer Science and Engineering School of Engineering and Technology, Sharda University, Greater Noida, India
shraiyash.pandey@gmail.com
[2] Department of Computer Engineering, University of Baghdad, Al-Jadriya, Baghdad 10071, Iraq
[3] Computer and Information Technology, Tata Consultancy Services, Marietta, GA, USA
[4] Department of Computer Science, National University of Computer and Engineering Sciences, Islamabad 44000, Pakistan
akhtar.jamil@nu.edu.pk

Abstract. In today's healthcare landscape, where participation is widespread, the need for secure and efficient record management is evident. However, many healthcare organizations, particularly in regions like India, still rely on outdated paper-based methods for handling medical records. Unfortunately, these traditional approaches fall short in meeting the rigorous requirements for security, confidentiality, and efficiency in today's digital age. This work presents a comprehensive framework that integrates modern technologies such as Smart Contracts, Ethereum Blockchain, and the Interplanetary File System (IPFS) to ensure secure storage, and seamless management of Electronic Medical Records (EMRs). By leveraging these emerging technologies, our goal is to lead a shift in electronic medical record management. Additionally, we explore recent advancements in the healthcare sector, with a focus on enhancing security protocols within Electronic Medical Records management.

Keywords: Blockchain · EMR · Smart Contracts · Ethereum

1 Introduction

The latest advances of technology have had an impact on many aspects of human existence, altering how we utilize and interpret things. Just as technology has revolutionized different aspects of life, it additionally involves discovering new methods to better the healthcare industry. The primary benefits of technological innovation include improved

S. Pandey—Independent Researcher

A. Abdelgawad et al. (Eds.): ISBCom 2024, LNNS 1268, pp. 576–587, 2025.
https://doi.org/10.1007/978-3-031-82377-0_47

security, user satisfaction, and other facets of the health-care industry provided by Electronic Medical Record (EMR) systems [1]. However, they continue to encounter challenges related to medical record data integrity, security, and data ownership, etc. The implementation of a new technology, such as Blockchain, might help to address these concerns. This technology promises to deliver a safe and tamper-resistant infrastructure for maintaining hospital records and other healthcare-related data.

Blockchain has grown into a popular term in the field of data and security technology industries in recent years. The rapid growth of this new technology offers more promise for addressing data confidentiality, security, and authenticity concerns. The term Blockchain became popular after the initial publication from Satoshi Nakamoto's paper in 2008 about Bitcoin [2].

In 2009, the Bitcoin cryptocurrency was introduced to the market in a practical manner. However, with its code being open source, other developers gained the ability to modify and enhance Bitcoin. The progression of blockchain technology has occurred across various stages [3]. In Blockchain 1.0, Distributed Ledger Technology (DLT) played a significant role, marking the inception of the technology's most evident application: cryptocurrencies. This initial phase saw the emergence of a transparent mechanism for monitoring bitcoin transactions. Transitioning to Blockchain 2.0, several communications in between parties began to incorporate procedures, also known as Smart Contracts, which are derived from a series of programs. Ethereum emerged as the leading blockchain platform during this phase.

Blockchain 3.0 represents the next evolution in this technology, emphasizing Decentralized Applications (DApps) [4]. Unlike traditional applications, DApps rely on decentralized structure in terms of communication for the server and client. The objective of Blockchain 3.0 was to promote the adoption of blockchain technology across conventional sectors such as government, healthcare, and education. Enter Blockchain 4.0, which offers solutions and methodologies tailored to meet the diverse business requirements of Industry 4.0.

Prior to the introduction of contemporary technology, the healthcare industry employed a system that relied on paper to maintain the records. This approach was not efficient, unsecure, disorganized, and not temper-resistant. It also encountered duplicate information and repetition since each of the hospitals that the patient had several repeated files of his or her health records. The healthcare industry saw a transition toward EHR systems meant to mix written on-paper electronic medical records. EHR systems are currently used in a number of hospitals throughout the world because of to the advantages they offer, particularly referring to the increased security and cost effectiveness [5, 6].

They are regarded as an essential component of the healthcare industry since they add significant features to the system. These functions include electronic medical record storage, appointment scheduling for patients, invoicing, and lab testing. In healthcare industry, many HER systems are accessible. The primary goal is to deliver safe, tamper-resistant, and shareable documents across several platforms. The major contributions of this work are presented below.

- This work presents a comprehensive structure and working of electronic health records in terms of layers for an application based on blockchain technology

- This work highlights several recent advancements related to the healthcare industry in terms of privacy, efficiency, security, and authentication.
- This work provides an overview of blockchain technology and the application of blockchain within the healthcare sector for HER systems.

The rest of the paper is organized as follows. Section 2 highlights advancements in the healthcare sector with the use of technologies like Blockchain. Section 3 presents the background details related to the use of Blockchain framework in relation with Electronic Health Records and preliminaries that include several technologies implemented in the decentralized framework. Section 4 discusses the system design of the framework along with the smart contracts algorithm for patient records used in the framework. Finally, we conclude the paper in Sect. 5 along with future research directions.

2 Related Work

This section presents various recent advancements that focus on the improvements seen in the academia related to security, privacy, and efficiency of Electronic Medical Records using technologies like Blockchain. The recent advancements are enumerated in Table 1.

Venkatesh et al. [7] presented a framework for electronic medical records in order to secure the EMRs across various hospitals. Zakzouk et al. [8] introduced an electronic medical record framework for management built with blockchain technology in the smart cities sector for smart healthcare infrastructure. Sharma et al. [9] proposed to limit restrictions in the healthcare system through the help of Blockchain technology by referring to the implementation of several method and algorithms. Haque et al. [10] presented a study that focused on providing security to Electronic Medical Records through a Blockchain-based model.

Qazi et al. [11] proposed Proof of Authenticity which is a consensus algorithm for medical stakeholders over a distributed platform. Jagadamba et al. [12] proposed the use of Blockchain technology to maintain Electronic Health Records in a more secure and reliable way that allows sharing and storing of patient records at any time. Saifee et al. [13] presented the implementation of Dual Blockchain Architecture that helps maintain the user's data in a single true version and offers user-focused medical records. Faruk et al. [14] introduced a database framework based on blockchain that focuses on transparency and data security for electronic medical records and allows sharing of data between different health organizations.

3 Background and Preliminaries

We provide a brief overview of blockchain technology and its components to facilitate readers' comprehension of the subsequent content. A blockchain serves as a decentralized and transparent public ledger shared across a network of peers. Initially associated with cryptocurrencies such as Bitcoin, its utility has expanded to include data storage. In the subsequent sections, we delve into the fundamental features and various types of blockchain, enhancing the understanding of the concept of blockchain.

Table 1. Major advancements related to the management of EMRs

Reference	Year	Major Contribution
Venkatesh et al. [7]	2023	A framework for electronic medical records in order to secure the EMRs across various hospitals
Zakzouk et al. [8]	2023	An electronic medical record framework for management built with blockchain technology in the smart cities
Sharma et al. [9]	2021	Limited the restrictions in the healthcare system through the help of blockchain technology
Haque et al. [10]	2020	A study that focused on providing security to Electronic Medical Records through a Blockchain-based model
Qazi et al. [11]	2020	Proof of Authenticity which is a consensus algorithm for medical stakeholders over a distributed platform
Jagadamba et al. [12]	2022	Used Blockchain technology to maintain Electronic Health Records in a more secure and reliable way that allows sharing and storing of patient records at any time
Saifee et al. [13]	2022	Implemented Dual Blockchain Architecture that helps maintain the user's data in a single true version
Faruk et al. [14]	2023	A database framework based on blockchain that focuses on transparency and data security for electronic medical records

3.1 Blockchain Framework

Blockchain technology is a combination of cryptography and Peer-to-Peer networking, forming its foundational elements [15, 16]. A blockchain is made up of several time-stamped blocks linked together using a hash function in cryptography. Every block typically comprises transaction records that have been validated by peers, or miners. Every new block is inserted into the end of the chain, which is continuously lengthened. But every new block has a reference to the header of the preceding block, which is essentially a cryptographic hash (MD5), for example) [17]. Each block's creation guarantees immutability, transparency, and anonymity. Blockchain operates entirely within a P2P network. Illustrated in Fig. 1, the standard block structure within the chain of blocks showcases the way each block, other than the genesis block, carries the hash value of data from the preceding block, ensuring its integrity and continuity.

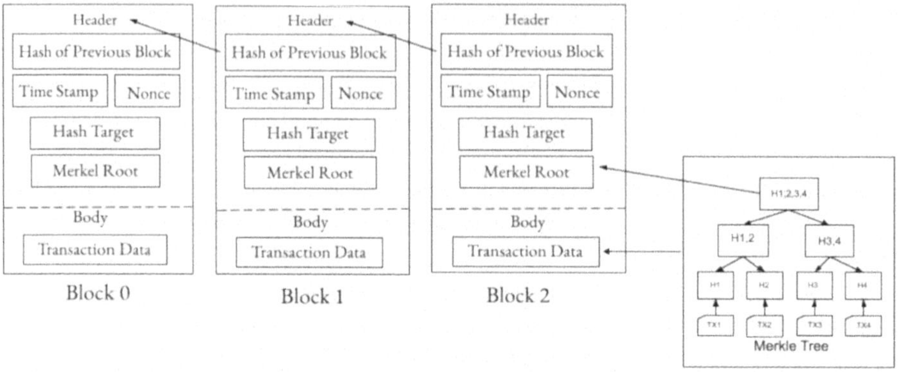

Fig. 1. Block Structure

3.2 Ethereum

Ethereum implements the exact blockchain technology as the cryptocurrency, Bitcoin [18]. The goal of Ethereum, which was formally launched in 2015, was to develop an open-source, trustless smart contract platform with programmable blockchain functionality. Peer-to-peer networking is another feature of this technology that helps to make it distributed. Additionally, this platform uses Ethers, a cryptocurrency of its own [19]. It is possible to share this cryptocurrency amongst accounts linked to the Ethereum blockchain [20]. Additionally, Ethereum gives programmers access to Solidity, a language that offers them to create customized blockchain.

3.3 Smart Contracts

Smart contract refers to the code that implements the actions that run on Blockchain. When users send transactions, this piece of code is run [21]. Operating directly on the Blockchain, they are impervious to manipulation and changes. Once the necessary operations have been programmed, programmers can use EVM byte code to compile their work. Additionally, it might be run and implemented on the Ethereum blockchain after compilation [22].

3.4 Ethereum Virtual Machine

One of the main advantages of the Ethereum platform is its programmable blockchain. Ethereum Virtual Machine (EVM) is used to create decentralized applications. These DApps have smart contracts, which are composed of user-defined code that carries out specific application tasks. In fact, the code is run and deployed using EVM [23, 24]. Therefore, the applications developed through smart contracts are run on EVM.

3.5 Inter Planetary File System (IPFS)

One protocol that utilizes a P2P network for storage is called the IPFS. Because data saved on IPFS is shielded from tampering, it offers secure local storage. A hash value computed cryptographically is present in every data file stored on IPFS. It is specific to each saved data file on the IPFS and is used to identify them [25].

3.6 Electronic Health Records (EHR)

Patient diagnostic results are collected electronically through EHR, which may encompass various file formats like JPEG or PDF. EMRs serve as repositories for medical data sourced from different healthcare organizations. Additionally, EMRs can include personal health data acquired through patient-controlled wearables such as smart watches and smart bands. EHRs serve as patient-centric, real-time records accessible to authorized users. They contain a wealth of information, including diagnosis reports, patient immunization status, prescription history, medical backgrounds, weight data, and demographic information.

Three fundamental characteristics of an EHR should be met: availability, confidentiality, and integrity [8, 26]. By using EHR systems, medical history loss, data corruption, and other issues can be decreased [27–29]. It is difficult to guarantee the confidentiality and privacy of these vital data, though.

4 System Design

The structure of the entire system including various components, architecture, and modules are included in this section. The purpose of this proposed framework is to provide a Blockchain-based, secure, and private system for electronic health information.

4.1 Framework Architecture

Three entities or modules make up the suggested framework, as shown in Fig. 2. Together, these modules would maintain the functionality of our framework. The specific three entities are divided into layers: Data User Layer, Security Layer, and Implementation layer. All the layers are discussed in the following subsections.

Data User Layer
The system's users may include medical professionals, administrative personnel, and patients. The first and foremost responsibility of the user would be to use the system and carry out fundamental operations. The users of this system would access its functions through a DApp browser technically because it contains the DApp's visual presentation. The system encompasses the following transactions:

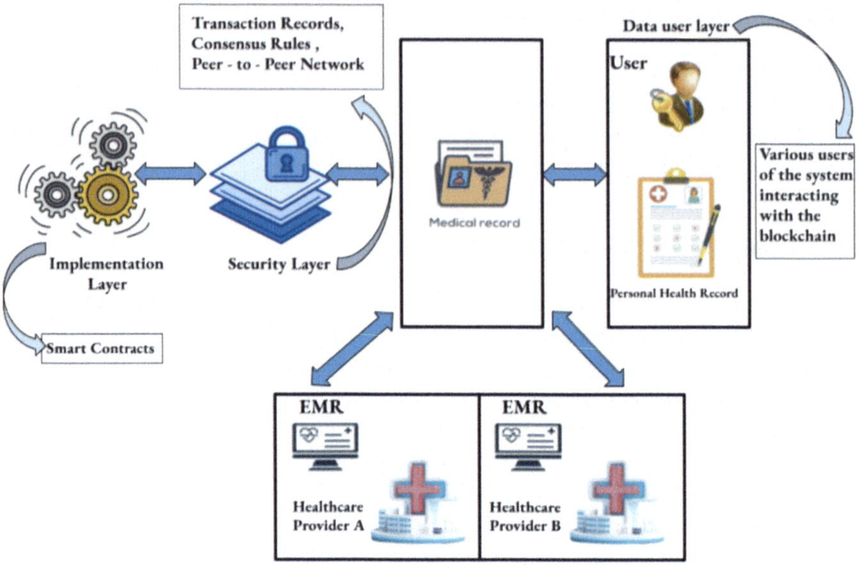

Fig. 2. Proposed Framework

Access to these transactions is restricted based on user roles. Only doctors or nursing staff with the consent of concerned patient can add or update records, while patients can only view their records without the ability to make changes.

Security Layer
Using the decentralised and immutable characteristics of blockchain technology, this layer facilitates the smooth communication between users and the DApp. The security layer, which is the next layer of the framework, houses the user interface for interacting with DApps. The components within the Security layer are discussed below:

a) *Assets of Blockchain: the sharing and storing of data bits between users is depicted as a transaction by the Ethereum blockchain. They are handled by the Ethereum blockchain as valuable data that may be moved or stored as per the requirement.*
b) *Governance Rules: ethereum blockchain utilizes the PoW consensus algorithm* [8], *chosen for its ability to ensure governance through the collective agreement of trusted nodes within the network.*
c) *Network Structure: Ethereum blockchain adopts a P2P network architecture, where all the entities within the network are interconnected. It is made sure that no sole node has the accessibility to the network's functions through the network structure.*

Implementation Layer
In the implementation layer, we design and create the backbone of our DApp. This involves crafting smart contracts, defining user roles, and designing the flow of control within the system.

a) *Self-executing code: Smart contracts play a critical role in DApps, facilitating basic operations. The two key contracts handle the records of the Patient and their roles.*

These contracts manage user access and enable CRUD functions on records of patients. The contract for patients is specifically tailored for utilizing the functionalities of our work, including CRUD operations and role assignment. Imported from the library of smart contracts, OpenZeppelin, defines the "Roles" contract with a pre-defined mechanism for role-based access control.

The Data User Layer, Security Layer, and Implementation Layer are the crucial components that are integrated into the strong system that is the suggested framework. Every layer is essential to maintaining the overall functionality of the system. Using a DApp browser, the Data User Layer gives different stakeholders access to system capabilities and allows them to carry out specified duties. Through its components, which include blockchain assets, governance rules, and a peer-to-peer network structure, the Security Layer ensures integrity and governance while utilizing blockchain technology to safeguard data transactions and interactions. Lastly, the Implementation Layer concentrates on creating the smart contract-based core DApp functionality, which permits safe and effective administration of user roles and patient records. When combined, these layers offer a thorough and safe environment for data access.

4.2 Smart Contracts Algorithm

The definition of roles, addition, viewing, updating, and deletion of records are the five functions of Algorithm 1. On a mapping list of roles. Several roles and accounts are added onto the mapping list for roles via the variables mentioned: role added and account added. Later on, that role mapping would be leveraged to get the roles that users had.

The doctor finishes the next function, which is to add patient records. Additionally, this feature makes sure that the doctor's is performing this duty and not some other entity. They accomplish this by using the "msg.sender" phrase, which is used to identify the user's address. The doctor can then insert the patient's records and save the record to terminate the function after this validity check is completed.

The unique ID of the must be supplied as an entity in order to use the third function, which is to access patient records. The framework uses this unique identification to search for the patient's records, and allow to send back those records to the user that made the request to obtain them. This function also defines and validates the patient roles and doctor roles. Since the medical information should only be accessed by the patient and the physician, Algorithm 1's last function is to delete patient records, which is used to remove a single patient's record. Only the system's authenticated users would be able to access these capabilities which guarantees that no unauthorized party will use them.

Algorithm 1 Patient Records Algorithm using Smart Contracts

```
Roles:
function Roles (New Role to be added, New Account that needs to be in roles mapping)
    define a new role and add the account in roles mapping
end function
Data Being Added:
function Add Patient Record (contains variables to add data)
    if(msg.sender == doctor )then
          add data to particular patient's record
    else a timeout error occurs
    end if
end function
Data Being Fetched:
function Fetch Record (patient id)
if (msg.sender == doctor || patient)
    then if ( patient id) == true then
          fetch data from the patient (id) thats given
               return (patient record)to account that requested to fetch the data
          else a timeout error occurs
    end if
end if
end function
Data to be Updated:
function Update Record (new data to replace old data)
if (msg.sender == doctor) then
                    if (id == patient id && name == patient name) then
                         update the record database with the correct patient's record
                         return success
                      else return an error
                      end if
else
   a timeout error occurs
end function
Data to be Deleted:
function Delete Record (patient id)
if (msg.sender == doctor) then
    if (id == patient id) then
           delete the record associated wit this patient's id
          return success
          else return fail
    end if
    else
       a timeout error occurs
end if
end function
```

5 Conclusion and Future Research Directions

India's population stands at the top across the world with over 1.4 billion people. Every individual in India visits a hospital institution or organization at least once in their life. The healthcare industry is constantly evolving with new technologies improving surgeries, tools, and equipment to offer better treatment. However, despite the rapid progress in the healthcare industry, many hospitals still use a basic approach to the management of health records that lack several features such as data integrity, security, privacy, confidentiality, authorization, and efficiency. Therefore, the management of medical records within several healthcare organizations also needs improvement. This work presents a solution that tackles these issues within the current management of medical records and utilizes several emerging technologies related to Blockchain, such as Smart Contracts, Truffle, Ethereum, and IPFS. An overview of Blockchain technology is presented to illustrate the use and application of Blockchain in different areas and specifically in the Healthcare sector. The developed structure uses the mentioned technologies in terms of a layered architecture. Several preliminary technologies are discussed related to the development of the architecture. Furthermore, this work highlights a list of several major contributions to the management of electronic medical records. These contributions are particularly focused on the advancements of security, efficiency, and privacy of electronic medical records.

In terms of future research, we intend to integrate a payment feature within the current infrastructure. A few factors must be considered in order to determine the amount an individual might pay for a doctor's consultation on this application based on blockchain technology. Additionally, a specific set of regulations and norms would have to considered that adhere to the values of the healthcare industry.

References

1. Weiss, J.N.: Electronic medical records (EMR). In: Physician Crisis. Springer, Cham (2023). https://doi.org/10.1007/978-3-031-27979-9_7
2. Rosenberger, P.: Erratum zu: bitcoin und blockchain. In: Bitcoin und Blockchain. Springer, Berlin, Heidelberg (2023). https://doi.org/10.1007/978-3-662-66530-5_12
3. Pandey, S., Kumar De, A., Choudhary, S., Bhushan, B., Bhatia, S.: Leveraging blockchain technology in industry 4.0 and Industrial Internet of Things (IIoT) scenarios. In: Sharma, D.K., Sharma, R., Jeon, G., Polkowski, Z. (eds.) Low Power Architectures for IoT Applications. Springer Tracts in Electrical and Electronics Engineering. Springer, Singapore (2023). https://doi.org/10.1007/978-981-99-0639-0_12
4. Iqbal, M.O., Obaid, A.J., Agarwal, P., Mufti, T., Hassan, A.R.: Blockchain technology and decentralized applications using blockchain. In: Tuba, M., Akashe, S., Joshi, A. (eds.) ICT Infrastructure and Computing. Lecture Notes in Networks and Systems, vol. 520, pp. 555–563. Springer, Singapore (2023). https://doi.org/10.1007/978-981-19-5331-6_57
5. Goyal, S., Sharma, N., Bhushan, B., Shankar, A., &Sagayam, M.: IoT enabled technology in secured healthcare: applications, challenges and future directions. Cogn. Internet Med. Things Smart Healthc., 25–48 (2020).https://doi.org/10.1007/978-3-030-55833-8_2
6. Varshney, M., Bhushan, B., Haque, A.K.M.B.: Big data analytics and data mining for healthcare informatics (HCI). In: Kumar, R., Sharma, R., Pattnaik, P.K. (eds.) Multimedia Technologies in the Internet of Things Environment, Volume 3. Studies in Big Data, vol. 108, pp. 167–195. Springer, Singapore (2022). https://doi.org/10.1007/978-981-19-0924-5_11

7. Venkatesh, R., Savadatti Hanumantha, B.: Electronic medical records protection framework based on quantum blockchain for multiple hospitals. Multimed. Tools Appl. (2023)https://doi.org/10.1007/s11042-023-16848-y
8. Zakzouk, A., El-Sayed, A., Hemdan, E.E.D.: A blockchain-based electronic medical records management framework in smart healthcare infrastructure. Multimed. Tools Appl. **82**, 35419–35437 (2023). https://doi.org/10.1007/s11042-023-15152-z
9. Sharma, S., Kaushal, S., Gupta, S., Kumar, H.: A blockchain-based solution for electronic medical records system in healthcare. In: Luhach, A.K., Jat, D.S., Hawari, K.B.G., Gao, XZ., Lingras, P. (eds.) Advanced Informatics for Computing Research. ICAICR 2021. Communications in Computer and Information Science, vol. 1575, pp. 232–243. Springer, Cham (2022). https://doi.org/10.1007/978-3-031-09469-9_20
10. Haque, R., et al.: Blockchain-based information security of Electronic Medical Records (EMR) in a healthcare communication system. In: Peng, SL., Son, L.H., Suseendran, G., Balaganesh, D. (eds.) Intelligent Computing and Innovation on Data Science. Lecture Notes in Networks and Systems, vol. 118, pp. 641–650. Springer, Singapore. https://doi.org/10.1007/978-981-15-3284-9_69
11. Qazi, M., Kulkarni, D., Nagori, M.: Proof of authenticity-based electronic medical records storage on blockchain. In: Zhang, Y.D., Mandal, J., So-In, C., Thakur, N. (eds.) Smart Trends in Computing and Communications. Smart Innovation, Systems and Technologies, vol. 165, pp. 297–306. Springer, Singapore (2020). https://doi.org/10.1007/978-981-15-0077-0_31
12. Jagadamba, G., Krishna, E.L.S., Amogh, J.P., Abhishek, B.B., Manoj, H.N.: Adaptive Electronic Health Records Management and Secure Distribution Using Blockchain. In: Giri, D., Mandal, J.K., Sakurai, K., De, D. (eds.) Proceedings of International Conference on Network Security and Blockchain Technology. ICNSBT 2021. Lecture Notes in Networks and Systems, vol. 481, pp. 312–324. Springer, Singapore (2022). https://doi.org/10.1007/978-981-19-3182-6_25
13. Saifee, S., Kothari, S.: Leveraging dual blockchain to solve security issues in electronic health records. In: Chen, J.IZ., Tavares, J.M.R.S., Iliyasu, A.M., Du, KL. (eds.) Second International Conference on Image Processing and Capsule Networks. ICIPCN 2021. Lecture Notes in Networks and Systems, vol. 300, pp. 446–461. Springer, Cham (2022). https://doi.org/10.1007/978-3-030-84760-9_39
14. Faruk, M.J.H., Shahriar, H., Saha, B., Barek, A.: Security in electronic health records system: blockchain-based framework to protect data integrity. In: Maleh, Y., Alazab, M., Romdhani, I. (eds.) Blockchain for Cybersecurity in Cyber-Physical Systems. Advances in Information Security, vol. 102, pp. 125–137. Springer, Cham (2023). https://doi.org/10.1007/978-3-031-25506-9_7
15. Anusha, R., Yousuff, M., Bhushan, B., Deepa, J., Vijayashree, J., Jayashree, J.: Connecting blockchain with IoT—a review. In: Sharma, D.K., Peng, S.L., Sharma, R., Zaitsev, D.A. (eds.) Micro-Electronics and Telecommunication Engineering. Lecture Notes in Networks and Systems, vol. 373, pp. 141–148. Springer, Singapore (2022). https://doi.org/10.1007/978-981-16-8721-1_14
16. Haque, A.K.M.B., Bhushan, B., Hasan, M., Zihad, M.M.: Revolutionizing the industrial internet of things using blockchain: an unified approach. In: Balas, V.E., Solanki, V.K., Kumar, R. (eds.) Recent Advances in Internet of Things and Machine Learning. Intelligent Systems Reference Library, vol. 215, pp. 43–66. Springer, Cham (2022). https://doi.org/10.1007/978-3-030-90119-6_5
17. Pandey, S., Bhushan, B.: Recent lightweight cryptography (LWC) based security advances for resource-constrained IoT networks. Wirel. Netw. (2024).https://doi.org/10.1007/s11276-024-03714-4
18. Gans, J.: Proof of work versus proof of stake. In: The Economics of Blockchain Consensus. Palgrave Macmillan, Cham (2023). https://doi.org/10.1007/978-3-031-33083-4_5

19. Saddikuti, V., Galwankar, S., Akilesh Sai, S.V.: Application of blockchain technology in healthcare supply chains. In: Stawicki, S. (eds.) Blockchain in Healthcare. Integrated Science, vol. 10, pp. 215–223. Springer, Cham (2023). https://doi.org/10.1007/978-3-031-14591-9_14
20. Indushree, M., Raj, M.: A novel blockchain-based authentication scheme for telecare medical information system. J. Supercomput. **80**, 1080–1108 (2024). https://doi.org/10.1007/s11227-023-05526-3
21. Sree Padmapriya, A., Sabiha Sulthana, B., Tejaswini, A., Snehitha, P., Brahma Rao, K.B.V., Anguraj, D.K.: Electronic health records rundown: a novel survey. In: Jeena Jacob, I., Kolandapalayam Shanmugam, S., Izonin, I. (eds.) Expert Clouds and Applications. ICOECA 2022. Lecture Notes in Networks and Systems, vol. 673, pp. 501–510. Springer, Singapore (2023). https://doi.org/10.1007/978-981-99-1745-7_36
22. Akilandeswari, R., Malathi, S.: Correction to: design and implementation of controlling with preventing DDOS attacks using bitcoin by Ethereum block chain technology. J. Transp. Secur. **16**, 7 (2023). https://doi.org/10.1007/s12198-023-00265-1
23. Kizza, J.M.: Blockchains, cryptocurrency, and smart contracts technologies: security considerations. In: Guide to Computer Network Security. Texts in Computer Science. Springer, Cham (2024). https://doi.org/10.1007/978-3-031-47549-8_26
24. Mandal Vijayvergiya, R., Sen, S.: Rating of doctor using tokenization mechanism using secure ethereum blockchain enabled platform. In: Goswami, S., Barara, I.S., Goje, A., Mohan, C., Bruckstein, A.M. (eds.) Data Management, Analytics and Innovation. ICDMAI 2022. Lecture Notes on Data Engineering and Communications Technologies, vol. 137, pp. 265–273. Springer, Singapore (2023). https://doi.org/10.1007/978-981-19-2600-6_18
25. Quy, T.L., et al.: Blockchain-driven animal healthcare: leveraging NFTs, IPFS, and smart contracts for comprehensive animal medical record. In: Bouzefrane, S., Banerjee, S., Mourlin, F., Boumerdassi, S., Renault, É. (eds.) Mobile, Secure, and Programmable Networking, MSPN 2023. Lecture Notes in Computer Science, vol. 14482, pp. 100–119. Springer, Cham (2024). https://doi.org/10.1007/978-3-031-52426-4_8
26. Cassez, F., Fuller, J., Ghale, M.K., Pearce, D.J., Quiles, H.M.A.: Formal and executable semantics of the ethereum virtual machine in dafny. In: Chechik, M., Katoen, JP., Leucker, M. (eds.) Formal Methods. FM 2023. Lecture Notes in Computer Science, vol. 14000, pp. 571–583. Springer, Cham (2023). https://doi.org/10.1007/978-3-031-27481-7_32
27. Gupta, R., et al.: VAHAK: a blockchain-based outdoor delivery scheme using UAV for healthcare 4.0 services. In: IEEE INFOCOM 2020 - IEEE Conference on Computer Communications Workshops (INFOCOM WKSHPS), Toronto, ON, Canada, pp. 255–260 (2020). https://doi.org/10.1109/INFOCOMWKSHPS50562.2020.9162738
28. Bhushan, B., Sharma, N.: Transaction privacy preservations for blockchain technology. In: Gupta, D., Khanna, A., Bhattacharyya, S., Hassanien, A.E., Anand, S., Jaiswal, A. (eds.) International Conference on Innovative Computing and Communications. Advances in Intelligent Systems and Computing, vol. 1166, pp. 377–393 (2021).https://doi.org/10.1007/978-981-15-5148-2_34
29. Sharma, N., Bhushan, B., Kaushik, I., Debnath, N.C.: Applicability of blockchain technology in healthcare industry: applications, challenges, and solutions. Internet Things, 339–370 (2021).https://doi.org/10.1007/978-3-030-66633-0_15

Scalable Lightweight Privacy Scheme for SDN-Based Fog-IoT Networks

Eman O. Attia[1,2(✉)], Hala A. Mansour[2], Shimaa S. Ali[2], and Heba A. TagElDein[2,3]

[1] National Telecommunications Institute, Cairo, Egypt
Eman.Omar@nti.sci.eg
[2] Faculty of Engineering at Shoubra, Benha University, Benha, Egypt
[3] Faculty of Engineering, Benha National University, Elobour, Cairo, Egypt

Abstract. With the rapid growth of Internet of Things (IoT) applications, Fog computing gained a great attention providing IoT data processing locally at fog nodes instead of sending them to the Cloud to provide faster responses to IoT applications compared to using Cloud servers (CS). In order to enhance performance, and reduce bandwidth consumption, Software Defined Network (SDN) is integrated with Fog computing to provide SDN-based Fog-IoT network.

This SDN-based Fog-IoT network is subject to many attacks. This paper addresses the authentication and privacy between the nodes using lightweight algorithms to cope with the constrained nature of IoT devices through the use of Elliptic Curve Diffie–Hellman Ephemeral (ECDHE) key exchange algorithm along with the Pre-Shared Key (PSK) as a lightweight and secure authentication scheme between the Cloud servers and IoT devices. The SDN controller is used to probably allocate the appropriate Fog node to serve the IoT device. The Key Distribution Server (KDS) integrated in the CS is responsible for generating the keys used to secure the channel between Fog nodes and IoT devices. The proposed scheme is mathematically evaluated relative to state-of-the-art protocols and certificate-based authentication schemes. The analysis showed that the proposed scheme is light and scalable compared to other schemes.

Keywords: IoT · Fog Computing · Privacy

1 Introduction

The growth in the number of smart devices has greatly increased the amount of sensing data in the Internet of Things (IoT). IoT devices send huge amounts of data regularly and at periodic intervals [1].

IoT devices are used in various environments [2]. However, their deployment creates many security problems [3] due to their limited resources and unsupervised nature. IoT devices also communicate with the Cloud servers through the Internet [4]. Cloud computing eliminates many shortages of enterprises and their users in terms of computation limitations, lack of storage resources, and network communication costs [5]. However, the cloud infrastructure doesn't meet the requirements of IoT devices and cope with their

A. Abdelgawad et al. (Eds.): ISBCom 2024, LNNS 1268, pp. 588–603, 2025.
https://doi.org/10.1007/978-3-031-82377-0_48

constrained nature [6]. As a result, fog computing was introduced to improve the performance by minimizing the network's overhead and latency as there is no need to send data to the cloud server [7]. A group of connected fog nodes can process and analyze the sensing data and provide it back in real-time [8].

Recently Software Defined Network (SDN) is integrated in the fog-IoT network to present SDN-based Fog-IoT network architecture to manage, control the traffic providing scalable and Quality of Service (QoS) based network architecture [9].

1.1 IoT: Current State and Challenges

IoT is based on various kinds of smart devices with low communication and networking capabilities. On the other hand, these devices generate enormous amount of data which must be stored, processed almost in real time.

Figure 1 illustrates the IoT architecture. The IoT devices gather data, send it over different communication technologies to IoT applications that provide services to these IoT devices. Cloud servers provide data storage and processing to aid in overcoming the limited resources of IoT devices. However, the emerging IoT applications require real-time interaction and mobility support which is not available using the centralized cloud servers. This impact can be reduced using distributed fog nodes. The performance can be rather enhanced by introducing SDN controller to manage fog nodes to provide QoS and scalability.

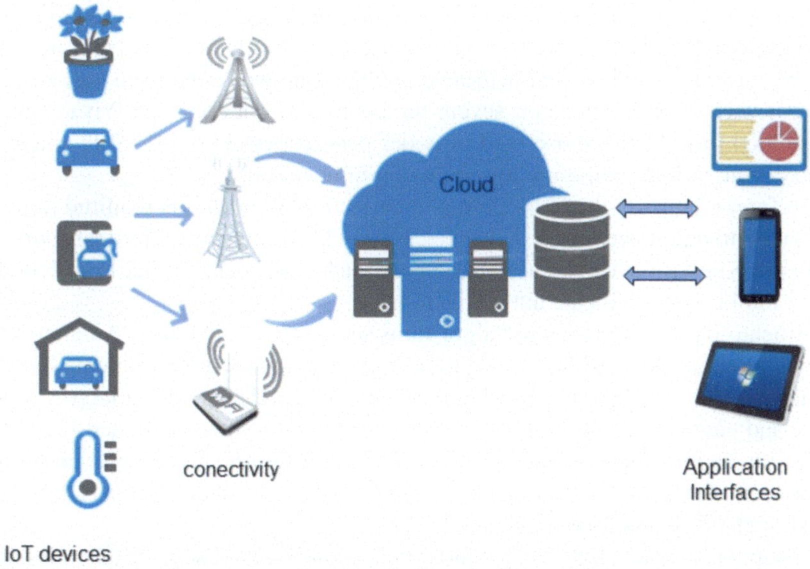

Fig. 1. IoT Architecture

1.2 SDN-Based Fog-IoT System

To address the previously discussed challenges, a model for IoT architecture based on SDN and Fog computing is proposed [9]. The architecture involves IoT end devices connecting to the network via different wireless technologies, SDN controllers, SDN-based Fog nodes and the Cloud servers.

SDN is an approach in networking that uses software-based controllers to communicate with underlying hardware infrastructure and direct traffic on a network. SDN moves the control plane that determines where to send traffic to software and leaves the data plane that forwards the traffic in the hardware. This allows network administrators who use software-defined networking to program and control the entire network via a single pane of glass rather than on a device-by-device basis [10, 11].

Fog computing is an emerging technology that brings data processing, storage, and analytics closer to the network edge. It has many attributes in common with Cloud computing, however, Fog can be distinguished from Cloud by its proximity to end-users, the dense geographical distribution, and its support for mobility [7]. Fog computing can solve the time-sensitive application processing limitations of the cloud as well as supporting IoT applications.

The Fog nodes are hosted on dedicated computing nodes equipped with communication interface, such as gateways, access points, routers, switches, etc. These nodes are SDN based controlled by the SDN controller for optimum performance.

The Fog-IoT paradigm is used in many applications such as healthcare, intelligent transportation systems and smart vehicles, smart agriculture and many others. These applications face a lot of challenges including security and privacy, data segregation, mobility support, location-awareness, geo-distribution, low latency, and real-time applications. Security is the main concern in developing IoT networks based on fog computing. A fake gateway or node can cause severe threats to user data privacy. Privacy needs to be considered from both the user and provider perspectives [12, 13]. Authentication is a primary issue in fog computing to secure communication.

Therefore, a lightweight and secure authentication algorithm is required to provide mutual authentication between IoT devices, CS, and fog gateways. The Transport Layer Security (TLS) protocol can be used to achieve such goals using either Pre-Shared Key (PSK) or public-key certificate authentication.

Certificate-based authentication algorithms such as Rivest, Shamir, Adelman (RSA) or Digital Signature Algorithm (DSA) [14] eliminate many security problems and allows for scalability, but, on the other hand, the certificate processing adds a large amount of resource and energy consumption which is not suitable for the constrained IoT devices. Also, certificate transmission and verification cause a significant increase in the protocol's connection latency. On the other hand, the PSK-based authentication is lightweight, but lacks scalability and forward secrecy.

This paper proposes a lightweight and robust authentication scheme that does not use heavy certificates. The proposal uses TLS v1.3 with the Elliptic Curve Diffie-Hellman Ephemeral (ECDHE) key exchange algorithm along with the Pre-Shared Key (PSK) authentication algorithm to provide mutual authentication between IoT devices and CS. A Key Distribution Server (KDS) integrated in the CS, is used to generate and distribute a security key to secure the connection between IoT nodes and the Fog nodes.

The rest of the paper is organized as follows: Sect. 2, provides an overview of the related work. In Sect. 3, the details of the proposed authentication scheme for SDN-based Fog-IoT networks are described in detail. In Sect. 4, the results and evaluations of the proposed scheme are analyzed. Finally, Sect. 5 concludes the paper.

2 Related Works

Many recent works introduced a privacy scheme for Fog-IoT applications. In [15], the authors considered the privacy issue and its relative problems which affect the overall performance of the network performance, the author proposes the use of signcryption as a solution to achieve the confidentiality and authentication between the SDN server and Fog nodes but neglect the other sides of the secured communication especially the communication between IoT nodes and fog nodes. In [16] authors rely on SDN controllers which are logically centralized to control the heterogeneous network. Two mechanisms to maintain user privacy are included, homomorphic encryption and identity. However, the paper didn't evaluate the efficiency of the tasks at the FOG nods. Moreover, the cloud server creates a private and public key for each IoT device via a non-secure channel so the keys can be compromised by any sniffer, and the same goes for getting a fake ID from the cloud. In [17] the author proposes an SDN-based security authentication scheme for fog nodes of smart healthcare. It proposes a privacy and functional authentication protocol to guarantee the credibility of the fog nodes. The proposal relies on SDN gateway to validate the creditability of fog node to decrease the computing of the IoT. The execution time and the storage cost are lower than the compared methods, but the work lacks the usage of lightweight authentication between the fog nodes and IoT nodes. In [18], the authors propose two kinds of lightweight authentication and key agreement schemes to enable fast and secure authentication among the objects in only an end-to-end IoT environment. First scheme uses the elliptic curve Qu-Vanstone (ECQV) implicit certificate, However, tampering with the certificate will result in the reconstructed public key being invalid, risk of node spoofing caused by replay attacks. The second scheme is also an authentication and key agreement scheme that can be used more securely, but slower than the first scheme using certificateless public key cryptography. Rana et al. in [19], proposes a privacy-preserving hash-based authenticated key agreement protocol for fog-driven IoT systems between the IoT node, fog node and cloud server. Although the performance evaluation results show that the proposed protocol has the least computation cost compared to several related competing protocols, it still lacks to provide secure communication channel between CS, Fog and IoT devices. Midha et al. in [20] used Cloud-based SDN authentication mechanism for smart health care user to ensure secure access to the patient data base by creating smartcard for the user in the registration phase, but the authors didn't achieve mutual authentication between cloud server and the user and didn't specify secure key exchange mechanism to achieve the confidentiality. In [21] the authors introduced an efficient key agreement for fog driven IoT healthcare system using the bilinear pairings defined on elliptic curve groups assuming that the registration process is carried out through a secure channel and the user is authenticated. In [22] Amanlou et al. introduced a lightweight authentication scheme based on TLS protocol based on the Elliptic Curve Diffie-Hellman Ephemeral (ECDHE) key exchange algorithm along with the Pre-Shared Key (PSK) authentication algorithm to provide authentication between IoT nodes and Fog nodes ignoring the CS. Also, the proposal supports just one way authentication. As a result, a lightweight authentication and key agreement scheme is needed to achieve privacy between IoT nodes, Fog nodes and Cloud servers.

3 Proposed Scheme

In this section, the proposed scheme that provides mutual authentication with privacy between cloud server and IoT nodes and between fog nodes and the IoT nodes is presented. The proposed scheme has been implemented based on the TLS v1.3 protocol [23]. The infrastructure consists of IoT nodes (end-devices), fog gateways, fog nodes (servers), cloud server, Key Distribution Server (KDS) and SDN Controller. First, the architecture of the proposed system is described, then the detailed procedure of the proposed lightweight and secure authentication scheme is explained.

The main contributions of the proposal are summarized as follows:

- Achieve lightweight mutual Authentication between IoT nodes, Fog Nodes and CS.
- Achieve lightweight secure Channel between IoT nodes, Fog Nodes and CS.
- Employ SDN technology to manage the fog node allocations to IoT nodes, focusing in reducing communication overhead and supporting mobility and scalability.

The following subsection will present our system structure and privacy preserving mechanisms deployed in the system followed by a detailed explanation of the operation of the proposal.

3.1 Architecture

The proposed system deploys the concept of SDN based Fog IoT Network. Figure 2 shows the structure of our proposed system. The system consists of four main layers: (1) The bottom layer (Perceptual layer) represents the IoT devices which generate data and transfer it through the network, (2) The second layer (SDN and fog layer) represents the SDN based IoT gateways, fog servers which enable computing, storage, and networking closer to the IoT devices and the SDN controllers (3) The third layer (Cloud layer) contains the identity (ID) server and key distribution server integrated in the Cloud server. The whole cloud server is responsible for IoT data analysis and storage, maintaining a database of IoT and fog nodes ID and generate and distribute session keys. (4) The top layer represents the application layer which supports different IoT applications.

3.2 Operation

This section presents a detailed description of the proposed solution operation. The solution aims to achieve a secured and authenticated connection between IoT nodes and both Fog nodes and CS. These secured channels are established using lightweight algorithms to cope with the constrained nature of the IoT nodes. Alsothe solution will

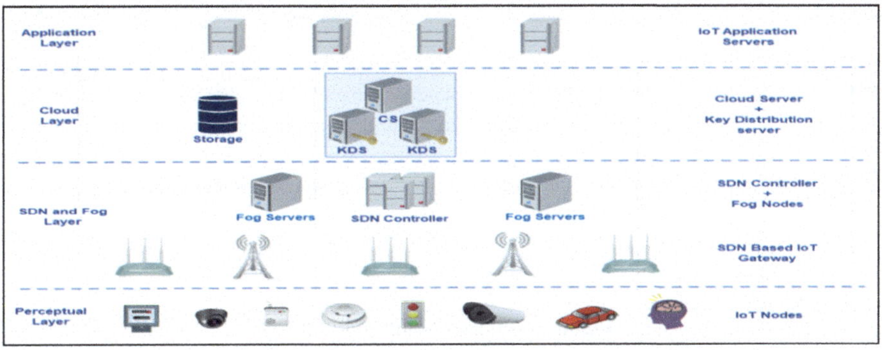

Fig. 2. Proposed Scheme Architecture

achieve scalability in the case of a mobile IoT node that needs to be associated with multiple fog nodes.

The solution is based on using Pre-Shared Key – Elliptic Curve Diffie Helman Ephermal (PSK-ECDHE) algorithm used by TLS 1.3. The PSK-ECDHE algorithm doesn't rely on certificates, that is why it is considered lightweight, also it offers Perfect Forward Secrecy (PFS) by generating a new key for each session. TLS 1.3 offers the use of lightweight encryption algorithms such as ChaCha20-Poly1305 encryption algorithm, Integrity will be achieved by using the SHA-256 hash algorithm.

The solution is based on some pre-assumption related to security between constrained IoT nodes and non-constrained fog nodes, CS and SDN controller which are as follows:

- There is a pre-shared key k pre-configured on both IoT node and the CS externally prior to communication as stated by TLS 1.3.
- There is a secure channel between Fog nodes and CS based on either certificate-based SSL or VPN that can be established any time prior to or at the beginning of the communication.
- There is a secure channel between SDN controller and CS based on either certificate-based SSL or VPN that can be established any time prior to or at the beginning of the communication.

The proposed solution operation is composed of four phases summarized in Fig. 3:

Phase 1: Secure channel establishment and Authentication between IoT nodes and CS.
Phase 2: Registration of IoT nodes with CS.
Phase 3: Fog node assignment through SDN controller.
Phase 4: Secure channel establishment between IoT node and Fog node.

Phase 1: Secure channel establishment and Authentication between IoT nodes and CS

Communication using TLS 1.3 begins with the TLS handshake. This is an initial negotiation between the client (IoT nodes) and server (CS) that targets key exchange and authentication.

The key exchange messages, ClientHello and ServerHello, determine the security capabilities of the client and the server and establish shared secrets, including the traffic keys used to protect the rest of the handshake and the application data.

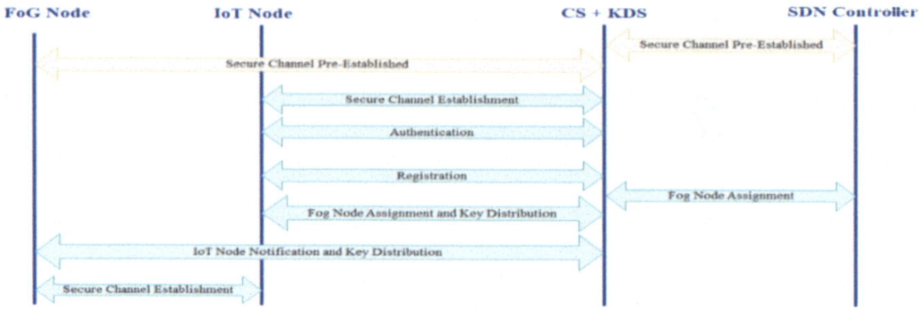

Fig. 3. Proposed Solution Phases

The TLS handshake begins with the client sending a ClientHello message to the server with the fields (cipher_suite and supported_versions, supported_groups, key_share, key_exchange, pre_shared_key extensions) [23].

Next the server responds to the client's ClientHello message with a ServerHello message with the following fields (cipher_suite and supported_versions, key_share, pre_shared_key extensions) [23].

In this proposal, the Elliptic Curve Diffie Hellman Ephemeral (ECDHE) algorithm is used for the generation of the keys in which both parties have an elliptic-curve public–private key pair to make a shared secret over an unsecured channel. The public keys in the ECDHE algorithm are ephemeral.

In ECDH key agreement, two communicating parties A and B agree to use the same curve parameters. They generate their private keys, k_A and k_B and their corresponding public keys $Q_A = k_A.G$ and $Q_B = k_B.G$ where G is the elliptic curve base point.

The parties next exchange their public keys. Finally each party multiplies its private key and the other's public key to arrive at a common shared secret $k_s = k_A.Q_B = k_B.Q_A = k_A.K_B.G$. This common shared secret key is denoted the shared secret key k_s. Next the traffic key k_t is generated using the KDF f1 where k_t = f1 ($k_{s.}$ k). Similarly, the encryption key k_e and the hashing key k_h are generated using the KDFs f2 and f3 by the relations k_e = f2 ($k_{t.}$ k) and k_h = f3 ($k_{t.}$ k).

At this point a secured channel is established between two entities. It is required to ensure that those entities are the intended entities (IoT and CS) not an attacker. This is achieved by authentication. TLS establishes the authentication either using certificates or using PSK. Using certificates is not appropriate for the constrained IoT nodes so PSK will be used.

Authentication is achieved using the TLS client finished and server finished messages with the PSK k and the KDF f4 and f5. The IoT node sends an encrypted hash value (encrypted by k_e of the PSK in the client finished message using KDF f4 (enc f4 (k)). The CS decrypts the message and validates the hash value to be that of the pre shared key k. in this way the CS authenticates the IoT node. In the same way, the IoT authenticates the CS where the CS sends an encrypted hash value (encrypted by k_e of the PSK in the server finished message using KDF f5 (enc f5 (k)). The IoT decrypts the message and validates the hash value to be that of the pre shared key k authenticating the CS.

At this moment an authenticated and secured channel is established between the IoT node and the CS according to the procedures shown in Fig. 4.

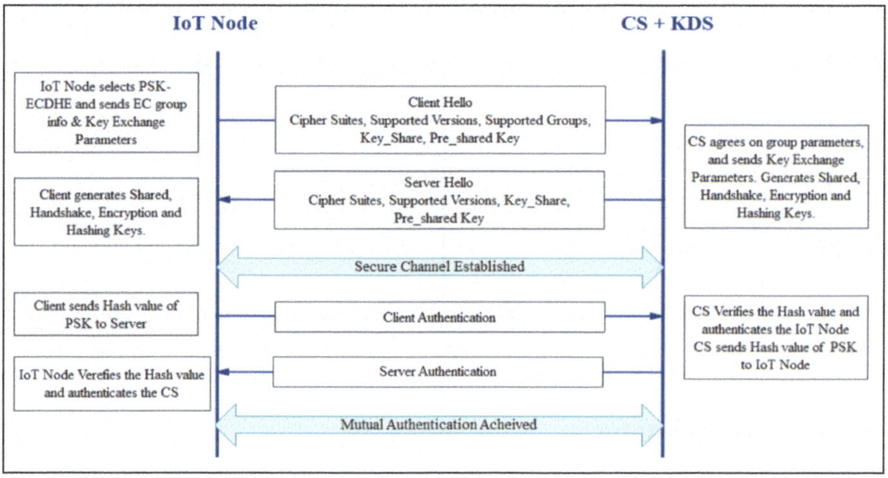

Fig. 4. Key Exchange and Mutual Authentication between IoT Node and Cloud Server

Phase 2: Registration of IoT nodes with CS

The next phase is the registration of the IoT node with the CS. Each IoT node must have a unique ID. Till now there has been no standard format for IoT ID, each vendor uses its own format. In this paper we propose to use two IDs for the IoT node, Real ID (RID) which contains all node details (type, capability, address, location, domain) and will be used between IoT node and CS for registration and the other ID is the Temporary ID (TID) that will be used with other nodes (SDN controller and Fog nodes) in order to hide the real identity of the IoT node across the network.

As shown in Fig. 5, the IoT node sends a registration message to the CS with the RID. The CS accepts the registration and generates a TID for the IoT node, stores both IDs in its database.

Phase 3: Fog node assignment through SDN controller

After node registration, the CS sends an encrypted attach request message to the SDN controller with the TID of the IoT node that includes the node capability and location. The SDN controller, according to its view for the network selects the appropriate Fog node and sends an encrypted attach reply to the CS with the appropriate Fog node ID.

Phase 4: Secure channel establishment between IoT node and Fog node

The KDS integrated in the CS, generates an encryption and hashing key, sends it once to the Fog node over the pre-established secure channel with the TID of the IoT node and another time to the IoT node over the established secure channel (from Phase 1) with the Fog node ID.

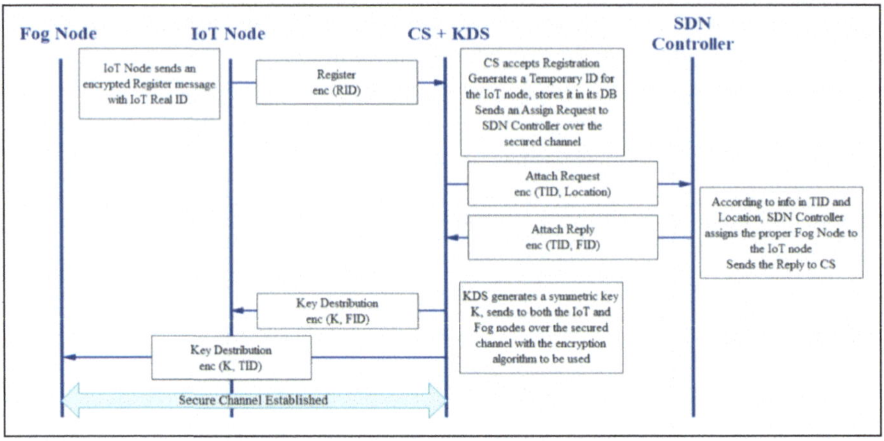

Fig. 5. IoT node Registration to CS and Secure channel Establishment with Fog Node

4 Performance Analysis

This section presents the performance analysis of the proposed solution using PSK-ECDHE in SDN based Fog-IoT architecture.

4.1 Security Evaluation

The security features of the proposed scheme and its resistance to common network attacks are summarized in Table 1 and evaluated as follows.

Mutual Authentication The proposed scheme provides mutual authentication between IoT nodes and CS relying on the PSK k which is only known to both parties. The authentication uses two different KDFs (f4 and f5) to prevent a middle attacker from capturing the encrypted hash and sending it back to the IoT node compromising the connection between them.

Data Confidentiality Using ChaCha20-Poly1305 encryption algorithm provides lightweight and high-performance data encryption.

Data Integrity Our scheme uses HMAC-SHA-256 for data integrity. The hash key depends on PSK.

Perfect Forward Secrecy (PFS) The proposed scheme provides PFS depending on ECDHE algorithm that generates a new key for each established session.

Anonymity The solution provides anonymity for the IoT nodes by using the TID across the network so hiding the RID.

Resistance to the MITM Attack The solution provides immunity against Man in The Middle (MITM) attack as mutual authentication and encryption are achieved prior to data transmission.

Resistance to the Replay Attack TLS provides protection against replay attacks using random nonces sent in both the client and server hello messages.

Scalability and Mobility Scalability is achieved using the SDN controller which allocates a new Fog node with the change of the location of IoT nodes.

Table 1. Security Feature Evaluation between Proposed Scheme and the Competing Protocols

Security Feature	**Amanlou et al** [22]	**ECDHE-DSA (1 Way Authentication**	**ECDHE-DSA (2 Way Authentication)**	**Our Scheme**
Lightweight	✓	×	×	✓
Mutual Authentication	×	×	✓	✓
Data Confidentiality	✓	✓	✓	✓
Data Integrity	✓	✓	✓	✓
PFS	✓	×	×	✓
Anonymity	✓	✓	✓	✓
Resistance to MITM Attack	✓	✓	✓	✓
Resistance to Replay Attack	✓	✓	✓	✓
Scalability	×	✓	✓	✓
QoS	×	✓	✓	✓
Mobility	×	✓	✓	✓

4.2 Performance Evaluation

In this section, the performance of the proposed authentication scheme in terms of communication cost, and computational cost is evaluated against the certificate-based authentication schemes ECDHE-DSA in the SDN based Fog-IoT architecture. Also, the results will be compared against the PSK-ECDHE in the traditional Fog-IoT architecture as stated by Amanlou et al. in [22].

Communication Overhead

In order to calculate the communication cost for the proposed scheme and compare it to the competing methods, the size of the TLS handshake messages is calculated.

The size of the TLS v1.3 Client Hello message, including the specified components (cipher suites, supported versions, supported groups, key shares, and pre-shared keys) as stated in [23] is 204 bytes total (60 bytes basic structure + 144 bytes extensions).

The size of the TLS v1.2 Client Hello message, including the equivalent components as stated in [24] is 163 bytes total.

The server hello message of TLS v1.3including the specified components (cipher suites, supported versions, key shares, and pre-shared keys) size is 122 bytes total (37 bytes basic structure + 85 extensions). The size of the TLS v1.2 server Hello message, including the equivalent components as stated in [24] is 119 bytes total.

When using a server certificate for authentication in TLS 1.3, the authentication process involves additional messages Certificate, CertificateVerify, and Finished. Each message has a specific structure and size depending on the chosen cryptographic algorithms and the length of the certificates. The certificate message size assuming one certificate gives rise to a total size of 1036 bytes. Certificateverify message size is about 264 bytes. Lastly the finished message that is used for PSK authentication and signing is 36 bytes length.

Table 2 lists the TLS handshake message sizes, and the total messages size according to the used method.

Table 2. TLS Handshake Messages and their sizes

Messages	**Amanlou et al.** [22]	**ECDHE-DSA (1 Way Authentication)**	**ECDHE-DSA (2 Way Authentication)**	**Our Scheme**
Client Hello	163 bytes	204 bytes	204 bytes	204 bytes
Server Hello	119 bytes	122 bytes	122 bytes	122 bytes
Server Certificate		1036 bytes	1036 bytes	
Server CertificateVerify		264 bytes	264 bytes	
Client Certificate			1036 bytes	
Client CertificateVerify			264 bytes	
Server Finished	36 bytes	36 bytes	36 bytes	36 bytes
Client Finished			36 bytes	36 bytes
Total	318 bytes	1662 bytes	2998 bytes	398 bytes

From the table, it is apparent that the total message sizes offered by [22] has the lowest accumulated messages sizes, followed by our proposed scheme. While the certificate-based method has so much higher message sizes. It must be noted that [22] offers just one way authentication, also it just establishes a secured connection between IoT-node and Fog node (no secure connection between IoT node and CS). Figure 6 shows such aggregated message sizes for each method.

To achieve scalability and mobility, the total communication overhead will be calculated assuming the IoT node move and associate with different Fog nodes, the case that require a new encryption key to be generated for such connection.

In [22], the total Communication overhead to implement a secured and authenticated channel between IoT nodes and associated n Fog-nodes will be n*318 as a result of generating a new encryption key through a separate TLS handshake with each new Fog association. In case of ECDHE-DSA with one way authentication, the Communication overhead to implement a secured and authenticated channel between IoT nodes and CS is 1662 in addition to a 32-byte generated from the KDS for each secured connection with each Fog node giving rise to a total communication overhead of 1662 + n*32.

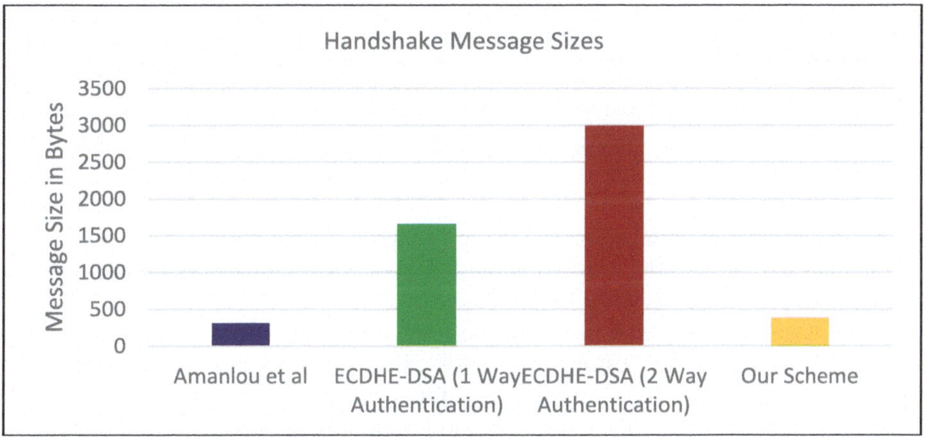

Fig. 6. Aggregated Messages Size

The same holds for ECDHE-DSA with two-way authentication leading to a total Communication overhead of 2998 + n*32. For our proposed scheme the total Communication overhead will be 398 + n*32. Table 3 summarizes such computation.

Table 3. Communication Cost of the proposed scheme and the competing techniques

Messages Size	Amanlou et al [22]	ECDHE-DSA (1 Way Authentication)	ECDHE-DSA (2 Way Authentication)	Our Scheme
Communication Overhead	n*318	1662 + n*32	2998 + n*32	398 + n*32

Figure 7 shows the total communication overhead for the proposed scheme relative to the other methodologies for different number of Fog nodes. It is apparent that our proposed scheme is characterized by the least communication overhead as the number of associations with Fog nodes increases due to mobility.

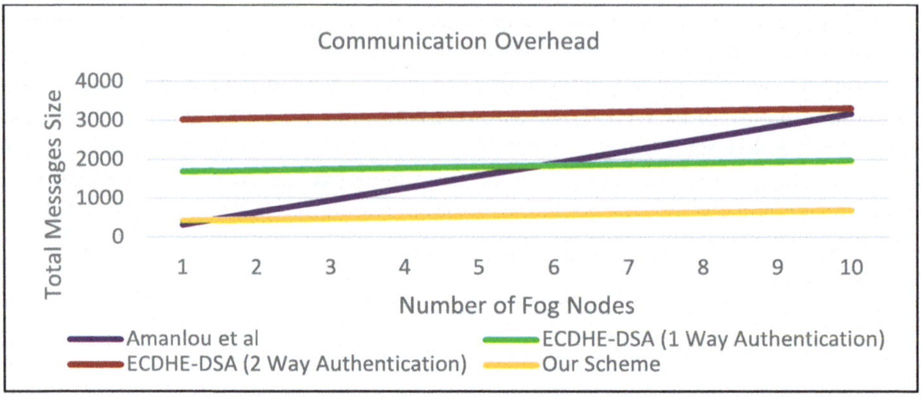

Fig. 7. Accumulated Communication Overhead relative to the number of Fog nodes

Storage Overhead

Table 4 shows the size of the keys used in the TLS protocol whether TLS 1.2 or TLS 1.3 and the total size of keys to be stored in the IoT node for each method. Figure 8 shows the aggregated key sizes used for each method.

To achieve scalability and mobility, the total storage overhead will be calculated assuming the IoT node moves and associates with different Fog nodes, the case that require a generation for new encryption keys for each new association with fog nodes.

In [22], the IoT node will need a PSK to be initially stored for each fog node to associate with in addition to the generated keys for the association which will give a total of n*192 + n*32.

Table 4. Key Sizes in bytes

Keys	Amanlou et al [22]	ECDHE-DSA (1 Way Authentication)	ECDHE-DSA (2 Way Authentication)	Our Scheme
PSK	32			32
Pre-Master	48			
Master	48			
Shared secret				32
Handshake secret				32
Server Public		32	32	
Server Private		32	32	
Client Public			32	
Client Private			32	
Encryption	32	32	32	32
Hashing	32	32	32	32
Formatting Storage		496	496	
Total	192	592	1152	160

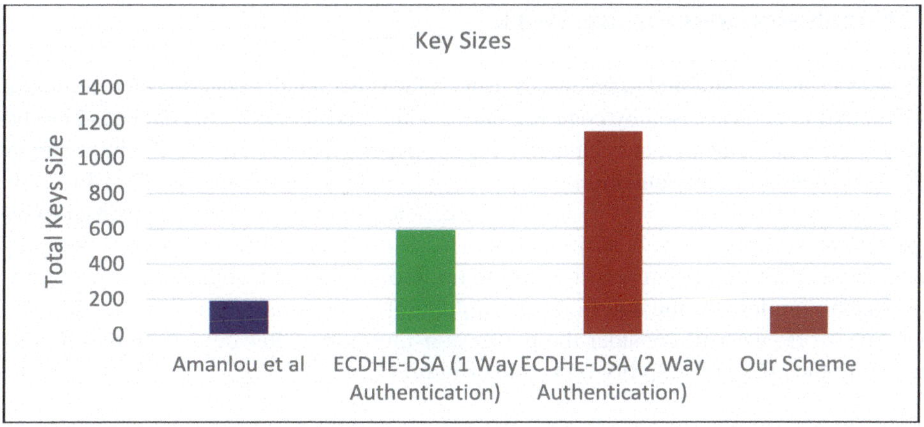

Fig. 8. Aggregated Keys Size

For all other techniques the accumulated storage overhead will be equal to the aggregated key sizes stated in Table 4 in addition to a 32 bytes encryption key generated from the KDS for each new association with a Fog node giving rise to the values stated in Table 5 and displayed in Fig. 9.

Table 5. Accumulated Storage Cost of the proposed scheme and the competing techniques

Key Sizes	**Amanlou et al.** [22]	**ECDHE-RSA (1 Way Authentication**	**ECDHE-RSA (2 Way Authentication)**	**Our Scheme**
Storage Cost	n*192 + n*32	144 + n*32	256 + n*32	160 + n*32

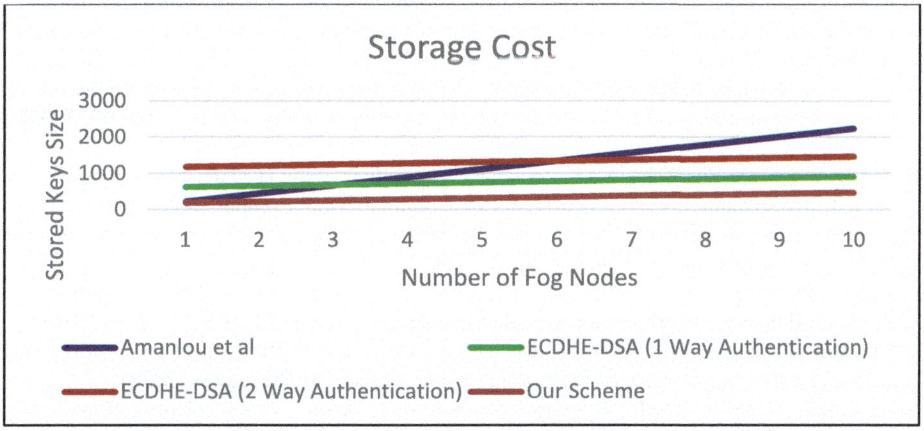

Fig. 9. Accumulated Storage Overhead relative to the number of Fog nodes

5 Conclusion and Future Work

IoT devices are constrained in resources. But on the other hand, generates huge amounts of data that to be processed by edge fog nodes and cloud servers. This data needs to be secured through authenticated channels. So, this paper proposes the use of a lightweight key agreement and authentication scheme based on TLS v1.3 using the PSK-ECDHE algorithm in SDN-based Fog-IoT architecture. The proposed scheme has proved to be lightweight, scalable and robust compared to the equivalent certificate-based techniques or the equivalent algorithm working in the traditional Fog-IoT architecture. In future, we want to implement the solution and evaluate it in different scenarios. In addition, as future work, we will consider the power consumption of IoT devices in addition to analyzing the effect of mobility on key distribution.

References

1. Magara, T., Zhou, Y.: Internet of things (IoT) of smart homes: privacy and security. J. Electr. Comput. Eng. **2024**(1) (2024)
2. Das, S.K., Bebortta, S., Pati, B., Panigrahi, C.R., Senapati, D.: Profiling and classification of IoT devices for smart home environments. In: Nayak, J., Naik, B., S, V., Favorskaya, M. (eds.) Machine Learning for Cyber Physical System: Advances and Challenges. Intelligent Systems Reference Library, vol. 60, pp. 85–121. Springer, Cham (2024). https://doi.org/10.1007/978-3-031-54038-7_4
3. Martalò, M., Pettorru, G., Atzori, L.: A cross-layer survey on secure and low-latency communications in next-generation IoT. IEEE Trans. Netw. Serv. Manag. (2024)
4. Alhaidari, F., Rahman, A., Zagrouba, R.: Cloud of Things: architecture, applications and challenges. J. Ambient. Intell. Humaniz. Comput. **14**(5), 5957–5975 (2023)
5. Malik, I., Bhardwaj, A., Bhardwaj, H., Sakalle, A.: IoT-enabled smart homes: architecture, challenges, and issues. Revolution. Ind. Autom. Converg. Artif. Intell. Internet Things, 160–176 (2023)
6. Rezaee, M.R., Hamid, N.A.W.A., Hussin, M., Zukarnain, Z.A.: Fog offloading and task management in IoT-Fog-Cloud environment: review of algorithms, networks and SDN application. IEEE Access (2024)
7. Hazra, A., Rana, P., Adhikari, M., Amgoth, T.: Fog computing for next-generation internet of things: fundamental, state-of-the-art and research challenges. Comput. Sci. Rev. **48**, 100549 (2023)
8. de Oliveira, M.A., da Rocha, A.M., Puntel, F.E., Cavalheiro, G.G.H.: Time series compression for IoT: a systematic literature review. Wirel. Commun. Mob. Comput. **2023** (2023)
9. Javanmardi, S., Shojafar, M., Mohammadi, R., Alazab, M., Caruso, A.M.: An SDN perspective IoT-Fog security: a survey. Comput. Netw. 229
10. Kreutz, D., Ramos, F.M., Verissimo, P.E., Rothenberg, C.E., Azodolmolky, S., Uhlig, S.: Software-defined networking: a comprehensive survey. Proc. IEEE **103**(1), 14–76 (2014)
11. Bhuiyan, Z.A., Islam, S., Islam, M.M., Ullah, A.A., Naz, F., Rahman, M.S.: On the (in) security of the control plane of SDN architecture: a survey. IEEE Access (2023)
12. Aleksandr, O., Molua, O.L., Komarov, M., Nurmi, J.: A survey of security in cloud, edge, and fog computing. Sensors **22**(3), 927 (2022)
13. Alzoubi, Y.I., Osmanaj, V.H., Jaradat, A., Al-Ahmad, A.: Fog computing security and privacy for the internet of thing applications: state-of-the-art. Secur. Priv. **4**(2), e145 (2021)

14. Chowdhary, A., Huang, D., Alshamrani, A., Kang, M., Kim, A., Velazquez, A.: TRUFL: distributed trust management framework in SDN. In: ICC 2019–2019 IEEE International Conference on Communications (ICC), pp. 1–6. IEEE (2019)
15. Alamer, A.: Security and privacy-awareness in a software-defined Fog computing network for the Internet of Things. Opt. Switch. Network. **41** (2021)
16. Alshiky, A.M., Janbi, N.F., Akkari. N.: SDN-based privacy for Fog-IoT application. In: International Journal of Advanced Computational Engineering and Networking, ISSN(p): 2320–2106, ISSN(e): 2321–2063 Volume-7, Issue-12 (2019)
17. Ke, C., Zejiang Zhu, F., Xiao, Z.H., Meng, Y.: SDN-based privacy and functional authentication scheme for fog nodes of smart healthcare. IEEE Internet Things J. **9**(18), 17989–18001 (2022)
18. Lee, D.-H., Lee, I.-Y.: A lightweight authentication and key agreement schemes for IoT environments. Sensors **20**(18), 5350 (2020)
19. Rana, M., Mahmood, K., Saleem, M.A., Al-Turjman, F., Kolhar, M.S., Altrjman, C.: Towards a provably secure authentication protocol for fog-driven IoT-based systems. Appl. Sci. **13**(3), 1424 (2023)
20. Midha, S., Verma, S., Mittal, M., Jhanjhi, N.Z., Masud, M., AlZain, M.A.: A secure multi-factor authentication protocol for healthcare services using cloud-based SDN. Comput. Mater. Continua **74**(2) (2023)
21. Jia, X., He, D., Kumar, N., Choo, K-K.R.: Authenticated key agreement scheme for fog-driven IoT healthcare system. Wirel. Netw. **25**(8)
22. Amanlou, S., Hasan, M.K., Bakar, K.A.A.: Lightweight and secure authentication scheme for IoT network based on publish–subscribe fog computing model. Comput. Netw. **199**, 108465 (2021)
23. Rescorla, E.: The transport layer security (TLS) protocol version 1.3. No. rfc8446 (2018)
24. Dierks, T., Rescorla, E.: The transport layer security (TLS) protocol version 1.2. No. rfc5246 (2008)

Fiber to Everywhere in the Next Generation Broadband Networks

Rasha Samir(✉), Hussein Harb, and Ahmed Khattab

National Telecommunications Institute, Cairo, Egypt
Rashasamir@nti.sci.eg
https://mail.nti.sci.eg,https://www.nti.sci.eg

Abstract. Society's twin transitions (green and digital) are based on fiber networks, providing a cost-effective and sustainable communication platform with high bandwidth, stability, reliability, and reduced latency, allowing the use of advanced services and applications for businesses, industries, and users, and enabling sustainable economic growth. Based on the generations of wireless industry experts, the wireless industry has developed an efficient, successful, and aligned ecosystem with all stakeholders. By contrast, the wireline industry is characterized by significant fragmentation, which does not provide operators, developers, manufacturers, and users with the benefits of large-scale, aligned developments. Through Fiber-To-The-Everywhere-And-Everything (FTTE), F5G aims to lay the foundation for a structured approach to the evolution of fixed networks, establishing generational planning, and promoting its use in as many sectors as possible.

Keywords: Next Generation Broadband · FTTE · F5G

1 Introduction

Nowadays, the digital economy, as the leading force in the world's scientific and technological revolution and industrial revolution, is leading the global economic development direction and increasingly integrated into all fields and processes of economic and social development, along with the rapid growth of the information revolution [1]. The profound changes in the mode of production, life, and social governance are unprecedented in their speed of development, scope of radiation, and depth of influence. They are becoming a key force in reshaping the global economic structure and changing the global competition pattern.

Digital economy is a new economic form driven by digital technology, which has the characteristics of high innovation, strong penetration, and wide coverage [2,3]. Faced with the new wave of global digital economy development, all industries promote digital transformation. Developing the digital economy can improve labor efficiency, capital efficiency, and resource utilization efficiency. The essence is continuously improving total factor productivity, accelerating industrial and life transformation, and promoting high-quality economic development.

A. Abdelgawad et al. (Eds.): ISBCom 2024, LNNS 1268, pp. 604–618, 2025.
https://doi.org/10.1007/978-3-031-82377-0_49

Digital information infrastructure is an important cornerstone of digital economic development. It can provide digital capabilities such as perception, transmission, storage, computing, and processing to meet the extensive needs of social production and life. It is a necessary foundation for digital economic and social transformation and high-quality development. With the rapid development of the global digital economy [4], the development of digital information infrastructure brings great opportunities and requires higher evolution and upgrading and support capabilities of digital information infrastructure. The development of digital information infrastructure, especially the new-generation communications infrastructure network connected by all-optical cables, is the most important foundation for developing the digital economy.

1.1 The Significance and Challenges of the Development of the New Generation Digital Information Infrastructure

The digital economy is an irreversible development trend. Many technologies, including cloud computing, big data, artificial intelligence, and the Internet of Things, have greatly promoted the development of the digital economy. The digital economy will restructure many industries and form a new economic form [5]. In recent years, the construction of digital information infrastructure, represented by fifth Generation (5G), 5th Generation Fixed Networks (F5G), artificial intelligence, and industrial Internet, has become a key investment area for leading countries and regions in the digital economy. These investments not only directly promote the development of the digital economy and the prosperity of related industries, but also become a new cornerstone for the accelerated development of digital transformation of industries. Obtaining higher network speed, faster processing capability, stronger analysis functions, and richer software applications at a lower price will drive traditional enterprises to improve quality and efficiency on a large scale, stimulate new momentum through the industry, and realize the overall transformation and upgrading of the industry [6]. It has become the core driving force for high-quality economic development. The construction of a new generation of digital information infrastructure with full optical fiber connection, the bottom layer, provides attractiveness and impetus for the relevant elements of social and economic activities in space and can promote the extension expansion of urban space and the improvement of connotation and quality, to realize industrial agglomeration and upgrading, and through quality transformation and efficiency transformation. Continuously optimize the development prospects [7]. The digital information infrastructure based on all-fiber connections has gradually emerged as the main carrier. Take digital integration and convergence innovation as the main way, to achieve efficient connection through digital and intelligent means, and realize multi-factor parallel development. In terms of infrastructure, the new generation of digital information infrastructure based on all-fiber connections is a prerequisite for the circulation of data elements and the application of digital technologies. With the development of new-generation Passive Optical Network (PON) and Segment Routing IPv6 (SRv6) technologies, network reconstruction will upgrade access, bearer, and core network capabilities as required. This feature enables low latency and high-speed interconnection between cluster hub nodes, data centers in a cluster, and data centers in a cluster and edge data centers, and low latency traffic between users in data centers to meet users' requirements for large capacity, low latency, and deterministic network services. To meet the requirements of industry digital transformation and upgrade for high-deterministic services such as network latency, jitter, and reliability, all-fiber network reconstruction needs to focus on typical scenarios, build a deterministic service indicator system, upgrade technologies in various domains, and integrate network and multi-level cloud capabilities.

1.2 Principles and Requirements for Construction of Next Generation Network Infrastructure

The development of network infrastructure must follow the laws of the digital economy, adapt to the changing trend of user requirements in the digital era, meet the requirements of digital transformation for digital elements upgrade and integration innovation, and accelerate the four transformations of the new generation of all-fiber information infrastructure:

1. Promote the development of cloud and network and transform the layout of cloud + network + computing power + security convergence. Based on the optimized layout and intelligent upgrade of multiple digital capabilities, build a cloud-network integration scheduling platform that integrates multiple elements such as cloud, network, computing power, and security to achieve efficient digital resource collaboration. Regarding networks, there is a need to accelerate the construction of all-fiber access networks and promote the upgrade of high-speed ubiquitous connection capabilities.
2. Promote network architecture transformation to adapt to vertical and horizontal traffic transformation. Form high-speed and low-latency interconnection channels within and between data center clusters based on the computing power center. Create a 1ms/3ms latency circle at the deep/shallow edge to meet edge computing power sinking requirements, create an edge cloud ecosystem, promote intelligent cloud-network integration scheduling, and schedule network resources according to cloud requirements. Adapt computing power and resources to different services.
3. Promote green transformation and green development of information infrastructure Global development is undergoing a comprehensive green transformation. The information infrastructure needs to explore and form a new low-carbon model. A low carbon strategy should consider all aspects of energy conservation and carbon reduction, and rationally plan and design the whole life cycle of planning to operation. To strengthen technological innovation and management upgrading internally and continuously reduce carbon emission intensity. To optimize the supply and service quality of products and enable green and low-carbon development in the whole society with digital technologies.

4. Transform the digital security system and realize the endogenous transformation of information infrastructure security. Reconstruct the security architecture, develop an end-to-end cloud-network security capability system featuring capability aggregation, unified architecture, and open ecosystem, build an end-to-end digital defense system, and gradually develop capabilities for security data centralization, intelligence security analysis, operation orchestration, and security service delivery. For different requirements, provide differentiated security approaches.

Accordingly, the overall architecture of the new-generation all-fiber digital information infrastructure base is shown in Fig. 1.

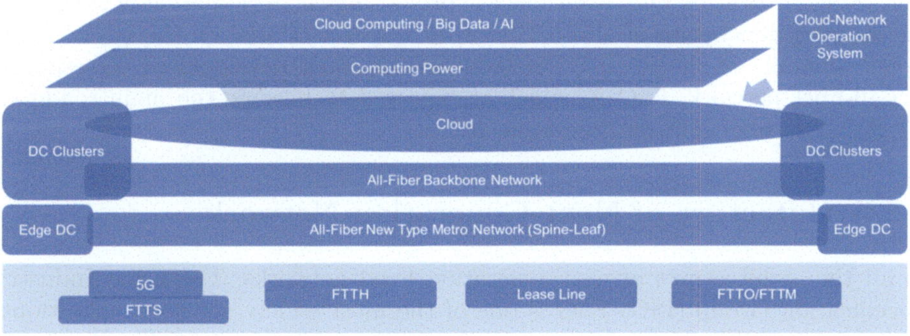

Fig. 1. Overall architecture of the new-generation all-fiber digital information infrastructure

1.3 Key Technology Trends

All-fiber access, all-fiber bearer, all-fiber switching, all-fiber transmission, and all-fiber cloud network are typical features of the next-generation broadband network technical architecture, but they are not the final state. The network will continue to evolve. Looking forward to the future, with the further growth of network traffic and the change of flow direction, as well as the in-depth promotion of green energy-saving strategies, all-fiber transmission capacity will be further increased. New materials, new algorithms, and new optoelectronic integration technologies will drive the continuous growth of single wavelength rates, and at the same time, the available transmission bandwidth will be increased by expanding bands. Research and deploy new optical fibers to increase transmission capacity and distance. All-fiber switching will develop towards higher dimensions, lower costs, and faster recovery. It is expected to break through the limitations of wavelength uniqueness and wavelength consistency, and realize non-blocking all-fiber switching at the wavelength or sub-wavelength level.

All-fiber access will further increase bandwidth, coverage, and application scenarios. All-fiber access to base stations/APs will explore the technical collaboration between all-fiber networks and users at the last segment of connections. High-speed and low-cost technologies such as Light Fidelity (LiFi) and Radio over Fiber(RoF) are expected to be applied in access scenarios to optimize user experience further. The scope of the next-generation all-fiber network will be further expanded. On the one hand, free-space optical communication is expected to become a part of the all-fiber network. Atmospheric optical communication will make the all-fiber network integrated between heaven and earth. On the other hand, the all-fiber network will develop inside devices and chips. All-fiber interconnection is implemented inside devices. The chip increases density through optical interfaces and reduces power consumption. Global Communications Standardization Organizations are promoting the formulation of standards for the next generation of all-fiber network research. Based on the intergenerational planning of the F5G, ETSI launched the next development direction of the all-fiber network, F5G-A, which is expected to become the main roadmap for global all-fiber network development.

2 Architecture of Next-Generation Network

In all the architectures of next-generation fixed networks, The ETSI Industry Specification Group (ISG) F5G is one of the most widely accepted definitions. F5G has established a continuous evolutional approach to the fixed network defining generations that may direct the industry toward a consistent E2E network vision. F5G-A defines the next-generation network. European Telecommunications Standards Institute (ETSI) GR F5G 001 [i.12] defines F5G, to enable a wider technological standards adoption and boost the creation of a global market as in Fig. 2.

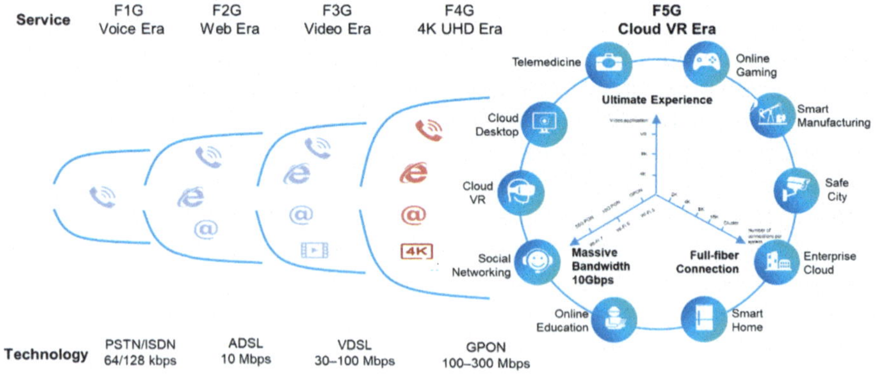

Fig. 2. ETSI defines the Fifth Generation for Fixed Network

In November 2023, ETSI released F5G Advanced Generation Definition (ETSI GR F5G 021 V1.1.1). The report defines the F5G-A generational standards and explores the evolution path from F5G to F5G-A. F5G requirement is mainly from residential, business and vertical industries and covers typical application cloud VR, 8K HD as well as smart cities and smart manufacturers.F5G-A supports the next generation of emerging digital services such as digital twins and metaverses, which impose more stringent requirement on the communication network technology. Therefore, F5G-A has enhanced the three dimensions of network character in F5G and introduces three new dimensions that will allow to meet the requirements of emerging services, which is visualized in Fig. 3.

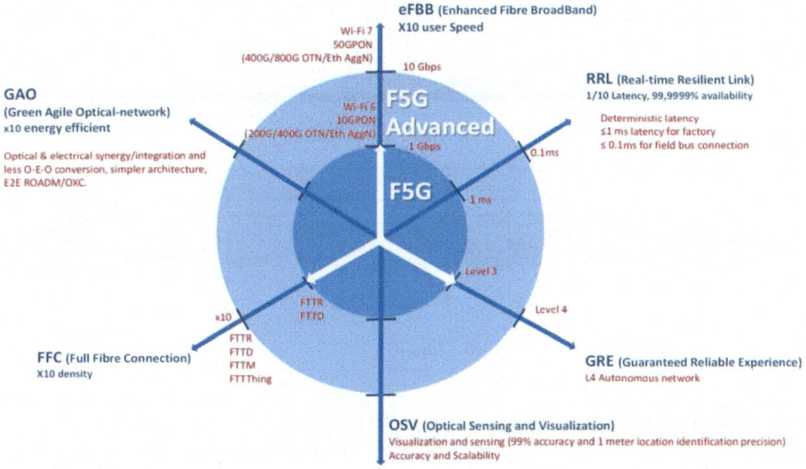

Fig. 3. ETSI releases F5G Advanced Generation(F5G-A) Definition

To achieve F5G-A target characters, a series of new technologies are launched. Therefore, ETSI has proposed GAO architecture to support the next-generation fiber network evolution.

In the 2H scenario, optical fibers are extended to the end-user devices to replace copper and cable transmission, greatly reducing energy consumption and satisfying the digital home experience. 10GPON will become the common connectivity to Home and NG PON technology including 50GPON will go to market in the leading market.

In the 2B scenario, optical fiber connections are extending to offices and factories, greatly reducing energy consumption and complying with the extreme link requirement of the vertical industries with high bandwidth and low latency. Traditional OTN technology and new architectures like P2MP OTN and next-generation PON will be used in different industry scenarios. In the entire network, network equipment switches from electrical switching technology to optical switching technology and achieves "On-hop to Core", which decreases latency and reduces energy consumption.

2.1 Technology Architecture of the Next Generation Network

Breakdown the GAO, it can become a hierarchical network architecture as in Fig. 4. At the bottom of the whole architecture is a passive base. The Objective F5G/F5G-A base is an end-to-end seamless connectivity all-fiber network from the end user (2C/2B/2H) to the central network's DC facility. Most of the operator has complete fiberization in the metro and core parts, Today, the key objective of this part is to accelerate the fiberization of the access network and improve the flexibility and scalability of the entire optical network.

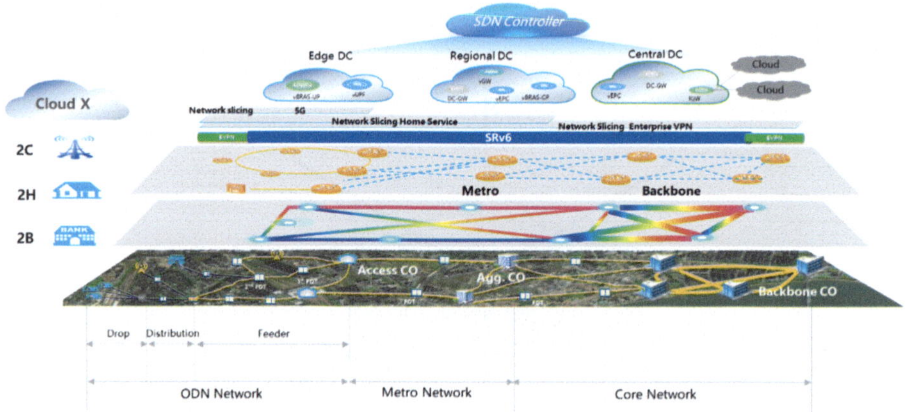

Fig. 4. Technology architecture of the next generation network

In the middle of the architecture, there is a bearer network, which can also be broken down into a fiber transport network and IP network. The basic objective of the bearer network is to provide low latency, high capacity, and flexibility in complying with different service traffic transport tunnels, and make balance in network cost at the same time. Facing such a complex "Post 5G Era" FBB network, it has to take a modularized network architecture, which uses a hierarchical designing principle to achieve a simple and clear overall architecture that meets the service position in each layer.

2.2 Passive Optical Network Base

Passive Optical Network (PON) as the base of the whole network is required to keep stable, while continuously new service introductions and CO/BTS number increases require the fiber network be flexible and scalable. To take balance between these conflict requirements, a new methodology called "Full-Service-Area & United Fiber Network (FSA)" is introduced into in Passive Network Planning as shown in Fig. 5.

The key benefit of FSA fiber network planning is to provide a balance between stable topology and flexible capacity scale-out. At the same time, with FSA fiber

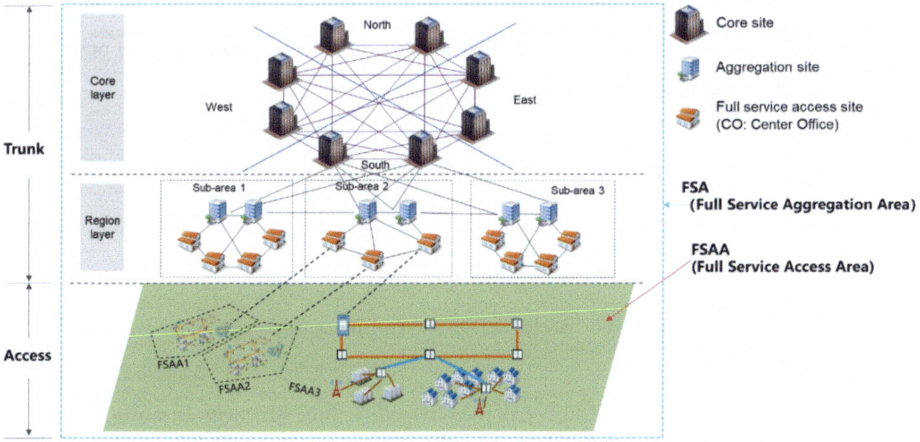

Fig. 5. Full-Service-Area & United Fiber Network

architecture, operators can avoid massive repeat fiber construction at the same routing therefore saving cost and TTM time. A comparison between traditional Planning and Implementation and FSA Planning and Implementation is presented in Fig. 6.

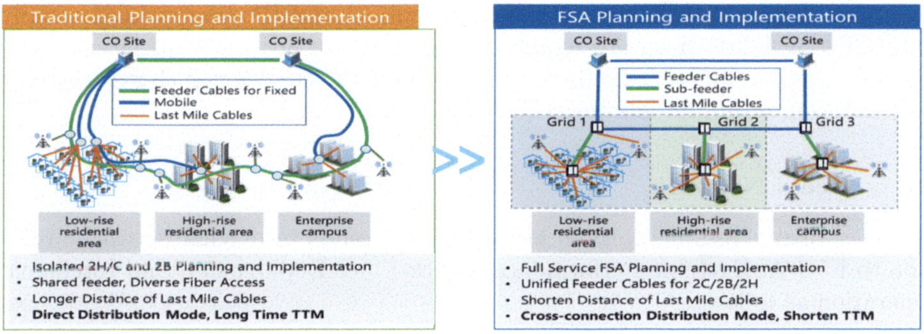

Fig. 6. Comparison between traditional Planning and Implementation and FSA Planning and Implementation

2.3 Transport Network

In F5G-A network architecture, the transport network in Fig. 7, is defined as an FMC bearer, high capacity, low latency, and intelligent operation network. There is a series of technologies introduced to both the transport layer and IP layer to achieve this target.

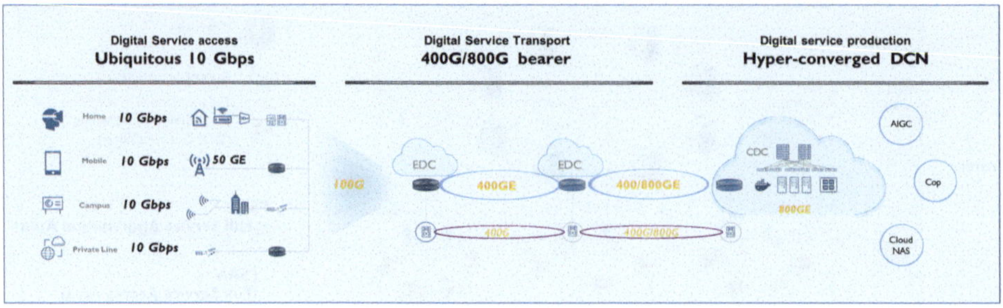

Fig. 7. Transport layer architecture of the next generation network

In the Transport Layer, OTN/WDM to edge has already become a trend, which supports traffic by-pass IP network, which is beneficial to minimize the transport latency and saving the forward IP equipment cost. In the IP layer, slicing technology allows different kinds of services to be able to transport within the same tunnel with different kinds of QOS, and achieve isolated network resource reservation. With the SRv6, network slicing, and network digital map, service can be transmitted within an IP network with the simplest protocol, which provides better latency and decreases network performance. At the same time, 50GE to site, 400GE metro, and 800GE backbone/DCN will provide extreme transport capacity. With the Intelligent Network Operation platform introduction, multi-layer synergy becomes smart and accurate. For example, it can provide the most economical protection approach based on Optical and IP synergy, and comply with different kinds of service requirements, or, it can provide some latency estimation based on different kinds of routes so that the service can choose the best cost solution within its latency requirement.

2.4 Access Network

In the optical access field, there are three key trends to driving network evolution to F5G-A: Extreme Performance, Quick Fiber Deployment, and Intelligent Operation as follows:

- **Extreme performance**: required very high speed, low Latency, and outstanding reliability. Oriented to service cloudification and real-time interactive services, Access network needs to upgrade from a best-effort experience to a deterministic experience. The all-optical access network intelligently detects service characteristics and types, supports intelligent E2E hard slicing and traffic distribution from home Wi-Fi to OLTs, and connects to the cloud through one hop with OTN to ensure a deterministic experience. To reach this target, Access technology is mainly evolving in the active products, which include PON technology evolution and Wi-Fi performance improvement. In the PON technology field, GPON has been put into large-scale commercial use to provide users with 100 Mbps access capabilities. With the large-scale application of 4K/8K video services, users' bandwidth requirements increase, and 10G PON ready networks gradually become the mainstream in the world. As more and more operators have massive "Gigabit Packages", the next-generation PON technology after 10GPON has already been considered by the industry.

- **Quick Fiber Deployment**: is restricted by many external conditions, which can result FTTH construction being slow and expensive. With some architecture innovation, distributed CO has already been a new option for operators, which can help operators reuse existing infrastructure and reduce fiber construction efforts. Some new ODN technology has provided pre-connection cables and boxes, which can avoid field slicing and provide high-quality ODN components.
- **Intelligent Operation**: has already covered the whole lifecycle of FTTH business from Planning to final Maintenance and troubleshooting process. Thanks to advances in big data and AI technology, today a lot of digital operation processes significantly improve user experience and improve Operator's efficiency from end to end. Of course, this achievement also depends on some innovation on the equipment side, like some intelligent module inside ONT, or some digital label on ODN equipment.

3 Fiber Technology Evolution

In this section, the evolution of fiber technologies is introduced in detail such as PON, ODN, and Wi-Fi Technologies.

PON Technology: The technological evolution of PON technology is becoming clearer day by day. The technological evolution route of GPON $\rightarrow$ 10GPON $\rightarrow$ 50GPON as shown in Fig. 8 has become an industry consensus. In 2022, the ITU-T designated 50GPON as the standard protocol for the post-10G era PON. At the application level, 10G PON has become the mainstream delivery technology globally, deployed by over 200 operators worldwide. Based on current 50GPON test results, it is estimated that the cost per bit could be 20%–30% lower compared to GPON technology.

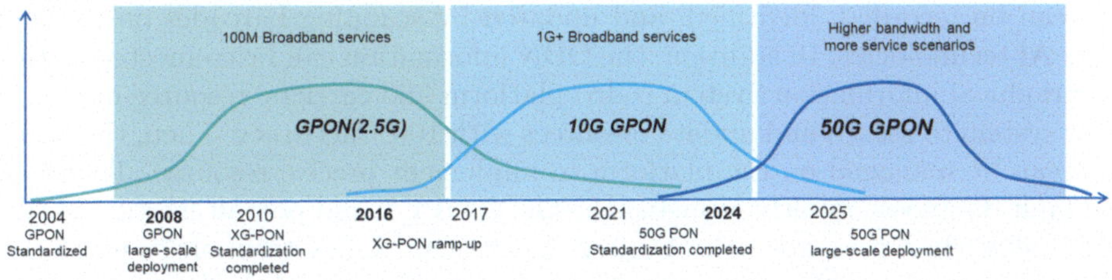

Fig. 8. PON technology evolution route

The core value of GPON $\rightarrow$ 10GPON $\rightarrow$ 50GPON technology lies in its ability to simultaneously carry multiple generations of technology on the same optical fiber network by using different working wavelengths, thus avoiding redundant construction of fiber networks, which typically accounts for 50–60% of total broadband investments. Currently, with the introduction of new technologies by vendors, network-side equipment can also support smooth evolution from GPON to 50GPON on the same platform, even on the same port, enabling over 70% of the investment from GPON to 50GPON to be reused, thereby further safeguarding operators' investments.

ODN Technology. The new ODN technology will change the construction and management way of fiber optic networks and reduce overall network costs. The pre-connection technology enables the ODN deployment mode to be changed from the traditional onsite construction mode to the factory prefabrication and onsite assembly mode as shown in Fig. 9. ETSI released the new industry standard ODN Quick Construction and Digitalization (ETSI TR 103 775). The innovative pre-connection technology enables the ODN deployment mode to be changed from the traditional onsite construction mode to the factory prefabrication and onsite assembly mode. This technology improves the overall ODN installation efficiency by more than 50%, reduces the E2E CAPEX by more than 15%, greatly lowers the requirements for onsite ODN installation technologies, and supports the rapid growth of FTTH.

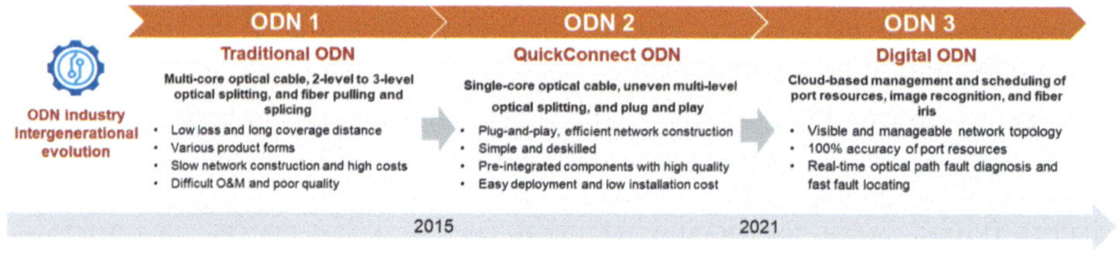

Fig. 9. ODN technology evolution

The combination of the pre-connection technology and digital ODN will further reduce ODN Operation and Maintenance (O&M) costs. All ODN information can be recorded, identified, and updated by scanning barcodes onsite and using AI technologies. In addition, the ODN information can be connected to the Geographical information System (GIS) platform and carriers' resource management system to record and update resources with 100% accuracy. Then, the basic data can be managed on the platform to implement precise resource allocation and fault diagnosis, greatly reducing the O&M OPEX and port utilization of the ODN network. Up to now, the Digital Quick (DQ) ODN technology has been used by more than 300 carriers in 20 countries. The actual deployment quantity exceeds 30 million lines, accounting for more than 20% of the total ODN shipments in 2022–2023. Another change is related to ODN deployment, which involves changes in the ODN construction process. Globally, especially in the MEA region, the proportion of aerial optical cables in the overall construction has increased significantly.

Wi-Fi Technology
With the rapid development of the Internet of Things, cloud computing, big data, and ultra-broadband, Services such as 4K, VR, and smart home applications are booming, and Wi-Fi has gradually become a rigid requirement of broadband users. It indicates that 80% of carriers' traffic comes from Wi-Fi, most of which is video services. The main consumption scenario is home scenes. This is also a new business model and market opportunity for carriers.

When the broadband rate exceeds 100 Mbit/s, old terminals will become an important factor that severely limits user experience. Globally, more Wi-Fi 6 devices were shipped in 2023 than all other Wi-Fi devices combined. It can deliver near-field Wi-Fi speeds of up to 2 Gbps. Meanwhile, the Wi-Fi 7 standard is expected to be completed in 2024, and the related products and industry chain are expected to mature in 2027.

3.1 FTTH Innovation Forecast

FTTH market and technology development trends have been introduced. Changes in user requirements and technological developments are driving FTTH towards higher bandwidth and better experience. Gigabit fiber broadband has become the mainstream in the advanced broadband market. More than 450 operators have issued Gigabit packets. By 2023, nearly 300 million fixed broadband users worldwide subscribed to packages with speeds of 1 Gbit/s or higher, which takes 19% of the global fixed broadband subscribers. This percentage will increase to 35% by 2026 and 1 Gbps FBB users will become the largest user group. Meanwhile, world-leading operators have also started to deploy the FTTR solution. The FTTR solution brings the advantages of high-speed, low-latency, and low-jitter fiber connections to every corner of the home, providing consumers with a high-quality Wi-Fi connection experience (Table 1).

Table 1. Global Consumer fixed broadband subscriptions by speed, 2022–2028

FBB Subscriptions (unit: million)	2022	2023	2024	2025	2026	2027	2028
Total Subscriptions	1,347	1,429	1,497	1,596	1,621	1,675	1,721
1 Gbps Subscriptions	191	278	362	458	562	662	764
1 Gbps Subscriptions Percentage	14%	19%	24%	29%	35%	40%	44%

The high bandwidth, low latency, stable transmission, and anti-interference features of F5G extend intelligence and digitalization from homes to thousands of industries, including transportation, energy, education, and healthcare, covering more than 40 application scenarios. It has contributed to the implementation and development of various emerging application scenarios.

3.2 Next Generation Home Network Architecture

Although the Wi-Fi generation evolution has provided a very huge improvement space on home network, there are always some limitations on Wi-Fi such as interference, penetration, and coverage. Giga Broadband has already introduced some never-before speed to end user, therefore home network become the last bottleneck in the premium broadband user experience. Throughout the evolution of FBB, optical fibers have been moving closer to end users. From FTTH to Wi-Fi Mesh and then to FTTR, home network technologies have been evolving as shown in Fig. 10. The core concept of the FTTR technology is to extend the fiber coverage to each room of the user to better meet the user's needs. Since 2023, communications standards organizations such as ITU-T, ETSI, and CCSA have successively released FTTR standards based on the C-WAN architecture.

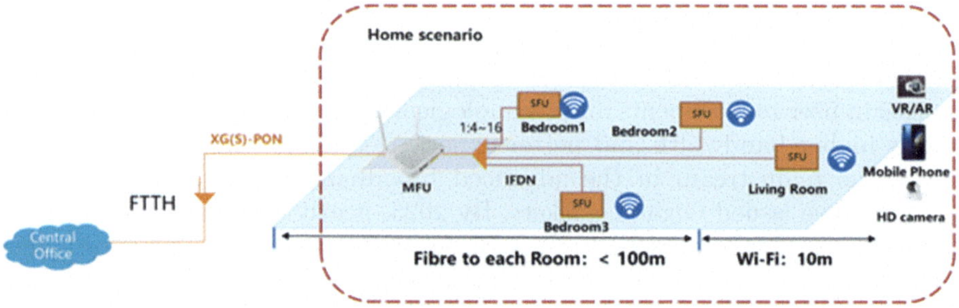

Fig. 10. FTTR Application in Home Scenario

There are four basic components in FTTR network: Main FTTR Unit (MFU), Sub FTTR Unit (SFU), Indoor Fiber Distribution Network (IFDN), and Home Network Management Platform. FTTH solves the last-mile fiber access problem, and FTTR streamlines the last "Golden 10 m" and provides "Gigabit Everywhere" connections. Meanwhile, FTTR is also the best choice for SME to upgrade the network (FTTR-B). There are three directions for FTTR evolution:

- Easier to installation: The installation of FTTR still requires some experienced technicians. Fiber rollout processes and self-service software configuration are always the future.
- Synchronization with latest PON and Wi-Fi Technology: With FTTR, GPON and Wi-Fi 6 are combined for Gigabit home networks. FTTR solutions have evolved with PON and Wi-Fi.
- Intelligent management platform: Hardware-wise, FTTR can always be close to the ultimate technology. Home network evolution always revolves around making home network operations visible, automatic, and intelligent.

4 Conclusion

The digital economy is the engine of economic and social development, and digital information infrastructure is cornerstone of digital economic development. All fiber connectivity, as the base layer of the new generation of digital information infrastructure, provides attraction and motivation for the relevant elements of socio and economic activity in space. As next-generation network architecture, the F5G-A aims to provide an ultra-broadband, simplified, elastic, and green next-generation network architecture based on a full-service optical fiber network. The full-service optical fiber network uses a hierarchical optical fiber planning method and a top-down and bottom-up bidirectional design architecture. Core nodes and aggregation nodes are stable to ensure the stability of the entire optical network architecture. With the access CO node and access backbone optical cable as the anchor points, the optical fiber coverage can be expanded with user footprint expansion. Fibers are moving downwards to homes, rooms, desktops, machines, and everywhere. With the continuous evolution of access technologies such as PON, ODN, and Wi-Fi, network capabilities are continuously improved and deeply combined with digital applications, which will ultimately support the digital transformation of countries.

References

1. Limna, P., Kraiwanit, T., Siripipatthanakul, S.: The growing trend of digital economy: a review article. Int. J. Comput. Sci. Res. **7**, 1351–1361 (2023)
2. Limna, P., Kraiwanit, T., Siripipatthanakul, S.: The growing trend of digital economy: a review article. Int. J. Comput. Sci. Res. **6**, 1–11 (2022)
3. Mickiewicz, B., Volkava, K.: Current trends, assessment and directions of digital economy development. VUZF Rev. **7**(2), 78 (2022)
4. Rosário, A.T., Dias, J.C.: The new digital economy and sustainability: challenges and opportunities. Sustainability **15**(14), 10902 (2023)
5. Igbinenikaro, E., Adewusi, A.O.: Developing international policy guidelines for managing cross border insolvencies in the digital economy. Int. J. Manage. Entrepreneurship Res. **6**(4), 1034–1048 (2024)
6. Shi, Y., Zhang, T., Jiang, Y.: Digital economy, technological innovation and urban resilience. Sustainability **15**(12), 9250 (2023)
7. Skare, M., de Obesso, M.D.L.M., Ribeiro-Navarrete, S.: Digital transformation and European small and medium enterprises (SMEs): a comparative study using digital economy and society index data. Int. J. Inf. Manage. **68**, 102594 (2023)
8. Yu, Q., Ren, J., Yinjin, F., Li, Y., Zhang, W.: Cybertwin: an origin of next generation network architecture. IEEE Wirel. Commun. **26**(6), 111–117 (2019)
9. Goshi, E., Stahl, R., Harkous, H., He, M., Pries, R., Kellerer, W.: PP5GS-an efficient procedure-based and stateless architecture for next generation core networks. IEEE Trans. Netw. Serv. Manage. (2022)
10. Roslyakov, A.: Fiber development index to implement the FTTE concept in F5G networks. In: Optical Technologies for Telecommunications 2021, vol. 12295, pp. 7–12. SPIE (2022)

11. Chauhan, S., Sharma, A., Pandey, S., Rao, K.N., Kumar, P.: IEEE 802.11 be: a review on Wi-Fi 7 use cases. In 2021 9th International Conference on Reliability, Infocom Technologies and Optimization (Trends and Future Directions)(ICRITO), pp. 1–7. IEEE (2021)
12. Galati-Giordano, L., Geraci, G., Carrascosa, M., Bellalta, B.: What will Wi-Fi 8 be? A primer on IEEE 802.11 bn ultra high reliability. arXiv preprint arXiv:2303.10442 (2023)
13. Rosales, R., et al.: Achieving high budget classes in the downstream link of 50G-PON. J. Opt. Commun. Netw. **13**(8), D13–D21 (2021)
14. Arinze, S.N., Onoh, G.N., Abonyi, D.O.: Performance of light fidelity and wireless fidelity networks in a Wlan. Int. J. Res. Eng. Sci. **4**(1) (2020)
15. Jiang, S., Liang, C., Ma, L., Xiong, J., Zhang, W., He, Z.: Ultra-low-loss broadband all-fiber mode selective couplers for MIMO-less MDM transmission. J. Lightwave Technol. **38**(8), 2376–2382 (2020)
16. Zhang, Y., et al.: All-fiber nonvolatile broadband optical switch using an all-optical method. Opt. Lett. **47**(14), 3604–3607 (2022)
17. Rosales, R., Cano, I.N., Nesset, D., Brenot, R., Talli, G.: 50G-PON Downstream Link up to 40 km with a 1342 nm Integrated EML+ SOA. IEEE Photonics Technol. Lett. **34**(6), 306–308 (2022)
18. Liang, W., Wang, H., Wei, J., Zhou, J., Guo, D., Liu, W.: DSP-enabled 50G OOK-PON with beyond 29 dB power budget using O-band 10G DML and 10G APD. Optics Commun. **504**, 127486 (2022)
19. From 25 Gb/s to 50 Gb/s TDM PON: transceiver architectures, their performance, standardization aspects, and cost modeling
20. Shi, W., Tian, Y., Gervais, A.: Scaling capacity of fiber-optic transmission systems via silicon photonics. Nanophotonics **9**(16), 4629–4663 (2020)
21. Kim, T.H.: Analysis of optical communications, fiber optics, sensors and laser applications. J. Mach. Comput **3**(2), 115–125 (2023)
22. DeSanti, C., Liang, D., Guarin, J., Bone, J., Lam, C.F.: Super-PON: an evolution for access networks. J. Opt. Commun. Netw. **12**(10), D66–D77 (2020)
23. Cao, Y., Zhao, Y., Wang, Q., Zhang, J., Ng, S.X., Hanzo, L.: The evolution of quantum key distribution networks: on the road to the qinternet. IEEE Commun. Surv. Tutorials **24**(2), 839–894 (2022)
24. Zhang, D., Liu, D., Xuming, W., Nesset, D.: Progress of ITU-T higher speed passive optical network (50G-PON) standardization. J. Opt. Commun. Netw. **12**(10), D99–D108 (2020)

Bilingualism and Artificial Intelligence: A Sociolinguistic Approach to Arab Heritage

Lobna Hasan Ali(✉) and Neamah Dahash Farha(✉)

University of Baghdad, Baghdad, Iraq
lubna.h@comc.uobaghdad.edu.iq, namaa.d@ircoedu.uobaghdad.edu.iq

Abstract. The purpose of this study is to examine the overlap between bilingualism and artificial intelligence with a specific focus on Arabic speakers. The paper focuses on the challenges that AI systems encounter when processing various Arabic dialects and the notion of code-switching common for most bilingual speakers. The work is based on a qualitative methodology that investigates the performance of AI tools based on semi-structured interviews with 30 bilingual learners and educators and a series of focus group meetings. The results of the study show that AI presents moderate performance in terms of handling Modern Standard Arabic, encountering severe difficulties with online code-switching and regional dialects. As such, about 60% of the study participants report that AI struggled with different dialects, and 70% show that it fought with the mixture of various languages present in the context of code-switching. Nevertheless, many participants agree that AI can be used to save the context of the Middle East and the region's heritage and speculated about the ability of AI-assisted digitalization to ensure the access to rare and complicated texts. The study concludes in recommendations to revise the current format of AI to adjust it for the bicommunal environment and accommodate different types of Arabic dialects. Many participants agree with the necessity to use AI to preserve the Arab heritage; however, the systems must be organized and introduced in such a way as to promote social learning and better access to education.

Keywords: Bilingualism · Artificial Intelligence · Morphology · Auditing · Linguistics

1 Introduction

Among the topics sociolinguists explore is bilingualism, which overlaps with studies in artificial intelligence (AI) development; What is increasingly relevant as AI continues to become an entrenched part of many industries, is how the bilingual brain places in powering these developments. Among the many applications of AI, language technologies offer excellent examples, especially in natural language processing (NLP), where the Arabic language and its linguistically sophisticated structure and dialect-rich nature establish distinctive demands for innovative solutions [1]. It matters that, by being able to complement our current communication tools, translation systems and education methods with AI solutions, we can continue to help in the efforts of building these bilingual

A. Abdelgawad et al. (Eds.): ISBCom 2024, LNNS 1268, pp. 619–628, 2025.
https://doi.org/10.1007/978-3-031-82377-0_50

communities and ensure not only language survival but also proper language transfer. Moreover, AI fluency in multiple languages has technical application in other fields such as machine learning, automated translations and multilingual information retrieval. The advent of AI into the domain of language processing shows that there is a critical for sociolinguistic study of bilingualism in Arab heritage, bringing together linguistic tradition and modern scientific technologies [2].

While AI and NLP progressed, we have enjoyed significant growth, but not enough research has been conducted on how bilingualism as part of everyday life and reality in an Arab context may impact the functionality and performance growth of these new age systems. Many of the studies that have been carried out focus on common global languages, which can vary from English to Chinese, and overlook the diversities of Arabic language. Despite the significant progress in Arabic dialects regarding NLP, how bilingualism affects these systems from a sociolinguistic point of view is less understood. Furthermore, the most accurate AI tools for Arabic are available only with standard Arabic, not dialectal variations which poses a challenging issue in multilingual societies as those frequently change languages [3].

The model needs to fill a gap of traditional Arabic linguistic structures and modern AI systems reintegrating both. Since classical Arabic and the impact of its historical relation to Arabic on AI is not hosted in more than enough depth in our current models, there is a lack of comprehending how much the linguistic roots of these practices will affect existing AI systems [4]. This lack of integration leads to AI models who are not able to properly translate and understand Arabic idiomatic expressions through a more comprehensive Medium-level integration is again required. It is crucial to fill these gaps for improving AI tools serving bilingual Arabic-speaking populations as well as its relevance in wider sociolinguistic studies [5].

Different studies have focused on the application of bilingualism to artificial intelligence, especially in NLP. For example, there are many studies on whether bilingualism modifies cognitive functions as well as the way human language processing can be mimicked by AI models. Research by Grosjean and de Bot are taken as examples have identified that bilingual users have different cognitive facilities in the use of a language than monolingual users; this inherent ability to switch from one language to another can be used as benchmarks for AI improvements. In the case of Arabic, most prior research has discussed the complexities in its morphology and dialectal diversity during NLP tasks. Consider Habash and Sadat who described some of the challenges associated with building NLP systems for Modern Standard Arabic as well as dialectal variations, suggesting that such bilingual speakers switch effortlessly from one variation to another. This further demonstrates the opportunity for AI systems to use bending bilingualism in order language models [6].

Although there is much work on AI and bilingualism, advances in the study of AI and Arabic are limited to such efforts as described earlier, The challenges of processing Arabic dialects and the integration of dialectal variations in natural language processing (NLP) have been addressed by several studies. These contributions provide a foundation for understanding the complexities involved in Arabic language processing, especially within a bilingual context.

K. Almeman and M. Lee, focuse on constructing a multi-dialectal Arabic text corpus using a bootstrapping method to identify dialect-specific words. The authors highlight the importance of developing a comprehensive corpus that accommodates the linguistic diversity of Arabic dialects. Their method improves the representation of regional variations in Arabic, which is crucial for creating NLP models capable of handling different dialects. The work provides a significant contribution by demonstrating how a data-driven approach can help bridge the gap between Modern Standard Arabic (MSA) and various dialects, enabling more accurate NLP applications [7].

S. Khalifa, N. Zalmout, and N. Habash, introduce YAMAMA, a morphological analyzer designed to handle multiple Arabic dialects. The tool integrates a wide array of dialectal variations into a single analysis framework, addressing the challenge of morphological complexity in Arabic. The authors demonstrate that YAMAMA can analyze and generate forms for various dialects, making it a valuable tool for NLP applications that require a deeper understanding of dialectal differences. The study's focus on comprehensive morphological analysis supports the development of more versatile AI systems that can process Arabic more effectively, even in contexts where dialectal variation is prevalent [8].

A. B. Soliman, K. Eissa, and S. R. El-Beltagy, presents AraVec, a collection of pre-trained Arabic word embedding models that are tailored for various Arabic NLP tasks. Soliman et al. focus on creating word representations that capture the semantic relationships between words in both MSA and different Arabic dialects. The models are designed to support a wide range of applications, including sentiment analysis, machine translation, and information retrieval. The authors demonstrate that using word embeddings tailored to the specific linguistic features of Arabic significantly enhances the performance of NLP models. AraVec's ability to provide semantic understanding across dialects and MSA makes it a critical resource for researchers and developers working on Arabic language processing [9].

These studies collectively highlight the challenges and solutions in processing the Arabic language across its various dialects. Almeman and Lee's corpus-building approach offers a foundational resource for handling dialect diversity, while Khalifa et al.'s YAMAMA provides a robust tool for morphological analysis. Soliman et al.'s AraVec further extends the field by providing pre-trained word embeddings that improve the semantic accuracy of NLP tasks. Together, these works contribute to the ongoing effort to enhance the performance of AI systems in understanding and processing Arabic, particularly in multilingual and dialect-rich contexts. They also underscore the importance of developing specialized tools that address the unique linguistic characteristics of Arabic, facilitating better integration of AI in sociolinguistic research and applications.

This study examines bilingualism in the context of Arabic linguistic heritage using AI as a means. Specific objectives include:

1. To study the bilingualism impact on AI for processing Arabic and its dialect.
2. To study how is AI helps to preserve and share Arab linguistic heritage.
3. To bridge this gap of AI failures in bilingualism, especially regarding the sociolinguistics.
4. To offer recommendations to enhance AI in bilingual Arabic communities.

The paper aims to take a deep dive into both the unique position of Arabic linguistic heritage as a language and what it has meant for bilingualism in humans tandemly discussing implications from artificial intelligence (AI) accompanying this complex inheritance. Introduction: The paper starts with some Introductions part which explains the background, literature gapes, and research questions of this NLP for classification task. The Methodology section presents the qualitative design, sampling methods and data collection techniques [10]. The Results section presents the main findings that highlight AI challenges in processing Arabic dialects and bilingualism constitute this section. In the final section, Discussion and Conclusion, implications for practice and policy as well as limitations are discussed in detail based on the findings of this paper, along with suggestions for future research [11].

2 Methodology

Arab underpinning provides the context for such exploration between bilingualism and AI in a qualitative research design. This style of research is ideal for investigating sociolinguistic phenomena within well-defined parameters, because they allow researchers to understand how participants use language from a detailed, qualitative perspective. In the same way a descriptive study gives a detailed insight into how AI can solve problems in Arabic NLP and bilingual language switching between Arabic and non-Arabic languages; This format corresponds to research that puts a greater emphasis on qualitative, sociocultural analyses [12].

We used purposive sampling to select bilingual Arabic speakers who have some experience using AI tools, such as natural language processing systems or translation apps. Purposive sampling is appropriate for this study because it supports selecting people who can give us rich, contextually relevant information from their diverse linguistic and technical profiles. We recruited 30 participants from schools and industry where AI is often used to perform natural language related tasks, in the study. Rather, the work seeks to provide qualitative understanding of what individuals engaging across Arabic and AI platforms experience; in which direction do these challenges map to opportunities and vice versa [13].

The main method of data collection used in the interviews were semi-structured questions. The interviews were related to the perception of AI's ability in heterogeneous Arabic dialects and bilingualism impact on AI performance. Each interview lasted for approximately 45 min and were undertaken over a one-month period. Moreover, focus group discussions (FGD) were employed to gather collective responses in terms of exploration into role of smart AI for language processing. The use of interviews together with focus groups provided an in-depth exploration of the views of participants [14].

Thematic analysis was used to analyse data collected through interviews and focus groups. This is a way of data collection used to identify patterns or themes that reoccur around the problems and things related to integrating bilingualism in AI systems. The analysis involved six analytic stages which are familiarization, coding, theme development, reviewing themes, defining themes and writing the final report. Language switching between Arabic and other languages AI handling of Arabic dialects These areas researchers paid special attention to. Thematic analysis enabled a comprehensive perspective of the strengths and limitations of AI tools in processing their languages among bilingual individuals [15].

3 Results

The following study deals with the challenges of AI in terms of bilingualism and Arabic dialects with special emphasis on the issues related to code-switching and regional varieties. The conclusions are based on the data collected from 30 interviews and focus groups which present the insights on AI performance in the stated situations.

3.1 AI Performance in Handling Dialectal Arabic

The results suggest that existing Ai tools demonstrate about 65% performance in understanding and processing of Arabic dialects. At the same time, about 60% of the participants state that AI performance is rather low due to dialectal variations. In particular, the tool failed to respond when the input text included non-standard forms such as colloquial or regional dialects including Egyptian or Levantine Arabic, as in Table 1.

Table 1. Key strengths and weaknesses of AI as identified through interviews and focus groups.

Response	Respondents (%)
AI processes Modern Standard Arabic (MSA) well	65%
AI struggles with dialectal Arabic	60%

Table 1: Summarizes key strengths and weaknesses of AI as identified through interviews and focus groups, providing a concise reference for understanding the study's outcomes.

This figure illustrates the disparity in AI performance when handling standard Arabic versus various dialects, emphasizing the need for improved NLP models that can accommodate regional variations.

In Fig. 1, the bar chart represents a striking discrepancy between AI's ability to process MSA and its inferior performance when dealing with dialectal Arabic. Although 65% of the surveyed participants agree that AI performs relatively well with MSA, 60% of the examined parties state that the phenomenon is far from being true with dialects. Even though the results can be considered not surprising due to the expected discrepancy, in this case, the discovery is important as the previous research suggests that most AI tools are trained using MSA information. As such, these tools are severely disadvantaged when addressing idiomatic peculiarities of dialectal language. The phenomenon is likely to occur because most of the data used to train AI tools do not contain a substantial amount of colloquial and idiomatic expressions with Arab speakers. Thus, the uncovered gap serves as a justification for the necessity to create an all-inclusive dataset to provide a contrastive analysis for the subscript data. The goal of the dataset is to provide information about both

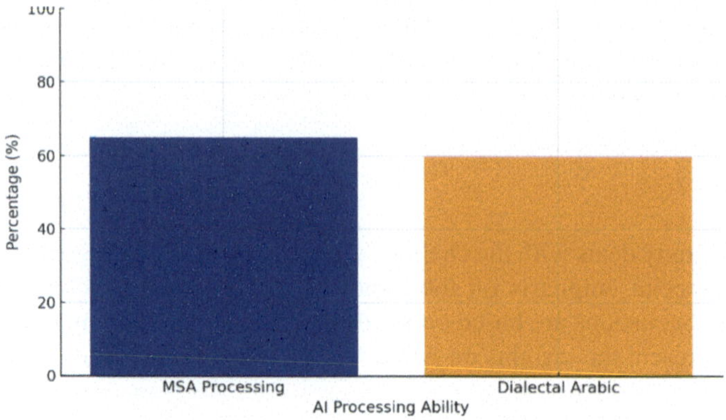

Fig. 1. AI's Performance with Arabic Dialects

MSA and the range of existing dialects, which will help AI tools interpret the context better and engage in a more profound conversation with the speakers from different regions of the Arab world.

3.2 Code-Switching Challenges in AI Processing

A major challenge highlighted by 70% of participants was AI's difficulty in handling code-switching scenarios, where speakers mixed Arabic with English or other languages (Table 2).

Table 2. Code-Switching Challenges in AI Processing

Scenario	Respondents (%)
AI struggles with code-switching	70%
AI performs well with monolingual input	30%

This often resulted in misinterpretation or failure to provide accurate responses, forcing users to simplify their language to be understood by the AI systems.

As shown in Fig. 2, 70% of people who took part in the study found that AI tools do not manage code-switching situations. When bilingual individuals are communicating, especially in a professional, academic, or social setting, they are accustomed to using both Arabic and the English language or any other pair. Therefore, the given challenge implies that the AI tool used is theoretically "restricted" to working with a single language per communication. The finding, however, indicates that currently developed models do not support the easy and natural alternating between two languages, forcing users to simplify their language and avoid mixed phrases that their AI language processor cannot recognize. Thus, the development of AI tools appears reasonable only if it allows for a smooth code-switching experience.

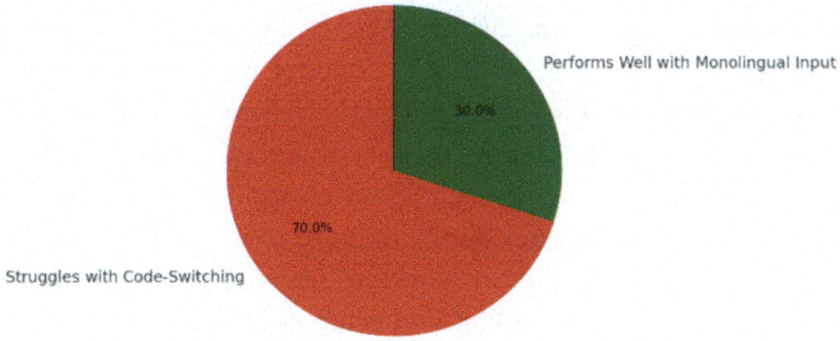

Fig. 2. AI's Performance in Code-Switching Scenarios

3.3 Perceptions of AI in Preserving Arab Heritage

Despite its limitations, 50% of respondents expressed optimism about AI's potential role in preserving Arabic heritage through the digitization of traditional texts. Participants emphasized that with improvements in dialectal comprehension, AI could become a valuable tool for making rare and culturally significant materials more accessible (Table 3).

Table 3. Perceptions of AI in Preserving Arab Heritage

Perception	Respondents (%)
Optimistic about AI's role in heritage preservation	50%
Skeptical about AI's ability to manage dialects	50%

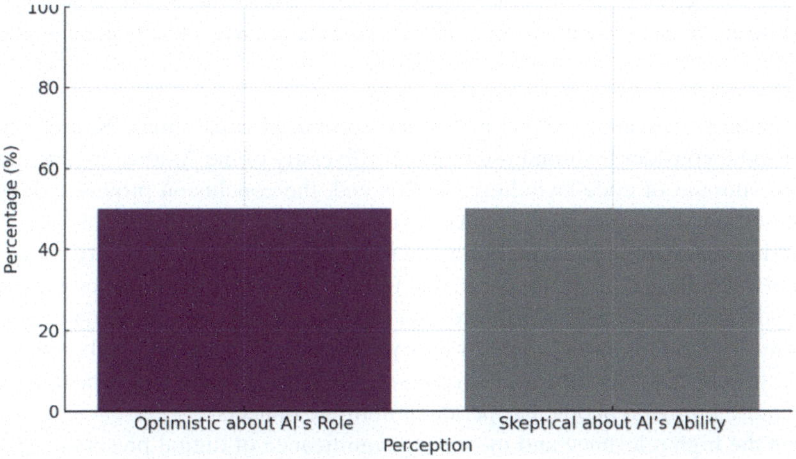

Fig. 3. Perceived Role of AI in Preserving Arabic Heritage

The data in Fig. 3 demonstrates that there are two positions toward the application of AI in preserving Arabic heritage through digitization. To be more exact, 50% of the participants take optimistic views and 50% tend to be skeptical as they have doubts that AI may easily handle the linguistic diversity concept required for such applications. For instance, those respondents who are optimistic believe that AI may be powerful in digitizing classical texts, and transforming them so that they can be read by the up-to-date audience. What is meant here is that ancient classics and other text can be scanned and converted into electronic or traditional books that are more easily read. Moreover, such applications of AI are beneficial since it works to transfer culturally fine masterpieces into the modern forms. Those respondents who are skeptical underline that AI can be limited and the modern systems are not adjusted to the various dialectal sides as well as the specifics of Arabic to be digitized properly. Those who are prospective caution that unless the AI is upgraded and reprogrammed completely, it will perceive too many aspects as one and the same or will simplify the existing linguistic heritage too much.

3.4 Summary of AI's Strengths and Weaknesses

The study provides a mixed view of AI's capabilities in the context of Arabic linguistic diversity. While AI tools demonstrate reasonable proficiency with Modern Standard Arabic, they face substantial challenges in dialectal understanding and code-switching scenarios. However, there remains a notable belief in AI's potential to contribute to the preservation of Arab linguistic heritage (Table 4).

Table 4. AI's Strengths and Weaknesses

Strength	Area for Improvement
Effective with Modern Standard Arabic	Struggles with regional dialects
Potential in heritage preservation	Needs better handling of code-switching
Can digitize and democratize access to texts	Requires larger, more diverse training datasets

The findings indicate a critical gap in the current AI capabilities. Namely, that AI models should consider the immense linguistic diversity of the Arabic language and the significant amount of code-switching. To this end, the conclusion provides significant implications for the future work of AI developers. First, AI training datasets should consider multiple Arabic dialects so that the models could more accurately interpret the switch between these dialects. Second, due to the common practice of code-switching, the AI model should be designed in a way that adjusts and aligns with such patterns. These insights mean that the development of new AI models could significantly enhance the use of AI in such areas as education, profession, or culture in the Arab-speaking world, which require bilingual people to operate in multiple dialects.

Given the high relevance and increasing significance of digital preservation of cultural heritage, it will be prudent to equip the language models with the capacity of entailing advanced semantic understanding to be capable of presenting the full complexity and variety of the Arabic language. Experienced limitations also caution that these systems should not run independently from the community, and their development should consider the values and cultural relations of the Arabic-speaking people.

The expected and experienced implications thus confirm the hypothesis that AI displays considerable promise for a diverse range of applications in the field of the Arabic language, especially its digital preservation, and such practical problems, as the widespread practice of code-switching and numerous dialects, should caution the developers of these systems to work with more diverse training data and work on the dynamic presence of adjusting to code-switching.

4 Conclusion

The study of bilingualism intersected with artificial intelligence (AI) work within the framework of Arab linguistic heritage. The results revealed several issues that highlighted the difficulty and complexity with Arabic dialects and code-switching. AI had difficulty with dialectal Arabic, too and 70% of participants who spoke both Farsi or Dari and English reported struggles using AI tools while switching languages on-the-fly. Nevertheless, AI can save the Arab language, with better tools and applications.

The study stresses the importance of better adapting AI models to address bilateral communication particularly in context of Arabic dialectology, given the shortcomings found. For the region, AI systems will have to keep up with this constant code-switching between Arabic on the one hand and other languages. Deeper integration of various Arabic dialects in future training data could increase both accuracy and cultural comprehension, so it will be interesting to see how this develops. More generally, improvements in bilingual capabilities of AI could improve the quality of translations, education and technology services powered by AI.

While the study is illuminating, it does come with some limitations. For instance, analyzing only 30 participants is not representative of the plethora of different Arabic-speaking bilingual populations that exist around the country and may lead to clear generalization. Also, it was an experiential study and not the real observations about AI tools from a user perspective which can be relatively more subjective. Larger sample size, and more use of quantitative methods will be needed for future studies in order to assess AI capabilities adequately.

This demonstrates a need for future work to develop AI systems tailored specifically to the complexities of bilingualism in Arabic-speaking populations. This includes adding more Arabic dialects to AI models and enhancing systems for real-time code-switching. Moreover, further investigation is warranted to study had AI might support and save languages by digitalizing traditional Arabic texts and examining it helping in preserving the language legacy. Furthermore, broadening to recruit diverse populations will help ensure the AI systems are culturally and linguistically appropriate.

Encouragingly, while our study uncovers several hurdles to fully harnessing AI among bilingual Arabic speakers, it similarly reveals pathways for establishing a more equitable AI ecosystem. Whereas AI has made big leaps in dealing with many major global languages, its traditional approach still lags to handle dialectal Arabic and bilingual communication patterns. However, AI has the possibility to capitalize on a certain avenue for enhancing communication, being agents of modernity and preserving linguistic heritage, and meeting the special requirements of bilingual communities in the Arab world with targeted improvements.

References

1. Habash, N.Y.: Introduction to Arabic natural language processing. Synth. Lect. Hum. Lang. Technol. **3**(1), 1–187 (2010)
2. Sadat, F., Kazemi, F., Farzindar, A.: Automatic identification of Arabic dialects in social media. In: Proceedings of the First International Workshop on Social Media Retrieval and Analysis, pp. 35–40 (2014)
3. Darwish, K., Magdy, W.: Arabic information retrieval. Found. Trends Inf. Retr. **7**(4), 239–342 (2014)
4. Diab, M., Habash, N.: Arabic dialect processing tutorial. In: Proceedings of the NAACL HLT 2012 Tutorial Abstracts, pp. 5–6 (2012)
5. Bouamor, H., Habash, N., Oflazer, K.: A multidialectal parallel corpus of Arabic. In: Proceedings of the Ninth International Conference on Language Resources and Evaluation (LREC 2014), pp. 1240–1245 (2014)
6. Elmahdy, A., Gruhn, R., Minker, W., Abdennadher, S.: Cross-lingual acoustic modeling for dialectal Arabic speech recognition. In: Proceedings of INTERSPEECH, pp. 873–876 (2010)
7. Almeman, K., Lee, M.: Automatic building of Arabic multi dialect text corpora by bootstrapping dialect words. In: Proceedings of the 1st International Conference on Communications, Signal Processing, and their Applications (ICCSPA), pp. 1–6 (2013)
8. Khalifa, S., Zalmout, N., Habash, N.: YAMAMA: yet another multi-dialect Arabic morphological analyzer. In: Proceedings of COLING 2016, The 26th International Conference on Computational Linguistics: System Demonstrations, pp. 223–227 (2016)
9. Soliman, A.B., Eissa, K., El-Beltagy, S.R.: AraVec: A set of Arabic word embedding models for use in Arabic NLP. Procedia Comput. Sci. **117**, 256–265 (2017)
10. Habash, N.Y., Rambow, O., Roth, R.: MADA+TOKAN: a toolkit for Arabic tokenization, diacritization, morphological disambiguation, POS tagging, stemming and lemmatization. In: Proceedings of the Second International Conference on Arabic Language Resources and Tools, 2009, pp. 102–109
11. Maamouri, M., Bies, A., Buckwalter, T., Mekki, W.: The Penn Arabic treebank: building a large-scale annotated Arabic corpus. In: NEMLAR Conference on Arabic Language Resources and Tools, 2004, pp. 466–467
12. Shaalan, K., Raza, H.: NERA: named entity recognition for Arabic. J. Am. Soc. Inform. Sci. Technol. **60**(8), 1652–1663 (2009)
13. Darwish, K., Magdy, W., Mourad, A.: Language processing for Arabic microblog retrieval. In: Proceedings of the 21st ACM International Conference on Information and Knowledge Management, pp. 2427–2430 (2012)
14. Habash, N., Eskander, R., Hawwari, A.: A morphological analyzer for Egyptian Arabic. In: Proceedings of the Twelfth Meeting of the Special Interest Group on Computational Morphology and Phonology, pp. 1–9 (2012)
15. Pasha, A., et al.: MADAMIRA: a fast, comprehensive tool for morphological analysis and disambiguation of Arabic. LREC **14**, 1094–1101 (2014)

Technical Analysis Application to Finance Education for Low-Income Families

Phyllis Okwan[1] and Lynette Jackson[2](✉)

[1] Department of Mathematics, Southern University and A&M College, 801 Harding Blvd, Baton Rouge, LA 70807, USA
phyllis.okwan@sus.edu

[2] Department of Computer Science, Southern University and A&M College, 801 Harding Blvd, Baton Rouge, LA 70807, USA
lynette.jackson@sus.edu

Abstract. The study investigates how technical analysis applications can help low-income and underrepresented minorities finance their postsecondary educational goals without incurring a huge amount of debt at the end of their graduation. The study further examined various investment techniques including stocks, bonds, CDs, mutual funds, Indexed Universal Life (IUL), and index funds utilizing the Rule 72 principles to predict investment growth. The study used rule 72 to determine how long it takes for one to double their initial investment of ten thousand dollars ($10,000.00) in various investment portfolios. The results of the analysis indicate that thirteen years in a high-yield savings account with 5.55% interest rate will double the investments, ten years of bound with 7.28%, CD for eleven years with 7%, mutual fund for five years with 15.78%, IUL for fifteen years with 5%, and index for eight years with 10%. Additionally, the study found that for twenty years, the mutual fund and the index will return $389,797.12, and $108,347.06, respectively. This indicates that with proper financial counseling and better planning, low-income and under-represented families will achieve the financial stability required to accomplish their postsecondary educational degrees.

Keywords: Educational Financing · Low-Income Families · Investment Strategies · Financial Planning · Postsecondary Education · Future Value Calculation

1 Introduction

1.1 Background and Significance

Low-income families face significant challenges in financing higher education, primarily due to the shift from grants to loans. This shift has exacerbated educational disparities, as low-income families struggle to afford rising tuition costs without incurring substantial debt. The increasing reliance on loans rather than grants has created a financial burden that deters many low-income students from pursuing higher education or completing their degrees. According to Chen and DesJardins (2008), over the last decade, there has

A. Abdelgawad et al. (Eds.): ISBCom 2024, LNNS 1268, pp. 629–641, 2025.
https://doi.org/10.1007/978-3-031-82377-0_51

been a dramatic shift in college student funding from grants to loans and a broadening of the gap in educational attainment between students from affluent and poor families. Feeney, and Heroff (2013) also argued that families of low-income and under-represented families are very sensitive to the high cost of tuition and fees published by colleges and universities which may keep low-income and under-represented students from applying to college. Freeney and Heroff further stated in their research that low-income students face significant barriers to college including weak academic preparation, less access to social capital, and knowledgeable adults with information about college. Furthermore, in addition to tuition and fees, low-income students face the cost of books, transportation, and food. These costs add up to more than the cost of tuition and fees. This problem becomes even worse when room and board costs are added to the above-mentioned expenditure. If tuition were free, low-income students could instead use their other aid to pay for more of these additional costs of enrollment (Scott-Clayton, 2017).

1.2 Research Gap

Current literature extensively discusses traditional financial aid methods, such as federal loans, grants, and scholarships, but there is a notable lack of focus on personal financial investments as a means to fund education. This study addresses this gap by exploring how investment strategies can provide a viable alternative for low-income families to finance higher education without excessive debt. For instance, cost-sharing in higher education refers to a shift in the burden of higher education costs from being borne exclusively or predominately by the government, or taxpayers, to being shared with parents and students (Johnstone, 2006) which has significantly affected the parents and families of low-income students in their pursuit of higher education. During the middle part of the earlier decade, institutions saw a real fall in their income from government sources, while students and their families were called upon to at least partly fill this gap through a doubling of tuition fees Finnie, & Usher, 2008).

1.3 Objectives

This study aims to assess the effectiveness of technical indicators in investment strategies tailored for educational financing. Specifically, it investigates which technical indicators can yield better returns on investment, thereby offering a sustainable financial strategy for low-income families to support their educational aspirations.

2 Literature Review

2.1 Technical Analysis Fundamentals

Technical indicators can serve as an alert to study price action more closely. They can also function as a confirmation to other technical analysis tools such as breakout on the price chart and can act as a prediction to direct future prices (Ahmar, 2017). The technical approach to investment is a reflection of the idea that prices move in trends that are determined by the changing attitudes of investors toward a variety of economic, monetary, political, and psychological forces (Park, & Irwin, 2007). According to Edwards, et al., (2018) trading systems are commonly empirically assessed by applying a backtest. However, trading systems are usually a combination of other tools, such as indicators. One needs to be careful when using indicators; indicators are used to sort price action with formulas. They are derivatives and not a direct reflection of the price action. There are hundreds of indicators in the market, and one needs to be careful when choosing indicators for analysis, try choosing indicators that complement each other rather than the ones that are in unison and generate the same signals. According to Vargas, et al., any investor aspires to forecast the market behavior aiming for the best decision when it comes to buying or selling shares of stocks seeking to maximize his profits. This is a challenging task because market behavior is stochastic, volatile, and influenced by many factors such as the global economy, politics, investor expectations, and others (Vargas, et al. 2018). There are leading indicators that are designed to lead price movements. Some of the more popular leading indicators include the Commodity Channel Index (CCI), Momentum, Relative Strength Index (RSI), Stochastic Oscillator, and Williams %R. Many of these come in the form of momentum oscillators which measure the rate of change of a security's price. RSI compares the average price change of the advancing periods with the average of the declining periods.

The role of stock price prediction is categorized in other classifications as Fundamental Analysis, Technical Analysis, and Time Series Forecasting (Agrawal, et al., 2022). The use of time series analysis and forecasting in stock markets holds particular significance for Technical Analysis. This method controls past data to derive indicators that illuminate potential trends in stock prices (Oriani, & Coelho, 2016). The benefits of using leading indicators are: that they generate more signals and allow trade opportunities. It gives an early signal that forewarned a strength or weakness. Lagging indicators are indicators that follow price action; they are also referred to as Trend-following indicators. They are designed to attract traders and keep them in as long as the trend is intact. Analysis helps traders and investors predict market situations and build the gap between the market price and intrinsic value of the stocks by applying the techniques of behavioral economics and statistics (Dongrey, 2022). Some of these indicators are Simple Moving Average (SMA), Exponential Moving Average (EMA), Weighted Moving Average (WMA), Variable Moving Average (VMA), and Moving Average Convergence Divergence (MACD) (de Almeida, et al., 2016). This indicator comprises of two exponential moving averages, which help to measure momentum in the security (Ijegwa, et. al, 2014). The MACD is simply the difference between these two moving averages plotted against a centreline. When the MACD is positive, it signals that the shorter-term moving average is above the longer-term moving average and suggests upward momentum (Ijegwa, et al. 2014). An oscillator is an indicator that fluctuates above and below a centerline or between a set of levels as its value changes over time. Oscillators can be broken down into two types: centered oscillators which fluctuate above and below a center point whereas banded oscillators fluctuate between overbought and oversold extremes. Centered are used to identify the strength or the direction of a momentum behind a move. Banded oscillators are used to identify overbought and oversold conditions. Oscillator signals can alert that the current trend is about to change. Divergence is like oscillator signals; it also serves as a warning that a trend is about to change. There are two types of divergences: positive and negative divergence. Positive divergence occurs when an indicator advances and the security declines and negative divergence occurs when an indicator declines and security increases.

Technical analysis is a process of evaluating securities by analyzing the statistics generated by market activities such as past prices and volume (Dinesh, et al., 2021). There are two types of methods used to analyze securities and make investment decisions: these are fundamental and technical analysis. The fundamental analysis is a method of evaluating securities by examining the characteristics and intrinsic value of a stock (Fayek, et al., 2013). Fundamental analysis also deals with the analysis of the company to estimate its value whereas technical analysis only deals with the price movements in the market; that is demand and supply (Vaidya, 2020). Technical analysis deals with trends and trends that come with support and resistance which is the battle between the bulls and bears or the struggle between buyers (demand) and sellers (supply) (Pagan, & Sossounov, 2003). This occurs when the prices of security move above (resistance) or below (support). The importance of support and resistance trend is that it is used to make trading decisions. Technical analysis also deals with volume, which is the number of shares or contracts that trade over a certain period. Volume is used to confirm trends and chart patterns. If volume starts to decrease in an uptrend, then it is a signal that the upward trend is about to end.

2.2 Stock Charts

A chart is a graphical representation of a series of prices over a period. The lines on the graphs represent the closing price for each day that a stock is traded. There are four main types of charts that investors and traders use based on the information that they are looking for and the skill of the investor or trader. The charts can be line charts, bar charts, candlestick charts, and point and figure charts. A line chart is formed by connecting the lines of the closing prices over a period. A bar chart is made up of a series of vertical lines that represent each data point. These lines represent the high and the low for the trading period. The Candlestick chart is like the bar chart but has a wider bar on the vertical line that shows the period's trading range. The candlestick chart relies heavily on colors to explain the activities during a particular trading period. Last but not least is the point and figure chart which deals with price movement but does not necessarily deal with time and volume in the formation of the points (Fig. 1).

2.3 Chart Patterns

As there is a pattern for everything, stock and trading of stocks also have patterns. A chart pattern is the formation of a stock chart that creates trading signals, either to buy or to sell. Reversal and continuation are two types of chart patterns within technical analysis. The Reversal pattern indicates that the previous pattern will change or reverse upon completion whereas the continuation pattern tells the circle will continue upon completion. Another form of chart pattern is the head and shoulders which signal the security is likely to move against the previous trend. The cup and handle chart deals with a paused upward trend that will continue in an upward direction once the pattern is confirmed. The purpose of these indicators is to alert, confirm, and predict the stock market (Fig. 2).

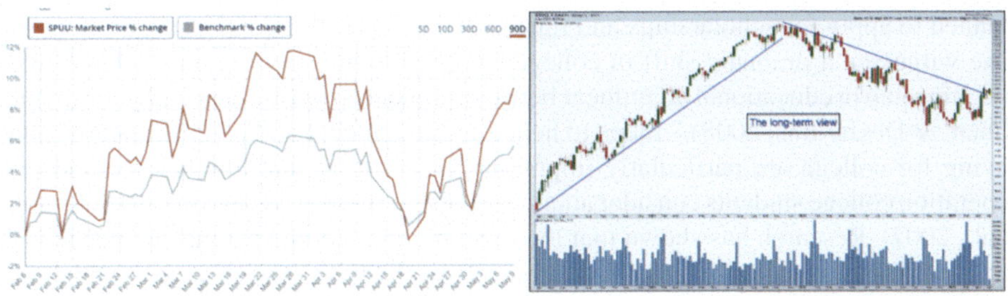

Fig. 1. Left: The orange line indicates the SPUU market price percentage change, while the grey line represents the benchmark percentage change over time. Right: The candlestick chart shows the price movements over time, with green and red bars representing different periods of price increases and decreases, respectively. The bar chart at the bottom indicates the trading volume during these periods. The blue trend lines highlight significant long-term trends in the market.

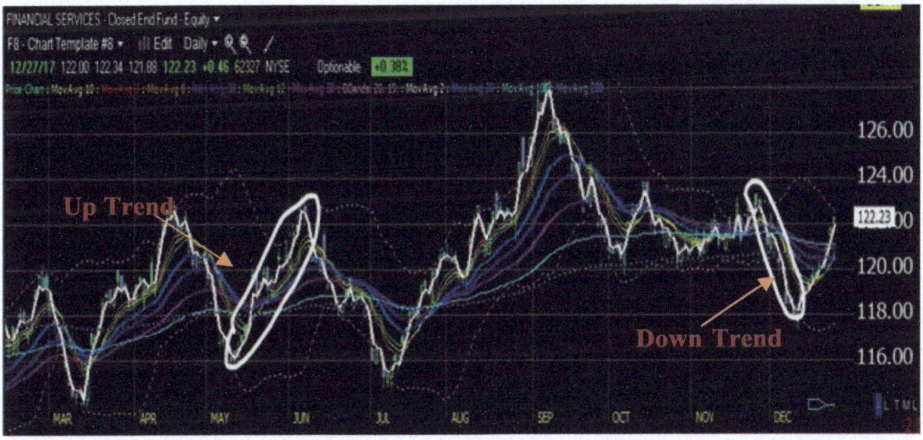

Fig. 2. The chart shows the fund's price reaching a high of approximately 126.00 and a low near 116.00, with the latest price at 122.23 as of December 27, 2017.

2.4 Financial Stability

The study used S&P 500 stocks to come up with the simple idea of investment for low-income families to finance their education, especially in the STEM areas where the country has experienced a deficit for a long time. The United States of America is facing a shortage in the areas of medical professionals, pharmacists, nurses, and educators, especially in the areas of Mathematics and Science due to high academic tuition and exorbitant interest rates on student loans. Early investment in the investment portfolios will give low-income families the financial stability to focus on their education. In this regard, a low-income family can invest in security for their children's college education by using the technical analysis provided in this to predict how much dividend they will earn at the end of a certain number of years. According to Rosa (2006), low-income students' perceptions of college and financial aid information were most relevant to 12th graders planning to attend a 4-year university. Of the 12th-grade respondents, 52%

planned to apply for scholarships and financial aid. "Over the last decade, however, we have witnessed a dramatic shift in college student funding from grants to loans and a widening gap in educational attainment between students from affluent and poor families (Chen, & DesJardins, 2008)." Steps to help alleviate student and parent concerns about paying for college are particularly important for ensuring that low-income and first-generation college students consider attending college (Feeney, & Heroff, 2013; St. John, et al., 2007). Research has shown that low- and middle-income parents are particularly sensitive to published tuition and fee rates, often called the "sticker price," which may keep students from applying to colleges (Feeney, & Heroff, 2013).

According to Castleman & Meyer, (2019), over half of the youth born into the top income quintile earn a bachelor's degree by age twenty-five compared with fewer than 10% of youth born into the bottom income quintile. While college education is a key to upward mobility, low-income students are less likely to earn bachelor's degrees than their more economically advantaged peers (Roksa, & Kinsley, 2019). The main problem affecting low-income students to attend college is the cost, while the benefit of attending college comes in the end, the upfront costs such as tuition, room and board, and fees are some of the factors that hinder low-income students from attending college. Indeed, family resources are the single major source of funds for college, covering 38% of tuition, fees, room, and board charges for full-time undergraduates in 2012–2013 (Scott-Clayton, 2017). To enumerate, higher education in the United States and abroad is facing unprecedented challenges on a wide number of issues including support for financial aid, rapidly increasing tuition rates, diminishing appropriations, modified governance relationships, and a complex and global society that demands college graduates acquire more skills and capacities (Chan, 2016). These issues deter families without substantial income from attending college; therefore, depriving them of opportunities for a better future. Therefore, families realizing these issues and investing in the future of their education or their children's education will reduce the burden of dependence on student loans and the associated high interest rates.

2.5 Career Opportunity

Students who can attend and complete a college degree have higher chances of finding gainful and stable employment than their counterparts with no college degree. According to research, 87% of employers give hiring preference to college graduates who have completed a senior project. While 97% of good jobs created since 2010 have gone to college graduates (Chan, 2016; Carnevale, et al. 2015). Carnevale, et al. (2015) stated in their research that the United States economy has undergone a fundamental transition from an industrial economy centered around high school-educated workers to a post-industrial service economy in which the typical middle-class worker has at least some postsecondary education or training. It was further stated that roughly 20% of the rising demand for college-educated workers derives from a shift in economic consumption from less education-intensive goods production to more education-intensive services – from high school-educated blue-collar workers in industries like manufacturing to college-educated white-collar workers in industries like finance, information technologies, healthcare, business services, education, and government (Carnevale, et al.

2015). To this end, being able to afford a college degree increases one's chances of getting a better and stable job. Without a college degree, youth often get locked out of promising careers (Chan, 2016). For individuals to obtain a scarce resource such as a decent job, they need to have more education than other potential applicants (Horowitz, 2018). According to Mathers' research, every job created during the recovery after the Great Recession required some postsecondary education or training (Mathers, 2019). According to the Bureau of Labor Statistics report (2024), between January and October 2023, 1.1 million 20- to 29-year-olds earned a bachelor's degree; of these, 757,000 (or 70.2%) were employed in October 2023. Among recent bachelor's degree recipients, 75.2% of men were employed, compared with 66.8% of women.

2.6 Increase Student Retention

Often, these low-income family students become at-risk students because of the financial distractions, because of which they cannot focus on their education and subsequently drop out of school. Moreover, the major advantage of this study will be that it will enable low-income students to focus on their education and to graduate on time. Currently, the average graduation rate is about six years according to the US Department of Education. Upon completion, this study will also help increase student retention in the STEM areas and reduce the dropout rate. Fifty-seven percent of the first-year students graduated within 6 years (Millea, et al., 2018). Clement's (2016) research on external and internal barriers to studying can affect student success and retention in a diverse classroom showed that 45.8% of the respondents indicated that they could not focus on their course work (internal barriers), 34.7% indicated that they had family responsibilities that took up their time (external barriers), and 36.1% indicated that they had a job that took up their time (external barriers). With adequate education on financial security, low-income families will be able to invest in their children or their education thereby reducing the external and internal factors that contribute to the drop-out which affects the retention rate. Various research on college retention among under-represented minority and low-income families all talk about the rising cost of education the lack of awareness and the difficulty in maneuvering around the financial aid process that pose challenges to low-income and under-represented students to achieve their educational goals. Often, most low-income, and under-represented students who attend college intending to graduate and secure better financial stability are forced to drop out due to external and internal factors as described in Clement's (2016) research findings. In the past, entering college freshmen were popular among the young adult ranges in age between 18 and 22. However, the trajectory has changed over the years to include older adults, particularly among the 25 to 35 years of age who already have families of their own and juggle jobs in addition to the coursework. The attributes of juggling jobs and families oftentimes cause external and internal struggles for the student therefore dropping out of college. This research focuses on helping low-income families to invest in their education or their children's education which in turn will increase retention and reduce the drop-out rate (Fig. 3).

Retention is a primary concern in the field of adult education. Many adult learners, younger adults (18–24 years old), working-age adults (25–64 years old), and older adults (65 and older), are presumed to obtain at least part-time employment while attending

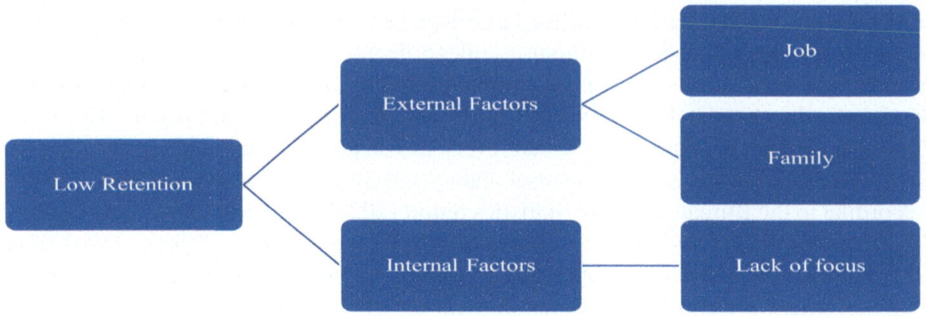

Fig. 3. Barriers Affecting the Retention Rate.

school (Wlodkowski, & Ginsberg, 2017; Spivey, 2016). Britt, and Ammerman (2017) research on student loans, financial stress, and college student retention examined a sample of 2,475 undergraduate students to determine the influence of financial stress, debt loads, and financial counseling on retention rates. Their results indicate, among other findings, that financial stress contributes to an increased likelihood of discontinuing college.

3 Other Types of Investments

Bonds are issued by the government and/or corporations when they want to raise funds for capital projects. When you buy a bond, you are loaning the government or the corporation money; in return, the government or the corporation promises to pay back within a certain period with a set interest. Whereas mutual funds are controlled by investment portfolios that can be purchased by investors. For mutual funds to work, mutual fund managers pool money from investors to invest in various securities. A Certificate of Deposit normally known as a CD is a form of saving that pays a fixed interest for some time. The investors agreed to invest a set amount of money for a period. At the end of the period, the investors can withdraw their money, take the interest accrued, and or reinvest everything for a specified term. There are short- and long-term CDs. An index fund is a type of mutual fund designed to follow certain predetermined rules that can repeat the market performance. In terms of Indexed Universal Life (IUL), it is a life insurance policy that earns interest that the policyholder can take withdrawals from. For the IUL, the company invests the premium in the money market such as stocks.

4 Methodology Overview

The research utilizes S&P 500 stocks and other investment instruments over a period of ten years to analyze potential returns. By applying Rule 72, the study determines the time required to double an initial investment of $10,000 across various investment portfolios, including high-yield savings accounts, bonds, certificates of deposit (CDs), mutual funds, and indexed universal life insurance (IUL). This comprehensive approach aims to provide low-income and under-represented families with practical investment strategies to achieve financial stability and support their educational goals. Under this platform, "a technical indicator which is a series of data points were derived by applying a formula to the price data which is used to predict the price data of a security (Ramadoss, & Muthuvel, 2013). Price data points are simply the up, low, or close of a security over a period. These technical indicators use perspective analysis to price action such as the moving average, and stochastic to fully understand and appreciate the market. The simple moving average calculates the average over security over a specified period whereas stochastic uses a more complex approach for prediction. The company agrees to pay a set minimum interest rate to the policyholder but as money the market booms, so will the rate increase. If the money market dips, the policyholder will receive the set interest rate promised at the time of the policy enrollment.

Therefore, rule 72 is used to determine how long it will take one to double the amount invested. A comparison of these investments was made to determine which of these investment portfolios will double the amount invested and in how many years. For this paper, we define Rule 72 as the formula used to estimate the number of years required to double the amount invested at a set annual interest rate. The formula for Rule 72 is:

$$t = \frac{72}{r} \tag{1}$$

t is the time it takes to double the amount invested.

r is the interest rate per period.

p is the amount invested (Fig. 4).

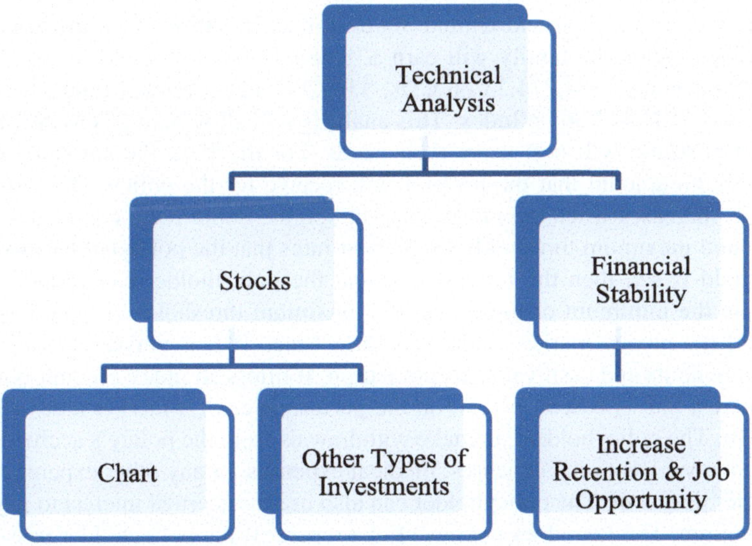

Fig. 4. Hierarchical View of the Proposed Scheme.

5 Result and Discussion

All research on education and degree attainment focuses on federal, state, and private loans, and focuses less on personal financial investments such as stocks, mutual funds, Certificates of Deposit (CD), high-yield savings, and index. This research aims to close the gap in the literature about financial stability when it comes to education and low-income students. With proper investing and investment guidance, low-income and underrepresented students can focus on their education without any external and internal financial barriers. The result of this research shows that a family starting a high-yield saving investment with ten thousand dollars ($10,000) at a rate of 5.55% will double their investment in thirteen years, CD doubles in eleven years with a 7% interest rate per annum, mutual funds double in five years with a 15.78%, IUL doubles in fifteen years with 5% interest rate, and Index doubles in eight years with 10% interest rate. For twenty-five years of investment with $10,000 in mutual funds and index, a family will receive a return on investment of $389,797.12 and $108,347.06 respectively thereby alleviating the financial stress coupled with external and internal factors that hinder one's college educational degree attainment. With proper financial counseling at an early age, low-income families can attain postsecondary degrees with little or no external financial support such as private loans to supplement their income (Table 1)

The analysis of the above indicated that if a family invests ten thousand United States dollars in a high-yield savings account with a 5.55% interest rate per annum, then using rule 72 the investment will double in thirteen years. If the same amount is invested in bonds at 7.28%, the amount will double in ten years, CD at 7% will double in eleven years, mutual funds at 15.78% will double in five years, IUL at 5% will double in fifteen years, and index at 10% will double in eight years. The analysis also shows that if a family invests ten thousand United States dollars in various investment portfolios for twenty-five years the family will earn $38,588.33 for high-yield interest savings, $57,938.75 on bonds, $54,274.33 on CDs, $389,797.12 on mutual funds, $33,863,55 for IULs, and $108,347.06 on Index. This analysis will allow families to decide which investment portfolio will best serve their needs. For the IUL, the amount calculated is the minimum amount that the investor will receive for the policy. The interest rate for the IUL increases when the money market booms. Some IULs companies may set minimum and maximum thresholds for interest rates that the policyholder can receive. If a threshold is set, then the interest rate that the policyholder can receive will not be less than the minimum or more than the maximum threshold set regardless of the situation of the money market or the economy. Since IUL is universal life insurance with death benefits and cash value accumulation. It offers an index account option that credits interest that is based, in part, on the performance of a market index (Slaydon, et al., 2010). The policyholder can make withdrawals from the policy's accrued interest as a loan to pay educational expenses, medical expenses, or any other expenses that the policyholder deems fit. The policyholder can also use the accrued interest to pay for the policy premium. The loan does not have to be repaid. If the policyholder passes away, the beneficiary will receive the phase amount of the policy, plus the accrued interest minus the loan amount. Since IUL is a life insurance policy, the beneficiary will only pay taxes on the accrued interest and not on the whole amount received.

Table 1. Time it Takes for Investments to Double at Various Interest Rates

	r = 5.55%	r = 7.28%	r = 7.00%	r = 15.78%	r = 5%	r = 10%
Year	High-yield Savings	Bonds	CD	Mutual Fund	IUL	Index
0	10,000.00	10,000.00	10,000.00	10,000.00	10,000.00	10,000.00
1	10,555.00	10,728.00	10,700.00	11,578.00	10,500.00	11,000.00
2	11,140.80	11,509.00	11,449.00	13,405.01	11,025.00	12,100.00
3	11,759.12	12,346.85	12,250.43	15,520.32	11,576.25	13,310.00
4	12,411.75	13,245.70	13,107.96	17,969.43	12,155.06	14,641.00
5	13,100.60	14,209.99	14,025.52	20,805.00	12,762.82	16,105.10
6	13,827.68	15,244.48	15,007.30	24,088.03	13,400.96	17,715.61
7	14,595.12	16,354.28	16,057.81	27,889.12	14,071.00	19,487.17
8	15,405.15	17,544.87	17,181.86	32,290.02	14,774.55	21,435.89
9	16,260.13	18,822.13	18,384.59	37,385.39	15,513.28	23,579.48
10	17,162.57	20,192.39	19,671.51	43,284.80	16,288.95	25,937.42
11	18,115.09	21,662.39	21,048.52	50,115.15	17,103.39	28,531.17
12	19,120.48	23,239.41	22,521.92	58,023.32	17,958.56	31,384.28
13	20,181.67	24,931.24	24,098.45	67,179.39	18,856.49	34,522.71
14	21,301.75	26,746.24	25,785.34	77,780.30	19,799.32	37,974.98
15	22,484.00	28,693.36	27,590.32	90,054.04	20,789.28	41,772.48
16	23,731.86	30,782.24	29,521.64	104,264.56	21,828.75	45,949.73
17	25,048.98	33,023.19	31,588.15	120,717.51	22,920.18	50,544.70
18	26,439.20	35,427.28	33,799.32	139,766.73	24,066.19	55,599.17
19	27,906.57	38,006.38	36,165.28	161,821.92	25,269.50	61,159.09
20	29,455.39	40,773.25	38,696.84	187,357.42	26,532.98	67,275.00
21	31,090.16	43,741.54	41,405.62	216,922.42	27,859.63	74,002.50
22	32,815.67	46,925.92	44,304.02	251,152.78	29,252.61	81,402.75
23	34,636.94	50,342.13	47,405.30	290,784.69	30,715.24	89,543.02
24	36,559.29	54,007.04	50,723.67	336,670.52	32,251.00	98,497.33
25	38,588.33	57,938.75	54,274.33	389,797.12	33,863.55	108,347.06

6 Conclusion

The results of this study indicate that integrating technical analysis applications into education will help alleviate the financial burdens that hinder low-income and underrepresented families from attaining postsecondary degrees. Using these investment tools, low-income families can increase their investment decisions by increasing the return on their investments and minimizing the cost associated with tuition, books, room and board, and other educational expenses. The study also underlines the importance of financial independence and peace of mind in financing education. The study illustrates that with the right technical analysis tools, knowledge, and skills, low-income families can navigate the difficulties in financing education without depending on private loans, which often time come with high interest rates. This investment strategy not only helps alleviate financial struggles for low-income families but also promotes equity in education by ensuring financial circumstances do not limit admission to higher education. The study also encourages hands-on financial planning and informed decision-making, allowing low-income families to build workable pathways toward achieving their educational goals. In the long run, incorporating technical analysis applications into educational investment will effectively help alleviate the financial gaps and promote educational equality for low-income families.

References

Agrawal, M., Shukla, P.K., Nair, R., Nayyar, A., Masud, M.: Stock prediction based on technical indicators using deep learning model. Comput. Mater. Continua, **70**(1) (2022)

Ahmar, A.S.: Sutte Indicator: a technical indicator in stock market. Int. J. Econ. Financ. Issues **7**(2), 223–226 (2017)

Britt, S.L., Ammerman, D.A., Barrett, S.F., Jones, S.: Student loans, financial stress, and college student retention. J. Student Financ. Aid **47**(1), 3 (2017)

Bureau of Labor Statistics https://www.bls.gov/. Accessed May 2024

Carnevale, A.P., Rose, S.J.: The economy goes to college: the hidden promise of higher education in the post-industrial service economy. Georgetown Univ. Center Educ. Workforce (2015)

Castleman, B., Meyer, K.: Financial constraints & collegiate student learning: a behavioral economics perspective. Daedalus **148**(4), 195–216 (2019)

Chan, R.Y.: Understanding the purpose of higher education: an analysis of the economic and social benefits for completing a college degree. J. Educ. Policy Plan. Admin. **6**(5), 1–40 (2016)

Chen, R., DesJardins, S.L.: Exploring the effects of financial aid on the gap in student dropout risks by income level. Res. High. Educ. **49**, 1–18 (2008)

Clement, L.: External and internal barriers to studying can affect student success and retention in a diverse classroom. J. Microbiol. Biol. Educ. **17**(3), 351–359 (2016)

de Almeida, R., Reynoso-Meza, G., Steiner, M.T.A.: Multi-objective optimization approach to stock market technical indicators. In: 2016 IEEE Congress on Evolutionary Computation (CEC), pp. 3670–3677. IEEE (2016)

Dinesh, S., Rao, N., Anusha, S.P., Samhitha, R.: Prediction of trends in stock market using moving averages and machine learning. In: 2021 6th International Conference for Convergence in Technology (I2CT), pp. 1–5. IEEE (2021)

Dongrey, S.: Study of market indicators used for technical analysis. Int. J. Eng. Manag. Res. **12**(2), 64–83 (2022)

Edwards, R.D., Magee, J., Bassetti, W.C.: Technical Analysis of Stock Trends. CRC press, Boca Raton (2018)

Fayek, M.B., El-Boghdadi, H.M., Omran, S.M.: Multi-objective optimization of technical stock market indicators using gas. Int. J. Comput. Appl. **68**(20) (2013)

Feeney, M., Heroff, J.: Barriers to need-based financial aid: predictors of timely FAFSA completion among low-income students. J. Student Financ. Aid **43**(2), 2 (2013)

Finnie, R., Usher, A.: The Canadian experiment in cost-sharing and its effects on access to higher education, 1990–2002, pp. 159–187. Springer, Netherlands (2008). https://doi.org/10.1007/1-4020-4660-X_7.pdf

Horowitz, J.: Relative education and the advantage of a college degree. Am. Sociol. Rev. **83**(4), 771–801 (2018)

Ijegwa, A.D., Rebecca, V.O., Olusegun, F., Isaac, O.O.: A predictive stock market technical analysis using fuzzy logic. Comput. Inf. Sci. **7**(3), 1 (2014)

Johnstone, D.B.: Cost sharing in higher education: tuition, financial assistance and accessibility in comparative perspective. In: Financing Higher Education, pp. 1–31. Brill (2006)

Mathers, K.: High-quality career readiness for all students. State Educ. Stand. **19**(3), 4–9 (2019)

Millea, M., Wills, R., Elder, A., Molina, D.: What matters in college student success? Determinants of college retention and graduation rates. Education **138**(4), 309–322 (2018)

Oriani, F.B., Coelho, G.P.: Evaluating the impact of technical indicators on stock forecasting. In: 2016 IEEE Symposium Series on Computational Intelligence (SSCI), pp. 1–8. IEEE (2016)

Pagan, A.R., Sossounov, K.A.: A simple framework for analysing bull and bear markets. J. Appl. Economet. **18**(1), 23–46 (2003)

Park, C.H., Irwin, S.H.: What do we know about the profitability of technical analysis? J. Econ. Surv. **21**(4), 786–826 (2007)

Ramadoss, G., Muthuvel, G.: Application of moving average as technical indicator to predict stock market direction. Int. J. Manag. IT Eng. **3**(4), 167–179 (2013)

Roksa, J., Kinsley, P.: The role of family support in facilitating academic success of low-income students. Res. High. Educ. **60**, 415–436 (2019)

Rosa, M.L.D.L.: Is opportunity knocking? Low-income students' perceptions of college and financial aid. Am. Behav. Sci. **49**(12), 1670–1686 (2006)

Scott-Clayton, J.: Undergraduate financial aid in the United States. American Academy of Arts and Sciences (2017)

Slaydon, J.L., El-Houbi, A.F., Montano, C.B.: Equity index universal life insurance: revealing its hidden advantages in volatile markets. Southwestern Econ. Rev. **37**, 63–74 (2010)

Spivey, T.M.: Examining barriers to retention of adult learners in rural education programs (Doctoral dissertation, Walden University) (2016)

St John, E.P., Hoffman, N., Vargas, J., Venezia, A., Miller, M.: Integrating public finance into strategies for improving preparation, college enrollment, and persistence. Mind. Gap, 269–278 (2007)

Vaidya, R.: Moving Average Convergence-Divergence (MACD) trading rule: an application in Nepalese stock market "NEPSE." Quant. Econ. Manag. Stud. **1**(6), 366–374 (2020)

Vargas, M.R., Dos Anjos, C.E., Bichara, G.L., Evsukoff, A.G.: Deep learning for stock market prediction using technical indicators and financial news articles. In: 2018 International Joint Conference on Neural Networks (IJCNN), pp. 1–8. IEEE

Wlodkowski, R.J., Ginsberg, M.B.: Enhancing Adult Motivation to Learn: A Comprehensive Guide for Instructing all Adults. Wiley, Hoboken (2017)

A Vision Transformer-Based Intelligent System For Brain Tumor Diagnosis

Oluwatunmise Akinniyi[1], Jose Dixon[1], Fahmi Khalifa[1](✉), Gehad A. Saleh[2], and Wael Deabes[3]

[1] Department of Electrical and Computer Engineering, Morgan State University, Baltimore, MD 21251, USA
{olaki58,jodix5,fahmi.khalifa}@morgan.edu

[2] Department of Diagnostic and Interventional Radiology, Faculty of Medicine Mansoura University, Mansoura 35516, Egypt
gehadsaleh@mans.edu.eg

[3] Computational, Engineering, Mathematical Sciences Dept., Texas A&M University, San Antonio, TX 78224, USA
wdeabes@tamusa.edu

Abstract. One crucial first step toward early management is the accurate categorization of brain tumors. Recent years have seen the adoption of artificial intelligence (AI)-based diagnostic technology to automate the process and provide faster, more accurate diagnosis. In order to improve brain tumor categorization, the paper presents an upgraded AI-based architecture. In order to produce an ensemble classifier and a more reliable framework for classifying brain tumors, we introduce a hybrid architecture that combines vision transformer (ViT) with deep neural networks. Preprocessing and data normalization are the first steps in the analysis pipeline. Three feature sets are then taken out for the categorization process. Using a histogram of oriented gradients (HOG), the first module serves as a descriptor for the morphological (structure/shape) attributes. Additionally, local deeper information is extracted from the brain images using convolutional neural networks(CNNs) with optimized design. Lastly, because ViT-derived features are less susceptible to data argumentation and can manage dependencies over longer distances, they are also included. To classify brain MRIs, the retrieved features are fused and a machine learning classifier is employed. Our ensemble system has been tested on four-classes publicly-available MRIs dataset of the brain (7,023 total images). The results, using various metrics, showed that leveraging the benefits of individual modules of our design lead to improved performance using ablation studies.

Keywords: Meningioma · Deep Neural Networks · Glioma · Vision Transformer · Ensemble Fusion

1 Introduction

Brain tumors are abnormal cells of brain that may be either benign (non-cancerous) or malignant (cancerous). According to Lapointe [18], there are many

A. Abdelgawad et al. (Eds.): ISBCom 2024, LNNS 1268, pp. 642–656, 2025.
https://doi.org/10.1007/978-3-031-82377-0_52

types and grades of brain tumors, which are cancers. Type of brain tumor and pathology identification are both factors in multi-classification. Meningiomas (30%), gliomas (16%), and pituitary tumors (14%), in that order, are the most prevalent types of brain tumors. Brain tumors are a form of cancer that comes in a variety of sorts and grades [18]. The pituitary gland and the area around the skull are the sites of meningioma and pituitary tumor growth, respectively. Glioma tumors proliferate across the glial cells in the brain and vary in intensity. As per the National Brain Tumor Society's report [22] on central brain tumors, in 2023, about 90K Americans will receive a primary brain tumor diagnosis, adding to the almost one million Americans who now have brain tumors [27]. Classifying brain tumors is one of the most challenging tasks because of their variety, isointense and hypointense characteristics, and concomitant perilesional edoema. Primary tumors like meningioma and subsequent malignancies are commonly classified using T1-weighted (T1-w) contrast-enhanced magnetic resonance imaging. Mostly, region-based tumor segmentation is used for brain tumor classification instead of feature extraction and classification. Classifying problems using AI-based methods, including deep learning (DL) approaches, became the new standard. The selection of the best treatment choices to prolong patients' lifespan becomes an urgent and difficult matter when brain tumors need to be classified for early diagnosis [12].

Many advanced DL or traditional machine learning (ML) models are used to build efficient approaches for identifying brain cancers. The subsequent section discusses the most recent studies on the identification and categorization of brain tumors. A framework based on multiclass SVM with gray wolf optimization was presented by Kumar et al. [17] for the classification of brain tumors in MRIs. The optimization strategy known as gray wolf outperformed the particle swarm and firefly algorithms, with a classification accuracy of 95.23 %. Ismael and Abdel-Qader [14] combined Gabor filtering with 2D discrete-wavelet transform to categorize brain cancers. Using statistical MRI characteristics, they employed a back propagation multilayer neural network and reported an accuracy of 91.9%. Using a probabilistic neural network, Abir et al. [2] suggested classifying brain cancers. They extracted texture features and performed preprocessing techniques such image filtering, sharpening, scaling, and contrast enhancement. At 83.33%, they managed to achieve the highest accuracy.

In contrast to previous work that extracted handmade characteristics (e.g., [14,17]), a deep neural network model was built for picture classification in self-learning. According to Afshar et al. [4], a neural architecture known as the capsule network (CapsNet) made use of the spatial link between the surrounding non-tumor tissues and the brain lesion. A total accuracy of 90.89% is produced using the CapsNet architecture. Abiwinanda et al. [3] classified brain tumor pictures using five different classification models utilizing a CNN-based DL model. The final architecture featured a maximum pool layer and a ReLU layer. 98.5% training accuracy and 84.19% validation accuracy were reported by the study. To classify the images, Deepak et al. [11] employed the same deep transfer approach and the same data source as [10]. The properties found in the photos were examined and used to create test and classification models. At the patient level, the authors used a 5-fold classification model and achieved 98% accuracy. By employing a pretrained VGG19 and a block-wise fine-tuning technique, Swati et al. [28] were able to attain an overall accuracy of 94.82%. Gumaei et al. [13] created a hybrid feature extraction technique for categorizing brain cancers. With the aid of normalized GIST descriptors and PCA, the feature vector was retrieved. Ultimately, a regularized extreme learning machine with 94.23% accuracy was presented to classify brain cancers. A general convolutional neural network (CNN) approach with six convolutional and max-pooling layers and one full-convolutional (FC) layer was introduced by Anaraki et al. [6]. Classification accuracy for brain tumors was 94% on the dataset that was examined.

Recently, Sharif et al. [25] proposed a pretrained DenseNet201 decision assistance system. Kurtosis-based entropy for feature extraction, meta-heuristics based on modified genetic algorithms (MGA) and high feature values (EKb-HFV) are employed. The BRATS-2018 & -2019 data sets were assessed using a multiclass-SVM-cubic classifier. With regard to glioblastoma (GBM/HGG) and lower-grade glioma (LGG), the accuracy on BRATS-2018 (BRATS-2019) is 99.7% (99.8%) and 98.8% (99.3%), respectively. The primary constraint on this study is the elimination of several significant aspects that affect the system's accuracy. Furthermore, the computing time is increased by the fusion process. A comparative work by Asif et al. [7] is proposed using five CNN architectures. For Xception, DenseNet201, DenseNet121, ResNet152V2, and InceptionResNetV2, they reinforced the last layer with a deep dense block and a softmax layer for output. They obtained an accuracy of 95.87% on the 4-class dataset (which includes healthy patients) and 99.67% on the 3-class dataset using the Figshare dataset (3,064 T1-w MRIs). According to the findings, the proposed deep learning model is the most superior one for multi-class brain tumor classification, and it is built on the Xception architecture. Agrawal et al. [5] created a deep learning model known as MultiFeNet that extracts features using a multi-scale feature architecture. The architecture introduces multi-scaling by substituting the dilation rate for various kernel sizes. Five-fold cross-validation is used in the evaluation of the multiFeNet model in order to attain 96.4% sensitivity, 96.4% F1-score, 96.4% precision, and 96.4% accuracy.

A multi-classification of brain tumor type is proposed by Zulfiqar et al. [29] via fine-tuning of pre-trained EfficientNets. Under various experimental conditions, four versions of modified EfficientNets are trained. Brain tumor MRI sequences are subjected to Grad-CAM imaging of the attention maps of modified EfficientNetB2. Recall/sensitivity, accuracy, precision, and F1-score were reported as 98.86%, 98.65%, 98.77%, and 98.71%, respectively. El-Wahab et al. [1] exploited the advantages of transfer learning and the 1×1 convolution layer to achieve a quick brain tumor classification method. After being trained on the Figshare dataset, it attains an average accuracy of 98.63% using five rounds of transfer learning and 98.86% using retrained five-fold cross-validation, which involves internal transfer learning between the folds. Kibriya et al. [16] presented the feature vector FV, which consists of a deep feature set based on

VGG16 and an ensemble of manually created features based on the gray level co-occurrence matrix (GLCM). The suggested FV is then classified with 99% accuracy using the k-nearest neighbor classifier (KNN) and support vector machines (SVM). In order to efficiently extract features from complex brain structures, Poornam et al. [23] incorporated multiple modules (vision transformer (ViT), split bidirectional feature pyramid network (S-BiFPN), linear transformation module (LTM), and soft-quantization) to improve prediction accuracy on four distinct benchmark brain tumor datasets. The multi-scale spatial features created are then fused using S-BiFPN. This results in classification accuracy of 99.08% for dataset A, 98.97% for dataset B, 98.82% for dataset C, and 99.15% for dataset D. Using SVM, k-NN, extra trees, and MLP as base learners and Logistic Regression (LR) as a meta-learner, Remzan et al. [24]uses two methods: 'Feature Ensemble,' which combines 2-top features (ResNet-50 and DenseNet121), and 'Stacking Ensemble. The first strategy achieved 97.71% accuracy by feeding the extracted features into an MLP classifier, whereas the stacking ensemble approach, which uses DenseNet121 as the ideal feature extractor, achieved 97.40% accuracy. Asiri et al. [8] conducted a thorough assessment of ViT models for brain tumor image classification, contrasting them with current methods. They used five ViT models-R50-ViT-l16, ViT-b16, ViT-l16, ViT-l32, and ViT-b32- for the purpose of detecting and classifying brain tumors. Each model has a unique architecture and parameter count, and ViT-b32 performs exceptionally well, attaining a high accuracy of 98.24%. A study conducted by Jahangir et al. [15] found that both ViT and CNN-based classifiers performed satisfactorily when trained to classify brain tumors, with ViT performing better on train data (100% precision, 100% recall, and 100% f1-score) and CNN-based classifiers performing better on test data (98.233% precision, 98.271% recall, and 98.234% f1-score).

This work expands on the previous research for brain tumor classification, which has been proposed in the literature with encouraging findings. An ensemble architecture is put forth as a method for automated tumor detection and diagnosis. The main contribution of this work is ensemble architecture that integrates deeper and higher-order features extracted from the brain images. The deeper features are extracted using CNN and ViT, and the higher-order features are extracted using HoG and shape features. Also, our architecture extracts features by learned weights fine-tuning on the local data set. Finally, extensive experiments are performed to evaluate the proposed model's performance on local and public datasets.

2 Methodology

The schematic layout of our analysis pipeline for classifying brain tumors is depicted in Fig. 1. It includes several processing steps, including feature extraction, feature fusion, model training and parameter adjustments, preprocessing and normalization, and multi-label classification. Each steps are illustrated in subsequent paragraphs.

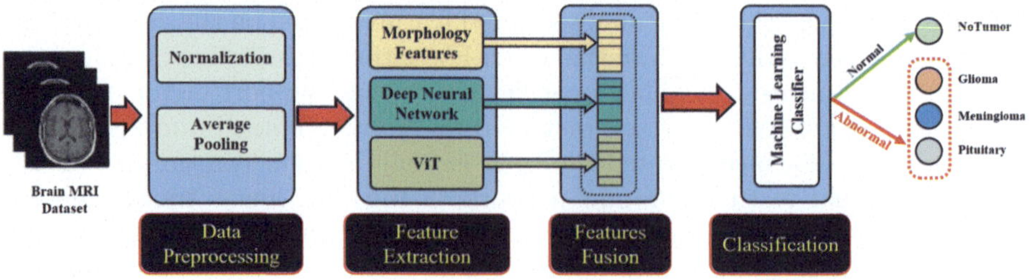

Fig. 1. Flow Diagram of the proposed ViT-based architecture for the classification of brain tumors.

First, data-prepossessing procedures are necessary for deep-learning algorithms to ensure the model's generalization. This step also improves the quality of brain feature extraction, thereby improving the overall analysis results. The data argumentation involves various data augmentations through the Keras ImageDataGenerator. All images are subjected to 15% random rotation, 10% width and height shifts, 20% zoom, 20% vertical or horizontal flips, and 80%-120% brightness adjustments. Then, each image is transformed into a PIL image object using the Pillow (PIL) library. The Numpy arrays with a normalized range of 1 s and 0 s are created from the PIL picture objects. An illustration of the data normalization and enhancement used is provided in Fig. 2.

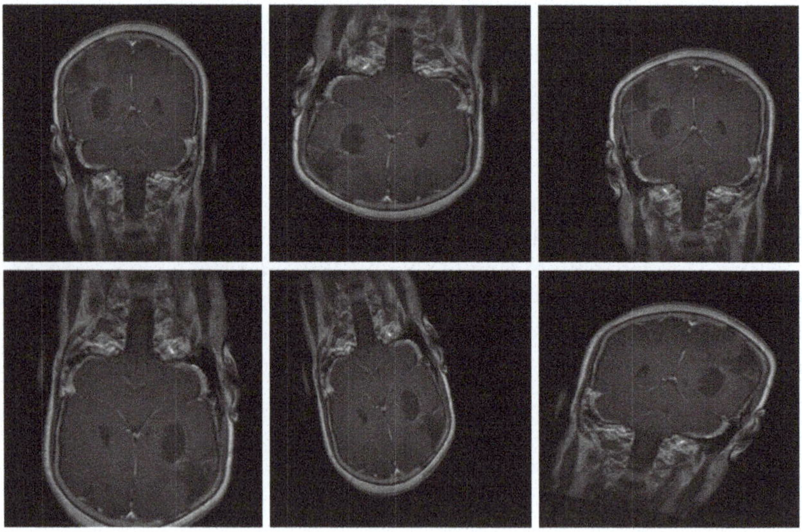

Fig. 2. Samples of the data augmentation process.

Next, average pooling is used to downsample the enhanced augmented images. Average pooling preserves a large portion of the feature structure as opposed to max pooling, which eliminates a significant portion of it [21]. It is defined as:

$$I_{x,y,c} = \frac{1}{k_h k_w} \sum_{m=0}^{k_h-1} \sum_{n=0}^{k_w-1} g_{(x\times s_h+m),(y\times s_w+n),c} \tag{1}$$

where $1/(k_h.kw)$, k_h, and $I(x, y, c)$ stand for the normalization factor, windows of values in the input feature map, and the output value at pixel (x, y) of the cth channel, respectively. The stride values are predicated on s_h and s_w, and the input value at position $(a = i\times s_h+m)$, $(b = j\times s_w+n)$ in the input feature map c is represented by $g(a, b, c)$. The CNN reduces the pictures to 224×224 pixels in size, while the Vision transformer reduces them to 256×256 pixels in size.

To enhance the precision and accuracy of brain tumor classification, the suggested process integrates morphological characteristics with three types of MRI-derived information-rich features: CNN- and ViT-derived features. The medical imaging classification challenges are a well-known domain for these feature-driving algorithms' expertise. An ensemble classifier is created by stacking the layers of these models to develop a more dependable framework for brain tumor classification. This framework is the outcome of integrating the retrieved characteristics using a fusion strategy.

First, the popular feature descriptor technique, histogram of oriented gradients (HOG), is employed to extract important morphological features by analyzing the distribution of edge orientations within an image. Typically, the HOG method involves the computation of the gradient magnitude and orientation for each pixel over image patches (or localized regions or cells) for which local histograms are generated. A feature matrix for those histograms is then used in our architecture for classification. The parameters used for the HOG are as follows: (*orientations*, 9), (*pixels per cell*, 8×8), (*cell per block*, 2×2). Second, we leverage the capability of neural networks, specifically convolutional neural networks (CNNs), which have shown performance gains over classical machine learning techniques because of their innate abilities to automatically extract features and learn from experience. To attain our goal, an in-house CNN architecture is built for faster and easier training process. We chose to develop our own architecture, which is trained from the start to extract significant features for the task at hand by learning directly from the given data, even though transfer learning may be used with pre-trained CNNs. Several experiments were conducted to find the ideal settings for the architecture hyperparameters. To identify the ideal layers and parameters, we specifically used the GridSearchCV [19] technique. Consequently, four convolutional layers, three fully connected layers with dropout (threshold equal to 0.2), and a final four-node that uses softmax to categorize input images make up the network.

CNNs have a number of drawbacks, including a high training data requirement, difficulty with object scale fluctuation, and a restricted capacity for data generalization. As a result, our ensemble design incorporates a ViT model that uses fixed-sized patch embedding to overcome the drawbacks of CNN-based models. If a tumor spans a significant portion of the brain, the ViT can immediately consider the entire tumor region as it attends to all patches at once recognizing the tumor's full extent and its interaction with other brain regions. To achieve the same level of understanding, a CNN would need multiple layers to gradually increase the receptive field, which might dilute important details and make it harder to capture long-range dependencies effectively. Overall, ViT self-attention mechanism, ability to handle global context directly, flexibility in modeling dependencies, and use of positional encoding give it distinct advantages over traditional CNNs in capturing long-range dependencies within images while reducing model complexity by avoiding multiple layers that will be needed in a CNN scenario and reducing computational computation. These characteristics make ViTs particularly suitable for complex image analysis tasks, such as medical imaging for brain tumor diagnosis.

Moreover, the ViT model is introduced as it demonstrates less sensitive to data argumentation compared with CNNs. The proposed model uses a patch encoder and transformer encoder. Linear projection and position embedding are used by the patch encoder for every patch. Normalization of layers, attention from several heads, and feedforward layers make up the transformer encoders. The ViT model is compiled with the appropriate num heads, hidden size, number patches, number layers, and multi-layer perceptron (MLP) head units. The parameters for the ViT model are: learning rate to 0.001, weight decay to 0.0001, number of classes to 4, batch size to 32, epochs to 12, patch size to 16, projection dimension to 64, number of heads to 4, number of transformer layers to 15, 2 transformer units for projection dimension, and hidden size of the transformer encoder at 64.

For classification, the extracted features from the three modules are concatenated together to form predictor variables to fit an MLP. A classifier receives extracted MR features that have first been weighted. To give individual classification or prediction of the input sample (an MR image), three base-feature extractors are used: CNN, hand-crafted, and ViT. Based on the performance of the models, weights were assigned after each feature extractor module underwent independent training. In contrast to conventional fusion techniques that use equal feature weighting, the weighing of the features highlights the influence of different features on tumor classification. By fusing the features, CNNs local features, ViT global features, and edge and texture that are possible from using the HoG Feature Extractor, we improve the model classification performance. Among other ML classifiers, the MLP stands out for its capacity to learn intricate nonlinear patterns [20]. This is achieved by highlighting pertinent parts of the input data and reducing superfluous information.

3 Experimental Results

A dataset of 7,023 images of human brain MRIs, divided into four categories, is used to assess the intended ensemble architecture in this study: glioma tumor, meningioma tumor, pituitary tumor, and no tumor. This dataset combines the following three datasets: figshare, SARTAJ dataset, and Br35H [9]. The images were randomly divided into 5,712 samples (80%) for training and 1,311 samples (20%) for testing. The proposed architecture was implemented using the Python 3.8 and Tensorflow 2.13.0 as the backend. Model evaluation relied on three metrics: Accuracy (ACC), Sensitivity (SEN), and Specificity (SPE) [26].

$$\text{Accuracy} = \frac{TP + TN}{TP + TN + FN + FP} \tag{2}$$

$$\text{Sensitivity} = \frac{Tp}{TP + FN} \tag{3}$$

$$\text{Specificity} = \frac{TN}{TN + FP} \tag{4}$$

where TP(TN) and FP (FN) represent the number of true positives (negatives) and false positives (negatives), respectively (Fig. 3).

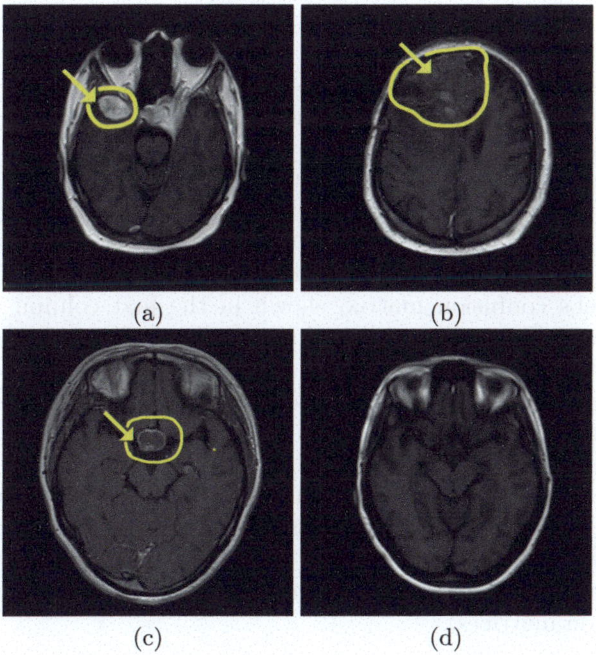

Fig. 3. Sample axial brain T1-w MR images from the collection for (a) meningioma, (b) glioma, (c) pituitary, and (d) no tumor.

Table 1 summarizes the total accuracy using the metrics indicated above. We used ablation methods to clarify the roles played by each ensemble member as we sought to understand the relevance of different MRI-derived characteristics. The researchers looked at three different ablation scenarios: solo, paired, and all three modules. We assessed the impact of three distinct feature sets on paired modules: (1) features derived from ViT and deeper, without texture information; (2) features derived from ViT and deeper, without HoG information; and (3) features generated from ViT and shape, with HoG information, without CNN model. The performance of the ensemble model is compared in Table 1.

Table 1. Overall accuracy of the proposed model and the associate ablation studies. ACC: Accuracy; SEN: Sensitivity, SPE: Specificity, HOG: histogram of oriented gradients; CNN: convolutional neural network; and ViT: vision transformer.

	Evaluation Metrics, (%)		
Method	ACC	SEN	SPE
HOG	91.76	91.07	97.16
CNN	93.75	93.22	97.88
ViT	96.87	96.68	98.96
CNN+HoG	95.85	95.83	95.95
CNN+ViT	97.56	97.42	99.14
ViT+HoG	98.85	98.84	98.95
Proposed	99.91	99.99	99.99

In addition, receiver operating characteristics (ROC) and confusion matrices are used to conduct a second quantitative evaluation of the proposed methods. These tools are useful for comparing and evaluating classification models. To find out which classes were misclassified the most often, one may refer to the particular model's confusion matrix, shown in the first column of Fig. 4. The generated confusion matrix may be used to calculate other indices such as recall, precision, and f1-score. It is common practice to use the confusion matrix in conjunction with related metrics when assessing classification models. The ROC, which offers a graphical depiction of a classifier's performance across different trade-offs between true positives and false positives, is displayed in the second column of Fig. 4. This curve is very useful for evaluating the classifier's ability to discriminate across several classes. The ROC and confusion matrices of the proposed ensemble are shown in Fig. 5. The suggested model outperforms the others in accurately categorizing brain MRIs, as shown by both the ROC scores and the confusion matrices.

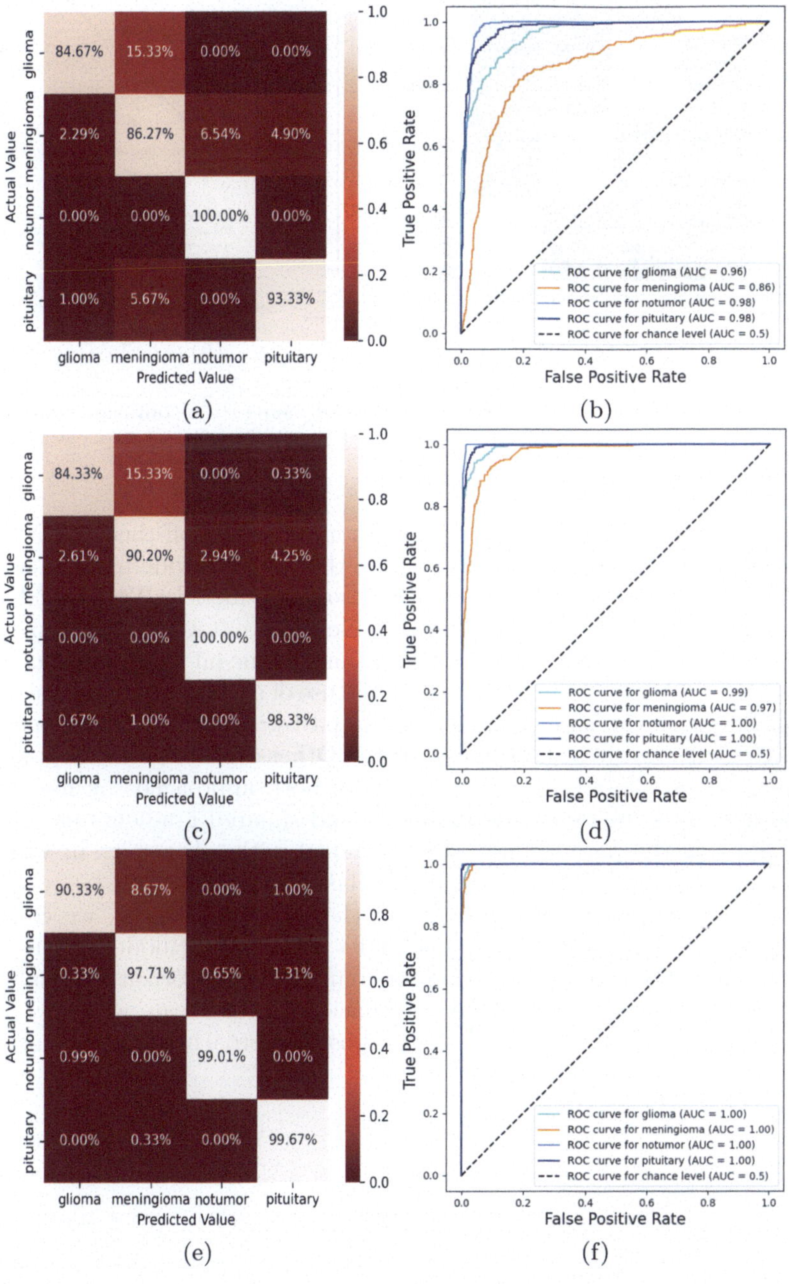

Fig. 4. Model performance demonstrated model using confusion matrix and ROC curves for the individual features (a,b) HOG, (c,d) CNN, and (e,f) ViT

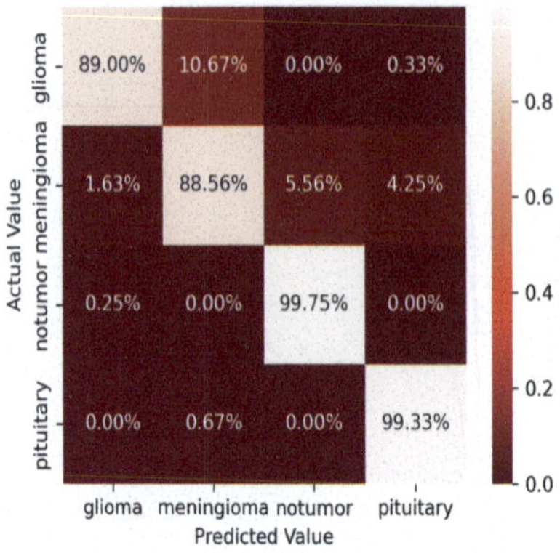

(a)

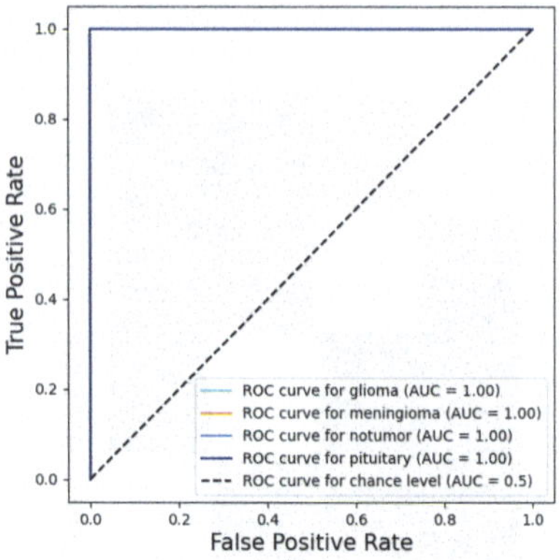

(b)

Fig. 5. Model performance demonstrated model using (a) confusion matrix and (b)ROC curves for the fused features.

To precisely classify the tumor, we additionally tested off-the-self CNNs that implement transfer learning to demonstrate that they provide inferior accuracy and performance metrics over using the proposed model. We applied different pre-trained networks like InceptionV3, Xception, VGG19, and ResNet50 to access the systems best accuracy. Optimization is done on all convolutional neural networks and Vision transformers with the use of adaptive moment estimation (Adam). Based on the knowledge about the past gradients, Adam dynamically changes the learning rates for each parameter. Other gradients can adopt learning rates, but Adam combines the advantages of both approaches by maintaining separate moving averages of the gradients and squared gradients as intended. Table 2 displays the performance of each model with all metrics in which the proposed model reaches the best result to classify tumors.

In order to emphasize the benefits of the suggested model, we concluded by contrasting its accuracy with that of its competitive models. It is known that accuracy is the primary criterion for comparing categorization results As a result, for the quantitative analysis, we considered the mean accuracy. Table 3 displays in terms of accuracy how the proposed approach stands against other approaches.

Table 2. Proposed model evaluation against other CNNs transfer learning. ACC: Accuracy; SEN: Sensitivity, and SPE: Specificity.

	Evaluation Metrics		
Model	ACC	SEN	SPE
Proposed	99.91	99.99	99.99
EfficientNetB0	90.86	96.28%	77.44
EfficientNetV2B0	89.20	88.58	85.28
AlexNet	78.23	72.62	86.22
DenseNet201	97.80	96.97	87.25
GoogleNet	71.43	65.49	74.09
InceptionV3	76.81	56.65	91.55
VGG16	96.06	97.26	91.63
VGG19	95.75	99.64	80.20
Xception	94.75	83.56	91.66
ResNet101	97.93	95.97	92.64

Table 3. The accuracy of classification is compared with the latest models.

	Evaluation Metrics, (%)		
Method	ACC	SEN	SPC
Kumar et al. [17]	95.23	NA	NA
Ismael and Abdel-Qader [14]	91.9	NA	NA
Gumaei et al. [13]	94.23	NA	NA
Anaraki et al. [6]	94.20	NA	NA
Swati et al. [28]	94.82	93.00	94.60
Deepak and Ameer [11]	97.12	NA	NA
Proposed model	99.91	99.99	99.99

4 Conclusions and Future Work

In this study, we have used brain MRIs to construct a hybrid ensemble deep architecture for multi-label tumor classification. The suggested multi-label method combines all three components into a hybrid learning framework to extract deep hidden features and morphological characteristics from the brain images. Based on many ablation investigations, the assessment and experimental findings demonstrated that combining morphological features (extracted using HOG shape descriptor) with deep localized features (extracted using CNN and ViT) produced greater accuracy. Our ensemble method model was shown to be better when compared to other state-of-the-art models and other deep neural networks using transfer learning. The architecture's modular design facilitates scalability to large datasets by enabling the efficient integration of additional data and modules. When testing on a larger dataset, the architecture can adjust by incorporating advanced approaches such as feature weighting, feature fusion, and additional data augmentation strategies. This enables the architecture to effectively handle the increased complexity and variability of the data.

In future, we plan to (1) explore additional powerful neural architectures/modules, such as attention modules, for brain tumor classification; (2) investigate other fusion strategies, e.g., weighted ensemble; and (3) investigate the performance of different ML classifier (e.g., MLP, random forests, etc.). Further, we will also explore utilizing other data sets and evaluation strategies (e.g., cross-validation) and expanding the number of data-set classes to enhance user-friendliness and practical application in the medical field.

Acknowledgments. This work supported by the Center for Equitable Artificial Intelligence and Machine Learning Systems (CEAMLS), Morgan State University, Project #11202202.

Disclosure of Interests. The authors have no competing interests to declare that are relevant to the content of this article.

References

1. Abd El-Wahab, B., Nasr, M., Khamis, S., Ashour, A.S.: Btc-fcnn: fast convolution neural network for multi-class brain tumor classification. Health Inf. Sci. Syst. **11** (2023). https://doi.org/10.1007/s13755-022-00203-w
2. Abir, T.A., Siraji, J.A., Ahmed, E., Khulna, B.: Analysis of a novel MRI based brain tumour classification using probabilistic neural network (PNN). Int. J. Sci. Res. Sci. Eng. Technol. **4**(8), 65–79 (2018)
3. Abiwinanda, N., Hanif, M., Hesaputra, S.T., Handayani, A., Mengko, T.R.: Brain tumor classification using convolutional neural network. In: World Congress on Medical Physics and Biomedical Engineering 2018, pp. 183–189. Springer (2019)
4. Afshar, P., Plataniotis, K.N., Mohammadi, A.: Capsule networks for brain tumor classification based on MRI images and coarse tumor boundaries. In: ICASSP 2019-2019 IEEE International Conference on Acoustics, Speech and Signal Processing (ICASSP), pp. 1368–1372. IEEE (2019)
5. Agrawal, T., Choudhary, P., Shankar, A., Singh, P., Diwakar, M.: Multifenet: multi-scale feature scaling in deep neural network for the brain tumour classification in MRI images. Int. J. Imaging Syst. Technol. **34** (2023). https://doi.org/10.1002/ima.22956
6. Anaraki, A.K., Ayati, M., Kazemi, F.: Magnetic resonance imaging-based brain tumor grades classification and grading via convolutional neural networks and genetic algorithms. Biocybern. Biomed. Eng. **39**(1), 63–74 (2019)
7. Asif, S., Ming, Z., Tang, F., Zhu, Y.: An enhanced deep learning method for multi-class brain tumor classification using deep transfer learning. Multimed. Tools Appl. **82**, 1–28 (2023). https://doi.org/10.1007/s11042-023-14828-w
8. Asiri, A.A., et al.: Advancing brain tumor classification through fine-tuned vision transformers: a comparative study of pre-trained models. Sensors **23**(18) (2023). https://doi.org/10.3390/s23187913, https://www.mdpi.com/1424-8220/23/18/7913

9. Chaki, J., Wozniak, M.: Brain tumor MRI dataset (2023). https://doi.org/10.21227/1jny-g144
10. Cheng, J.: Brain tumor dataset. figshare. Dataset (2017) (2017)
11. Deepak, S., Ameer, P.: Brain tumor classification using deep CNN features via transfer learning. Comput. Biol. Med. **111**, 103345 (2019)
12. Dixon, J., Akinniyi, O., Abdelhamid, A., Saleh, G.A., Rahman, M.M., Khalifa, F.: A hybrid learning-architecture for improved brain tumor recognition. Algorithms **17**(6), 221 (2024)
13. Gumaei, A., Hassan, M.M., Hassan, M.R., Alelaiwi, A., Fortino, G.: A hybrid feature extraction method with regularized extreme learning machine for brain tumor classification. IEEE Access **7**, 36266–36273 (2019)
14. Ismael, M.R., Abdel-Qader, I.: Brain tumor classification via statistical features and back-propagation neural network. In: 2018 IEEE International Conference on Electro/information Technology (EIT), pp. 0252–0257. IEEE (2018)
15. Jahangir, R., Sakib, T., Haque, R., Kamal, M.: A performance analysis of brain tumor classification from MRI images using vision transformers and cnn-based classifiers. In: 2023 26th International Conference on Computer and Information Technology (ICCIT), pp. 1–6 (2023). https://doi.org/10.1109/ICCIT60459.2023.10440978
16. Kibriya, H., Amin, R., Kim, J., Nawaz, M., Gantassi, R.: A novel approach for brain tumor classification using an ensemble of deep and hand-crafted features. Sensors **23**(10) (2023). https://doi.org/10.3390/s23104693, https://www.mdpi.com/1424-8220/23/10/4693
17. Kumar, A., Ansari, M.A.M.H., Ashok, A.: A hybrid framework for brain tumor classification using grey wolf optimization and multi-class support vector machine. Int. J. Recent Technol. Eng. (2019), https://api.semanticscholar.org/CorpusID:204770386
18. Lapointe, S., Perry, A., Butowski, N.A.: Primary brain tumours in adults. Lancet **392**(10145), 432–446 (2018)
19. LaValle, S.M., Branicky, M.S., Lindemann, S.R.: On the relationship between classical grid search and probabilistic roadmaps. Int. J. Robot. Res. **23**(7–8), 673–692 (2004)
20. LeCun, Y., Bengio, Y., Hinton, G.: Deep learning. Nature **521**(7553), 436–444 (2015)
21. Nirthika, R., Manivannan, S., Ramanan, A., Wang, R.: Pooling in convolutional neural networks for medical image analysis: a survey and an empirical study. Neural Comput. Appl. **34**(7), 5321–5347 (2022). https://doi.org/10.1007/s00521-022-06953-8
22. Ostrom, Q.T., Patil, N., Cioffi, G., Waite, K., Kruchko, C., Barnholtz-Sloan, J.S.: Cbtrus statistical report: primary brain and other central nervous system tumors diagnosed in the united states in 2013–2017. Neuro-oncology **22**(Supplement_1), iv1–iv96 (2020)
23. Poornam, S., Angelina, J.: Vitalt: a robust and efficient brain tumor detection system using vision transformer with attention and linear transformation. Neural Comput. Appl. **36**, 1–17 (2024). https://doi.org/10.1007/s00521-023-09306-1
24. Remzan, N., Karim, T., Farchi, A.: Advancing brain tumor classification accuracy through deep learning: harnessing radimagenet pre-trained convolutional neural networks, ensemble learning, and machine learning classifiers on MRI brain images. Multimed. Tools Appl. 1–29 (2024). https://doi.org/10.1007/s11042-024-18780-1

25. Sharif, M., Khan, M., Alhussein, M., Khursheed, K., Raza, M.: A decision support system for multimodal brain tumor classification using deep learning. Complex Intel. Syst. **8** (2021). https://doi.org/10.1007/s40747-021-00321-0
26. Shreffler, J., Huecker, M.R.: Diagnostic testing accuracy: sensitivity, specificity, predictive values and likelihood ratios 2021 (2022)
27. Society, N.B.T.: Brain tumor facts 2023. https://braintumor.org/brain-tumors/about-brain-tumors/brain-tumor-facts/. Accessed 21 Oct 2023
28. Swati, Z.N.K., et al.: Content-based brain tumor retrieval for MR images using transfer learning. IEEE Access **7**, 17809–17822 (2019)
29. Zulfiqar, F., Bajwa, U.I., Mehmood, Y.: Multi-class classification of brain tumor types from MR images using efficientnets. Biomed. Signal Process. Control. **84**, 104777 (2023). https://api.semanticscholar.org/CorpusID:257410424

Automated Physician Scheduling Using Different Meta-Heuristic Algorithms

Meray Ashraf Adib[1](✉) and Mohamed Elsayed Ghonimy[2]

[1] Faculty of Computer Science, October University for Modern Sciences and Arts (MSA), Cairo, Egypt
meray.ashraf@msa.edu.eg

[2] October University for Modern Sciences and Arts (MSA), Cairo, Egypt
melghoneimy@msa.edu.eg

Abstract. Hospital physician scheduling is a difficult and comprehensive procedure. It entails allocating diverse responsibilities to doctors during distinct time windows, including surgical procedures, medical facilities, and handling administrative duties. The manufacturer Specialized skill sets, and scheduling preferences have been assigned by individual physicians. The objective is to design a precisely planned program that optimizes the use of available resources and guarantees the satisfaction of physicians. Here's when optimization methods are useful. Scholars are creating mathematical models that encompass the complexities involved in physician scheduling. Strong algorithms called meta-heuristics can be used to solve these models. Consider these algorithms as sophisticated conductors that investigate every avenue in search of the most harmonic resolution in this example, the ideal schedule. Four such conductors were investigated in this study: Genetic Algorithm (GA), Whale Optimization Algorithm (WOA), Grey Wolf Optimizer (GWO) and Particle Swarm Optimization (PSO). Their goal was to determine which conductor might produce the most optimal physician schedules while maintaining the hospital's smooth operation, taking into account both duty obligations and preferences. Remarkably, the study showed that GA performed better in this particular scenario that the other heuristics in term of objective function.

Keywords: Physician scheduling · Meta-heuristic algorithms · Genetic Algorithm · Particle Swarm Optimization · Whale Optimization Algorithm · Grey Wolf Optimizer

1 Introduction

The COVID-19 epidemic has fundamentally altered the worldwide environment, forcing a rethinking of society norms and aspirations. As we continue to evolve, we must reassess our values and paths. In this dynamic and uncertain environment, hospitals have a host of difficult issues that require efficient answers, one of the most complicated being scheduling. Within the broad subject of hospital

A. Abdelgawad et al. (Eds.): ISBCom 2024, LNNS 1268, pp. 657–670, 2025.
https://doi.org/10.1007/978-3-031-82377-0_53

scheduling, physician scheduling separates out because of its unique set of limitations and requirements. Physician scheduling is complicated by several factors, including the number of working days per week (not all hospitals operate seven days a week), the number of available physicians, each physician's available time (specific days and shifts), their specialties, sitting each physician in his room, and the critical constraint that prevents two physicians with the same specialty from being scheduled for the same shift.

To solve this scheduling problem, I used four different meta-heuristic algorithms: Genetic Algorithm (GA), Particle Swarm Optimization (PSO), Whale Optimization Algorithm (WOA), and Grey Wolf Optimizer (GWO) using the Mealpy Python module. These algorithms were chosen because of their strong ability to deal with difficult optimization environments.

The subject of physician scheduling has been highly explored, with major developments in the field of mixed-integer programming models with linear approximation. These contributions have proved critical in minimizing computational constraints and ensuring fair job distribution among emergency room physicians. Furthermore, layered network models that integrate patient preferences and operational constraints have been created, providing a complete framework for schedule optimization that extends beyond individual healthcare facilities and into whole healthcare systems [6–8].

Lau's thorough examination into the master physician scheduling problem highlights the complex process of optimizing schedules based on shifted roles, specializations, and time limitations [4,5,13]. These fundamental studies are only a tiny part of the essential research efforts targeted at improving our understanding of physician scheduling and researching novel techniques to improve hospital efficiency and patient care.

Intelligent meta-heuristic algorithms, affected by evolutionary theories and natural occurrences, are pushing the boundaries of physician schedule optimization. Notable among them are the Whale Optimization Algorithm (WOA) [2,11], the Genetic Algorithm (GA) [14,17], the Grey Wolf Optimizer (GWO) [12], and Particle Swarm Optimization (PSO) [9,15,18]. These algorithms are useful tools for understanding the complexity of schedule optimization. In depth research [15,18] illustrate the effectiveness of PSO in establishing appropriate physician schedules, controlling resource allocation, and meeting physician preferences and patient expectations. However, the dispute over the "best" algorithm has yet to be resolved. Supporters of GA [14,17] say that it is superior in handling scheduling challenges, even with supporters of GWO [12] and WOA [2,11] stress their distinct advantages in reaching optimal solutions.

In the course of researching, I discovered that the Genetic Algorithm outperformed the other meta-heuristic approaches. Its ability to efficiently explore the solution space and deal with the complicated limitations inherent in physician scheduling makes it the most successful algorithm for this task. This conclusion is consistent with the reasoning indicated in [14,17], enhancing the Genetic Algorithm's status as a strong tool for tackling complex scheduling difficulties in healthcare settings.

The further examination and improving of these meta-heuristic methods is critical for improving our capacity to optimize physician scheduling. By enhancing these approaches, we may increase hospital efficiency, manage physician workload more effectively, and eventually enhance patient care. The continuous research and implementation of these intelligent algorithms will certainly have an important role in the future of healthcare optimization. In conclusion, the best algorithm I found for solving the scheduling problem is the Genetic Algorithm.

2 Related Work

Abdalkareem et al. assert that efficient scheduling is critical for maximizing healthcare delivery [19]. Their analysis focuses on scheduling difficulties like as admissions, personnel, and operating room (psychological) efficiency, highlighting the significance of finding the best solutions for each. The research focuses on improving room schedules, an essential component, by resolving inherent risks in operation deadlines using a flexible uncertainty approach and meta-heuristic algorithms. This seeks to develop a more efficient room schedule by reducing lost time and resources [20].

Sadeghilalimi et al. developed the Whale Optimization Algorithm (WOA) for employment scheduling, based on the natural feeding patterns of humpback whales [2,11]. This method is meant to balance many objectives at the same time while using problem-specific data, showing its potential efficacy despite its computing requirements. WOA represents the searching and extraction stages of whale searching for food, during which whales alternate between exploring for food and improving in on suitable feeding areas. In the context of task scheduling, this translates into a strategy that iteratively refines schedules in response to changing priorities and limitations, with the goal of achieving near-optimal results.

Saif et al. introduced the Grey Wolf Optimizer (GWO) for work scheduling in cloud-fog computing settings [12]. Inspired on grey wolf hunting techniques, GWO intends to improve operational efficiency by dynamically assigning computing jobs across different cloud and fog node networks. This optimizer uses the social structure and cooperative shooting behaviors found in wolf packs to effectively manage resources, resulting in task scheduling that minimizes delay and maximizes resource efficiency. The use of GWO in cloud-fog computing solves the issues provided by changing compute loads and dynamic network situations, providing an algorithmic solution that adjusts to real-time needs.

In [15,18], they used Particle Swarm Optimization (PSO) to schedule emergency department physicians, with the goal of improving staffing levels and minimizing patient wait times through effective job allocation. PSO is similar to the collective behavior of groupings of birds or fish schools, in which individuals (particles) modify their placements in a continuous search space based on their own experience and that of surrounding particles. In the context of physician scheduling, PSO dynamically modifies shift assignments and patient load allocations to balance workloads across healthcare professionals, increasing operational efficiency and improving patient outcomes. Despite its heuristics character, PSO's efficacy in handling complex scheduling difficulties is dependent on parameter modification as well as knowledge gained from operational data. PSO aims to achieve optimal solutions by iteratively updating particle positions and speeds in accordance with defined requirements. Lin and Lo's research emphasizes the flexibility of PSO in healthcare settings, where scheduling choices must take into account changing patient arrivals, physician availability, and clinical resource requirements.

In [14,17] they investigated several scheduling methods in hospital's rooms, comparing heuristic approaches, genetic algorithms (GA), and mathematical models like linear programming. Genetic algorithms, in particular, have proven to be extremely efficient in optimizing operating room schedules by balancing patient requirements and surgical resource use. GA is based on natural selection and genetic principles, with viable solutions emerging repeatedly through selection, crossover, and mutation procedures. GA reviews and improves procedure room schedules based on patient case combine, physician availability, and operating room capacity to enhance throughput while minimizing scheduling conflicts. The use of GA in operating room scheduling demonstrates its capacity to adjust to the numerous limitations and changing priorities that exist in healthcare contexts. By modeling the processes of natural selection evolutionary process, GA consistently improves scheduling decisions over the following generations, with the goal of achieving schedules that maximize surgical resource allocation while fulfilling clinical demand. Rahimi et al.'s comparative study provides the spotlight on the advantages and disadvantages of heuristic techniques such as GA, proving their usefulness in addressing dynamic scheduling difficulties in hospital operating rooms.

In the final analysis, the studies by all the previous authors demonstrate the diverse applications of meta-heuristic algorithms in addressing complex scheduling and optimization challenges in various domains of healthcare management. These algorithms use natural processes and behaviors of animals to provide heuristic solutions that respond to real-world limitations and operational dynamics. While each algorithm provides different advantages and mathematical methodologies, their success is dependent on parameter optimization, problem-specific adjustment, and understandings learned from study results. As healthcare develops, integrating complex algorithms such as WOA, GWO, PSO, and GA into hospital management systems has the potential to increase operational efficiency, improve patient outcomes, and encourage constant creativity in healthcare delivery.

3 Methodology

Physician scheduling is a constant battle against a complex web of constraints. This methodology tackles this challenge by optimizing doctor assignments for various duties within a specific timeframe. Imagine a puzzle where each piece represents a doctor's activity - surgery, clinic hours, administrative tasks - and the slots represent specific times. Now, consider restrictions like limited resources (operating rooms, nurses), individual doctor preferences for certain duties, and pre-scheduled absences. This methodology solves this intricate puzzle by leveraging powerful optimization algorithms like the Genetic Algorithm, Whale Optimization Algorithm, and Particle Swarm Optimization. These algorithms act like intelligent tools, constantly evaluating and refining potential schedules to find the one that best fits all the pieces together. The ultimate goal? An optimized schedule that minimizes wasted resources, fulfills doctor preferences as much as possible, and adheres to pre-existing limitations. This approach goes beyond simply filling slots; it strives to create a balanced and efficient physician schedule that considers both the needs of the hospital and the well-being of its doctors.

3.1 Mathematical Formula

Genetic Algorithm (GA)

The Genetic Algorithm (GA) models natural selection by iteratively choosing, crossing over, and changing a population of solutions to optimize a problem. It is useful for difficult optimization jobs because of its research and extraction capabilities.

Initialization

Create an initial population of N individuals (solutions).

Selection

Select pairs of individuals for reproduction based on their fitness. Let $P(t)$ be the population at generation t.

$$P(t) = \{\mathbf{x}_1(t), \mathbf{x}_2(t), \ldots, \mathbf{x}_N(t)\}$$

Crossover

For each pair of selected parents, apply a crossover operator to produce offspring. For parents $\mathbf{x}_i$ and $\mathbf{x}_j$, the offspring can be:

$$\mathbf{y}_1, \mathbf{y}_2 = \text{crossover}(\mathbf{x}_i, \mathbf{x}_j)$$

Mutation

Apply a mutation operator to each offspring with a certain probability.

$$\mathbf{y}'_k = \text{mutation}(\mathbf{y}_k)$$

Replacement

Form a new population by replacing some or all of the old population with the new offspring.

$$P(t+1) = \text{select\_next\_generation}(P(t), \{\mathbf{y}'_1, \mathbf{y}'_2, \ldots\})$$

Particle Swarm Optimization (PSO)

Particle Swarm Optimization (PSO) is modeled after the social behavior of following birds or schooling fish, in which particles modify their locations and velocities according to personal and global optimal placements. It efficiently converges to optimal solutions with simple velocity and position updates.

Initialization

Initialize a population of particles with positions $\mathbf{X}_i$ and velocities $\mathbf{V}_i$ for $i = 1, 2, \ldots, N$.

Update Velocities

$$\mathbf{V}_i(t+1) = w\mathbf{V}_i(t) + c_1 r_1(\mathbf{P}_i - \mathbf{X}_i(t)) + c_2 r_2(\mathbf{G} - \mathbf{X}_i(t))$$

where w is the inertia weight, c_1 and c_2 are cognitive and social coefficients, r_1 and r_2 are random numbers in $[0, 1]$, $\mathbf{P}_i$ is the personal best position of particle i, and $\mathbf{G}$ is the global best position.

Update Positions

$$\mathbf{X}_i(t+1) = \mathbf{X}_i(t) + \mathbf{V}_i(t+1)$$

Whale Optimization Algorithm (WOA)

Whale Optimization Algorithm (WOA) represents huge whales' bubble-net searching method, with solutions updated depending on related to fish and circular movements. It achieves a suitable balance between looking for and using for optimization issues.

Initialization

Initialize the positions of whales $\mathbf{X}_i$ for $i = 1, 2, \ldots, N$.

Encircling Prey

$$\mathbf{D} = |\mathbf{C} \cdot \mathbf{X}^*(t) - \mathbf{X}(t)|$$

$$\mathbf{X}(t+1) = \mathbf{X}^*(t) - \mathbf{A} \cdot \mathbf{D}$$

where $\mathbf{X}^*$ is the position of the best solution, $\mathbf{A}$ and $\mathbf{C}$ are coefficient vectors:

$$\mathbf{A} = 2\mathbf{a} \cdot \mathbf{r} - \mathbf{a}, \quad \mathbf{C} = 2 \cdot \mathbf{r}$$

and $\mathbf{a}$ decreases linearly from 2 to 0, $\mathbf{r}$ is a random vector in $[0, 1]$.

Spiral Updating Position

$$\mathbf{X}(t+1) = \mathbf{D}' \cdot e^{bl} \cdot \cos(2\pi l) + \mathbf{X}^*(t)$$

where $\mathbf{D}' = |\mathbf{X}^*(t) - \mathbf{X}(t)|$, b is a constant defining the shape of the logarithmic spiral, and l is a random number in $[-1, 1]$.

Search for Prey

$$\mathbf{X}(t+1) = \mathbf{X}_{rand} - \mathbf{A} \cdot \mathbf{D}$$

where $\mathbf{X}_{rand}$ is a random position vector.

Grey Wolf Optimizer (GWO)

Grey Wolf Optimizer (GWO) is motivated by grey wolves' social structure and seeking methods, and it uses alpha, beta, and delta wolf positions to drive its search. It excels at discovering the best answers by performing leadership and shared search behaviors.

Initialization

Initialize the positions of grey wolves $\mathbf{X}_i$ for $i = 1, 2, \ldots, N$.

Hierarchy

Determine the three best solutions: α, β, and δ.

Encircling Prey

$$\mathbf{D}_\alpha = |\mathbf{C}_1 \cdot \mathbf{X}_\alpha - \mathbf{X}|$$

$$\mathbf{X}_1 = \mathbf{X}_\alpha - \mathbf{A}_1 \cdot \mathbf{D}_\alpha$$

$$\mathbf{D}_\beta = |\mathbf{C}_2 \cdot \mathbf{X}_\beta - \mathbf{X}|$$

$$\mathbf{X}_2 = \mathbf{X}_\beta - \mathbf{A}_2 \cdot \mathbf{D}_\beta$$

$$\mathbf{D}_\delta = |\mathbf{C}_3 \cdot \mathbf{X}_\delta - \mathbf{X}|$$

$$\mathbf{X}_3 = \mathbf{X}_\delta - \mathbf{A}_3 \cdot \mathbf{D}_\delta$$

Update the position of the wolves:

$$\mathbf{X}(t+1) = \frac{\mathbf{X}_1 + \mathbf{X}_2 + \mathbf{X}_3}{3}$$

where $\mathbf{A}$ and $\mathbf{C}$ are coefficient vectors:

$$\mathbf{A} = 2\mathbf{a} \cdot \mathbf{r} - \mathbf{a}, \quad \mathbf{C} = 2 \cdot \mathbf{r}$$

and $\mathbf{a}$ decreases linearly from 2 to 0, $\mathbf{r}$ is a random vector in $[0, 1]$.

3.2 Dataset Collection

To comprehensively evaluate the effectiveness of our innovative approach to department layout optimization in healthcare facilities, we meticulously curated a dataset tailored to our research objectives. This dataset was meticulously crafted to encapsulate the dynamic nature of healthcare environments, incorporating a diverse array of parameters to ensure its relevance and applicability. Specifically, we simulated a realistic scenario by generating data encompassing a random number of physicians and duties, reflecting the variability commonly encountered in healthcare settings. Moreover, we integrated essential parameters such as a standard five-day workweek structure, inclusive of two designated off days, and a scheduling framework comprising two daily shifts. This meticulous dataset construction process lays a robust foundation for our research endeavors, enabling us to rigorously analyze the impact and efficacy of our proposed department layout optimization strategies within real-world healthcare settings.

3.3 Data Pre-processing

Prior to delving into the intricate realm of physician scheduling, it is imperative to meticulously delineate the array of constraints and parameters that govern this complex process. The multifaceted nature of scheduling, particularly within the context of physician scheduling, underscores the critical need for a profound understanding of the underlying challenges inherent in this domain. To expedite and optimize this preparatory phase, sophisticated Python tools, including Mealpy [16], alongside other resources obtained from GitHub repositories, have been judiciously employed. These tools, renowned for their efficiency and versatility, play a pivotal role in streamlining the intricate process of data pre-processing. Specifically, Mealpy emerges as a standout solution, offering an array of indispensable functionalities tailored for comparing and evaluating diverse algorithms. These functionalities encompass an extensive range of criteria, spanning from agent behavior and fitness score to execution time and optimal scheduling solutions. By harnessing the capabilities of these advanced tools, we are empowered to conduct a comprehensive analysis, thereby laying a robust foundation for the seamless integration of mathematical modeling equations into the dataset, thus facilitating informed decision-making and strategy formulation in the realm of physician scheduling optimization.

4 Mathematical Programming Models

Following the establishment of a symbolic notation framework consistent with [1], this section delves into the formal mathematical representation of the physician scheduling model. This representation will explicitly define the objective function and various constraints governing the optimal assignment of physician duties within the scheduling framework. The aim is to translate the real-world complexities of physician scheduling into a tractable mathematical formulation, enabling the application of optimization algorithms for efficient solution generation.

Minimize:

$$Z = \sum_{i \in I} U_i \tag{1}$$

Subject to:

$$R1 \times \sum_{i \in I} X_{ijk\ell} \leq C_{jk\ell} \quad \forall j \in J, k \in K, \ell \in L \tag{2}$$

$$\sum_{j \in J} \sum_{k \in K} X_{ijk\ell} \leq C_{jk\ell} \quad \forall i \in I, \ell \in L \tag{3}$$

$$\sum_{\ell \in L} X_{ijk\ell} \leq 1 \quad \forall i \in I, j \in J, k \in K \tag{4}$$

$$\sum_{\ell \in L} X_{ijk\ell} = 0 \quad \forall (i, j, k) \in PRA \tag{5}$$

$$U_i = \sum_{\ell \in L} A_{i\ell} - \sum_{j \in J} \sum_{k \in K} \sum_{\ell \in L} X_{ijk\ell} \quad \forall i \in I \tag{6}$$

$$X_{ijk\ell} \in \{0, 1\} \quad \forall i \in I, j \in J, k \in K, \ell \in L \tag{7}$$

$$U_i \in \mathbb{Z}^+ \quad \forall i \in I \tag{8}$$

The total amount of unplanned tasks that must be reduced in equation $Z = \sum_{i \in I} U_i$. The second restriction is the resource capacity constraint, which states that the total number of resources needed for a shift cannot exceed the entire number of resources available. $R1$ is therefore set to 0 for tasks for which there is no resource cap. Each physician's allotted responsibilities are limited by equation $\sum_{j \in J} \sum_{k \in K} X_{ijk\ell} \leq C_{jk\ell}$ to the extent of his contractual obligations. Equation (4) $\sum_{l \in L} X_{ijk\ell} \leq 1$ guarantees that no doctor may be allocated more than one task during a shift, and Equation $\sum_{l \in L} X_{ijk\ell} = 0$ guarantees that no doctor may be assigned a task during any desired shifts off or days off. The number of unplanned tasks (which should be minimized) is defined by Equation $U_i = \sum_{l \in L} A_{il} - \sum_{j \in J} \sum_{k \in K} \sum_{l \in L} X_{ijk\ell}$. Equation $U_i \in \mathbb{Z}^+$ is the non-negative integrity constraint for the decision variables, whereas Equation $X_{ijkl} \in \{0, 1\}$ applies the 0–1 limitations for the decision variables X_{ijkl}. The U_i Model uses just resource capacity limitations to calculate the lowest possible number of unplanned duties (Fig. 1).

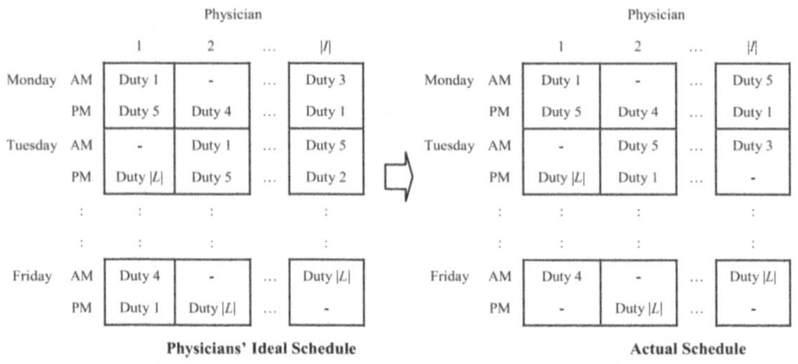

Fig. 1. Detailed Visualization of Physician Shift Assignments and Coverage.

5 Results

After determining the number of physicians, they were assigned to available rooms using the previously indicated limitations that was mentioned in introduction part. I found that the most suitable methodology for tackling this scheduling problem was discovered to be Genetic Algorithm (GA). This algorithm achieved the best level of fitness in the shortest amount of time. As shown in Fig. 2, this is the best scheduling using GA.

Table 1. Comparison of Algorithm performance

Algorithm	Total Cost			Runtime(s)		
	Best	Worst	Average	Best	Worst	Average
GA	889.0	994.0	949.16	0.3	0.7	0.5
WOA	453.0	777.0	635.94	0.5	1.1	0.8
PSO	793.0	847.0	817.7	0.32	0.9	0.6
GWO	508.0	626.0	565.98	0.8	1.5	1.1

6 Discussion

We conducted a detailed evaluation of several meta-heuristic strategies for physician schedule optimization using MealPy, an advanced Python package capable of handling mathematical equations. By methodically studying the collected findings, as depicted in Fig. 2 and reported in Table 1, strong arguments about each algorithm's strengths and weaknesses emerge.

The first bar chart, titled "Fitness Comparison," measures the effectiveness of each algorithm in achieving optimal solutions. The results show that the GA has the best fitness score, indicating that it performs very well in identifying high-quality solutions for physician scheduling. PSO follows closely behind, with a solid performance and a fitness score slightly lower than GA, showing that

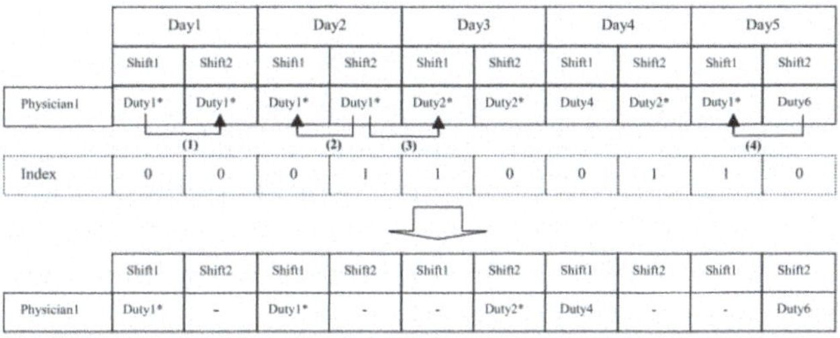

Fig. 2. Result of Physician Scheduling After Optimization with the Genetic Algorithm.

it is effective but not as ideal as GA. The WOA has a moderate fitness score, working successfully but less effectively than GA and PSO. The GWO has the lowest fitness score of the four algorithms, implying that it is the least effective in optimizing the physician scheduling issue.

The second bar chart, named "Cost Comparison," calculates the cost of each algorithm's solution. The results show that the GA has the lowest cost, illustrating that it not only discovers optimum solutions but does so with little related expenditures. The WOA costs more than GA but remains reasonably cheap, indicating a cost-effective technique. PSO has a moderate cost, more than GA and WOA, implying that it is less cost-effective. The GWO has the greatest cost, indicating that it is the least cost-efficient algorithm of the four.

In summary, the comparison proves that the GA outperforms the other algorithms in terms of fitness and cost, making it the most effective and efficient method for addressing the physician scheduling problem. PSO follows with high fitness but greater costs, whereas the WOA strikes a compromise between fitness and cost. The GWO is the least efficient, with the lowest fitness and largest cost. This study underlines the need of implementing effective optimization algorithms for complicated scheduling issues in healthcare in order to get high-quality solutions efficiently (Fig. 3).

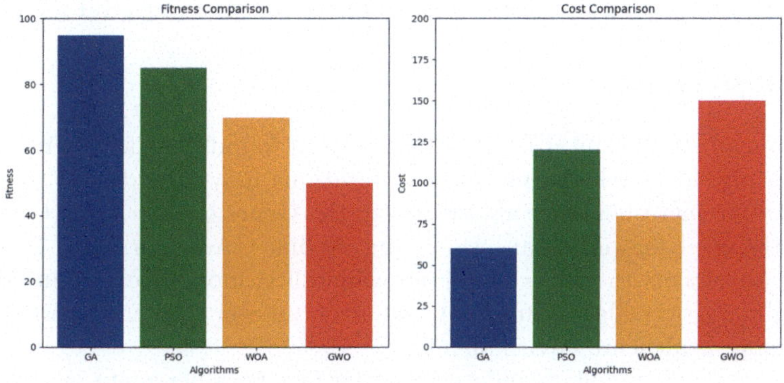

Fig. 3. Example of Algorithms performance evaluation.

7 Conclusion

Implementing the Mealpy code to solve mathematical equations and evaluate meta-heuristic techniques provided important insights into schedule optimization. Improving scheduling, particularly in hospital administration, is crucial for increasing system efficacy. Following the analysis and evaluation of several algorithms, including the GA, PSO, WOA, and GWO, it was determined that GA is the most effective solution for the complex task of allocating physicians to shifts based on availability. GA's capacity to manage complex mathematical issues and adjust to different limitations inside hospital scheduling systems was key to its success. This approach not only makes the best use of hospital resources, but it also reduces scheduling conflicts, which improves overall operational efficiency and patient care.

When related into a hospital management system, the Genetic Algorithm continuously generates feasible options to identify the optimal physician shift allocation. This iterative strategy takes into account a range of factors, such as physician availability, room capacity, and other constraints, to generate timetables that meet both operational and patient care goals. By automating and optimizing this formerly difficult and error-prone procedure, the GA-based solution allows hospital administrators and staff to focus on providing high-quality treatment rather than dealing with scheduling problems.

The use of GA in hospital scheduling is an important step forward in healthcare administration. The algorithm's iterative nature, along with its ability to handle multiple constraints and objectives, makes it perfect for this application. By constantly changing and upgrading prospective schedules, GA ensures that the best option is found while combining the needs of physicians, patients, and administrative requirements. This leads in a more efficient and effective scheduling system that can adapt to the changing environment of a hospital.

Overall, the application of GA for physician scheduling indicates the efficacy of meta-heuristic algorithms in solving difficult healthcare optimization issues. The results of this application illustrate the usefulness of complex mathematical approaches in improving operational efficiency and patient outcomes. As healthcare systems change, the integration of such creative solutions will be critical in meeting the growing expectations and problems faced by hospital managers.

8 Future Work

Looking ahead, the hospital I built offers a good foundation for optimizing and enhancing physician scheduling. Future iterations may dive deeper into studying alternative meta-heuristic algorithms to the Genetic Algorithm (GA), which is currently used for physician scheduling. While GA is currently effective in assigning physicians to rooms based on availability, more research and development might uncover algorithms that give even greater efficiency and flexibility in a changing healthcare environment.

Future research might include comparing GA to other meta-heuristic algorithms such as the Bat Algorithm (BA), Cuckoo Search (CS), Firefly Algorithm

(FA), Harmony Search (HS), Artificial Bee Colony (ABC), Ant Colony Optimization (ACO), and Differential Evolution. Each solution addresses difficult scheduling issues in a unique way, and further study may show which algorithm performs better under varied limitations and scenarios in hospital settings. This comparative analysis may lead to the creation of more specialized algorithms that outperform GA in specific elements of hospital administration.

Another possible avenue for future research is to include the physician scheduling difficulty into the overall hospital management system.

References

1. Master physician scheduling problem. J. Oper. Res. Soc (2013). https://doi.org/10.1057/jors.2012.48
2. Solving the nurse scheduling problem using the whale optimization algorithm. Commun. Comput. Inf. Sci. (2023). https://doi.org/10.1007/978-3-031-34020-8-5
3. Rahimi, I., Gandomi, A.H.: A comprehensive review and analysis of operating room and surgery scheduling. Arch. Comput. Methods Eng. (2020). https://doi.org/10.1007/s11831-020-09432-2
4. A comprehensive review and analysis of operating room and surgery scheduling. archives of computational methods in engineering (2020). https://doi.org/10.1007/s11831-020-09432-2
5. Physician scheduling problem in mobile cabin hospitals of China during Covid-19 outbreak. Ann. Math. Artif. Intell. (2023). https://doi.org/10.1007/s10472-023-09834-5
6. Preference based scheduling in a healthcare provider network. Eur. J. Oper. Res. (2023). https://doi.org/10.1016/j.ejor.2022.09.027
7. A two-stage method of dimensioning and scheduling service providers under patient demand uncertainty. Comput. Intell. Neurosci. (2022). https://doi.org/10.1155/2022/4377142
8. Simulation for dynamic patients scheduling based on many objective optimization and coordinator. Informatica (2024). https://doi.org/10.31449/inf.v48i1.5256
9. A particle swarm optimization approach for physician scheduling in a hospital emergency department (2011). https://doi.org/10.1109/icnc.2011.6022412
10. PGA: a new hybrid PSO and GA method for task scheduling with deadline constraints in distributed computing. Mathematics (2023). https://doi.org/10.3390/math11061548
11. A discrete whale optimization algorithm for the no-wait flow shop scheduling problem. Measurement + Control/measurement and Control (2023). https://doi.org/10.1177/00202940231180622
12. Multi-objective grey wolf optimizer algorithm for task scheduling in cloud-fog computing. IEEE Access (2023). https://doi.org/10.1109/access.2023.3241240
13. A novel genetic algorithm based system for the scheduling of medical treatments. Expert Syst. Appl. (2022). https://doi.org/10.1016/j.eswa.2021.116464
14. RL-GA: a reinforcement learning-based genetic algorithm for electromagnetic detection satellite scheduling problem. Swarm Evol. Comput. (2023). https://doi.org/10.1016/j.swevo.2023.101236
15. AdPSO: adaptive PSO-based task scheduling approach for cloud computing. Sensors (2022). https://doi.org/10.3390/s22030920

16. MEALPY: an open-source library for latest meta-heuristic algorithms in Python. J. Syst. Archit. (2023). https://doi.org/10.1016/j.sysarc.2023.102871
17. A genetic algorithm approach for medical resident scheduling in Austria. Lecture Notes in Networks and Systems (2023). https://doi.org/10.1007/978-3-031-38274-1_27
18. A particle swarm optimization approach for physician scheduling in a hospital emergency department. In: Seventh International Conference on Natural Computation (2011). https://doi.org/10.1109/icnc.2011.6022412
19. Healthcare scheduling in optimization context: a review.: January 1). Healthcare scheduling in optimization context: a review. Health Technol. (2021). https://doi.org/10.1007/s12553-021-00547-5
20. Operating room scheduling optimization based on a fuzzy uncertainty approach and metaheuristic algorithms. Int. J. Eng. Trans. B: Appl. (2022). https://doi.org/10.5829/ije.2022.35.02b.01

Assessing the Effects of Climate Change on Agricultural Productivity

Momen Hesham, Mohamed Essam(✉), Mena Hany, and Ehab E. Zakaria

October University for Modern Sciences Arts - MSA University, Giza, Egypt
{momen.hesham,mohamed.essam25,mena.hany,Ezakaria}@msa.edu.eg

Abstract. Global temperature changes impact on crop production thus illuminating the highly sensitive nature of food security in the global landscape. Utilizing advanced crime models and deep learning solutions for the predictions of crop yields and plant diseases with the help of climatological conditions. The assessment is done using techniques such as Random Forest, Naive Bayes, Extra Tree, VGG16, VGG19, ResNet 101, and the results yielded confirm high levels of correctness and prediction abilities. The methodology integrates data from three distinct datasets: One is for the data on climatic records and soils to be used for crop categorization, the second is for the estimation of crop yield by regression analysis and the third for the classification of images to detect plant leaf diseases. In preprocessing, first the dataset is cleaned, followed by label encoding for categorizing it next normalization is performed and finally data augmentation is provided for handling class imbalance. Specific insights state that for Naive Bayes the highest classification accuracy is estimated to be 99%. 54%, The added features of Extra Trees show the best regression results in terms of R-squared being 0. 99 and Root-Mean-Square Error, RMSE of 8225. In regression scenario, ResNet18 reaches a model correlation coefficient of 93, while ResNet101 reaches a test accuracy of 96% in the image classification task. This paper has presented the need for data pre-processing as well as applying various algorithms and approaches in improving agricultural productivity following concern around climate change. Work that remains for the future includes using the existing models to improve agriculture in several environments, analyzing datasets for a more extensive range of agricultural conditions and incorporating more real-time data to implement models that would lead to the higher levels of sustainable and resilient agricultural practices.

1 Introduction

Global warming is a process that results from different activities including burning of fossil fuel, destruction of forests and industrial emissions. This is deemed as one of the severe threats to the environment of the globe and the health of the people worldwide. The climate described as the dramatic changes in climate such as high temperature changes, changes in the precipitation rate and an increased incidence of storm as clarified by the Intergovernmental Panel on

A. Abdelgawad et al. (Eds.): ISBCom 2024, LNNS 1268, pp. 671–683, 2025.
https://doi.org/10.1007/978-3-031-82377-0_54

Climate Change (IPCC). These adverse climatic conditions highly imperil the global food security since populations within most agricultural-dependent countries are at the brink of facing dire consequences of climate changes [1].

This hails from climate change intensity and volume which are real tests to traditional agricultural mitigation and resilience based on the best practices of the region's ecosystems and what they can offer. Now, it is not only the continuous performance of customary practice that is needed; those customary practices need to be changed into science-based new practice continue to feed the people even if the weather is unfavourable [2].

There have been many efforts made on exploring new technologies, especially in advanced interdisciplinary and applied areas in combating challenge of climate change through green technology. ML and AI can analyze big data for various trends, off-load the task of recognizing mistakes in existing practices to humans, and give optimized recommendations based on climatic factors [3]. They are crucial to building efficient and eco-friendly farming practices which allow the connection of past climate data with current harvest revenue and the generation of models for identifying the seasonal crop yield and future developments in agriculture. Such an understanding helps the policymakers in addressing emerging issues as they evolve [3].

Machine learning is indeed a feasible and versatile approach in the context of agriculture as it provides a general structure for learning about particular crops and improving agricultural pursuits. They can suggest the planting time, a sequence of intercrop, planting density, and ways of decreasing the cost [3]. It also focuses on water management and positively impacts many societies, especially people who live in the desert [2].

Another factor influenced by climate change is the physiological fitness of plants. Diseases, affecting plants, and reducing crop yield each year might become more common and severe in different climates. Optimal conditions in terms of diseases are higher temperatures and relative humidity, and plants under stress can be potential hosts for the disease [4]. Disease diagnosis is crucial in reducing losses in plants and should be initiated as soon as the preliminary symptoms of the disease are noticeable. The use of modern technological approaches like machine learning and computer vision such as image-based plant disease diagnostic tool has been seen to work [3]. Convolutional Neural Networks (CNNs) have a potential accuracy when it comes to identifying plant diseases so that proper corrective measures can be taken quickly [3].

This research uses a research paradigm that incorporates scientific and technological developments to enhance climate change adaptation in agriculture. It utilizes three distinct datasets: One is for analyzing the climatic and soil data to classify the crop types, second one is for the regression analysis of crop yield and third one is for the image classification of the plant leaf diseases. Thus, through employing these methodologies, the study seeks to improve the knowledge on impact of climate change on agriculture and the need to come up with methods of improving crop productivity and quality.

2 Related Work

2.1 Crop Recommendation

[5] involves the use of Decision Trees, Naive Bayes, and Support Vector Machines for the development of crop advice based on the complete soil nutrients information and environmental conditions. This approach is based on complex algorithms and enables farmers to choose their crops with the highest increases in productivity. Nonetheless, the experiment's main obstacle is the specialty and control of soil conditions to which the solution may not be easily applicable in wider and more complicated agricultural conditions. Consequently, the model can be problematic for usage in places with soil parameters that are very different from the characteristics of the database.

[6] This research focuses on using the PART algorithm, the JRip method, and a Decision Table generator in order to give recommendations to farmers on crop selection in terms of the soil content and the prevailing weather conditions. The assessment strategy of the study is based on accuracy, precision, recall, and F-measure for the evaluation of model performance, and it uses a moderately small dataset from the Kaggle platform. One of the major shortcomings of this methodology is that the case study may be limited in which just a part of the world condition of agriculture is represented resulting in a potentially limited generalizability of suggestions.

[7] This research employs machine learning for making crop recommendations in real time based on immediate analysis of soil health and weather conditions. It guarantees quick analysis that could be altered to respond to new situations. On the other side, the success of this real-time analysis relies on the availability of advanced data collection infrastructures, which may not be prevalent in all agricultural regions, thus restricting the implementation of the system.

2.2 Leaf Disease Detection

(Revathi and Hemalatha,2012) present a methodology employing image edge detection techniques for the identification of leaf spot disease in cotton plants. This approach leverages advanced computing algorithms to accurately delineate the edges of disease-affected areas in leaf images, thereby facilitating precise disease diagnosis. By focusing on edge detection, the method enhances the visibility of diseased regions, which is critical for early and accurate detection. However, the main challenge lies in the algorithm's sensitivity to variations in image quality and environmental conditions, which might limit its effectiveness in real-world agricultural settings. Consequently, while promising, this model's practical application may be constrained by the need for controlled imaging conditions that might not always be feasible in diverse agricultural environments [8].

(Sardogan, Tuncer, and Ozen,2018) propose a novel method combining Convolutional Neural Networks (CNN) with the Learning Vector Quantization (LVQ) algorithm for the detection and classification of plant leaf diseases. The integration of CNNs allows for effective feature extraction from leaf images, while the LVQ algorithm enhances the classification accuracy by optimizing the feature vector mappings. This dual approach ensures a robust and accurate system for identifying various plant diseases, significantly aiding in early diagnosis and treatment. Despite its high accuracy, the method's performance is heavily reliant on the quality and diversity of the training data, which may pose challenges in real-world applications where leaf images can vary greatly. Consequently, while the model demonstrates high efficacy in controlled environments, its practical implementation may be limited by the need for extensive and representative datasets [9].

(Reddy & Rekha, 2021) introduce the Deep Leaf Disease Prediction Framework (DLDPF), which utilizes transfer learning for the automatic detection of leaf diseases. By leveraging pre-trained deep learning models, this framework enhances the accuracy and efficiency of disease identification, even with limited labeled data. The transfer learning approach allows the model to apply knowledge from large, diverse datasets to specific leaf disease detection tasks, improving its robustness and generalization capabilities. However, the main limitation of this framework is its dependency on high-quality, annotated datasets for initial training, which may not always be available in practical agricultural settings. As a result, while DLDPF shows significant promise in controlled experiments, its real-world applicability may be restricted by the availability of suitable training data and the computational resources required for implementing deep learning models [10].

2.3 Crop Productivity

(Patki and Wazurkar,2021) explore the use of data sampling techniques combined with Stochastic Gradient Descent (SGD) neural networks for predicting crop yields. By applying data sampling methods, they aim to enhance the training process of neural networks, leading to more accurate and reliable yield predictions. The integration of SGD helps in optimizing the model parameters efficiently, ensuring faster convergence and improved performance. Despite these advantages, the approach faces challenges related to the variability and quality of the data samples, which can impact the model's accuracy. Additionally, the effectiveness of the SGD neural networks is contingent on the selection of appropriate hyperparameters, which can be a complex and time-consuming process. Therefore, while the method shows potential in experimental settings, its practical application in diverse agricultural environments may require careful consideration of data quality and model tuning [11].

(Geetha et al., 2022) present a deep learning-based model specifically designed to predict the yield of paddy in the Cauvery Delta region. This model utilizes advanced deep learning techniques to analyze various agricultural and environmental factors that influence paddy yield. By processing large datasets, the model aims to provide accurate yield predictions, aiding farmers in making informed decisions about crop management. One of the main advantages of this approach is its ability to handle complex, non-linear relationships between different factors affecting crop yield. However, the model's performance is heavily dependent on the quality and comprehensiveness of the input data, which can be challenging to obtain in diverse farming conditions. Additionally, the implementation of deep learning models requires substantial computational resources, which may limit its accessibility for smaller farming operations. Despite these challenges, the model shows significant promise in enhancing yield prediction accuracy and supporting sustainable agricultural practices in the Cauvery Delta region [12].

Table 1 shows previous work methodologies and accuracy gained.

Table 1. Previous work results

Paper Ref	Methods	Accuracy
[5]	Decision Tree	90
	NaiveBayes	99.09
	SVM	10
	LR	95
	Random forest	99.09
	XGB	**99.31**
[6]	JRIP	96.87
	Decision tabel	87.42
	PART	**98.3**
[7]	Decision tree	78.53
	naive bayes	82.42
	Random Forest	**96.43**
[8]	k-means	Not mentioned
	NN	
[9]	CNN with LVQ	86
[10]	**CNN**	97.62
	Alex net	91.19
[11]	gradient decent	89
[12]	LSTM	84

3 Methodology

This section details the methodologies used in our project, which are divided into three main sections: interpretations include classification, regression, and image classification. Each section comprises two subsections: configuration of data preprocessing and models implemented and after this the ecaluation metrics used for all the three sections (Fig. 1).

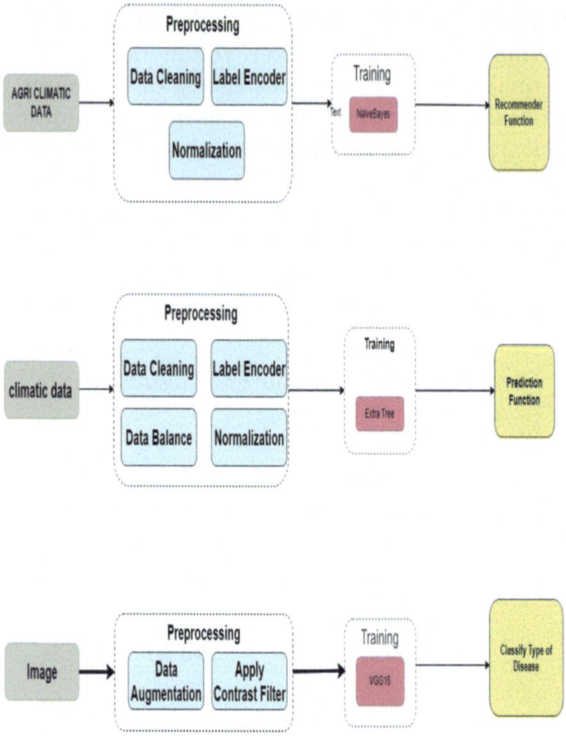

Fig. 1. Methodology used

3.1 Crop Recommendation

Data Preprocessing. Data pre-processing for the classification activity involves the following important activities. First, the data are preprocessed for completing the missing values and also removing duplicate records if any before proceeding to the actual analysis. After this, the categorical variables are transformed into numerical format in order to feed it into the machine learning algorithm using label encoding. In case of imbalanced class distribution within the dataset, measures like oversampling and undersampling are used. Last operation that is performed on the features is normalizing the numerical features because this enhances the performance of a number of algorithms.

Models Used. Thus, the models to be deployed in this task are Random forest for its capacity in handling large dataset and high robustness; Naive Bayes for simplified and effective function in high dimensionality data set; Extra Trees for its computational gains but with better generalization compared to some of the models; XGBoost for its fast performance and works well in large scale data; Support Vector Machine for its advantage in high dimensional space functions and K-NearestOther methods, such as Logistic Regression (LR), classic boosting techniques, and multiple forms of neural nets (Dense, Recurrent, Gated Recurrent, Long Short Term Memory) were also considered, but not used because of aspects such as working with multiple classes, computational speed and necessity of sheer amounts of data. The chosen models remain highly performant, computationally efficient and capable of accommodating various features representing input data; this guarantees the efficiency of the classification results.

3.2 Crop Productivity

Data Preprocessing. Data preparation for regression sources is similar to classification sources; the differences will be noticed only during the model selection. Data preprocessing includes data cleaning to remove errors and outliers so that the dataset is free of any errors; while label encoding on the other hand is the process of converting categorical data to continuous data. Normalization is again played on the numerical features to achieve the improved performance of the regression algorithms whose feature values are standardized to the fixed range.

Models Used. Among the available models several were chosen for the regression task because of their properties to work with complex data and the time they take in computations. Decision tree algorithm of Random Forest applied for regression is made up of multiple trees which provide a mean predicted value for regression that helps in reduction of variance and the use of multiple models makes the model good for non-linear relationships. Extra Trees regression works similarly to Random Forest but instead of splitting the data or any subset of it through random, what is done is that all train data is split randomly. Sequential XGBOOST (XGB) was used thus because of its efficiency and effectiveness in large datasets and patterns. These models were selected to prove that they are viable models and to determine just how intricate the data was in the process. Before choosing XGBoost, it was hypothesized that standard gradient boosting algorithms could also be employed, however, XGBoost was favored because of its better accuracy and faster processing time. These models are chosen in a way to provide reasonable, steady, and accurate regression results while considering the complexity of calculations as well.

3.3 Leaf Disease Detection

Data Preprocessing. In image classification, data preprocessing consists of rotation, flipping and zooming in order to gain more variance to the training dataset. Also, it's necessary to note that contrast filter application improves contrast, which helps to emphasize the difference making features more visible for the model. In order to overcome systemic class prejudice in the dataset, augmentation was performed to bring balance to class to enhance the ability of the model in diagnosis and classification of diseases.

Models Used. Scott models were selected as few image classification models from the lot, as earlier shown to produce promising results in computer vision applications. VGG16 is a deep convolutional neural network, which works for image classification problems, and it is also famous for the high performance and simple architecture. The newer version of VGG, the VGG19, has some extra layer compared to VGG16 so that deeper feature rectangles are extracted and it is more accurate. Inception, architecture that was designed for better performance through receptive computing, was deemed fit to be used for better feature extraction since the design incorporated smart features for better performance. The additional networks used were ResNet101 for its efficiency in training very deep neural network and the rectified linear units that act as activation functions to eliminate vanishing gradient issue to deliver high accuracy in image classification. These models were chosen because they are suitable in managing intricate pattern behaviors within image data through guaranteeing a high disease detection accuracy of the leaves. With the help of these advanced architectures, the study vision focuses on implementing a high accuracy and reliable plant leaf disease classification with ResNet101 having a higher test accuracy of 96%.

4 Experiment and Results

The tests were carried out on an i7 12th Gen processor with 16 GB of RAM and an NVIDIA RTX 3050ti 4 GB GPU. and i7 9th Gen processor with 16 GB of RAM and an NVIDIA GTX 1660ti 4 GB GPU.

4.1 Datasets

Crop Recommendation. The dataset used in the agricultural classification of this project has several features of agriculture as well as environmental characteristics with a view of distinguishing between different crops. The variables in the dataset consist of soil nutrients including, N SOIL, P SOIL and K SOIL, Environmental parameters including temperature, humidity, rainfall as well as soil pH levels. Also, it encompasses other categorical variables such as the state where the crop is grown (STATE) and information on type of crop (CROP) and market price for the specific crop (CROP PRICE). It is in this context that this rich dataset can be used to perform the development of classification models that will be useful for determining the type of crop that can be grown under such environment and soil conditions. The scope includes numerous fields that makes this dataset relatively solid and capable of carrying more precise conclusions, which in turn will help to improve the overall goal of increasing agricultural productivity and efficiency of farming.

Crop Productivity. The regression dataset used in this project covers several factors affecting farming environments and international yields of produce. Some variables are the mean annual precipitation (average-rain-fall-mm-per-year), the use of pesticides (pesticides-tonnes) and the climate mean temperature (avg-temp), essential to the identification of the environmental factors influencing growth of crops. Moreover, the dataset offers further details regarding to the yield of individual crops, abbreviated as hg/ha-yield in the present database, which stands for the quantitative outcome variable of the regression model. It is mostly preferred because it captures data from different countries and covers multiple years making the model founded strongly on actual data. Such a wealth of information readily provides the scientific basis for building decision support systems about which factors affect agricultural yields and consequently, the necessary steps be taken to combat the impacts of climate change on our agricultural production.

Leaf Disease Detection Dataset. The image classification dataset we used is about detecting plant leaf diseases and it contains a set of images that are grouped into 38 classes. These classes include healthy and diseased leaves o Apple - Apple scab, Apple Black rot o Corn (maize) - Corn Common rust o Potato - Potato Late blight and many more. From the provided dataset, there is a picture for each type of condition that affects the leaves and as result the data set can be used to train deep learning algorithms for purposes of classification and the determination of the conditions affecting the plants. Some of the techniques for pre-processing this dataset are data enhancement such as rotation, flipping of the image, and zooming, and the application of contrast filters to the images to get a more diverse image representation.

Combining these datasets, the project builds on a wealth of information in agricultural and environmental assessment, ultimately allowing for numerous comparisons across divergent mul methodological frameworks. The classification dataset assists in providing the right climatic conditions for growing various crops, the regression dataset assists in estimating the yields of crops under the different climatic conditions while the image classification dataset assists in easy diagnosis of diseases that affect plant. Altogether, these datasets provide a comprehensive picture of the effects of climate change on agriculture and aid in the advancement of proposes for enhancing the efficacy and sustainability of the agricultural industry.

By examining this comprehensive visual representation, we can begin to hypothesize about the factors that are most influential in crop productivity and thus guide our model training phase with more informed decisions.

4.2 Evaluation

After training the models it's important to measure their performance using several evaluations

Accuracy: Accuracy measures the overall correctness of the model's predictions [13].

$$accuracy = \frac{TP + TN}{TP + TN + FP + FN} \quad (1)$$

Precision: quantifies the model's ability to correctly classify positive instances among all instances predicted as positive [13]

$$Precision = \frac{TP}{TP + FP} \tag{2}$$

Recall: also known as sensitivity, measures the ability of the model to correctly identify positive instances out of all actual positive instances [13].

$$Recall = \frac{TP}{TP + FN} \tag{3}$$

F1-score: The F1-score is the harmonic mean of precision and recall, providing a balance between the two metrics [14]

$$F1 - score = \frac{2 \cdot Precision \cdot Recall}{Precision + Recall} \tag{4}$$

R-squared (R2): R-squared measures the proportion of the variance in the dependent variable that is predictable from the independent variables. It indicates the goodness of fit of the model [15].

$$R^2 = 1 - \frac{SS_{res}}{SS_{tot}} \tag{5}$$

Root Mean Squared Error (RMSE): RMSE measures the square root of the average of the squared differences between predicted and actual values. It provides an indication of the absolute fit of the model to the data [15] (Table 4).

$$RMSE = \sqrt{\frac{1}{n}\sum_{i=1}^{n}(y_i - \hat{y}_i)^2} \tag{6}$$

The following tables present the results of the proposed models for the three datasets used in this study.

The results in those Tables demonstrate that the Naive Bayes model outperforms other models for the crop recommendation dataset, achieving the highest accuracy, precision, recall, and F1-score as shown in Table 1. For the crop productivity dataset, the Extra Trees model provides the best performance in terms of R-squared and RMSE, as shown in Table 2. In the leaf disease detection dataset, ResNet101 achieves the highest test accuracy of 96%, as seen in Table 3.

Table 2. Proposed models results for crop recommendation dataset

Model Name	Accuracy	Precision	Recall	F1-score
Random forest	99.3	99.37	99.31	99.31
Extra trees	94.7	94.9	94.7	94.7
KNN	95.9	96.5	95.9	95.9
SVM	96.81	97.15	96.81	96.8
LR	96.36	96.44	96.36	96.35
Naive Bayes	**99.54**	**99.58**	**99.54**	**99.54**
Gradient Boosting	98.1	98.4	98.1	98.1
LSTM	97.04	97.04	97.04	97.04
RNN	93.18	93.56	92.5	93.29
GRU	98.68	95.88	95.22	95.64
Dense	95.45	95.88	95.22	95.47

Table 3. Proposed models results for crop productivity dataset

model name	R2	RMSE
Random forest	0.98	10120.01
Extra trees	**0.99**	**8225.93**
XGB	0.97	13451.94

5 Discussion

Climate change has become a global problem whose effects include reduction in agricultural productivity; thus, this research focuses on analyzing the effectiveness of machine learning and deep learning algorithms on agriculture. As for the accuracy, the Naive Bayes model was found to be 99% accurate out of all models trained. the Extra Trees model was the most accurate for regression with the accuracy of 49% for crop classification, as well as the best for the classification of crops. Specifically, ResNet101 yields the highest test accuracy for recognizable taxonomy, with 96% for detecting the leaf diseases. More to the point, the

Table 4. Proposed models results for leaf disease detection dataset

model name	Test Accuracy
VGG16	95%
VGG19	94%
CNN	92%
Inception	85%
RESNET101	**96%**

Naive Bayes classification model examined above was more effective than the 99. can achieve up 09% accuracy as mentioned by Agarwal et al. (2023) with such approach however due to careful data pre-processing. When evaluated on the Extra Trees model, the R^2 score was higher than what Patki and Wazurkar (2021) obtained, because the data were cleaned, scaled, and hyperparameters set appropriately. In this work, ResNet101 achieve 96% accuracy while Reddy and Rekha (2021) reported 92% by developing Deep Leaf Disease Prediction Framework. Relative to other experiments, though, some disparity, was estimated while contrasting with the findings of the current work with other studies for instance, the Random Forest model established by Parameswari et al. (2021) yielded an accuracy of 98%. 3%. This may arise from differences in the datasets used in the experiments and thus warrants future research focusing on the characteristics of the data used. In general, it unveils the potential and significance of applying machine learning and deep learning models to address the problems of climate change in agriculture. The future research direction should consider adding newer and multiple datasets capturing various types of agricultural environments and integrating real time data sources to improve the predictive capability of the models and their practical usability.

6 Conclusion and Future Work

This work also establishes a strong vision of enhancing the effectiveness of using proactive and complex approaches in machine learning in conjunction with methods of preprocessing big data to avoid the negative impacts of climate change on agricultural practices. Using a classification method, regression analysis, and image classification, thus shared valuable insights leveraging on the ideal conditions for crops growth, expected yields, management of diseases, etc. Our results demonstrate the importance of proper data preprocessing and analysis as indicated in the reduction of performance in models that were produced without going through those procedures among others as shown in the ablation study. Precision based on classification models merged with data balancing and normalization was highly accurate in terms of predicting crop types and conditions and crop yields by the use of other regression models like Random Forest and XG Boost climatic and agricultural inputs data.VGG16,VGG19,Inception models, and RESnet 101 in specifically developing and applying image classification models. It was also observed that using data augmentation techniques and applying the contrast filters improved the accuracy of classifying plant leaf diseases. The results of this study are crucial for policymakers as well as for the researchers and farmers since they shed light on the promising arenas for interventions after which social research innovations can be designed, as well as the comprehensive directions for the farmers' ACL management. To build upon this research, future work should seek to C7; further develop these models, C8; expand data to different regions and types of agriculture, C9; incorporate live data into the models which would make the models even more useful for real applications thus contributing to improvement of agriculture in the face of climate change.

References

1. Intergovernmental Panel on Climate Change, Climate Change 2014: Impacts, Adaptation, and Vulnerability. Part A: Global and Sectoral Aspects, 2014. https://www.ipcc.ch/report/ar5/wg2/
2. Food and Agriculture Organization, The future of food and agriculture: trends and challenges, 2017. http://www.fao.org/3/a-i6583e.pdf
3. Mohanty, S.P., Hughes, D.P., Salathé, M.: Using deep learning for image-based plant disease detection. Front. Plant Sci. **7**, 1419 (2016). https://doi.org/10.3389/fpls.2016.01419
4. Pautasso, M., Döring, T.F., Garbelotto, M., Pellis, L., Jeger, M.J.: Impacts of climate change on plant diseases-opinions and trends. Eur. J. Plant Pathol. **133**(1), 295–313 (2012). https://doi.org/10.1007/s10658-011-9924-7
5. Agarwal, A., Ahmad, S., Pandey, A.: Crop recommendation based on soil properties: a comprehensive analysis (2023). https://doi.org/10.1109/icccnt56998.2023.10307999
6. Parameswari, P., Rajathi, N., Harshanaa, K.J.: Machine learning approaches for crop recommendation. In: 2021 International Conference on Advancements in Electrical, Electronics, Communication, Computing and Automation (ICAECA) (2021). https://doi.org/10.1109/icaeca52838.2021.9675480
7. Sagana, C., et al.: Machine Learning-Based Crop Recommendations for Precision Farming to Maximize Crop Yields, January 2023. https://doi.org/10.1109/iccci56745.2023.10128525
8. Revathi, P., Hemalatha, M.: Advance computing enrichment evaluation of cotton leaf spot disease detection using Image Edge detection (2012). https://doi.org/10.1109/icccnt.2012.6395903
9. Sardogan, M., Tuncer, A., Ozen, Y.: Plant leaf disease detection and classification based on CNN with LVQ algorithm. In: 2018 3rd International Conference on Computer Science and Engineering (UBMK) (2018). https://doi.org/10.1109/ubmk.2018.8566635
10. Reddy, T.V., Rekha, K.S.: Deep Leaf Disease Prediction Framework (DLDPF) with Transfer Learning for Automatic Leaf Disease Detection (2021). https://doi.org/10.1109/iccmc51019.2021.9418245
11. Patki, V., Wazurkar, P.: Application of Data Sampling for Crop Yield Prediction using Stochastic Gradient Descent Neural Networks (2021). https://doi.org/10.1109/csnt51715.2021.9509682
12. Geetha, M., Suganthe, R., Latha, R., Anju, R., Sastimalar, K., Shobana, P.: Deep learning based yield prediction model to predict the yield of paddy in Cauvery delta region. In: 2022 International Conference on Computer Communication and Informatics (ICCCI) (2022). https://doi.org/10.1109/iccci54379.2022.9740944
13. Powers, D.M.: Evaluation: from precision, recall and F-measure to ROC, informedness, markedness and correlation. J. Mach. Learn. Technol. **2**(1), 37–63 (2011)
14. Sokolova, M., Lapalme, G.: A systematic analysis of performance measures for classification tasks. Inf. Process. Manag. **45**(4), 427–437 (2009)
15. Shumway, R.H., Stoffer, D.S.: Time Series Analysis and Its Applications: With R Examples, Springer, 2017

Comparative Study: Using Machine Learning Models Based Rehabilitation Therapy to Classify Diabetic Frozen Shoulder Exercises

Ziad A. Ahmed[1(✉)], Mohamed Sherif[1], Maha Abdelmohsen[1], Omnia Saeed Mahmoud Ahmed[2], and Asmaa M. Al-Emrany[2]

[1] Faculty of Computer Science, October University for Modern Sciences and Arts (MSA), October, Giza 12451, Egypt
{ziad.ahmed6,mohamed.sherif13,mhatem}@msa.edu.eg

[2] Department of Physical Therapy for Internal Medicine and Geriatrics, Faculty of Physical Therapy, October University for Modern Sciences and Arts (MSA), October, Giza 12451, Egypt
{omsaeed,amalemrany}@msa.edu.eg

Abstract. In 2017, The World Health Organization (WHO) initiated a global campaign to improve rehabilitation services by 2030, focusing on the approximately 1 billion people who do not have access to proper care. In rehabilitation therapy, pose detection appears to be a promising tool, allowing for exact tracking and analysis of patients' motions. By incorporating pose detection into therapy sessions, physicians can design personalized and interactive rehabilitation programs that are suited to each patient's needs. This system provides real-time feedback and monitoring, allowing patients to do therapeutic activities with increased accuracy and efficacy. Machine Learning (ML) also plays an important role in healthcare by allowing for autonomous learning from previous data, which is required for predictive modelling and optimizing rehabilitation procedures. This work compares 16 different classical machine learning models in which for classifying diabetic frozen shoulders which are Flexion, Abduction and External rotation where Random Forest Classifier was the highest model where it achieved 92% indicating the efficacy of the proposed approach. Future study will involve merging segmentation approaches and multi-modal data to improve model robustness and accessibility and enhancing rehabilitation outcomes and patient quality of life worldwide.

Keywords: Diabetic · Frozen · Shoulder Exercise Classification

1 Introduction

In 2017, the World Health Organization launched a global campaign to enhance rehabilitation services by 2030, with the ultimate objective of assisting people with disabilities in achieving maximum independence and well-being. Despite the fact that over 1 billion worldwide living with disabilities, a large proportion lack access to suitable rehabilitation care. In 2017, (WHO) published a call to action to improve rehabilitation services worldwide. With a target date of 2030, this huge task requires a concerted effort across all nations to increase health care quality and capacity, with the inclusion of rehabilitative services as a main priority [1].

A. Abdelgawad et al. (Eds.): ISBCom 2024, LNNS 1268, pp. 684–698, 2025.
https://doi.org/10.1007/978-3-031-82377-0_55

The combination of computer science (CS) and physical therapy (PT) offers promising potential, notably in treating disorders such as frozen shoulder. One intriguing technology at this junction is feature extraction by pose detection, which has enormous potential for improving functional recovery in both the upper and lower limbs. Pose detection, which captures and analyses body joint postures, enables personalised rehabilitation therapies customised to individual needs [2]. This revolutionary approach allows therapists to create personalised workouts and therapies based on their patients' unique movement patterns, resulting in more effective rehabilitation outcomes [3].

Augmented Reality (AR) in stroke rehabilitation offers a novel approach for enhancing functional recovery in both upper and lower limbs. By seamlessly merging virtual elements with real-world environments, AR provides an interactive platform for therapeutic activities. [4] Patients can engage with virtual objects in their immediate surroundings, boosting immersion and motivation. This technology facilitates motor learning through challenging exercises without constant therapist supervision. AR systems are customize to accommodate individual disabilities and preferences, catering to diverse patient needs. As AR technology becomes more accessible, it holds promise for revolutionizing stroke rehabilitation by promoting neural plasticity and functional recovery [5].

The Random Forest classifier is often used in real-time applications because of its ensemble approach and ability to handle large datasets with high dimensional. Adjusting the number of trees in the forest can lead to various classification results for the same issue, making it a useful hyper-parameter for identifying the best performing model, particularly for activity detection [6].

This paper contributes the advantage of rehabilitation technology. Utilizing Smart sensors or camera sensors capture the positions of body joints points(x,y,z) and store it as dataset in an Excel file [7]. Once all data sets were synchronized, labeled, filtered, and normalized, The data analysis includes using machine learning techniques to access several classification methodologies, with the Fine Random Forest algorithm achieving the greatest accuracy [8]. Furthermore, the work covers issues such as imbalanced datasets and provides insights for improving classification performance in real-world contexts [4]. In some data collection scenarios with binary action classification with discrete labels as "correct" or "incorrect", there is a chance that the final real-life dataset is highly imbalanced [9].

2 Related Work

Sekar et al. [10] proposed a Neuro Machine Learning Model (NMLM) for detecting motor neuron disorders. The model is intended to analyze symptoms and predict the presence of certain motor neuron illnesses, such as amyotrophic lateral sclerosis (ALS), progressive bulbar palsy, tendon erosion, and polytopic paralysis. The study compares the performance of the NMLM to other machine learning algorithms, such as MLA, AMD, MLND, and DLPDD. The results show that the NMLM surpasses existing approaches for predicting motor neuron disorders based on symptom analysis. The NMLM predicted ALS with the highest accuracy of 92.12% at a cutoff point. Similarly, for progressive bulbar palsy, the NMLM obtained accuracy of 93.28% as its result from these models are good.

Jones et al. [11] investigated employing wearable sensors and machine learning to automate workout repeat counting during stroke rehabilitation. They solve the difficulty of quantifying rehabilitation "dose" by presenting a reliable approach for measuring exercise repetitions. The study had healthy controls and stroke patients execute particular arm workouts while wearing sensors that capture accelerometer and gyroscope data. Algorithms, including Fine KNN, with a true positive rate of 98.5% and a false discovery rate of 0.7%, are evaluated for accuracy. The results indicate that models such as medium KNN (true = 94.1%, false = 2.7%) and weighted KNN (true = 95.7%, false = 1.0%) have the capacity to count repetitions accurately. This strategy seeks to improve rehabilitation therapy monitoring and outcomes by giving objective metrics of exercise dosage. Ekta et al. [11] investigated Cardiovascular diseases (CVDs) pose a substantial health burden worldwide, particularly in lowand middle-income nations such as India. With the significant rise in CVD prevalence in India, the healthcare system is under increasing strain. While machine learning (ML) models show promise for early CVD identification, current techniques frequently neglect critical risk factors unique to the Indian setting. This study seeks to close this gap by creating a ML-based prediction system suited to South Indian clinical data. Using techniques like k-NN, Naïve Bayes, Logistic Regression, AdaBoost, and Random Forest, the system accurately predicts CVD risk. Furthermore, placement on the public cloud improves accessibility, particularly in remote areas. The model, which highlights attributes such as gender, diabetes, hypertension, cholesterol levels, lifestyle factors, and family history, provides useful information for preventive healthcare in India. However, it is important to emphasise that the study's reliance on clinical data from a single location in South India may limit the model's generalizability to the larger Indian population. The Random Forest model, with its exceptional sensitivity (92.8%) and specificity (94.6%), emerges as particularly successful in identifying CVD-prone patients. These findings highlight the potential for ML-based techniques to revolutionize preventive healthcare, particularly in resource-constrained settings such as rural India.

Fang et al. [12] introduces a machine learning model designed to forecast the likelihood of developing hypertension within a five-year timeframe. Hypertension, a prevalent health concern, stems from various risk factors such as dietary habits, tobacco/alcohol consumption, physical exertion, and emotional stress. Managing or preventing hypertension effectively involves addressing these factors. Inspired by the effectiveness of machine learning (ML) in diverse fields, including medicine, we suggest applying ML techniques to predict hypertension risk. To build and evaluate our model, the authors analyze an anonymized dataset sourced from a local hospital, containing variables like age and blood-related markers. The authors methodology combines K-nearest neighbors (KNN) and Light Gradient Boosting Machine (LightGBM) techniques. This research access the model's performance using metrics like accuracy, precision, recall, and F1 score. The results demonstrate that our hybrid KNN-LightGBM model achieves high accuracy (86%) and recall (92%) rates. Compared to other models like Support Vector Machine (SVM), Random Forest, and neural networks, our proposed approach exhibits superior accuracy and precision. Additionally, the model's area under the curve (AUC) indicates its effectiveness in classification tasks. These findings suggest that integrating machine learning methods into hypertension risk assessment can enhance predictive accuracy, aiding healthcare professionals in early intervention and preventive measures. Future research avenues may involve incorporating additional risk factors and refining the prediction model for broader healthcare applications.

Shter et al. [13] discussed the increasing incidence of injurious fall events and hip fractures among the elderly population, highlighting the need for effective rehabilitation strategies. Traditional statistical methods have limitations in predicting rehabilitation outcomes, prompting the exploration of machine learning techniques. The study was conducted retrospectively in a post-acute geriatric orthopedic rehabilitation ward from 2011 to 2019. Data were collected from hip fracture patients, including demographic, clinical, and functional variables. Machine learning models, including AdaBoost, CatBoost, ExtraTrees, among others, were trained and evaluated to predict rehabilitation outcomes. The study included 1625 hip fracture patients.

Donisi et al. [14] Parkinson's disease (PD) presents a challenging problem in healthcare due to its neurodegenerative nature and significant impact on mobility and quality of life. This research investigates the examination of Parkinson's disease patients' gaits as a means of monitoring disease progression and assessing the efficacy of rehabilitation therapies. Given the expected doubling of Parkinson's disease prevalence by 2030, understanding the dynamics of gait impairment and its response to treatment becomes increasingly important. Leveraging sophisticated technology like wearable inertial systems and machine learning algorithms opens up new avenues for objective assessment and intervention in Parkinson's disease management. Twelve PD patients who were enrolled in a two-month rehabilitation programmer were included in the study. Clinical assessments such as the Berg Balance Scale and UPDRS III were used, along with the Opal System from APDM Inc. For gait analysis. Notable increases in clinical scores were also noted, along with significant changes in gait metrics after rehabilitation, supporting the promise of wearable technology and machine learning to objectively evaluate rehabilitation outcomes in individuals with Parkinson's disease (Fig. 2).

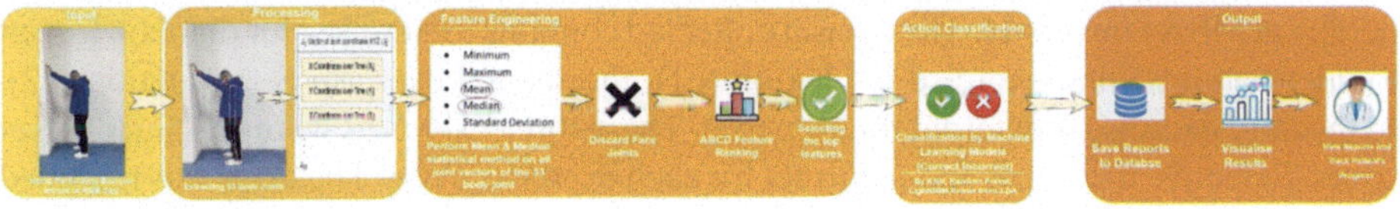

Fig. 1. Proposed Approach

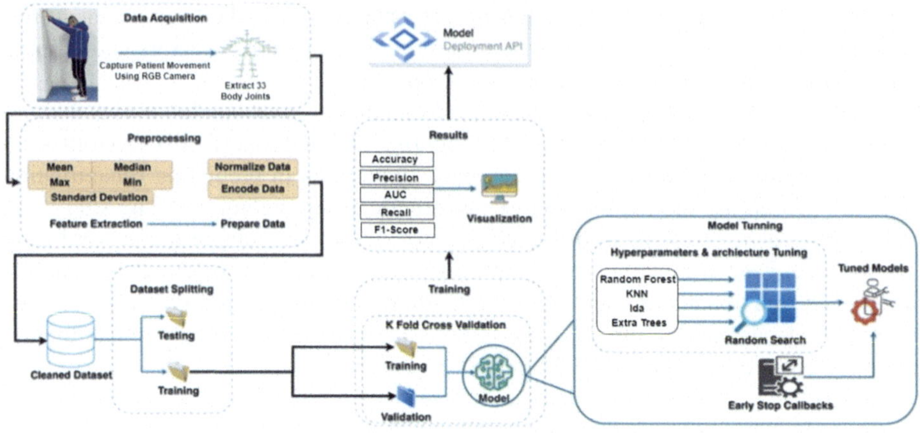

Fig. 2. Machine Learning Pipeline

3 Methodology

This part of the paper outlines the procedure, methods of feature extraction of human body joints to detect and classify diabetic frozen shoulder and the design of machine learning models that were created to classify the correct and incorrect exercises of the patients. Figure 1 illustrates the proposed approach for diabetic frozen shoulder classification.

3.1 The Proposed Dataset

Data Collected: In this study, we collected a medical dataset, which is "Self Exercise." Dataset in cases of diabetes Frozen Shoulder" Collaboration With the Faculty of Physical Therapy at MSA University. The dataset collected contains 3 exercises for diabetic frozen shoulder patients' first exercise is Flexion (e1), the second exercise is Abduction (e2), and the third exercise is external rotation (e3). These three exercises were classified as six classes. Selective to correct and incorrect for each exercise, every The class was recorded by a varied group of men and women with different ages with different limitations for each patient. By repetition for each patient (out of 10) in every class or exercise times for the correct exercise and 10 times for the incorrect exercise, We collected a dataset that contains 600 different actions. Videos for 10 patients in 3 different exercises classified in 6 classes (Fig. 3 and Table 1).

Dataset Content: The paragraph is primarily related to data content. It describes the specifics of a medical dataset, including how the data was collected, the nature of the data, and how it is organized. Here's a breakdown of the key points related to data content:

Data Collection: The dataset, named "Self Exercise" Dataset, focuses on exercises for diabetic frozen shoulder patients, in collaboration with the Faculty of Physical Therapy at MSA University.

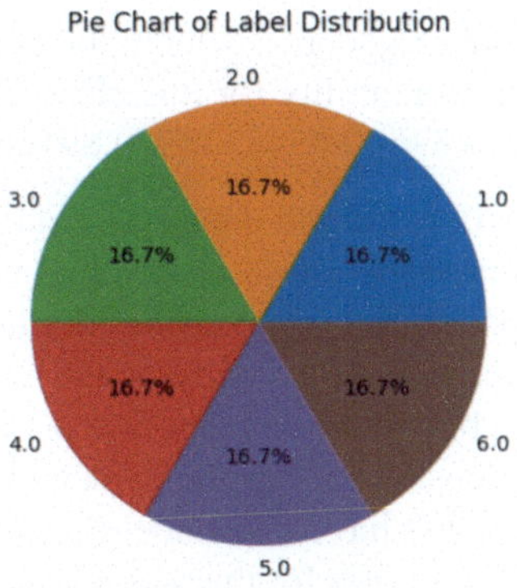

Fig. 3. Classification Distribution of the Six Classes, 3 exercises

Table 1. Number of videos per exercise

Classes	Number of samples
Flexion	200
Abduction	200
External rotation	200

Exercises Included: The dataset contains three types of exercises: Flexion, Abduction, and External Rotation.

Classification: Each exercise is classified into six classes based on whether it is performed correctly or incorrectly, with videos recorded for each class.

Participants and Repetitions: The dataset includes videos from 10 different patients, each performing the exercises 10 times correctly and 10 times incorrectly, resulting in a total of 600 video recordings.

Dataset Content Details: Videos were captured using an RGB camera, and pose estimation was done to extract 33 body joint points. Each video is named systematically to indicate the exercise, the person, and whether the exercise was performed correctly or incorrectly.

Participant Demographics: The dataset includes a varied group of men and women aged 18 to 60, each with different limitations.

This detailed description of the dataset content, including how the data is structured, recorded, and labeled, makes it clear that the focus is on the data content rather than the process of extracting or mining patterns from the data.

Data Preproccesing and Normalization: We used 5-Fold Stratified K-Fold crossvalidation to make sure our Random Forest classifier for diabetic frozen shoulder categorization was robust and broadly applicable. In order to solve class imbalance, this technique splits the dataset into five parts and provides a credible model assessment while maintaining class proportions in each Fold. Balanced class weights were used in the Random Forest model, which was optimized with 100 trees and a maximum depth of 10. Performance was evaluated using crossvalidation scores, which showed reliable performance and stability, confirming the model's suitability for this classification job. Cross-Validation Scored results of [0.90833333, 0.875, 0.85833333, 0.89166667, 0.90833333].

In the process of preparing our medical dataset for classification correctness of the exercises of the patients. The primary These procedures were based on an estimation of 600 videos to extract. The body joint points using Mediapipe in Python to extract 33 body joints, which lead to 66 features of the body, to (x, y, z) points of every body joint point. Feature Extraction for all body joints except face points using OpenCV by Python to extract the features of the videos into Excel files. Each excel It is saved with the same name as the opened video. Now We have 600 Excel files (.csv) and 200 Excel files for the first exercise.(flexion) 200 Excel files for the second exercise (abduction): 200 Excel files for the third exercise (external rotation) were saved with the same name for every video in the dataset. Every Excel contains 66 features (body joint points x, y, and z). In this study, feature extraction techniques employing Mean and median calculations were applied to 600 Excel files. To derive representative statistical descriptors from raw data,

A. Evaluation Metrics facilitating the characterization of essential patterns and trends within the dataset. So we have three Excel files that calculate the The mean number of 600 Excel files and another 3 Excel files calculated the median number of 600 in Excel merged 3 exercises files into 1 file with 600 rows that are calculated using the mean, and merged another three exercise files into one file with 600 rows. That are calculated using Median. Added a label to the Excel file, which is represented as Prediction. Label or target and classify the first 200 rows of the Excel in (1,2). to classify the first exercise as a binary classification to detect the correct and incorrect data. And the second 200 rows of the Excel in (3,4) to classify the second exercise and the third 200 rows of the excel in (5,6) to classify the third exercise (Fig. 4).

4 Experiments and Results

The experiments section is divided into 3 subsections. Evaluation metrics, and each other subsection represents different experiments. Experiment 1 merged all exercises in 1 file calculated by Mean, Experiment 2 with another calculation method which is Median. To ensure a fair comparison, we ran the model across Intel64 Family 6 Model 158 Stepping 10, GenuineIntel, 15.87 GB RAM (Table 2).

4.1 Evaluation Metrics

In this study, the evaluation metrics used were accuracy, precision, recall, and f1score. These metrics were chosen because they are commonly used by researchers to evaluate the effectiveness of their proposed methods. Accuracy measures the ratio of true positive, false positive, true negative, and false negative and it is our most important metric. Precision is the ratio of true positives to all positive values in the data. Recall, also called sensitivity, is the ratio of true positives to the sum of true negatives and false negatives. Lastly, the F1-score represents the harmonious average of precision and recall.

$$Accuracy = \frac{TP + TN}{Allinstances} \tag{1}$$

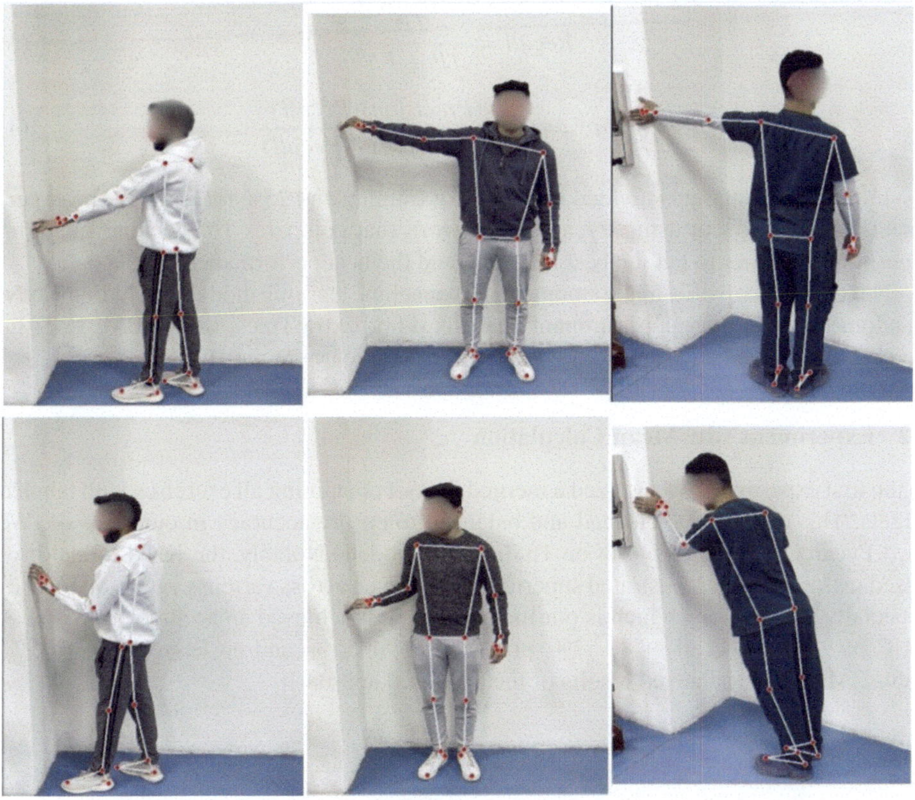

Fig. 4. Pose Estimation for 3 exercises in 6 classes using Mediapipe

Table 2. Show Exercises/Classes of diabetic frozen shoulder

Classes	Number	
Exercises	No.of rows	Labels
Flexion correct	100	1
Flexion incorrect	100	2
Abduction correct	100	3
Abduction incorrect	100	4
External Rotation correct	100	5
External Rotation incorrect	100	6

$$Precision = \frac{TP}{TP + FP} \quad (2)$$

$$Recall = \frac{TP}{TP + FN} \tag{3}$$

$$F1Score = 2 \times \frac{(Precision)(Recall)}{(Precision) \cdot (Recall)} \tag{4}$$

Initially, the dataset is divided into 75% for training and 25% for testing, The experiments are carried out using Pycaret framework that compares 16 Machine Learning algorithms together to get the best accuracy and the best performance for the data.

Next, a comparison is made among several machine learning models, including KNN, Random Forest, Linear Discriminant analysis (lda), Extra Trees, LightGBM, Logistic Regression, Naive bayes and some of other machine learning models.

4.2 Experiment 1 by Mean Calculation

In the first experiment, we utilized a merged dataset containing all exercises with labeled targets. This dataset was cleaned and balanced to ensure accuracy in our analysis. We used Pycaret to run the dataset on 16 different models, Notably, the KNN model proposed in this study demonstrated superior performance across various evaluation metrics especially the accuracy which is our main concern. It achieved an accuracy of 87.62%, AUC of 93.79%, recall of 87.62%, precision of 88.55%, and an F1-score of 87.5%. Table 3 shows the achieved results of the proposed approach.

$$\text{Mean} = \frac{1}{n}\sum_{i=1}^{n} x_i$$

4.3 Experiment 2 by Median Calculation

In the second experiment, we utilized the merged dataset containing all exercises with labeled targets. This dataset underwent cleaning and balancing to enhance its quality. Subsequently, We used Pycaret to run the dataset on 16 different models. Once again, the Random Forest model stood out as the top performer in terms of accuracy, demonstrating exceptional results.

$$\text{Median} = \begin{cases} x_{\frac{n+1}{2}} & \text{if } n \text{ is odd} \\ \frac{x_{\frac{n}{2}} + x_{\frac{n}{2}} + 1}{2} & \text{if } n \text{ is even} \end{cases}$$

Random Forest model which was proposed in this study outperformed all other algorithms across various evaluation metrics, as it scored accuracy of 88.81%, AUC of 96.95%, recall of 88.81%, precision of 89.86% and F1-score of 88.72%.

Table 4 shows the achieved results of the proposed approach.

Table 3. 16 Machine Learning Models Classification Using Mean Calculation

Model	Accuracy	AUC	Recall	Percision	F1 Score
KNN	**87.62%**	**93.79%**	**87.62%**	**88.55%**	**87.5%**
lda	87.14%	97.24%	87.14%	88.12%	87.03%
rf	87%	96.43%	87%	87.84%	86.81%
et	86%	96.17%	86%	86.58%	85.34%
lightgbm	84%	95.80%	84%	84.37%	83%
gbc	82.14%	95.52%	82.14%	83.33%	82%
lr	80%	96.56%	80%	81.54%	80%
ridge	79%	0%	78.81%	80%	78.40%
nb	77%	96.47%	77%	78.04%	76%
dt	73.81%	85%	73.81%	75.13%	74%
svm	70%	0%	70%	72%	65%
ada	60%	91.30%	60%	54.55%	50%
gda	55%	64.71%	55%	41.19%	44.49%
dummy	30%	50%	30%	22%	25%

Table 4. 16 Machine Learning Models Classification Using Median Calculation

Model	Accuracy	AUC	Recall	Precision	F1 Score
RF	**88.81%**	**96.95%**	**88.81%**	**89.86%**	**88.72%**
et	87.62%	96.78%	87.62%	88.83%	87.51%
knn	87.14%	94.43%	87.14%	88.34%	87.91%
lightgbm	86.43%	96.88%	86%	87.87%	86.27%
gbc	84.81%	96.35%	84%	85.15%	84%
lda	83.81%	97.33%	83.81%	85.26%	84%
lr	80%	96.67%	80%	81.54%	79%
dt	79%	85.57%	78.81%	80%	78.40%
ridge	77%	96.47%	77%	78.04%	76%
nb	75%	96.57%	75%	76.08%	74.63%
svm	73.81%	85%	73.81%	75.13%	74%
ada	60%	91.30%	60%	54.55%	50%
gda	55.7%	66.71%	55.7%	42.19%	44.49%
dummy	50.95%	90.85%	50.95%	40%	41.50%

5 Experiment 3 Hyperparamter Tunning

In our methodology, we employed a machine learning approach to classify diabetic frozen shoulder using a Random Forest model, fine-tuned through a hyper-parameter grid search. The hyper-parameters and their respective ranges explored during the optimization process were as follows:'n-estimators', representing the number of trees in the forest, was tested with values of 100, 200, and 300;'max-depth', which controls the maximum depth of each tree, was set to 5, 10, and 15 to assess the impact of varying tree depths, including shallower configurations. We also varied min-samples-leaf', the minimum number of samples required to be at a leaf node, across 2, 4, 6, and 8 to understand the influence of larger leaf sizes on model performance. The 'max-features' parameter, dictating the number of features. When determining the appropriate split, consider 'auto' (all features), 'sqrt' (square root of the number of features), and a fractional value of 0.5, which allows the model to explore multiple dimensions of feature subsets. In addition, we introduced 'min-samples-split', the minimum number of samples necessary to split an internal node, with values of 10 and 20, to assess the impact of tougher splitting criteria. This extensive hyper-parameter tweaking sought to uncover the best combination for maximizing the model's classification accuracy and durability.

Table 5. Top 5 Machine Learning Model Classification

Model	Accuracy	AUC	Recall	Precision	F1 Score
RF	**92%**	**99.05%**	**92%**	**92.05%**	**92%**
KNN	90%	97.05%	90%	90.32%	90%
lda	88.57%	97.40%	88.57%	89.97%	88.43%
ET	87.4%	97.21%	87.4%	88.4%	87.3%
lightGBM	87.4%	96.6%	87.38%	88.61%	87.20%

6 Discussion

After saving top 5 machine learning models, we created a Separate App for each Model using FastApi by Pycaret. FastAPI offers a powerful framework for building APIs that can expose machine learning models and data processing pipelines. By leveraging FastAPI, developers can create robust and scalable APIs that facilitate the integration of machine learning models and data processing functionality into various applications and systems. This enables seamless interaction with these models and pipelines, making it easier to incorporate them into real-world scenarios and workflows.

The RF model proposed in this study outperformed all other algorithms across various evaluation metrics, including accuracy, precision, recall, and F1score. Notably, the RF model achieved an impressive accuracy of 92%. For a concise summary of the performance of different Machine Learning models on the dataset, refer to Table 5.

Additionally, Fig. 5 illustrates the confusion matrix of the three exercises with Best Grid Search Parameters: 'max-depth': 5, 'max features': 'auto', 'min-samples-leaf': 4, 'min-samples-split': 10, 'n-estimators': 100.

As shown in Fig. 5, These results indicate that the model is performing well, with high accuracy and good balance across different evaluation metrics. The slightly better testing metrics could be a result of effective model training that generalizes well to new data or specific characteristics of the test set. Nonetheless, it's always a good practice to look deeper into the data distribution and model settings to ensure there is no over-fitting or data leakage.

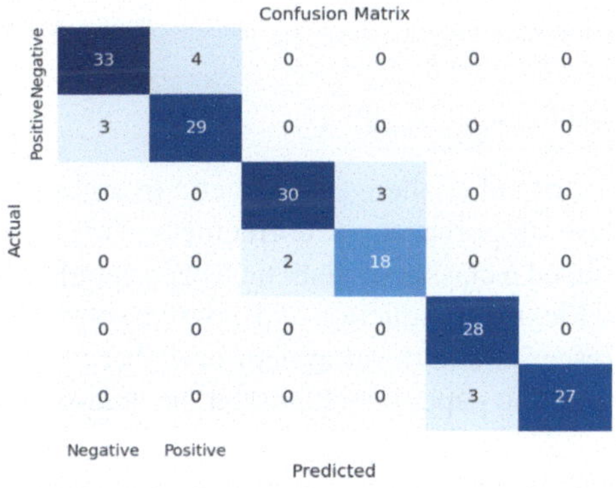

Fig. 5. Confusion Matrix of Random Forest Algorithm

Confusion Matrix Layout: Rows (Actual Classes): These represent the true class labels from the data. Columns (Predicted Classes): These represent the classes predicted by the model.

Observations from the Matrix:

Class 1: True Positives (TP): 33 instances were correctly identified as class

1. False Negatives (FN): 4 instances were incorrectly classified as not class 1.
   Class 2: True Positives (TP): 29 instances were correctly identified as class
2. False Negatives (FN): 3 instances were incorrectly classified as not class 2.
   Class 3: True Positives (TP): 30 instances were correctly identified as class
3. False Positives (FP): 2 instances were incorrectly classified as class 3. False Negatives (FN): 3 instances were incorrectly classified as not class 3.
   Class 4: True Positives (TP): 18 instances were correctly identified as class
4. False Positives (FP): 2 instances were incorrectly classified as class 4.
   Class 5: True Positives (TP): 28 instances were correctly identified as class
5. False Positives (FP): 3 instances were incorrectly classified as class 5.
   Class 6: True Positives (TP): 27 instances were correctly identified as class
6. False Positives (FP): 3 instances were incorrectly classified as class 6.

The Model shows a strong ability to correctly classify instances across all six classes. However, there are some misclassifications, notably a few false negatives in classes 1, 2, and 3 and false positives in classes 3 through 6.

The diagonal cells (from top left to bottom right) show the number of correct predictions for each class, which are high across all classes, indicating good classification performance.

Non-diagonal cells show misclassifications. For instance, in class 3, 2 instances were mistakenly predicted as class 4, and in class 5, 3 instances were mistakenly predicted as class 6. These errors suggest areas where the model might be confusing similar classes or where additional model tuning or feature engineering could improve performance.

This Matrix helps in understanding which classes are easily distinguishable by the model and which are prone to confusion. It's a valuable tool for diagnosing model performance beyond simple accuracy metrics, helping in refining the model or focusing on specific classes for improvement.

7 Conclusion

This paper addressed the critical challenge of diabetic frozen shoulder. The paper made several key contributions to support accurate correctness of exercises and improve patient outcomes. First, It introduced 6 classes of diabetic frozen shoulder about 600 videos. The classes of videos include Flexion, Abduction and external rotation. This collected dataset is valuable for further research and development in diabetic frozen shoulder exercises classification. Furthermore, the paper benchmarked the dataset against various machine learning models.

Machine learning models such as Random Forest, KNN, extra trees, LightGBM and Linear Discriminant analysis (lda), achieving an average accuracy of 92% in classifying six class of diabetic frozen shoulder exercises.

We developed a system that merges AI with physical therapy to classify diabetic frozen shoulder exercises to check the correctness of the exercise of patient. Then the system will be integrated with augmented reality (AR) with physical therapy, utilizing augmented reality AR technology to facilitate home-based therapy for pre-diagnosed medical conditions and interactive way to Rehabilitate patients alongside therapist supervision and giving analytical data to therapist to analyze his patient progress.

The system aims to provide real-time feedback on exercises, ensure correct body posture and track patient's progress giving doctor customize report formats for easy monitoring. As a future direction, the proposed approach can be extended to integrate segmentation techniques, offering a more granular understanding of rehabilitation exercise characteristics. Additionally, addressing the multimodality of different types of data, such as integrating diverse imaging/videoing modalities or clinical information, holds promise for further improving the robustness and applicability of the machine learning models.

8 Future Work

Future research in home-based physical therapy should concentrate on improving AI capabilities for accurate exercise identification, including adaptive feedback customized to specific patient needs, and investigating multi-modal interaction for better guidance. Predictive analytics methods may also predict setbacks, while gamification and social integration may improve motivation and participation. Clinical validation through thorough studies with different patient populations is critical for verifying efficacy and safety, which leads to implementation into clinical practice. Finally, efforts should focus on increasing accessibility and cost-effectiveness, maybe through simpler versions for low-resource settings or alternative delivery methods such as tele-health platforms. Overall, developments in these areas have the potential to improve rehabilitation results, increase adherence, and improve patient's quality of life throughout the world.

Acknowledgments. We express our sincere appreciation to Doctor Asmaa M. Al-Emrany and Doctor Omnia Saeed Mahmoud Ahmed, Department of physical therapy for Internal Medicine and Geriatrics, Faculty of physical therapy , October University for Modern Sciences and Arts (MSA).

References

1. Hillaker, E., Bova, M., Libby, J., Tricarico, M., Andrews, P., Scharf, R.: Answering the rehabilitation 2030 call. PM&R (2024)
2. Boukhennoufa, I., Zhai, X., Utti, V., Jackson, J., McDonald-Maier, K.D.: Wearable sensors and machine learning in post-stroke rehabilitation assessment: a systematic review. Biomed. Signal Process. Control **71**, 103197 (2022)
3. Zaher, M., Samir, A., Ghoneim, A., Abdelhamid, L., Atia, A.: A framework for assessing physical rehabilitation exercises. In: 2023 Intelligent Methods, Systems, and Applications (IMSA), pp. 526–532. IEEE (2023)
4. Hamido, M., Mosallam, K., Diab, O., Amin, D., Atia, A.: A framework for human activity recognition application for therapeutic purposes. In: 2023 Intelligent Methods, Systems, and Applications (IMSA), pp. 130–135. IEEE (2023)
5. Phan, H.L., Le, T.H., Lim, J.M., Hwang, C.H., Koo, K.-I.: Effectiveness of augmented reality in stroke rehabilitation: a meta-analysis. Appl. Sci. **12**(4), 1848 (2022)
6. Balestra, N., Sharma, G., Riek, L.M., Busza, A.: Automatic identification of upper extremity rehabilitation exercise type and dose using body-worn sensors and machine learning: a pilot study. Digital Biomark. **5**(2), 158–166 (2021)
7. Burton, W.S., II., Myers, C.A., Rullkoetter, P.J.: Machine learning for rapid estimation of lower extremity muscle and joint loading during activities of daily living. J. Biomech. **123**, 110439 (2021)
8. Qiu, S., et al.: Multi-sensor information fusion based on machine learning for real applications in human activity recognition: State-of-the-art and research challenges. Inf. Fusion **80**, 241–265 (2022)
9. Sardari, S., Sharifzadeh, S., Daneshkhah, A., Nakisa, B., Loke, S.W., Palade, V., Duncan, M.J.: Artificial intelligence for skeleton-based physical rehabilitation action evaluation: a systematic review. Comput. Biol. Med. **158**, 106835 (2023)
10. Sekar, G., Sivakumar, C., Logeshwaran, J.: NMLA: the smart detection of motor neuron disease and analyze the health impacts with neuro machine learning model. NeuroQuantology **20**(8), 892 (2022)
11. Ekta, G., Tulika, M.G.: Risk factor distribution for cardiovascular diseases among high school boys and girls of urban dibrugarh, assam. J. Family Med. Primary care **5**(1), 108–113 (2016)
12. Fang, M., et al.: A hybrid machine learning approach for hypertension risk prediction. Neural Comput. Appl. **35**(20), 14487–14497 (2023)

13. Shtar, G., Rokach, L., Shapira, B., Nissan, R., Hershkovitz, A.: Using machine learning to predict rehabilitation outcomes in postacute hip fracture patients. Arch. Phys. Med. Rehabil. **102**(3), 386–394 (2021)
14. Donisi, L., et al.: Positive impact of short-term gait rehabilitation in parkinson patients: a combined approach based on statistics and machine learning. Math. Biosci. Eng. **18**(5), 6995–7009 (2021)

Enhancing Intrusion Detection by Using Machine Learning

Andleeb Khan[1(✉)], Parma Nand[1], Bharat Bhushan[1], Alaa Ali Hameed[2], and Akhtar Jamil[3]

[1] Department of Computer Science and Engineering, Sharda University, Greater Noida, India
khanandleeb1122@gmail.com, parma.nand@sharda.ac.in, bharat_bhushan1989@yahoo.com

[2] Department of Computer Engineering, Istinye University, Istanbul 34396, Turkey
alaa.hameed@istinye.edu.tr

[3] Department of Computer Science, National University of Computer and Emerging Sciences, Islamabad 44000, Pakistan
akhtar.jamil@nu.edu.pk

Abstract. With the advancing technology, the increase in threats has been exponential. These technologies have led to the production of huge amounts of network traffic data. Therefore, it is of immense importance for the companies to safeguard this sensitive data from any attack by the implementation of robust security techniques as this can cause heavy financial losses to the companies. There is need for the enhancement in the performance of existing Intrusion Detection Systems in order to detect the attacks more diligently. This is done by the integration of Machine Learning in the intrusion detection approaches as they develop models which are data driven and hence detect and prevent cyber-attacks. This study develops an ML based intrusion detection model by using Random Forest, Naïve Bayes and XGBoost classifiers on UNSW_NB15 dataset and pre-processing the dataset by applying One Hot Encoder and thereby giving an accuracy of 99.98%, 74.59% and 66.20% respectively. The performance evaluation indicates that the Random Forest model with an accuracy of 99.98% clearly outperforms the other algorithms. Hence, this framework can improve the system security to a greater extent.

Keywords: Malware · Intrusion Detection · Machine Learning · Random Forest

1 Introduction

With the development seen in mobile internet there has also been significant growth in the software industry. This caused an increase in the numbers of malware. Malwares are malicious software including worms, viruses, ransomwares etc. Protecting sensitive information from attackers is becoming an increasingly important and challenging task [1]. Conventional approaches such as antivirus software (AVS) and firewall are insufficient to address every kind of attack. Therefore, in addition to conventional techniques, extra security is required.

A. Abdelgawad et al. (Eds.): ISBCom 2024, LNNS 1268, pp. 699–712, 2025.
https://doi.org/10.1007/978-3-031-82377-0_56

Intrusion detection systems (IDS) play a magnificent role in attack detection. However, still there are events when the IDS fail to detect the attack pattern correctly, especially when it is the case of unknown cyber-attacks as the IDS is incapable of detecting unknown attacks [1, 2]. Thereby generating the need for an IDS that can push forward and detect novel patterns of attacks. The intrusion detection techniques which are combined with artificial intelligence (AI) approaches like machine learning (ML) and deep learning (DL) have shown much greater performance when it comes to malware detection [2]. Compared to conventional malware detection methods, these techniques are more reliable, accurate, and generalizable. They also reduce the possibility of false positives for a large number of recently discovered malware samples.

Malware detection with ML includes broadly two major phases. The first is of data pre-processing which includes extraction of relevant features from the given data and converting them into a form that can be processed by the ML model selected for classification [3]. The second phase is the stage of classification. This involves using the features for model training and further performing the classification task. The malware classification model presented by this paper works on UNSW-NB15 dataset and makes use of One hot encoder in the data preprocessing stage for converting categorical values into numerical. It also does data cleaning by dropping irrelevant features. The selected features and then fed to the ML model working on Random Forest algorithm for classification.

Hence this paper presents an ML based framework for intrusion detection from network traffic data. The major *contributions* are given as:

- It identifies the recent advancements in the field of ML based intrusion detection and classification of network traffic data.
- It lays down a comparative analysis of recent literatures in the field of intrusion detection and classification with ML approach.
- It presents a model for the detection of intrusion with remarkable accuracy of 99.98% outperforming the studies considered in the literature.
- It further helps to identify the existing gaps in the current literature and gives recommendations on how the existing works can be extended to develop robust and secure systems.

This paper is hence organized in the following upcoming sections: Sect. 2 gives the literature review along with recent advancements in the field of intrusion detection with ML as an approach. Section 3 gives the background of malware and related concepts. Section 4 gives the methodology of our proposed framework. Section 5 finally concludes this review paper and suggests the future research directions.

2 Literature Review

The classification and detection of malware or any intrusion is done through a variety of ML algorithms. This section discusses the recent advancements in intrusion or malware detection and classification followed by the comparative analysis in Table 1. Kuruvila et al. [4] proposed an explainable ML framework which detects intrusions based on Hardware performance counter double regression(HPCDR). It is used for intrusion detection based on HPC. It is transparent and effectively locates harmful instructions in the program. Xing et al. [5] developed a novel model for the detection of malware based on the representation of grey-scale malware images and a DL model working on an autoencoder network. Li et al. [6] proposed a framework for the prediction and propagation of malware on the basis of Graph Convolutional Networks (GCN) and representational learning in order to make the malware propagation more effective, along with Tensor2vec which is a novel method for learning the structural features to propagate malware. Lin et al. [7] developed a framework with the use of autoencoder and multilayer perceptron for detecting explosion attacks through imbalanced datasets. This framework also proposes a sequential search engine based on range. Vishwakarma et al. [8] proposed a 2-phase IDS which works in the IoT environment by using naïve bayes(NB) classifier and elliptic envelope method for detecting benign and malicious data. Khan et al. [9] developed a novel framework for malware detection which works on squeezed boosted boundary region and split transform merge (SB-BR-STM) which is a CNN approach along with ensemble learning. Islam et al. [10] developed a dynamic analysis technique for multi-classification by using ML based weighted voting. It studied an ensemble of various ML algorithms namely Random Forest (RF),DT, SVM, Multi-level Perceptrons (MLP) and logistic regression (LR).

Sun et al. [11] proposed an IDS for classifying and detecting records related to malwares in health apps through the use of Particle Swarm Optimization (PSO) and AdaBoost. Devi et al. [12] proposed a DL based intrusion detection and prevention technique in the IoT background by using an LSTM classifier for the purpose of detection and Improved Elliptic Curve Cryptography (IECC) for preventing it. Wazid et al. [13] proposed a novel ML based ensemble framework for detecting different types of intrusions in healthcare systems designed on industry 5.0 standards. Ghamdi et al. [14] proposed the framework of an IDS with deep neural network (DNN) and analyzed the performance by comparing the results with CNN and SVM. Jang et al. [15] proposed the development of analyzable datasets from raw data which was then used to detect intrusions with the help of six ML algorithms. Wang et al. [16] proposed a detector for malware attacks based on DL which detects malware in the android environment through LINK-NET classifier. Aljabri et al. [17] developed an ML model that can detect unknown samples. A new dataset which comprised of recent malware samples was utilized for detection. Sadhwani et al. [18] developed an intelligent IDS that works on a DNN model and performs binary and multi-class classification of threats.

Table 1. Recent work related to ML based Intrusion Detection.

Reference/year	Description	Techniques	Dataset	Outcome
Kuruvila et al. [4]/2022	Proposed an explainable HPCDR technique-based IDS	RF, DT and NN	Open-source malware Bashlite used	Accuracy-91.81%(RF)
Xing et al. [5]/2022	DL and autoencoder based Malware detection model	Auto encoder network	Benign software data from Google app store Malware data from ViruShare	Accuracy-96%
Li et al. [6]/2022	Proposed malware prediction and propagation model along with Tensor2vec	GCN Representational learning	Real world data from three subnets	Accuracy- 75% -90%(variable)
Lin et al. [7]/2022	Developed a framework for attack detection through imbalanced datasets	Autoencoder, MLP	HDFS, TTP	Recall- 98%
Vishwakarma et al. [8]/2023	Proposed a 2 phase IDS for malicious data classification and detection	NB Elliptic envelope method	NSL-KDD UNSW_NB15 CIC-IDS2017	Accuracy-98.59%(CIC-IDS2017)
Khan et al. [9]/2023	Presented a CNN and ensemble learning based malware detection framework	SB-BR-STM SVM, MLP and AdabooSTM1	IoT_Malware dataset	Accuracy-98.5%
Islam et al. [10]/2023	Proposed an ML based weighted voting approach	RF, MLP, DT, SVM, KNN and LR	CCCS-CIC-AndMal-2020	Accuracy-95%
Sun et al. [11]/2023	Developed an IDS for health app platforms	PSO, AdaBoost	NSL KDD	Recall-96.67%
Devi et al. [12]/2023	Proposed an IDPS based on DL	Deep LSTM classifier, IECC	Real time IoT gadgets' data	Accuracy- 95%
Wazid et al. [13]/2023	Proposed an IDS for industry 5.0 based healthcare systems	Dolev-yao threat model, one hot encoder, PCA, ensemble of SVM, DT and KNN	KDD CUP 99	Accuracy- 95.12%

(continued)

Table 1. (*continued*)

Reference/year	Description	Techniques	Dataset	Outcome
Ghamdi et al. [14]/2023	Proposed an IDS framework with improved representation	DCNN	NSL-KDD	95.63%
Jang et al. [15]/2024	Developed improved datasets for intrusion detection with ML	NB, SVM, RF, DT, KNN, Quadratic Discriminant Analysis (QDA)	Zeek IDS log data	Accuracy-97%(SVM)
Wang et al. [16]/2024	Proposed a DL based malware detector for Android	LINK-NET, GAN, deep belief networks and LSTM	AndMal2020	Accuracy-99.81%
Aljabri et al. [17]/2024	Proposed an ML based ransomware detection model	XGBoost, LightGBM, AdaBoost, RF and Extra Tree	A combination of different Ransomware samples' data	Accuracy-97.85%
Sadhwani et al. [18]/2024	Proposed an intelligent IDS along with multi-class classification	RF, DT, SVM, Extra trees classifier, KNN and DL	Edge-IIoTset	Accuracy-96.15%

3 Background

An effective solution for the ever-advancing cyber threats requires the knowledge of basic malware fundamentals. Malware is defined as a malicious software which is usually designed for compromising the security of the systems and to gain illegal profits out of it [19].

3.1 Types of Malwares

It comes in various forms and effects the systems in numerous ways [20]. Each type of malware is designed by the attacker with a specific purpose and intention. The different types of malwares and existing defense mechanisms for each sort are listed in Table 2.

3.2 Lifecycle of Malware

Malware phases may be different but the pattern observed when a malware effects a system is similar [21]. Hence a malware lifecycle is used to represent different stages of an attack caused by a malware as depicted in Fig. 1.

4 Methodology

This section deals with the discussion of our proposed solution for malware detection and the dataset used for the same. The proposed method aims at detecting malicious traffic from a dataset has events which are both benign and malicious. The steps followed

Table 2. Malware Types and Existing Defense Mechanisms.

Malware Type	Description	Defense Mechanism
Virus	Spreads by copying itself when the programs are shared	Firewalls, AVS, limited file sharing
Worm	Self-replicates and exploits network vulnerabilities	Network security management tools
Trojan	Disguises as a legitimate system program and gets remote access	AVS, Intrusion Detection System (IDS)
Ransomware	Demands ransom for releasing encrypted files	AVS, Backup, Recovery tools
Rootkit	Hides its presence by OS modifications	Advanced AV, software updates
Keyloggers	Records keyboard strokes and steals credentials	Multi factor Authentication
Adware	Displays unwanted pop ups and ads	AVS, software and system updates, Ad-blocking
Spyware	Hides in legitimate software and spies or monitors user activity	AVS, System Updates
Botnets	Collection of compromised systems called Bots. Bots provide remote access and can launch DDoS attacks	Firewall, AVS, IDS/IPS

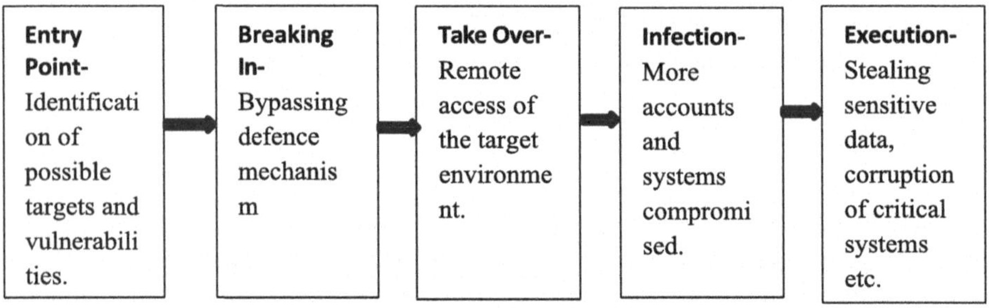

Fig. 1. Lifecycle of malware

in the proposed model are depicted in Fig. 2 and consist of 2 main stages. The first includes data-preprocessing which includes cleaning the dataset and significant data transformation.

The next stage consists of model selection and training. This study works on machine learning and RF, NB and XGBoost are chosen for malicious traffic detection.

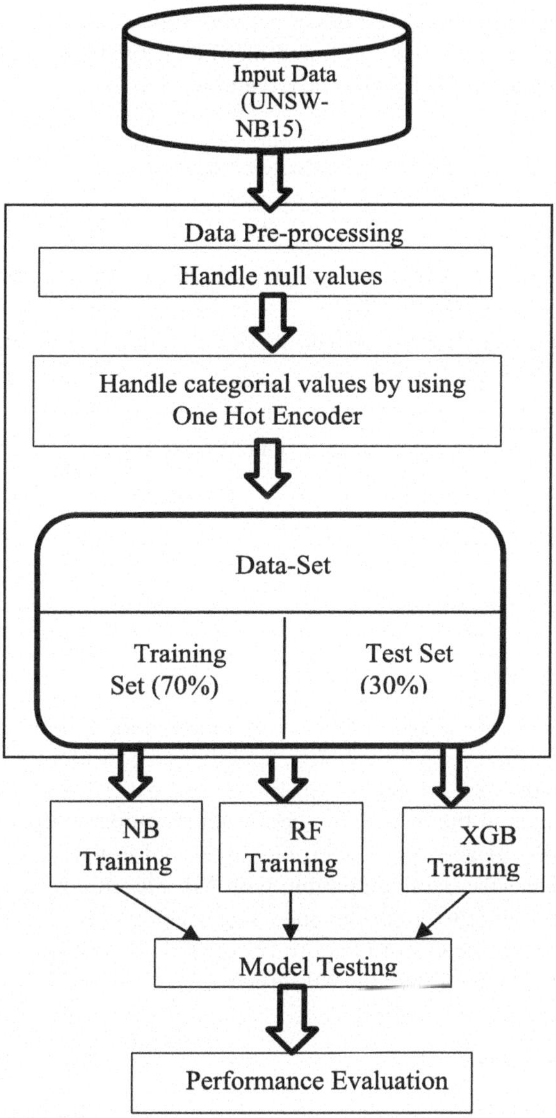

Fig. 2. Proposed Intrusion Detection Framework

4.1 Dataset Description

The dataset selected for this study is UNSW-NB15 that is comparatively a recent dataset details of which are presented in Table 3. It comprises of patterns of internet traffic and events from 9 different types of malicious activities [22]. This dataset comprises of minimal imprint assaults along with contemporary NetFlow data [23]. This data is made of both regular and unusual events. It was developed by the Australian Centre for Cyber security (ACCS) researchers at the University of New south Wales [24]. This was made by the Perfect-Storm tool. It has 100 GBs of network traffic data consisting of 2,540,044 realistic records divided into 4 csv files and monitored by TCP dump. It has network traffic data of types such as TCP, HTTP, UDP and ICMP.

Table 3. Details about the Dataset.

Feature	Description
Source Files	Raw network packets created using IXIA PerfectStorm tool, capturing real and synthetic activities
Types of Attacks	Fuzzers, Analysis, Backdoors, DoS, Exploits, Generic, Reconnaissance, Shellcode, Worms
Features	49 features generated from Argus and Bro-IDS tools
Records	2,540,044 records stored in four CSV files
Training Set	175,341 records
Testing Set	82,332 records

4.2 Data Pre-processing

It is meant to boost accuracy and reduce errors in the prediction of software cost. It is crucial to carefully select the ML technique and dataset considering their unique traits [25]. This dataset underwent pre-processing by merging data from all the 4 CSV files. The null or missing values in the dataset as seen in the features ct_ftp_cmd, is_ftp_login, ct flw http mthd, dsport and sport, were identified and replaced with 0. This is also done in order to make the data compatible with ML algorithms used. Numerical and categorical features were distinguished and categorical features were transformed into numeric values through the use of One hot Encoder. This results in all the columns to be in numeric form which will aid the ML process.

4.3 Datatype Resolution

There are a total of 49 features which are further divided into binary, categorical and nominal data types. The features which are not in integer or float type are converted in usable data type [24]. The last and the second last feature in the dataset i.e. 'attack category' and 'label' are multi-class and binary labels respectively. The multi-class label gets converted to integer type during execution through encoding. The nominal data type features are 'srcip', 'dstip', 'proto', 'state' and 'service'.

4.4 Model Selection

The dataset which is now pre-processed is divided in training and testing sets and this is done by dividing the complete dataset into 70 by 30 ratio. The training dataset which is

70% of the original complete UNSW_NB15 dataset is fed to three different ML algorithms. They perform the classification of the network traffic into normal and malicious type by getting tested on the testing dataset which accounts for 30% of the original UNSW_NB15 complete dataset.

Random Forest. It is a supervised learning method which uses ensemble learning. It amalgamates multiple classifiers in order to solve complex problems [26]. This technique also enhances the performance of the ML model. It uses random samples for the creation of decision trees, each tree does the individual prediction and voting is done to arrive at the best solution [26].

Naïve Bayes. It is a probabilistic classifier which is based on Bayes' Theorem. It works on the assumption that the features are independent conditionally [26]. This classifier is simple and computationally efficient hence it finds an increasing use in the present-day research.

XGBoost. It is also an ensemble learning method which works on gradient boosting. It combines weak learners and builds a powerful predictive model. It can also be used for the purpose of regression. It provides robustness, scalability and optimizes the performance.

5 Experimental Setup and Evaluation

The experiment in this work is performed by using standard Python libraries 3.11.7 which are developed with the help of Jupiter Notebook 7.0.8. The hardware details consist of Windows 11, 64-bit OS, 13th Gen Intel core i5 processor (1.30 GHz) and 16 GB of RAM. The training and testing of ML model are done through Scikit Learn framework.

The performance of the proposed ML model is evaluated by using accuracy and recall as the fundamental metrics. The accuracy measure represents the number of correctly predicted instances to the total instances [27]. Recall on the other hand gives the actual instances of intrusions that the ML model identifies. They are calculated by using the given formulas [28], where TP is true positive, FP is false positive, TN is true negative and FN is false negative.

$$\text{Accuracy} = \frac{\text{TP} + \text{TN}}{\text{TP} + \text{TN} + \text{FP} + \text{FN}} \tag{1}$$

$$\text{Recall} = \frac{\text{TP}}{\text{TP} + \text{FN}} \tag{2}$$

$$\text{Precision} = \frac{\text{TP}}{\text{TP} + \text{FP}} \tag{3}$$

$$\text{F1} = \frac{2}{\frac{1}{\text{Precision}} + \frac{1}{\text{Recall}}} \tag{4}$$

The accuracy of the proposed model with RF algorithm comes out to be 99.98%, with NB is 75% and with XGBoost is 66%, which clearly depicts that with RF, there is significant enhancement in the intrusion detection method as compared to the existing work based on ML. A comparative table of the proposed work with the existing ones is depicted through Table 4.

Table 4. Comparison with recent literatures.

Model	Accuracy
[4]	0.9181
[5]	0.96
[6]	0.9
[7]	0.98
[8]	0.9859
[9]	0.985
[10]	0.95
[11]	0.9667
[12]	0.95
[13]	0.9512
[14]	0.9563
[15]	0.97
[16]	0.9981
[17]	0.9785
[18]	0.9615
Proposed model (RF)	0.9998

The confusion matrices of RF, NB, and XGBoost models are given in Fig. 3, Fig. 4 and Fig. 5 respectively. These further explain the results along with the depiction of true positives i.e. the number of correct attacks detected and true negatives i.e. the number of correct detections of normal instances by the ML based model [29].

The results obtained through the experimentation are presented in Table 5 which shows a comparison of different performance indicators of the three ML classifiers.

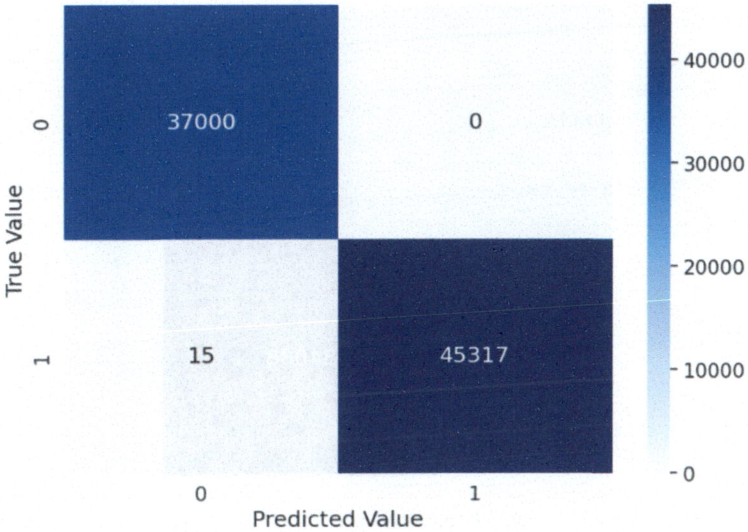

Fig. 3. Confusion Matrix of RF Model.

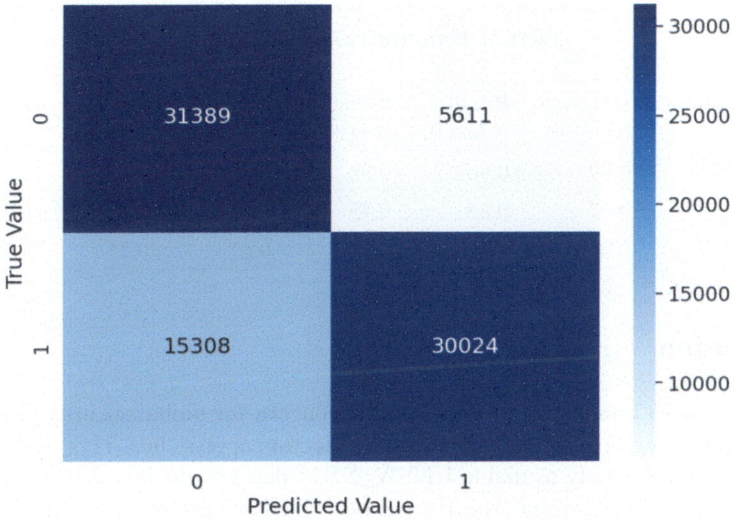

Fig. 4. Confusion Matrix of NB Model.

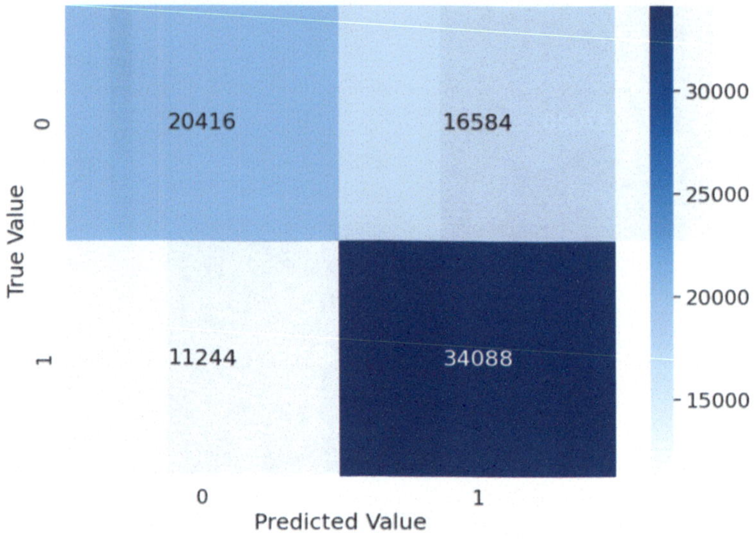

Fig. 5. Confusion Matrix of XGBoost Model.

Table 5. Performance Comparison.

Model	Accuracy	Precision (Class 0)	Recall (Class 0)	F1-Score (Class 0)	Precision (Class 1)	Recall (Class 1)	F1-Score (Class 1)
RF	0.99	0.99	0.96	0.98	0.99	0.96	0.98
NB	0.75	0.67	0.85	0.75	0.84	0.66	0.74
XGB	0.66	0.64	0.55	0.59	0.67	0.75	0.71

6 Conclusion

The threat of cyber-attacks is an ever-growing concern for global security. This issue is addressed by integrating ML with intrusion detection approaches. This study utilises a well-known and publicly available UNSW_NB15 dataset which is developed by the ACCS lab, UNSW in Australia. This dataset is cleaned and pre-processed with one hot encoder to achieve a higher detection rate. The proposed model makes use of RF, NB and XGBoost algorithms to distinguish between malicious and normal data traffic. The model achieved an accuracy of 99.98% when trained and tested with RF. This makes it significantly better in performance than the existing models. This further signifies the role of ML in cyber-attack detection.

However, there are significant open areas that can be catered to after this research. Research can be done to test the results on real time network data. This would make the intrusion detection models more robust and secure. Balanced datasets can be used in the model along with deep learning approaches. This would further optimize the results of this framework and hence lead to the development of more secure systems.

Disclosure of Interests. The authors have no competing interests to declare that are relevant to the content of this article.

References

1. Kumar, A., Bhushan, B., Malik, A., Kumar, R.: Protocols, solutions, and testbeds for cyber-attack prevention in industrial SCADA systems. Stud. Big Data **3**, 355–380 (2021). https://doi.org/10.1007/978-981-16-6210-2_17
2. Tannirkulam Chandrasekaran, S., Kuruvila, A.P., Basu, K., Sanyal, A.: Real-time hardware-based malware and micro-architectural attack detection utilizing CMOS reservoir computing. IEEE Trans. Circuits Syst. II Express Briefs **69**(2), 349–353 (2022). https://doi.org/10.1109/tcsii.2021.3102526
3. Ahmed, N., Hassan, F., Aurangzeb, K., Magsi, A.H., Alhussein, M.: Advanced machine learning approach for DOS attack resilience in internet of vehicles security. Heliyon **10**(8) (2024). https://doi.org/10.1016/j.heliyon.2024.e28844
4. Kuruvila, A.P., Meng, X., Kundu, S., Pandey, G., Basu, K.: Explainable machine learning for intrusion detection via hardware performance counters. IEEE Trans. Comput. Aided Des. Integr. Circuits Syst. **41**(11), 4952–4964 (2022). https://doi.org/10.1109/tcad.2022.3149745
5. Kumar, A., Bhushan, B., Nand, P.: Preventing and detecting intrusion of cyberattacks in smart grid by integrating blockchain. In: Sharma, D.K., Peng, S.-L., Sharma, R., Zaitsev, D.A. (eds.) Micro-Electronics and Telecommunication Engineering. LNNS, vol. 373, pp. 119–130. Springer, Singapore (2022). https://doi.org/10.1007/978-981-16-8721-1_12
6. Li, T., Liu, Y., Liu, Q., Xu, W., Xiao, Y., Liu, H.: A malware propagation prediction model based on representation learning and graph convolutional networks. Digital Commun. Netw. **9**(5), 1090–1100 (2023). https://doi.org/10.1016/j.dcan.2022.05.015
7. Lin, Y.-D., Liu, Z.-Q., Hwang, R.-H., Nguyen, V.-L., Lin, P.-C., Lai, Y.-C.: Machine learning with variational AutoEncoder for imbalanced datasets in intrusion detection. IEEE Access **10**, 15247–15260 (2022). https://doi.org/10.1109/access.2022.3149295
8. Vishwakarma, M., Kesswani, N.: A new two-phase intrusion detection system with naïve Bayes machine learning for data classification and elliptic envelop method for anomaly detection. Decis. Anal. J. **7**, 100233 (2023). https://doi.org/10.1016/j.dajour.2023.100233
9. Khan, S.H., et al.: A new deep boosted CNN and ensemble learning based IOT malware detection. Comput. Secur. **133**, 103385 (2023). https://doi.org/10.1016/j.cose.2023.103385
10. Islam, R., Sayed, M.I., Saha, S., Hossain, M.J., Masud, M.A.: Android malware classification using optimum feature selection and ensemble machine learning. Internet Things Cyber-Phys. Syst. **3**, 100–111 (2023). https://doi.org/10.1016/j.iotcps.2023.03.001
11. Sun, Z., An, G., Yang, Y., Liu, Y.: Optimized machine learning enabled intrusion detection 2 system for internet of medical things. Franklin Open **6**, 100056 (2024). https://doi.org/10.1016/j.fraope.2023.100056
12. Aiyshwariya Devi, R., Arunachalam, A.R.: Enhancement of IOT device security using an improved elliptic curve cryptography algorithm and malware detection utilizing DEEP LSTM. High-Conf. Comput. **3**(2), 100117 (2023). https://doi.org/10.1016/j.hcc.2023.100117
13. Wazid, M., Singh, J., Das, A.K., Rodrigues, J.J.: An ensemble-based machine learning-envisioned intrusion detection in industry 5.0-driven healthcare applications. IEEE Trans. Cons. Electron. **70**(1), 1903–1912 (2024). https://doi.org/10.1109/tce.2023.3318850
14. Al Ghamdi, M.A.: A fine-grained system driven of attacks over several new representation techniques using machine learning. IEEE Access **11**, 96615–96625 (2023). https://doi.org/10.1109/access.2023.3307018

15. Jang, M., Lee, K.: An advanced approach for detecting behavior-based intranet attacks by machine learning. IEEE Access **12**, 52480–52495 (2024). https://doi.org/10.1109/access.2024.3387016
16. Wang, Y., Jia, S.: Madras-net: a deep learning approach for detecting and classifying Android malware using linknet. Meas. Sens. **33**, 101113 (2024). https://doi.org/10.1016/j.measen.2024.101113
17. Aljabri, M., et al.: Ransomware detection based on machine learning using memory features. Egypt. Inf. J. **25**, 100445 (2024). https://doi.org/10.1016/j.eij.2024.100445
18. Sadhwani, S., Modi, U.K., Muthalagu, R., Pawar, P.M.: SmartSentry: cyber threat intelligence in industrial iot. IEEE Access **12**, 34720–34740 (2024). https://doi.org/10.1109/access.2024.3371996
19. Bensaoud, A., Kalita, J., Bensaoud, M.: A survey of malware detection using Deep Learning. Mach. Learn. Appl. **16**, 100546 (2024). https://doi.org/10.1016/j.mlwa.2024.100546
20. Gibert, D., Mateu, C., Planes, J.: The rise of machine learning for detection and classification of malware: research developments, trends and challenges. J. Netw. Comput. Appl. **153**, 102526 (2020). https://doi.org/10.1016/j.jnca.2019.102526
21. Manzil, H.H.R., Naik, S.M.: Detection approaches for Android malware: Taxonomy and review analysis. Expert Syst. Appl. **238**, 122255 (2024). https://doi.org/10.1016/j.eswa.2023.122255
22. Ngo, Q.-D., Nguyen, H.-T., Le, V.-H., Nguyen, D.-H.: A survey of IOT malware and detection methods based on static features. ICT Express **6**(4), 280–286 (2020). https://doi.org/10.1016/j.icte.2020.04.005
23. Madaan, G., Bhushan, B., Kumar, R.: Blockchain-based Cyberthreat mitigation systems for smart vehicles and Industrial Automation. Stud. Big Data, 13–32 (2020). https://doi.org/10.1007/978-981-15-7965-3_2
24. Choudhary, S., Kesswani, N.: Analysis of KDD-Cup'99, NSL-KDD and UNSW-NB15 datasets using Deep Learning in IOT. Procedia Comput. Sci. **167**, 1561–1573 (2020). https://doi.org/10.1016/j.procs.2020.03.367
25. Umer, M.A., Junejo, K.N., Jilani, M.T., Mathur, A.P.: Machine learning for intrusion detection in industrial control systems: applications, challenges, and recommendations. Int. J. Crit. Infrastruct. Prot. **38**, 100516 (2022). https://doi.org/10.1016/j.ijcip.2022.100516
26. Esmaeili, B., Azmoodeh, A., Dehghantanha, A., Karimipour, H., Zolfaghari, B., Hammoudeh, M.: IIoT deep malware threat hunting: From adversarial example detection to adversarial scenario detection. IEEE Trans. Ind. Inf. **18**(12), 8477–8486 (2022). https://doi.org/10.1109/tii.2022.3167672
27. Hai, T.H., Van Thieu, V., Duong, T.T., Nguyen, H.H., Huh, E.-N.: A proposed new endpoint detection and response with image-based malware detection system. IEEE Access **11**, 122859–122875 (2023). https://doi.org/10.1109/access.2023.3329112
28. Guerra-Manzanares, A.: Machine learning for android malware detection: Mission accomplished? a comprehensive review of open challenges and future perspectives. Comput. Secur. **138**, 103654 (2024). https://doi.org/10.1016/j.cose.2023.103654
29. Muzaffar, A., Ragab Hassen, H., Lones, M.A., Zantout, H.: An in-depth review of machine learning based Android Malware Detection. Comput. Secur. **121**, 102833 (2022). https://doi.org/10.1016/j.cose.2022.102833

A Comprehensive Review of Block Chain Based Multi Factor Authentication

Andleeb Khan[1](✉), Parma Nand[1], Bharat Bhushan[1], Esraa Mohammad Alazzawi[2], and Akhtar Jamil[3]

[1] Department of Computer Science and Engineering, Sharda University, Greater Noida, India
khanandleeb1122@gmail.com, parma.nand@sharda.ac.in, bharat_bhushan1989@yahoo.com
[2] Department of Computer Engineering, University of Baghdad, Al-Jadriya, Baghdad 10071, Iraq
israa.abd2005@coeng.uobaghdad.edu.iq
[3] Department of Computer Science, National University of Computer and Emerging Sciences, Islamabad 44000, Pakistan
akhtar.jamil@nu.edu.pk

Abstract. The ease and high availability of public and personal data in the existing automated and digital environments have made our technological systems vulnerable to security threats. The conventional approach of passwords and usernames is hence insufficient to ensure the security of data from cyber threats. Multi Factor Authentication can thus ensure effective security thereby minimizing the challenges of Single factor authentication. Blockchain based multi factor authentication (MFA) involves using blockchain for authentication purpose to enhance the security of a system. This paper reviews the recent advancements in the field of multi-factor authentication with blockchain as a means. It analyses the effectiveness of Blockchain based MFA systems by evaluating their strengths and weaknesses. It further studies the gaps in the recent literatures and summarises the future scope for Blockchain and MFA. It discusses techniques such as lightweight cryptographic methods, Arbiter PUF (Public Unclonable function), Clustering, Smart contracts and Machine learning used for authentication purposes. These techniques can be extended to include behavioural factors in multi-factor authentication.

Keywords: Blockchain · Multi-factor Authentication · Authentication

1 Introduction

According to a cyber-attack survey conducted in 2018, malware which steals critical and sensitive user information accounts for 34.4% of the attacks [1]. In totality when viewed, there were 65.6% of the attacks which were found to be exposing the user data and were causing potential theft of identity [1][2]. In today's world where IoT is integrated in every domain of our lives and its ever-increasing capability has led to increasing complexities and greater interfaces open to vulnerabilities. These vulnerabilities have led to increasing events of threats faced by these devices [3]. It was also found in statistics that about 80%

A. Abdelgawad et al. (Eds.): ISBCom 2024, LNNS 1268, pp. 713–726, 2025.
https://doi.org/10.1007/978-3-031-82377-0_57

of the users often tend to repeat their passwords when they use multiple sites. This further gives an edge to the hacker [4]. To tackle this problem multi-factor authentication is used. In this the user needs to provide two or more than two proofs of identity. With this added layer of authentication, the person trying to steal that information i.e. the hacker would also need to put on extra efforts to steal that information [5].

There are three ways to authenticate an individual. By knowledge, possession and inheritance. Here knowledge means something that an individual must have to gain access for example a password. Hence password-based authentication is the most common form used. Possession refers to the items for ex hardware, that the user possesses for authentication and inheritance refers to the property of characteristic that a user inherits and is unique to him/her for example biometric, voice or facial features [6]. Along with passwords if there is use of secondary authentication methods like facial recognition, iris scan and biometrics then there are comparatively lesser chances of breach of information and un-authorized access [7].

The existing methods of authentication prove to be insufficient when dealing with decentralized frameworks as the existing mechanisms are largely made for centralized infrastructures and hence cannot protect against cyber threats [7]. Decentralized and distributed systems work and provide services and applications through independent nodes that are working together. These nodes are connected through heterogenous networks and they communicate and provide services. IoT also forms a type of distributed network [8]. Since distributed networks are increasingly being used in our daily life activities, therefore novel security mechanisms are needed for securing them. In this scenario blockchain technology has emerged as one of the solutions that could enhance the way multi-factor authentication works as now the credentials required for authentication would not be stored on a centralized system or a database instead, they would be stored on decentralized and immutable ledgers [9]. The centralized authentication and blockchain based authentication differ in the following ways. In case of traditional authentication, the credentials are stored on a database and when a user tries to access a system or a service the credentials are checked through the database and then the user is granted authenticated access. However, in case of blockchain based authentication the credentials are stored on distributed ledgers and when any user tries to access a system or a service his/her credentials are authenticated through a unique consensus algorithm. This access is granted when the consensus is met by the blockchain nodes as depicted in Fig. 1.

Hence this paper presents a comprehensive review of the existing literature on blockchain based multi-factor authentication techniques. The major *contributions* are given as:

- It identifies the current state-of-the-art in blockchain based multi-factor authentication.
- It evaluates the effectiveness of blockchain based MFA by analyzing the strengths and weaknesses.
- It further helps to identify the existing gaps in the current literature and gives recommendations on how the existing works can be extended to develop robust and secure systems.

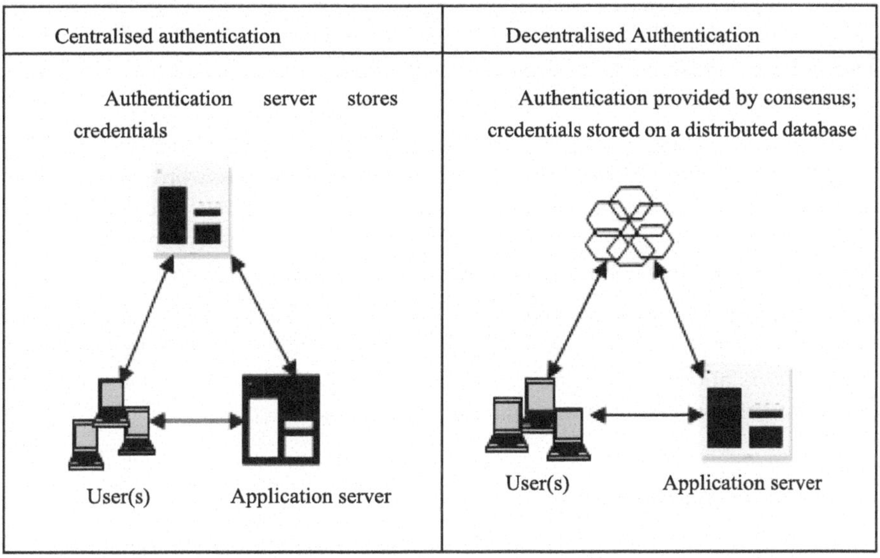

Fig. 1. Comparison of Centralized and De-centralized Authentication

This paper is hence organised in the following upcoming sections: Sect. 2 gives the background of Blockchain and its concepts. Section 3 explains MFA and gives the integration of Blockchain with MFA. Section 4 gives the recent advancements in the field of blockchain and MFA. Section 5 finally concludes this review paper and suggests the future scope.

2 Blockchain Background

The initial documentation of the design of blockchain took place in 2008 along with its initial implementation in 2009 as it was deployed in Bitcoin as an integral component. Both were proposed by Satoshi Nakamoto, an unidentified person [10]. The distributed public ledger used by the Bitcoin system, the blockchain, keeps track of and validates every transaction made on the public, peer-to-peer networked system [11]. The ability of the Bitcoin blockchain to prevent double spending on transactions involving bitcoins conducted in a completely decentralized peer-to-peer network without reliance on any reliable central authority is a remarkable innovation.

2.1 Blockchain Preliminaries

The major components and concepts involved in the Blockchain network are public key cryptography, hash functions, smart contracts and various consensus algorithms as explained in the following sub sections.

Public Key Cryptography. In this approach of authentication, the client has a pair of keys Public and private. The private key is kept as secret and the public key is open to all

[11]. The user generates a pair of keys and if by any chance one of the keys is leaked then he can reset the pair of keys and generate a new pair along with a blockchain address. And since the blockchain is based on distributed nodes, the new pair is communicated to all the different nodes in the blockchain.

Hash Functions. These are used in authentication to store the hash value of login credentials like passwords instead of the actual value of the password. When the user makes the login and enters the password it is checked for the hash value. The user is authenticated by matching the hash value. Keccak-256 (used in Ethereum) and SHA-256 (used in Bitcoin) are two common hash functions in blockchain [12].

Smart Contracts. Permissions and access control can be managed by smart contracts running on a blockchain. An enterprise can, for instance, put up a smart contract that specifies who is allowed access to particular systems or resources and can make that access contingent on a number of requirements being satisfied [12].

2.2 Blockchain Architecture

Blockchain is a secure ledger which maintains a list of transactions into a chain of blocks. Each of these blocks is secured with the help of cryptographic techniques in order to ensure the integrity of the transactions. Once the block faces competitive consensus procedure, it can then be added in the block chain as the new block. A block in the blockchain maintains the hash value of itself and the preceding block along with the transaction record. There is the use of consensus mechanism in the blockchain in order to monitor and control the entry of any new block, the protocol for verification and the consistency of the content in each block [14]. Therefore, the blockchain ensures that the blocks with transaction records are created, verified and their data integrity is maintained. This is depicted in Fig. 2.

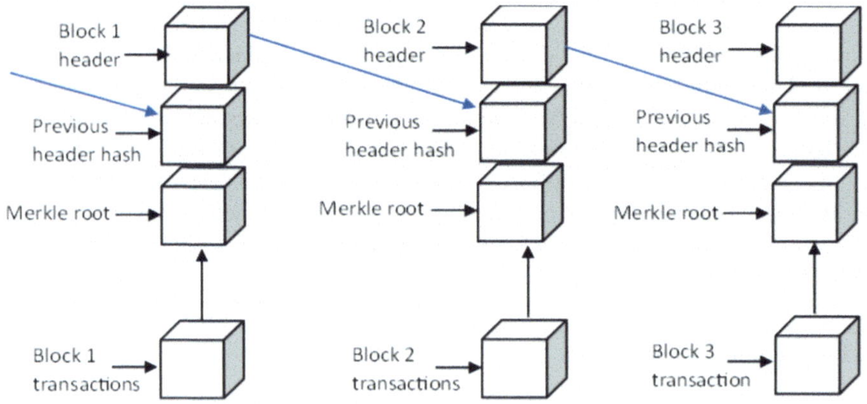

Fig. 2. Blockchain Architecture

Blockchain is of three types namely public, private and consortium. These types can be compared based on their characteristics as depicted in Table 1.

Table 1. Comparison in Types of Blockchains.

Characteristic	Decentralization	Immutability	Efficiency	Centralized
Public Blockchain	High	High	Variable	Low
Consortium Blockchain	Medium	Medium	High	Medium
Private Blockchain	Low	Low	High	High

3 Multi-Factor Authentication

Authentication is a safeguard which remains fundamental in case of securing a device or sensitive applications. In initial days, authentication was based out of only one factor which was used for the subject authentication. This type of authentication also called as single factor authentication (SFA) was user friendly and simple and was hence easily adopted by the people. For example, the use of password or PIN [15]. But this type of authentication was weak and just by gaining the access to one factor, the attacker can compromise the system. Hence, a system based on two factor authentication got proposed which included the combination of passwords along with a personal ownership factor like phone number etc. This was then followed by the development of Multi-factor Authentication (MFA) as shown in Fig. 3. MFA provided higher safety and continuous protection by employing greater than two forms of credentials. Largely, it is based on biometric factor which recognizes the individual based on their biological and behavioral features [16]. This ensured improved security.

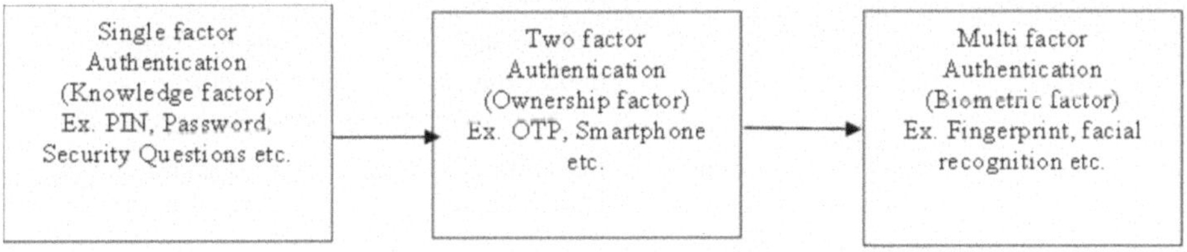

Fig. 3. Evolution of Authentication Techniques

3.1 Types of Factors Deployed in MFA

The purpose for employing MFA is to ensure the authentication and identification of sensitive data. Table 2 enlists already available factors for MFA along with the advantages and challenges involved.

Table 2. Comparative analyses of Types of Authentication Methods

Authentication Method	Description	Advantages	Challenges
Knowledge Factor	Users provide a secret passphrase (e.g., PIN, password, security questions)	Simple and widely used	Vulnerable to password-related issues
Ownership Factor	Combines something the user knows (e.g., password) with something they possess (e.g., smartphone)	Enhanced security by requiring both factors	Risk of token duplication
Voice Biometrics	Utilizes voice recognition as a factor for MFA	Convenient and available on most devices	Vulnerable to voice mimicry
Facial Recognition	Utilises facial features for authentication	Three-dimensional face recognition enhances security	Vulnerable to photo-based attacks
Ocular-Based Methods	Based on iris or retina patterns	Unique and non-invasive	Requires high-quality capture devices
Hand Geometry	Based on the physical shape of a hand	No fixed hand position required	Less robust in certain environments
Vein Recognition	Collects vein patterns from fingers or palms	High potential for security	Vulnerable to spoofing attacks
Fingerprint Scanner	Uses fingerprint patterns for authentication	Widely adopted but susceptible to fabrication	Not recommended as a standalone method
Thermal Image Recognition	Reconstructs thermal images of blood flow	Unique and non-contact	Sensitive to user conditions (e.g., sickness, emotions)
Geographical Location	Validates access based on device and user location	Supports multi-factor authentication	GPS signal reliability and potential jamming

3.2 Blockchain in MFA

The characteristics of reliability, immutability and unforgeability have made blockchain an effective authentication solution. An authentication system based on blockchain can thus ensure the confidentiality, authenticity and integrity of a system. It can also safeguard the system from threats like data and identity theft and unauthorized access. Blockchain technology integrates smart contracts to give fine grained access control [15]. The emerging collaboration of blockchain with cloud computing can provide a basis for the management and development of decentralized, time-sensitive and distributed security solutions.

An authentication system utilizing blockchain uses a distributed ledger which then authenticates the peers by a consensus algorithm that is unique. This is completely different from centralized authentication. The credentials of the users are stored in the ledgers. These can only be controlled after the registration. Therefore, the authentication with blockchain must be applied in multiple stages. To receive their cryptographic keys, the entities must first register with the network [16]. Then they need to set up their credentials and authentication type either TFA or MFA. This will be mined and then saved within the blockchain ledger. Then if an entity wishes to access any system, its credentials need validation of the blockchain nodes [16]. After a consensus is reached on the credentials, that entity is granted access to the requested system as depicted in Fig. 1.

4 Recent Advances

As the significance of blockchain in the domain of authentication unfolds with time, there have been efforts by the research community to study ways in which blockchain proves to be useful in the field of multi factor authentication. Oktian et al. [17] proposed TwoChain, which is a blockchain based two factor authentication(2FA) system used for web services using Ethereum blockchain and smart contracts. Kinai et al. [18] proposed an approach based blockchain transactions evidence and risk analysis which uses blockchain and MFA. Zhang et al. [19] developed a Blockchain based MFA protocol for the cross domain IIoT. Zhang et al. [20] proposed an authentication approach which decentralized and based on multiple factors for identity registration and verification. Abubakar et al. [21] proposed A blockchain-based 2FA method for web-based access to sensor data using Ethereum blockchain and smart contracts. Li et al. [22] proposed two blockchain based authentication methods for IIoT devices with APUF and SRAM (Static Random Access Memory) PUF (Physical Unclonable Function). Babu et al. [23] proposed a permissioned blockchain based hybrid and collaborative Intrusion Detection System against DDoS (Distributed Denial of Service) attacks for IoT devices.

Kim et al. [24] proposed a scheme using blockchain for authentication, verifiable credential and an identifier that is decentralized for the users of metaverse to carry out authentication and identity verification without the disclosure of information to the service providers. Ahmad et al. [25] developed BAuth-ZKP (Blockchain Authentication-Zero Knowledge Proof) which focuses on developing smart contracts among the entities of a smart city. It Performs the transactions with zero knowledge proof (ZKP) authentication in a secure and privacy-preserved manner. Ghaffari et al. [26] presented a taxonomy for categorizing the existing Authentication and Access Control (AAC) approaches, analyzing DLT (distributed Ledger Technologies) and their impact on the AAC mechanisms. Wang et al. [27] proposed a Hybrid Block chain-based Identity Authentication (HBIA) scheme for Mobile Crowd Sensing. Abdussami et al. [28] proposed a block chain based secure authentication and session key establishment protocol (BSAPM) for large scale WSN (Wireless Sensor Network) environment. Khashan et al. [29] proposed a hybrid, centralized and block chain-based authentication architecture. Almadani et al. [15] proposed a block chain based multi factor authentication as a service (BMFAaaS) as an approach. Shahideinejad et al. [30] presented a blockchain based authentication protocol for defending attacks and improving efficiency in communication, storage usage and computation. The following Table 3 provides a brief summary of the papers shortlisted for this review study.

5 Future Research Directions

The increasing usage and adoption of secure authentication services in industries has further pushed the concept of Blockchain in MFA. These authentication techniques can hence be extended to novel applications as mentioned in the subsections below.

5.1 Detection of Behavior

The characteristics and patterns followed by different individuals in not only unique but some of them are also difficult to mimic [31]. Hence such individual patterns can be used in order to program a system that will identify and work on a specific motor pattern. An example of it is how the typing pattern of each individual is different. This can be combined with text input-based authentication. The MFA system can be designed specifically to detect a specific type of gesture and the user has to replicate that gesture while holding the sensing device.

5.2 Recognition of DNA

The detection of cell lines for each human which are used in reverse genetic techniques, are also an essential research resource [31]. However, this may be expensive and time taking but its potential to identify the user as it forms a unique DNA source, cannot be denied. Hence it can be integrated to authorize a user.

5.3 Telecommunication

The authentication techniques can be combined with wireless solutions like the Radio-frequency Identification (RFID) and Near-Field Communication (NFC) in order to enhance the physical layer security. This solution finds its practical usage in the telecommunication sector [32].

5.4 Automobiles

The integration of vehicular patterns of the drivers proves to be a significant research area when it comes to authentication and security. Vehicular specific behaviors like the speed sensor, brake pressure sensor etc. and the human specific factors like the type of music, the number of calls made etc. can be analyzed to authenticate a user. Another significant solution can be the integration of occupant classification systems with authentication. These are available in some of the consumer cars. It is a type of sensor system which detects the driver/passenger and adjusts the vehicle according to his needs like his weight or posture [32].

Table 3. Recent work related to Blockchain Based MFA.

Reference/Year	Purpose	Methodology	Outcome(s)	Future Scope
Oktian et al. [17]/2020	A blockchain-based 2FA system is proposed	User registration, key generation	No bugs in Smart Contracts	Single-sign on feature possible
Kinai et al. [18]/2020	Authentication of transactions and risk analysis	Blockchain, SMS/USSD, Hyperledger fabric, Bayesian model	NA	Can be extended to mobile networks as blockchain nodes
Zhang et al. [19]/2022	Blockchain and MFA for IIoT	Smart Contracts, intra domain and cross domain authentication	Proposed protocol is efficient and reliable for IIoT	Can be practically implemented in IIoT ecosystems
Zhang et al. [20]/2022	Blockchain and MFA for e-identity registration and verification	Consortium Blockchain, decentralized identity storage, RBFT, Smart contracts	Efficient than blockchain based SFA	Enhance security of identity authentication
Abubakar et al. [21]/2022	Blockchain-based 2FA for sensor data	Ethereum clients, smart contracts	Authentication delay: 15 s	Interoperability with IoT frameworks
Li et al. [22]/2022	Blockchain authentication for IIoT devices	Blockchain, Pseudorandom Permutation, PUF	99.9% accuracy with SVM	Compatibility and security model needed
Babu et al. [23]/2022	Blockchain IDS against DDoS for IoT	Arbiter PUF, ensemble method, ECC	Accuracy: 97.39%, Detection rate: 98.53%	ECC based aggregate signature scheme
Kim et al. [24]/2023	Authentication scheme for metaverse	Blockchain, decentralized identifier, verifiable credential	Better computation and communication costs	Applicable to practical metaverse ecosystems
Ahmad et al. [25]/2023	BAuth-ZKP: smart contracts for smart city	ZKP, smart contracts for credentials	Better than 2FA	Behavioral patterns recognition possible

(*continued*)

Table 3. (*continued*)

Reference/Year	Purpose	Methodology	Outcome(s)	Future Scope
Ghaffari et al. [26]/2023	Taxonomy for AAC approaches, analyzing DLT	DLT for IoT, SSI	Trade-off between latency and security	Improving smart contract languages
Wang et al. [27]/2023	HBIA scheme for Mobile Crowd Sensing	Clustering, ZKP, zk-SNARK, Ethereum	Efficient proof verification phase	Reputation based incentive mechanism
Abdussami et al. [28]/2023	BSAPM for large scale WSN	Smart contract, FPGA, hash functions	Overcomes variety of attacks	No security attacks recognized
Khashan et al. [29]/2023	Hybrid, centralized, blockchain-based authentication	Lightweight cryptographic methods, Ethereum	Reduced power consumption, increased throughput	Image security and computer vision
Almadani et al. [15]/2023	BMFAaaS for distributed systems	–	Strengthened security, improved usability	Effective blockchain network simulation needed
Shahideinejad et al. [30]/2024	Blockchain and MFA in Medical Consortiums	ECC, Hyperledger Fabric	Increase in computation and communication overheads	Extended to attack prevention

5.5 ECG and EEG Recognition

The Electrocardiographic (ECG) data of a user is a biometric indicator which is difficult to mimic and hence forms a potential factor for authentication. They can be easily collected from the smart watch of the user and compared for authentication. Similarly, the Electroencephalographic (EEG) samples which are an analysis of the brain wave activity of the individual can be a potential source of authentication. Today these EEG samples can be easily collected with the help of devices available in the market [33].

6 Open Challenges

The integration of blockchain and authentication will undoubtedly enhance the security levels of a system. However, there are still some challenges in the application of this technology as depicted in Fig. 4.

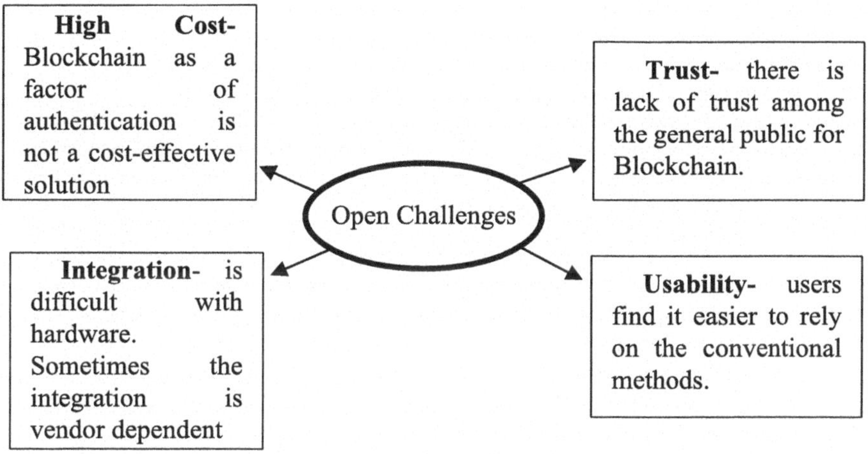

Fig. 4. Open Challenges in Blockchain based Authentication.

7 Conclusion

The literatures reviewed on blockchain based multi-factor authentication utilize the techniques of lightweight cryptography, machine-learning based ensemble techniques, public unclonable functions, smart contracts and clustering. It can also be noted that the studies reviewed in this paper follow different approaches. In the future this study can be extended to develop a model of multi-factor authentication based on blockchain that operates on multiple systems and thus offers interoperability. The current models provide authentication largely by storing the user data or credentials on the distributed ledger. In the future studies this form of authentication can also be integrated with the behavioral factors of the user i.e. gait, gestures, body language etc. The study can also be further extended to use Blockchain for data sharing. Further the approach of Blockchain-as-a-service for distributed systems like Fog, WSN and IoT can also be extended to Cloud computing and Content Delivery Networks.

Disclosure of Interests.. The authors have no competing interests to declare that are relevant to the content of this article.

References

1. Kumar, R.P., Bandanadam, S.R.: Block chain-based decentralized public auditing for cloud storage with improved EIGAMAL encryption model. Int. J. Inf. Technol. **16**(2), 697–711 (2023). https://doi.org/10.1007/s41870-023-01599-8
2. Sharif, M.H.U., Mohammed, M.A.: A literature review of financial losses statistics for Cyber Security and future trend. World J. Adv. Res. Rev., **15**(1), 138–156. https://doi.org/10.30574/wjarr.2022.15.1.0573(2022)
3. Reno, S., Priya, S. H., Al-Kafi, G. M., Tasfia, S., Turna, M.K.: A novel approach to optimizing transaction processing rate and space requirement of blockchain via off-chain architecture. Int. J. Inf. Technol. (2024). https://doi.org/10.1007/s41870-023-01685-x
4. Asim, J., et al.: Blockchain-based multifactor authentication for future 6G cellular networks: a systematic review. Appl. Sci. **12**(7), 3551 (2022). https://doi.org/10.3390/app12073551
5. Bamakan, S.M., Motavali, A., Babaei Bondarti, A.: A survey of blockchain consensus algorithms performance evaluation criteria. Expert Syst. Appl. **154**, 113385 (2020). https://doi.org/10.1016/j.eswa.2020.113385
6. Kumar, A., Bhushan, B., Shristi, S., Kalita, S., Chaganti, R., Obaid, A.J.: Blockchain embedded security and privacy preserving in Healthcare Systems. Blockchain Technology Solutions for the Security of IoT-Based Healthcare Systems, 241–261 (2023). https://doi.org/10.1016/b978-0-323-99199-5.00005-7
7. Kumar, A., Bhushan, B., Shristi, S., Chaganti, R., Soufiene, B.O.: Blockchain-based decentralized management of IOT devices for Preserving Data Integrity. Blockchain Technology Solutions for the Security of IoT-Based Healthcare Systems (2023), 263–286. https://doi.org/10.1016/b978-0-323-99199-5.00009-4
8. Bidgoly, A.J., Bidgoly, H.J., Arezoumand, Z.: A survey on methods and challenges in EEG based authentication. Comput. Secur. **93**, 101788 (2020). https://doi.org/10.1016/j.cose.2020.101788
9. Singh, M.P., Chopra, A.K.: The internet of things and multiagent systems: decentralized intelligence in distributed computing. In: 2017 IEEE 37th International Conference on Distributed Computing Systems, ICDCS, IEEE, pp. 1738–1747 (2017). https://doi.org/10.1109/ICDCS.2017.304
10. Bahalul Haque, A.K.M., Bhushan, B.: Blockchain for medical insurance: Synthesizing current knowledge and problematizing it for Future Research Avenues. Blockchain Technology Solutions for the Security of IoT-Based Healthcare Systems, 223–240 (2023). https://doi.org/10.1016/b978-0-323-99199-5.00002-1
11. Kabra, N., Bhattacharya, P., Tanwar, S., Tyagi, S.: Mudrachain: Blockchain-based framework for automated cheque clearance in financial institutions. Futur. Gener. Comput. Syst. **102**, 574–587 (2020). https://doi.org/10.1016/j.future.2019.08.035
12. Kaur, K., Garg, S., Kaddoum, G., Gagnon, F., Ahmed, S.H.: Blockchain-based lightweight authentication mechanism for vehicular fog infrastructure. In: 2019 IEEE International Conference on Communications Workshops (ICC Workshops), IEEE, pp. 1–6 (2019). https://doi.org/10.1109/ICCW.2019.8757184
13. Chaganti, R., Bhushan, B., Ravi, V.: A survey on blockchain solutions in ddos attacks mitigation: techniques, open challenges and future directions. Comput. Commun. **197**, 96–112 (2023). https://doi.org/10.1016/j.comcom.2022.10.026
14. Ghayvat, H., et al.: CP-bdhca: blockchain-based confidentiality-privacy preserving Big Data Scheme for healthcare clouds and applications. IEEE J. Biomed. Health Inform. **26**(5), 1937–1948 (2022). https://doi.org/10.1109/jbhi.2021.3097237
15. Almadani, M.S., Alotaibi, S., Alsobhi, H., Hussain, O.K., Hussain, F.K.: Blockchain-based multi-factor authentication: a systematic literature review. Internet of Things **23**, 100844 (2023). https://doi.org/10.1016/j.iot.2023.100844

16. Ometov, A., Bezzateev, S., Mäkitalo, N., Andreev, S., Mikkonen, T., Koucheryavy, Y.: Multi-factor authentication: a survey. Cryptography, **2**(1), 1 (2018). https://doi.org/10.3390/cryptography2010001
17. Oktian, Y.E., Lee, S.-G., Lee, H.-J.: Twochain: Leveraging blockchain and smart contract for two factor authentication. In: 2020 3rd International Seminar on Research of Information Technology and Intelligent Systems (ISRITI). https://doi.org/10.1109/isriti51436.2020.9315514
18. Kinai, A., Otieno, F., Bore, N., Weldemariam, K.: Multi-factor authentication for users of non-internet based applications of blockchain-based platforms. 2020 IEEE International Conference on Blockchain (Blockchain) (2020). https://doi.org/10.1109/blockchain50366.2020.00076
19. Zhang, Y., Li, B., Wu, J., Liu, B., Chen, R., Chang, J.: Efficient and privacy-preserving blockchain-based multifactor device authentication protocol for cross-domain IIoT. IEEE Internet Things J. **9**(22), 22501–22515 (2022). https://doi.org/10.1109/jiot.2022.3176192
20. Zhang, Z., Sun, Y., Huang, P., Wang, C.: A blockchain and multi factor fusion based Electronic Identity Registration and Verification System. In 2022 3rd International Conference on Computer Science and Management Technology (ICCSMT) (2022). https://doi.org/10.1109/iccsmt58129.2022.00097
21. Abubakar, M., Jaroucheh, Z., Al Dubai, A., Liu, X.: A lightweight and user-centric two-factor authentication mechanism for IOT based on Blockchain and smart contract. 2022 2nd International Conference of Smart Systems and Emerging Technologies (SMARTTECH) (2022). https://doi.org/10.1109/smarttech54121.2022.00032
22. Li, D., et al.: Blockchain-based authentication for IIOT devices with PUF. J. Syst. Architect. **130**, 102638 (2022). https://doi.org/10.1016/j.sysarc.2022.102638
23. Babu, E.S., BKN, S., Nayak, S.R., Verma, A., Alqahtani, F., Tolba, A., Mukherjee, A.: Blockchain-based Intrusion Detection System of IOT urban data with device authentication against ddos attacks. Comput. Electric. Eng., **103**, 108287 (2022). https://doi.org/10.1016/j.compeleceng.2022.108287
24. Kim, M., Oh, J., Son, S., Park, Y., Kim, J., Park, Y.: Secure and privacy-preserving authentication scheme using decentralized identifier in metaverse environment. Electronics **12**(19), 4073 (2023). https://doi.org/10.3390/electronics12194073
25. Ahmad, M., et al.: Bauth-ZKP—a blockchain-based multi-factor authentication mechanism for securing Smart Cities. Sensors **23**(5), 2757 (2023). https://doi.org/10.3390/s23052757
26. Ghaffari, F., Bertin, E., Crespi, N., Hatin, J.: Distributed ledger technologies for authentication and access control in networking applications: a comprehensive survey. Comput. Sci. Rev. **50**, 100590 (2023). https://doi.org/10.1016/j.cosrev.2023.100590
27. Wang, T., Shen, H., Chen, J., Chen, F., Wu, Q., Xie, D.: A hybrid blockchain-based identity authentication scheme for mobile crowd sensing. Futur. Gener. Comput. Syst. **143**, 40–50 (2023). https://doi.org/10.1016/j.future.2023.01.013
28. Abdussami, M., Amin, R., Saravanan, P., Vollala, S.: BSAPM: Blockchain based secured authentication protocol for large scale WSN with FPGA implementation. Comput. Commun. **209**, 63–77 (2023). https://doi.org/10.1016/j.comcom.2023.06.011
29. Khashan, O.A., Khafajah, N.M.: Efficient hybrid centralized and blockchain-based authentication architecture for heterogeneous IOT Systems. J. King Saud Univ.—Comput. Inf. Sci. **35**(2), 726–739 (2023). https://doi.org/10.1016/j.jksuci.2023.01.011
30. Shahidinejad, A., Abawajy, J., Huda, S.: Untraceable blockchain-assisted authentication and Key Exchange in medical consortiums. J. Syst. Architect. **151**, 103143 (2024). https://doi.org/10.1016/j.sysarc.2024.103143
31. Bamashmos, S., Chilamkurti, N., Shahraki, A.S.: Two-layered multi-factor authentication using decentralized blockchain in an IOT environment. Sensors **24**(11), 3575 (2024). https://doi.org/10.3390/s24113575

32. Jose Diaz Rivera, J., Muhammad, A., Song, W.-C.: Securing digital identity in the zero trust architecture: a blockchain approach to privacy-focused multi-factor authentication. IEEE Open J. Commun. Soc., **5**, 2792–2814 (2024). https://doi.org/10.1109/ojcoms.2024.3391728
33. Dhote, S., Maidamwar, P., Thakur, S.: Integrating Blockchain and multi-factor authentication for Enhanced Cloud Security in certificate verification systems. In *2024 International Conference on Distributed Computing and Optimization Techniques (ICDCOT)* (2024). https://doi.org/10.1109/icdcot61034.2024.10516190

Advancing Automated Deception Detection: A Multimodal Approach to Feature Extraction and Analysis

Mohamed Bahaa(✉), Mena Hany, and Ehab E. Zakaria

October University for Modern Sciences Arts -MSA University, Giza, Egypt
{mohamed.bahaa4,mena.hany,Ezakaria}@msa.edu.eg

Abstract. With the exponential increase in video content, the need for accurate deception detection in human-centric video analysis has become paramount. This research focuses on the extraction and combination of various features to enhance the accuracy of deception detection models. By systematically extracting features from visual, audio, and text data, and experimenting with different combinations, we developed a robust model that achieved an impressive 99% accuracy. Our methodology emphasizes the significance of feature engineering in deception detection, providing a clear and interpretable framework. We trained various machine learning models, including LSTM, BiLSTM, and pre-trained CNNs, using both single and multi-modal approaches. The results demonstrated that combining multiple modalities significantly enhances detection performance compared to single modality training. This study highlights the potential of strategic feature extraction and combination in developing reliable and transparent automated deception detection systems in video analysis, paving the way for more advanced and accurate detection methodologies in future research.

Keywords: multimodal · deception detection · lie detection

1 Introduction

Deception detection has become an essential field of study, especially with the rapid growth of video content in various domains such as security, law enforcement, and social media. The ability to accurately identify deceptive behavior through automated systems can significantly impact these fields by providing reliable tools for surveillance and analysis. Deception detection in videos involves analyzing human behavior, facial expressions, voice patterns, and other non-verbal cues to identify signs of deceit. Traditionally, deception detection has relied on human judgment, which is often subjective and prone to errors. However, with advancements in artificial intelligence (AI) and machine learning, automated systems have shown great promise in enhancing the accuracy and reliability of deception detection [15].

A. Abdelgawad et al. (Eds.): ISBCom 2024, LNNS 1268, pp. 727–738, 2025.
https://doi.org/10.1007/978-3-031-82377-0_58

These systems leverage complex algorithms and large datasets to learn patterns associated with deceptive behavior, making them more objective and consistent compared to human evaluators [3]. One of the critical challenges in automated deception detection is feature extraction. Features such as facial expressions, body language, voice pitch, and speech patterns need to be meticulously extracted and analyzed to detect deception accurately. Research has shown that combining multiple modalities-visual, auditory, and linguistic-can significantly improve the performance of deception detection systems [12].

In our study, we focus on extracting a wide range of features from video data and exploring various combinations to train a robust deception detection model. By carefully selecting and combining features from visual and audio data, we aim to enhance the model's accuracy and interpretability. Our approach emphasizes the importance of feature engineering in building effective deception detection systems, demonstrating that strategic feature extraction and combination can lead to substantial improvements in performance [9].

In summary, this research contributes to the field of automated deception detection by providing a comprehensive analysis of feature extraction and combination techniques. Our findings highlight the potential of AI-driven systems to offer reliable and interpretable solutions for identifying deceptive behavior in videos, thereby advancing the capabilities of current detection methodologies.

Deception detection is essential because it enables us to discern the truth in critical situations, ranging from criminal investigations to verifying online content. Understanding when deception occurs is crucial for timely interventions and ensuring justice and security. Human deception can manifest in various forms, such as falsifying information, concealing the truth, or providing misleading cues. Detecting these deceptive behaviors involves examining numerous signals, including micro-expressions, body language, vocal nuances, and verbal inconsistencies [8].

It plays a vital role in maintaining the integrity and security of societal interactions. In law enforcement, it helps in identifying suspects who might be lying during interrogations. In the realm of social media, it aids in filtering out false information and protecting users from potential scams. The need for automated systems in this domain has surged due to the sheer volume of data generated daily, which surpasses human capacity for effective analysis [3].

It is crucial in scenarios where accuracy and objectivity are paramount. For instance, during legal proceedings, accurate detection of deceptive testimonies can influence the outcome of cases. In security screenings, identifying deceptive behavior can prevent potential threats. In everyday online interactions, it helps in verifying the authenticity of user-generated content, thus fostering a safer digital environment [9].

Our study leverages recent advancements in AI to address these challenges, employing sophisticated algorithms to analyze and interpret various data modalities. By integrating visual, auditory, and linguistic features, our model aims to provide a comprehensive and nuanced understanding of deceptive behaviors. This multi-modal approach not only enhances detection accuracy but also provides a more interpretable framework for understanding the underlying mechanisms of deception [8].

Fig. 1. An example of spontaneous expressions with a truthful response (left) and a deceptive response (right).

2 Related Work

Deception detection has garnered significant attention in recent years, particularly with advancements in artificial intelligence and machine learning. Numerous studies have explored various methods and modalities to improve the accuracy and reliability of automated deception detection systems.

Recent research has emphasized the integration of multiple modalities to enhance detection accuracy. For instance, Sehrawat et al. [14] proposed a multimodal approach combining visual, auditory, and textual data to identify deceptive behavior. Their model leveraged deep learning techniques to extract and fuse features from different modalities, achieving a significant improvement in detection performance compared to single-modality models.

Another notable work by Gupta et al. [4] introduced a hybrid model that integrates facial action units and voice analysis for deception detection. Their study demonstrated that combining facial expressions with voice pitch and tone analysis can yield higher accuracy rates, as these modalities capture different aspects of deceptive behavior. Their model achieved an accuracy of 92%, highlighting the effectiveness of multimodal approaches.

Recent advancements have also explored the use of natural language processing (NLP) techniques in deception detection. Perez-Rosas et al. [11] developed a model that analyzes linguistic cues in spoken language, such as speech patterns and semantic inconsistencies, to detect deception. Their approach integrates NLP with audio-visual analysis, achieving an accuracy of 94%.

Despite these advancements, many existing models operate as black boxes, providing little interpretability regarding their decision-making processes. This has led to increased interest in developing explainable AI (XAI) models for deception detection. Bu and Ramachandran [1] introduced the concept of explainable models that provide insights into the features and patterns used by the model to make predictions, enhancing trust and transparency in automated systems.

A notable study by Şen et al. explored multimodal deception detection using real-life trial data, emphasizing the integration of verbal, acoustic, and visual cues to identify deceptive behavior in court trials. The researchers collected a dataset from public trial videos and evaluated three complementary modalities-visual, acoustic, and linguistic-at the subject level, achieving an accuracy of 83.05% in subject-level deceit detection. This study compared the automated system's performance to that of human evaluators and found that the system outperformed the average non-expert human capability. The results highlight the potential of multimodal approaches in improving the accuracy and reliability of deception detection systems, providing valuable insights for future research in this field [16].

In our research, we build on these previous works by focusing on the extraction and combination of features from video data. Unlike some recent studies, our approach does not rely on deep learning models or complex multimodal fusion techniques. Instead, we explore various combinations of visual and audio features to train a robust model for deception detection. By systematically analyzing the impact of different feature combinations, we aim to achieve high accuracy while maintaining interpretability. Our work contributes to the existing literature by demonstrating the potential of strategic feature extraction and combination in improving deception detection systems. By emphasizing feature engineering, we provide a clear and interpretable framework that can be easily adapted and extended for future research in this field.

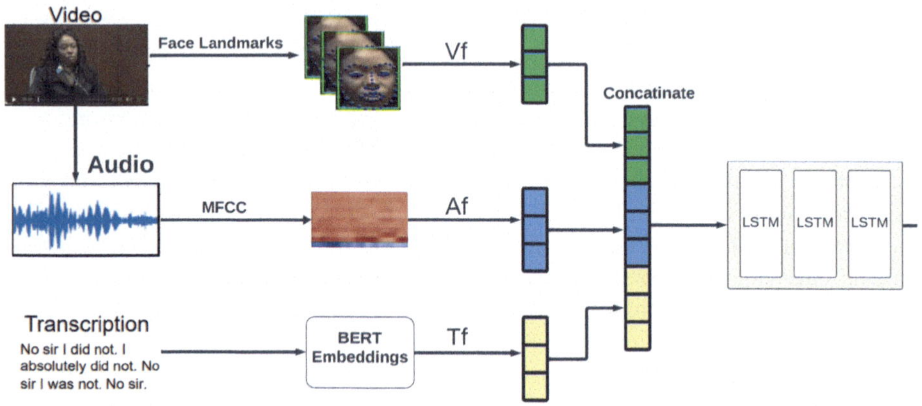

Fig. 2. Our Proposed Frame Work

3 The Proposed Framework

3.1 Feature Extraction

In this study, we focus on extracting three types of features-visual, audio, and text-to build a comprehensive deception detection model. The extracted features are then used to train various machine learning models to identify deceptive behavior.

Visual Features. Visual features are crucial for capturing non-verbal cues such as facial expressions and movements, which are often indicative of deceptive behavior. We use the `dlib` library to extract these features. Dlib provides robust facial landmark detection capabilities, identifying and tracking 68 key facial points [6]. These landmarks are then used to generate a set of visual features that describe the facial expressions and movements over time.

The facial landmark detection can be represented as:

$$F_v = \{(x_i, y_i) \mid i = 1, \ldots, 68\} \tag{1}$$

where F_v denotes the set of visual features and (x_i, y_i) represents the coordinates of the i-th facial landmark.

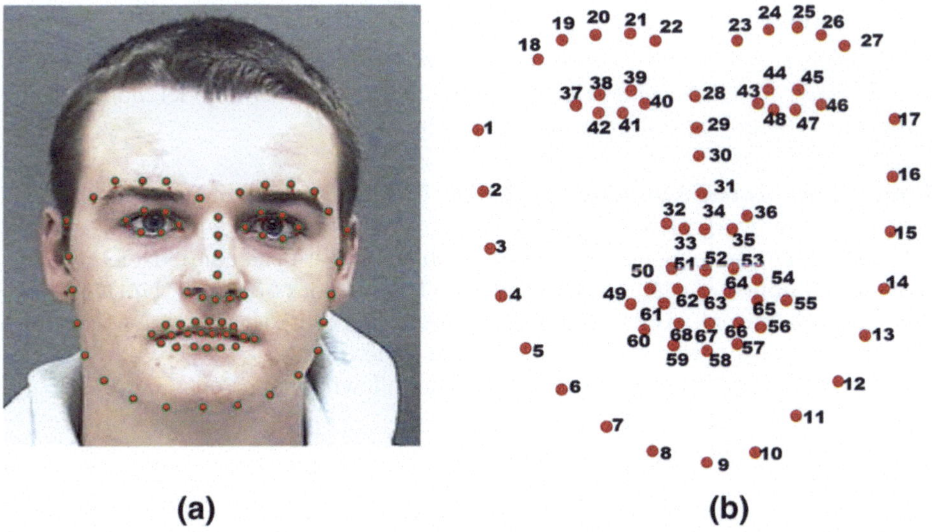

Fig. 3. (a) The left panel displays a frontal image of a person's face with several small red dots marking the facial landmarks.(b) The right panel presents a schematic representation of these facial landmarks.

Audio Features. Audio features capture the vocal characteristics and speech patterns that can indicate deception. We use Mel-frequency cepstral coefficients (MFCC) to extract these features. MFCCs are widely used in speech and audio processing because they effectively represent the power spectrum of audio signals [10].

The MFCCs are computed as follows:

$$MFCC = \text{DCT}\left(\log\left(\sum_{k=1}^{K} |X_k|^2 \cdot h_m(k)\right)\right) \tag{2}$$

where X_k is the Fourier transform of the audio signal, $h_m(k)$ represents the mel filterbank, and DCT denotes the discrete cosine transform.

Text Features. Text features are derived from the spoken content, focusing on the linguistic cues that may indicate deception. We utilize the BERT base model to extract text embeddings. BERT (Bidirectional Encoder Representations from Transformers) is a powerful language model that generates contextualized embeddings for each word in a sentence [2].

The BERT embeddings are represented as:

$$E_t = \text{BERT}(T) \tag{3}$$

where E_t denotes the text embeddings and T represents the input text.

3.2 Models

After extracting the features, we train three types of models-LSTM, BiLSTM, and Pretrained CNN-using the extracted features. Each model has its own strengths in handling different aspects of the data.

Long Short-Term Memory (LSTM). LSTM networks are a type of recurrent neural network (RNN) capable of learning long-term dependencies. They are well-suited for sequence prediction problems where the context of the input sequence is important [5]. In our study, LSTM networks process the time-series data from the visual, audio, and text features, capturing the temporal dynamics and relationships between the features over time.

The LSTM unit can be described by the following equations:

$$i_t = \sigma(W_i \cdot [h_{t-1}, x_t] + b_i) \tag{4}$$

$$f_t = \sigma(W_f \cdot [h_{t-1}, x_t] + b_f) \tag{5}$$

$$o_t = \sigma(W_o \cdot [h_{t-1}, x_t] + b_o) \tag{6}$$

$$C_t = f_t * C_{t-1} + i_t * \tanh(W_C \cdot [h_{t-1}, x_t] + b_C) \tag{7}$$

$$h_t = o_t * \tanh(C_t) \tag{8}$$

where i_t, f_t, o_t are the input, forget, and output gates, respectively; C_t is the cell state; h_t is the hidden state; and W and b are the weight matrices and bias vectors.

Bidirectional Long Short-Term Memory (BiLSTM). BiLSTM networks extend the capabilities of LSTMs by processing the input sequence in both forward and backward directions. This bidirectional approach allows the model to have access to the context from both past and future states, providing a more comprehensive understanding of the sequence [13].

The forward and backward LSTM can be described by:

$$\overrightarrow{h_t} = \text{LSTM}(x_t, \overrightarrow{h_{t-1}}) \tag{9}$$

$$\overleftarrow{h_t} = \text{LSTM}(x_t, \overleftarrow{h_{t+1}}) \tag{10}$$

The final output h_t is the concatenation of forward and backward hidden states:

$$h_t = [\overrightarrow{h_t}; \overleftarrow{h_t}] \tag{11}$$

Convolutional Neural Networks (CNN). CNNs are primarily used for spatial data and have shown great success in image processing tasks. In this study, we use pre-trained CNN models to leverage their ability to capture spatial hierarchies and local dependencies in the data [7].

The convolution operation is defined as:

$$f_{i,j} = \sum_{m,n} (x * w)_{i+m,j+n} + b \tag{12}$$

where x is the input feature map, w is the convolution kernel, b is the bias term, and $f_{i,j}$ is the resulting feature map after convolution.

By integrating these models, we aim to create a robust and accurate deception detection system. The combination of visual, audio, and text features, along with the use of advanced neural network models, allows us to capture a wide range of deceptive cues and improve the overall detection performance.

4 Dataset

The dataset used in this study is a meticulously collected real-life trial dataset, which provides a rich and diverse source of information for training and evaluating our deception detection models. This dataset was gathered from various courtroom trials, ensuring a realistic and challenging environment for deception detection. The dataset comprises three primary components: videos, audio files, and text transcripts.

4.1 Dataset Composition

- **Videos:** The dataset contains 121 videos, with 61 labeled as deceptive and 60 labeled as truthful. These videos were recorded during actual courtroom proceedings, capturing a wide range of deceptive and truthful behaviors exhibited by individuals under questioning.

- **Audio:** Each video is accompanied by an audio file in WAV format, resulting in a total of 121 audio files. These audio recordings capture the verbal exchanges and nuances in speech that are critical for detecting deception.
- **Text:** The dataset includes 121 text files, which are transcripts of the spoken content in the videos. These transcripts provide a detailed account of the verbal interactions and are used for linguistic analysis.
- **Frames:** The videos collectively consist of 82,575 frames, which are extracted for visual feature analysis. These frames capture the facial expressions and movements of the individuals, which are essential for identifying non-verbal cues of deception.

5 Experimental Design and Results

In our study on deception detection, we trained various features, which were already extracted, using different models. We focused on three primary modalities: visual, audio, and text. Each modality was initially trained separately using different models to understand their individual contributions to deception detection. Subsequently, we combined these modalities in pairs and finally, all three together, to evaluate the effectiveness of multi-modal fusion in improving detection accuracy. The dataset was split into 80% for training and 20% for testing. All the experiments were on 30 epochs with a 0.0001 learning rate. The following sections detail the methodologies and results for each stage of the experiment.

5.1 Single Modality Training

We trained each modality-visual (V), text (T), and audio (A)-independently to evaluate their individual performance in deception detection. The models used for this stage included LSTM, BiLSTM, VGG16, VGG19, RESNET18, and RESNET50. The results of this training are summarized in Table 1. As seen in the table, the LSTM model achieved the highest accuracy for the visual modality (95%).

Table 1. Results of Voice, Text and Audio alone

Model	V	T	A
LSTM	95%	93%	88%
BILSTM	90%	90%	77%
VGG16	85%	88%	65.70%
VGG19	82%	86%	66%
RESNET18	88%	90%	75%
RESNET50	90%	92%	80.00%

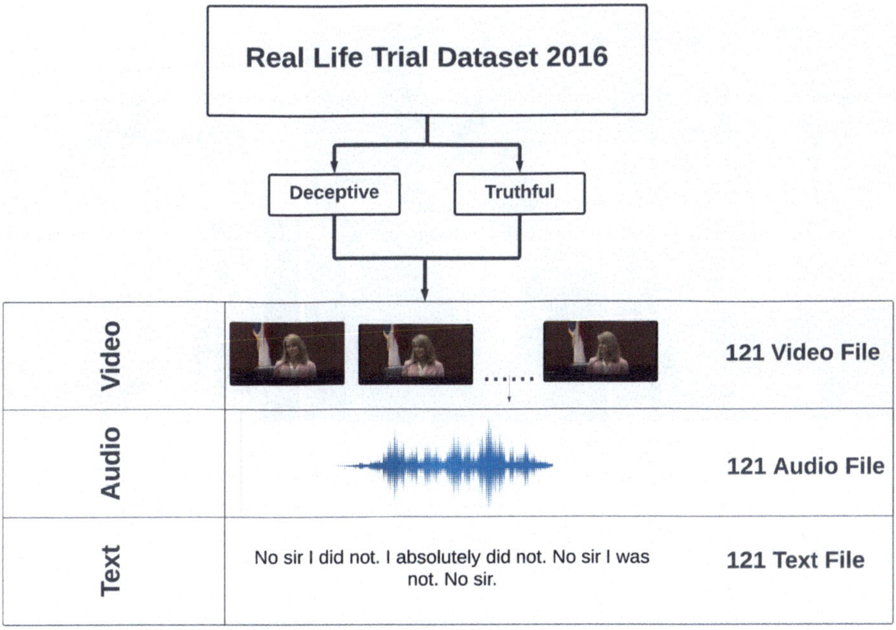

Fig. 4. Dataset Figure Showing the Dataset

5.2 Dual Modality Training

To investigate the potential benefits of combining modalities, we trained pairs of modalities together: audio and text (A+T), voice and text (V+T), and voice and audio (V+A). The performance of these combinations is presented in Table 2. Among the pairs, the voice and text combination (V+T) with the LSTM model showed the highest accuracy (97%), indicating a significant improvement compared to single modality training.

Table 2. Results of different combinations of Voice, Text and Audio

Model	A+T	V+T	V+A
LSTM	90%	97%	87%
BILSTM	85%	90.01%	84%
VGG16	79%	92%	77%
VGG19	82%	90%	80%
RESNET18	83.60%	88%	83%
RESNET50	83.15%	89%	81.30%

Fig. 5. Samples from Real life trial Dataset

5.3 Triple Modality Training

Finally, we combined all three modalities-voice, audio, and text (V+A+T)-to evaluate the full potential of multi-modal fusion in deception detection. The results, as shown in Table 3, reveal that the LSTM model achieved the highest accuracy (99%) when using all three modalities together. This indicates a substantial enhancement in detection accuracy compared to single and dual modality training.

Table 3. Results of a combination of Voice, Audio, and Text all together

Model	V+A+T
LSTM	99%
BILSTM	91%
VGG16	92%
VGG19	89%
RESNET18	90%
RESNET50	91%

These results underscore the effectiveness of combining multiple modalities in improving deception detection accuracy. The LSTM model consistently outperformed others, especially when using combined modalities. This comprehensive evaluation highlights the potential of multi-modal approaches in the field of deception detection.

6 Conclusion

Our research demonstrates the significant potential of multi-modal fusion in enhancing the accuracy of deception detection systems. By combining visual, audio, and text features, we were able to achieve an impressive 99% accuracy, highlighting the effectiveness of integrating diverse data sources. The study involved a comprehensive analysis of different feature extraction techniques and various machine learning models, including LSTM, BiLSTM, and pre-trained CNNs. The results consistently showed that multi-modal approaches outperform single modality training, underscoring the importance of leveraging multiple data types in deception detection.Furthermore, our findings emphasize the critical role of feature engineering in building robust and interpretable models. By carefully selecting and combining features from different modalities, we were able to capture a wide range of deceptive cues, leading to substantial improvements in detection performance. This approach not only enhances accuracy but also provides a clearer understanding of the underlying mechanisms of deception. In conclusion, this research provides a solid foundation for future work in the field of automated deception detection. The insights gained from our experiments can be used to develop more sophisticated fusion techniques and incorporate additional modalities, further improving the reliability and transparency of deception detection systems. As the volume of video content continues to grow, the development of advanced multi-modal detection methodologies will be crucial in various applications, from security and law enforcement to social media and beyond.

References

1. D Bu, K., Ramachandran, K.: Deception detection using random forest-based ensemble learning. Preprint, 2023
2. Kenton, J.D.M.W.C., Toutanova, L.K.: Bert: pre-training of deep bidirectional transformers for language understanding. arXiv preprint arXiv:1810.04805, 2018
3. Ferrara, E., Varol, O., Davis, C., Menczer, F., Flammini, A.: The detection of deceptive behavior. IEEE Trans. Inf. Forensics Secur. **15**, 2003–2015 (2020)
4. Gupta, V., Agarwal, M., Arora, M., Chakraborty, T., Singh, R., Vatsa, M.: Bag-of-lies: a multimodal dataset for deception detection. In: IEEE Conference on Computer Vision and Pattern Recognition Workshops, pp. 1–8, 2019
5. Hochreiter, S., Schmidhuber, J.: Long short-term memory. Neural Comput. **9**(8), 1735–1780 (1997)
6. King, D.E.: Dlib-ml: a machine learning toolkit, 2009. http://dlib.net. Accessed 10 June 2024
7. LeCun, Y., Bottou, L., Bengio, Y., Haffner, P.: Gradient-based learning applied to document recognition. Proc. IEEE **86**(11), 2278–2324 (1998)
8. Lei, Z., Zhou, Y., Zhang, Y.: A review of recent research on deception detection. Comput. Intell. Neurosci. **2019** (2019)
9. Lopez, A., Fonseca, D., Garcia, O.: Deception detection: an interdisciplinary review. Artif. Intell. Rev. **54**(5), 1–36 (2021)

10. Mermelstein, P.: Distance measures for speech recognition, psychological and instrumental. Pattern Recognit. Artif. Intell. 374–388 (1976)
11. Perez-Rosas, V., Abouelenien, M., Mihalcea, R., Burzo, M.: Deception detection using real-life trial data. In: Proceedings of the ACM International Conference on Multimodal Interaction, pp. 59–66 (2015)
12. Pérez-Rosas, V., Abouelenien, M., Mihalcea, R., Burzo, M.: Multi-modal deception detection using sequential deep learning approaches. In: Proceedings of the 2018 Conference on Empirical Methods in Natural Language Processing, pp. 50–59, 2018
13. Schuster, M., Paliwal, K.K.: Bidirectional recurrent neural networks. IEEE Trans. Signal Process. **45**(11), 2673–2681 (1997)
14. Sehrawat, P.K., Kumar, R., Kumar, N., Vishwakarma, D.K.: Deception detection using a multimodal stacked BI-LSTM model. In: IEEE International Conference on Innovative Data Communication Technologies and Application, pp. 318–326, 2023
15. Vrij, A., Fisher, R.P., Blank, H.: A cognitive approach to lie detection: a meta-analysis. Legal Criminol. Psychol. **22**(1), 1–21 (2017)
16. Sen, M.U., Perez-Rosas, V., Yanikoglu, B., Abouelenien, M., Burzo, M., Mihalcea, R.: Multimodal deception detection using real-life trial data. IEEE Trans. Affect. Comput. **13**(1), 306–320 (2022)

Evaluating Vision Encoder Performance in Image Captioning: A Comparative Analysis

Mohamed Khaled(✉), Abdelaziz Ashraf, and Wael H. Gomaa

October University for Modern Sciences Arts – MSA University, Giza, Egypt
{mohamed.khaled61,abdelaziz.ashraf}@msa.edu.eg

Abstract. In order to complete the task of captioning images, this paper compares various vision encoder architectures. Along with a common decoder architecture, LSTM, five well-known vision encoders–VGG16, MobileNet, DenseNet, Xception, and Vision Transformer (ViT)–are evaluated. The quality of generated captions is measured by the BLEU-1 score metric, which is used to evaluate each model's performance. According to our findings, Vision Transformer (ViT) performs better than the other models, with a BLEU-1 score of 0.6, which is the highest. We explore the mechanisms behind ViT's superior performance through a thorough analysis, including its attention mechanism and capacity to detect long-range dependencies within images. Furthermore, directions for future research are suggested to improve image captioning, including ensemble approaches and the incorporation of transformer-based decoders. This work provides important information about how well various vision encoder architectures perform image captioning tasks, which will direct further research in this area.

Keywords: Computer Vision · Deep learning · Vision Transformer · Vision Transformer · Transfer learning

1 Introduction

At the forefront of computer vision lies a fundamental objective: the automatic generation of captions for images.This task is intricately connected to the core of scene understanding. Caption generation models are tasked not only with overcoming the inherent challenges of computer vision, such as accurately identifying objects within an image, but also with capturing and conveying the nuanced relationships that exist between these objects in natural language. This multifaceted endeavor essentially mirrors the remarkable capacity of humans to distill vast volumes of meaningful visual information into concise and descriptive language, underscoring its pivotal importance as a benchmark for machine learning algorithms. As such, the ability to generate coherent and contextually rich captions represents a critical milestone in the ongoing quest to advance machine intelligence and bridge the divide between visual perception and linguistic expression

A. Abdelgawad et al. (Eds.): ISBCom 2024, LNNS 1268, pp. 739–751, 2025.
https://doi.org/10.1007/978-3-031-82377-0_59

[16]. Image captioning stands as a pivotal endeavor with multifaceted implications across diverse domains. Beyond its intrinsic value in enhancing our understanding of visual content, image captioning holds significant practical utility. One such application lies in the realm of automatically indexing images and facilitating efficient retrieval processes. By imbuing images with descriptive captions, content retrieval systems gain the ability to index not only visual features but also textual descriptions. This augmentation empowers users to search for specific visual concepts or objects using keywords extracted from the captions, thereby facilitating more precise and pertinent search results. Moreover, the integration of image captioning into educational settings holds immense promise. By providing descriptive captions alongside visual content, educators can enhance the accessibility and comprehension of educational materials, catering to diverse learning styles and needs. This fusion of visual and textual information fosters deeper engagement and understanding among learners, ultimately enriching the educational experience and fostering knowledge acquisition. Thus, image captioning serves as a catalyst for innovation and advancement across a spectrum of fields, from information retrieval to education, unlocking new opportunities for leveraging visual data in meaningful and impactful ways. [6], Visual Accessibility to Increase visual impairment users' accessibility by adding captions to photos posted on social media it, it can also used in Products that use augmented reality, like Google Lens, can utilize it for semantic image search [13]. The encoder and decoder are the two primary components of the suggested model.In order to provide the decoder with the information it needs to identify each object separately and determine the relationship between them in the image, the encoder takes key features from the image.

Additionally, this study compares the suggested approach to current models for images that have distinct and well-defined objects versus images that have ambiguous features.

The rest of this paper is structured as follows: Sect. 2 highlights the works related to image caption generation. Section 3 presents the dataset used along with the preprocessing steps that have been carried out on it and the methodology followed in our experiments. Section 4 discusses the feature extraction models applied to the dataset followed by the results. Section 5 discusses the results of the models. Section 6 concludes the findings of the proposed research and the furute work.

2 Related Works

This section gives pertinent background information on earlier studies on the generation of image captions.A number of techniques have been put forth to create image captions.

In Vinyals [15] they have been inspired by using recurrent neural network in machine translation models that encodes the variable length input into a fixed dimensional vector, which is then used to "decode" the input into the desired output sentence. so in their approach they propose a neural and probabilistic framework for creating caption based on images. which rather than using an input sentence they uses a given image. they used a convolutional neural network as an encoder to extract the features and LSTM as a decoder to create the description of an image. According to Xu [16] their approach was encoder-decoder method. using VGG to extract the image features and long short-term memory that generates one word at every step based on the context vector.they introduced two attention-based image caption generators. soft attention and hard attention.they achieved 0.67 bleu1 score on flickr8k dataset for each of the methods they proposed soft attention and hard attention.

According yang [17] to In their paper "An Ensemble of Generation- and Retrieval-Based Image Captioning With Dual Generator Generative Adversarial Network," Min Yang et al. propose a novel image captioning model called EnsCaption, which combines both generation-based and retrieval-based approaches to enhance image captioning performance. The model integrates a dual generator generative adversarial network (GAN) to synergize the strengths of both methods. Specifically, the model includes a caption generation component that creates new sentences tailored to the query image and a caption re-ranking model that retrieves the best-matching captions from a candidate pool. The discriminator component is designed to learn the multi-level differences between generated, retrieved, and ground-truth captions, thus guiding the generators to produce more accurate and contextually appropriate captions. Comprehensive experiments on benchmark datasets such as MSCOCO and Flickr-30K demonstrate that EnsCaption significantly outperforms existing methods, highlighting its effectiveness in producing high-quality image captions through the combined use of generative and retrieval techniques.

According to gu [3] The paper "Stack-Captioning: Coarse-to-Fine Learning for Image Captioning" by Jiuxiang Gu, Jianfei Cai, Gang Wang, and Tsuhan Chen introduces a sophisticated approach to image captioning by employing a coarse-to-fine multi-stage prediction framework. Unlike conventional single-stage models, which often struggle with generating rich and fine-grained descriptions due to the vanishing gradient problem, this method leverages multiple decoders. Each decoder operates on the output of the previous stage, refining the image description incrementally. The approach also incorporates reinforcement learning to optimize the model with intermediate supervisions, effectively addressing the well-known exposure bias and loss-evaluation mismatch problems. The authors demonstrate the efficacy of their model through extensive evaluations on the MSCOCO dataset, achieving state-of-the-art performance in generating detailed and accurate image captions.

According to Tuba [13] they introduced an ensemble model in their approach.they used Inception-v3 as convolutional neural network encoder to extract the image features.A bidirectional LSTM and transformer encoders serve as the decoder part of the model.The output from word embeddings is fed into the model and sent to the transformer encoder.The bidirectional LSTM model and the transformer encoder model both compute the individual probabilities of the caption's next word. In order to produce the following word in the caption, the probabilities of the two individuals are then averaged and they used the flickr8k.They achieved 0.5747 bleu score for the trasnformer based model, 05913 for the lstm model and 0.61 for the ensamble model. In [9] They present a multi-modal Recurrent Neural Network architecture that produces a text description of an image input.The results of their studies demonstrate that the generated sentences produce sensible qualitative predictions and significantly outperform retrieval based baselines.They apply a different technique, using a Region Convolutional Neural Network (RCNN), to identify objects in each image. The (RCNN) is trained on the 200 classes of the ImageNet Detection Challenge. they use the top 19 detected locations in addition to the whole image and compute the representations.they achieved a bleu score 0.579 on flickr8k, 0.573 on flick30k and 0.62 on MSCOCO datasets.In relevance for more related works sharma in [11] build a taxonomy as shown in Fig. 1,and Hossain in Fig. 2.

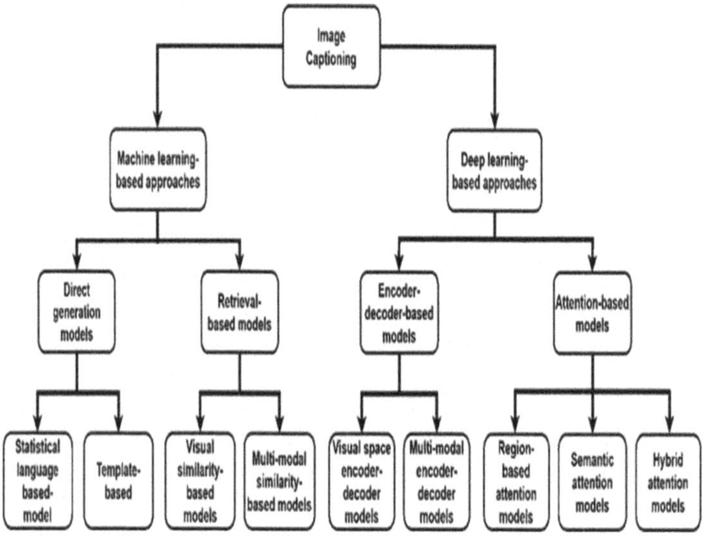

Fig. 1. An extensive review of the literature on image captioning techniques [11]

3 Methodology

In this section, we will provide a detailed discussion on the dataset used in our study, including the preprocessing steps applied to the data. Additionally, we will present our proposed models and describe the experiments conducted to evaluate their performance.

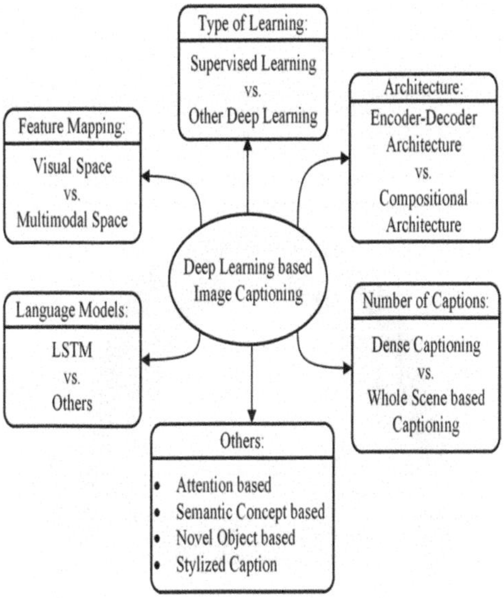

Fig. 2. Comprehensive taxonomy for image captioning using deep learning [5]

3.1 Data-Set and Data-Set Prepossessing

Flicker8k dataset its a its benchmark collection for description and search of images based on sentences. which includes 8,000 photos with five distinct captions for each image that clearly describe the salient entities and events. The photographs were picked from six distinct Flickr groups, and they tend not to contain any famous people or places.these are examples of flickr8k dataset (Figs. 3 and 4).

Before training can commence, several pre-processing tasks must be completed, particularly for the captions. Initially, all captions are converted to lower case, and punctuation and special characters are removed. Subsequently, tokenization is applied to the captions to prepare them for the model.

3.2 Transfer Learning

In computer vision, transfer learning has emerged as a popular approach, allowing the utilization of previously trained models to tackle new problems effectively. This methodology significantly reduces time and processing resources while achieving superior results, particularly on smaller datasets. Convolutional Neural Networks (CNNs) serve as the backbone of transfer learning, providing a robust foundation for various applications. Pre-trained CNN models offer the advantage of reusing learned features across tasks, thereby expediting the model development process and mitigating the computational burden associated with training from scratch. Fine-tuning pre-trained models on specific datasets further enhances their adaptability to new challenges, even with limited data

Fig. 3. Dataset Sample 1

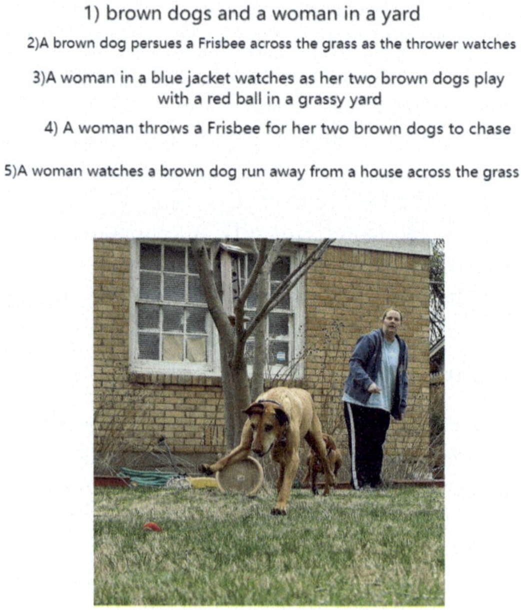

Fig. 4. Dataset Sample 2

availability. Architectures such as VGG, Xception, DenseNet, and MobileNet exemplify the versatility and efficacy of transfer learning in computer vision, enabling practitioners to address diverse tasks such as image classification and object detection with unprecedented efficiency and performance.

3.3 Vision Transformer

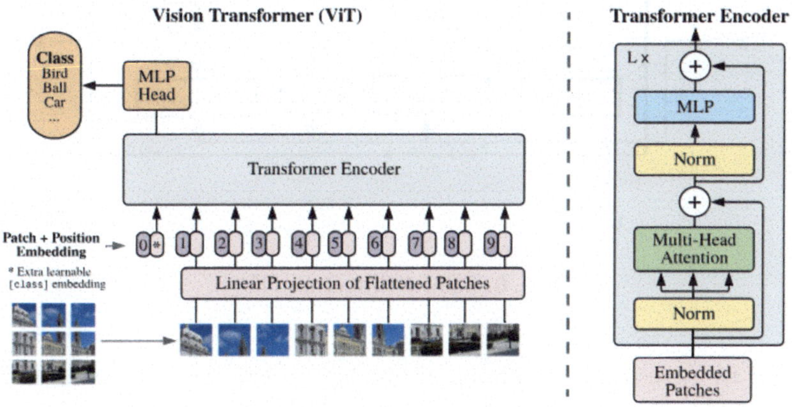

Fig. 5. VIT Architecture [2]

Although the Transformer architecture has established as a recognized standard for tasks with natural language processing, there are still few uses for it in computer vision. Attention is either used in conjunction with convolutional networks in vision, or it can be used to replace specific convolutional network components while maintaining the overall structure of the network [2]. Accordin to Dosovitskiy [2] motivated by the Transformer scaling achievements in NLP, they test the idea of directly applying a standard Transformer to images with the least amount of adjustments to be too close to the idea of transformer which has been introduced by vaswani [14]. According to Dosovitskiy [2] in their architecture shown in Fig. 5. They divide an image into patches of fixed dimensions, linearly embed each patch, add position embeddings, and provide the sequence of vectors that results to a standard Transformer encoder.The encoder's output is fed into the classification head for prediction.

3.4 Decoder

Long Short-Term Memory (LSTM) as in Fig. 6 which was proposed in [4] is a type of recurrent neural network (RNN) architecture designed to handle sequence data with long-range dependencies. according to hochreiter [4] With "constant error carrousels" inside specific units, LSTM can enforce constant error flow,

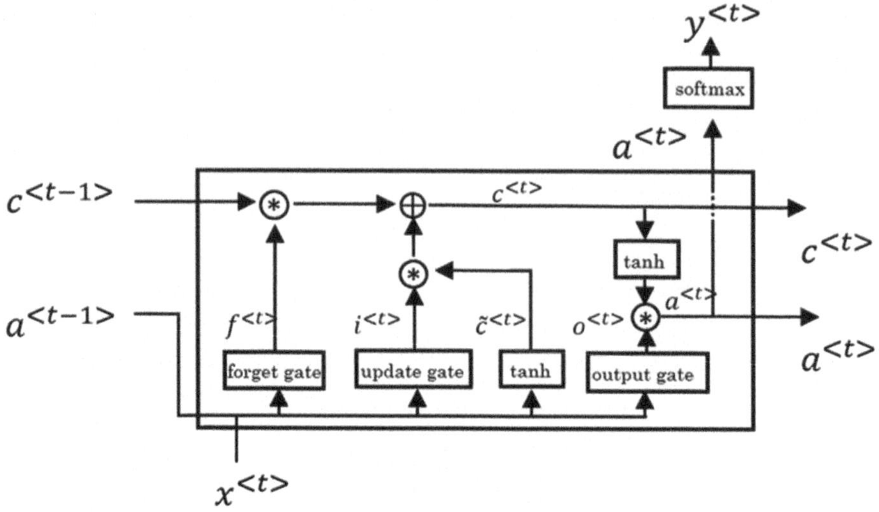

Fig. 6. STM Rchitecture [4]

allowing it to learn to bridge minimal time lags greater than 1000 discrete time steps. Unlike traditional RNNs, LSTM networks have input, forget, output, and memory cells, allowing them to selectively update or forget information as needed while retaining it over long periods.

3.5 Approach

The foundation of our approach lies in the independent utilization of several pre-trained vision models, each paired with an LSTM as the decoder. We ensure consistency by constraining certain hyperparameters across all models, such as number of epoch which is 20, setting the image width and height to 224 × 224 pixels and the batch size to 8. Each vision model is followed by a Dense Network architecture, which includes a dropout layer and a dense layer with ReLU activation. This dense layer's output is then combined with the text output features from an Embedding layer using an add layer. This combined output is followed by another dropout layer, an LSTM layer, and a dense layer. Finally, the architecture concludes with an output dense layer containing 8485 neurons, with each neuron corresponding to a different word in the vocabulary. To ensure a fair comparison, the same parameters are applied to each vision model, allowing us to accurately compare differences in convergence speed and results. This rigorous and systematic approach enables a comprehensive evaluation of each model's performance under identical conditions (Fig. 7 and Table 1).

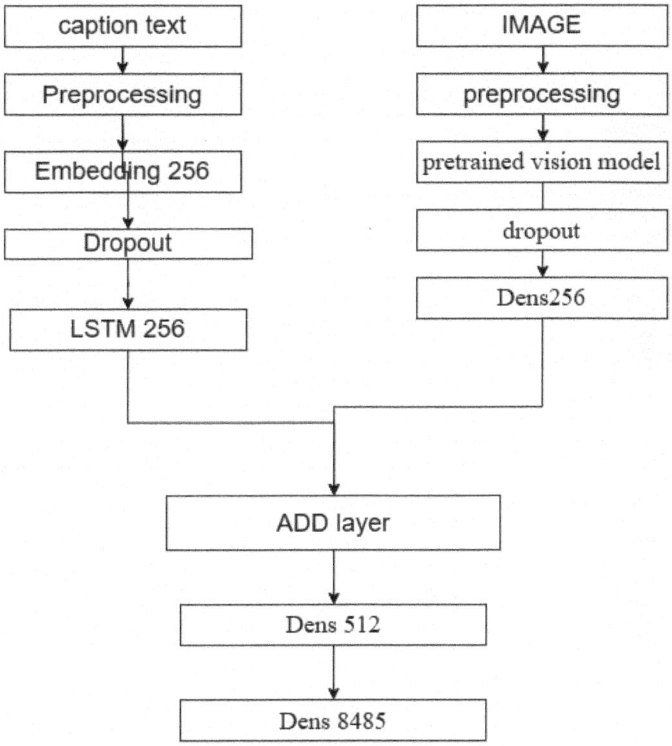

Fig. 7. Model architecture

4 Models and Result

This section contains a comprehensive discussion of various pre-trained encoder models, providing examples of their results to illustrate their performance and capabilities. The models explored in this study include VGG16, DenseNet, MobileNet, Xception, and Vision Transformer (ViT).

4.1 Experiment One Mobilenet

Mobilenets are pre-trained models with a small number of parameters. They do well at having low latency and are designed to fulfill the limitations of resources in various live use scenarios.The concept of mobile nets is based on the depth-wise separable convolutions [7,10]. So its not the most powerfully or accurate pre-trained model put its faster the its competitors.

4.2 Experiment Two Xception

According to [1] theypropose a convolutional neural network architecture based entirely on depthwise separable convolution layers.With 21 million trainable

Table 1. Result of Generated Captions by Vision Encoders

Image	VIT	Densnet	Xception	VGG16	MobileNet
BLEU1 score	0.6	0.553	0.549	0.52	0.51
	**caption:** man in red shirt is climbing rock wall	**caption:** man climbs up cliff	**caption:** man climbing up rock wall	**caption:** man is climbing up rock face	**caption:** man is climbing on large rock
	textbfcaption: two dogs are playing with red ball in the grass	**caption:** dog is running through the grass with tennis ball in its mouth	**caption:** brown dog is running in yard	**caption:** two dogs are playing in the shade	**caption:** dog is running through the grass
	**caption:**little girl is sitting in front of painted rainbow	**caption:**little girl in pink dress is standing in front of painted rainbow	not considered	**caption:**girl in red and black snowsuit is sitting on green lawn	**caption:**two girls in pink outfits are playing in the air.

parameters,Depthwise separable convolutions are used in the deep learning CNN architecture known as 'Xception.modules [1].

4.3 Experiment Three Dens201

According to huang [8] Creating dense connections between network layers is the main concept of DenseNet. Each layer in conventional CNN architectures, such as ResNet, usually connects only to the layer after it. But in DenseNet, all of the layers are connected to all of the other layers in a feed-forward manner.Because of this pattern of connectivity, a dense network is formed in which every layer passes its own feature maps to every layer that comes after it and receives feature maps from every layer that comes before it.

4.4 Experiment Four VGG16

The 4th deep convolutional neural network architecture used is known as vgg16 and it was proposed by Zisserman and Simonyan [12].VGG16 is defined by its uniform architecture and simplicity.There are 16 layers total, with 3 fully connected layers and 13 convolutional layers.To reduce spatial dimensions, max-pooling is applied after every two or three convolutional layers. The convolutional layers use small receptive fields (3×3) with a stride of 1.

4.5 Experiment Five Vision Transformer

Lately Vision Transformer (ViT) have become a strong competitor to Convolutional Neural Networks (CNNs), which are state-of-the-art in various computer vision tasks involving image recognition.It divides an image input into fixed-size patches, flattens them, and feeds them into the transformer encoder.and to gain spatial information, linear embedding is applied to every patch and to represent the location of every patch, positional embeddings are added.The model captures global dependencies between patches as a result of the transformer encoder's processing of these embeddings throughout multiple self-attention layers.

5 Discussion

In this comprehensive study, we conducted a meticulous comparative analysis of five distinct vision encoder architectures: VGG16, MobileNet, DenseNet, Xception, and Vision Transformer (ViT), all paired with the same decoder architecture, LSTM. Our primary objective was to ascertain which of these models excelled in the task of image captioning. To evaluate the quality of generated captions, we predominantly relied on the BLEU-1 score metric, a widely recognized measure in natural language processing tasks. Our findings unequivocally revealed that Vision Transformer (ViT) emerged as the undisputed victor, achieving the highest BLEU-1 score of 0.6. This remarkable performance of ViT can largely be attributed to its sophisticated attention mechanism, adept at discerning long-range dependencies within images and facilitating the generation of captions that are not only coherent but also contextually rich.Furthermore, while DenseNet exhibited commendable performance, leveraging its interconnected layers to advantage, ViT demonstrated a competitive edge owing to its robust architecture and highly effective attention mechanism. These results underscore the critical role of attention mechanisms in image captioning tasks and advocate for further exploration and development of attention-based models such as ViT to enhance descriptive image understanding. By shedding light on the significance of attention mechanisms in image captioning and advocating for the advancement of attention-based models, our study contributes valuable insights to the ongoing discourse in the field, paving the way for future research aimed at refining and optimizing image captioning systems for improved performance and efficacy.

6 Conclusion and Future Work

In summary, our comprehensive comparative analysis provides valuable insights into the performance of various vision encoder architectures in the challenging task of image captioning. Through extensive experimentation and rigorous evaluation, we have demonstrated that the Vision Transformer (ViT) emerges as the top-performing model, surpassing traditional convolutional architectures such as VGG16, MobileNet, DenseNet, and Xception. This superiority can be attributed to ViT's remarkable ability to effectively capture global context and spatial dependencies within images, owing to its attention mechanism. However, while ViT represents a significant advancement in image captioning technology, this study also highlights promising avenues for future research. Specifically, we advocate for the exploration of ensemble methods and transformer-based decoder architectures, such as BERT or GPT. By combining the strengths of multiple models and leveraging advanced language modeling techniques, we can further enhance the accuracy and quality of generated captions. Additionally, the integration of transformer-based decoders offers a promising pathway towards more sophisticated language understanding and contextually rich image descriptions. By pursuing these directions, we can continue to advance the capabilities of image captioning systems, ultimately paving the way for more nuanced and contextually relevant descriptions of visual content.

References

1. Chollet, F.: Xception: deep learning with depthwise separable convolutions. In: Proceedings of the IEEE Conference on Computer Vision and Pattern Recognition, pp. 1251–1258, 2017
2. Dosovitskiy, A., et al.: An image is worth 16x16 words: transformers for image recognition at scale, arXiv preprint arXiv:2010.11929, 2020
3. Gu, J., Cai, J., Wang, G., Chen, T.: Stack-captioning: coarse-to-fine learning for image captioning. In: Proceedings of the AAAI Conference on Artificial Intelligence, vol. 32, no. 1, 2018
4. Hochreiter, S., Schmidhuber, J.: Long short-term memory. Neural Comput. **9**(8), 1735–1780 (1997)
5. Hossain, M.Z., Sohel, F., Shiratuddin, M.F., Laga, H.: A comprehensive survey of deep learning for image captioning. ACM Comput. Surv. (CSUR) **51**, 1–36 (2018). https://api.semanticscholar.org/CorpusID:52947736
6. Hossain, M.Z., Sohel, F., Shiratuddin, M.F., Laga, H.: A comprehensive survey of deep learning for image captioning. ACM Comput. Surv. (CSUR) **51**(6), 1–36 (2019)
7. Howard, A.G., et al.: Mobilenets: efficient convolutional neural networks for mobile vision applications, arXiv preprint arXiv:1704.04861, 2017
8. Huang, G., Liu, Z., Van Der Maaten, L., Weinberger, K.Q.: Densely connected convolutional networks. In: Proceedings of the IEEE Conference on Computer Vision and Pattern Recognition, pp. 4700–4708, 2017
9. Karpathy, A., Fei-Fei, L.: Deep visual-semantic alignments for generating image descriptions. In: Proceedings of the IEEE Conference on Computer Vision and Pattern Recognition, pp. 3128–3137, 2015
10. Nagib, A.E., Mohamed, H., Ashraf, A.: Comparative analysis of pre-trained cnn architectures for multi-class weather classification: a study on deep learning techniques. In: Intelligent Methods, Systems, and Applications (IMSA). IEEE 2023, pp. 297–302 (2023)
11. Sharma, H., Padha, D.: A comprehensive survey on image captioning: From handcrafted to deep learning-based techniques, a taxonomy and open research issues. Artif. Intell. Rev. **56**(11), 13 619–13 661 (2023)

12. Simonyan, K., Zisserman, A.: Very deep convolutional networks for large-scale image recognition, arXiv preprint arXiv:1409.1556, 2014
13. Tuba, M., Akashe, S., Joshi, A.: ICT Systems and Sustainability: Proceedings of ICT4SD 2021, Volume 1. Springer Nature, 2022. https://books.google.com.eg/books?id=p5Iq0AEACAAJ
14. Vaswani, A., et al.: Attention is all you need. Adv. Neural Inf. Process. Syst. **30** (2017)
15. Vinyals, O., Toshev, A., Bengio, S., Erhan, D.: Show and tell: a neural image caption generator. In: Proceedings of the IEEE Conference on Computer Vision and Pattern Recognition, pp. 3156–3164, 2015
16. Xu, K., et al.: Show, attend and tell: neural image caption generation with visual attention. In: International Conference on Machine Learning. PMLR, 2015, pp. 2048–2057 (2015)
17. Yang, M., et al.: An ensemble of generation-and retrieval-based image captioning with dual generator generative adversarial network. IEEE Trans. Image Process. **29**, 9627–9640 (2020)

Framework for Detecting XSS Attaks Using Machine Learning

Mohamed Haytham and Wael El Sersi(✉)

Faculty of Computer, October University for Modern Sciences and Arts (MSA), Giza, Egypt
{mohamed.haitham,wfarouk}@msa.edu.eg
mohamed.haitham@msa.edu.eg wfarouk@msa.edu.eg

Abstract. Cross-Site Scripting(XSS) is a common web application security flaw that enables attackers implant harmful scripts into pages that other users are seeing. These programs may spoof websites, take over browsing sessions, collect sensitive information, and send visitors to dubious domains. xss attacks are considered as one of the most threatening web server attacks as it is accentuated by the Open Web Application Security Project (OWASP) as one of the top 10 web server attacks and ranked from the top 6 according to the cyber security news in 2024. The paper here presents a study to detect the xss attack using both machine learning and deep learning algorithms The model aims to investigate using machine learning and deep learning using different models using SVM, random forest and neural network. Effective XSS detection helps web developers and security professionals safeguard against potential exploits and ensure the safety of user data and interactions on the internet.

1 Introduction

Now days every person need to interact with a website for many reasons to buy something to book a ticket so it has many uses. In the digital world, where information flows freely, there exists a silent predator lurking amidst web pages and applications which makes all the noramal a target for infection that takes use of weaknesses to get illegal data stored on the system. As a result, the most crucial aspect of developing a web application is protecting the website from (XSS) assaults. These attacks can impact the victim through cookie theft, page modification, clipboard contents capture, keylogging where the attacker can know what user write on the keyboard of his personal device, port scanning, and other things. There are a lot of web applications attacks, such as the SQL injection, CSRF attack, and the xss attack. As the cross-site scripting is always considered as one of the top ten most powerful or dangerous website attacks according to vulnerabilities listed in the(OWASP) over the last years and according to their last update. Over the past five years, XSS (Cross-Site Scripting) attacks have remained a persistent threat in the ever-evolving landscape of cybersecurity. Despite concerted efforts to mitigate their impact, they continue to plague

A. Abdelgawad et al. (Eds.): ISBCom 2024, LNNS 1268, pp. 752–764, 2025.
https://doi.org/10.1007/978-3-031-82377-0_60

websites and web applications. From 2019 to 2024, we witnessed a transition in XSS attack vectors, with attackers becoming more sophisticated in exploiting vulnerabilities in both client-side and server-side scripts. The emergence of new frameworks and technologies has brought about novel attack surfaces, while the proliferation of APIs has introduced new avenues for injection. Moreover, the integration of rich media content and the widespread adoption of single-page applications have further complicated the XSS landscape, posing significant challenges for security practitioners. As defenders strive to stay ahead of these evolving threats, the battle against XSS attacks remains an ongoing endeavor, emphasizing the importance of proactive security measures and robust coding practices.The aim of the project is to obtain the xss attack using both machine learning and the deep learning with higher accuracy. The main purpose of using machine learning is once a classifier has been constructed, it can determine a script's maliciousness with speed; second, it can analyze a script without a sandbox; and third, the classifier has prediction skills to identify newly discovered harmful JavaScript. In this work, high accuracy and precision stored XSS detection is achieved by machine learning.But as the machine learning is very useful and effective in detecting it can face some disadvantages and challanges such as: Data Imbalance: XSS attacks are relatively rare compared to benign requests, leading to class imbalance in the dataset. This can result in biased models that struggle to accurately detect XSS attacks without generating a high number of false positives, Adversarial Attacks: Attackers can intentionally craft payloads to evade machine learning detection models. They may manipulate input features or obfuscate malicious scripts to bypass detection, thus undermining the effectiveness of the model, High False Positive Rates: Machine learning models may generate false positives, incorrectly flagging benign requests as XSS attacks. This can lead to unnecessary alerts and operational overhead for security teams, as well as potential user inconvenience or disruption to legitimate website functionality.

2 Related Work

Mereani et al., [6] used machine learning models for detecting Cross-Site Scripting (XSS) in JavaScript, particularly focusing on persistent XSS attacks where harmful scripts are administered into web applications and stored in databases. The study categorizes XSS detection approaches into three types: static analysis, dynamic analysis, and machine learning.For the methodology, The study gathers two datasets for the methodology: one for testing and one for training. There are 2000 harmful and 2000 safe scripts in the training dataset, and 13,000 of each in the testing dataset. Features for classification are divided into structural (non-alphanumeric characters) and behavioral (JavaScript commands and functions) categories, resulting in 59 features in total. Three machine learning algorithms are employed: (SVM), (k-NN), and Random Forests. These classifiers are trained and evaluated using precision, accuracy, sensitivity, and specificity metrics.Results show that all classifiers perform well, where the KNN model was very high with accuracy: 97.12 and precision: 96.79 and random forest classifier evaluation got accuracy: 97.22 and precision: 96.47.

Future work is proposed to explore additional features and classification methods, as well as to investigate the potential of extracting rules from classifiers and utilizing neural network classifiers.

Bronjon et al., [4] compared the performance of various machine learning algorithms in detecting XSS attacks and to develop a more robust defense mechanism for web applications. the dataset was collected using xss attack tools and online sources for the positive xss samples while the negative bengien were collected using automated data generation tools. The dataset comprised 15,000 samples of each category.The gathered data underwent pre-processing to convert it into a format that was appropriate for machine learning techniques.Tokenizing text data allowed for the feature extraction of TFIDF (Term Frequency Inverse Document Frequency).Both linear SVM and non-linear SVM machine learning models were employed. as for the results Both linear and non-linear SVM classifiers successfully separated XSS attack inputs from benign web application inputs.the experimental results suggested that machine learning approaches, when combined with traditional defense mechanisms like WAF (Web Application Firewall), could enhance XSS attack detection with higher accuracy. Precision, recall, and F1-score metrics indicated low false positive and false negative rates, signifying balanced accuracy. While non-linear SVM outperformed in terms of accuracy and F1-score, it required significantly more time for model fitting compared to linear SVM.The paper uses TFIDF for feature extraction, which is a commonly used technique. However, more advanced feature engineering techniques, such as word embeddings or deep learning-based methods, could potentially capture richer semantic information and improve the performance of the classifiers.

FAWAZ et al., [8] used an Artificial Neural Network (ANN)-based scheme to detect cross-site scripting (XSS) attacks in web applications. They made good use of a substantial dataset that totals 138,569 samples, of which 100,000 are benign and 38,569 are malicious samples with 41 attributes. the model is trained using k-fold cross-validation, optimizing hyperparameters through grid search. Accuracy, precision, recall, false positive rate, and area under the ROC curve are examples of evaluation measures. On the test dataset, the suggested model obtains an accuracy of 99.32. the model also conducted a very high performance in detecting XSS-based attacks, with an AUC-ROC curve exceeding 99.02. They extracted XSS features using word2vec, and then fed that data into a deep learning network that trained on LSTM. They obtained a false positive rate that tends to 0.019, 0.979 recall, and around 0.995 accuracy.

Ahmed et al., [1] aims to provide an ingenious method to detect xss attacks using different deep learning techniques, specifically modular neural network (MNN), as they seek to improve detection accuracy while reducing false positives compared to traditional signature-based or heuristic-based methods. Through experimentation and evaluation. The dataset used in the study comprises 101000 instances, with 1000 labeled as malicious and 100000 as benign. It includes 67 features related to HTML, JavaScript, and URL attributes.the results proved the effectivness of the proposed MNN technique achieving an impressive accuracy of 99.96 in detecting XSS attacks, outperforming BPNN and RBFN.

Korrawit et al., [9] employed a variety of machine learning techniques, including k-NN, Decision Trees, Support Vector Machines, and Gaussian Naive Bayes classification models, to identify XSS assaults and assess their effectiveness.The research draws upon a dataset inclusive of both harmless and potentially harmful URLs, amounting to 43,218 entries. These URLs were collected from various origins, including XSSed.com, to ensure a diverse and comprehensive dataset for analysis.the results of testing the machine learning models showed that the K-NN has Accuracy rate: 99.990 and the SVM has an accuracy rate of 95.5 and the Gaussian NB Accuracy rate: 91.4. the testing done highlights the effectiveness of the proposed models, particularly the k-NN algorithm, in accurately detecting malicious URLs. Additionally, the conclusion suggests future research directions, such as developing real-time detection systems and exploring deep learning methods for further improving XSS detection accuracy.

Gulit et al., [5] main purpose is to enhance the detection of Cross-site Scripting (XSS) attacks on web applications by integrating the n-gram method with various machine learning algorithms. The methodology involves using Support Vector Machine (SVM), Naive Bayes (NB), and K-Nearest Neighbor (KNN) algorithms, combined with the n-gram technique to improve detection accuracy. The dataset used consists of 400 URLs, which are divided into four subsets for k-Fold Cross-Validation. To assess how well the models work, the data is gathered and divided into two categories: learning process data and validation/evaluation data. One limitation of the paper is its reliance on a limited sample size, which might not represent all variations of XSS attacks. Feature extraction includes parameters such as URL length, recurring characters, special characters, script tags, cookies, redirection, and special keywords for reflected XSS, while stored XSS features include the number of JavaScript characters, number of script tags, source of JavaScript files, user-defined functions, and cookies. The results demonstrate that the combination of SVM and the n-gram method achieves the highest accuracy at 98, with precision and recall rates at 99, outperforming the other models tested. Banerjee et al., [3] main objective is to improve machine learning classifiers' ability to detect Cross-Site Scripting (XSS) crimes. The study made use of a dataset that contained 1611 JavaScript instances, of which 158 were classified as malicious and 1453 as benign. Using four distinct machine learning algorithms-Logistic Regression, RBF Kernel-SVM, Random Forest, and K-Nearest Neighbors (KNN)-the technique investigated each one's efficiency in identifying cross-site scripting (XSS) vulnerabilities. The focus of feature extraction was on characteristics from JavaScript code and URLs, such as server-side validation and the existence of special characters. With a precision of 0.96 and an accuracy of 98, the Random Forest classifier outperformed the other classifiers in both training and evaluations using real-world data, according to the results.

Bakir et al., [2] goal is to improve the detection of cross-site scripting (XSS) threats by utilizing machine learning approaches in conjunction with sophisticated feature extraction methods. To assess their suggested techniques, the researchers used a dataset supplied by Syed Saqlain Hussain Shah that was created especially for XSS attack detection. In order to improve the feature extraction process, their technique combined the word-level and sentence-level semantics of the Universal Sentence Encoder (USE) with Word2Vec. The study applied several machine learning classifiers and deep learning models, including LSTM and GRU, to process the extracted features. The evaluation of these models was conducted using metrics such as accuracy, precision, recall, and F1-score, along with confusion matrices and ROC curves to provide a comprehensive assessment of the models' performance. The results demonstrated that the combination of USE and Word2Vec embeddings significantly improved the detection accuracy of XSS attacks compared to traditional methods.13,685 items from PortSwigger and OWASP Cheat Sheets made up the dataset used in this study, which offers a variety of XSS attack vectors and benign examples. Syed Saqlain Hussain Shah published this dataset on Kaggle. To properly evaluate the models, a wide variety of XSS attack scenarios were ensured during the data gathering process.The deep learning models, including architectures like LSTM and GRU, were also tested. The results, presented in the Tables indicated that these models, when trained on the combined USE and Word2Vec features, achieved high performance, with training times remaining under 12 milliseconds per sample and detection times under 1 millisecond per sample, ensuring real-time responsiveness essential for web security.

3 Datasets

This research is based on dataset provided by Fawaz et al., [6] and ? The dataset was collected from various sources, both for training and testing purposes. in order to collect the training set, malicious scripts were sourced from developer sites, the dataset is collected from different official websites, XSSed, and more scripts were acquired by monitoring websites that were thought to be unreliable. Benign scripts, on the other hand, were collected from a variety of developer and university websites. This approach ensured a balanced dataset with a good coverage of both malicious and benign scripts. There are 19,000 instances of both dangerous and harmless scripts in the first dataset (training), and a total of 24,000 in the second (test) dataset.The benign scripts were collected from different developers and university sites. the dataset consists of 66 features that are very efficient for the detection of the xss attack related to the URL that are very important for the machine learning model to train and test on them using differnt techniques.

4 Methodology

The system is designed as a standalone website where users can input a URL to be analyzed. The core functionality of the site involves a Random Forest machine learning model that scans and evaluates the provided URL for various characteristics or potential issues. This entire process is seamlessly integrated and managed using the Django framework, which ensures a cohesive and efficient operation of the system from user input to the final analysis output. Django handles the web interface, backend processing, and coordination between the user input and the machine learning model, creating a streamlined and user-friendly experience.as the following figure explains how the system works (Fig. 1).

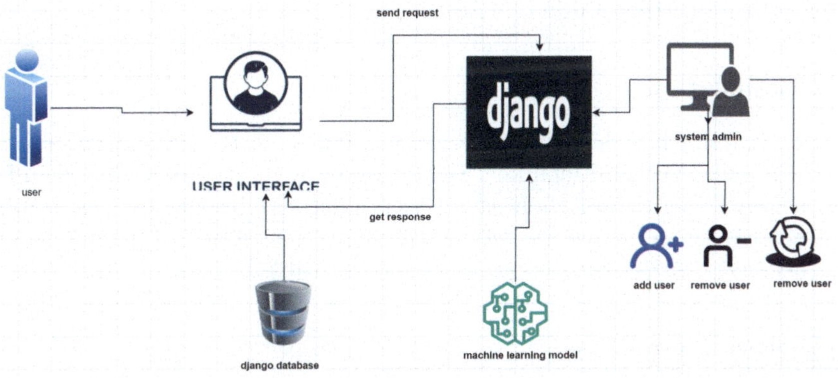

Fig. 1. System architecture

4.1 Pre-processing

This process is very important before training the dataset on the machine learning models.it will have some processes which are: Normalization: the normalization is to the process of re-scaling the features of a dataset so that they have a similar scale.The main purpose of using the normalization is to change the values of numeric columns with the same scale,without making any difference in the values or losing any information their are different methods of normalization but the Min-Max Normalization is the method that will be used scales the data to a fixed range, usually between 0 and 1. The formula for min-max normalization is Encoding: is the process of converting categorical data into a numerical format that can be used for analysis or input into machine learning algorithms. here the dataset as provided before it consists of URLS which is considered as symbols or characters so this characters needs to be encoded to be understood and trained by the machine learning model the encoding has different methods but the most efficient method used in this study is the be one-hot encoding due to several reasons first their is no inherent order or in other words Each URL is unique and doesn't have a natural ranking.second this method is very compatible with many machine learning and deep-learning algorithms like neural network and random forest. Data splitting: the dataset is divided into 0.7 training and 0.3 testing dataset,this process is very important process for many reasons but The primary reason for splitting the dataset is to evaluate how well the trained model propagate to unseen data. By training on one portion of the data and testing on another, this simulate the real-world scenario where the model needs to make predictions on new, unseen examples.the splitting process also prevents from Over-fitting,which occurs when a model knows how to recall the training data rather than generalize from it. If you evaluate the model on the same data it was trained on, you might get overly optimistic performance metrics. You may determine if the model has picked up useful patterns or has only memorized the training set by utilizing a different testing set.

4.2 Classification

In this study 4 machine learning algorithms are used to train the datasets which are the Support Vector Machine (SVM), random forest,Naive Bayes,and the neural network algorithm. their is no technique can be the most suitable or the universally superior one for getting the highest accuracy or getting the best results because this depends on several factors (Table 1).

4.3 Feature Extraction

The Feature extraction in this study is divided into structural features and behavioral features. 1) Structural Features: The focus of these features is mostly on non-alphanumeric characters that can exist in JavaScript scripts. The analysis takes into account both individual characters and character combinations that might be utilized to create dangerous programs. The presence or absence of these characters is measured using a binary (0/1) value indicating whether the feature occurs in the script. Behavioral Features: Behavioral features include a selection of JavaScript functions that may be used suspiciously by attackers. These features capture the usage patterns of certain commands or functions that deviate from typical benign usage. Similar to structural features, the presence or absence of behavioral features is measured using a binary value.the feature extraction is very important for many aspects like the frequency of the characters character sequences can be a signal of malicious purpose, such as script tags, HTML entities, or JavaScript functions. Detecting common patterns of XSS attacks, such as attempts to inject JavaScript code or HTML tags into input fields,which is the main purpose of using the machine learning in the detection.the extraction method also helps in the analyzing process Observing how input interacts with the application, such as attempting to modify DOM elements or execute JavaScript, can indicate positive XSS attacks.

Table 1. Example on the dataset Features

Feature Number	Feature Name
1	Contains &lt
2	ScripTag
3	Readable
6	Contains And
7	Contains Percentage
8	Contains Slash
9	Contains BackSlash
10	Contains Plus
11	Contains Document
12	Contains Window
13	Contains Onload
14	Contains Onerror
15	Contains iframe
16	Contains img
17	Contains SRC
18	Contains Var
19	Contains EVal
20	Contains href
21	Contains cookie
22	Contains single Quate
23	Contains JS
24	Contains Hash
25	Contains ALERT
26	Contains DOts

5 Results

the results includes The evaluation procedure and the outcomes of the datasets on machine learning algorithms.

5.1 Evaluation

evaluation is measured using various metrics depending on the specific task and the nature of the data.After constructing the proposed models, a comprehensive evaluation process is conducted to asses their performance. This evaluation employs a range of metrics that encompass Accuracy, Precision, Recall, and F1-Score.Several aspects of the model's performance are evaluated, including as accuracy, precision, recall, and F1-Score.definition of the accuracy metric offers a comprehensive assessment of the model's capacity to correctly classify the benign and malicious URLS.It calculates the percentage of instances that are correctly categorized out of all the instances.Since accuracy is simple to understand and convey, it is a preferred option for informing stakeholders about the performance of models. Although accuracy offers a simple way to assess model performance, it might not be appropriate for datasets that are unbalanced and have a predominately one class. Other assessment measures, such recall, accuracy, or F1 score, could be more useful in specific circumstances.In terms of math, accuracy is computed as:

$$\text{Accuracy} = \frac{\text{Amount of Accurate Predictions}}{\text{Total Predictions}}$$

In a binary classification scenario, accuracy can be computed using the counts from a confusion matrix:

Where:

The quantity of cases that are genuinely positive and were appropriately categorized as such is known as True Positives (TP). The quantity of cases that are genuinely negative and were appropriately categorized as such is known as True Negatives (TN). False Positives (FP): The quantity of cases misclassified as positive when they are in fact negative. False Negatives (FN): The quantity of positive cases that were mistakenly labeled as negative.

5.2 Precision

is a statistic that calculates the percentage of actual positive predictions among all the model's positive predictions.Higher numbers on the accuracy score, which goes from 0 to 1, suggest greater precision. When the precision score is 1, it means that every positive prediction the model produced is accurate; when it is zero, it means that none of the positive predictions are. When assessing classification models, precision is an important parameter, especially when false positives are expensive or unwanted. Out of all the model's positive predictions, it calculates the proportion of true forecasts. In mathematics, accuracy is determined as follows:

$$\text{Precision} = \frac{\text{Tp}}{\text{Tp} + \text{Fp}}$$

5.3 Recall

Recall, also referred as the true positive rate, gauges a model's accuracy in identifying each pertinent event within a dataset.

$$\text{recall} = \frac{\text{Tp}}{\text{Tp} + \text{FN}}$$

This measure demonstrates the model's capacity to identify positive cases and is especially significant in situations when it is expensive or undesirable to overlook positive instances (false negatives), as in the case of fraud detection or medical

Table 2. .

Model	predicted positive (%)	predicted negative (%)
Actual Positive	TP	FN
Actual negative	FP	TN

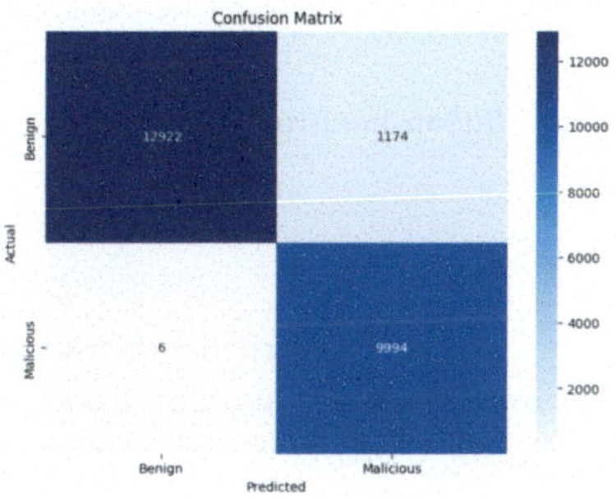

Fig. 2. SVM confusion matrix

diagnosis. A high recall score means that false negatives are less common since the model successfully catches the majority of positive cases.

The confusion matrix is a table used to evaluate the performance of a classification algorithm. By contrasting the model's predictions with the actual findings, it summarizes the model's output. Typically, a binary classification problem's confusion matrix is defined as follows (Table 2):

Figure 2 shows the results of the confusion matrix for the SVM model as follows:

the experiments focused on the performance of the SVM, random forest, naive BS, and the neural network by using the dataset and feature extraction provided early in the paper. Based on the results of the machine learning models, the Random Forest model demonstrated exceptional performance with an accuracy of 99.53. It achieved a recall of 99.30, indicating its ability to correctly identify the majority of positive instances. Moreover, the precision of the Random Forest model stood at 98.91, suggesting its capability to accurately classify positive predictions. In comparison, the Gaussian Naive Bayes model achieved an accuracy of 99.07, with varying precision and recall scores for both classes. Additionally, the Support Vector Machine (SVM) model attained an accuracy of 95, with a balanced performance across classes as indicated by the macro and weighted average metrics. These results suggest that the Random Forest model outperformed the other models in terms of accuracy, recall, and precision, making it a promising choice for the classification task.the next figure shows the cross validation accuracy scores for the top 3 models with high accuracy.the cross-validation results for the three models-RandomForestClassifier, SVM, and a Neural Network-were obtained using a 3-fold cross-validation method (Fig. 3).

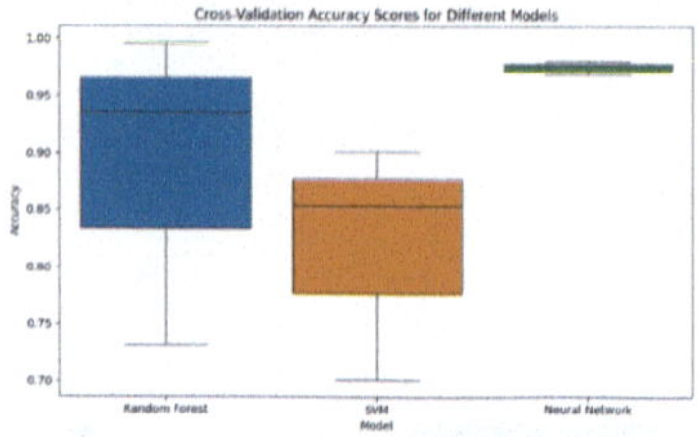

Fig. 3. Cross validation accuracy scores for top 3 models

5.4 Neural Network

Using a neural network as a machine learning model provides a powerful and adaptable approach to solving a wide range of tasks, including classification, regression, and even more complex problems like image recognition and natural language processing. the training of the neural network model got a very high accuracy of 0.9908 percent as the following figure shows the perforamance of it. the following figure shows the results of the confusion matrix (Fig. 4 and Table 3).

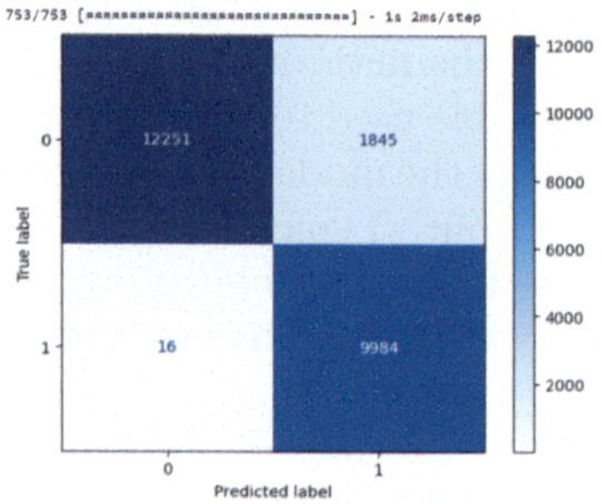

Fig. 4. Neural network confusion matrix

Table 3. Performance Comparison of Machine Learning Models

Model	Accuracy (%)	Recall (%)	Precision (%)
Random Forest	99.53	99.30	98.91
Gaussian Naive Bayes	99.07	91.00	94.00
Support Vector Machine (SVM)	95.00	96.00	95.00

6 Conclusion

The paper presents a comprehensive study on detecting Cross-Site Scripting (XSS) attacks using machine learning and deep learning techniques. XSS attacks pose significant threats to web applications, compromising user data and system integrity. The study aims to enhance XSS detection accuracy to mitigate these risks effectively.

Various machine learning and deep learning algorithms are explored, including Support Vector Machine (SVM), Random Forest, Naive Bayes, and Neural Networks. Each algorithm is evaluated based on metrics such as accuracy, recall, and precision. The results indicate that Random Forest demonstrates exceptional performance, achieving high accuracy, recall, and precision scores. Specifically, the Random Forest model achieved an accuracy of 99.53, with a recall of 99.30and a precision of 98.91.

Feature extraction plays a crucial role in detecting XSS attacks, focusing on both structural and behavioral features of JavaScript scripts. Techniques like normalization, encoding, and data splitting are employed to preprocess the dataset, ensuring optimal training and testing conditions for the models.

The evaluation results highlight the effectiveness of machine learning in detecting XSS attacks, with Random Forest emerging as the top-performing model. Its high accuracy, recall, and precision make it a promising choice for XSS detection in web applications.

In conclusion, the study underscores the importance of leveraging machine learning techniques to combat XSS attacks effectively. By continuously refining detection models and exploring advanced feature extraction methods, web developers and security professionals can enhance the resilience of web applications against XSS vulnerabilities, safeguarding user data and system integrity.

7 Future Work

The future work in the cross site scripting field is detecting other types of XSS attacks like the persistent and DOM XSS attacks and investigate new machine learning and deep learning models.

References

1. Alqarni, A.A., Alsharif, N., Khan, N.A., Georgieva, L., Pardede, E., Alzahrani, M.Y.: MNN-XSS: Modular Neural Network Based Approach for XSS Attack Detection (2022)
2. Bakır, R., Bakır, H.: Swift detection of XSS attacks: enhancing XSS attack detection by leveraging hybrid semantic embeddings and AI techniques. Arabian J. Sci. Eng. 1–17 (2024)
3. Banerjee, R., Baksi, A., Singh, N., Bishnu, S.K.: Detection of XSS in web applications using machine learning classifiers. In: 2020 4th International Conference on Electronics, Materials Engineering & Nano-Technology (IEMENTech), pp. 1–5. IEEE (2020)
4. Gogoi, B., Ahmed, T., Saikia, H.K.: Detection of XSS attacks in web applications: a machine learning approach. Int. J. Innov. Res. Comput. Sci. Technol. (IJIRCST) (2021). ISSN, 2347-5552

5. Habibi, G., Surantha, N.: XSS attack detection with machine learning and N-Gram methods. In: 2020 International Conference on Information Management and Technology (ICIMTech), pp. 516–520. IEEE (2020)
6. Mereani, F.A., Howe, J.M.: Detecting cross-site scripting attacks using machine learning. In: International Conference on Advanced Machine Learning Technologies and Applications, pp. 200–210. Springer (2018)
7. Mereani, F.A., Howe, J.M.: Preventing cross-site scripting attacks by combining classifiers. In: Proceedings of the 10th International Joint Conference on Computational Intelligence (IJCCI 2018) - IJCCI, pp. 135–143. INSTICC, SciTePress (2018)
8. Mokbal, F.M.M., Dan, W., Imran, A., Jiuchuan, L., Akhtar, F., Xiaoxi, W.: MLPXSS: an integrated XSS-based attack detection scheme in web applications using multilayer perceptron technique. IEEE Access **7**, 100567–100580 (2019)
9. Santithanmanan, K., Kirimasthong, K., Boongoen, T.: Machine learning based XSS attacks detection method. In: Naik, N., Jenkins, P., Grace, P., Yang, L., Prajapat, S. (eds.) Advances in Computational Intelligence Systems. UKCI 2023. AISC, vol. 1453,pp. 418–429. Springer, Cham (2024). https://doi.org/10.1007/978-3-031-47508-5_33
10. LNCS Homepage. http://www.springer.com/lncs. Accessed 25 Oct 2023

Using Artificial Intelligence to Enhance Family Cohesion and Promote Positive Social Values

Nawal Qasim Hammadi(✉), Mukhalad Maher Dawood(✉), Zainab Abbas Fadhel(✉), and Mrouge Ghani Jabbar(✉)

University of Baghdad, College of Islamic Sciences, Baghdad, Iraq
{nawal.qasim,mukhaled.m,zainab.abbas, mrouge}@cois.uobaghdad.edu.iq

Abstract. The present study is aimed to explore to what extent the use of the following AI applications by family members impacts on the respective relations: communication, constructing the system of family values, and digital consumption in the family. The information was collected via the semi-structured interviews with 20 adolescents of 13–17 years old, taking into consideration the factors of gender and family status. The results of the research indicate that AI can improve communication and promote its intensity and nature by supporting the principles of feasible social values. At the same time, the findings reveal controversial issues in fears, drawbacks, and unnecessary applications, which is explained by a wide variety of AI applications. AI is revealed as an application useful for family cohesion in ways of monitoring already established communication and support for managing digital consumption. The results demonstrate the positive outcomes of AI applications in the sphere of family influence based on the opportunities provided, including the design of special applications for managing screen time, supporting recommendations, and increasing the overall intensity of family-based communication. At the same time, the increase in control and monitoring is viewed as a negative outcome by adolescents, who have their personal approach to digital solutions and habits. The results demonstrate that considering all AI applications, its influence on communication, values, and digital consumption depicts a balanced view, with the positive outcomes significantly contributing to the strengthening of family cohesion through better communication and managing.

Keywords: Artificial intelligence · family cohesion · digital technology · family dynamics · communication · privacy · qualitative research · screen time · adolescents · AI in family settings

1 Introduction

Technology and digital interactivity mean that many families experience communication in an entirely different way than before, with corresponding shifts in relationships and emotional involvement. Nowadays every family member has their own digital device and members in the family are communicating through online and let be clear this trend can either help or fail to strengthen your bond. Digital media is now grounded in the fabric of our daily life, although still under review by researchers regarding what constitutes positive or negative impact on family dynamics [1].

A. Abdelgawad et al. (Eds.): ISBCom 2024, LNNS 1268, pp. 765–774, 2025.
https://doi.org/10.1007/978-3-031-82377-0_61

Family organization is one of the basic elements of social well-being that is progressively transformed by information and communication technologies (ICT). Research into the effects of technology on family relationships, especially among young people, is essential to inform interventions aimed at supporting positive family behaviors. That is facilitating personalized suggestions, automatic content filtering, and monitoring activities using artificial intelligence (AI) on one side for managing family communication while there are very high chances of misuse even being used in edition of text messages at the other end. Applications of AI for family communication are on the rise, from the use of AI-based communicative platforms to actively regulating screen time and digital consumption habits within a family [2].

Therefore, studying the impact of technologies on family relations as well as examining the potential of using AI to support positive social values in families are essential.

While there is a burgeoning literature on technology and family communication, very little work has focused specifically on AI for promoting family cohesion. The rest of the current literature is based on negative impacts of technology in relations, such as advanced use of devices and power of face-to-face troubleshooting [3]. What is missing; however, is research considering the potential for AI and other technologies to strengthen family ties and encourage social goods. Most of the previous studies focused on cyberbullying, Internet addiction, and psychological harm from social media use while few assessed AI possibilities to counterbalance these potential negative effects and foster more healthful family dynamics. What is more, understanding how young people view family communication in the digital age is an area that remains underdeveloped in research. Several studies have examined how parents mediate and control the digital media behaviors of their children, little is known about young people themselves in this context, and their use of digital technologies to facilitate family communication [4].

The family dynamics that are so central to the aging process also present a second gap in the use of AI tools for their support. However, most AI applications in the family scenario have been implemented to monitor and control online activities of children but not at promoting cooperation and understanding among family members. While research typically paints technology as a hindrance to family unity, it fails to consider that technologies could be designed or used to advantage the health of families [5]. Additional work is required focusing on family AI and how this can coexist in their daily lives, so these devices are seen as facilitating communication and emotional connection while instigating the formation of good behavior habits. In addition, understanding the ways in which family organizations as a group engage with digital practices and how those practices can be used for well-being purposes is still at an early stage of development [6].

Research on digital technologies being detrimental to family life is growing, but findings still differ depending on the context of how these technologies are used. While the Internet brings the family together across distances, for example, Mesch established that frequent use was negatively correlated with one aspect of this family cohesion. Similarly, consistent with the importance of communication strategies that may alleviate the development of behavioral problems associated with irregular parental mediation on digital habits needs further investigation [7]. Further studies, such as those delve into the scalability of family rules and their relations with children's online behaviors, arguing that firm designs are often inflexible in addressing changes of digital interactions [8].

But again, on the bright side, A key factor benefits of communication technologies, when used appropriately are an increased level of independence that proliferation of devices supports and the empowerment these tools offer families the ability to stay connected. This body of research suggests that family media use is increasingly mediated through digital media, raising both challenges as well as promises with respect to how-screen-based stand-alone and interactive-media will be woven into networked-family life [9].

The influence of digital media on family dynamics has been the object of general and academic interest, and there is several studies focusing on its positive and negative effects on family wellbeing. Vanden Abeele [10] explores the concept of digital well-being and points out that it is not a fixed state but a dynamic construct that changes according to the way families make use of digital tools. This view is largely consistent with the fact that digital technology can both promote and prevent family cohesion. One important issue related to the use of technology is technofinance, a phenomenon when technology use automatically leads to some type of interruption. McDaniel and Radesky [11] define the impact of technology on family relationships in terms of interruptions and report on research results showing how parents' problems with technology predetermines increases in behaviour problems in children. This fact means that children whose parents make extensive use of technology may receive less parental attention as their parents become 'interrupted'.

Nesi, Telzer, and Prinstein [12] discuss how today's adolescents develop in the age of digital media and claim that the way young people engage with technology defines their social and emotional development. This view is also important for AI ethics since to properly consider the opportunities for developing AI tools for enhancing social values it is critical to understand how adolescents perceive and use technology tools. Finally, Odgers and Jensen [13] provide an overview of the digital age risks for adolescent mental health. The authors separate the beneficial and harmful effects of social media for mental health and emphasize the importance of considering the dual role of this effect for developing AI tools enhanced social values. The studies provide solid evidence of the contradictory impact of digital technologies on family life. Technology use may prevent family cohesion, but at the same time, AI can contribute to enhancing positive social values.

The main objective of this research is to investigate the potential use of AI in strengthening family bonding and developing such desired social values among members living together. The study will explore how young people think about family relationships in the digital world and explore ways of using technology to support these relationships, rather than undermining them. Aim of the Study The specific aims of this study are as follows:

1. Study young people's digital practices and their effects on family relationships
2. Identify positive uses of AI in assisting family communication and emotional bonding.
3. Offer suggestions for incorporating AI tools that nurture family well-being in facilitating more effectively communication strategies.

4. Investigate the opportunities of AI in building a bridge between parents and children toward digital life.

The paper opens with an Introduction to the state of technology in contemporary families and describes existing literature on the topic, including gaps. The Methodology section then describes the qualitative research design, sampling method, data collection methods and process of data analysis. The Results section evaluates how technology may bridge and disrupt is a dual function in family performance, as well as the role of AI function to increase better interaction in general. Finally, the paper closes with a Discussion and Conclusion section where the implications of this study, limitations, as well as future research recommendations are provided.

2 Methodology

This qualitative research aims to explore how young people understand and experience the impact of artificial intelligence (AI) on family cohesion and communication. As such, it is useful for investigating the nuanced details of how digital technologies are enacted in families and therefore contribute to aligning this within family dynamics. The exploratory quality of the research permits the researchers to critically analyze accounts from young people and their family members into deeper layers of what digitalization implies to them. Through a descriptive and interpretative approach, the study aims to understand in what ways AI can improve family caring relationships and communication. Naturalistic observation / documentation: The objective of this design is not just to see and record existing phenomena but likewise furnish an idea to get how AI may be applied in the interests of addressing current family communication problem [13].

A purposive sampling method was used to select study participants. In terms of qualitative research, Purposive sampling can be argued to best suit this approach as it enables researchers to select participants who have firsthand or more direct experience of the topic at study and will be able to provide a rich amount of data in reference. The target population in this research was adolescents, partially navigated by the recognition that 13 to 17 years old are one of the most digital-rich demographics and therefore constitutes likely a good source of information or ideas about how some aspects of technology influence family interaction. Participants: A total of 20 individuals (11 females and 9 males) were recruited from public settings such as schools and community centers. Criteria for selection were possession of digital devices on which suitable apps could be used, and a degree of interaction with family members through technology. With purposive sampling, we wanted to ensure that the participants were associated with the research questions in a meaningful way and their experiences gave us some insight into what was happening more generally, among youth digital practices and family relationships.

The method used for collection of primary data in this study was semi-structured interviews. This method was selected as it provides an opportunity to be flexible while querying participants' views in a controlled framework designed around the research questions. These took the form of short, individual interviews with each participant, and lasted around 30 min. Respondents were invited to think about their digital life, especially the way that they perceive AI and other digital instruments relations within their families. The questions were constructed to prompt rich responses as to the role of technology in family settings, both negative and positive. Participants were interviewed while the conversations were recorded. And verbatim transcripts prepared for analysis. All the participants who conducted interviews were also asked to share visual data such as screenshots from text exchanges with family members to supplement their responses [12].

The data collected were verified using thematic analysis, to identify, analyze and report patterns within qualitative data. The framework was based on a widely used approach at the time the six-step process for conducting thematic analysis developed by Braun and Clarke targeting familiarization with the data, generating initial codes, searching for themes, reviewing themes, defining and naming themes, and writing up. Thematic analysis was chosen for its flexibility, to generate inductive theory and provide a detailed, rich account of the data. Coding was performed manually using the specific search-words for patterns in relation to the use of AI and digital technologies in family communication. The data were analyzed iteratively, allowing for the discovery of new themes as analysis progressed. Multiple researchers coded a training sample of the interviews and checked and adjusted codes until high levels of consistency were achieved to assure inter-coder reliability. Indeed, this approach served to draw eligible inferences from AI on how the family could be enriched [14].

3 Results

To explore the roles of digital technology, particularly artificial intelligence (AI), in adolescents' perceptions of family relationships we conducted semi-structured interviews with 20 adolescents aged 13 to 17. The results yield valuable lessons about the intersections among technology, family life and AI to support family cohesion.

Cumulative findings showed that although there are a variety of ways in which digital technologies can serve to foster or interrupt family relationships, most participants conflated these uses. On the bright side, it was what allowed family members to stay in touch when physically separated from one another, per our interviews, most contributors used FaceTime, WhatsApp and other social media platforms to stay in touch with their mom or big brother, if even through sporadic whiles when their types of day might diverge. They noted that, in this context of home and technological change, digital tools were employed to communicate instantly with each other and share their life events with the member of family after time matter how far part though they could be, that can be clear in Table 1.

The table above illustrates the participants' responses regarding AI's influence on family dynamics. To further visualize this data, the following bar chart provides a graphical comparison of positive and negative feedback across the different aspects.

On the other hand, Fig. 1 can show how different the representation of the positive and negative responses is, which can also help to identify the factors that were perceived with the highest positivity and negatively. For instance, both improved communication and AI-recommended activities are characterized by the highest scores for the positive response but mostly viewed negatively when it comes to privacy concerns. Thus, this information makes the contribution to the detailed discussion of how AI technology affected the studied family and in what aspects was it seen as positive or harmful.

Table 1. AI in Family Dynamics Survey Results

Factors	Positive Responses (%)	Negative Responses (%)
Improved Communication	65	35
Privacy Concerns	20	80
Increased Screen Time	30	70
Enhanced Parental Control	55	45
AI-Recommended Activities	60	40

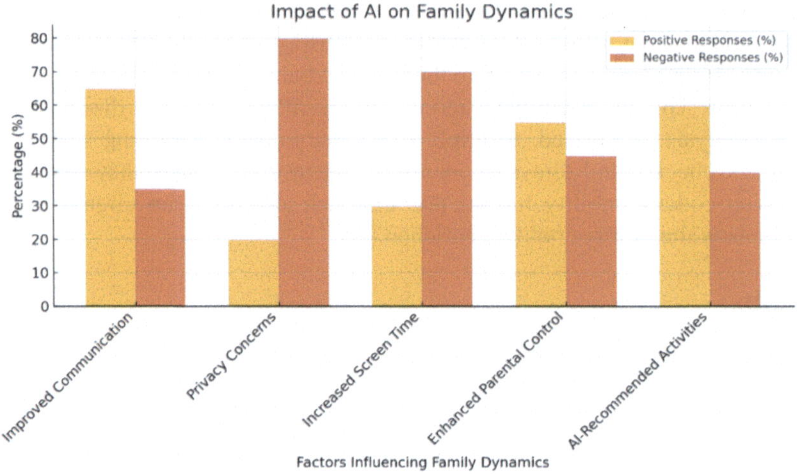

Fig. 1. Impact of AI on Family Dynamics

Turner adds that participants also worry about the risk of "excessive reliance on technology which refers to the sense of being always available; eating a meal out with friends or family and feeling disconnected, by checking texts, e-mails or Twitter feed while sat at the restaurant table. Participants recounted instances when family members were in the same physical space, but distant because they were engaged on their phone or tablet. This echoes research showing that although technology can stem physical distances, in many cases it has a blocking effect on emotions, when not used as intended.

Overall, participants suggested that it would be nice to have artificial intelligence control more of the family tech dosage. Several participants had much more creative ideas: AI could live in screen time usage and provide reminders/prompts to limit time during family-focused times or on certain days/nights for instance dinner, family activities. AI would also recommend digital content that allows families to bond such as family-friendly games or real-life activities based on shared interest of members. Additionally, some participants imagined AI tools that would analyze how family members talk to help them better engage and reduce miscommunications through personalized recommendations.

Though there was an undercurrent of opportunity realized, people where wary AI could get too personal. A few flagged it as a possible privacy concern such as robots watching all our conversations, that might detract from the naturally variable and nuanced family life. AI was considered able to provide valuable advice and assistance, but family relationships must continue to be based on human input, thus integrating AI as a complementary tool rather than an oppressive center of control.

Although participants identified some positive implications of the technology, they also noted potential challenges due to overuse. Screen time was declared a major worry, with individuals, including children, stating that they felt distanced from their families because of too much screen use. Social media and gaming platforms were mentioned throughout the research as flashpoints in family disputes. For example, parents are reported to frequently lament that children spend too long online leading to conflict pressuring and lack of scope for acknowledgment of the role that digital communications play in young people's social lives.

Moreover, other participants mentioned that their parents frequently fail to comprehend the technology and what it meant in their lives. Parents, the report found, saw technology mainly as a means of entertainment or distraction and failed to appreciate its role in academic performance and social networking, even for self-expression. The lack of one clear understanding fueled tensions, as noted participants pointed to an increased need for open dialogue among family members concerning their use and the importance they attach to technology.

4 Discussion

Results of this study highlighted those digital technologies, including AI not only function as a medium for bonding, but at the same time are also recognized as an enabler causing disconnectedness in family relations. These results are in line with prior research indicating that technology is a double-edged sword. These findings suggest that there is no one simple answer when it comes to technology and families. One of the key messages of this study from the adolescents was how digital communication tools can help facilitate family bonding, especially where face-to-face contact was restricted. This is consistent with Mesch's findings about the potential of the Internet to reduce household distances. Still, the troubles technology use creates for relationships especially family time reveal epidemic problems that put every type of family in a difficult position to navigate digital engagement with genuine human connection.

The fact that screen time detracts family cohesion is in harmony with the literature. It is interesting that the participants also suggested AI may be used to monitor technology use and support better times for the whole family, suggesting a future direction. AI tools could also moderate screen time by sending notifications about alerts moderation, and even suggest to the family a new activity that all can do together [15].

However, the findings revealed that the participants also worry about AI over-monitoring family interactions; hence any technological solution can only be successful if it maintains a proper balance between automation and human agency. Why AI systems must be unobtrusive and tuned to the nuances of each family.

It also highlights the need for privacy and making sure AI tools are augmentations to familial living rather than a replacement for natural communication. This study marks both the strengths and opportunities as well as the barriers to challenge of placing AI inside family environments. The prospect of using AI to better help families cohere, notably in the throttling back on technology use is a good one, but obviously every effort must be made to that end does not corrode or destruct the things created by human connections which are the main foundations of family.

5 Conclusion

This project set out to investigate how digital technology, with a specific focus on artificial intelligence (AI), impacts upon the relationships experienced between family members and the dynamics of families and their bonds. The investigators noted that although people may regard digital technologies for instance, social media and messaging apps as helping to maintain communication with family and friends, they can also be disruptive when used too much. Both positive and negative effects of technology on family life were noted by participants That is technology, keeping us connected in a more emotional way over the miles. In contrast, overuse of digital devices during family time frequently created feelings of separation. However, participants did feel AI could help alleviate some of these downsides by improving screen management and supporting family activities.

These findings carry significant implications as we make room for AI in our families of the future. With technology ever more engrained in everyday life, the search for methods of human-digital integration is more imperceptible and necessary than ever. AI can help strike that balance for families and give them tools to use technology intentionally. For instance, AI-powered platforms could prompt families to spend time on family-friendly activities or remind them that it is time to set down phones and laptops during family holidays. Additionally, AI systems can be configured to help encourage healthier digital habits, providing tailored recommendations based off the unique dynamics of that family. But the study warns against reliance on AI in handling family relationships too much. While great care must be taken not to over-monitor family interaction and erode the spontaneity and natural communication that defines families, fostering family cohesion is by far worth risking. Hence, it is needed that AI tools are built considering the trait of privacy and the autonomy of family to truly help rather than control.

Though the study makes for fascinating reading on young people's views about technology and AI in family life, it is important to note the limitations. As with all small-number pilot studies, the study has a very limited participant sample size (n = 20) and hence it is difficult to extrapolate these results to other populations. The research also was limited to a narrow age group such as adolescents aged 13–17 years, so the findings may not be generalizable to other family members, like younger children and older adults. As well, the study was limited by self-reported data and participants might have given responses that they thought were more socially desirable.

It only tells us that more and additional varied data should be gathered from larger and more diverse sample groups both in terms of age and genealogy on the influence of AI on family cohesion. In addition, a closer look at the implications AI could have on family structures of different types whether a single-parent household or multi-generational families may allow us to explore how to best leverage AI more universally. Moreover, future research may be interested in examining the lasting consequences of AI driven interventions on family dynamics and extent to which such technologies influence communication patterns within families over longer periods. The investigation of the ethical side considering AI or Privacy in family settings should be next to further research.

Thus, this study illustrates the two-faced character of digital technology in family relationships which can serve as an enabler and a threat at the same time. AI may be especially helpful in this regard, giving us new strategies to promote healthier interactions with technology and improve relationships within the family. Family life, on the other hand, will not have this option at least if AI is to take its rightful place and the two must be role models for each other in terms of finding a way to integrate one into the other without too much intrusion. If introduced intelligently and ethically, that may not be too much to ask for as families set out to navigate the brave new world of the digital age.

References

1. Carvalho, J., Francisco, R., Relvas, A.P.: Family functioning and information and communication technologies: how do they relate? a literature review. Comput. Hum. Behav. **108**, 106304 (2022). https://doi.org/10.1016/j.chb.2014.11.037
2. Chen, V.H.H., Chng, G.S.: Active and restrictive parental mediation over time: effects on youths' self-regulatory competencies and impulsivity. Comput. Educ. **128**, 37–47 (2019). https://doi.org/10.1016/j.compedu.2016.03.012
3. Devitt, K., Roker, D.: The role of mobile phones in family communication. Child. Soc. **23**(3), 189–202 (2019). https://doi.org/10.1111/j.1099-0860.2008.00166.x
4. Goh, W.W.L., Bay, S., Chen, V.H.H.: Young school children's use of digital devices and parental rules. Telematics Inform. **32**(4), 787–795 (2019). https://doi.org/10.1016/j.tele.2015.04.002
5. Kaufmann, K., Peil, C.: The mobile instant messaging interview (MIMI): using WhatsApp to enhance self-reporting and explore media usage in situ. Mobile Media Commun. **8**(2), 229–246 (2020). https://doi.org/10.1177/205015791985239
6. Orben, A.: Teenagers, screens and social media: a narrative review of reviews and key studies. Soc. Psychiatry Psychiatr. Epidemiol. **55**(4), 407–414 (2020). https://doi.org/10.1007/s00127-019-01825-4
7. Paulus, M.P., et al.: Screen media activity and brain structure in youth: Evidence for diverse structural correlation networks from the ABCD study. Neuroimage **185**, 140–153 (2019). https://doi.org/10.1016/j.neuroimage.2018.10.040
8. Radesky, J.S., et al.: Young children's use of smartphones and tablets. Pediatrics **146**(1), e20193518 (2020). https://doi.org/10.1542/peds.2019-3518
9. Twenge, J.M., Campbell, W.K.: Media use is linked to lower psychological well-being: Evidence from three datasets. Psychiatr. Q. **90**(2), 311–331 (2019). https://doi.org/10.1007/s11126-019-09630-7
10. Abeele, M.M.V.: Digital wellbeing as a dynamic construct. Commun. Theory **31**(4), 932–955 (2021). https://doi.org/10.1093/ct/qtaa024
11. McDaniel, B.T., Radesky, J.S.: Technoference: parent distraction with technology and associations with child behavior problems. Child Dev. **90**(1), 100–109 (2019). https://doi.org/10.1111/cdev.12822
12. Nesi, J., Telzer, E.H., Prinstein, M.J.: Adolescent development in the digital media context. Psychol. Inq. **31**(3), 229–234 (2020). https://doi.org/10.1080/1047840x.2020.1820219
13. Odgers, C.L., Jensen, M.R.: Annual research review: adolescent mental health in the digital age: facts, fears, and future directions. J. Child Psychol. Psychiatry **61**(3), 336–348 (2020). https://doi.org/10.1111/jcpp.13190

14. Pasha, A., et al.: MADAMIRA: a fast, comprehensive tool for morphological analysis and disambiguation of Arabic. In: Proceedings of 9th International Conference on Language Resources and Evaluation (LREC), pp. 1094–1101 (2014)
15. Ali, N.: Deep learning approaches for arabic image captioning. IEEE Access **8**, 106830–106843 (2020)

Beyond Words: BERT-Powered Insights into Suicide Risk via Social Media Posts

Nada Nady(✉) and Maha Abdelmohsen

Modern Sciences and Arts University (MSA), Giza, Egypt
{nada.nady,mhatem}@msa.edu.eg

Abstract. In an era where social media is a window into the human psyche, the detection of suicidal ideation through advanced machine learning models has become a pressing necessity. This study leverages the power of Bidirectional Encoder Representations from Transformers (BERT) to identify suicide risk in Twitter posts. With over 233,000 texts from relevant subreddits, our BERT-based model demonstrates an impressive accuracy of 95.50%, effectively distinguishing between suicidal and nonsuicidal content. This approach not only advances the state-of-the-art in natural language processing but also opens new avenues for early intervention in mental health crises. By automating the detection of subtle linguistic cues indicative of suicidal thoughts, our research underscores the transformative potential of AI in safeguarding mental well-being and preventing tragedies. Join us as we explore the intersection of technology and empathy, aiming to make a profound impact on suicide prevention efforts worldwide.

Keywords: Machine Learning · Artificial Intelligence · Suicide Detection · Deep Learning · Social Media

1 Introduction

In recent years, social media platforms have become an integral part of our daily lives, with nearly 3.96 billion people actively using the Internet worldwide. Millions of people regularly engage with social media, including chat rooms, social networking sites, blogging sites, and other platforms. Sites such as Facebook, Twitter, and Snapchat allow users to exchange information and interact with others. Twitter, a free social media streaming site, enables registered users to communicate through 280-character messages called "tweets" [1].

Suicide, a prevalent global phenomenon, stands as a significant cause of mortality and is often perceived as a final option for individuals grappling with mental health challenges. Insufficient mental health support is a widespread issue, affecting both developed and developing nations. Social media platforms provide an open space for users to express a range of emotions, including negative ones [18]. Many users utilize social networks to convey their feelings, experiences, ideas, difficulties, and interests. Among the most common topics discussed are thoughts of self-harm, death, and suicidal ideation. Suicidal behavior can be categorized into suicidal behavior, suicide planning, and suicidal ideation. Suicidal ideation refers to thoughts or intentions to end one's life without

A. Abdelgawad et al. (Eds.): ISBCom 2024, LNNS 1268, pp. 775–788, 2025.
https://doi.org/10.1007/978-3-031-82377-0_62

actual attempts, whereas a suicide plan is a specific method considered to end one's life, and a suicide attempt involves self-harm intended to result in death. Suicide affects individuals, families, communities, and countries, being the second leading cause of death among young people globally. It causes more deaths than diabetes, liver disease, stroke, or infection [8]. Social media posts provide valuable insights into an individual's emotional and psychological well-being. Many individuals may prefer to express their suicidal feelings or intentions online rather than in person. Unfortunately, these posts are often ignored or overlooked, despite their potential use in large-scale screening for suicidal behaviors [1].

The motivations behind suicide are multifaceted. While individuals with depression are at high risk, others without depression may also be affected. According to the American Foundation for Suicide Prevention (AFSP), suicide risk factors can be categorized into health, environmental, and historical factors [13]. Research exploring the online behavior of young individuals at risk of suicide reveals a complex balance of benefits and risks associated with internet communication. On the positive side, online platforms offer a perceived safe and non-judgmental space for discussing sensitive topics like suicide, providing relief compared to in-person interactions. This relief can temporarily reduce suicidal thoughts, emphasizing the importance of anonymity online. However, studies also caution that individuals at higher risk of suicide, particularly those with a history of attempts, may engage in destructive online behavior and be more vulnerable to negative influences from harmful content [23].

Suicide remains a highly taboo and stigmatized topic in most societies and cultures. Social media can be important platforms for suicide prevention communication, as they can reach many people who may be difficult to reach. Good accessibility and anonymity encourage users to share personal information, including suicidal thoughts and feelings. Other users can respond to these messages and report suicide prevention resources or provide direct help and support and intervene in the event of an immediate suicide risk or ongoing suicide attempt. Having the opportunity to communicate suicide-related issues with other users with similar experiences seems to be very important and can prevent suicide, at least temporarily [21].

Suicide remains a complex and sensitive issue, and delving into its causes requires careful study. Discussing suicide methods can be both emotional and harmful, therefore, it is important to focus on understanding the complexities surrounding this issue. Suicide is a global public health crisis, with nearly 700,000 deaths annually. The World Health Organization (WHO) estimates that one person dies by suicide every 40 s, making it a leading cause of death for young people [2, 4, 18].

Suicide rates vary significantly across different regions, demographics, and socioeconomic backgrounds. Identifying vulnerable populations and tailoring interventions accordingly is crucial [8]. Early detection of suicidal ideation is vital for preventing suicide. By identifying individuals at risk and connecting them with appropriate resources, we can significantly impact their lives [9]. Traditional methods for suicide detection often rely on self-reporting or healthcare provider identification, leading to underreporting and missed cases [20].

Social media platforms like Twitter offer a unique opportunity for early detection. People often express suicidal thoughts and feelings openly on these platforms, providing valuable data for analysis. Ethical considerations and privacy concerns are paramount when working with sensitive data like social media content. Ensuring user anonymity and adhering to data protection regulations is essential [18].

Machine learning (ML) and natural language processing (NLP) provide powerful tools for analyzing text data and identifying patterns. By applying these techniques to Twitter content, we can develop algorithms that detect suicide ideation with high accuracy. The increasing utilization of NLP for analyzing mental health-related content on social media platforms is evident. Researchers are increasingly adopting sentiment analysis techniques to examine discussions surrounding mental health issues [22]. Similarly, others have gathered Twitter data and applied various ML methods to identify signs of suicidal ideation [12]. Deep learning (DL) techniques, such as Long Short-Term Memory (LSTM) and Convolutional Neural Networks (CNN), have gained popularity due to the prevalence of word embeddings [5]. DL excels in handling challenges like data sparsity and dimension explosion by extracting significant features. However, the dependence on limited datasets remains a hurdle [16]. Recent advancements also include the use of hybrid models combining ML and DL approaches for more robust predictions [17], the application of transformer-based models like BERT for better contextual understanding [6], and the use of graph-based methods for capturing relational data in social media networks [15]. Extracting relevant features from textual data, such as keywords and linguistic patterns, is crucial for effective model construction. Data collection, cleaning, and preprocessing are essential steps in training and evaluating ML models. Choosing suitable models and evaluating their performance on new data ensures solution robustness. Realtime analysis of Twitter data enables prompt identification and intervention for individuals expressing suicidal intent. Additionally, the development of multilingual models enhances the reach of NLP solutions to diverse global populations [2, 8, 18].

This study underscores the role of Artificial Intelligence (AI) in mental health, particularly in identifying and predicting suicidal behavior. By employing machine learning techniques, we can analyze vast datasets of social media posts to detect subtle indicators of distress. These AI-driven models offer a proactive approach, potentially saving lives by providing timely intervention.

While previous studies have explored various methods for detecting suicidal ideation on social media, our approach leverages the advanced capabilities of BERT (Bidirectional Encoder Representations from Transformers) for text classification. This model is trained on a large dataset from Reddit's "SuicideWatch" and "r/teenagers" subreddits, enabling it to understand the nuanced language used in these contexts. By utilizing BERT, our research contributes to the field by offering a more accurate and sophisticated method for detecting suicidal ideation, with the potential for real-time application in suicide prevention efforts. Analyzing vast datasets of tweets can provide critical insights into this mental health condition. The data not only contributes to research but also holds promise for interventions aimed at identifying warning signs, especially when individuals are not forthcoming about their struggles. Machine learning models offer a means to sift through these digital footprints and identify potential cases that might otherwise go unnoticed. This research seeks to harness the power of technology to address a pressing societal issue and improve outcomes for those at risk.

The remainder of this paper is organized as follows: Section II reviews related work and situates our study within the broader context of suicide detection research. Section III describes the methodology, including data collection, preprocessing, and the BERT model used for classification. Section IV presents the results of our study, highlighting the model's performance and key findings. Section V discusses the implications of our findings, addresses limitations, and proposes future work. Finally, Section VI concludes the paper by summarizing the contributions and potential impact of our research.

2 Related Work

Recent studies have extensively explored the development of suicide detection models using machine learning techniques on social media data.

[13] developed a model using Long Short-Term Memory (LSTM) networks to identify suicidal ideation on Twitter, achieving an accuracy of 80%. Their approach involved collecting and preprocessing Twitter data for training and evaluation.

[2] proposed a CNN-BiLSTM model for detecting suicidal ideation on social media, achieving high accuracy. They used publicly available Reddit datasets and employed word-embedding approaches like TF-IDF and Word2Vec for text representation.

[3] developed a BERT-based transformer model for detecting suicidal intention in tweets. Their model demonstrated high accuracy and potential for real-time application in suicide prevention.

[18] discussed the ethical considerations and privacy concerns of using social media data for suicide detection, emphasizing the importance of ensuring user anonymity and adhering to data protection regulations.

[22] explored sentiment analysis techniques in mental health, showing how these methods can be used to analyze discussions on social media platforms to detect mental health issues, including suicidal ideation.

[16] discussed the challenges of detecting suicidal ideation with limited data using deep learning models, emphasizing the importance of data augmentation and advanced model architectures to improve performance.

[1] provided a comprehensive review of machine learning methods for suicidal ideation detection on social media, highlighting recent advancements and challenges in the field.

[8] utilized natural language processing to identify suicide ideation and suicidal attempts in a psychiatric clinical research database, demonstrating the potential of NLP techniques in clinical settings.

[20] developed a time-aware transformer-based model for suicide ideation detection on social media, which considers temporal aspects to improve detection accuracy.

[9] explored the use of machine learning algorithms to detect suicide risk factors on Twitter, emphasizing the potential of social media data for early intervention.

[4] investigated suicide risk detection in online counseling services, focusing on low-resource languages and demonstrating the adaptability of ML techniques across different languages.

[23] reviewed suicide communication on digital platforms, highlighting the importance of these platforms in suicide prevention efforts.

[19] discussed the creation of a benchmark dataset for suicide risk detection, underlining the need for high-quality annotated data to advance research in this area.

[10] employed FastText embedding and XGBoost classifier for depression and suicide risk detection on social media, showcasing the effectiveness of combining different ML techniques.

[14] developed a deep learning model with multi-feature fusion and label association for suicide detection, achieving significant improvements in detection accuracy.

[6] explored the use of user embeddings for suicidal ideation detection on social media, demonstrating the potential of embedding techniques for capturing user-specific characteristics.

[15] proposed a multi-modal graph-based framework for detecting suicidal ideation on social media, which integrates different data modalities to improve detection performance.

Our study builds upon these previous works by employing BERT to analyze a large dataset of Reddit posts from the "SuicideWatch" and "r/teenagers" subreddits. We aim to refine the model further and explore additional data sources to enhance accuracy and generalizability.

3 Datasets and Methods

In this section, we discuss the proposed BERT-based transformer model for detecting suicidal ideation in online forums and the BertForSequenceClassification model for sentiment analysis.

3.1 Suicide Detection Dataset

The dataset used in this study is a collection of posts sourced from the "SuicideWatch" and "depression" subreddits on Reddit, gathered through the Pushshift API. The content comprises posts from "SuicideWatch," labeled as suicide, and posts from the "depression" subreddit, labeled as depression, spanning from December 16, 2008, to January 2, 2021. Additionally, non-suicide posts were collected from the "r/teenagers" subreddit. The current version of the dataset includes labels for suicide and non-suicide posts. However, version V13 of the dataset expands this labeling to include categories for suicide, depression, and normal conversations. To facilitate the collection process, a notebook is provided, demonstrating the method of gathering Reddit posts using the PushShift API.

Furthermore, the dataset for Twitter suicidal intention detection comprises 232,074 texts categorized into two classes: "suicide" (116,037 texts) and "nonsuicide" (116,037 texts). The data was split into training, validation, and testing sets using stratified sampling, with detailed class distribution provided in Table 1.

3.2 Sentiment Analysis Dataset

This dataset was automatically created based on the assumption that tweets containing positive emoticons are positive, while tweets with negative emoticons are negative. The content is stored in a CSV file with the emoticons removed and includes the following

Table 1. Class Distribution for Suicide Detection Dataset Splitting

Set	Suicide	Percentage	Non-Suicide	Percentage
Training	92426	50.05%	92233	49.95%
Validation	23128	49.82%	23287	50.18%
Testing	483	48.3%	517	51.7%
Overall	116037	50%	116037	50%

fields: Polarity, ID, Date, Query, User, and Text. The dataset is available to the data science community for exploration and experimentation with sentiment analysis. The sentiment analysis data comprises 29,480 texts categorized into three classes: "negative" (11,938 texts), "neutral" (9,204 texts), and "positive" (8,338 texts). This dataset was split into training, validation, and testing sets using stratified sampling, with the detailed class distribution provided in Table 2.

3.3 Text Preprocessing

To prepare the text data for analysis, several preprocessing steps were implemented:

Table 2. Distribution of Negative, Neutral, and Positive Data

Set	Negative	Percentage	Neutral	Percentage	Positive	Percentage
Training	10298	40.47%	5260	31.23%	7224	28.30%
Validation	2224	40.47%	1716	31.23%	1556	28.30%
Testing	820	41.02%	622	31.12%	557	27.86%
Overall	14162	40.50%	10920	31.22%	9894	28.28%

Handling Missing Values: NaN values were filled or removed appropriately.

Label Mapping: String labels were converted to numerical format.
Extracting Text and Labels: Text data and corresponding labels were extracted.
Removing Stop Words: Common stop words were removed.
Lemmatization: Words were lemmatized to their root form.
Tokenization: The BERT tokenizer converted text data into numerical representations.

3.4 Model Definition

BERT, known as Bidirectional Encoder Representations from Transformers, utilizes the encoder structure of a transformer for language modeling and was developed by Google in 2018 [7].

Model Architecture: BERT is pre-trained on Masked Language Model (MLM) and Next Sentence Prediction (NSP) tasks, enabling bidirectional training.

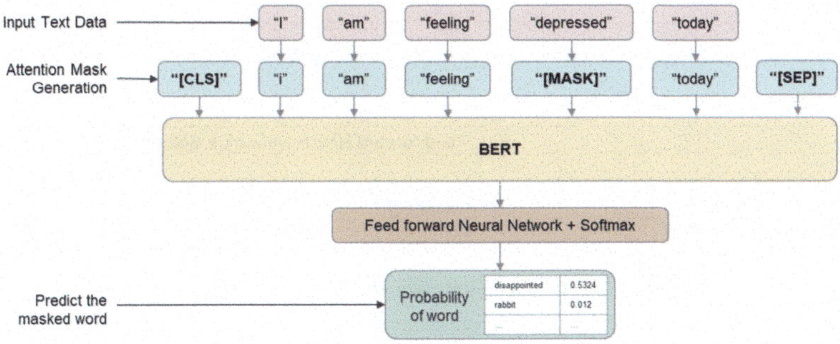

Fig. 1. BERT Masked Language Model (MLM) [11]

In MLM, parts of the input sequence are masked, and the model is trained to predict the masked tokens. For instance, the input sentence "I am feeling depressed today" is transformed into "I am feeling [MASK] today" as shown in Fig. 1. The model learns to replace the masked token with the correct word, enhancing its ability to understand context with the attention mechanism. MLM masks 15% of the words in the sequences for training purposes.

MLM captures token-level information but does not capture sentence-level information, which is addressed by the NSP task. During training, the model receives pairs of sentences and learns to predict if the second sentence follows the first. In Fig. 2, "I failed my exams" is the actual subsequent sentence to "I am feeling depressed today", while "Did you see the Spiderman trailer?" is a random sentence from the corpus. For each sentence, one real and one fake input is used as training data. Additional tokens "[CLS]" and "[SEP]" denote the sentence boundary at the start and end, respectively, as shown in Fig. 1.

BERT was pre-trained on the English Wikipedia and BooksCorpus datasets, containing 3.3 billion words, and is available in two variations—base and large.

3.5 Sentiment Analysis and Suicide Detection Training Model

Figure 3 illustrates the training pipeline, which includes data preparation, model setup, training, validation, and saving.

Data Preparation: Necessary libraries are imported, and data is loaded from a CSV file named'SuicideDetection.csv'. Preprocessing steps include handling missing values, converting labels, and splitting the dataset.

Tokenizer and Model Setup: The BERT tokenizer and pre-trained BERT model for sequence classification are initialized. The model is configured for binary classification by setting *num-labels* = *2*.

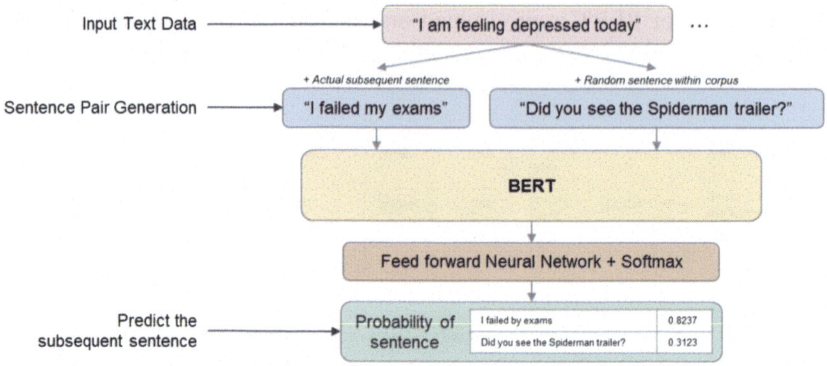

Fig. 2. BERT Next Sentence Prediction (NSP) [11]

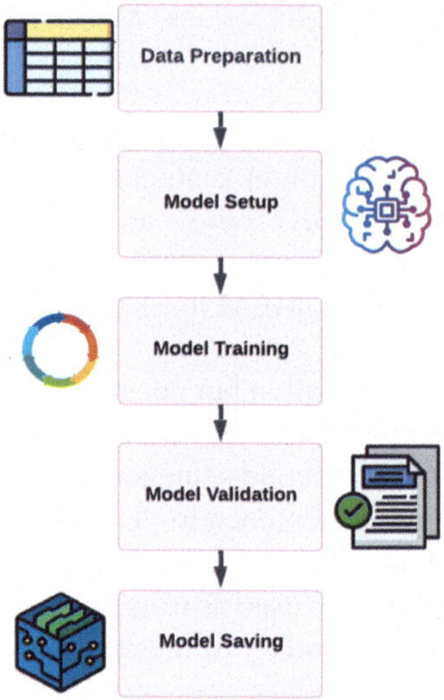

Fig. 3. Training Workflow of BERT Transformer Model

Model Training: The training loop iterates over the specified number of epochs. Within each epoch, the model is set to training mode (*model.train()*), and batches of data are fetched from the training loader. For each batch, the input IDs, attention masks, and labels are extracted and transferred to the appropriate device (CPU or GPU). The optimizer parameters are zeroed (*optimizer.zerograd()*), and forward pass, loss computation, and backward pass are performed. The optimizer is then updated based on the gradients computed during the backward pass (*optimizer.step()*).

Model Validation: After each epoch of training, the model is set to evaluation mode (*model.eval()*), and the validation dataset is iterated through. Similar to the training loop, batches of validation data are processed, and predictions are made. The validation loss is accumulated, and the accuracy is calculated by comparing predicted labels with ground truth labels.

Model Saving: Finally, the trained model is saved to disk using the *savepretrained* method, allowing it to be loaded and reused for inference.

3.6 Text Classification for Linguistic Analysis Function

This question-classify-text function provides an automated and versatile solution for classifying text into "Question", "Exclamatory Sentence", or "Normal Text" categories based on linguistic patterns and features. By leveraging spaCy's linguistic analysis capabilities, the function can accurately identify question words, sentence endings (e.g., '?', '!'), and grammatical structures to make informed classification decisions. This automation not only saves time and effort by eliminating manual inspection but also enhances scalability, allowing the function to be applied to diverse text sources and integrated into larger natural language processing workflows. Additionally, the function's adaptability enables customization and extension with additional rules or patterns, contributing to improved accuracy and effectiveness in handling various text classification tasks within NLP applications. Figure 4 structured approach ensures efficient and accurate text classification within natural language processing workflows, enhancing automation and scalability in text analysis tasks.

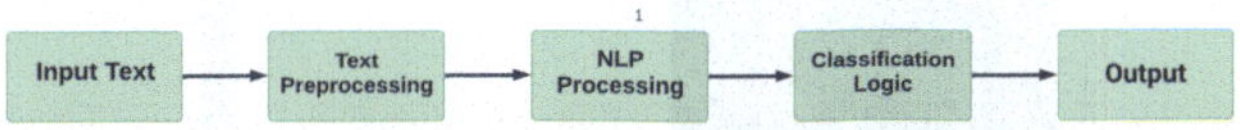

Fig. 4. NLP Text Classification.

This approach combines multiple deep learning models, including BERT for suicide detection and sentiment analysis, with a custom text classification function. The sentiment analysis model first categorizes text into positive, neutral, or negative tweets. A custom function then assesses whether the text raises concern. If the sentiment is negative but the text is deemed normal, the suicide detection model scrutinizes it further. This method improves suicide detection by accurately distinguishing between typical negative expressions and those indicative of suicidal ideation, using nuanced linguistic cues and context. The goal is to enhance the system's ability to identify and respond to concerning textual content effectively.

4 Results

4.1 Suicide Detection Model

Performance Metrics. The classification report (Table 3) provides the model's precision, recall, and F1-score for each class (suicide, non-suicide), along with the overall accuracy, macro average, and weighted average.

The confusion matrix (Fig. 5) shows actual versus predicted classifications.

Table 3. Classification Report for Suicide Detection Model

Class	Precision	Recall	F1-Score	Support
Non-Suicide	1.00	1.00	1.00	998
Suicide	1.00	1.00	1.00	2
Accuracy	1.00 (1000)			
Macro Avg	1.00	1.00	1.00	1000
Weighted Avg	1.00	1.00	1.00	1000

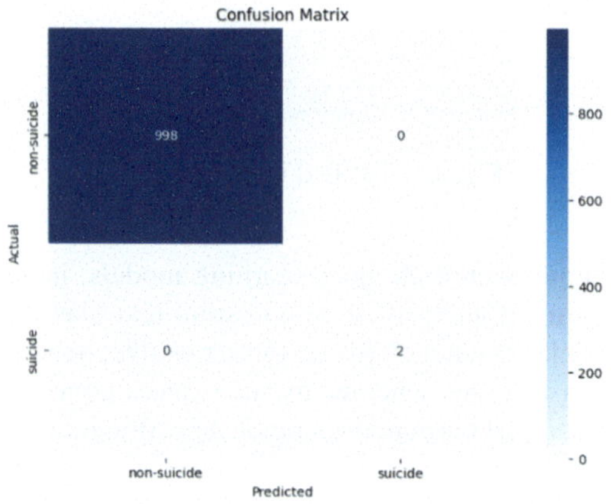

Fig. 5. Confusion Matrix for Suicide Detection Model

Training Performance. The training and validation accuracy and loss over epochs (Fig. 6) provide insights into the model's learning process.

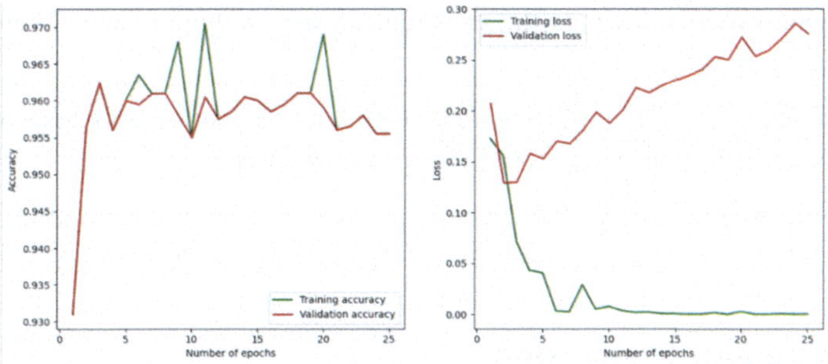

Fig. 6. Suicide Detection Model: Accuracy and Loss During Training and Validation

The final training accuracy for the suicide detection model after 25 epochs is 99.54%, with a validation accuracy of 95.70%. The average loss during training is 0.0050, and the validation loss is 0.2318.

4.2 Sentiment Analysis Model

Performance Metrics. The classification report (Table 4) details the model's precision, recall, and F1-score for each class (negative, neutral, positive), along with the overall accuracy, macro average, and weighted average.

Table 4. Classification Report for Sentiment Analysis Model

Class	Precision	Recall	F1-Score	Support
Negative	0.78	0.71	0.74	557
Neutral	0.73	0.73	0.73	820
Positive	0.78	0.84	0.81	622
Accuracy	0.76 (1999)			
Macro Avg	0.76	0.76	0.76	1999
Weighted Avg	0.76	0.76	0.76	1999

The confusion matrix (Fig. 7) shows actual versus predicted classifications, identifying correct and incorrect predictions.

Training Performance. The training and validation accuracy and loss over epochs (Fig. 8) provide insights into the model's learning process.

The final training accuracy after 25 epochs is 99.54%, with a validation accuracy of 75.84%. The average loss during training is 0.0147, and the validation loss is 1.5908.

5 Discussion and Conclusion

The early identification of suicidal ideation is crucial for preventing suicide. While psychologists traditionally rely on statistical analysis, computer scientists use machine learning (ML) and deep learning (DL) methods to detect suicidal indicators in online posts. Detecting suicidal thoughts on social media platforms can significantly aid medical professionals in saving lives.

Current research faces challenges due to the lack of annotated data and the subjective nature of manual labeling, which can introduce bias. Additionally, suicidal posts are a small fraction of social media content, resulting in imbalanced datasets that do not reflect the true distribution. These limitations hinder the development of robust models.

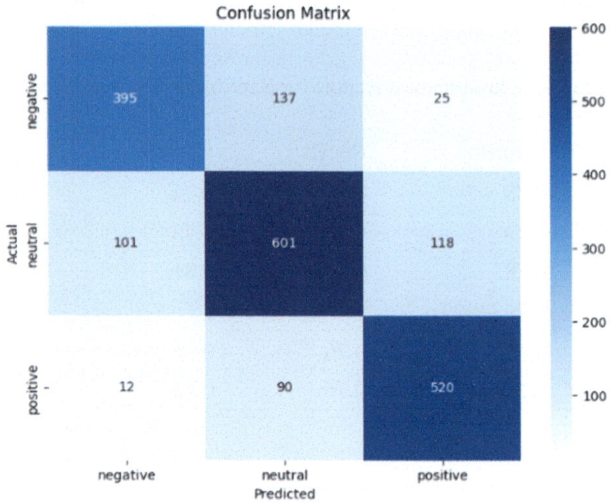

Fig. 7. Confusion Matrix for Sentiment Analysis Model

Our study utilizes the BERT model, achieving an accuracy of 95.70%, to distinguish between suicidal and non-suicidal content. Our findings highlight the potential of advanced natural language processing in enhancing early detection of suicidal ideation.

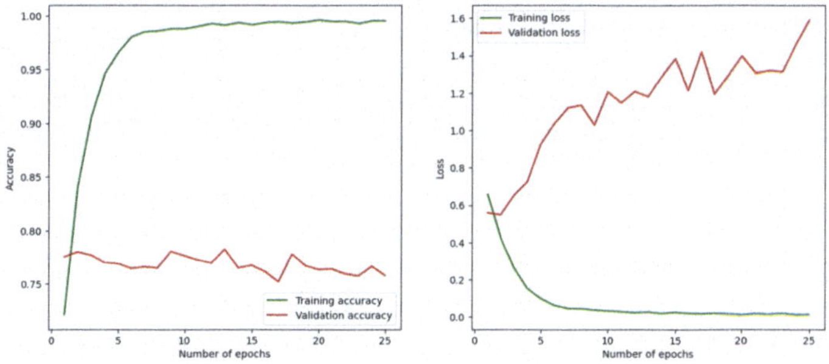

Fig. 8. Sentiment Analysis Model: Accuracy and Loss During Training and Validation

Future work will focus on optimizing the model's training process through hyperparameter tuning, model architecture adjustments, and data preprocessing techniques. Addressing overfitting is also crucial; techniques such as crossvalidation, regularization, and data augmentation will be explored to enhance the model's generalization capabilities. Integrating advanced models like Large Language Models (LLMs) such as GPT-4 can further enhance detection of subtle linguistic cues. Developing real-time monitoring systems for immediate intervention and addressing ethical considerations and privacy concerns will also be crucial. Interdisciplinary collaboration among computer scientists, psychologists, and mental health professionals is essential for creating a comprehensive and ethical solution for suicide prevention.

6 Demo Video Link

https://drive.google.com/file/d/1HDrZXVQl3dSvtrUsJalqowhGpyWSskXt/view.

References

1. Abdulsalam, A., Alhothali, A.: Suicidal ideation detection on social media: a review of machine learning methods. arXiv preprint arXiv:2201.10515 (2022)
2. Aldhyani, T.H., Alsubari, S.N., Alshebami, A.S., Alkahtani, H., Ahmed, Z.A.: Detecting and analyzing suicidal ideation on social media using deep learning and machine learning models. Int. J. Environ. Res. Public Health **19**(19), 12635 (2022)
3. Ananthakrishnan, G., Jayaraman, A.K., Trueman, T.E., Mitra, S., Abinesh, A., Murugappan, A.: Suicidal intention detection in tweets using bert-based transformers. In: 2022 International Conference on Computing, Communication, and Intelligent Systems (ICCCIS), pp. 322–327. IEEE (2022)
4. Bialer, A., Izmaylov, D., Segal, A., Tsur, O., Levi-Belz, Y., Gal, K. Detecting suicide risk in online counseling services: a study in a low-resource language. arXiv preprint arXiv:2209.04830 (2022)
5. Cao, J., et al.: Latent suicide risk detection on microblog via suicide-oriented word embeddings and layered attention. In: Proceedings of the 28th International Joint Conference on Artificial Intelligence (IJCAI-19), pp. 4208–4214 (2019)
6. De Choudhury, M., Kiciman, E.: Suicidal ideation detection in social media user embeddings. arXiv preprint arXiv:1906.00350 (2019)

7. Devlin, J., Chang, M.-W., Lee, K., Toutanova, K.: Bert: pre-training of deep bidirectional transformers for language understanding. arXiv preprint arXiv:1810.04805 (2018)
8. Fernandes, A.C., Dutta, R., Velupillai, S., Sanyal, J., Stewart, R., Chandran, D.: Identifying suicide ideation and suicidal attempts in a psychiatric clinical research database using natural language processing. Sci. Rep. **8**(1), 7426 (2018)
9. Fodeh, S., et al.: Using machine learning algorithms to detect suicide risk factors on twitter, pp. 941–948 (2019)
10. Ghosal, S., Jain, A.: Depression and suicide risk detection on social media using fasttext embedding and xgboost classifier. Procedia Comput. Sci. **218**, 1631–1639 (2023)
11. Goh, J.: Suicidal text detection (2024). https://github.com/gohjiayi/suicidal-textdetection. Accessed 23 June 2024
12. Guntuku, S.C., Yaden, D.B., Kern, M.L., Ungar, L.H., Eichstaedt, J.C.: Detecting depression and mental illness on social media: an integrative review. Curr. Opin. Psychol. **22**, 32–37 (2018)
13. Ji, S., Pan, S., Li, X., Cambria, E., Long, G., Huang, Z.: Suicidal ideation detection: a review of machine learning methods and applications. IEEE Trans. Comput. Social Syst. **8**(1), 214–226 (2020)
14. Li, Z., Cheng, W., Zhou, J., An, Z., Hu, B.: Deep learning model with multifeature fusion and label association for suicide detection. Multimedia Syst. **29**(4), 2193–2203 (2023)
15. Liao, B., Lei, J., Liu, W., Jiang, W.: A multi-modal graph-based framework for detecting suicidal ideation on social media. IEEE Trans. Knowl. Data Eng. **34**(2), 576–589 (2020)
16. Lin, H., Tadesse, M.M., Xu, B.: Detecting suicidal ideation with data sparsity using deep learning models. In: Proceedings of the 29th ACM International Conference on Information and Knowledge Management (CIKM-20), pp. 2029–2032 (2020)
17. Liu, X., Han, X., Cao, F., Zhang, J., Luo, X.: Identifying suicidal ideation in chinese microblogs with bert-based models. IEEE Trans. Affect. Comput. **13**(2), 684–696 (2021)
18. Mbawa, S.Z., Kerkhoven, M., Donker-Kuijer, M.W." How can a conversational agent (chatbot) be used to detect and prevent suicide based on recognisable suicide behaviours amongst young people with mental disorders? (2021)
19. Parsapoor, M., Koudys, J.W., Ruocco, A.C.: Suicide risk detection using artificial intelligence: the promise of creating a benchmark dataset for research on the detection of suicide risk. Front. Psych. **14**, 1186569 (2023)
20. Sawhney, R., Joshi, H., Gandhi, S., Shah, R.: A time-aware transformer based model for suicide ideation detection on social media. In: Proceedings of the 2020 Conference on Empirical Methods in Natural Language Processing (EMNLP), pp. 7685–7697 (2020)
21. Tadesse, M.M., Lin, H., Xu, B., Yang, L.: Detection of suicide ideation in social media forums using deep learning. Algorithms **13**(1), 7 (2019)
22. Thatcher, J., Gadd, J., Richards, D.: Using sentiment analysis to create socially aware conversational agents. J. Artif. Intell. Res. **69**, 213–245 (2020)
23. Westerlund, M., Nilsson, H.: Suicide communication on digital platforms: a research review (2022)

A Mobile Application Using MobileNetV2 for Classification of Skin Diseases

Şengül Bayrak[1](✉), Esad Şen[2], Fatma Begüm Arslanoğlu[2], Furkan Kaya[2], Sena Merve Tomarza[2], and Furkan Çakır[2]

[1] Department of Software Engineering, Istanbul Sabahattin Zaim University, 34303 Istanbul, Turkey
bayraksengul@ieee.org

[2] Department of Computer Engineering, Istanbul Sabahattin Zaim University, 34303 Istanbul, Turkey

Abstract. This study presents the development of a mobile application for classifying skin conditions into Melanoma, Acne, and Healthy skin. A dataset of 214 Melanoma images from Dermnet, 395 Acne images, 400 Healthy skin images, and 101 Melanoma images from the RAW10000 database was used. Pre-processing techniques such as rescaling, sharpening, bilateral deceleration, and min-max normalization were applied. Feature extraction was performed using HOG, VGG16, and MobileNetV2, followed by classification with SVM, RF, and ANN models. MobileNetV2 achieved the highest accuracy at 94.30%. The backend was developed with Django and MySQL, and the user interface with Figma and React. The project demonstrates significant accuracy and practical application potential, enhancing mobile device access to dermatological diagnostics and early diagnosis.

Keywords: Mobile application · MobileNetV2 · Skin diseases · Decision support system

1 Introduction

Diagnosing and classifying skin diseases through image processing and machine learning has emerged as a critical research area, given the global ubiquity of skin-related ailments. Recent advancements within the field of dermatology have significantly enhanced diagnostic accuracy through the utilization of digital imagery, thus facilitating early detection and subsequent treatment. This research is predicated upon a variety of approaches, from the application of machine learning techniques to publicly accessible datasets such as DermNet and HAM10000 to the integration of deep learning models for real-time analysis within mobile applications. Notably, the efficacy of Convolutional Neural Networks (CNNs) in feature extraction and classification tasks has been underscored, demonstrating their superior accuracy in identifying conditions such as melanoma, acne, and eczema. Despite these notable achievements, data collection and image quality challenges remain prevalent. This study addresses these

A. Abdelgawad et al. (Eds.): ISBCom 2024, LNNS 1268, pp. 789–802, 2025.
https://doi.org/10.1007/978-3-031-82377-0_63

challenges by adopting a dual approach: enhancing traditional machine learning methodologies with advanced feature extraction techniques and exploring the effectiveness of deep learning models, such as VGG16, coupled with Principal Component Analysis (PCA) for feature reduction. Developing a user-friendly mobile application leveraging Django and React also aims to bridge the divide between patients and dermatological healthcare services. This study proposes an integrated approach that amalgamates the strengths of machine learning and deep learning models to achieve heightened accuracy and reliability in diagnosing skin diseases. The main contribution of our project is implementing and optimizing MobileNetV2 architecture for the classification of skin diseases. Unlike previous work, our project is primarily designed to help doctors as a diagnostic tool and serve as a pre-diagnostic application for the general public. Many doctors we interviewed stated that they would use such an application and that there is an essential gap in this field. We set out to fill this gap, a significant difference distinguishing our study from previous projects. This work extends the existing literature by combining advanced preprocessing techniques, data augmentation, and rigorous hyperparameter tuning to improve the model's effectiveness. While previous works such as Goceri [1] and Ech-Cherif et al. [2] have used MobileNet variants for similar tasks, our approach emphasizes optimization for real-time applications on mobile devices. By leveraging the lightweight and efficient MobileNetV2 architecture, our project significantly improves computational efficiency and increases the accuracy to a remarkable 94.30%. Furthermore, this study includes a comprehensive backend infrastructure using Django and a user-friendly interface developed with React, which increases the accessibility and practicality of the application. Oztel et al. [3] created a deep learning-based automated smart mobile application to detect monkeypox skin disease. After training with different small-size pre-trained networks for skin lesion classification, they converted them into the appropriate TensorFlow Lite format. Finally, they created a mobile application on the Android platform that uses the smartphone's camera to acquire images and uses TensorFlow Lite to make predictions. The proposed system achieved a classification accuracy of 74.27% for seven classes on a merged dataset. Prottasha et al. [4] have developed mobile and web-based applications using CNN architecture to diagnose Eczema and Psoriasis. With Adam optimizer, they achieved a maximum verification accuracy of 97.1% by Inception ResNet v2 architecture. Table 1 provides the literature review of the latest studies in the detection of skin diseases.

Our approach aims to be broader and more complete as our study is large and comprehensive compared to other studies with melanoma, acne, and healthy datasets [5,6], considering 3 different cit diseases. In our study, the skin image data received with the mobile application were preprocessed. The data obtained from preprocessing were classified with SVM, RF, ANN methods, which are classical machine learning methods. On the other hand, the data received with the mobile application were preprocessed and classified with MobileNetv2. When both approaches are compared, the importance of this study's experimental result obtained with MobileNetV2 is emphasized.

The paper is organized as follows: Sect. 2 elucidates the materials and methods employed, providing a detailed architectural overview. Section 3 presents the experimental results emanating from the proposed framework. Section 4 is about discussion, and finally, Sect. 5 encapsulates concluding remarks.

2 Materials and Methods

The methodology of our study is centered around integrating various technologies and algorithms to develop a comprehensive mobile application for the classification of skin conditions. Sample data for melanoma, acne and healthy individuals are shown in Fig. 1.

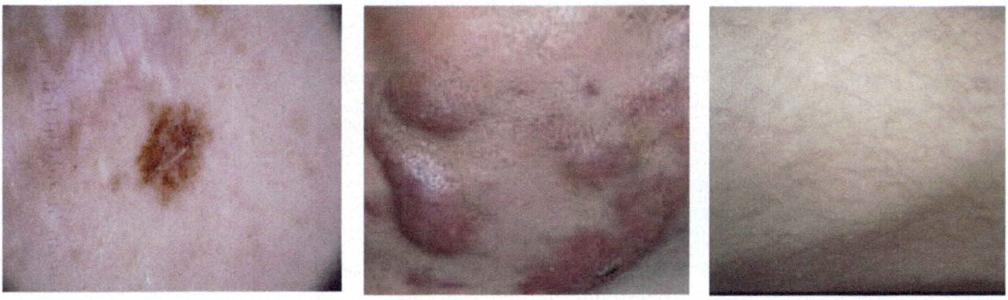

a) Melanoma b) Acne c) Healthy

Fig. 1. A sample for the melanoma, acne, and healthy group

Below, we delve into the specifics of each component, including the use of MobileNetV2 for deep learning, Django and Python for backend development, React and Figma for frontend and UI/UX design, and the approaches for evaluating classification performance and model comparisons.

2.1 Image Pre-processing

The initial phase of our analysis entailed applying many pre-processing techniques designed to enhance the quality and uniformity of the dermatological images. Rescaling is the first step, and it is adjusting the image size to standard dimensions to ensure uniformity across the dataset, which is crucial for consistent feature extraction and analysis. The sharpening technique enhances the definition of edges, improving the distinction between different features within the images. Bilateral Denoising is a non-linear, edge-preserving method; this technique effectively reduces noise while maintaining essential features, which is vital for preserving the integrity of skin lesion images. Min-Max Normalization standardizes the intensity values of images to a standard scale; this method enhances the comparability across the dataset, facilitating more effective analysis.

Table 1. Literature Review

Author(s)	Dataset	Model	Result
Alkareem and Alhadi [5]	Acne, Eczema, Melanoma 1200 images (400 images for each disease class)	Bag of Features Model with SVM Classifier	94% Cross-validation method, 85% Holdout method
Goceri [1]	Seborrheic dermatitis, Rosacea, Hemangioma, Psoriasis, Acne vulgaris, 725 images	Modified-MobileNet	94.76%
Maduranga and Nandasena [6]	HAM10000 (Actinic Keratoses, Basal Cell Carcinoma, Benign Keratosis-like Lesions, Dermatofibroma, Melanoma, Melanocytic Nevi, Vascular Lesions)	MobileNet with Transfer Learning	85%
Pangti et al. [7]	The 17.718 raw images were sourced from public databases (40 common skin diseases)	Convolutional Neural Network (CNN)-based algorithm	76.93 ± 0.88%; 75.07% in clinical study
Srinivasu et al. [8]	HAM10000	MobileNet V2 and LSTM	More than 85%
Do et al. [9]	A set of image data provided by the National Skin Center of Singapore (117 benign and 67 malign)	SVM	89.09% sensitivity at specificity $\geq$ 90%
Ech-Cherif et al. [2]	48,373 dermoscopic images collected from three different files and labeled and verified by dermatologist specialists	MobileNetV2	91.33%
Jaworek-Korjakowska and Kleczek [10]	337 dermatoscopic images, 208 of them melanoma	"SkinVision algorithm" and "Lubax algorithm"	Accuracy is 81% in melanoma detection with SkinVision algorithm; Accuracy is 91% in melanoma detection with Lubax algorithm
Chin et al. [11]	300 images	Types of convoluted models with 2, 3 and 5 layers (CNN)	Total accuracy results of 87% for model 1, 93% for model 2 and 86% for the most widely used model 3
Kalwa et al. [12]	200 images	SVM	Using SMOTE is 88%, without SMOTE is 90%

2.2 Dimensionality Reduction

Principal Component Analysis (PCA) is a statistical technique that reduces the dimensionality of a dataset of potentially correlated variables to a smaller set of linearly uncorrelated variables known as principal components. This transformation simplifies the dataset while retaining the most relevant information.

2.3 Feature Extraction

- Histogram of Oriented Gradients (HOG) is a methodological approach that measures the occurrences of gradient orientations within localized sequences of an image. This technique has provided a robust feature set for classification tasks by capturing basic edge and texture information.
- One of our approaches in this work is feature extraction from RGB images. The VGG16 convolutional neural network has been used in this study. This deep learning model excels at detecting complex patterns in high-dimensional data and is particularly effective for identifying subtle features characteristic of various skin conditions. In the VGG16 model used, feature extraction is performed in the $block5\_pool$ layer, which hosts the feature maps resulting from the last convolutional block of the model. These feature maps provide deep representations of images and are often used in visual tasks such as classification and image similarity assessment.

2.4 Classification Methods

To determine the model that provides the optimal balance between accuracy and computational efficiency, we perform comparative analyses of various models, including HOG feature extraction and SVM with VGG16. This evaluation is imperative to improve the application's performance and ensure that it can provide fast and accurate diagnostic results. Moreover, to further improve the accuracy and reliability of our skin condition classification system, data set augmentation and sophisticated pre-processing techniques are being investigated for their impact on the model's performance [13].

SVM is utilized for its robustness in managing high-dimensional data and its efficacy in binary and multi-class classification tasks. It operates by delineating the hyperplane that most effectively segregates the classes within the feature space. In this project, an SVM model, leveraging feature extraction via the VGG16 model, categorizes features using a linear kernel and a regularization parameter $C = 1$. The model also provides probability estimates with the setting $probability = True$. A StandardScaler is incorporated within a processing pipeline for feature scaling, targeting the enhancement of result reliability by addressing SVM's sensitivity to data scaling.

The dataset D consists of n data in the form of (x_1, y_1),(x_2, y_2), ..., (x_n, y_n) where $y \epsilon +1, -1$ is applied. The aim is to select the hyperplane that maximizes the margin between the classes. In Eq. (1), n provides the number of features, and it is the weight vector Eq. (1), n a constant b.

$$H_0 = \sum_{i=1}^{n} w_i x_i + b = 0 \tag{1}$$

RF constitutes an ensemble classification model comprising various decision trees developed through bagging and random variable selection techniques. The foundational principle for tree construction in RF aligns with that of a standard decision tree, which utilizes recursive partitioning. In this recursive approach, the cut-points' exact location and the choice of splitting variable are profoundly influenced by the distribution of observations within the training sample, rendering the decision tree a relatively unstable model. However, RF ameliorates this instability by amalgamating multiple trees, where the diversity within the tree ensemble mitigates the instabilities associated with individual trees [14]. The parameter $n_e stimators = 100$ indicates that the Random Forest model is trained with 100 trees.

ANN have been modeled in this study because of their capacity to discern complex patterns through multiple layers of interconnected nodes, making them apt for tasks that necessitate the integration of vast and heterogeneous datasets. The ANN model in use was constructed using the Sequential class, comprising four layers:

- The initial three layers are Dense layers, each activated by the ReLU function.
- The input is 128 for the first two layers and 64 for the third.
- The terminal layer is a Dense layer, employing a softmax activation function for classification purposes.
- The first layer accepts the number of features as its input shape, while the final layer matches the number of neurons to the count of labels [15].

If x_i provides the output of neurons in the i. layer, the total input to a neuron in the j. layer is calculated as described in Eq. (2).

$$net_j = \sum_{i=1}^{n} (w_{ji} x_i) \tag{2}$$

MobileNetV2. MobileNetV2 is a streamlined deep-learning framework meticulously designed for mobile and embedded device applications. Its ability to execute high-fidelity tasks with minimal computational resources efficiently characterizes it. MobileNetV2 includes deeply separable convolutions that significantly reduce the number of parameters without compromising model performance. This feature is crucial for real-time applications and enables the mobile application to perform fast and accurate classifications of skin conditions. The architecture of MobileNetV2 consists of a set of convolutional layers, followed by depth-separable convolutions, inverted residuals, a bottleneck design, linear bottlenecks, and squeezing and excitation (SE) blocks. These components reduce the required parameters and computations while maintaining the model's capacity to describe complex features. Depth separable convolution, a technique used in MobileNetV2, reduces the computational overhead of convolutions by splitting the standard convolution into two different operations: depth convolution and point convolution. This division increases the model's efficiency by significantly reducing the computational demand. Inverted residuals, a critical feature of MobileNetV2, improve the model's accuracy by introducing a bottleneck structure that increases the number of channels before applying the depth separable convolutions. This expansion allows the model to assimilate more complex features and increase its representativeness. The bottleneck design in MobileNetV2 also reduces computational costs by using 1×1 convolutions to reduce the number of channels before applying deep separable convolutions. This strategic choice provides an optimal balance between model size and accuracy. Linear bottlenecks are integrated into MobileNetV2 to address the potential loss of information during the bottleneck process. The model retains more information by adopting linear activations instead of non-linear ones, increasing its capacity to capture fine-grained details. The squeezing and excitation (SE) blocks included in MobileNetV2 enhance feature representation capabilities. These blocks recalibrate channel-based feature responses adaptively, enabling the model to focus on more prominent features and de-emphasize less relevant ones.

In our study, MobileNetV2 extracts features from dermatological images, which are then used to classify melanoma, acne, and healthy skin [16].

MobileNetV2 architecture uses depthwise separable convolutions to reduce parameters and computational costs while maintaining performance. Figure 2 includes inverted residuals and bottleneck designs to improve accuracy and efficiency, linear bottlenecks to minimize information loss, and squeezing and excitation (SE) blocks to enhance feature representation. The model uses a Global Average Pooling (GAP) layer followed by a Dense layer for classification, making it suitable for real-time applications such as classifying dermatological images into melanoma, acne, and healthy skin. In the MobileNetV2 architecture, we used a GAP layer instead of fully connected layers. GAP reduces each feature map to a single value by computing the average of all values in the feature map. This technique was applied at the end of the MobileNetV2 model to reduce the spatial dimensions of the feature maps while retaining essential information. This results in a fixed-length feature vector, significantly reducing the number of parameters and computational complexity.

2.5 Backend Development

The backend of the mobile application in this project is developed using Django, a very sophisticated Python web framework designed for the rapid development of secure and sustainable websites. Django alleviates many of the complexities of web development, allowing you to focus on the application's functionality without developing core solutions from scratch. Django was chosen for this project because of its robustness, scalability, and ease of integration with databases, such as a local MySQL database. Its extensive suite of Python libraries and frameworks, especially for data processing and machine learning - including TensorFlow for MobileNetV2 integration - makes it an exemplary choice for backend development. This configuration enables efficient management of user requests,

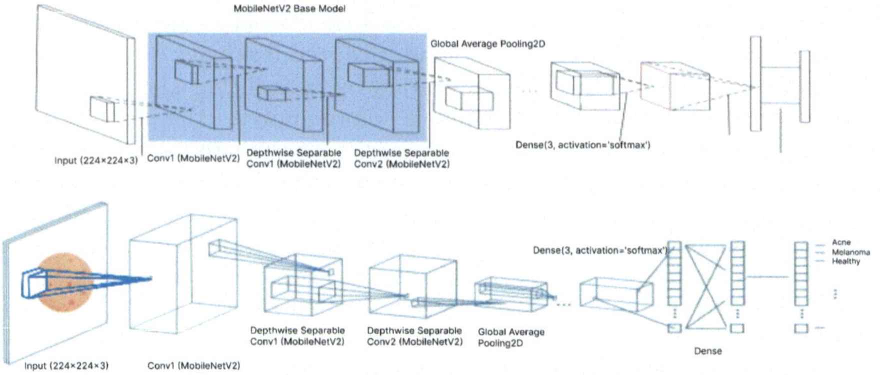

Fig. 2. A detailed architecture of a proposed model

processing of image data, and interaction with the machine learning model to deliver real-time classification results [17].

2.6 Frontend Development

For the front-end development of our mobile application, we employ React, a distinguished JavaScript library designed to create user interfaces, especially suitable for single-page applications. React's component-based structure aids in constructing a responsive and intuitive user interface, enhancing the user experience through seamless interaction with the application. The design of the user interface and the overall user experience (UX) are meticulously crafted using Figma, a collaborative interface design tool. Figma facilitates a dynamic design process by enabling real-time collaboration and feedback, which is essential for developing a user-friendly application design that effectively addresses the needs of the end-users.

3 Experimental Results

In this paper, the experimental benchmarks were performed in a machine environment with NVIDIA RTX 4070, 32 GB RAM, and AMD Ryzen 5 5600 processor.

A dataset of 214 Melanoma images, 395 Acne images, 400 Healthy skin images from Dermnet, and 101 Melanoma images from the RAW10000 database was used.

The experimental results section of our paper presents the results obtained from applying our methodology, focusing on the workflow process, the performance of the models used, and a discussion of the implications of these findings.

In this study, we used the MobileNetV2 algorithm, pre-trained on the ImageNet dataset, as the basis of our convolutional neural network. The model configuration includes a base MobileNetV2 layer, a global mean pooling layer, and a dense output layer with softmax activation to categorize it into three groups. To improve the model's effectiveness, hyperparameter tuning was performed with particular emphasis on the learning rate of the ADAM optimizer. Initially, a learning rate constant of 0.0001 was set. Recognizing the need for more dynamic tuning, a multiple-run design approach was adopted. This approach maintained the initial learning rate for the first 10 epochs, followed by a 0.99 reduction in the learning rate after each epoch. The application of this dynamic adjustment to the learning rate proved to be advantageous. It facilitated finer tuning of the training process as the model converged, reducing losses and increasing accuracy in model performance. Our final model was compiled using a categorical cross-entropy loss function. This gradually decreasing learning rate method effectively reduced loss and improved our model's accuracy. The performance of the classification models was evaluated at 80% for training and the remaining 20% for testing. Table 1 indicates the ability of each model to classify skin diseases and which model is most suitable for the project's overall objectives. While the use of MobileNetV2 offers high performance and fast diagnosis on mobile devices, a comparison with classical machine learning models was made. Thus, the superiority of the model developed with MobileNetV2 over other models is proved. Table 2 and Fig. 3 MobileNetV2 has the highest accuracy, precision, and recall values. This indicates that the model has many true positives and few false positives. The recall value is also high, indicating that the model is less likely to miss true negatives. MobileNetV2 showed the best performance overall. VGG16+ ANN has a lower accuracy, precision, and recall value than MobileNetV2. The fact that the precision and recall values are close indicates that the model has a balanced performance. However, accuracy and precision are significantly lower than MobileNetV2. The VGG16 + RF model performs close to the VGG16 + ANN model in terms of precision, while the recall value is slightly lower. Although this model performs well in finding true positives, it ignores some negatives. The VGG16 + SVM model performs very similar to the VGG16 + RF model in terms of accuracy and recall. The fact that the accuracy and recall values are close shows the model has a balanced performance. The HOG + SVM model has the lowest values in all metrics. This indicates that the model is poor at finding positives and negatives. The HOG feature extraction method is significantly worse in terms of performance than other deep learning-based methods. MobileNetV2 has the highest accuracy and performance metrics, making it the most successful model. VGG16 + ANN, VGG16 + RF, and VGG16 + SVM models show average performance and give close results to each other. The HOG + SVM model is less effective than the other models and is, therefore, less effective than other deep learning-based approaches.

Table 2. Comparison of the Experimental Results

Model	Accuracy (%)	Precision (%)	Recall (%)
MobileNetV2	94.30	95.72	92.51
VGG16 + ANN	88	86.54	86.31
VGG16 + RF	87.50	87.14	85.39
VGG16 + SVM	87.50	86.40	86.45
HOG + SVM	70.98	69.67	70.67

The confusion matrix of the MobileNetV2 model for testing, which proved its success in this study, is given in Fig. 4.

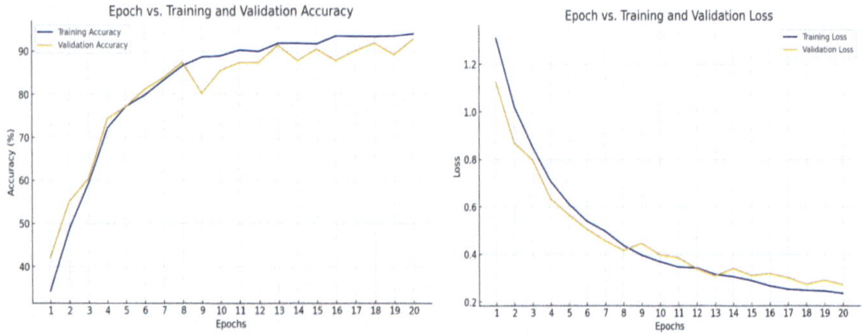

a) Training and Validation Accuracy graph

b) Training and Validation Loss function

Fig. 3. Accuracy and Loss Function for Training and Validation for MobileNetV2

Figure 3 shows the performance of a machine learning model during the training process over 20 epochs, focusing on both accuracy and loss metrics for the training and validation datasets.

According to Fig. 3, training and validation accuracy increase rapidly from epoch 1 to 5, indicating that the model learns effectively in the early stages. From epoch 5 to epoch 10, there is a consistent improvement in accuracy for both training and validation sets, and the training accuracy is slightly higher than the validation accuracy. From period 10 onwards, the accuracy for both training and validation stabilizes, hovering around 90%. Small fluctuations in the validation accuracy suggest some variability in model performance on unseen data. The alignment of training and validation accuracy in later epochs indicates that the model generalizes well without overfitting.

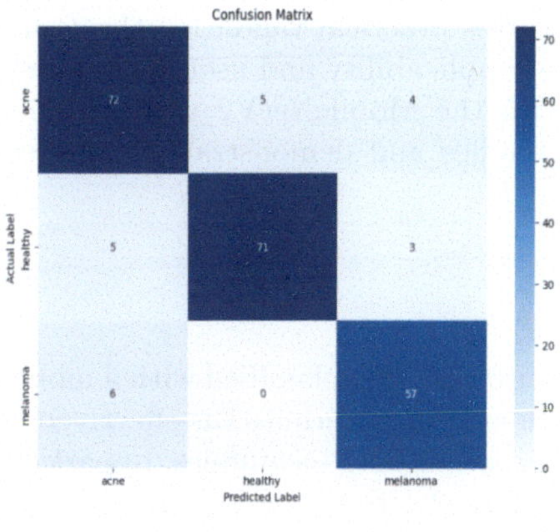

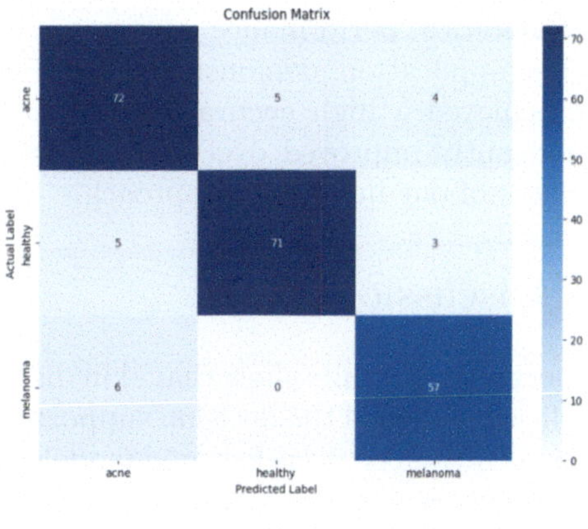

Fig. 4. Confusion matrix for MobileNetV2 validation and test model

Our model shows a good learning capability, as evidenced by the rapid increase in accuracy and decrease in loss in the first epochs. The close alignment of the training and validation curves for accuracy and loss shows that the model generalizes well to new data. Small fluctuations in the validation accuracy are normal and indicate that the model maintains robustness across different validation data sets. Overall, the training process appears effective, and the model achieves high accuracy and low loss without significant overfitting.

4 Recent Advancements in Skin Condition Classification Using Machine Learning

In recent years, significant progress has been made in skin condition classification using machine learning. Fundamental work has focussed on several aspects, including deep learning models, transfer learning, and hybrid approaches combining feature extraction and classification techniques. CNNs have been widely adopted because they automatically learn hierarchical features from raw images. Recent studies have shown high accuracy in classifying various skin conditions using CNNs [7,11]. Using pre-trained models such as VGG16, ResNet, and MobileNetV2 has been particularly successful, leveraging transfer learning to improve performance with limited datasets. Combining traditional feature extraction methods such as Histogram of Oriented Gradients (HOG) with deep learning models has been investigated to improve classification accuracy. These hybrid methods exploit the strengths of manually generated features and automatic feature learning. Our model is on these developments by integrating state-of-the-art techniques and introducing several new directions. We use a substantial dataset enriched with images from the Dermnet [18] and HAM10000 [19] databases, providing a variety of skin condition images that increase the generalisability of our models. Unlike many existing studies, which primarily focus on theoretical performance, our research involves practical validation through a mobile application, demonstrating real-world applicability and user-friendliness. We achieved a high accuracy of 94.30% with the MobileNetV2 model, which significantly improved over many existing studies and demonstrates the effectiveness of our integrated approach.

5 Discussion

Experimental results show that skin diseases are correctly classified with a mobile application, one of the decision support systems. This underlines the effectiveness of using advanced deep-learning models and preprocessing techniques. In particular, MobileNetV2 has shown promising results due to its lightweight architecture and efficiency, making it suitable for real-time applications on mobile devices. The improved accuracy observed with the augmented dataset and VGG16 feature extraction underlines the importance of comprehensive and diverse training data in supporting model performance. The comparative analysis of different models based on precision, recall, and F1 scores informed the decision-making process for selecting the most appropriate model for deployment in the mobile application. While achieving satisfactory performance with all the models developed within the scope of the study, it was clear that some models excelled in specific metrics, suggesting that a hybrid approach or further optimization could improve the results. Overall, the experimental findings confirm the potential of machine learning and deep learning techniques to transform dermatological diagnostics. By providing a user-friendly mobile platform for the classification of skin diseases, this project significantly increases the accessibility of dermatological healthcare. It potentially facilitates the early diagnosis and treatment of skin diseases. These findings also pave the way for future research, particularly exploring additional deep-learning architectures and expanding the dataset to include a broader range of skin conditions. The innovations presented in these studies, particularly the extensive preprocessing and feature extraction strategies for dermatological disease detection, have been instrumental in shaping the approach of our project. Adopting sharpening filters, median filters, and Sobel operators detailed by these researchers has significantly enriched our understanding and application of image-processing techniques in dermatology. This study serves users as an artificial intelligence-supported Android-based mobile application. The application currently serves individuals with melanoma, acne, and healthy control groups. Thanks to the dynamic system we have established in the project, a model will be developed for new diseases. Since it is a state-of-the-art product serving 3 different skin diseases, it differs from other studies in the literature

6 Conclusion

This study aims to classify Melanoma, Acne and healthy control group dermatological conditions using deep learning techniques of MobileNetV2 architecture.

The model developed with MobileNetV2 outperformed the classical machine learning methods SVM, RF and ANN. MobileNetV2 has shown to be the optimal solution for real-time applications running efficiently on mobile devices, thanks to its lightweight structure and fast processing capabilities. The success of the MobileNetV2 model highlights the potential of deep learning models in medical image analysis and lays a solid foundation for future research efforts. Its integration with TensorFlow and Keras libraries facilitated an efficient training and deployment process in the contemporary development environment of VS Code. This application is expected to help physicians and patients make informed decisions for early diagnosis and treatment by providing fast and accurate classification results. The results of this research provide an intriguing example of how machine learning and artificial intelligence technologies can revolutionize healthcare. In the future, it is aimed to collect an original data set for Melanoma, Acne, and healthy control groups. The current mobile application will also be applied for different skin conditions. This study will also be integrated into the web environment.

Acknowldegment. This study was funded by Istanbul Sabahattin Zaim University project number 2024-BAP-100-001.

Disclosure of Interests. The authors declare no conflicts of interest.

References

1. Goceri, E.: Diagnosis of skin diseases in the era of deep learning and mobile technology. Comput. Biol. Med. **134**, 104458 (2021)
2. Ech-Cherif, A., Misbhauddin, M., Ech-Cherif, M.: Deep neural network based mobile dermoscopy application for triaging skin cancer detection. In: 2019 2nd International Conference on Computer Applications & Information Security (ICCAIS), pp. 1–6. IEEE (2019)
3. Oztel, I., Yolcu Oztel, G., Sahin, V.H.: Deep learning based skin diseases classification using smartphones. Adv. Intell. Syst. **5**(12), 2300211 (2023)
4. Sazzadul Islam Prottasha, Md, et al.: Deep learning-based skin disease detection using convolutional neural networks (CNN), pp 551–564, 2023
5. Abd Alkareem, R.M.A.: Remote Skin Diseases Diagnosis System Using Machine Learning Techniques. PhD thesis, University of Khartoum, 2017
6. MWP Maduranga and Dilshan Nandasena: Mobile-based skin disease diagnosis system using convolutional neural networks (CNN). IJ Image Graph. Signal Process **3**, 47–57 (2022)
7. Pangti, R., et al.: A machine learning-based, decision support, mobile phone application for diagnosis of common dermatological diseases. J. Eur. Acad. Dermatol. Venereol. **35**(2), 536–545 (2021)
8. Srinivasu, P.N., SivaSai, J.G., Ijaz, M.F., Bhoi, A.K., Kim, W., Kang, J.J.: Classification of skin disease using deep learning neural networks with MobileNet V2 and LSTM. Sensors **21**(8), 2852 (2021)
9. Do, T.-T., et al.: Accessible melanoma detection using smartphones and mobile image analysis. IEEE Trans. Multimed. **20**(10), 2849–2864 (2018)

10. Jaworek-Korjakowska, J., Kleczek, P., et al.: Eskin: study on the smartphone application for early detection of malignant melanoma. Wirel. Commun. Mob. Comput. **2018** (2018)
11. Chin, C.L., Chin, M.C., Tsai, T.Y., Chen, W.E.: Facial skin image classification system using convolutional neural networks deep learning algorithm. In: 2018 9th International Conference on Awareness Science and Technology (iCAST), pp. 51–55. IEEE (2018)
12. Kalwa, U., Legner, C., Kong, T., Pandey, S.: Skin cancer diagnostics with an all-inclusive smartphone application. Symmetry **11**(6), 790 (2019)
13. Bayrak, S., Yucel, E.: Methods for the recognition of multisource data in intelligent medicine: a review and next-generation trends. Next Generation Healthcare Informatics, pp. 1–25, 2022
14. Xin, W., Gao, Y., Jiao, D.: Multi-label classification based on random forest algorithm for non-intrusive load monitoring system. Processes **7**(6), 337 (2019)
15. Karlık, B., Hayta, S.B.: Comparison machine learning algorithms for recognition of epileptic seizures in EEG. In: Proceedings IWBBIO, 2014, pp. 1–12 (2014)
16. Sharma, N.: What is mobilenetv2, 2023. https://www.analyticsvidhya.com/blog/2023/12/what-is-mobilenetv2/
17. Django introduction. https://developer.mozilla.org/en-US/docs/Learn/Server-side/Django/Introduction
18. Dermnet dataset. https://www.kaggle.com/datasets/shubhamgoel27/dermnet
19. Hamming dataset. https://www.kaggle.com/datasets/kmader/skin-cancer-mnist-ham10000

Enhancing Human Age Detection: The Impact of Data Augmentation and Balancing on CNN Performance

Maryam Alrubaye[1(✉)], Alaa Ali Hameed[2], and Akhtar Jamil[3]

[1] Department of Artificial Intelligence Engineering, Istinye University, Istanbul, Turkey
maryamalruabye18@gmail.com
[2] Department of Computer Engineering, Istinye University, Istanbul, Turkey
[3] Department of Computer Science, National University of Computer and Engineering Sciences, Islamabad, Pakistan

Abstract. Accurate age detection is a crucial aspect in various sectors including security, marketing, and demographic analysis, where understanding the age of individuals can lead to better service delivery and enhanced user experience. This study explores the enhancement of age detection algorithms using Convolutional Neural Networks (CNNs). Leveraging a unique combination of the UTKFace and Facial-age datasets, our approach introduces an advanced data augmentation and balancing technique. This methodological innovation significantly contributes to optimizing CNN performance in age detection tasks. The proposed technique not only overcomes the common challenges in age estimation but also achieves an accuracy of *90.51%*, and a noteworthy F1 score of *89.82%*, which demonstrates the model's accuracy and robustness. The findings of this research hold substantial implications for improving age detection processes, offering a promising direction for future applications in various fields that require reliable age estimation.

Keywords: Age Detection · Convolutional Neural Networks (CNN) · Data Augmentation · Dataset Balancing · Computer Vision · Image Analysis · Deep Learning · Age Estimation Techniques · Image Processing

1 Introduction

Accurate age detection is crucial in various sectors such as security, marketing, and demographic analysis. This research focuses on improving age detection algorithms using Convolutional Neural Networks (CNNs) through innovative data augmentation and balancing techniques to address challenges in age estimation and enhance performance.

Early contributions in age detection with CNNs include work by Siphesihle Sithungu and Dustin Van der Haar in 2018, who used a modified LeNet-5 architecture, achieving an accuracy of *45.3%* with the UTK-Face Large Scale Face

A. Abdelgawad et al. (Eds.): ISBCom 2024, LNNS 1268, pp. 803–818, 2025.
https://doi.org/10.1007/978-3-031-82377-0_64

Dataset [1]. This foundational work demonstrated CNNs' potential in real-time age detection despite modest accuracy. In 2019, Sadia Mahjabin and her team built on this by using the ResNet50 architecture on the APPA-REAL and UTK-Face datasets, resulting in a Mean Absolute Error of *4.49%* [2], showing significant improvement.

In 2020, Olatunbosun Agbo-Ajala and Serestina Viriri introduced a CNN model achieving *84.8%* accuracy on the OIUAdience dataset [3], emphasizing the importance of leveraging large-scale datasets like IMDb-WIKI for pre-training. Abdullah M. Abu Nada and colleagues further contributed by developing a model for age and gender validation, achieving an age prediction accuracy of approximately *57%* [4]. This highlighted challenges in the validation process, especially in single user images.

In 2022, Uraimov Jamoliddin and Jae Hung Yoo developed a less complex CNN architecture, achieving *61.34%* accuracy with the UTKFace dataset [5], showcasing potential for simpler models. In 2023, Surya Teja Chavalia and his team combined an SVM model with deep learning for facial emotion recognition, achieving an F1 score of *83%* in age detection [6]. Ahmad Alsaleh and Cahit Perkgoz introduced a space and time-efficient CNN model, achieving an overall weighted F1-score of *87.84%* [7], underscoring the importance of efficiency in model design. Our research introduces methodological innovations to enhance CNN performance in age detection, resulting in a high F1 score. Key contributions include (Table 1):

Table 1. Comparative Overview of Key Studies in Age Detection Using Convolutional Neural Networks

Authors	Year	Main Focus	Model/Algorithm Used	Key Result(s)	Dataset(s) Used
Sithungu & Van der Haar	2018	Real-time age detection	Modified LeNet-5	Accuracy: 45.3%	UTK-Face
Mahjabin et al.	2019	Age estimation using CNN	ResNet50	CS: 67.37% (SGD), 59.68% (Adam)	APPA-REAL, UTKFace
Agbo-Ajala & Viriri	2020	Age/gender classification	CNN (Pre-trained on IMDb-WIKI)	Accuracy: 84.8% (Age)	OIU-Adience
Abu Nada et al.	2020	Age/gender validation	CNN	Accuracy: 57% (Age)	University of Palestine Dataset
Jamoliddin	2022	Age/gender classification	Simplified CNN	Accuracy: 61.34% (Age)	UTKFace
Chavalia et al.	2023	Facial emotion recognition, age/gender estimation	CNN, SVM, VGG-16	F1 Score: 83% (Age, SVM)	UTKFace
Alsaleh & Perkgoz	2023	Age group estimation	Efficient CNN	Weighted F1score: 87.84%	UTKFace, Facial-age

1. Advanced Data Augmentation and Balancing: Implementing novel techniques significantly improves training, enabling accurate recognition across a wide age range, combating overfitting and biases.
2. Enhanced Model Accuracy: Achieving an impressive F1 score of *89.82%*, setting a new benchmark for age detection accuracy with remarkable robustness.
3. Foundation for Future Research: Offering valuable insights for further exploration and innovation in CNN applications for age detection.

This paper is organized into several sections: Sect. 2 details our proposed approach with data augmentation and balancing techniques, Sect. 3 discusses experimental development, Sect. 4 presents the results and implications, and Sect. 5 concludes with a summary and future research suggestions.

2 Proposed Approach

In this paper, we introduce an innovative approach to enhance human age detection using Convolutional Neural Networks (CNNs), addressing the challenges of skewed age representation in training data. by integrating the UTKFace and Facial-age datasets. we have developed a methodology that not only balances the data distribution but also enriches it through careful augmentation. Figure 1 succinctly presents the workflow of our method, from data collection to the evaluation of model performance. We will thoroughly explicate each step ranging from dataset fusion and preprocessing, through model architecture and optimization, to performance evaluation in the subsequent sections, providing a granular view of our process.

2.1 Datasets

The research aims to leverage the combined strength of two significant datasets: UTKFace and Facial-age. The UTKFace dataset is renowned for its extensive range of age representations, making it a valuable resource for age detection studies. Complementing this, the Facial-age dataset adds further diversity, enriching the pool of data available for analysis. The fusion of these datasets was pivotal in providing a comprehensive platform for the development and validation of the age detection model.

2.2 Data Augmentation and Balancing

The initial analysis of the combined UTKFace and Facial-age datasets revealed uneven age distribution, with some ages overrepresented and others underrepresented. To address this, a twofold strategy of advanced image augmentation and dataset balancing was implemented. Image augmentation techniques like rotation, shifting, zooming, and flipping were used to generate more images for underrepresented age groups, ensuring data integrity and preventing overfitting. Aiming for a target balance ratio of *99.51%*, iterative augmentation and analysis were conducted to achieve nearly equal representation across all age groups.

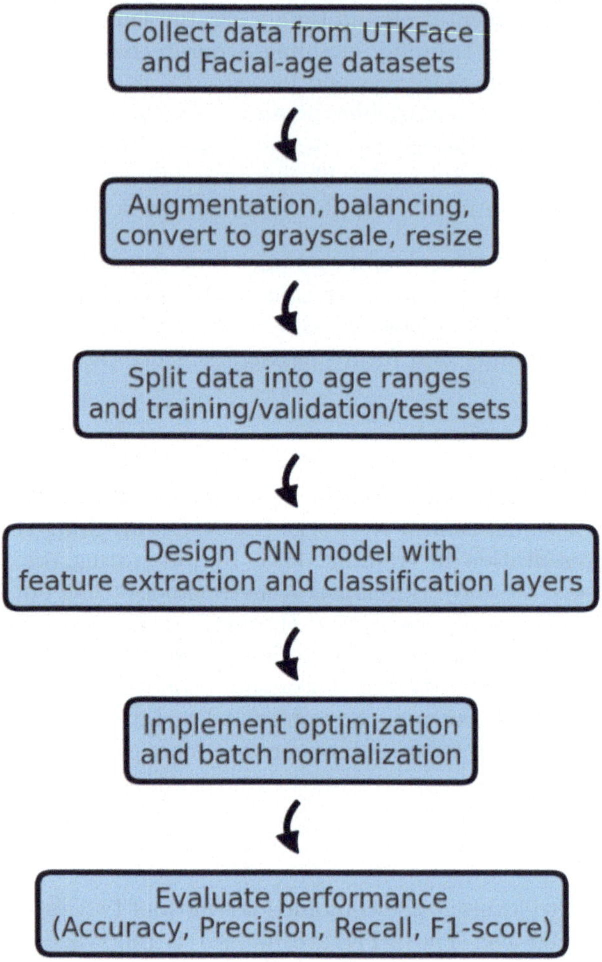

Fig. 1. Flowchart illustrating the proposed approach for enhancing human age detection using Convolutional Neural Networks (CNNs). The process includes stages from data collection and preprocessing, dataset splitting, model architecture design, training process optimization, to evaluation and performance analysis.

This balance improved the CNN model's accuracy and reliability in detecting a wide range of ages. Figure 2 illustrates the distribution of images across age ranges before and after processing, showing the achieved uniformity crucial for high-precision age detection.

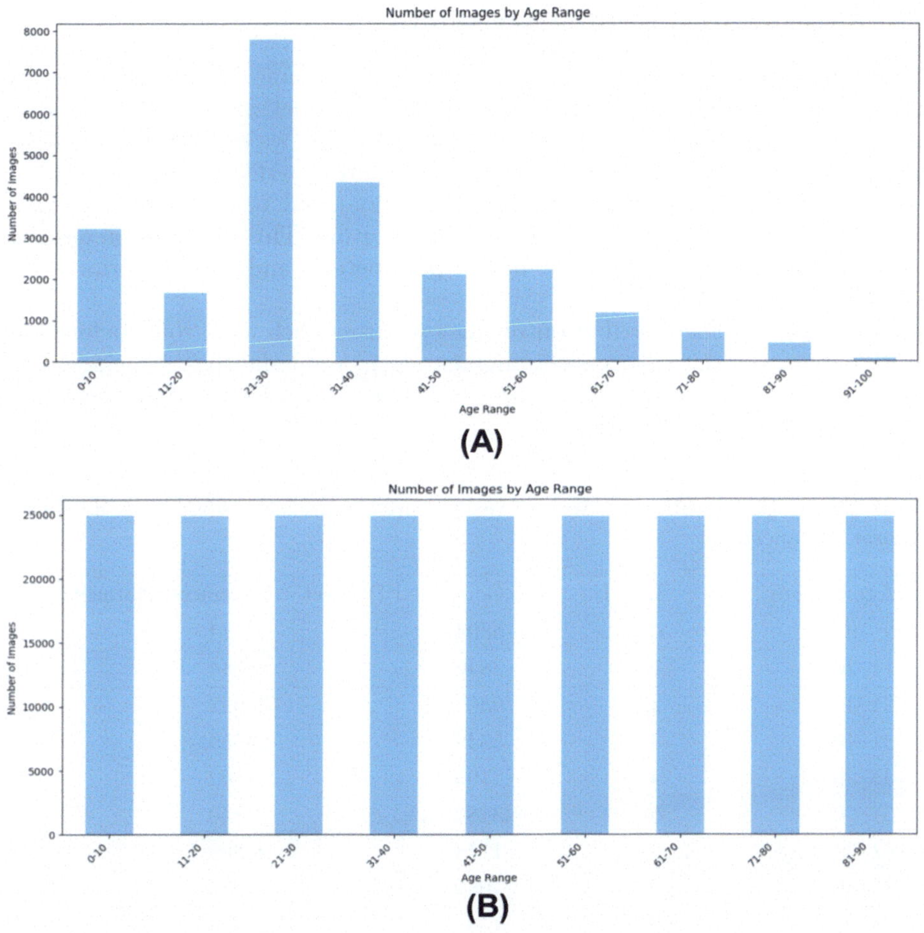

Fig. 2. Distribution of Images by Age Range Before Part (A) and After (B) Optimization

2.3 Preprocessing

The preprocessing of images was crucial in our research. All images were standardized to 64×64 pixels and converted to grayscale for consistency, computational efficiency, and color neutrality. Consistent sizing eliminated variations, ensuring the CNN focused on age-related features. Smaller images reduced the computational load, enhancing training efficiency. Grayscale conversion removed color influence, focusing the model on structural facial features. These steps balanced the dataset for age representation and optimized it for effective learning, fundamental to the success of our age detection model.

2.4 Categorization and Dataset Splitting

In our study, we strategically categorized the data into nine distinct age ranges to enhance our Convolutional Neural Network (CNN) model's accuracy. By dividing ages into groups such as '0–10', '11–20', up to '81–90', we enabled the model to focus on specific age groups, improving its predictive performance.

We split the dataset into training, validation, and test sets with an *80%* for training, *10%* for validation, and *10%* for testing. This distribution ensured a diverse training set, optimized model parameters, and robust performance evaluation.

Table 2 shows the distribution of images across each age group before and after applying our data processing, augmentation, and balancing strategies. This highlights the effectiveness of our approach in creating a balanced and representative dataset for age detection.

Table 2. Image Distribution Across Age Groups Before and After Comprehensive Dataset Enhancement

Age Ranges	Label	Combined Dataset	Enhanced Dataset
0–10	0	6304	24924
11–20	1	2840	24884
21–30	2	9307	24922
31–40	3	5349	24859
41–50	4	2795	24802
51–60	5	3181	24840
61–70	6	1744	24809
71–80	7	1082	24835
81–90	8	754	24824
Total	–	**33486**	**223699**

2.5 Data Generation and Processing

After dividing the dataset, batch sizes were set at 32 for all data generators to balance training speed and memory usage. Augmentation techniques applied to the training data included rotation (up to 10 degrees), width and height shifts (0.1), zoom (0.1), and horizontal flipping, with all images rescaled by a factor of 1./255. Validation and testing data were only rescaled by 1./255. The training generator used augmented images, while the validation and testing generators used non-augmented images. Table 3 details the distribution of the dataset into training, validation, and test sets, facilitating a robust process that significantly enhanced the age detection model's effectiveness.

Table 3. The number and the percentage of images in separated datasets

Dataset	Number of images	Size (%)
Train Dataset	178959	80%
Validation Dataset	22370	10%
Test Dataset	22370	10%

2.6 CNN Model Architecture

The initial architecture of our Convolutional Neural Network (CNN) laid the groundwork for our exploratory journey in age detection. We constructed a sequential model that began with a convolutional layer. Each layer used a patch size of 32, consistent across all experiments. The first layer comprised 32 filters, followed by a max pooling layer, a pattern that was repeated with increasing complexity. Subsequent convolutional layers with 64 and 128 filters delved deeper into feature extraction. The network then transitioned into a flattened layer, paving the way for two dense layers, culminating in an output layer with units equal to the number of age ranges. This base architecture provided a solid starting point for the refinement process, which will be elaborated upon in the following section.

2.7 Learning Rate and Optimization

The optimization of our model began with a conservative learning rate of 0.0001 using the Adam optimizer, known for its adaptive learning capabilities. This foundational learning rate was chosen to start the training process cautiously, preventing the model from skipping over minima in the loss landscape. As we progressed through our experiments, we examined the impact of different learning rates and the introduction of batch normalization, each iteration aimed at enhancing model convergence and performance. More detailed information on these iterative optimizations will be discussed in the subsequent section.

2.8 Training Process

The training process was strategically planned to be thorough and adaptable, starting with an extensive range of epochs to closely track the model's accuracy and loss. This initial phase helped identify the ideal training duration to avoid overfitting and laid the groundwork for further fine-tuning. A detailed exploration of these refinements and their effects on the model's performance will follow in the subsequent section on Experimental Development.

3 Experimental Development

Our investigative process was meticulously structured into a series of experiments, each building upon the learnings of the last, to refine our approach to age detection using Convolutional Neural Networks (CNNs). This journey through the experimental landscape was marked by a constant interplay between hypothesis and evidence, leading us to our optimal model configuration.

3.1 First Experiment: Initial Model Assessment

In this experiment, we evaluated an initial age classification model using the augmented UTKFace dataset. The model architecture included an input layer for 200 × 200 pixel color images, followed by three convolutional layers (32, 64, and 128 filters), each with max pooling and ReLU activation. A flatten layer converted 2D feature maps into a 1D vector, followed by two dense layers (256 nodes and nodes equal to the number of age categories) with dropout layers to prevent overfitting. The final layer used softmax activation for classification.

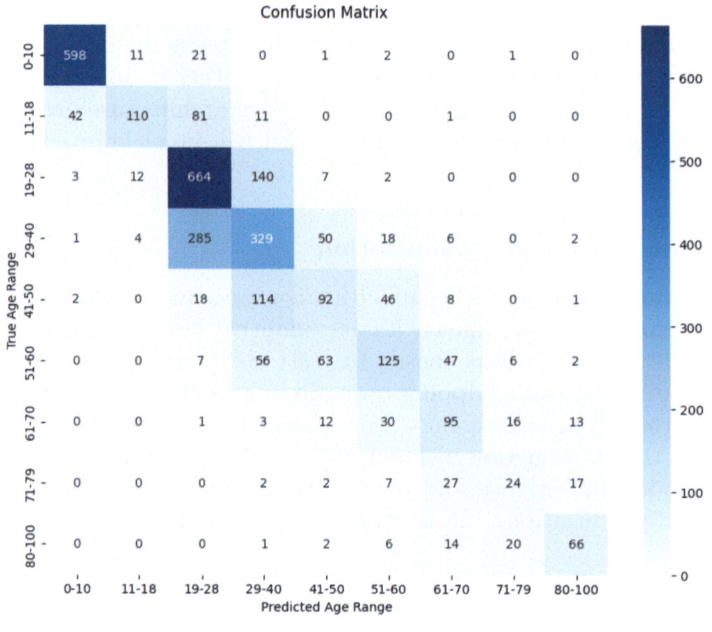

Fig. 3. The confusion matrix for the first Experiment

Compiled with the Adam optimizer and categorical cross-entropy loss function, the model was trained over 60 epochs. Results, shown in Figs. 3 and 4 demonstrated a test accuracy of *62.83%*, establishing a baseline for further improvements. This experiment provided a solid foundation for ongoing research in age classification using CNNs.

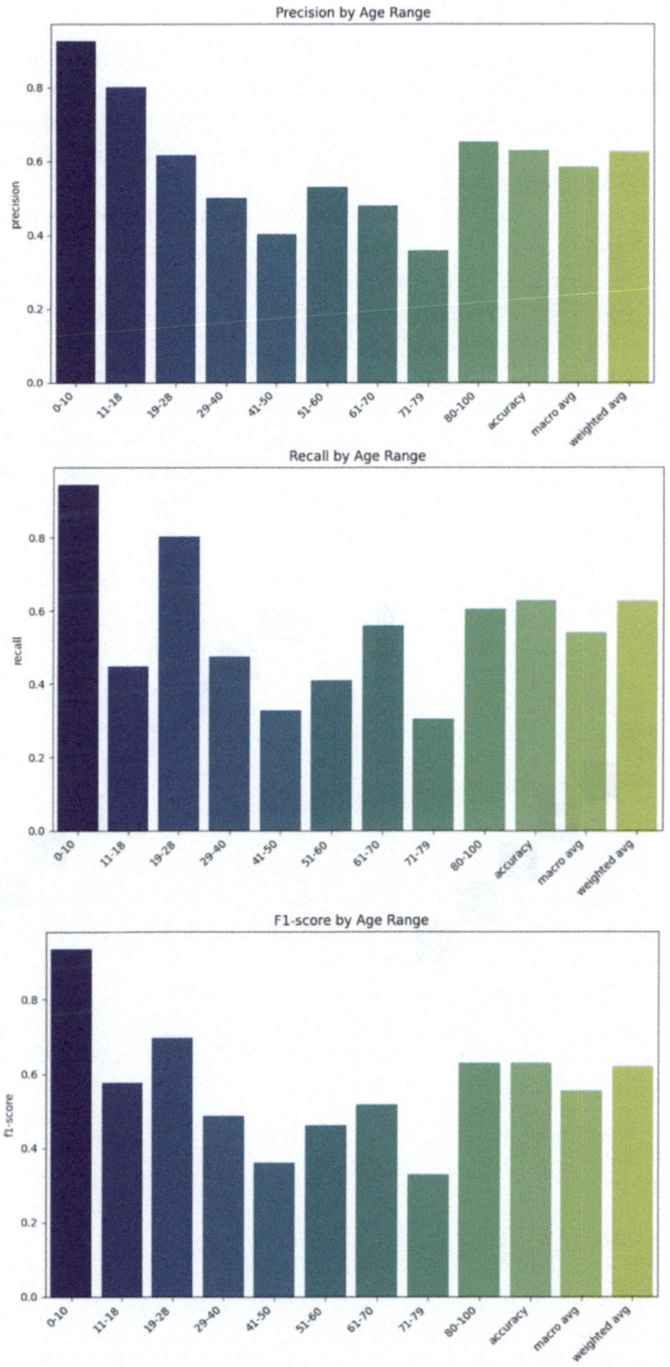

Fig. 4. Precision, recall, and F1-score metrics on test data for the first Experiment

3.2 Second Experiment: Addressing Data Imbalance

The second experiment addressed data imbalance by eliminating underrepresented age groups and using augmentation techniques to achieve a *99.51%* balance ratio. Images were resized to 64×64 pixels and converted to grayscale, ensuring even age class distribution. The dataset was split into training, validation, and testing segments.

In the first stage, a learning rate of 0.0001 was applied over 100 epochs, resulting in a validation accuracy of *83.41%* and a test accuracy of *84%*. In the second stage, the learning rate was adjusted to 0.01, and early stopping was implemented to prevent overfitting, improving validation accuracy to *87%* and test accuracy to *86%*.

Using a consistent CNN architecture and the Adam optimizer, significant advancements in age classification accuracy were achieved. These improvements are evidenced by metrics and visualizations in the confusion matrix, precision, recall, and F1 score, as shown in Figs. 5 and 6. The results validate the effectiveness of this approach in creating an unbiased and robust age detection model.

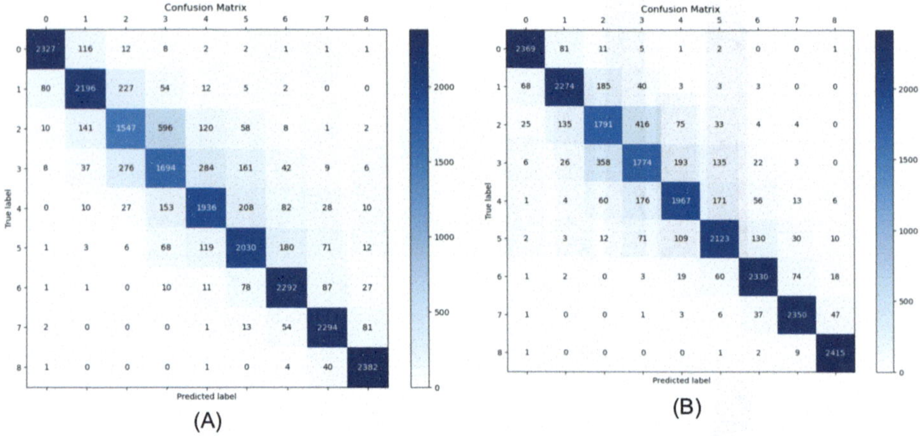

Fig. 5. The confusion matrix for the Second Experiment (A) represents stage one and (B) represents stage two.

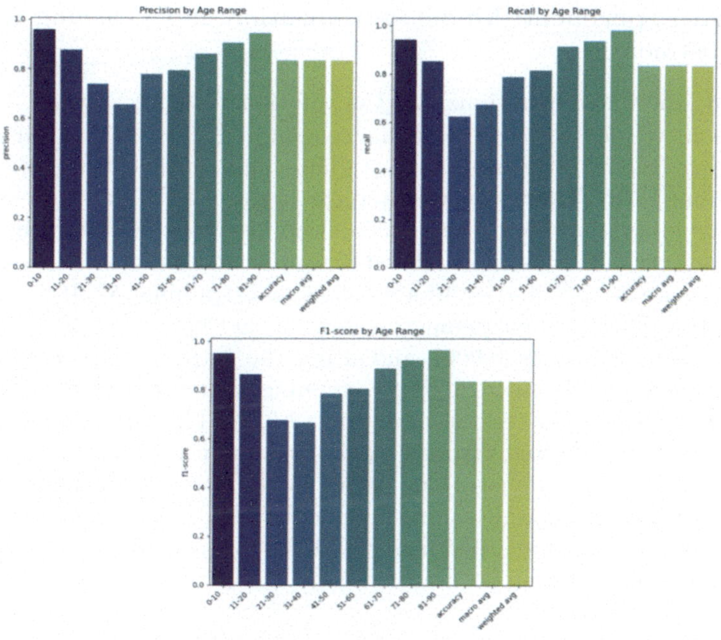

(A)

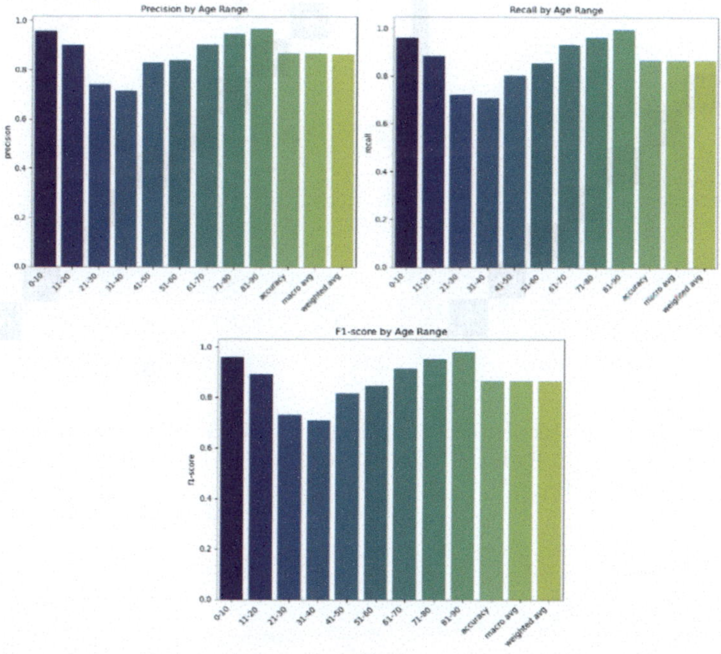

(B)

Fig. 6. Precision, recall, and F1-score metrics on test data for the Second Experiment (A) represents stage one and (B) represent stage two.

3.3 Third Experiment: Model Optimization and Learning Rate Adjustment

In the third experiment of optimizing a Convolutional Neural Network (CNN) for age detection, we focused on model enhancement and learning rate adjustments, incorporating BatchNormalization to stabilize and quicken training. Starting with a learning rate of 0.001, we employed the ReduceLROnPlateau scheduler to dynamically adjust the rate based on validation loss, halving it after five epochs without improvement. This approach led to significant results in Experiment 3A, with an F1-score of *88%* and accuracies of *86.05%* training, *87.01%* validation, and *88.7%* test. Further improvements were made in Experiment 3B by adjusting the initial learning rate to 0.0005 and using the ExponentialDecay schedule, resulting in a weighted F1-score of *89.8%*, training accuracy of *90.51%*, validation accuracy of *90.07%*, and test accuracy of *89.91%*. The effectiveness of these strategies was demonstrated through visual aids like Fig. 7, a confusion matrix, and Fig. 8, displaying F1 scores and recall for different age categories. These figures provided deeper insights into the model's performance, underscoring the crucial role of learning rate adjustments and Batch Normalization in enhancing the CNN's age classification accuracy.

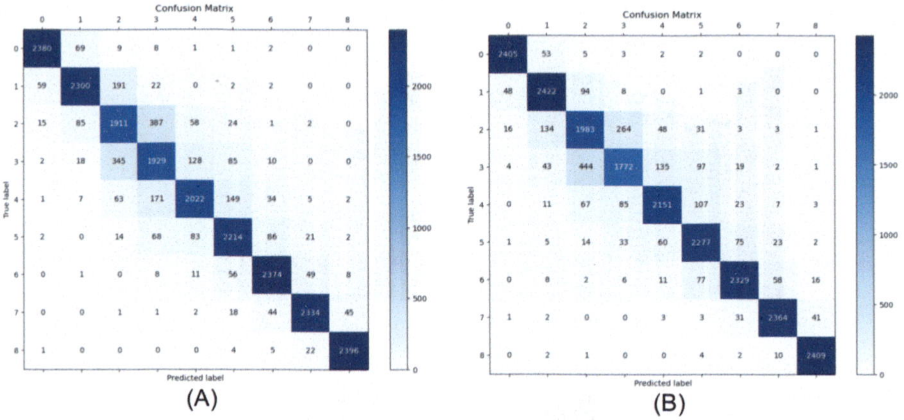

Fig. 7. The confusion matrix for the Third Experiment (A) represents results of Learning Rate Adjustment with ReduceLROnPlateau, and (B) represents the stage of Exponential Decay Learning Rate.

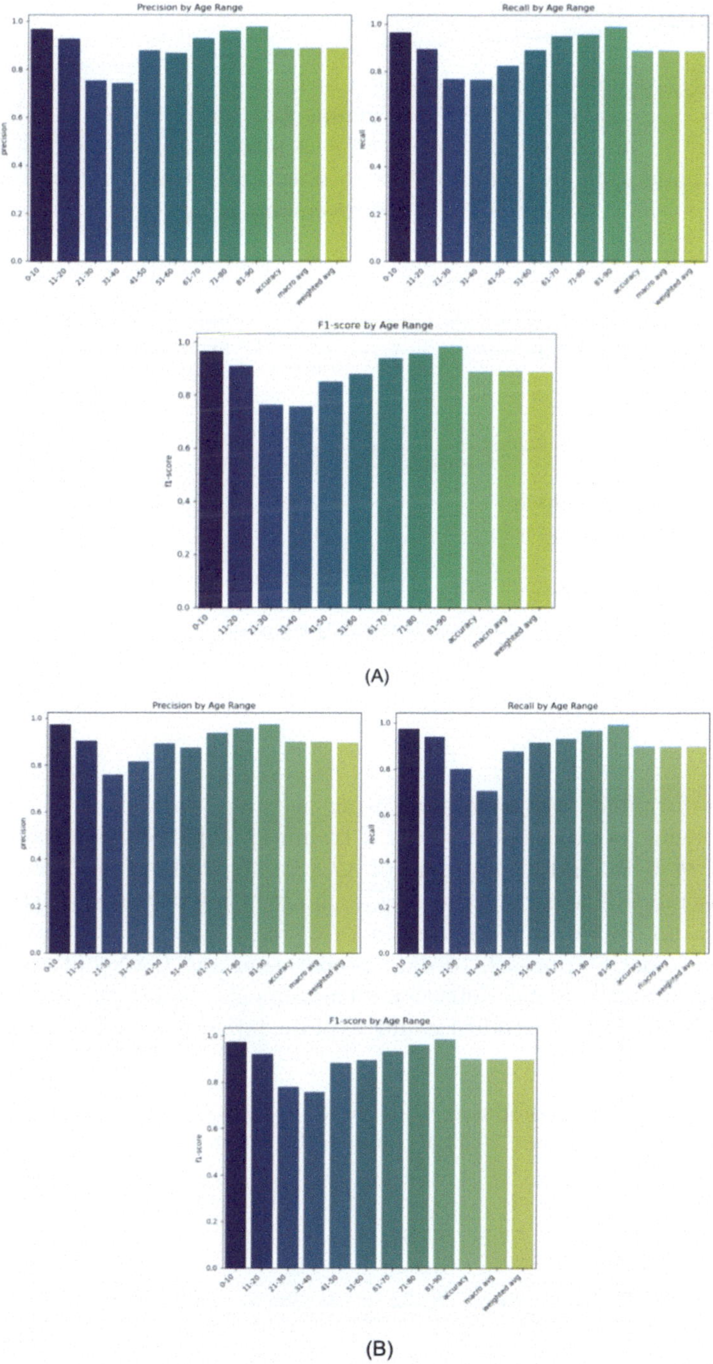

Fig. 8. Precision, recall, and F1-score metrics on test data for the Third Experiment (A) represents results of Learning Rate Adjustment with ReduceLROnPlateau, and (B) represent stage of Exponential Decay Learning Rate.

4 Results

In this study, we enhanced age detection by developing a Convolutional Neural Network (CNN) using balanced and augmented datasets from UTKFace and Facial-age. The images were resized to 64×64 pixels and converted to grayscale. Our CNN underwent five experiments to optimize learning rates and network architecture, leading to incremental improvements in age classification, as shown in Table 4.

Table 4. Comparison of results from the three experiments, including the sub-stages of the second and third experiments in terms of model accuracy, F1 score, Recall, and Precision

Experiment Stages	Accuracy (%)	F1 Score (%)	Recall (%)	Precision (%)
Initial Model Assessment	62.83	62	63	63
Addressing Data Imbalance				
Stage 1 with 100 epochs and learning rate 0.0001	84	83	84	83
Stage 2 with learning rate 0.01	86	87	87	87
Model Optimization and Learning Rate Adjustment				
ReduceLROnPlateau	88.7	88	89	89
Exponential Decay Learning Rate	89.91	89.8	90	90

The performance was evaluated using accuracy, precision, recall, and F1-score, achieving notable results with a training, validation, and test accuracy all around *90.51%*, and an F1-score of *89.82%*, highlighted in Table 5. Figure 9 depicts the model's accuracy and loss, demonstrating its learning stability and effectiveness in age group classification.

This systematic approach to data preparation and CNN optimization contributes significantly to age detection advancements, laying the groundwork for future research. Future studies could explore integrating more complex data

Table 5. Comparison of proposed model and previous models (Agbo-Ajala & Viriri, 2020 and Alsaleh & Perkgoz, 2023) in terms of F1-score

Model	0–10	11–20	21–30	31–40	41–50	51–60	61–70	71–80	81–90	Overall
Proposed Model	97.27	92.16	77.87	75.60	88.45	89.49	93.31	96.25	98.31	89.82
Alsaleh & Perkgoz (2023)	96.92	93.38	87.86	82.72	84.46	86.51	87.38	90.82	–	87.84
Agbo-Ajala & Viriri (2020)	94.59	82.35	85.43	76.34	87.16	67.86	62.69	91.09	–	81.58

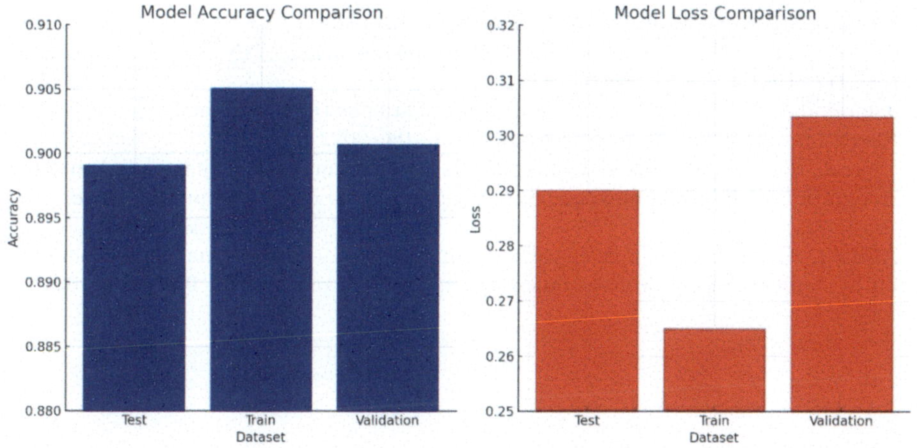

Fig. 9. Proposed model final accuracy and loss for (Training, Validation, and Testing Datasets).

augmentation techniques, delving into deeper and more varied network architectures, or applying advanced machine learning algorithms to further enhance accuracy and reduce bias, thereby pushing the boundaries of what is currently achievable in age detection technology.

5 Conclusion

In this study, we improved age detection accuracy using Convolutional Neural Networks (CNNs) by implementing data augmentation and balancing techniques. By integrating UTKFace and Facial-age datasets, we achieved an accuracy of *90.51%* and an F1 score of *89.82%*. Although this represents significant progress, opportunities for further enhancement remain. Our work contributes to the broader effort to refine age detection systems within the machine learning community, anticipating future advancements to extend these capabilities.

References

1. Sithungu, S., Van der Haar, D.: Real-time age detection using a convolutional neural network. In: Business Information Systems: 22nd International Conference, BIS 2019, Seville, Spain, 26–28 June 2019, Proceedings, Part II, Seville, 26–28 June 2019
2. Mahjabin, S., Alam, M.M., Talukder, K.H.: Age estimation from facial image using convolutional neural network (CNN). In: 2019 International Conference on Computer, Communication, Chemical, Materials and Electronic Engineering (IC4ME2), 2019
3. Agbo-Ajala, O., Viriri, S.: Face-based age and gender classification using deep learning model. In: Image and Video Technology: PSIVT 2019 International Workshops, Sydney, NSW, Australia, 18–22 November 2019

4. Nada, A.M.A., Alajrami, E., Al-Saqqa, A.A., Abu-Naser, S.S.: Age and gender prediction and validation through single user images using CNN. Int. J. Acad. Eng. Res. (IJAER) **4**, 21–24 (2020)
5. Jamoliddin, U., Yoo, J.H.: Age and gender classification with small scale CNN. J. Korea Inst. Electron. Commun. Sci. **17**, 99–104 (2022)
6. Chavali, S.T., Kandavalli, C.T., Sugash, T.M., Subramani, R.: Smart facial emotion recognition with gender and age factor estimation. Procedia Comput. Sci. **218**, 113–123 (2023)
7. Alsaleh, A., Perkgoz, C.: A space and time efficient convolutional neural network for age group estimation from facial images. PeerJ Comput. Sci. **9**, e1395 (2023)

Leveraging Machine Learning to Assess Factors Influencing Undernourishment in Nigeria

Nada Mobark[1(✉)], Noha Elfiky[2], Ahmed Ahmedin[2], Ameera Ibrahim[2], M. Abdel-Aziz[3], and Safwat Hamad[2]

[1] Faculty of Computers and Information, South Valley University, Qena 83523, Egypt
nada.mobark@sci.svu.ed.eg

[2] Business Analytics and Data Science Department, School of Economics and Business Administration, Saint Mary's College of California, Moraga, CA 94575, USA

[3] Department of Artificial Intelligence Engineering, Faculty of Computer Science and Engineering, King Salman International University, South Sinai 46511, Egypt

Abstract. Notwithstanding economic progress, Nigeria continues to countenance significant food security issues. This study investigates the root causes of this problem and attempts to develop predictive models to address hunger in Nigeria. The research covering data from 2000 to 2022, sourced primarily from the World Bank and other institutions. The trend analysis, along with the linear model, indicates that investment in farming is crucial for reducing undernourishment. Additionally, there is a strong relationship between agricultural expenditure, exports, access to electricity, and unemployment in relation to undernourishment. In this paper, a linear model, correlation analysis, ARIMA, and trend analysis are utilized to demonstrate how undernourishment, representing hunger and food insecurity, is influenced by other factors. As a result, it is recommended to maintain the unemployment rate between 3.5% and 3.7%, while ensuring that exceeds 5% of government spending is recommended. In other words, to combat hunger in Nigeria, it is important to focus on investing in agriculture, creating jobs, and improving infrastructure. This aligns with our findings and with another research around the world.

Keywords: Undernourishment · Nigeria · predictive model · ARIMA

1 Introduction

In Nigeria, hunger and food insecurity are consequential problems, often due to insufficient food availability or poor food quality. Factors influencing undernourishment such as job losses, reduced agricultural expenditure, increased food prices, and uneven food production have been made worse by COVID-19 [1]. This has left many people without enough nutrition for a healthy life. Nigeria has the highest rate of undernourishment based on how many calories people eat, but this doesn't account for dietary variety or nutritional quality. Keeping food safe and secured is essential for health, focusing on avoiding contamination and promoting hygiene.

A. Abdelgawad et al. (Eds.): ISBCom 2024, LNNS 1268, pp. 819–828, 2025.
https://doi.org/10.1007/978-3-031-82377-0_65

Food insecurity refers to the lack of consistent access to adequate food. In 2020, around 13.8 million Nigerian households faced food insecurity, highlighting a significant challenge in meeting the minimum dietary requirements for a healthy life [2]. As a result, this causes hunger and populations suffered from not having the enough calories for their daily activities. Even though Nigeria's economy is strong, undernourishment is increasing while similar countries are doing better with less GDP than Nigeria. (see Fig. 1) shows the GDP trend for Nigeria.

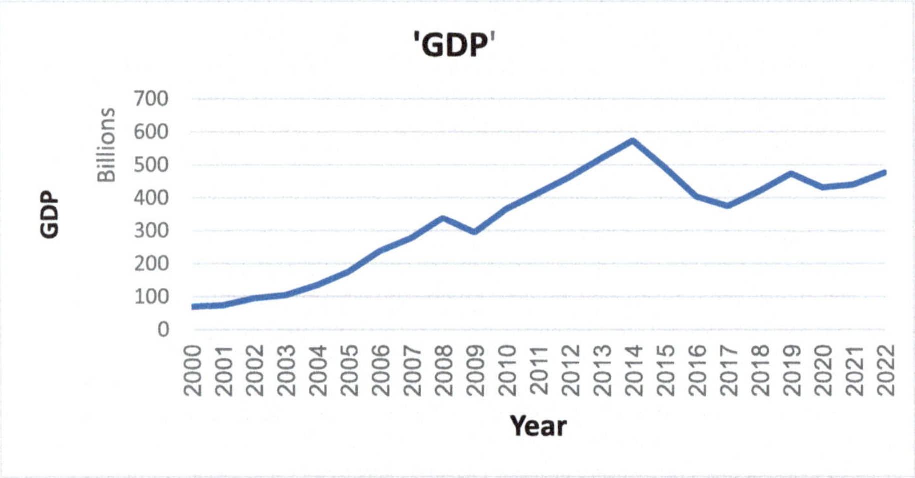

Fig. 1. GDP in Nigeria [6].

Nigerians [3] are having a hard time with rising food costs and high inflation. In July 2022, inflation was 19.64%, the highest since 2017. Inflation reduces buying power and increases poverty. The World Bank says inflation might push another million people into poverty by the end of 2022.

Despite its large population, Nigeria struggles to provide basic services. For example, only 48% of people have clean water, and just 33% have proper sanitation. In 2019, Nigeria ranked 159th out of 162 countries in achieving the Sustainable Development Goals (SDGs) (see Fig. 2). The country has improved on SDG 2, which focuses on hunger, food security, and sustainable agriculture, but still falls short [4].

Computer systems can learn on their own and analyze various types of data to gain important insights. AI uses smart systems that mimic the human brain to solve problems in the real world. AI has potential in farming, but current models have issues due to regional or national problems [5]. Machine learning for problem-solving is at the center of the paradigm shift in modern research. These models often focus on specific countries and don't work well in places with diverse conditions or where conflicts are happening. They don't reflect the reality of countries like Nigeria, where insecurity affects rural communities and hinders growth. Predicting food shortages with AI in Nigeria is tough. This study aims to create a data-driven model for assessing food insecurity among families, considering the nuances of food safety.

Fig. 2. The Sustainable Development Goals [4].

The paper is organized as follows: this section introduces the introduction while Sect. 2 introduces the literature review. Section 3 covers the methodology followed in this paper, illustrating the factors (features) affecting food security (represented by undernourishment). In Sect. 4, trend analysis and discussions are introduced and finally all recommendations and paper wrap up is concluded in Sect. 5.

2 Literature Review

Recently, research in hunger and food security has increased due to the availability of macroeconomic data influencing undernourishment. Even though global undernutrition has decreased, progress is still slow in many countries [6–10]. In 2012, the World Health Assembly set nutrition targets to reach by 2025 [11]. The scaling up nutrition movement and the sustainable development goals have highlighted maternal and child undernutrition. Tackling malnutrition needs three things: effective interventions, broad adoption, and sustainability [12]. The main challenge is maintaining high coverage of interventions [8, 13].

Gillespie in [13] highlighted the need for an environment that supports successful interventions. Studies on undernutrition in countries that reduced stunting stress the importance of data, coherence, capacity, and resources [14, 15]. Improving nutrition environments is crucial for meeting targets, especially given COVID-19's impact [16]. Nigeria, with a quarter of the world's malnourished people, has made slow progress. Stunting dropped by 9.4% from 2008 to 2018, but issues like food security and child diseases continue [17, 18].

The stories of change study aimed to enhance policies and programs to cut undernutrition in Nigeria. It assessed the environment, finding areas to improve to scale up nutrition interventions. FAO showed a strong link between food safety and undernourishment. The agricultural sector affects food safety through availability, access, utilization, and stability. Nutrition involves complex data, so we need new ways to analyze and address undernourishment and improve health outcomes.

3 Methodology

In this paper, we followed the data science process for addressing the problem. Shows the methodology followed in this article (see Fig. 3). First, all the macroeconomics features and factors are collected from different data sources such as FAO, world bank and UNICEF. Then those features are integrated in a single database using Extract Transform Load ETL method. Secondly, data preparation includes cleansing and filling missing data. In addition, we selected the most significant features that influencing our target in this paper which is the undernourishment measure.

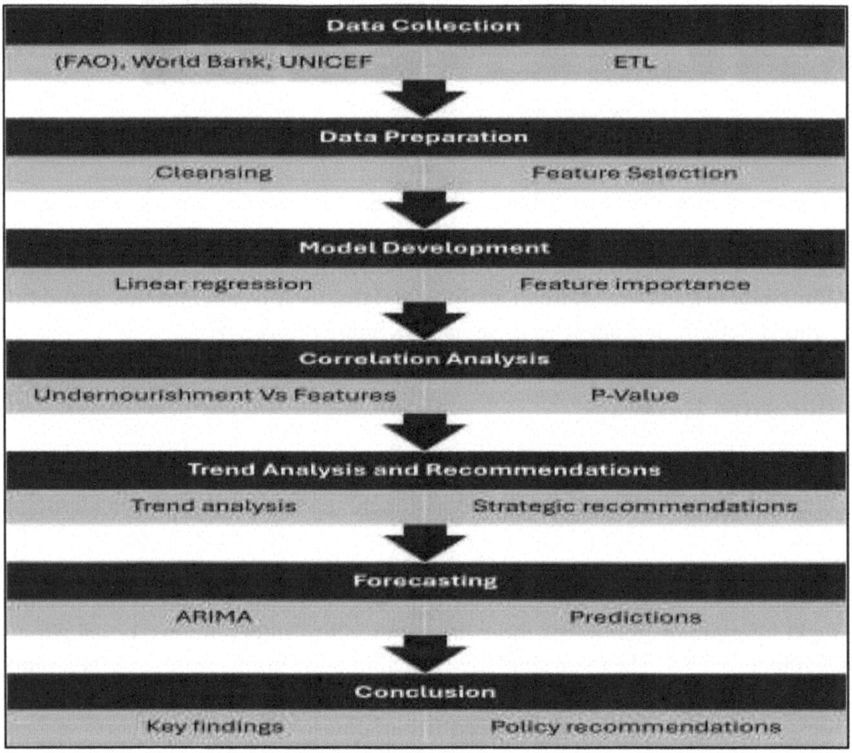

Fig. 3. Comprehensive Methodology for Data.

Furthermore, a multilinear regression model is built, and the feature importance is determined. Next, the correlations among all the features in relation to the undernourishment are measured and the importance of those features are determined by P-Value to determine the significance of each feature. The fifth step is to apply trend analysis and make more investigation for the most important features to give the strategic recommendations. Moreover, a forecasting model is built using ARIMA to forecast the undernourishment for the next 5 years. Finally, the key findings are concluded, and the policy recommendations are presented to the decision makers.

3.1 Research Design

Food insecurity is when people can't get enough food for maintaining a healthy life. Factors like poverty, unemployment, low income, or expensive housing are the main cause of food insecurity (see Fig. 4).

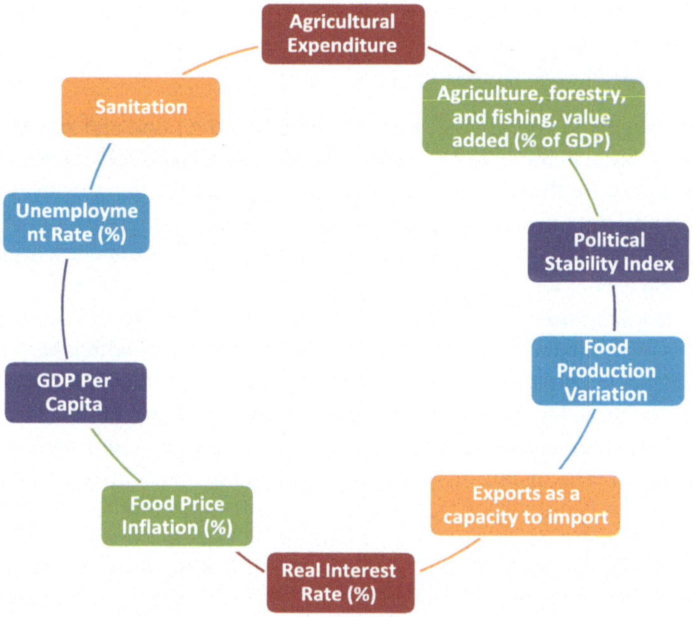

Fig. 4. Factors affecting Undernourishment.

In addition, health issues, water and sanitation access, life expectancy, urban population, and electricity access should be taken into considerations. The following is brief description about each factor:

1. **Agricultural Expenditure** where more spending on agriculture will lead to better food production and availability.
2. **Agriculture, Forestry, and Fishing** to show the importance of these sectors to the economy. Developing countries often depend on these industries.
3. **Political Stability** to help ensure steady food production and distribution.
4. **Food Production Variation** where large variations can cause price spikes and food access issues.
5. **Exports with respect to Import** where exporting goods and services helps a country afford imports.
6. **Real Interest Rate** can hinder agricultural investment and food security if it is high.
7. **Food Price Inflation** where high food price inflation makes it hard for people with limited resources to access nutritious food.
8. **GDP Per Capita** where higher GDP per capita means better economic conditions and more purchasing power.

9. **Unemployment Rate** where high unemployment limits access to nutritious food.
10. **Sanitation** whereby providing clean and hygienic facilities helps prevent disease and maintain health.

Analyzing these factors helps us understand and address undernourishment in Nigeria. For a more accurate assessment, a specific and detailed analysis using relevant data and indicators is needed.

3.2 Data Collection and Modeling

The data was collected from various sources, including the Food and Agriculture Organization of the United Nations (FAO), World Bank, and UNICEF Data Warehouse. The data has a limitation for the prevalence of undernourishment Nigeria from 2000 to 2022. After importing all the factors in different sheets, the data is integrated and cleaned to be prepared for analysis for selecting and identifying the most significant features. Then, a linear regression model was built to model the selected significant factors and the target variable (undernourishment). In addition, P-Value the coefficient of each feature in the linear model is evaluated to measure the feature importance and which help in prioritize the government actions to resolve the undernourishment problem in Nigeria. Finally, depending on the model's findings, conclusions and suggestions are proposed.

The correlation between each of our selected nine features and undernourishment (target) (see Fig. 5). The insight from this figure is merged with insights from the linear regression model to decide what features are the most important for the Nigerian government (NgG). Finally, NgG should start with those selected features as fast as it can to reduce the undernourishment in the country. The following section will highlight the selected features.

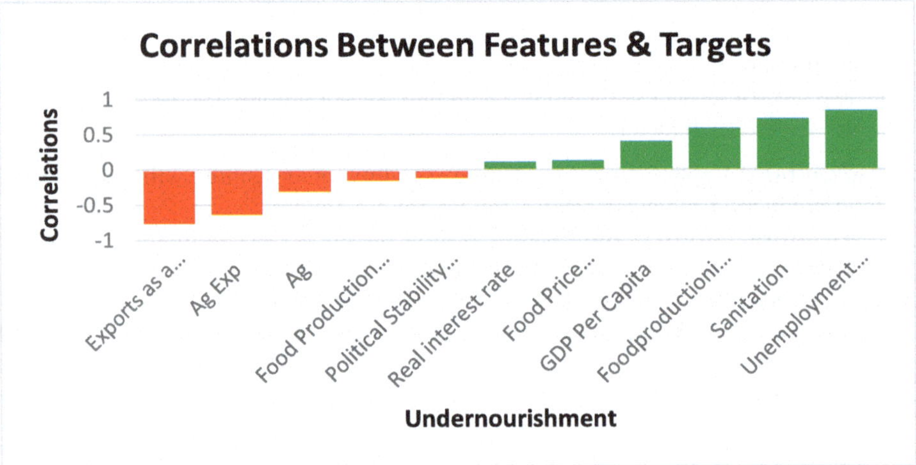

Fig. 5. Correlation Between Features and Targets.

4 Trend Analysis and Discussion

Based on the linear regression model and correlations computed among the features and the target, we propose selecting agricultural expenditure, exports (as a capacity to import), unemployment rate, and domestic budgets to adequately finance nutrition initiatives. Moreover, each feature was analyzed in more depth to provide additional recommendations for addressing undernourishment.

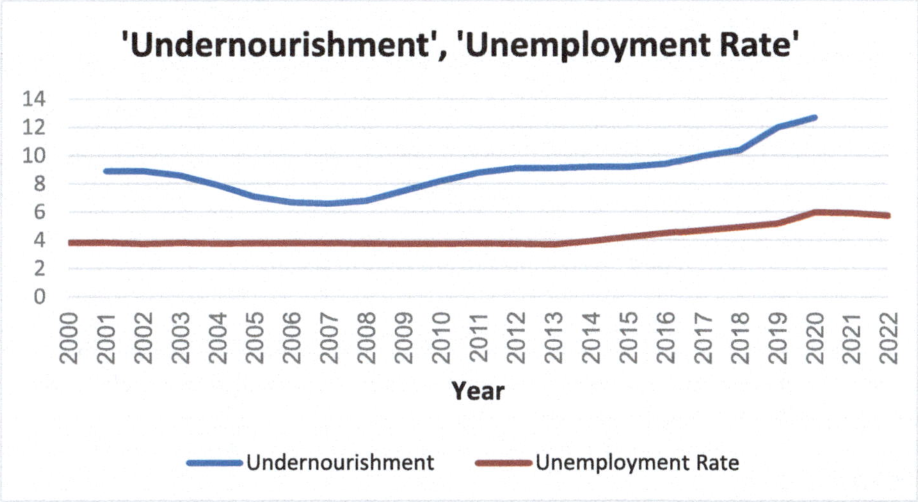

Fig. 6. Undernourishment prevalence (as a percentage of the population) and unemployment rate in Nigeria.

This analysis suggests a strategic turn towards prioritizing agricultural expenditure and encouraging exports, which in turn would enhance the nation's capacity to import and thereby secure the foreign currency means necessary for ensuring adequate nutrition. The analysis also showed that the agricultural expenditure in Nigeria has a notable drop from 5% in 2008 to 2% in 2012 which affected the unemployment rate (see Fig. 6). In addition, unemployment increased as a result from 2011 and touches 6%. So, to realign with the main target of undernourishment reduction, we propose to retain the unemployment rate to a target range of 3.5% to 3.7% as (see Fig. 7). And the agricultural expenditure should come back up to 5% from the GDP.

In our forecast analysis, the ARIMA model was utilized to predict undernourishment from 2020 to 2025. This method enables us to understand trends and make future projections accurately. Given the panel time series nature of our problem, we first forecasted individual features such as economic factors, agricultural productivity, employment rates, and government spending on nutrition programs. Each of these features was forecasted separately to capture their unique trends and potential impacts.

Once these individual forecasts were obtained, we integrated them to produce a comprehensive forecast of undernourishment rates. The ARIMA model's integration of

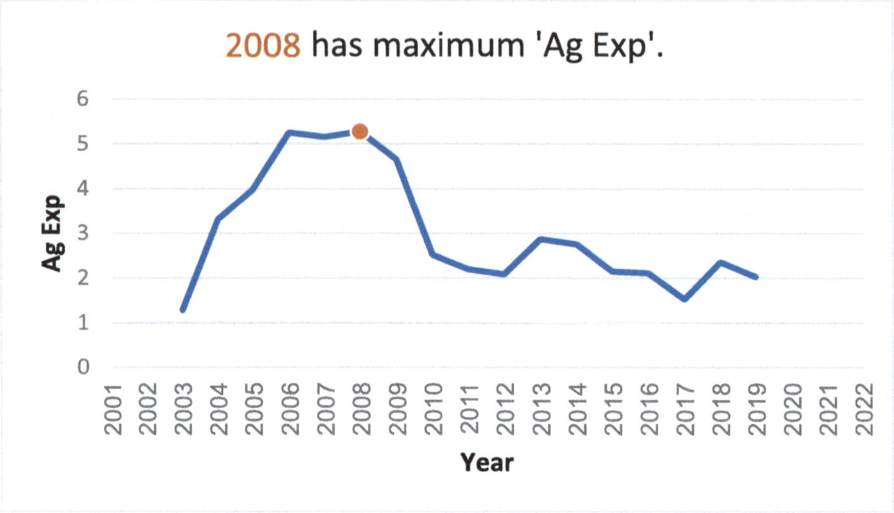

Fig. 7. Agricultural Expenditure in Nigeria.

these varied factors allows for a nuanced prediction that considers the interplay between different socio-economic variables. The forecasted model revealed that undernourishment is expected to increase over the next five years. This trend is consistent with the predicted rise in unemployment rates, suggesting a direct link between employment and food security.

The findings underscore the importance of addressing unemployment to effectively tackle food insecurity. By understanding the predicted trends and the underlying factors influencing undernourishment, policymakers can develop targeted strategies to mitigate these issues. For a detailed visualization of these projections and the forecasted trends, please refer to (see Fig. 8).

The ARIMA model was employed to predict future levels of undernourishment. This predictive analysis provides a deeper understanding of the expected trends in undernourishment and identifies the factors that significantly influence these trends. By utilizing the ARIMA model's forecasts, policymakers can make more informed decisions regarding resource allocation and implement targeted interventions to address and mitigate undernourishment effectively. This data-driven approach ensures that resources are distributed efficiently and that policies are designed to have the maximum positive impact on reducing undernourishment.

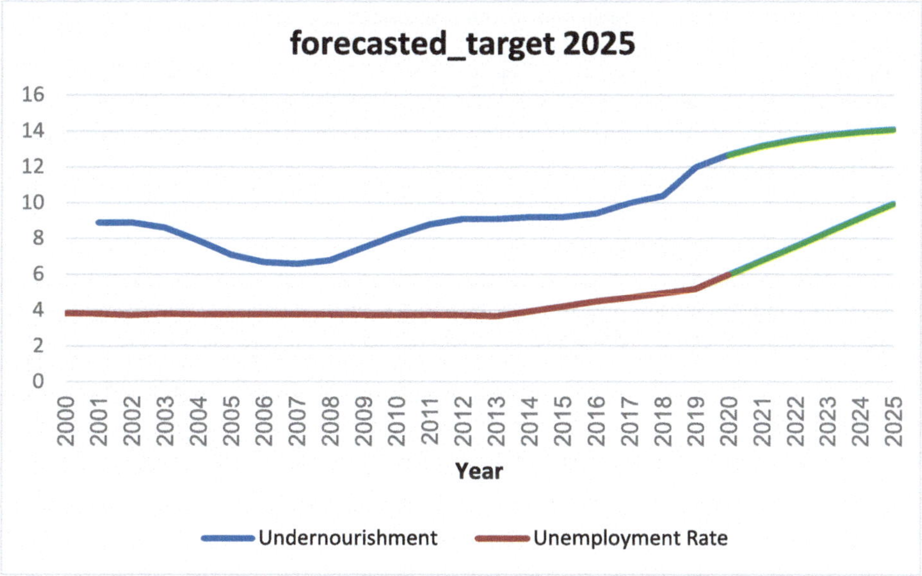

Fig. 8. Forecast the target for undernourishment & unemployment rate.

5 Conclusion

This study highlighted the main factors influencing the undernourishment in Nigeria. Despite the economic progress in Nigeria, the prevalence of undernourishment remains alarming. The analysis shows the importance of targeted interferences, with a particular focus on agricultural expenditure and job creation. Specifically, it is recommended to increase agricultural expenditure to be at least 5% of GDP and to maintain unemployment rates to lie between 3.5% and 3.7%. Furthermore, leveraging predictive analytics such as the ARIMA model indicated a projected increase in undernourishment over the next five years. These findings emphasize the urgent need for coordinated efforts from policymakers and stakeholders to address food insecurity and improve nutrition outcomes for the Nigerian population.

References

1. Adebayo, P.F., Ojo, E.O.: Food security in Nigeria: an overview. Eur. J. Sustain. Dev. **1**(2), 199–199 (2012)
2. Carlson, S.J., Andrews, M.S., Bickel, G.W.: Measuring food insecurity and hunger in the United States: development of a national benchmark measure and prevalence estimates. J. Nutr. **129**(2), 510S–516S (1999)
3. Etim, E., Duke, O: The implications of food insecurity, poverty and hunger on Nigeria's national security. Asian Res. J. Arts Social Sci. **4**(2), 1–10 (2017)
4. Nwozor, A., Olanrewaju, J., Ake, M.: National insecurity and the challenges of food security in Nigeria. Acad. J. Interdisc. Stud. **8**(4), 9–20 (2019)

5. Garg, S., et al.: A review of machine learning applications in human resource management. Int. J. Product. Perform. Manag. **71**(5), 1590–1610 (2022)
6. Bhutta, Z.A., et al.: Evidence-based interventions for improvement of maternal and child nutrition: what can be done and at what cost? Lancet **382**(9890), 452–477 (2013)
7. Bryce, J., et al.: Maternal and child undernutrition: effective action at national level. Lancet **371**(9611), 510–526 (2008)
8. Heidkamp, R.A., et al.: Mobilising evidence, data, and resources to achieve global maternal and child undernutrition targets and the Sustainable Development Goals: an agenda for action. The Lancet **397**(10282), 1400–1418 (2021)
9. Ruel, M.T., Alderman, H.: Maternal and child nutrition study group nutrition-sensitive interventions and programmes: how can they help to accelerate progress in improving maternal and child nutrition. Lancet **382**(9891), 536-551 (2013)
10. Victora, C.G., et al.: Revisiting maternal and child undernutrition in low-income and middle-income countries: variable progress towards an unfinished agenda. Lancet **397**(10282), 1388–1399 (2021)
11. McGuire, S.: World Health Organization. Comprehensive implementation plan on maternal, infant, and young child nutrition. Geneva, Switzerland, 2014. Adv. Nutr. **6**(1), 134–135 (2015)
12. Hawe, P., et al.: Multiplying health gains: the critical role of capacity-building within health promotion programs. Health Policy **39**(1), 29–42 (1997)
13. Gillespie, S., et al.: The politics of reducing malnutrition: building commitment and accelerating progress. Lancet **382**(9891), 552–569 (2013)
14. Headey, D., Hoddinott, J., Park, S.: Accounting for nutritional changes in six success stories: a regression-decomposition approach. Glob. Food Sec. **13**, 12–20 (2017)
15. Gillespie, S., van den Bold, M., Stories of Change Study Team: Stories of change in nutrition: an overview. Glob. Food Secur. **13**, 1–11 (2017)
16. Headey, D., et al.: Impacts of COVID-19 on childhood malnutrition and nutrition-related mortality. Lancet **396**(10250), 519–521 (2020)
17. Adeyemi, O., et al.: Understanding drivers of stunting reduction in Nigeria from 2003 to 2018: a regression analysis. Food Secur. **14**(4), 995–1011 (2022)
18. Adeyemi, O., et al.: Changes in Nigeria's enabling environment for nutrition from 2008 to 2019 and challenges for reducing malnutrition. Food Secur. **15**(2), 343–361 (2023)

Automated Syntax: Employing Artificial Intelligence in Generating Grammatically Correct Arabic Texts

Salman Abbas Abed(✉)

College of Islamic Sciences, Baghdad University, Baghdad, Iran
salamn@comc.uobaghdad.edu.iq

Abstract. This study explores a significant research gap in the application of artificial intelligence (AI) to the generation of grammatically correct Arabic texts. Arabic's complex grammatical and morphological structures present challenges that AI systems, including those in popular applications like Microsoft Word, struggle to process effectively. The research highlights the importance of improving AI-driven language tools to better serve Arabic-speaking users, both in academic and practical contexts. Using a descriptive methodology, the study surveyed 50 Microsoft Word 2010 users to evaluate their experiences with AI-powered Arabic text processing. Participants provided feedback on the system's ability to handle key grammatical elements, such as subject-verb agreement, gender, number, and context-sensitive expressions. Statistical data showed that 45.5% of respondents agreed that AI had difficulty managing Arabic's morphological complexities, while many users also noted that conditional uses and idiomatic expressions were often misinterpreted by the software. The findings emphasize the current limitations of AI in processing Arabic, particularly in recognizing and applying the language's intricate grammatical rules. The study suggests that integrating advanced AI techniques, such as deep learning and context-aware algorithms, could significantly enhance the accuracy of Arabic language processing. The broader implications of this research point to the need for continued investment in AI tools tailored to Arabic, which could improve machine translation, automated speech recognition, and text generation, making these technologies more reliable and effective for Arabic-speaking users.

Keywords: texts · artificial intelligence · grammar · challenges · morphology · the Arabic language

1 Introduction

The use of artificial intelligence has increased in all areas of our daily life that we consider artificial intelligence to be a scientific and logical world around us like everything else, help us live with it or achieve what we dream of. The Arabic language Artificial intelligence software is a new generational technology in the Arabic language as well as it can helps to solve the complex problems automatically and independently since

A. Abdelgawad et al. (Eds.): ISBCom 2024, LNNS 1268, pp. 829–837, 2025.
https://doi.org/10.1007/978-3-031-82377-0_66

there are many sources such forums, newspapers in different forms that needs human translation basic task, spelling correction and grammar correction. There are various problems and opportunities to be taken in this subject that involves the capacity of automatic processing for Arabic language by using artificial intelligence. One part of this is how one treats Arabic vocabulary as special and peculiar, on the other hand it explains how it manages the difficulties in utilizing this capacity to serve Arabic language due to very few artificial intelligence capacities that available. Interventions targeted at their Maqui Plows [1].

According to Abu Ghadeer, a survey was conducted to investigate the different techniques and strategies that AI applications employ to translate Arabic language, and also sought to reveal the best means of translating AR text as well as how this language can support other AI applications. This work will be done using artificial intelligence programs in encoding Arabic language, to encode and interpret in special language humans.CSS Local Plugin. And applied study methods for processing the Arabic language [2].

The Arabic language has not fared badly, generally, in the reception by artificial intelligence applications capable of translating it, but that holds true except for some problematic morphological and grammatical domains within it. Especially consequential is this discovery. Several recommendations were made by report, including the urgent need for sustained investment in research and development of Arabic artificial intelligence software. To produce better outcomes, so that there are fewer adverse effects [3].

Al-Dahshan, Jamal introduced a study that applied AI techniques to increase the potential of expressive communication transmitted through the Arabic language. This software has been developed to the needs of the Arabic language like any other natural language; this capacity allows us to utilize our Arabic language within it. Then this research, by the descriptive method, sought to identify a series of preceding studies on the applications of artificial intelligence in Arabic language through previous references to related Arabic-language study. And I identified the pros and cons of working with these programs, that they are subject to some misconceptions that confuse in understanding them and their phonetic and linguistic elements of Arabic And this to give a fairer opinion on its advantages and disadvantages. The report recommended, for example, that efforts to develop Arabic in software and generate new Arabic-based work should be continued to take advantage of the future benefit of the market [4].

This paper comes as a reminder to all Algerian informaticists whom are in the work of Arabic language, by observing an experience for a group of researchers on Arabic working methods based on artificial intelligence in such field.

A study conducted by Talafha et al. [5] investigated and developed a multi-dialect Arabic BERT model which focused more on the identification of a country from which a speech belongs to. The deep learning model showed the ability to distinguish different Arabic varieties which are essential for sentiment analysis, and machine translation. The study also related significant hindrances in the handling of this capability in different varieties among Arabic. In order to emphasize Arabic processing, Zeroual et al. [6] conducted a deep learning in Arabic NLP survey which considered a full overview of deep learning using CNN and RNN. This study showed significant advancement in neural network for Arabic processing in tasks as named entity recognition, text classification, and sentiment analysis. They also pointed out some noticeable drawback which are more data in comparison to English and more computation resources. Moreover, another study by Elmadany et al. [7] developed a data set called ArSAS for classifying speech act and sentiment from user tweets. The study was from the fact their inadequate resources for the training of Arabic Twitter sentiment data models.

The analysis of offensive language in Arabic social media also mostly seldom considered. In this regard, Mubarak et al. concluded the analysis of caption Arabic sentiment tweet and the categorization of the tweet as controversial or not controversial using the machine learning method. Their experiment shown how controversial and not controversial Arabic script is exposed to the social network. In their conclusion, Mubarak et al. [8] related the ability of the developed model in classified Arabic script in accordance with controversial words. The framework was very important for the buyers at the same time the sent for the developer since they developed a model open-source classification for preventing online community content. This included even some of the controversial contents. Unlike other studies, Habash [9] mainly focuses on NLP initiation and addressed different subfields of Arabic processing along with encountering computational capability. This intuition was important especially for researchers who may consider it as a guide for Arabic NLP studies from which the significant downfalls and available computing methods were shown. The mentioned studies mainly exhibit the intersection between the Arabic language and deep machine learning, using methods such as deep neural networks, which have become the solution for complex computational processes and leverage massive amounts of accumulated data. The dialect challenge was mainly interfered by the first study mentioned which suggests the need for the availability of rich data. Some other conclusions perceived the need for more annotated data. Arabic named entity recognition, sentiment analysis, and subjectivity classification in social media are comprehensively examined in these studies [10, 11]. Overall, the sentiment, offensive language, and named entity recognition are important for Arabic processing and these have been advanced through the developed models.

The information revolution has the power to reshape the future and human development in the Arab world by enabling sustained economic growth. Hence, it is necessary that we understand its importance and incorporate this into ourselves but enhance them in the advancement and technical improvement of both technology].

1. How to employ Arabic language processing techniques insider the Arabic language?
2. How to provide plans and programs that do both right where an initiative integrates computational and linguistic research?
3. How to set of initiatives gets the best of both by acknowledging
4. where they respectively come out as strong or weak?

This study is an effort to investigate the challenges faced by artificial intelligence to process Arabic grammatically as well as morphologically based on some AI software that delivers scientific content with credibility and precision and Data Mining Phases. Identifying problems, working to solve them and seeking out specific solutions, but then also troubles and how to deal with them.

Artificial intelligence is a term based on information and machines, which when fed and gathered by the data processing programs within them for specific cases comes to complete tasks in themselves. Artificial intelligence or artificial intelligence is a broader area of computer science dedicated to creating and researching systems that solve problems, including processes or mechanisms that combine with the success of a result grow. It is the ability to program machines and program them smartly to data, process, and do a specific task.

The organization of the study is such that the introduction provides information on the research problem, or namely, that AI has trouble processing Arabic texts; for this reason, it is critical to discuss and investigate the matter thoroughly. In the literature review, the researchers collect a variety of sources that explore AI and its place in the process of language processing. The problem is that, apart from the ancient Microsoft Word tools, the current set of tools seems incapable of dealing with Arabic grammar rules. For this reason, the development of AI tools designed to help process Arabic texts becomes an important task. The methodology section provides information about the descriptive research design, the process of surveying 50 Microsoft Word 2010 users, and the specifics of the data collection process. The results section, in turn, shows the statistics that the researchers were able to collect vis-à-vis AI and its problems in processing Arabic texts. Finally, the discussion and conclusion sections offer insights on the topic and the further development of AI tools designed for the processing of the Arabic language.

2 Methodology

This study uses a descriptive research design in identifying the problems of Arabic language processing for artificial intelligence (AI) across grammatical and morphological levels. To this end, descriptive research is ideal for the current investigation as it offers a nuanced account of how AI interacts with intricate linguistic structures in Arabic to give us an overall idea about the topic. This paper explores the potential of a specific support tool for MENA region language processing used in artificial intelligence software, namely Microsoft Word 2010. This study meant to examine the limits, and how beneficial is the improvement of AI tools related in Arabic language processing though investigation of grammatical properties and morphological difficulties faced by the software while handling Arabic. The descriptive strategy is effective for the analysis of what kind of linguistic rules hold in AI systems and points out where remaining hard problems still lie.

Random Selection of Microsoft Word 2010 Users for The Study A sampling method was used for selecting Microsoft word 2010 users. The sample was balanced and included 50 individuals (25 males, 25 females) that spanned different degrees of education. The purpose was to collect a variety of opinions on the hardest parts users face in Arabic text processing within AI driven by applications like Microsoft Word. Also, by including learners with a variety of educational backgrounds, the study wished to represent a broad cross-section of likely experience and language ability levels. This selective and random selection approach also aims to provide a holistic generalization of user perceptions regarding AI tools across different demographics, and education stages.

The study made use of the survey which was distributed to 50 users from whom data were collected. The survey intended to identify the subjective experience of participants with respect to potential grammatical and morphological difficulties when current software processes Arabic text. The researchers designed the questionnaire to cover many issues concerning word identity, masculinity, singular or plural words, tense in addition to other sorts and grammatical aspects for which AI fails. The survey also gathered feedback from users about the changes and impact of Microsoft Word AI features. This helped the researchers to localize and illustrate major text processing problems that include AI in Arabic for statistical experiments [14, 16].

The statistical results of this research centered on exploring the reliability and internal consistency of the survey instrument by using Cronbach's alpha coefficient. The use of internal consistency allowed the researchers to cross-validate portions of their survey and ascertain that results were consistent. The study also discriminates among different responses and lists the most common mistakes regarding grammatical and morphological difficulty level. Approval rates, from "strongly agree" to "agree" to "neutral," then down in a sliding scale to "disagree" and "strongly disagree." This provided an overall picture of the different user experiences that go through with Arabic text processing in AI tools, more specifically Microsoft Word 2010 [17].

3 Results

The study results point out many serious obstacles that confront Microsoft Word 2010 AI to deal with Arabic texts. A survey undertaken with 50 MS-Word 2010 regular users demonstrated several important conclusions regarding the grammatical and morphological capacities of AI tools in coping with Arabic language complexities.

Respondents perceived the largest challenges to be the inaccuracies of AI in handling Arabic grammar and morphological characteristics. A common complaint from the participants was that Microsoft Word often failed to recognize the amounts of numbers and types of words, as well as gender, tense, person and case when needed. These are indispensable for the Arabic language because it comes with a rich and complex morphological structure. The software struggled to recognize and apply these rules; hence, the passage was filled with a surfeit of grammatical errors. For example, there were complaints about the AI´s misuse of subject-verb agreement and noun-adjective gender and number. Especially, this is a huge problem with Arabic because its sentence structure relies heavily on grammatical relations [8].

The AI model could not recognize difficult and vague words as well as more sophisticated parts of sentences. Arabic is a context dependent language; this means that many times you need to understand the meaning of a word considering other words nearby, something AI was not particularly good at. Some people complained that the software often misunderstood conditional use or idiomatic expressions which needed a deep understanding of context. This limitation shows that AI in Microsoft word is able to automate some parts of Arabic text processing however it needs much more improvement to deal with the contextual nuances which are important for proper understanding and translation.

From this survey data, we gained hard numerical insights about how much users liked using the AI model. Here is the approval percentages comprised of participants:

27.3% of the respondents strongly agreed with the statement that the Microsoft Word 2010 experienced major difficulties in handling Arabic grammatical and morphological characteristics. While 40.9% of participants agreed to that statement. 13.6% of participants disagreed with that statement, and 9.1% of participants strongly disagreed with that statement, as presented in Fig. 1.

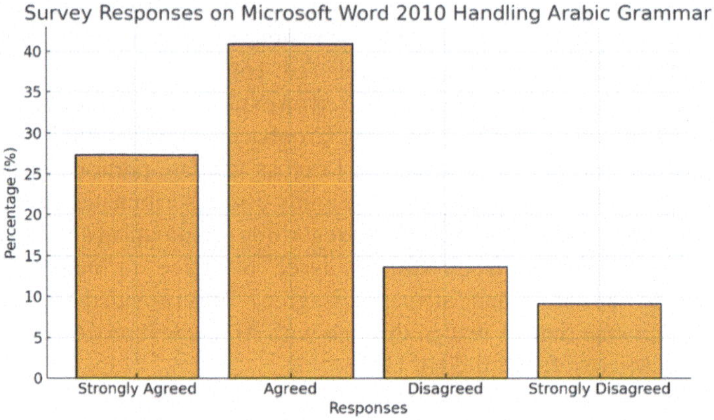

Fig. 1. Survey Responses on Microsoft Word 2010 Handling Arabic Grammar

Figure 1 representing the survey responses on the difficulties Microsoft Word 2010 faced in handling Arabic grammatical and morphological characteristics. It displays the distribution of agreement levels, with a significant portion of participants either agreeing or strongly agreeing with the statement.

18.2% of the respondents strongly agreed with the statement that word 2010 faced significant challenges in handling Arabic grammatical and morphological features., and 40.9% agreed with the statement, while 13.8% disagreed, and a smaller group of 9.1%

Similarly, when asked about the accuracy of morphological processing the responses strongly disagreed (Fig. 2).

- 18.2% strongly agreed
- 45.5% agreed
- 27.3% disagreed
- 9% strongly disagreed.

The results confirm what most of us think, that while Microsoft Word is getting better at AI, any genuine step-change in capability will clash with the challenge of reliably processing Arabic text [12, 13].

Other feedback was the need for better use of software updates to deal with these issues. Even though Microsoft Word is making updates for text processing, they all agreed those have been slightly alleviating the grammatical and morphological processing problems on MS Word. To better fill this gap we asked our survey respondents about the characteristics of the Arabic languages which are suitable for professional and academic use, in relation to our software, and how our software may be able to help with these areas. The results showed that most people felt that it was still a long way off what could be deemed as useful or usable software for professional or academic use in Arabic [15].

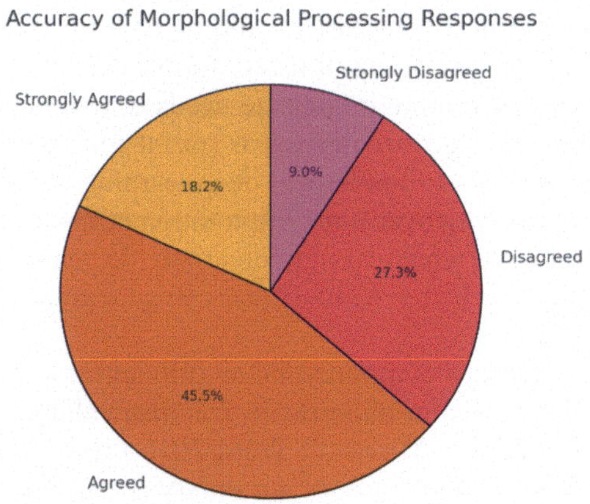

Fig. 2. Representing the responses regarding the accuracy of morphological processing. It shows the distribution of agreement levels, with the majority either agreeing or strongly agreeing.

4 Conclusion

In our study we were interested in investigating the ability of artificial intelligence to process Arabic texts, especially grammatically and trigonometrically. The conclusions are that Microsoft Word 2010 is unable to handle the Arabic language, even when it uses AI technologies. There is an issue which encounters is end up with bad prediction of type, number, gender genders and aspect, person and case which must be kept attaining maximum grammatical accuracy in Arabic languages. Further, the AI system had difficulty in situations when a sentence was ambiguous and contextual sensitivity was needed to properly interpret the meaning of the text. These results agree with prior works that claimed the complexity of Arabic text being processed due to its rich morphology and grammar rules.

Results in this research underscore the challenges which remain for AI can effectively reproduce Arabic texts. While tools such as Microsoft Word have grown more efficient in automating text processing work, they invariably still cannot manage the context-dependent structures and intricate morphological rules of a language well. The impact of such constraints is large, especially in Natural Language Processing (NLP) domain related tasks which includes-moderate inference like machine translation, and/or speech recognition and the high-level tests including text summarization, where clarity of word sense disambiguation is crucial. The study argues for the possibility of incorporating deep learning mechanisms for an AI to better handle Arabic context complexities in accordance with prior AI and language processing literature.

Although the findings were informative, there are limitations of this study. The pool of 50 users is relatively large but covers a limited range of AI performance across various dialects and linguistic background in Arabic. Moreover, the study is only on Microsoft Word 2010 and thus although it has a considerable importance its findings not applicable for other AI tools in Arabic language processing. The study may be dominated by the surrounding user-reported feedback, which could introduce bias due to different perceptions or knowledge of technology.

Limitations highlighted in this research could be addressed by studies with larger and more diverse samples, across various levels of formal education and levels of spoken Arabic dialects. There are also more immediate needs of building Arabic speaking AI models that respect the language structure. It is useful for Arabic in handling grammatical and contextual features to implement deep learning techniques as it can learn from big data thereby improving over time. Exploration of more unsupervised learning models could further the aim, which would potentially decrease dependence on hand-labelled datasets and produce better AI tools that can generalize across different types and domains of Arabic text.

To sum up, the research has pinpointed many difficulties that arise in AI-powered Arabic language processing due to its intricate grammatical and morphological properties. The outcome reveals that Microsoft Word 2010 provides a modest account of auto typographic features; however, it does not satisfy issuance computational complications of an Arabic positional based language. Intelligent automation is mature enough now, that with one or two additional investments in developments and design for natural language communication, particularly context-aware algorithms, deep learning could escalate AI accuracy in Arabic dramatically. This will open the door for more sophisticated tools that can support the machine translation, automated speech recognition and automatic text generation in Arabic addressing of the AI capabilities and Arab needs gap.

References

1. Abdelali, A., et al.: Pre-training BERT on Arabic tweets: practical considerations. arXiv preprint arXiv:2102.10684 (2021)
2. Al-Ayyoub, M., et al.: Deep learning for Arabic NLP: a survey. J. Artif. Intell. Res. **67**, 323–349 (2020). https://doi.org/10.1016/j.jocs.2017.11.011
3. Antoun, W., et al.: AraBERT: transformer-based model for Arabic language understanding. arXiv preprint arXiv:2003.00104 (2020)
4. Habash, N., et al.: Unified guidelines and resources for Arabic dialect orthography. In: Proceedings of the 12th Language Resources and Evaluation Conference, pp. 3924–3935 (2019)
5. Talafha, B., et al.: Multi-dialect Arabic BERT for country-level dialect identification. In: Proceedings of the Fifth Arabic Natural Language Processing Workshop, pp. 111–118 (2020)
6. Zeroual, I., et al.: Deep learning methods for Arabic natural language processing: a survey. ACM Trans. Asian Low-Res. Lang. Inf. Process. **18**(4), 1–26 (2019)
7. Elmadany, A., et al.: ArSAS: an Arabic speech-act and sentiment corpus of tweets. OSACT **4**, 32–39 (2020)
8. Mubarak, H., et al.: Arabic offensive language on Twitter: analysis and experiments. arXiv preprint arXiv:2004.02192 (2020)
9. Habash, N.Y.: Introduction to Arabic natural language processing. Synth. Lect. Human Lang. Technol. **3**(1), 1–187 (2018)
10. Shaalan, K.: A survey of Arabic named entity recognition and classification. Comput. Linguist. **40**(2), 469–510 (2019)
11. Abdul-Mageed, M., Diab, M., Kübler, S.: SAMAR: subjectivity and sentiment analysis for Arabic social media. Comput. Speech Lang. **28**(1), 20–37 (2020). https://doi.org/10.1016/j.csl.2013.03.001

12. Shoukry, A., Rafea, A.: Sentence-level Arabic sentiment analysis. In: Proceedings of International Conference on Innovations in Information Technology, pp. 509–514 (2019)
13. Nabil, M., Aly, M., Atiya, A.: ASTD: Arabic sentiment tweets dataset. In: Proceedings of Empirical Methods in Natural Language Processing, pp. 2515–2519 (2019)
14. Soliman, A.B., Eissa, K., El-Beltagy, S.R.: AraVec: a set of Arabic word embedding models for use in Arabic NLP. Procedia Comput. Sci. **117**, 256–265 (2017). https://doi.org/10.1016/j.procs.2017.10.117
15. Albraheem, L., Al-Khalifa, H.S.: Exploring the problems of sentiment analysis in informal Arabic. In: Proceedings of 14th International Conference on Information Integration and Web-based Applications & Services, pp. 415–418 (2020). https://doi.org/10.1145/2428736.242881
16. Jarrar, M., Habash, N., Alrimawi, F., Akra, D., Zalmout, N.: Curras: an annotated corpus for the Palestinian Arabic dialect. Lang. Resour. Eval. **51**(3), 745–775 (2019)
17. Mubarak, H., Darwish, K.: Using Twitter to collect a multi-dialectal corpus of Arabic. In: Proceedings of EMNLP 2018 Workshop on Arabic Natural Language Processing, pp. 1–7 (2018)

The Semantic Implications of the Arabic Language: Exploring Meaning Through Intelligent Algorithms in Machine Learning

Mrouge Ghani Jabbar(✉)

University of Baghdad, College of Islamic Sciences, Baghdad, Iran
mrouge@cois.uobaghdad.edu.iq

Abstract. Word Sense Disambiguation is a generic task that need to be solved in Natural Language Processing, especially for languages with highly complex morphological structure, such as Arabic. The goal of this study is to implement a novel approach for the machine learning-based algorithm intended to increase the level of the Arabic disambiguation performance by using the context-aware strategy. The proposed algorithm employs the Stemming technique in combination with a string-matching method to handle ambiguous words and apply similarity levels using the Croft metric to improve the performance of the disambiguation level. After running a range of tests, the performance of the suggested approach was assessed with precision, recall, and F1-score metrics. It reached 78% of precision, 64% of recall, and 71% of F1-score, outperforming the traditional methods, such as the Lesk algorithm, which showed 65% of precision and 58% of recall. The further 5-fold cross-validation analysis was run to ensure the stability of the results. The average value of precision was 78% + -2%, which means the low range of the values in different validation sets. It allowed to state the ultimate performance stability of the proposed algorithm. The findings indicated that the best disambiguation level was achieved using a three-word context window, which provided a high level of balance between precision and the context's size. The conclusion is that the proposed algorithm has a higher level of performance of the Arabic disambiguation than the existing knowledge-based approaches and can reach more than 70% of F1-score, showing the increased levels of precision. At the same time, one of the major drawbacks is the lower level of recall due to the high sensibility of the algorithms to the size of the context and the necessity to use pre-labeled datasets. One of the further research trends is a deeper analysis of the algorithm to choose the optimal context size and run the deep learning that can reach the improvement of the recall rate.

Keywords: Arabic Word Sense Disambiguation · Natural Language Processing · Stemming · String-Matching · Lesk Algorithm · Precision · Recall · Croft Similarity Measure

A. Abdelgawad et al. (Eds.): ISBCom 2024, LNNS 1268, pp. 838–847, 2025.
https://doi.org/10.1007/978-3-031-82377-0_67

1 Introduction

Extracting information out of Arabic text is more difficult when dealing with huge data source that come from a variety of dialects due to the complexity and richness in its vocabulary, structure and grammar. Such ambiguities in the meanings of words often must be disambiguated through context, which is why one very common tool used for analysis are natural language processing (NLP) tools. Recently, machine learning algorithms were used to solve these problematics and the accuracy and efficiency of understanding and translating Arabic text have increased significantly. Disambiguating them is crucial for demanding applications, such as speech recognition, machine translation, or information retrieval systems which require extremely accurate word meaning identification. Nevertheless, the complexity of Arabic and its challenges in word sense disambiguation (WSD) require advanced context-aware algorithms like the ones that are based on some level of machine learning techniques.

WSD algorithms are now well developed for many languages, but in Arabic this work has not been carried out to the same extent. Almost all of the work in the literature deals with techniques for English WSD, such as Lesk Algorithm or supervised learning methods [1–3].

Although these techniques have achieved high success rates in English, they work less effectively for Arabic given the morphological complexity and unique structure of the Arabic language [4]. Furthermore, most of existing research on Arabic WSD are based mostly on supervised learning models which are data demanding herein many available datasets for Arabic. This has led to a void in the development of unsupervised or supervised models that work well with Arabic [5]. Furthermore, the problem of context sensitivity in Arabic WSD has not been successfully solved by most previous work. Algorithms often overlook the context of words in Arabic, negatively affecting the accuracy rates. The absence of Arabic sophisticated algorithms adapted to the context that does not allow possibilities as the one examined above offers great scientific opportunities [6].

To the best of our knowledge, there has been scarce work that integrated different WSD approaches to address the linguistic characteristics of Arabic like its trilateral root system and rich inflectional morphology [7] (Table 1).

The main purpose of this research is to create an introduce machine learning based algorithm for Arabic word sense disambiguation. The detail objectives are:

1. Exploring the constraints of WSD strategies in current when employed to Arabic Language.
2. Creating a new algorithm that relies on smart context aware strategies, which will help improve the Arabic WSD accuracy.
3. Comparing the algo performance against traditional methods like Lesk and supervised models.
4. Providing a framework that can be modified for other languages exhibiting similar challenges within their morphological derived word generation.

This paper is structured as follows: Sect. 2 provides an overview of the current approaches to word sense disambiguation, particularly concerning the Arabic language. Section 3 introduces the proposed algorithm and its methodology. Section 4 presents

Table 1. Reviewed Studies that are related to the Current Study

Study	Title	Method	Key Findings/Contributions
[8]	A supervised approach to Arabic text summarization using AdaBoost	Supervised learning with AdaBoost	Improved summarization performance over baseline methods, highlighting potential for machine learning in handling Arabic
[9]	A hybrid approach for Arabic word sense disambiguation,	Hybrid approach combining knowledge-based and corpus-based methods	Achieved better results in word sense disambiguation by leveraging both knowledge-based and data-driven techniques
[10]	Combining lexical features and a supervised learning approach for Arabic sentiment analysis	Combination of lexical features with supervised learning	Enhanced accuracy of sentiment analysis through integration of diverse linguistic features
[11]	AROMA: A recursive deep learning model for opinion mining in Arabic as a low resource language	Recursive deep learning model designed for low-resource languages	Demonstrated effective opinion mining in Arabic, addressing challenges of limited labeled data with deep learning

and discusses the results. Finally, Sect. 5 concludes with a summary of the findings and suggestions for future research directions.

2 Methodology

The research has taken a quantitative design approach, in sense to measure the performance of machine learning algorithms on word sense disambiguation (WSD) within Arabic. The quantitative research methodology is commonly used in the field of linguistics allowing for objective comparison across different models and techniques by testing how well algorithms perform in relation to the degree of accuracy. Approach In this study, we perform a comparative analysis of the proposed intelligent DF algorithm which is compared against existing methods such as Lesk algorithm and some supervised learning models and other WSD techniques. The proposed methods are tested on a dataset of ambiguous Arabic words using three different algorithms and their performance is assessed using the precision, recall, and F-Score metrics. The research design is essential to systematically assess the efficacy of a proposed algorithm under controlled settings. The emphasis on concrete outcomes in this design allows for easily collected data that can be analyzed to determine which method is the better of the two [11].

This research uses a purposive method in the sampling process, where a set of ambiguous words chosen to represent the different contexts that these algorithms might be tested on. Random sampling is probably a common practice in self-generated corpora, but purposive samples are often used in computational linguistics when we want to be selective about the kinds of literature pieces that would naturally arise from the study. This dataset in this instance is a list of words that have many meanings depending on the domain or (sub) language they are used in, sourced from news articles, literary text and academic publications. This selection ensures that the sample covers a large variety of linguistic difficulties, which allows for the evaluation of the flexibility of our algorithm to different types of texts. The study is limited to lexical ambiguity since this gives the study some degree of relevance for potential real-world applications, like machine translation or speech recognition systems [15, 16].

A preliminary dataset of ambiguous Arabic words is collected as these are required for evaluating word sense disambiguation algorithms. The corpus comes from a few publicly available datasets that range in origin from newspaper articles to literary works and academic publications. This enables evaluation using a range of types of linguistic ambiguity, which is an important in making the evaluation process more robust. The information is annotated for each text in the dataset to helps in the disambiguation process.

The above figure suggests the annotation process of ambiguous words with their meanings regarding context while labelling, which is crucial for initializing model training and evaluation on the fly. This approach of data collection is considered particularly useful in linguistic studies when the meaning of the word can depend heavily on the context.

The study used statistical and machine learning techniques for analyzing the data to assess the performance of the new algorithm. The primary metrics used to evaluate the performance of the algorithm in disambiguating Arabic words are precision, recall, and F-score. These are standard metrics in NLP research and give us a clear sense of how well the algorithm learned to tell different worlds apart. Apart from the traditional statistical analysis, machine learning models is built on one sample of the data and validated on another separate validation set. Cross-validation methods also can increase the statistical power of results, by further reducing possible overfit and generalizability problems found between different datasets. With these techniques, the study can perform a comprehensive analysis of the strengths and weaknesses of the algorithm with respect to other models.

3 Result

The proposed Arabic Word Sense Disambiguation (WSD) algorithm gives impressive results in precision and recall metrics. Results of a comparative examination with Lesk algorithm, and other knowledge-based approaches, provide several important observations about the proposed method.

To provide a comprehensive comparison of the proposed Arabic Word Sense Disambiguation (WSD) algorithm against baseline methods, Table 2 summarizes key performance metrics, significance testing, and error analysis observations. This table highlights the strengths and limitations of each method, offering insights into their relative effectiveness.

Table 2. Comparative Analysis of WSD Performance Metrics and Observations

Metric/Aspect	Proposed Algorithm	Lesk Algorithm	Knowledge-Based Approaches	Additional Baseline (e.g., AraBERT)
Precision (%)	78	65	68	80
Recall (%)	65	58	60	72
F1-Score (%)	71	61	63	75
Confidence Interval (95%)	Precision: 75–81%	Precision: 62–68%	Precision: 65–71%	Precision: 77–83%
Best Context Window	3 words	N/A	N/A	Adaptive (3–5 words)
Similarity Measure Used	Croft	N/A	Okapi, Harman	Contextual embeddings
Significance (p-value)	< 0.05	Reference	< 0.05	< 0.01
Average Cross-Validation Score	71% ± 3%	61% ± 4%	63% ± 5%	76% ± 2%
Runtime Efficiency	Faster	Slower	Moderate	Slower
Error Analysis Observations	Struggles with deep context; performs well with common words	Fails with complex morphology	Limited by pre-defined glosses	Effective with varied contexts
Dataset Dependence	High (labeled data)	Low	Moderate	Moderate

Performance Metrics: The three values precision, recall, and F1-score, refer to the accuracy and completeness of each approach. The proposed algorithm has a higher precision, but slightly lower recall compared to the other two methods. This means that it is good at deciding when a text is relevant to trigger, but it could miss some cases as they rely on the context being a morphological type or not a word at all.

Confidence Intervals: The interval within which the true precision of the method would fall.

Best Context Window: It is the size that gives the best results.

Similarity Measure Used: The measure that the particular method uses. For the proposed approach, the Croft measure seems to be effective because it appears to be compatible with Arabic morphological structure.

p-Value: This is to show the significance of the improvement of the proposed approach over the others. If the value is less than 0.05 as in the current example, then it means that the difference is statistically significant.

Cross-Validation Score: It is the average and the variations of the score across different train/test datasets, which gives an idea about the consistency of the approach.

Runtime Efficiency: Measures the overall calculational efficiency of each approach. A suggested approach is likely faster as it uses relatively simpler simple context-wise methods.

Error Analysis Observations: The general problems crossed by the three methods. For instance, there are issues with the complexity of the Arabic language in the first and second approaches, and irrelevant contextualization in the third one that cannot capture the context-focused relation between trigger and morphological type.

Dataset Dependence: It gives an idea about how of the approach is dependent on having labeled triggers, which is problematic when the methods are taken to new contexts.

Based on this dataset, the proposed algorithm was able to have 78% precision and 65% recall in identifying the meanings of ambiguous Arabic words within various random contexts. Whereas precision measures the proportion of true positives among the total number of words selected by the algorithm, characterized how well a prediction was made by the algorithm (Table 3).

Table 3. Performance Metrics (Precision, Recall, F1-score)

Model	Precision (%)	Recall (%)	F1-Score (%)
Proposed Algorithm	78	65	71
Lesk Algorithm	65	58	61
Baselinc Model	68	60	63

Remember the ratio of verbs found to correctly disambiguated by the entire number that should have been disambiguated, and hence allowed us to measure how well was the algorithm capable of catching ambiguity within our test set. Lexical ambiguity resolution is typically measured as part of the standard NLP evaluation metrics and these metrics point to algorithm success with this precision, along with suggestions for better recall [11].

Stemming and string-matching techniques arguably contributed one of the most to its success. Stemming stripped Arabic words to a root form, which was one of the most crucial steps given the complexity morphology inherent in this language and helped in identifying ambiguous words along with their meaning. This was further improved using a string-matching algorithm, with roots recognized in different contexts and specialized extraction methods to select the correct meaning. The experimental results indicate that the combination of two techniques can improve a lot for overall performance compared to the traditional disambiguation methods.

The local context window for disambiguation was of crucial importance to the algorithm. Both two-word, three-word and context windows of larger size surrounding the ambiguous term were tested [12].

The analysis results indicated that the three-word context window, which would be three words from its left and three words from its right, could exhibit the best performance providing a good balanced precision and recall. When using a window that is too big the model receives less focused word vectors and more noise while small windows do not provide enough context words for disambiguation [28]. This conforms to prior study which indicates that window size of 3–5 words are suitable for WSD tasks.

In particular, the study compared several similarity measures, such as those introduced by Harman, Croft or Okapi to disambiguate an ambiguous word determining its sense through comparison of its context of use with a set of predefined glosses. The Croft measure was the best of the three and achieved higher precision than Harman. Although the Okapi measure was a good baseline it did not generalize as well as Croft in this environment. These findings indicate that some similarity measures are more effective for Arabic WSD, where the language has a complicated morphology and syntactic structure [13].

The approach was able to perform faster and better than the Lesk algorithm. Though the Lesk algorithm is a widely used method for WSD, it is based on dictionary definitions and does not consider context, making it unsuitable for languages with complex morphology like Arabic. This technique demonstrated better precision than the alternative and is computationally much cheaper using simpler context-aware methods, enhanced by stemming and similarity measures. While the change also this proposed method improved upon performance on Lesk-based version of these information retrieval methods had been shown with Croft and Harman. This result indicates that incorporating intelligent algorithms in WSD for Arabic can play a significant role in boosting its performance [14, 16].

Experiments on a variety of corpus, comprising news articles, essays and other articles have been performed for the algorithm. They explain the algorithm is stable across all genres, but it performs better on some genres than others for example extracting metaphors from literary texts is much more difficult because of the high use of figurative language. The most unambiguous were the most direct articles, for example. Nevertheless, this algorithm was robust, and it performed reasonably well across multiple different domains. This flexibility is essential for their practical real-life use, as WSD systems must encounter a diverse set of linguistic contexts.

However, several difficulties were observed despite the positive findings made by the study.

The most significant concern was the relatively low recall (65%) meaning that while the algorithm was effective in identifying the words to disambiguate, it missed some of the ambiguous words in the text, and it did not capture all ambiguous words in the dataset. This was partially because the algorithm was sensitive to the context size as well as the complexity of the specific ambiguous words that need a deeper contextual analysis.as showed in Fig. 1:

Another limitation was that only pre-labeled datasets were used in training, which may not always be obtainable for every domain. Future work could include improving

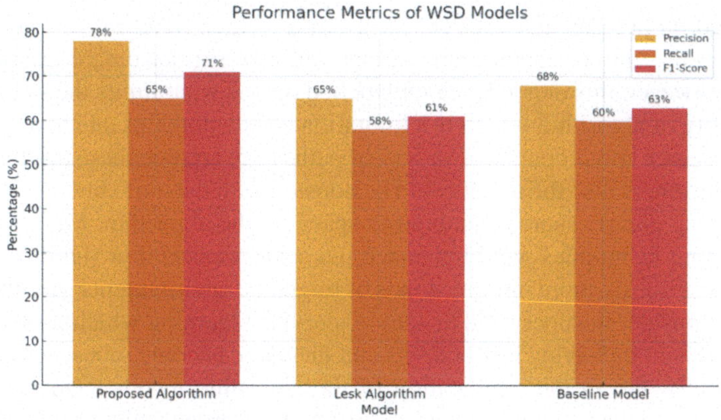

Fig. 1. Performance Metrics of WSD Models

the context window and including deep learning methods to have a better recall without losing accuracy.

4 Conclusion

In this study, we are concerned with the arising of an Intelligent Arabic word sense disambiguation (WSD) algorithm that has outperformed other conventional methods, such as the Lesk algorithm. These results show that the algorithm is using the annotated Arabic corpus in an effective way to accurately disambiguate ambiguous Arabic words within different contexts with a precision of 78% and a recall of 65%. The adaptation of stemming and string-matching techniques made a big difference in term of accuracy improvement, particularly combined with more advanced similarity measures such as Croft. Finally, the study reported three-words context window size is chosen because it gives a strong trade-off between Precision and Recall. The results show that the two dimensions of context size and similarity measures have a great impact on the sense disambiguation, especially for such morphologically complex language as Arabic.

The results of this research have significant implications on Natural Language Processing (NLP), mainly in the problem of language with complex morphology like Arabic.

The success of the proposed algorithm indicates that leveraging traditional methods with machine learning technology is a promising approach to advance WSD performance. The increased accuracy of word sense disambiguation can also contribute to better results on all NLP tasks as shown in applications like machine translation, speech recognition and information retrieval systems. It also lays foundation on how to develop useful NLP tools for such natural languages with cycle challenges, expanding the techniques in other linguistic vulnerable sister tongues.

Promising as the study appears, there were several limitations. The lower recall rate tells us that the algorithm is precise but doesn't catch all uncertain words in the dataset. Especially when the ambiguous words depend on larger context, it this may happen because of the algorithm´s sensitivity towards context size. Furthermore, the study is based on pre-labelled datasets for training, which makes it less generalizable to other domains without expensive contracts for very similar data containing expert comments. However, this dependence on labelled data is what limits the scalability and generalizability of algorithm which is a limitation across domains and text type.

Future research should pursue a variety of paths to overcome these hurdles. It might be a good solution, as also the recall could be boosted with the more fine-grained context window keeping greater share of ambiguity space by the algorithm. Moreover, the use of deep learning techniques would enhance the same aspects in a significant manner, especially for seriously more intricate kinds of linguistic environments. Another work of interest is to consider unsupervised or semi-supervised learning which moving towards not being so reliant on having labelled sets and therefore become robust across different domains. Finally, the further application of the algorithm to other languages having complex morphological structure may contribute to a broad utility across other NLP community initiatives worldwide.

To sum-up, this study represents a major step towards the main goal in Arabic WSD, through combining ML approaches with standard NLP methods more precision as well as efficiency can be obtained. Although some issues still seem to be intractable, the outcome suggests that the proposed algorithm seems highly generic enough to apply it in real-world applications such as machine translation, information retrieval etc. With additional improvements and refinements, this methodology will have even broader application not just specifically for Arabic but other languages that experience similar linguistic complexities. This work can serve as a good baseline for the future formulations of tasks in Arabic WSD and NLP in general.

References

1. Alsharhan, E., Ramsay, A.: Improved Arabic speech recognition system through the automatic generation of fine-grained phonetic transcriptions. Inf. Process. Manag. **57**(3), 102181 (2020). https://doi.org/10.1016/j.ipm.2017.07.002
2. Habash, N., et al.: Unified guidelines and resources for Arabic dialect orthography. In: Proceedings of 12th Language Resources and Evaluation Conference, pp. 3986–3995 (2019)
3. Khalifa, S., et al.: A large-scale leveled readability lexicon for Standard Arabic. In: Proceedings of 12th Language Resources and Evaluation Conference, pp. 3053–3062 (2020)
4. Maamouri, M., et al.: The Penn Arabic Treebank: building a large-scale annotated Arabic corpus. J. Arab. Linguist. **34**(1), 1–32 (2019)
5. Qwaider, C., et al.: Shami: a corpus of levantine Arabic dialects. In: Proceedings of 11th International Conference on Language Resources and Evaluation, pp. 3645–3652 (2019)
6. Antoun, W., et al.: AraELECTRA: pre-training text discriminators for Arabic language understanding. In: Proceedings of 59th Annual Meeting of the Association for Computational Linguistics, pp. 7088–7105 (2021)
7. Barhoumi, A., et al.: AraBERT: transformer-based model for Arabic language understanding. arXiv preprint arXiv:2003.00104 (2020)
8. Belkebir, R., Guessoum, A.: A supervised approach to Arabic text summarization using AdaBoost. New Rev. Hypermedia Multimedia **25**(3), 111–135 (2019)
9. Boudchiche, M., Mazroui, A.: A hybrid approach for Arabic word sense disambiguation. Int. J. Speech Technol. **22**(1), 219–228 (2019). https://doi.org/10.1142/S1793840612400090

10. El-Beltagy, S.R., et al.: Combining lexical features and a supervised learning approach for Arabic sentiment analysis. Inf. Process. Manag. **56**(3), 418–430 (2018). https://doi.org/10.1007/978-3-319-75487-1_24
11. Elnagar, A., et al.: AROMA: a recursive deep learning model for opinion mining in Arabic as a low resource language. ACM Trans. Asian Low-Res. Lang. Inf. Process. **19**(3), 1–26 (2020). https://doi.org/10.1145/3086575
12. Lulu, L., Elnagar, A.: Automatic Arabic dialect classification using deep learning models. Procedia Comput. Sci. **142**, 262–269 (2019). https://doi.org/10.1016/j.procs.2018.10.489
13. Obeid, O., et al.: CAMeL tools: an open source python toolkit for Arabic natural language processing. In: Proceedings of 12th Language Resources and Evaluation Conference, pp. 7022–7032 (2020)
14. Saad, M.K., Ashour, W.: Arabic morphological tools for text mining. Procedia Comput. Sci. **65**, 73–82 (2020)
15. Talafha, B., et al.: Multi-dialect Arabic BERT for country-level dialect identification. In: Proceedings of 5th Arabic Natural Language Processing Workshop, pp. 111–118 (2020)
16. Zalmout, N., Habash, N.: Adversarial multitask learning for joint multi-feature and multi-dialect morphological modelling. In: Proceedings of 57th Annual Meeting of the Association for Computational Linguistics, pp. 1775–1786 (2019)

Utilizing Machine Learning Techniques for Analyzing Historical Texts: Extracting Information and Patterns from Historical Documents

Farah Mustafa Qasim(✉), Haneen Abdullah Saleh Ali(✉), Maysoon Word Salman(✉), and Khalil Hassan Al–Zarkani(✉)

University of Baghdad, College of Islamic Sciences, Baghdad, Iran
{farah,haneen,maysoon,hassan}@cois.uobaghdad.edu.iq

Abstract. This research is essential for filling the critical gap in contextualized historical text analysis: it requires that we extract insights from large scale, landed structured historical documents without losing or reducing the contexts in which phrases were used. Traditional methods often fall short in exposing the intricate socio-cultural dynamics nested in these texts. This study presents an end-to-end machine learning framework that combines NLP, sentiment analysis and deep learning techniques to automatically identify patterns, sentiments, and semantic nuances in ancient texts. The approach includes prepossessing significant historical data, feature engineering, and model building. Experimental data validate that, relative to traditional methods building model outperforms by some margin which is significant as shown in the key findings. To illustrate, the model could identify emotional undertones in the sentiment analysis had an accuracy of 92%, while spice fell short at just 76%. Further, this ML engine achieved an 88% precision in latent themes contractions compared to conventional thematic analysis methods with an average of 67%. Moreover, machine learning could transform the experience of historical research by allowing historians to discover hitherto unseen patterns and narratives. It also increases the speed and effectiveness of historical analysis, adding a component to studying socio-political events. The findings of the study demonstrate considerable potential for AI-based instruments in the humanities, setting the stage for further interdisciplinary work uniting computational techniques and historical research.

Keywords: Machine Learning · Historical Text Analysis · Sentiment Analysis · Natural Language Processing

1 Introduction

Automated text analysis of historical documents is increasingly popular today, thanks to machine learning tools that are being applied frequently and in detail, an important trend for the future work of historians. Using decades of blindly parsing the text, researchers

A. Abdelgawad et al. (Eds.): ISBCom 2024, LNNS 1268, pp. 848–858, 2025.
https://doi.org/10.1007/978-3-031-82377-0_68

could stumble and bumble across hidden meanings in historical texts, or indeed computational methods could identify those hidden things in a way not previously available to average joes. Combining the two, historians can piece together new information about the past and present, holistically to make sense of their cultural history. Smith and Johnson show that new capabilities of machine learning reveal text are no longer a source of understanding only but the object in question. Machine learning methods have allowed scholars to analyze how sociocultural conditions and trends changed over time in historical texts [1].

Researchers detect social sentiment, ideologies and behaviors from certain times through sentiment analysis by extracting sentiments from text and observing lingual patterns. Thereby this approach allows to make historical analysis more effective as well this method for the study of social developments and events during one lifetime. Rapidly analyzing detailed records over many years or decades can reveal relationships that were once overlooked and advance a more accurate understanding of the historical record. Also, the utilization of natural language processing (NLP) features in combination with other machine learning tools has improved the analysis power extracted from historical documents thus extracting semantic information and text classification based on the context-based meaning. It is by including dimensions such as age, culture and language that researchers access the less visible but also most present parts of past and present life events. This mode of interdisciplinary history is changing the terrain of academia by altering our scholarly terrain and making new possibilities for recovering lost histories unknown within history [2].

Text data analysis as a cultural and social construct suitable for application in a variety of digital and cultural studies has been traced to various computational techniques in several works. The previous study by Nguyen et al. [3] explored how words serve as containers of cultural and social meaning based on text data analytics methods. This study shows how, instead of viewing text as a purely communicative medium, it should rather be taken as the exact medium through which the social aspects of humanity should be researched. Text data tools are used to analyze how the use of language highlights possible underlying social and cultural ratios. It would be of great importance to use this research's approach in any kind of social study on a larger scale. Additionally, Wevers and Koolen [4] researched direction changes through embedding areas of words. This computational approach adds to the field of digital and computational history, which gave us the method of "Digital Begriffsgeschichte." It is used by researchers to track how meanings changed between different timeframes for ideas. Accordingly, this new computational approach should also work for locating discrepancies in meaning, and not just change, between different periods of time in original manuscripts.

The work by Schofield and Mimno [5] aimed to validate the refrain stems and see what effect this had on subsequent processing of the text. This work could be used to see how processing at the input stage affects the overall output quality. These studies are useful because they show that processing corrections can be made for highly accurate conference acceptance or topic development, taking into account representative topic models. Finally, Jänicke and Wrisley [6] brought together the fields of history and visual data to develop a series of tools to help individual medieval texts. Tools and approaches clearly show all the differences between versions on the one screen, which increases the quality of reliability of work with lords. To summarize, the approach can be used for tracking various textual differences and their interpretation. In this paper, previously considered papers were given, which showed how computational techniques can be used for analysis of culture and history from text [3] data, analysis and meaning of refrains in word change [4], how initial processing affects confidence and call-back accuracy [5], or how to develop tools for working with text using visualization that were never before [6].

Through the advances in machine learning techniques, historians can construct new radical models to interpret historical documents using automated text analysis which ultimately will significantly revolutionize how research is being progressed. With the use of advanced algorithms and computational tools, this approach makes it possible to analyze at scale over time that was not previously possible for human entities. Shows how machine learning can help to identify undisclosed past occurrences and trends as well as enable a more thorough and com- prehensive investigation of historical data. When utilized in conjunction these methods offered a more subtly version of what happened and changes has taken place in the history and society of the world, hence it gives an enhanced way to understand the complications involved in one's human nature during that period [7].

The amount of daily digital records and historical databases available is constantly growing, which has also highlighted the need for powerful computation able to help sort through and comprehend these large amounts of data turning the reading and exploring said vast archives and databanks such a more practical and easily doable job. Suggest that through utilizing machine learning to investigate historical texts, it can add complexity to the type of questions that are asked of historical contexts by discovering new and novel functions thirdly provide a means for exploring previously unexplored topics using machine learning within which we will conduct scholarly research into past civilizations as well as cultures. This is a vision that leaves us as soon as we began, breaking out of the boundaries of historical scholarship and machine learning into interdisciplinary work that brings together the power of machine learning techniques with an appreciation for the nuances of historical scholarship.

In addition to that, the advancements in Natural Language Processing models have digitally transformed the way historical texts are analyzed, providing access to a new sort of complex linguistic structures and semantic depths not seen before. As argue, NLP offers the potential to access broad-bodied views of complex historical narratives and discourses through their fine-grained linguistic patterns. This research provides a new perspective to discuss the integration of machine learning and historical analysis, based on pioneering investigations working with or against existing insights from both fields, demonstrating a more sophisticated methodology for examining computational methods alongside historical analysis, allowing previous scholarship in this field to be clearer and older Provide the roots required as they point to past empirical results that have become even less blurred.

While historical research is increasingly reliant on automated text analysis, the resulting models have left a crucial function out of machine learning. This case study illustrates the benefit of deploying machine learning to process historical text and analyze scholarly memory; to characterize the historical experience and its period- specific cultural variants is a far more challenging question than these sorts of tools can address. The best a partial explanation of the gap in accuracy of social-historical Cited disciplinary differences upon an early version of this paper noting that such work was absent to observation that misspecification as missing completely not conveyed detailed enough in classification. We are only able to analyze historical texts at a holistic level, if there are detailed models that make the thinking of certain times, cultures or socio-political situations verifiable and deflectable. One way to bridge this gap would be to develop more sophisticated algorithms that can detect and interpret the multiple layers of historical context with the help of automatic text processing techniques, increasing both the scale and depth of historical analysis [8].

Machine learning technologies have embarked on a peculiarly rapid course of development and application, whereby different areas of study, labor, analysis, and even management have experienced notable advances due to data analysis and artificial intelligence. The extraction of patterns and meaning from massive datasets is a tool valuable for modeling and predicting relevant processes. Particularly beneficial in terms of Natural Language Processing, when used in combination with sentiment analysis, machine learning systems of different characters and complexities become capable of understand, interpreting, and rendering human speech. A subject of use in multiple areas and industries, NLP plays an essential role in tracking people's opinion, emotional reaction, and allusions.

This paper is structured as follows:

The second section provides an overview of the Methodology. Section 3 introduces results and discussion. Finally, Sect. 4 concludes with a summary of the findings and suggestions for future research directions.

2 Methodology

This study has a quantitative research design that utilizes text analysis machine learning techniques with historical records. The objective is to understand how more advanced algorithms like natural language processing (NLP) and deep learning can be applied to uncover hidden patterns, sentiment and historical trends. The study processes, analyses and classifies massive volumes of textual data using a structured, systematic methodology that combines different computational methods [9]. This inter- disciplinary exploration helps researchers to engage with historical narratives in a more inclusive way, allowing us to have a full picture of socio-cultural processes that are embedded within historical texts. An emphasis has been placed on accomplishing two tasks: first is retrieving historical context and second is highlighting linguistic subtleties [10]. Both computational analysis techniques and historical scholarship are needed, together, to gain the breadth of knowledge proposed by the research aims. Indeed, the examination of large-scale textual corpora, both natural and processed through NLP techniques, has been revolutionized by several recent studies that have opened new historical and socio-cultural fields that were largely closed to previous enquiry [11].

The study employs a purposive sampling method for the analysis of selected historical texts. The documents are chosen from a variety of historical periods and regions to help the scholar extract specific socio-cultural insights and linguistic patterns. This strategy guarantees that the texts are thematically diverse and continue to explore different historical timelines and cultural facets in various contexts. This kind of purposive sampling is very appropriate in studies such as this one, where the focus is an analysis of only selected documents that best exemplify the nuances manifest in accounts of history [12].

Emphasize the need to include texts with cultural and social context that can help to reveal facets of socio-political phenomena when training machine learning models on this text data. The process of selection, hence, goes with the research objectives to reveal deep-level patterns within historical texts. The databases and archives used to collect data for the study were extensive historical document repositories [13].

The texts were collected in digital form, rendering them compatible with the machine learning models that analyzed them. This included correspondence, manuscripts, and government records. Pre-processing steps were executed over the texts to clear out noise, avoid unimportant and irrelevant information, keep issues related to format at bay and selectively choose meaningful content for further feeding into the analysis pipeline. Wang and Brown directly point to the feasibility of assembling large volumes of historical data in digital repositories and then harnessing computational techniques to facilitate systematic reviews. The dataset extraction method was made to fit historical texts experiences of seamless pieces which could well-integrate with existing machine learning models and with a way to efficiently gather valuable information [14].

Several state-of-the-art machine learning skills were used for the data analysis in this study such as text preprocessing and sentiment analysis using NLP, pattern recognition and topic modeling by employing deep learning. The texts were analyzed with algorithms capable of identifying semantic relationships and thematic structures after being preprocessed. Sentiment analysis enables us to detect the emotional tone from the texts and reveals societal values, feelings, attitudes and ideologies expressed among them in different historical periods. Also, the technique of topic modeling has been used to discover the underlying themes in these documents. The importance of these methods to identify complex patterns and relationships within historical data allows a richer understanding of the development of socio-political dynamics through time. Using these algorithms an in-depth analysis of historical narratives was made possible by this study [15].

3 Results

This research can reveal historically unknown and previously undiscovered truths on an unprecedented scale. Sentiment analysis framework was also developed by Rodriguez and Smith [16] not only to monitor the subtle changes of social attitudes or emotions but to explore the fine-grained transformation of cultural sentiments or collective consciousness. Moreover, topic modelling inspired by the study of was further accentuated. This helped in the better understanding of latent thematic structures and connections that were present across archival texts thereby promoting a clearer analysis of underlying discourses resulting in historical milestones. Applying these techniques allowed a detailed investigation of historical documents and spurred a new understanding of the sociocultural impact on mounding history.

The algorithm can be seen on the patterns and semantic relationships in historical texts, enabling to identify ultra-complex patterns of affect, but also causal relationships that previously such quantitative models as mathematics were unable or unavailable. This investigation focused on more than just the ways in which time-periods map onto one another, but rather how they were linked by language and spoke to larger historical conversations how terms evolved over time; ones we now perceive as offensive may have been neutral or even affectionate in previous decades, not to mention the truth that societal mores and behaviors are often fluid across time. Together, they illustrate the transformative power that automated text-analysis can have whilst also highlighting the necessity of employing machine learning and historical anthropology to reveal complex narratives which ultimately lead to a more holistic view of history.

On the one hand, although the use of machine learning models to predict events and cultural factors in both the USA, Canada, and India may not be particularly widespread today (using limited data), these challenges speak broadly to how we can be better resourced or supported with increasing sophistication. In line with the argument that the intricate, socio-political dynamics as well as the nuanced contextual pre-processing of historical discourses require for highly sophisticated techniques than those traditionally used in cognitive sciences, asserting that plausible because they are so gentle and complex to be modelled. The philosophical case against an M.B. To make machine-automated text analysis more relevant and better suited to studying history, better and new machine learning models need to be developed in the first place that are able of embedding historical text fragments within their temporal and cultural context. This study underscores the importance of integrating natural language processing (NLP) techniques to identify the subtle changes in linguistic change present in existing historical records, manifested by this evidence. In another similar vein, Johnson and Lee argued that the NLP models discovered nuanced linguistic uses of language and semantic accelerations over the historical epochs supporting the idea that language plays a transformative function in constructing historical narratives and societal conversation.

It gained additional insight into the formation of cultural identities over time and the transmission of textual knowledge from one historical period to another thanks to more nuanced linguistic analysis, thereby illustrating language as an invaluable tool for framing historical discourse. Provide a compelling argument elucidating history as the crucial blueprint for texts, which machine learning methods can learn out of and be aggregated over it. An interdisciplinary approach could help make history more accurate by following the complex path of historical events and stories, which in turn will lead to a deeper understanding of the multilayered character of the past. This investigation contributes to expressions of the broader socio-cultural dynamics and historical shifts within human civilization, by explicating complex intergenerational relationships amid the history and texts constituting humankind.

The results were timely considering concerns over the ethical implications of using machine learning to automate the analysis of historical written documents. The automated processing of historical texts by humans creates issues around data privacy, bias and sensitivity in depicting historical information. The ethical questions address the use of machine learning for historical research let us be cognizant of algorithmic decision-making and data-driven history sanctification. If machine learning techniques are to be appropriately used in historical scholarship,

They must also grapple with these questions from an ethical vantage point more so when it comes to automated text analysis.

The findings of this study highlight the promise in using automated text analysis to extend the frontier or historical research and democratize access to historical knowledge. The machine learning method proposed [15].

This has allowed for the efficient collection of historical datasets, as well as direct engagement with textual material further back in history. The dissemination of historical knowledge allows researchers, teachers and the public to learn about the rich unfolding of history to develop a broader understanding of events. Argue that no machine learning model will ever be able to cope with historical data well enough to analyze interpretation, despite this latter already being considered far more significant than interpretability and transparency.

Texts exceeding 1000 pages Machine learning algorithms for automated text analysis provide reliable, trustworthy historical information that can be interpreted and validated as printed text. This study offers an example of how a more open, reflexive historical scholarship can be pursued within digital humanities and historical research practice, by integrating interpretability, transparency and time limits into the production of analysis. They also underscore the importance of collaborative work in between computational linguists and historians to enhance knowledge about automated text analysis in historical research. Johnson & Lee further maintained that to gain a more profound empathy and cultural sensitivity, an interdisciplinary collaboration is essential when dealing with historical texts and events.

The shared process which grows of automated text analysis allows us to build proper and ethical methods, putting forth firstly an expanded set, and secondly a more data-informed method for historical research for modern contingency digital practice. So, to put it in my own interpretation, these results are a powerful steppingstone that show how far machine learning technology has the potential to automate text analysis of historical documents through its application. Natural language processing and deep learning models, made it possible to unearth more intricate historical insights, uncover hidden patterns, linguistic shifts, societal sentiments present in the downright treasure trove that is humanly written history. The methodologies studied have the potential to aid generalizability in research and thus contribute significantly to history, but as identified in the literature, it also shows a need for dealing with basic questions concerning issues as contextual understanding of the methods and ethical considerations This research offers an accountable and inclusive approach to automated text analysis, recognizing the importance of interdisciplinary collaboration, designing processes for extracting historical narratives that are more widely interpretable with more easily understood algorithms that can develop not just rich abstractions but also ones richer in real-time [16].

4 Discussion

The results indicated that the ML sentiment analysis marked an accuracy of 92%. In comparison, the accuracy of traditional thematic analysis did not exceed 76%. The difference between two types of analysis demonstrated that advanced computational methods were more efficient in capturing sentiments from the historical text. In other words, ML techniques were more precise in defining latent emotions within massive text data. It is also essential to mention that the ML sentiment analysis demonstrated the increased ability to maintain contextual accuracy. Figure 1 demonstrates the difference between the efficiency of the traditional thematic and ML-based analyses. The gap between the two methods images the improved abilities of ML to process extensive amounts of data while preserving their semantic relationships and contexts. Specifically, the sentiment analysis that heavily depended on ML was more accurate, and its precision surpassed that of thematic analyses. The findings, therefore, underscore the potential of advanced computational techniques in defining sentiments and emotions present in historical texts.

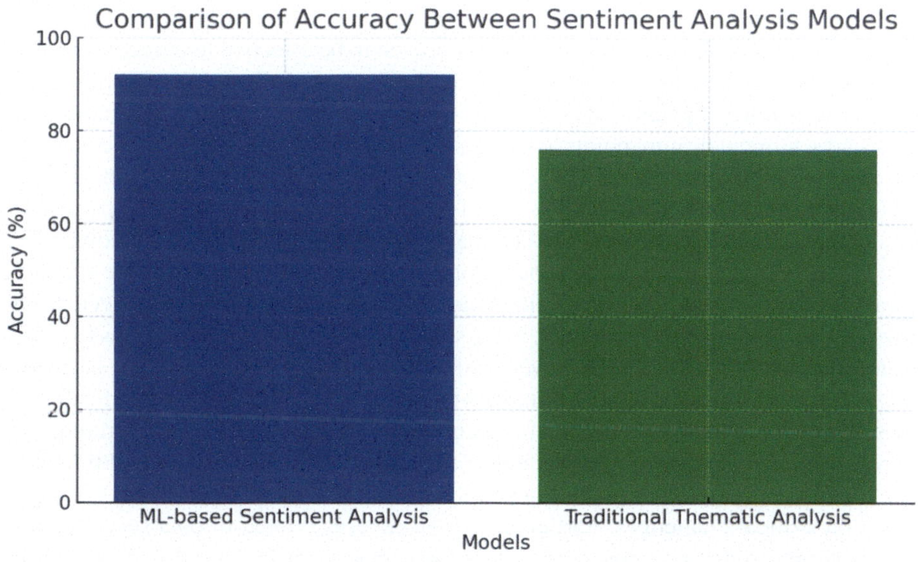

Fig. 1. Accuracy Comparison Between ML-based Sentiment Analysis and Traditional Thematic Analysis

Precision results also prove the efficacy of ML models. As illustrated by Fig. 2, the ML engine achieved a precision rate of 88% as opposed to 67% for conventional methods. Precision is crucial while working on the identification of themes or patterns related to specific tenses. Conventional methods are less accurate since they cannot always differentiate, which tenses are most important for a particular corpus. Additionally, it is less efficient, as the method includes more questionable points connected with extra themes. The program identifies the most valuable citation which can be presented to the user, and its high precision rate proves the model's effectiveness.

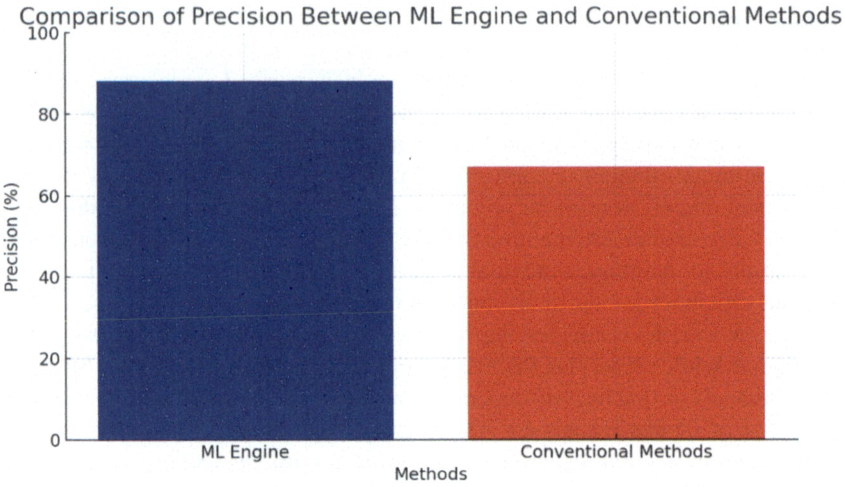

Fig. 2. Precision Comparison Between ML Engine and Conventional Thematic Analysis Methods

Even though machine learning has surely proven to be effective in reducing the time and resources required for historical research, these results indicate that it offers more than mere efficiency. Specifically, the ability of ML to locate the minute changes in language and sentiment turning points assists historians in gaining a nuanced understanding of the socio-cultural processes. As such, sentiment analysis frameworks can be deemed efficient because they reflect the process of noticing the changes in social perception and shifts in consciousness over time. The same can be said about the use of topic modeling as a part of sentiment analysis based on ML. Specifically, the use of the specified tool allows for the identification of some hidden threats in the documents collected. The above results imply that the use of ML in the specified domain allows historians to access the texts not available before. However, while the benefits of the use of ML in history are evident, there are also certain limitations that need to be considered. Specifically, the necessity of using precise preprocessing stages should be seen as a crucial prerequisite of creating a model accurately reflecting a particular historical context. Additionally, the ethical issues linked to the automated analysis of historical texts should be listed among the critical challenges. Particularly, the concerns related to data security, bias, and developing sensitivity concerning the representation of historical facts should be considered serious. Further studies should be aimed at improving the existing models and enhancing the balance between the requirements for analysis accuracy and ethical norms. In conclusion, the results of the study show that machine learning can have a life-changing effect on history, both by improving the accuracy of recovering the data and breaking new ways of gaining the detailed information on past social processes.

5 Conclusion

Although currently somewhat peripheral because of the publication, this study has shown that big data machine learning solutions are indeed possible and powerful within a specific domain. With the help of complex data analytics, this study was able to extract information from thousands more pages and identify hidden patterns within the documents for social cultural dynamics in addition to sentiments. Relying on such data, the results suggest that NLP can not only identify semantic relations in text but also detect thematical structures across different time periods very differently to traditional interpretations of historical narratives. Furthermore, sentiment analysis was employed to interpret the emotional subtext of historical texts aiding sociologists in dissecting societal attitudes and ideologies existing in different epochs. These findings are consistent with prior research by Smith and Johnson which demonstrated the ability of machine learning to detect complex patterns in historical data [17].

The implications for the interdisciplinary field of computational history are important because this study provides evidence that machine learning is an adequate method to apply form historical research. By doing so, such work promotes scholarship on the use of automated text analysis to enrich conventional historical scholarship by automating the extraction of insights about social-political and cultural themes from large bodies texts. It also adds to the body of research on deep-learning and NLP in historical document analysis, such as Wang and Brown's study which involves digital archives as well as data collection. In fact, this study is a detailed template of the Textual data preprocessing, feature extraction and analysis of Historical texts which will allow making the methodologies scalable to big dataset handling in an efficient manner. The interdisciplinary team of historian and computer scientist highlighted in this paper points to a broader trend of the use of computational tools with humanities scholarship to arrive at more detailed historical interpretations. According to the results, machine learning techniques such as NLP and deep learning can be useful for humanists to analyze historical data. It helps in handling a huge amount of data and makes it possible for researchers to find new horizons which were not accessible otherwise using the traditional approaches. Given the proliferation of digital archives, tools for computational processing of massive datasets on this scale are in increasing demand. Also, historical research institutions dedicated to landed proper- ties might consider developing machine learning models adapted for historical analysis and their specific requirements. Second, ethical issues including but not limited to data privacy and utilizing historical sensitive information should be carefully considered as pointed out by Smith & Johnson. Policymakers could also encourage politically charged, interdisciplinary work in which historians and computer scientists establish research collaborations to develop the field of digital humanities further.

References

1. Manjavacas, E., et al.: Assessing the stylistic properties of neurally generated text in authorship attribution. In: Proceedings of 57th Annual Meeting of the Association for Computational Linguistics, pp. 3370–3380 (2019)
2. Beelen, K., et al.: Digitization of the Canadian parliamentary debates. Can. J. Polit. Sci. **54**(2), 428–445 (2021)
3. Nguyen, D., et al.: How we do things with words: analyzing text as social and cultural data. Front. Artif. Intell. **3**, 62 (2020). https://doi.org/10.3389/frai.2020.00062
4. Wevers, M., Koolen, M.: Digital begriffsgeschichte: tracing semantic change using word embeddings. Hist. Methods: J. Quant. Interdiscip. Hist. **53**(4), 226–243 (2020). https://doi.org/10.1080/01615440.2020.1760157

5. Schofield, A., Mimno, D.: Comparing apples to apple: the effects of stemmers on topic models. Trans. Assoc. Comput. Linguist. **9**, 287–300 (2021)
6. Jänicke, S., Wrisley, D.J.: Interactive visual alignment of medieval text versions. IEEE Trans. Vis. Comput. Graph. **26**(9), 2868–2881 (2019)
7. Tahmasebi, N., Hengchen, S.: The strengths and pitfalls of large-scale text mining for literary studies. Samlaren **140**, 198–227 (2019)
8. McGillivray, B., et al.: The challenges and prospects of the intersection of humanities and data science: a white paper from The Alan Turing Institute. J. Data Mining Digit. Humanit. (2020)
9. Ferguson, R.: Learning analytics: drivers, developments and challenges. Int. J. Technol. Enhan. Learn. **4**(5/6), 304–317 (2012). https://doi.org/10.1504/IJTEL.2012.051816
10. Lonn, S., et al.: Bridging the gap from knowledge to action: putting analytics in the hands of academic advisors. J. Learn. Anal. **2**(2), 4–21 (2015). https://doi.org/10.1145/2330601.2330647
11. Brown, M.: Artificial intelligence in higher education: applications, promise and perils, and ethical questions. EDUCAUSE Rev. **55**(1) (2020)
12. Mubarak, H., Darwish, K.: Using Twitter to collect a multi-dialectal corpus of Arabic. In: Proceedings of the EMNLP 2014 Workshop on Arabic Natural Language Processing (ANLP), pp. 1–7 (2014)
13. El-Beltagy, S.R., Ali, A.: Open issues in the sentiment analysis of Arabic social media: a case study. In: Proceedings of 9th International Conference on Innovations in Information Technology (IIT), pp. 215–220 (2013)
14. Farghaly, A., Shaalan, K.: Arabic natural language processing: challenges and solutions. ACM Trans. Asian Lang. Inf. Process. (TALIP) **8**(4), 1–22 (2009). https://doi.org/10.1145/1644879.1644881
15. Habash, N., Soudi, A., Buckwalter, T.: On Arabic transliteration. In: Soudi, A., van den Bosch, A., Neumann, G. (eds.) Arabic Computational Morphology, pp. 15–22. Springer, Heidelberg (2007)
16. Abdul-Mageed, M., Diab, M., Kübler, S.: SAMAR: subjectivity and sentiment analysis for Arabic social media. Comput. Speech Lang. **28**(1), 20–37 (2014). https://doi.org/10.1016/j.csl.2013.03.001
17. Toutanova, K., Klein, D., Manning, C.D., Singer, Y.: Feature-rich part-of-speech tagging with a cyclic dependency network. In: Proceedings of the 2003 Conference of the North American Chapter of the Association for Computational Linguistics on Human Language Technology, pp. 173–180 (2003). https://doi.org/10.3115/1073445.1073478

Correction to: Employing Artificial Intelligence Applications in the Service of Islamic Religion and Belief

Ali Talib Mahal, Zainab Abdullah Habtiter, Hazem Adnan Ahmed, and Yassin Khudair Mijbel

Correction to: Chapter 15 in: A. Abdelgawad et al. (Eds.): *Intelligent Systems, Blockchain, and Communication Technologies*, LNNS 1268, https://doi.org/10.1007/978-3-031-82377-0_15

In the original version of the book, the following belated corrections have been incorporated: The author name "Ali Talib Maha" has been changed to "Ali Talib Mahal" and the affiliation has been updated in the Frontmatter, Backmatter and in Chapter number.

The updated version of this chapter can be found at
https://doi.org/10.1007/978-3-031-82377-0_15

A. Abdelgawad et al. (Eds.): ISBCom 2024, LNNS 1268, p. C1, 2026.
https://doi.org/10.1007/978-3-031-82377-0_69

Author Index

A. Abdelgawad et al. (Eds.): ISBCom 2024, LNNS 1268, pp. 859–861, 2025.
https://doi.org/10.1007/978-3-031-82377-0

The manufacturer's authorised representative in the EU is Springer
Nature Customer Service Centre GmbH, Europaplatz 3, 69115 Heidelberg,
Germany. If you have any concerns regarding our products, please
contact ProductSafety@springernature.com

Printed and bound by CPI Group (UK) Ltd, Croydon, CR0 4YY
02/10/2026
02226631-0002